SECOND EDITION

PRINCIPLES OF ELECTRIC CIRCUITS
Electron Flow Version

THOMAS L. FLOYD

MERRILL PUBLISHING COMPANY
Columbus Toronto London Melbourne

49305

Published by
Merrill Publishing Company
Columbus, Ohio 43216

This book was set in Times Roman.

Administrative Editor: Dave Garza
Production Coordinator: Rex Davidson
Developmental Editor: Carol Thomas
Art Coordinator: Peter A. Robison
Cover Designer: Cathy Watterson
Text Designer: Cynthia Brunk
Cover Photo: Morton & White Photographic, Rod Joslin, Clay White

Library of Congress Catalog Card Number: 89–64076
International Standard Book Number: 0–675–21292–8
Printed in the United States of America
1 2 3 4 5 6 7 8 9—94 93 92 91 90

To Debbie, Cyndi, and Sissy

MERRILL'S INTERNATIONAL SERIES
IN ELECTRICAL AND ELECTRONICS TECHNOLOGY

FLOYD	*Digital Fundamentals. Fourth Edition,* 21217-0
	Electric Circuits Fundamentals, 20756-8
	Electronic Devices, Second Edition, 20883-1
	Electronics Fundamentals: Circuits, Devices and Applications, 20714-2
	Essentials of Electronic Devices, 20062-8
	Principles of Electric Circuits, Third Edition, 21062-3
	Principles of Electric Circuits, Electron Flow Version, Second Edition, 21292-8
GAONKAR	*Microprocessor Architecture, Programming, and Applications with the 8085/8080A, Second Edition,* 20675-8
	The Z80 Microprocessor: Architecture, Interfacing, Programming, and Design, 20540-9
GILLIES	*Instrumentation and Measurements for Electronic Technicians,* 20432-1
HUMPHRIES	*Motors and Controls,* 20235-3
KULATHINAL	*Transform Analysis and Electronic Networks with Applications,* 20765-7
LAMIT/LLOYD	*Drafting for Electronics,* 20200-0
LAMIT/WAHLER/HIGGINS	*Workbook in Drafting for Electronics,* 20417-8
MARUGGI	*Technical Graphics: Electronics Worktext,* 20311-2
McINTYRE	*Study Guide to accompany Electronic Devices, Second Edition,* 21145-X
	Study Guide to accompany Electronics Fundamentals, 20676-6
MILLER	*The 68000 Microprocessor: Architecture, Programming, and Applications,* 20522-0
MONACO	*Introduction to Microwave Technology,* 21030-5
	Laboratory Activities in Microwave Technology, 21031-3
NASHELSKY/BOYLESTAD	*BASIC Applied to Circuit Analysis,* 20161-6
QUINN	*The 6800 Microprocessor,* 20515-8
REIS	*Electronic Project Design and Fabrication,* 20791-6
ROSENBLATT/FRIEDMAN	*Direct and Alternating Current Machinery, Second Edition,* 20160-8
SCHOENBECK	*Electronic Communications: Modulation and Transmission,* 20473-9
SCHWARTZ	*Survey of Electronics, Third Edition,* 20162-4
SORAK	*Linear Integrated Circuits: Laboratory Experiments,* 20661-8
STANLEY, B.H.	*Experiments in Electric Circuits, Third Edition,* 21088-7
STANLEY, W.D.	*Operational Amplifiers with Linear Integrated Circuits, Second Edition,* 20660-X
TOCCI	*Electronic Devices: Conventional Flow Version, Third Edition,* 20063-6
	Fundamentals of Electronic Devices, Third Edition, 9887-4
	Fundamentals of Pulse and Digital Circuits, Third Edition, 20033-4
	Introduction to Electric Circuit Analysis, Second Edition, 20002-4
WEBB	*Programmable Controllers: Principles and Applications,* 20452-6
WEBB/GRESHOCK	*Industrial Control Electronics,* 20897-1
YOUNG	*Electronic Communication Techniques, Second Edition,* 21045-3
ZANGER	*Fiber Optics: Communications and Other Applications,* 20944-7

PREFACE

This second edition of *Principles of Electric Circuits, Electron Flow Version,* is a result of a very thorough review process and a painstaking effort to ensure thoroughness, clarity, accuracy, and effectiveness in the presentation of the material.

A great effort has been made to enhance the quality of the text by making certain changes in organization and by expanding the coverage of certain topics. The major new features and modifications that make this text a more effective teaching tool are as follows:

1. A new full chapter on magnetism and electromagnetism (Chapter 10).
2. A separate chapter covering phasors and complex numbers (Chapter 12) and more emphasis on phasor applications throughout ac.
3. More illustrations—over 1300.
4. Standard resistor values are used in most circuits.
5. More end-of-chapter problems, including those of a more challenging nature.
6. Calculator sequences in selected examples.
7. For continuity, transformer coverage (Chapter 15) now follows the chapter on inductors.
8. More emphasis on troubleshooting and applications.
9. Test instruments are introduced throughout the text at appropriate points so that an instrument's operation and use can be related to the topic being covered.

This second edition features the popular pedagogy of *Principles of Electric Circuits,* third edition (conventional flow version). Each chapter begins with a *list of objectives,* a chapter *overview,* and a *list of sections.* A *section review* follows each section within a chapter to reinforce key concepts and to provide frequent feedback on the student's comprehension. Answers to these section reviews are given at the end of each chapter.

At the end of each chapter, there is a *summary* and a *list of formulas*. The *self-test* at the end of each chapter (with solutions at the end of the book) allows students to check their comprehension of the chapter. The *problem sets* are grouped by section for easy reference and assignment. Answers to selected odd-numbered problems appear at the end of the book.

SUPPLEMENTS

A lab manual, *Experiments in Electric Circuits, 3/E* by Brian Stanley contains 26 dc experiments and 25 ac experiments that are cross-referenced to this text. Like the text, this lab manual has been extensively revised based on user feedback. Procedural steps have been added where appropriate, and while the experiments do not require theory beyond that presented in the text, original thought by the student is encouraged. The manual has its own Solutions Manual.

A Solutions Manual for this text provides completely worked-out solutions to all end-of-chapter problems. Also provided is a transparency package that includes transparency masters and two-color overlay transparency acetates. New to this edition is a testbank of questions available in printed or computerized (for the IBM) form. Contact your local Merrill representative for copies of these supplements.

SUGGESTIONS FOR USE

Principles of Electric Circuits, Electron Flow Version, can be used in several ways to accommodate a variety of scheduling requirements. Some suggestions are as follows:

☐ A two-semester dc/ac sequence can cover most of the book, although some selectivity may be necessary, depending on class requirements. Chapters 1 through 10 are covered in the first semester. Capacitors and inductors (Chapters 13 and 14) can be covered at the end of the first term, if desired, by delaying coverage of the ac topics in Sections 13–6, 13–7, 14–6, and 14–7 until the second semester.

☐ A two-quarter dc/ac sequence can cover much of the book by light treatment or omission of selected topics, depending on the course objectives and emphasis. Since program requirements vary greatly, it is difficult to make specific suggestions for selective coverage. However, a few general recommendations are as follows:
   1. Assign Chapter 1 for outside study.
   2. Omit computer analysis sections.
   3. Lightly cover most of Chapters 8 and 9.
   4. Lightly cover Chapters 10, 19, 20, and 21.
   5. Omit Chapter 22.

Suggestions for light treatment or omission do not imply that the topics are less important than others, but only that, in the context of a specific program, they may not require as much emphasis as the more fundamental topics. Obviously, requirements vary from program to program so that content decisions must be tailored to fit *your* program.

☐ A one-term dc/ac course can cover the more fundamental topics by selectively omitting a number of areas as follows:

1. Omit Chapter 1 or assign as outside reading.
2. Omit Chapters 8 and 9 (except Thevenin's theorem perhaps).
3. Omit all of Chapter 10 except topics necessary for the study of inductance.
4. Lightly cover transformers (Chapter 15).
5. Omit Chapters 19, 20, and 21.
6. Omit computer analysis sections.
7. Other topics throughout may be lightly covered or omitted, depending on program requirements.

ACKNOWLEDGMENTS

I wish to express thanks to all the people at Merrill Publishing Company who had a part in making this revision a reality. In particular, the efforts of Dave Garza, Steve Helba, Carol Thomas, Pete Robison, and Rex Davidson are greatly appreciated. While space does not permit acknowledgment of all those who have submitted suggestions or comments for this new edition, I do wish to thank the following reviewers for their invaluable input: Jack Braun, Houston Community College; Dave Buchla, Yuba College; Li Chang-Qi, Civil Aviation Institute of China; Steve Harsany, Mt. San Antonio College; Samuel Kraemer, Oklahoma State University; Harvey Laabs, North Dakota State School of Science; Greg McBride, Heald Institute of Technology—San Francisco; Roman Ozarka, Dekalb College; Fred Schoenfeld, Nassau Community College; and Ulrich Zeisler, Salt Lake City Community College. In addition I am grateful to Harvey Laabs and Greg McBride for submitting material for use in the problem sets. Finally, I thank the following reviewers for checking the accuracy of this text: Chuck Donahue, Pensacola Jr. College; Ken Ferguson, Midlands Technical College; Frank Gergelyi, Metropolitan Technical Institute; John Kvartek, Pensacola Jr. College; Aaron B. Loggins, Technical-Vocational Institute, Montoya Campus; and James Yu, San Jose State University. To those who are using *Principles of Electric Circuits, Electron Flow Version,* for the first time and those who used it in the previous edition, I hope that you will like what we have done with this edition and that it will serve you well. Again, as with all my books, my wife, Sheila, has helped greatly with her love and support during the preparation of this text.

Thomas L. Floyd

CONTENTS

SIX
PARALLEL RESISTIVE CIRCUITS

173

SEVEN
SERIES-PARALLEL CIRCUITS

225

EIGHT
CIRCUIT THEOREMS AND CONVERSIONS

277

THIRTEEN
CAPACITORS

475

FOURTEEN
INDUCTORS

531

FIFTEEN
TRANSFORMERS

571

SIXTEEN
RC CIRCUIT ANALYSIS   **609**

SEVENTEEN
RL CIRCUIT ANALYSIS   **673**

EIGHTEEN
RLC CIRCUITS AND RESONANCE   **719**

NINETEEN
FILTERS
773

TWENTY
CIRCUIT THEOREMS IN ac ANALYSIS
805

TWENTY-ONE
PULSE RESPONSE OF REACTIVE CIRCUITS
843

TWENTY-TWO
POLYPHASE SYSTEMS IN POWER APPLICATIONS
887

ONE

INTRODUCTION

This chapter presents a brief history of the fields of electricity and electronics and discusses some of the many areas of application. Also, to aid you throughout the book, the basics of scientific notation and metric prefixes are reviewed, and the quantities and units commonly used in electronics are introduced. The use of the computer as an aid in circuit analysis using the BASIC language is discussed.

In this chapter, you will learn:

☐ A brief history of electricity and electronics.
☐ Some of the important areas in which electronics technology is applied.
☐ How to recognize some important electrical components and measuring instruments.
☐ The electrical quantities and their units.
☐ How to use scientific notation (powers of ten).
☐ The metric prefixes and how to use them.
☐ Some aspects of BASIC.

1–1 HISTORY OF ELECTRICITY AND ELECTRONICS

One of the first important discoveries about static electricity is attributed to William Gilbert (1540–1603). Gilbert was an English physician who, in a book published in 1600, described how amber differs from magnetic loadstones in its attraction of certain materials. He found that when amber was rubbed with a cloth, it attracted only light-weight objects, whereas loadstones attracted only iron. Gilbert also discovered that other substances, such as sulfur, glass, and resin, behave as amber does. He used the Latin word *elektron* for amber and originated the word *electrica* for the other substances that acted similarly to amber. The word *electricity* was used for the first time by Sir Thomas Browne (1605–82), an English physician.

Another Englishman, Stephen Gray (1696–1736), discovered that some substances conduct electricity and some do not. Following Gray's lead, a Frenchman named Charles du Fay experimented with the conduction of electricity. These experiments led him to believe that there were two kinds of electricity. He called one type *vitreous electricity* and the other type *resinous electricity*. He found that objects charged with vitreous electricity repelled each other and those charged with resinous electricity attracted each other. It is known today that two types of electrical *charge* do exist. They are called *positive* and *negative*.

Benjamin Franklin (1706–90) conducted studies in electricity in the mid-1700s. He theorized that electricity consisted of a single *fluid,* and he was the first to use the terms *positive* and *negative*. In his famous kite experiment, Franklin showed that lightning is electricity.

Charles Augustin de Coulomb (1736–1806), a French physicist, in 1785 proposed the laws that govern the attraction and repulsion between electrically charged bodies. Today, the unit of electrical charge is called the *coulomb*.

Luigi Galvani (1737–98) experimented with current electricity in 1786. Galvani was a professor of anatomy at the University of Bologna in Italy. Electrical current was once known as *galvanism* in his honor.

In 1800, Alessandro Volta (1745–1827), an Italian professor of physics, discovered that the chemical action between moisture and two different metals produced electricity. Volta constructed the first battery, using copper and zinc plates separated by paper that had been moistened with a salt solution. This battery, called the *voltaic pile,* was the first source of steady electric current. Today, the unit of electrical potential energy is called the *volt* in honor of Volta.

A Danish scientist, Hans Christian Oersted (1777–1851), is credited with the discovery of electromagnetism, in 1820. He found that electrical current flowing through a wire caused the needle of a compass to move. This finding showed that a magnetic field exists around a current-carrying conductor and that the field is produced by the current.

The modern unit of electrical current is the *ampere* (also called *amp*) in honor of the French physicist André Ampère (1775–1836). In 1820, Ampère measured the magnetic effect of an electrical current. He found that two wires carrying current can attract and repel each other, just as magnets can. By 1822, Ampère had developed the fundamental laws that are basic to the study of electricity.

One of the most well-known and widely used laws in electrical circuits today is *Ohm's law*. It was formulated by Georg Simon Ohm (1787–1854), a German teacher, in 1826. Ohm's law gives us the relationship among the three important electrical quantities of resistance, voltage, and current.

Although it was Oersted who discovered electromagnetism, it was Michael Faraday (1791–1867) who carried the study further. Faraday was an English physicist who believed that if electricity could produce magnetic effects, then magnetism could produce electricity. In 1831 he found that a moving magnet caused an electric current in a coil of wire placed within the field of the magnet. This effect, known today as *electromagnetic induction,* is the basic principle of electric generators and transformers.

Joseph Henry (1797–1878), an American physicist, independently discovered the same principle in 1831, and it is in his honor that the unit of inductance is called the *henry*. The unit of capacitance, the *farad,* is named in honor of Michael Faraday.

In the 1860s, James Clerk Maxwell (1831–79), a Scottish physicist, produced a set of mathematical equations that expressed the laws governing electricity and magnetism. These formulas are known as *Maxwell's equations*. Maxwell also predicted that electromagnetic waves (radio waves) that travel at the speed of light in space could be produced.

It was left to Heinrich Rudolph Hertz (1857–94), a German physicist, to actually produce these waves that Maxwell predicted. Hertz performed this work in the late 1880s. Today, the unit of frequency is called the *hertz*.

THE BEGINNING OF ELECTRONICS

The early experiments in electronics involved electric currents flowing in glass tubes. One of the first to conduct such experiments was a German named Heinrich Geissler (1814–79). Geissler found that when he removed most of the air from a glass tube, the tube glowed when an electrical potential was placed across it.

Around 1878, Sir William Crookes (1832–1919), a British scientist, experimented with tubes similar to those of Geissler. In his experiments, Crookes found that the current flowing in the tubes seemed to consist of particles.

Thomas Edison (1847–1931), experimenting with the carbon-filament light bulb that he had invented, made another important finding. He inserted a small metal plate in the bulb. When the plate was positively charged, a current flowed from the filament to the plate. This device was the first *thermionic diode*. Edison patented it but never used it.

The electron was discovered in the 1890s. The French physicist Jean Baptiste Perrin (1870–1942) demonstrated that the current in a vacuum tube consists of negatively charged particles. Some of the properties of these particles were measured by Sir Joseph Thomson (1856–1940), a British physicist, in experiments he performed between 1895 and 1897. These negatively charged particles later became known as *electrons*. The charge on the electron was accurately measured by an American physicist,

Robert A. Millikan (1868–1953), in 1909. As a result of these discoveries, electrons could be controlled, and the electronic age was ushered in.

PUTTING THE ELECTRON TO WORK

A vacuum tube that allowed electrical current to flow in only one direction was constructed in 1904 by John A. Fleming, a British scientist. The tube was used to detect electromagnetic waves. Called the *Fleming valve*, it was the forerunner of the more recent vacuum diode tubes.

Major progress in electronics, however, awaited the development of a device that could boost, or *amplify*, a weak electromagnetic wave or radio signal. This device was the *audion*, patented in 1907 by Lee de Forest (1873–1961), an American. It was a triode vacuum tube capable of amplifying small electrical signals.

Two other Americans, Harold Arnold and Irving Langmuir (1881–1957), made great improvements in the triode tube between 1912 and 1914. About the same time, de Forest and Edwin Armstrong, an electrical engineer, used the triode tube in an *oscillator circuit*. In 1914, the triode was incorporated in the telephone system and made the transcontinental telephone network possible.

The tetrode tube was invented in 1916 by Walter Schottky, a German. The tetrode, along with the pentode (invented in 1926 by Benjamin D. H. Tellegen, a Dutch engineer), provided great improvements over the triode. The first television picture tube, called the *kinescope*, was developed in the 1920s by Vladimir Zworykin, an American researcher.

During World War II, several types of microwave tubes were developed that made possible modern microwave radar and other communications systems. In 1939, the *magnetron* was invented in Britain by Henry Boot and John Randall. In the same year, the *klystron* microwave tube was developed by two Americans, Russell Varian and his brother Sigurd Varian. The *traveling-wave* tube was invented in 1943 by Rudolf Komphner, an Austrian-American.

THE COMPUTER

The computer probably has had more impact on modern technology than any other single type of electronic system. The first electronic digital computer was completed in 1946 at the University of Pennsylvania. It was called the Electronic Numerical Integrator and Computer (ENIAC). One of the most significant developments in computers was the *stored program* concept, developed in the 1940s by John von Neumann, an American mathematician.

SOLID STATE ELECTRONICS

The crystal detectors used in the early radios were the forerunners of modern solid state devices. However, the era of solid state electronics began with the invention of the *transistor* in 1947 at Bell Labs. The inventors were Walter Brattain, John Bardeen, and William Shockley. Figure 1–1 shows these three men, along with the notebook entry describing the historic discovery.

(a)

(b)

FIGURE 1–1

(a) Nobel Prize winners Drs. John Bardeen, William Shockley, and Walter Brattain, shown left to right, with apparatus used in their first investigations that led to the invention of the transistor. The trio received the 1956 Nobel Physics award for their invention of the transistor, which was announced by Bell Laboratories in 1948. (b) The laboratory notebook entry of scientist Walter H. Brattain recorded the events of December 23, 1947, when the transistor effect was discovered at Bell Laboratories. The notebook entry describes the event and adds, "This circuit was actually spoken over and by switching the device in and out a distinct gain in speech level could be heard and seen on the scope presentation with no noticeable change in quality." [(a) and (b) courtesy of AT&T Bell Laboratories]

In the early 1960s, the integrated circuit was developed. It incorporated many transistors and other components on a single small *chip* of semiconductor material. Integrated circuit technology continues to be developed and improved, allowing more complex circuits to be built on smaller chips. The introduction of the microprocessor in the early 1970s created another electronics revolution: the entire processing portion of a computer placed on a single, small, silicon chip. Continued development brought about complete computers on a single chip by the late 1970s.

SECTION REVIEW 1–1

1. Who developed the first battery?
2. The unit of what electrical quantity is named after André Ampère?
3. What contribution did Georg Simon Ohm make to the study of electricity?
4. In what year was the transistor invented?
5. What major development followed the invention of the transistor?

1–2 APPLICATIONS OF ELECTRICITY AND ELECTRONICS

Electricity and electronics are diverse technological fields with very broad applications. There is hardly an area of our lives that is not dependent to some extent on electricity and electronics. Some of the applications are discussed here in a general way to give you an idea of the scope of the fields.

COMPUTERS

One of the most important electronic systems is the digital computer; its applications are broad and diverse. For example, computers have applications in business for record keeping, accounting, payrolls, inventory control, market analysis, and statistics, to name but a few.

Scientific fields utilize the computer to process huge amounts of data and to perform complex and lengthy calculations. In industry, the computer is used for controlling and monitoring intricate manufacturing processes. Communications, navigation, medical, military, and home uses are a few of the other areas in which the computer is used extensively.

The computer's success is based on its ability to perform mathematical operations extremely fast and to process and store large amounts of information.

Computers vary in complexity and capability, ranging from very large systems with vast capabilities down to a computer on a chip with much more limited performance. Figure 1–2 shows some typical computers of varying sizes.

COMMUNICATIONS

Electronic communications encompasses a wide range of specialized fields. Included are space and satellite communications, commercial radio and television, citizens' band and

(a) (b) (c)

FIGURE 1–2
Typical computer systems. (a) Large computer system (courtesy of Unisys Corp.).
(b) PDP-11V23 computer (courtesy of Digital Equipment Corp.). (c) Personal computer
(courtesy of Tandy Corp.).

amateur radio, data communications, navigation systems, radar, telephone systems, military applications, and specialized radio applications such as police, aircraft, and so on. Computers are used to a great extent in many communications systems. Figure 1–3 shows a telephone switching system as an example of electronic communications.

FIGURE 1–3
A telephone switching system (GTD-5® EAX. Photo courtesy of AG Communication Systems Corp.).

AUTOMATION

Electronic systems are employed extensively in the control of manufacturing processes. Computers and specialized electronic systems are used in industry for various purposes, for example, control of ingredient mixes, operation of machine tools, product inspection, and control and distribution of power. Figure 1–4 shows an example of automation in a manufacturing facility using robots.

FIGURE 1–4
Robots on an automobile assembly line (courtesy of Ford Motor Co.).

MEDICINE

Electronic devices and systems are finding ever-increasing applications in the medical field. The familiar *electrocardiograph* (ECG), used for the diagnosis of heart and other circulatory ailments, is a widely used medical electronic instrument. A closely related instrument is the *electromyograph,* which uses a cathode ray tube display rather than an ink trace.

The *diagnostic sounder* uses ultrasonic sound waves for various diagnostic procedures in neurology, for heart chamber measurement, and for detection of certain types of tumors. The *electroencephalograph* (EEG) is similar to the electrocardiograph. It records the electrical activity of the brain rather than heart activity. Another electronic instrument used in medical procedures is the *coagulograph.* This instrument is used in analysis of blood clots.

Electronic instrumentation is also used extensively in intensive-care facilities. Heart rate, pulse, body temperature, respiration, and blood pressure can be monitored on a continuous basis. Monitoring equipment is also used a great deal in operating rooms. Some typical medical electronic equipment is pictured in Figure 1–5.

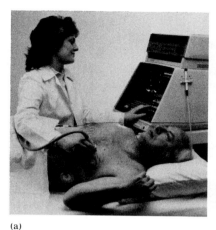
(a)

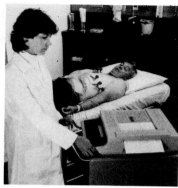
(b)

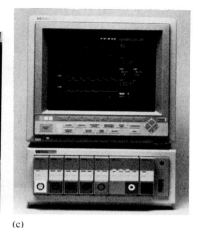
(c)

FIGURE 1–5

Typical medical instrumentation. (a) Doppler enhancement provides quantified measurements of blood-flow direction and velocity through the heart. (b) PageWriter cardiographs can provide analysis reports, measurements, and/or interpretation at the bedside in under 90 seconds. (c) Cardiac output module computes cardiac output and measures continuous pulmonary artery blood pressure (courtesy of Hewlett-Packard Co.).

CONSUMER PRODUCTS

Electronic products used directly by the consumers for information, entertainment, recreation, or work around the home are an important segment of the total electronics market. For example, the electronic calculator and digital watch are popular examples of consumer electronics. The small personal computer is used widely by hobbyists and is also becoming a common household appliance.

Electronic systems are used in automobiles to control and monitor engine functions, control braking, provide entertainment, and display useful information to the driver.

Most appliances such as microwave ovens, washers, and dryers are available with electronic controls. Home entertainment, of course, is largely electronic. Examples are television, radio, stereo, and recorders. Also, many new games for adults and children incorporate electronic devices.

SECTION REVIEW 1–2

1. Name some of the areas in which electronics is used.

2. The computerization of manufacturing processes in an example of _____.

1-3   CIRCUIT COMPONENTS AND MEASURING INSTRUMENTS

In this text you will study many types of electrical components and measuring instruments. This coverage of dc/ac circuit fundamentals provides the foundation for understanding electronic devices and circuits. Several types of electrical and electronic components and instruments that you will be studying in detail later in this and in other courses are introduced briefly in this section.

RESISTORS

These components *resist*, or limit, the flow of electrical current in a circuit. Several common types of resistors are shown in Figure 1–6.

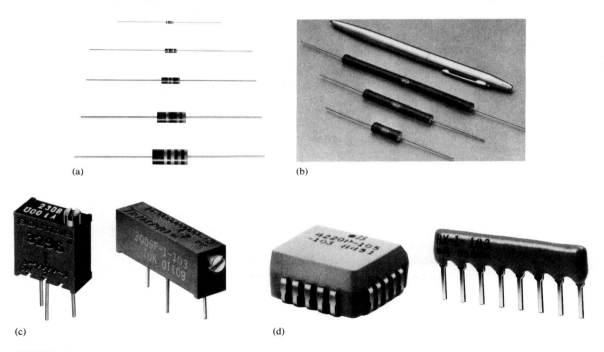

(a)   (b)

(c)   (d)

FIGURE 1–6
Typical fixed and variable resistors. (a) Carbon-composition resistors with standard power ratings of ⅛W, ¼W, ½W, 1W, and 2W (courtesy of Allen-Bradley Co.). (b) Wirewound resistors (courtesy of Dale Electronics, Inc.). (c) Variable resistors or potentiometers (courtesy of Bourns, Inc.). (d) Resistor networks (courtesy of Bourns, Inc.).

CAPACITORS

These components *store* electrical charge; they are found in a variety of applications. Figure 1–7 shows several typical capacitors.

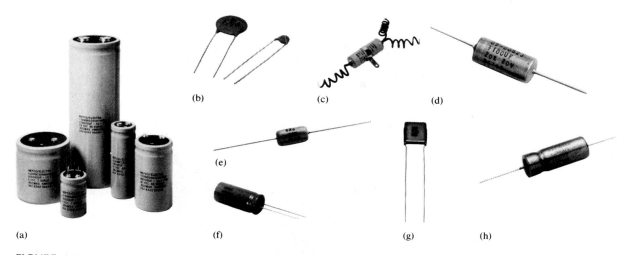

FIGURE 1–7
Typical capacitors [(a,d,e,h) courtesy of Philips Components. (b) courtesy of Murata Erie, North America. (c,f,g) courtesy of Sprague Electric Co.].

INDUCTORS

These components, also known as *coils*, are used to store energy in an electromagnetic field; they serve many useful functions in an electrical circuit. Figure 1–8 shows several typical inductors.

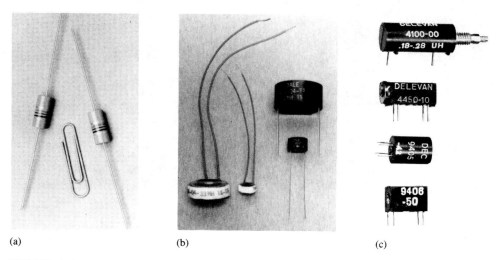

FIGURE 1–8
Typical inductors [(a) and (b) courtesy of Dale Electronics, Inc. (c) courtesy of Delevan.].

TRANSFORMERS

These components are sometimes used to couple ac voltages from one point in a circuit to another, or to increase or decrease the ac voltage. Several types of transformers are shown in Figure 1–9.

FIGURE 1–9
Typical transformers (courtesy of Dale Electronics, Inc.).

SEMICONDUCTOR DEVICES

Several varieties of diodes, transistors, and integrated circuits are shown in Figure 1–10.

(a)   (b)

FIGURE 1–10
A grouping of typical semiconductor devices (courtesy of Motorola Semiconductor Products, Inc.).

ELECTRONIC INSTRUMENTS

Figure 1–11 shows a variety of instruments that are discussed throughout the text. Typical instruments include the power supply, for providing voltage and current; the voltmeter, for measuring voltage; the ammeter, for measuring current; the ohmmeter, for measuring resistance; the wattmeter, for measuring power; and the oscilloscope for observing and measuring ac voltages.

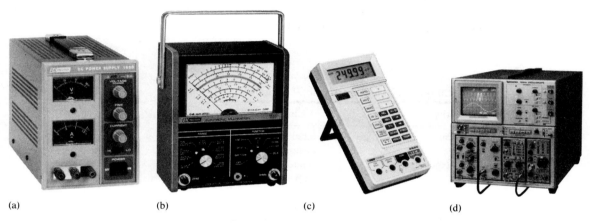

(a) (b) (c) (d)

FIGURE 1–11
Typical instruments. (a) dc power supply (courtesy of B&K Precision). (b) Analog multimeter (courtesy of B&K Precision). (c) Digital multimeter (courtesy of B&K Precision). (d) Oscilloscope (courtesy of Tektronix, Inc.).

SECTION REVIEW 1–3

1. Name three types of common electrical components.
2. What instrument is used for measuring electrical current?
3. What instrument is used for measuring resistance?

1–4   ELECTRICAL AND MAGNETIC UNITS

In electronics work, you must deal with measurable quantities. For example, you must be able to express how many volts are measured at a certain test point, how much current is flowing through a wire, or how much power a certain amplifier produces.

In this section you will be introduced to the units and symbols for many of the electrical and magnetic quantities that are used throughout the book. Definitions of the quantities are presented as they are needed in later chapters.

Symbols are used in electronics to represent both quantities and their units. One symbol is used to represent the name of the quantity, and another is used to represent the unit of measurement of that quantity. For example, P stands for *power,* and W stands for *watt,* which is the unit of power. Table 1–1 lists the most important electrical quantities, along with their SI units and symbols. The term *SI* is the French abbreviation for *International System (Système International* in French). Table 1–2 lists magnetic quantities, along with their SI units and symbols.

TABLE 1–1
Electrical quantities and units with SI symbols.

Quantity	Symbol	Unit	Symbol
capacitance	C	farad	F
charge	Q	coulomb	C
conductance	G	siemen	S
current	I	ampere	A
energy	W	joule	J
frequency	f	hertz	Hz
impedance	Z	ohm	Ω
inductance	L	henry	H
power	P	watt	W
reactance	X	ohm	Ω
resistance	R	ohm	Ω
time	t	second	s
voltage	V	volt	V

TABLE 1–2
Magnetic quantities and units with SI symbols.

Quantity	Symbol	Unit	Symbol
flux density	B	tesla	T
magnetic flux	ϕ	weber	Wb
magnetizing force	H	ampere-turns/meter	At/m
magnetomotive force	F_m	ampere-turns	At
permeability	μ	webers/ampere-turn-meter	Wb/Atm
reluctance	$\mathcal{R}$	ampere-turns/weber	At/Wb

SECTION REVIEW 1–4

1. What does *SI* stand for?

2. Without referring to Table 1–1, list as many electrical quantities as possible, including their symbols, units, and unit symbols.

3. Without referring to Table 1–2, list as many magnetic quantities as possible, including their symbols, units, and unit symbols.

1–5   ## SCIENTIFIC NOTATION

In electronics work, you will encounter both very small and very large numbers. For example, it is common to have electrical current values of only a few thousandths or even a few millionths of an ampere. On the other hand, you will find resistance values of several thousand or several million ohms. This range of values is typical of many other electrical quantities also.

POWERS OF TEN

Scientific notation uses *powers of ten,* a method that makes it much easier to express large and small numbers and to do calculations involving such numbers.

Table 1–3 lists some powers of ten, both positive and negative. The power of ten is expressed as an exponent of the base 10 in each case. The exponent indicates the number of places that the decimal point is moved to the right or left in the original number. If the power is *positive,* the decimal point is moved to the *right.* For example,

$$10^4 = 1 \times 10^4 = 1.0000. = 10,000$$

If the power is *negative,* the decimal point is moved to the *left.* For example,

$$10^{-4} = 1 \times 10^{-4} = .0001. = 0.0001$$

TABLE 1–3
Some positive and negative powers of ten.

$1,000,000 = 10^6$	$0.000001 = 10^{-6}$
$100,000 = 10^5$	$0.00001 = 10^{-5}$
$10,000 = 10^4$	$0.0001 = 10^{-4}$
$1,000 = 10^3$	$0.001 = 10^{-3}$
$100 = 10^2$	$0.01 = 10^{-2}$
$10 = 10^1$	$0.1 = 10^{-1}$
$1 = 10^0$	

EXAMPLE 1–1

Express each number using a positive power of ten:
(a) 200   (b) 5000   (c) 85,000   (d) 3,000,000

Solution:
In each case there are many possibilities for expressing the number in powers of ten. We do not show all possibilities in this example but include the most common powers of ten used in electrical work.
(a) $200 = 0.0002 \times 10^6 = 0.2 \times 10^3$
(b) $5000 = 0.005 \times 10^6 = 5 \times 10^3$
(c) $85,000 = 0.085 \times 10^6 = 8.5 \times 10^4 = 85 \times 10^3$
(d) $3,000,000 = 3 \times 10^6 = 3000 \times 10^3$

EXAMPLE 1–2

Express each number using a negative power of ten:
(a) 0.2   (b) 0.005   (c) 0.00063   (d) 0.000015

Solution:
Again, all the possible ways to express each number as a power of ten are not given. The most commonly used powers are included, however.
(a) $0.2 = 2 \times 10^{-1} = 200 \times 10^{-3} = 200,000 \times 10^{-6}$
(b) $0.005 = 5 \times 10^{-3} = 5000 \times 10^{-6}$
(c) $0.00063 = 0.63 \times 10^{-3} = 6.3 \times 10^{-4} = 630 \times 10^{-6}$
(d) $0.000015 = 0.015 \times 10^{-3} = 1.5 \times 10^{-5} = 15 \times 10^{-6}$

EXAMPLE 1–3

Express each of the following powers of ten as a regular decimal number:
(a) 10^5 (b) 2×10^3 (c) 3.2×10^{-2} (d) 250×10^{-6}

Solution:
(a) $10^5 = 1 \times 10^5 = 100,000$ (b) $2 \times 10^3 = 2000$
(c) $3.2 \times 10^{-2} = 0.032$ (d) $250 \times 10^{-6} = 0.000250$

CALCULATING WITH POWERS OF TEN

The advantage of scientific notation is in addition, subtraction, multiplication, and division of very small or very large numbers.

Rules for Addition   The rules for adding numbers in powers of ten are as follows:

1. Convert the numbers to be added to the *same* power of ten.
2. Add the numbers directly to get the sum.
3. Bring down the common power of ten, which is the power of ten of the sum.

**EXAMPLE
1–4**

Add 2×10^6 and 5×10^7.

Solution:
1. Convert both numbers to the same power of ten:

$$(2 \times 10^6) + (50 \times 10^6)$$

2. Add $2 + 50 = 52$.
3. Bring down the common power of ten (10^6), and the sum is 52×10^6.

Rules for Subtraction The rules for subtracting numbers in powers of ten are as follows:

1. Convert the numbers to be subtracted to the *same* power of ten.
2. Subtract the numbers directly to get the difference.
3. Bring down the common power of ten, which is the power of ten of the difference.

**EXAMPLE
1–5**

Subtract 25×10^{-12} from 75×10^{-11}.

Solution:
1. Convert each number to the same power of ten:

$$(75 \times 10^{-11}) - (2.5 \times 10^{-11})$$

2. Subtract $75 - 2.5 = 72.5$.
3. Bring down the common power of ten (10^{-11}), and the difference is 72.5×10^{-11}.

Rules for Multiplication The rules for multiplying numbers in powers of ten are as follows:

1. Multiply the numbers directly.
2. Add the powers of ten algebraically (the powers do not have to be the same).

**EXAMPLE
1–6**

Multiply 5×10^{12} and 3×10^{-6}.

Solution:
Multiply the numbers, and algebraically add the powers:

$$(5 \times 10^{12})(3 \times 10^{-6}) = 15 \times 10^{12+(-6)} = 15 \times 10^6$$

Rules for Division   The rules for dividing numbers in powers of ten are as follows:

1. Divide the numbers directly.
2. Subtract the power of ten in the denominator from the power of ten in the numerator.

EXAMPLE 1–7

Divide 50×10^8 by 25×10^3.

Solution:

The division problem is written with a numerator and denominator as

$$\frac{50 \times 10^8}{25 \times 10^3}$$

Dividing the numbers and subtracting 3 from 8, we get

$$\frac{50 \times 10^8}{25 \times 10^3} = 2 \times 10^{8-3} = 2 \times 10^5$$

SECTION REVIEW 1–5

1. Scientific notation uses powers of ten (T or F).
2. Express 100 as a power of ten.
3. Do the following operations:
   (a) $(1 \times 10^5) + (2 \times 10^5)$ **(b)** $(3 \times 10^6)(2 \times 10^4)$
   (c) $(8 \times 10^3) \div (4 \times 10^2)$

1–6   METRIC PREFIXES

In electrical and electronics work, certain powers of ten are used more often than others. The most frequently used powers of ten are 10^9, 10^6, 10^3, 10^{-3}, 10^{-6}, 10^{-9}, and 10^{-12}.

It is common practice to use *metric prefixes* to represent these quantities. Table 1–4 lists the metric prefix for each of the commonly used powers of ten.

TABLE 1–4
Metric prefixes and their symbols.

Power of Ten	Value	Metric Prefix	Metric Symbol	Power of Ten	Value	Metric Prefix	Metric Symbol
10^9	one billion	giga	G	10^{-6}	one-millionth	micro	μ
10^6	one million	mega	M	10^{-9}	one-billionth	nano	n
10^3	one thousand	kilo	k	10^{-12}	one-trillionth	pico	p
10^{-3}	one-thousandth	milli	m				

USE OF METRIC PREFIXES

Now we use examples to illustrate use of metric prefixes. The number 2000 can be expressed in scientific notation as 2×10^3. Suppose we wish to represent 2000 watts (W) with a metric prefix. Since $2000 = 2 \times 10^3$, the metric prefix *kilo* (k) is used for 10^3. So we can express 2000 W as 2 kW (2 kilowatts).

As another example, 0.015 ampere (A) can be expressed as 15×10^{-3} A. The metric prefix *milli* (m) is used for 10^{-3}. So 0.015 becomes 15 mA (15 milliamperes).

EXAMPLE 1–8

Express each quantity using a metric prefix:
(a)   50,000 V   (b)   25,000,000 Ω   (c)   0.000036 A

Solution:
(a) $50,000 \text{ V} = 50 \times 10^3 \text{ V} = 50 \text{ kV}$
(b) $25,000,000 \text{ } \Omega = 25 \times 10^6 \text{ } \Omega = 25 \text{ M}\Omega$
(c) $0.000036 \text{ A} = 36 \times 10^{-6} \text{ A} = 36 \text{ } \mu\text{A}$

ENTERING NUMBERS WITH METRIC PREFIXES ON THE CALCULATOR

To enter a number expressed in scientific notation on the calculator, use the EXP key (the EE key on some calculators). The following example shows how to enter numbers with metric prefixes on a typical scientific calculator such as the Sharp EL506 or EL525. Consult your user's manual for your particular calculator.

EXAMPLE 1–9

(a)   Enter 3.3 kΩ (3.3×10^3 Ω) on the calculator.
(b)   Enter 450 μA (450×10^{-6} A) on the calculator.

Solution:
(a)   Step 1:   Enter $\boxed{3}$ $\boxed{.}$ $\boxed{3}$. The display shows 3.3.
        Step 2:   Press $\boxed{\text{EXP}}$. The display shows 3.3 00.
        Step 3:   Enter $\boxed{3}$. The display shows 3.3 03.
(b)   Step 1:   Enter $\boxed{4}$ $\boxed{5}$ $\boxed{0}$. The display shows 450.
        Step 2:   Press $\boxed{\text{EXP}}$. The display shows 450 00.
        Step 3:   Press $\boxed{+/-}$. Enter $\boxed{6}$. The display shows 450 − 06.

SECTION REVIEW 1–6

1. List the metric prefix for each of the following powers of ten: 10^6, 10^3, 10^{-3}, 10^{-6}, 10^{-9}, and 10^{-12}.

2. Use an appropriate metric prefix to express 0.000001 ampere.

1–7	**METRIC UNIT CONVERSIONS**

It is often necessary or convenient to convert a quantity from one metric prefix unit to another, such as from milliamperes (mA) to microamperes (μA). A metric prefix conversion is accomplished by moving the decimal point in the quantity an appropriate number of places to the left or to the right, depending on the particular conversion. The following examples illustrate a few conversions.

EXAMPLE 1–10

Convert 0.15 milliampere (0.15 mA) to microamperes (μA).

Solution:
Move the decimal point three places to the right:

$$0.15 \text{ mA} = 150 \ \mu\text{A}$$

EXAMPLE 1–11

Convert 4500 microvolts (4500 μV) to millivolts (mV).

Solution:
Move the decimal point three places to the left:

$$4500 \ \mu\text{A} = 4.5 \text{ mA}$$

EXAMPLE 1–12

Convert 5000 nanoamperes (5000 nA) to microamperes (μA).

Solution:
Move the decimal point three places to the left:

$$5000 \text{ nA} = 5 \ \mu\text{A}$$

EXAMPLE 1–13

Convert 47,000 picofarads (47,000 pF) to microfarads (μF).

Solution:
Move the decimal point six places to the left:

$$47,000 \text{ pF} = 0.047 \ \mu\text{F}$$

EXAMPLE 1–14

Convert 0.00022 microfarad (0.00022 μF) to picofarads (pF).

Solution:
Move the decimal point six places to the right:

$$0.00022 \ \mu\text{F} = 220 \text{ pF}$$

| EXAMPLE 1–15 | Convert 1800 kilohms (1800 kΩ) to megohms (MΩ). |

Solution:
Move the decimal point three places to the left:

$$1800 \text{ k}\Omega = 1.8 \text{ M}\Omega$$

When adding (or subtracting) quantities with different metric prefixes, first convert one of the quantities to the same prefix as the other as the next example shows.

| EXAMPLE 1–16 | Add 15 mA and 8000 µA. |

Solution:
Convert 8000 µA to 8 mA and add:

$$15 \text{ mA} + 8000 \text{ µA} = 15 \text{ mA} + 8 \text{ mA} = 23 \text{ mA}$$

SECTION REVIEW 1–7

1. Convert 0.01 MV to kV.
2. Convert 250,000 pA to mA.
3. Add 0.05 MW and 75 kW.

1–8 THE COMPUTER AS AN ANALYSIS TOOL

The computer is a very useful tool in the areas of circuit analysis and problem solving.

Many of the chapters in this book provide computer programs as examples of programming for the analysis of electrical circuits. These sections on computer analysis are optional and can be omitted without affecting the flow of material.

All of the programs are patterned after selected example problems that are presented in the text. The programs are limited to fundamental program statements so that few if any changes are necessary for use on various machines.

THE BASIC LANGUAGE

A computer program is a *series of instructions* that tell the computer, step-by-step, what to do. Generally, in BASIC, each instruction must have a *line number*. Instruction words such as CLS, PRINT, LET, INPUT, GOTO, FOR/NEXT, and IF/THEN tell the computer what operation to perform at each step in the program.

Mathematical operators such as $+$, $-$, $*$, $/$, $\uparrow$, and $=$ are used for addition, subtraction, multiplication, division, exponentiation, and equating, respectively.

These instruction words and operators, of course, do not represent the extent of the BASIC language, but most of the programs found throughout this text are limited to these. Table 1–5 lists each of the BASIC language components mentioned above and gives a brief description of each. For a detailed and thorough coverage of BASIC, refer to your computer instruction manual or a BASIC language textbook.

TABLE 1–5
Some BASIC instruction words and mathematical operators.

Instruction Word	Description	Example
CLS	Clears video screen	CLS
PRINT	Causes the computer to display or print out any message that follows in quotes or the value of a designated variable or the result of a specified calculation.	PRINT "HELLO"→HELLO PRINT X→Value of X PRINT 2 + 3→5
LET	Assigns a value to a variable. Can be omitted in some versions of BASIC for simplicity.	LET X = 8 or simply X = 8 LET Y = X/2 or simply Y = X/2
INPUT	Allows data to be entered into computer, such as variable values.	INPUT X (The computer stops and waits for you to enter a value for X from the keyboard.)
GOTO	Causes the computer to branch from its place in the program to a specified line number and skip everything in between.	GOTO 50
FOR/NEXT	These two instruction words are used together to set up loops.	A value for Y is calculated for each of three values of X (1, 2, and 3). 10 FOR X = 1 TO 3 20 Y = X∗2 30 NEXT X
IF/THEN	These two instruction words are used together to set up conditional statements.	IF X = 6 THEN GOTO 50 or IF X = 6 THEN 50

Mathematical Operator	Description	Example
=	Equals	X = 3 (X equals 3)
+	Addition	Y = X + 5 (X plus 5)
−	Subtraction	A = 10 − X (10 minus X)
∗	Multiplication	Z = 2∗X (2 times X)
/	Division	W = Y/4 (Y divided by 4)
↑ ([)(∗∗)	Exponentiation	X↑2 (X squared)

SECTION REVIEW 1–8

1. List seven BASIC instruction words or sets of words.

2. List six BASIC mathematical operators.

SUMMARY

1. Areas of electronics applications include computers, communications, automation, medicine, military, and consumer products.

2. Resistors limit the flow of electrical current.

3. Capacitors store electrical charge.

4. Inductors store energy electromagnetically.

5. Inductors are also known as *coils*.

6. Transformers magnetically couple ac voltages.

7. Semiconductor devices include diodes, transistors, and integrated circuits.

8. Power supplies provide current and voltage.

9. Voltmeters measure voltage.

10. Ammeters measure current.

11. Ohmmeters measure resistance.

12. Metric prefixes are a convenient method of expressing both large and small quantities.

SELF-TEST

Solutions appear at the end of the book.

1. List the units of the following electrical quantities: current, voltage, resistance, power, and energy.

2. List the symbol for each unit in Question 1.

3. List the symbol for each quantity in Question 1.

4. Express the following using metric prefixes:
   (a) ten milliamperes   **(b)** five kilovolts
   (c) fifteen microwatts   **(d)** twenty megohms

5. Express each of the following metric quantities as a decimal number with the appropriate nonprefixed unit:
   (a) $8 \ \mu A$ **(b)** 25 MW   **(c)** 100 mV

PROBLEMS

Section 1–5

1–1   Express each of the following numbers in scientific notation:
   (a) 3000   **(b)** 75,000   **(c)** 2,000,000

1–2   Express each number in scientific notation:
   (a) 1/500   **(b)** 1/2000   **(c)** 1/5,000,000

1–3   Express each of the following numbers in three ways, using 10^3, 10^4, and 10^5:
   (a) 8400   **(b)** 99,000   **(c)** 0.2×10^6

1–4   Express each of the following numbers in three ways, using 10^{-3}, 10^{-4}, and 10^{-5}:
   (a) 0.0002   **(b)** 0.6   **(c)** 7.8×10^{-2}

1–5   Express each power of ten in regular decimal form:
   (a) 2.5×10^{-6} **(b)** 50×10^2 **(c)** 3.9×10^{-1}

1–6   Express each power of ten in regular decimal form:
   (a) 45×10^{-6} **(b)** 8×10^{-9} **(c)** 40×10^{-12}

1–7   Add the following numbers:
   (a) $(92 \times 10^6) + (3.4 \times 10^7)$ **(b)** $(5 \times 10^3) + (85 \times 10^{-2})$
   (c) $(560 \times 10^{-8}) + (460 \times 10^{-9})$

1–8   Perform the following subtractions:
   (a) $(3.2 \times 10^{12}) - (1.1 \times 10^{12})$ **(b)** $(26 \times 10^8) - (1.3 \times 10^9)$
   (c) $(150 \times 10^{-12}) - (8 \times 10^{-11})$

1–9   Perform the following multiplications:
   (a) $(5 \times 10^3)(4 \times 10^5)$ **(b)** $(12 \times 10^{12})(3 \times 10^2)$
   (c) $(2.2 \times 10^{-9})(7 \times 10^{-6})$

1–10   Divide the following:
   (a) $(10 \times 10^3) \div (2.5 \times 10^2)$ **(b)** $(250 \times 10^{-6}) \div (50 \times 10^{-8})$
   (c) $(4.2 \times 10^8) \div (2 \times 10^{-5})$

1–11   Convert each of the following numbers to another number having a multiplier of 10^{-6}:
   (a) 2.37×10^{-3} **(b)** 0.001856×10^{-2} **(c)** 5743.89×10^{-12}
   (d) 100×10^3

1–12   Perform the following indicated operations:
   (a) $(2.8 \times 10^3)(3 \times 10^2)/2 \times 10^2$ **(b)** $(46)(10^{-3})(10^5)/10^6$
   (c) $(7.35)(0.5 \times 10^{12})/[(2)(10 \times 10^{10})]$ **(d)** $(30)^2(5)^3/10^{-2}$

1–13   Perform each operation:
   (a) $\sqrt{49 \times 10^6}/\sqrt{4 \times 10^{-8}}$ **(b)** $(3 \times 10^3)^2/1.8 \times 10^3$
   (c) $(1.5 \times 10^{-6})^2(4.7 \times 10^6)$ **(d)** $1/(2\pi\sqrt{(5 \times 10^{-3})(0.01 \times 10^{-6})})$

Section 1–6

1–14 Express each of the following as a quantity having a metric prefix:
    (a) 31×10^{-3} A   **(b)** 5.5×10^3 V   **(c)** 200×10^{-12} F

1–15 Express the following using metric prefixes:
    (a) 3×10^{-6} F   **(b)** 3.3×10^6 Ω **(c)** 350×10^{-9} A

1–16 Express the following quantities as powers of ten:
    (a) 258 mA   **(b)** 0.022 μF   **(c)** 1200 kV

1–17 Express each of the following quantities as a metric unit:
    (a) 2.4×10^{-5} A   **(b)** 970×10^4 Ω **(c)** 0.003×10^{-9} W

1–18 Complete the following operations:
    (a) (50 μA)(6.8 kΩ) = _____ V   **(b)** 24 V/1.2 MΩ = _____ A
    (c) 12 kV/20 mA = _____ Ω

1–19 Complete the following operations and express the results using metric prefixes:
    (a) $(100 \ \mu A)^2$(8.2 kΩ) = _____ W   **(b)** $(30 \ kV)^2$/2 MΩ = _____ W

Section 1–7

1–20 Perform the indicated conversions:
    (a) 5 mA to microamperes.   **(b)** 3200 μW to milliwatts.
    (c) 5000 kV to megavolts.   **(d)** 10 MW to kilowatts.

1–21 Determine the following:
    (a) The number of microamperes in 1 milliampere.
    (b) The number of millivolts in 0.05 kilovolt.
    (c) The number of megohms in 0.02 kilohm.
    (d) The number of kilowatts in 155 milliwatts.

ANSWERS TO SECTION REVIEWS

Section 1–1

1. Volta.   **2.** Current.
3. He established the relationship among current, voltage, and resistance as expressed in Ohm's law.
4. 1947.   **5.** Integrated circuits.

Section 1–2

1. Computers, communications, automation, medicine, and consumer products.
2. Automation.

Section 1–3

1. Resistors, capacitors, inductors, and transformers.
2. Ammeter.   **3.** Ohmmeter.

Section 1-4
1. It is the abbreviation for Système International.
2. Refer to Table 1-1 after you have compiled your list.
3. Refer to Table 1-2 after you have completed your list.

Section 1-5
1. T.   **2.** 10^2.   **3. (a)** 3×10^5;   **(b)** 6×10^{10};   **(c)** 2×10^1.

Section 1-6
1. Mega (M), kilo (k), milli (m), micro (μ), nano (n), and pico (p).
2. 1 μA (one microampere).

Section 1-7
1. 10 kV.   **2.** 0.00025 mA.   **3.** 125 kW.

Section 1-8
1. CLS, PRINT, LET, INPUT, GOTO, FOR/NEXT, IF/THEN.
2. =, +, −, *, /, ↑ or **.

TWO

VOLTAGE, CURRENT, AND RESISTANCE

Three basic electrical quantities are presented in this chapter: voltage, current, and resistance. No matter what type of electrical or electronic equipment you may work with, these quantities will always be of primary importance.

To help you understand voltage and current, the basic structure of the atom is discussed, and the concept of charge is introduced. The basic electric circuit is studied, along with techniques for measuring voltage, current, and resistance.

In this chapter, you will learn:

☐ The concept of the atom.
☐ That the electron is the basic particle of electrical charge.
☐ What voltage, current, and resistance are.
☐ How voltage causes current.
☐ How resistance restricts current.
☐ Various types of voltage sources.
☐ Various types of fixed and variable resistors.
☐ How to determine resistance value by color code.
☐ What a basic electric circuit consists of.
☐ The concepts of closed and open circuits.
☐ Various types of switches and how they are used.
☐ Various types of fuses and circuit breakers.
☐ How to measure voltage, current, and resistance.
☐ How to read both analog and digital meters.

2–1   ATOMS

An atom is the smallest particle of an element that still retains the characteristics of that element. Different elements have different types of atoms. In fact, every element has a unique atomic structure.

Atoms have a planetary type of structure, consisting of a central nucleus surrounded by orbiting electrons. The nucleus consists of positively charged particles called *protons* and uncharged particles called *neutrons*. The electrons are the basic particles of negative charge.

Each type of atom has a certain number of electrons and protons that distinguishes the atom from all other atoms of other elements. For example, the simplest atom is that of hydrogen. It has one proton and one electron, as pictured in Figure 2–1(a). The helium atom, shown in Figure 2–1(b), has two protons and two neutrons in the nucleus, which is orbited by two electrons.

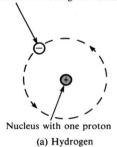

One electron orbiting the nucleus

Nucleus with one proton

(a) Hydrogen

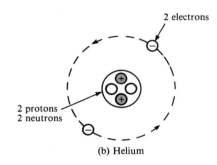

2 electrons

2 protons
2 neutrons

(b) Helium

FIGURE 2–1
Hydrogen and helium atoms.

ATOMIC WEIGHT AND NUMBER

All elements are arranged in the periodic table of the elements in order according to their atomic number, which is the number of protons in the nucleus. The elements can also be arranged by their atomic weight, which is approximately the number of protons and neutrons in the nucleus. For example, hydrogen has an atomic number of one and an atomic weight of one. The atomic number of helium is two, and its atomic weight is four.

In their normal, or neutral, state, all atoms of a given element have the same number of electrons as protons. So the positive charges cancel the negative charges, and the atom has a net charge of zero.

THE COPPER ATOM

Since copper is the most commonly used metal in electrical applications, let us examine its atomic structure. The copper atom has 29 electrons in orbit around the nucleus. They do not all occupy the same orbit, however. They move in orbits at varying distances from the nucleus. The orbits in which the electrons revolve are called *shells*. The number of electrons in each shell follows a predictable pattern.

The first shell of any atom can have up to 2 electrons, the second shell up to 8 electrons, the third shell up to 18 electrons, and the fourth shell up to 32 electrons. A copper atom is shown in Figure 2–2. Notice that the fourth or outermost shell has only 1 electron, called the *valence* electron.

FIGURE 2–2
The copper atom.

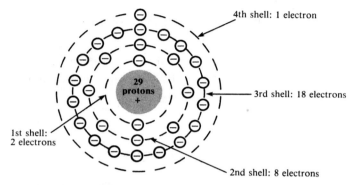

FREE ELECTRONS

When the electron in the outer shell of the copper atom gains sufficient energy from the surrounding medium, it can break away from the parent atom and become what is called a *free electron*. The free electrons in the copper material are capable of moving from one atom to another in the material. In other words, they drift randomly from atom to atom within the copper. As you will see, the free electrons make electrical *current* possible.

Three categories of materials are used in electronics: *conductors, semiconductors,* and *insulators*.

CONDUCTORS

Conductors are materials that allow electrons to flow easily. They have a large number of free electrons in their structure. Most metals are good conductors. Silver is the best conductor, and copper is next. Copper is the most widely used conductive material because it is less expensive than silver.

SEMICONDUCTORS

These materials are classed below the conductors in their ability to carry current because they have fewer free electrons in their structure than do conductors. However, because of their unique characteristics, certain semiconductor materials are the basis for modern electronic devices such as the diode, transistor, and integrated circuit. Silicon and germanium are common semiconductor materials.

INSULATORS

Insulating materials are poor conductors of electric current. In fact, they are used to *prevent* current where it is not wanted. Compared to conductive materials, insulators have very few free electrons.

SECTION REVIEW 2–1

1. What is the basic particle of negative charge?
2. Define *atom*.
3. What does a typical atom consist of?
4. Do all elements have the same types of atoms?
5. What is a free electron?

2–2

ELECTRICAL CHARGE

As you saw in the previous discussion of the structure of an atom, there are two types of charge: *positive* and *negative*. The electron is the smallest particle that exhibits negative electrical charge. When an excess of electrons exists in a material, there is a net negative electrical charge. When an excess of protons exists, there is a net positive electrical charge. The charge of an electron and that of a proton are equal in magnitude. Electrical charge is symbolized by Q.

Static electricity is the presence of a net positive or negative charge in a material. Everyone has experienced the effects of static electricity from time to time, for example, when attempting to touch a metal surface or another person or when the clothes in a dryer cling together.

Materials with charges of opposite polarity are attracted to each other, and materials with charges of the same polarity are repelled, as indicated in Figure 2–3. There is a force acting between charges, as evidenced by the attraction or repulsion. This force, called an *electric field*, consists of invisible lines of force, as shown in Figure 2–4.

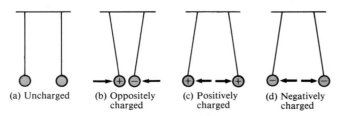

FIGURE 2–3
Attraction and repulsion of electrical charges.

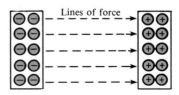

FIGURE 2–4
Electric field between oppositely charged surfaces.

THE COULOMB: THE UNIT OF CHARGE

Electrical charge is measured in *coulombs,* abbreviated C. *One coulomb is the total charge possessed by 6.25 × 10[18] electrons.* A single electron has a charge of 1.6 × 10^{-19} C. This unit is named for Charles Coulomb (1736–1806), a French scientist.

HOW POSITIVE AND NEGATIVE CHARGES ARE CREATED

Consider a neutral atom—that is, one that has the same number of electrons and protons and thus has no net charge. If a valence electron is pulled away from the atom by the application of energy, the atom is left with a net positive charge (more protons than electrons) and becomes a *positive ion.* If an atom acquires an extra electron in its outer shell, it has a net negative charge and becomes a *negative ion.*

The amount of energy required to free a valence electron is related to the number of electrons in the outer shell. An atom can have up to eight valence electrons. The more complete the outer shell, the more stable the atom and thus the more energy is required to release an electron. Figure 2–5 illustrates the creation of a positive and a negative ion when sodium chloride dissolves and the sodium atom gives up its single valence electron to the chlorine atom.

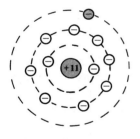

Sodium atom
(11 protons, 11 electrons)

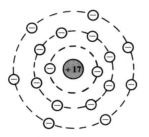

Chlorine atom
(17 protons, 17 electrons)

(a) The sodium atom has a single
valence electron.

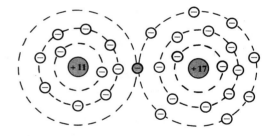

(b) The atoms combine by sharing
the valence electron to form
sodium chloride (table salt).

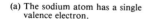

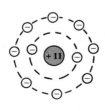

Positive sodium ion
(11 protons, 10 electrons)

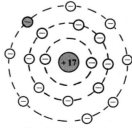

Negative chlorine ion
(17 protons, 18 electrons)

(c) When dissolved, the sodium atom gives up the valence electron to become a positive
ion, and the chlorine atom retains the extra valence electron to become a negative ion.

FIGURE 2–5
Example of the formation of positive and negative ions.

EXAMPLE 2–1

How many coulombs do 93.75×10^{16} electrons represent?

Solution:

$$Q = \frac{\text{number of electrons}}{\text{number of electrons in one coulomb}}$$

$$= \frac{93.75 \times 10^{16} \text{ electrons}}{6.25 \times 10^{18} \text{ electrons/C}} = 0.15 \text{ C}$$

SECTION REVIEW 2–2

1. What is the symbol for charge?
2. What is the unit of charge, and what is its symbol?
3. How much charge, in coulombs, is there in 10×10^{12} electrons?

2–3   VOLTAGE

As you have seen, there is a force of attraction between a positive and a negative charge. A certain amount of energy must be exerted in the form of work to overcome the force and move the charges a given distance apart. All opposite charges possess a certain potential energy because of the separation between them. The difference in potential energy of the charges is the *potential difference*. For example, consider a water tank that is supported several feet above the ground. A given amount of energy must be exerted in the form of work to pump water up to fill the tank. Once the water is stored in the tank, it has a certain potential energy which, if released, can be used to perform work. For example, the water can be allowed to fall down a chute to turn a water wheel.

Potential difference in electrical terms is more commonly called *voltage (V)* and is expressed as energy or work *(W)* per unit charge *(Q):*

$$V = \frac{W}{Q} \tag{2–1}$$

where W is expressed in joules (J) and Q is in coulombs (C).

The unit of voltage is the *volt*, symbolized by V. *One volt is the potential difference (voltage) between two points when one joule of energy is used to move one coulomb of charge from one point to the other.*

EXAMPLE 2–2

If 50 joules of energy are available for every 10 coulombs of charge, what is the voltage?

Solution:

$$V = \frac{W}{Q} = \frac{50 \text{ J}}{10 \text{ C}} = 5 \text{ V}$$

SOURCES OF VOLTAGE

The Battery   A voltage source is a source of potential energy that is also called *electromotive force* (emf). The battery is one type of voltage source that converts chemical energy into electrical energy. A voltage exists between the electrodes (terminals) of a battery, as shown by a voltaic cell in Figure 2–6. One electrode is positive and the other negative as a result of the separation of charges caused by the chemical action when two different conducting materials are dissolved in the electrolyte.

Batteries are generally classified as *primary cells,* which cannot be recharged, and *secondary cells,* which can be recharged by reversal of the chemical action. The amount of voltage provided by a battery varies. For example, a flashlight battery is 1.5 V and an automobile battery is 12 V. Some typical batteries are shown in Figure 2–7.

FIGURE 2–6
A voltaic cell converts chemical energy into electrical energy.

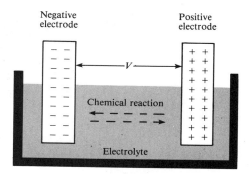

FIGURE 2–7
Typical batteries (courtesy of SAFT America, Inc.)

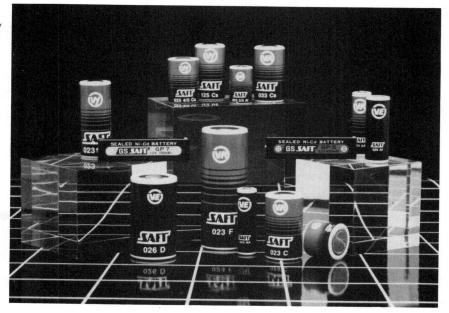

The Electronic Power Supply   These voltage sources convert the ac voltage from the wall outlet to a constant (dc) voltage which is available across two terminals. Typical commercial power supplies are shown in Figure 2–8.

(a)   (b)

FIGURE 2–8
Electronic power supplies [(a) courtesy of B&K Precision. (b) courtesy of B&K Precision. (top); courtesy of Heath Company (bottom)].

The Solar Cell   The operation of solar cells is based on the principle of *photovoltaic action,* which is the process whereby light energy is converted directly into electrical energy. A basic solar cell consists of two layers of different semiconductive materials joined together to form a junction. When one layer is exposed to light, many electrons acquire enough energy to break away from their parent atoms and cross the junction. This process forms negative ions on one side of the junction and positive ions on the other, and thus a potential difference (voltage) is developed. Construction of a basic solar cell is shown in Figure 2–9.

FIGURE 2–9
Construction of basic solar cell.

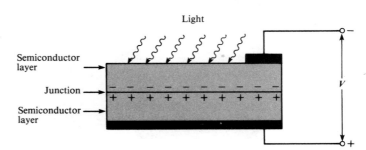

The Generator   Generators convert mechanical energy into electrical energy using a principle called *electromagnetic induction* (to be studied later). A conductor is rotated through a magnetic field, and a voltage is produced across the conductor. A typical generator is pictured in Figure 2–10.

FIGURE 2–10
Cutaway view of dc generator (courtesy of Pacific Scientific Motor & Control).

SECTION REVIEW 2–3

1. Define *voltage*.

2. How much is the voltage when there are 24 joules of energy for 10 coulombs of charge?

3. List four sources of voltage.

2–4   CURRENT

As you have seen, there are free electrons available in all conductive and semiconductive materials. These electrons drift randomly in all directions, from atom to atom, within the structure of the material, as indicated in Figure 2–11.

FIGURE 2–11
Random motion of free electrons in a material.

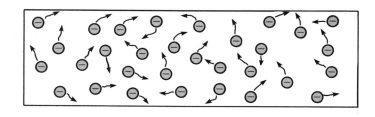

Now, if a voltage is placed across the conductive or semiconductive material, one end becomes positive and the other negative, as indicated in Figure 2–12. The repulsive force of the negative voltage at the left end causes the free electrons (negative charges) to move toward the right. The attractive force of the positive voltage at the right end pulls the free electrons to the right. The result is a net movement of the free electrons from the negative end of the material to the positive end, as shown in Figure 2–12.

FIGURE 2–12
Electrons flow from negative to positive when a voltage is applied across a conductive or semiconductive material.

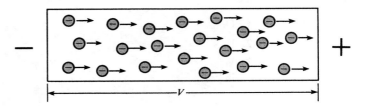

The movement of these free electrons from the negative end of the material to the positive end is the *electrical current*, symbolized by *I*. *Electrical current is defined as the rate of flow of electrons in a conductive or semiconductive material*. It is measured by the number of electrons (amount of charge, *Q*) that flow past a point in a unit of time:

$$I = \frac{Q}{t} \tag{2–2}$$

where *I* is current, *Q* is the charge of the electrons, and *t* is the time.

THE AMPERE: THE UNIT OF CURRENT

Current is measured in a unit called the *ampere* or *amp* for short, symbolized by A. It is named after André Ampère (1775–1836), a French physicist whose work contributed to the understanding of electrical current and its effects.

One ampere (1 A) is the amount of current when a number of electrons having one coulomb (1 C) of charge move past a given point (P) in one second (1 s). (See Figure 2–13.) One coulomb is the charge carried by 6.25×10^{18} electrons.

FIGURE 2–13
Illustration of one ampere of current in a material (1 C/s).

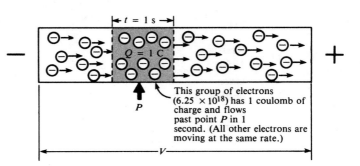

This group of electrons (6.25×10^{18}) has 1 coulomb of charge and flows past point *P* in 1 second. (All other electrons are moving at the same rate.)

**EXAMPLE
2–3**

Ten coulombs of charge flow past a given point in a wire in 2 seconds. How many amperes of current is this?

Solution:

$$I = \frac{Q}{t} = \frac{10 \text{ C}}{2 \text{ s}} = 5 \text{ A}$$

SECTION REVIEW 2–4

1. Define *current* and state its unit.

2. How many electrons make up one coulomb of charge?

3. What is the current in amperes when 20 coulombs flow past a point in a wire in 4 s?

2–5

RESISTANCE

When there is current in a material, the free electrons move through the material and occasionally collide with atoms. These collisions cause the electrons to lose some of their energy, and thus their movement is restricted. The more collisions, the more the flow of electrons is restricted. This restriction varies and is determined by the type of material. The property of a material that restricts the flow of electrons is called *resistance,* designated *R. Resistance is the opposition to current.* The schematic symbol for resistance is shown in Figure 2–14.

FIGURE 2–14
Resistance/resistor symbol.

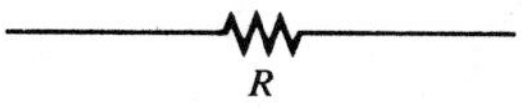

     When there is current through any material that has resistance, heat is produced by the collisions of electrons and atoms. Therefore, wire, which typically has a very small resistance, becomes warm when there is current through it. A discussion of how the size of wire affects its resistance is given in Appendix A.

THE OHM: THE UNIT OF RESISTANCE

Resistance, *R,* is expressed in the unit of *ohms,* named after Georg Simon Ohm (1787–1854) and symbolized by the Greek letter omega (Ω). *There is one ohm (1 Ω) of resistance when there is one ampere (1 A) of current in a material when one volt (1 V) is applied across the material.*

Conductance   Later, in circuit analysis problems, you will find conductance to be useful. Conductance, symbolized by *G,* is simply the *reciprocal* of resistance, and is a measure of the *ease* with which current is established. The formula is

$$G = \frac{1}{R} \qquad\qquad (2\text{--}3)$$

The unit of conductance is the siemen, abbreviated S.

RESISTORS

Components that are specifically designed to have a certain amount of resistance are called *resistors*. The principal applications of resistors are to limit the current and, in certain cases, to generate heat. Although there are a variety of different types of resistors that come in many shapes and sizes, they can all be placed in one of two main categories: *fixed* or *variable*.

Fixed Resistors This kind of resistor is available with a large selection of ohmic values that are set during manufacturing and cannot be changed easily. Fixed resistors are constructed using various methods and materials. Several common types are shown in Figure 2–15.

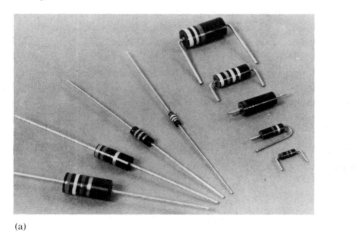

(a)

(b)

(c)

FIGURE 2–15
Typical fixed resistors [(a) and (b) courtesy of Stackpole Carbon Co. (c) courtesy of Bourns, Inc.].

One common fixed resistor is the *carbon-composition* type, which is made with a mixture of finely ground carbon, insulating filler, and a resin binder. The ratio of carbon to insulating filler sets the resistance value. The mixture is formed into rods, and lead connections are made. The entire resistor is then encapsulated in an insulated coating for protection. Figure 2–16 shows the construction of a typical carbon-composition resistor.

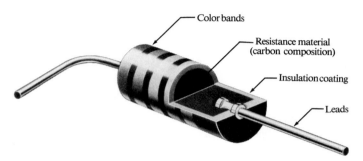

FIGURE 2–16
Cutaway view of carbon-composition resistor (courtesy of Allen-Bradley Co.).

Other types of fixed resistors include carbon film, metal oxide, metal film, metal glaze, and wirewound. In film resistors, a resistive material is deposited evenly onto a high-grade ceramic rod. The resistive film may be carbon (carbon film), nickel chromium (metal film), a mixture of metals and glass (metal glaze), or metal and insulating oxide (metal oxide). In these types of resistors, the desired resistance value is obtained by removing part of the resistive material in a helical pattern along the rod using a *spiraling* technique. Very close tolerance can be achieved with this method.

Wirewound resistors are constructed with resistive wire wound around an insulating rod and then sealed. Normally, wirewound resistors are used because of their relatively high power ratings.

Some typical fixed resistors are shown in the construction views of Figure 2–17.

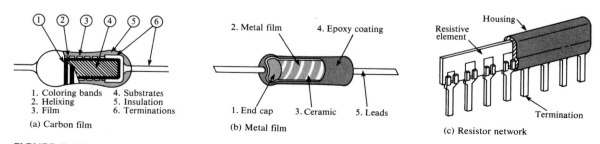

FIGURE 2–17
Construction views of typical fixed resistors [(a) and (b) courtesy of Stackpole Carbon Co. (c) courtesy of Bourns, Inc.].

Resistor Color Codes   Fixed resistors with value tolerances of 5%, 10%, or 20% are color coded with four bands to indicate the resistance value and the tolerance. This color-code band system is shown in Figure 2–18, and the color code is listed in Table 2–1.

FIGURE 2–18
Color-code bands on a resistor.

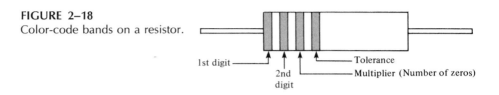

1st digit ⎯ Tolerance
2nd digit ⎯ Multiplier (Number of zeros)

TABLE 2–1
Resistor color code.

	Digit	Color
	0	Black
	1	Brown
	2	Red
	3	Orange
Resistance value, first three bands	4	Yellow
	5	Green
	6	Blue
	7	Violet
	8	Gray
	9	White
Tolerance, fourth band	5%	Gold
	10%	Silver
	20%	No band

The color code is read as follows:

1. Beginning at the banded end, the first band is the first digit of the resistance value.
2. The second band is the second digit of the resistance value.
3. The third band is the number of zeros, or the *multiplier*.
4. The fourth band indicates the tolerance.

For example, a 5% tolerance means that the *actual* resistance value is within ±5% of the color-coded value. Thus, a 100-Ω resistor with a tolerance of ±5% can have acceptable values as low as 95 Ω and as high as 105 Ω.

For resistance values less than 10 Ω, the third band is either gold or silver. Gold represents a multiplier of 0.1, and silver represents 0.01. For example, a color code of red, violet, gold, and silver represents 2.7 Ω with a tolerance of ±10%. Standard resistance values are given in Appendix B.

**EXAMPLE
2–4**

Find the resistance value in ohms and the percent tolerance for each of the color-coded resistors shown in Figure 2–19.

FIGURE 2–19

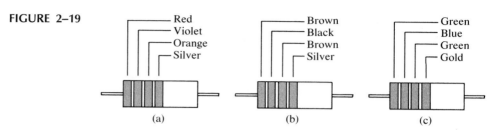

(a)   (b)   (c)

Solution:

(a) First band is red = 2, second band is violet = 7, third band is orange = 3 zeros, fourth band is silver = 10% tolerance.

$$R = 27,000 \ \Omega \ \pm 10\%$$

(b) First band is brown = 1, second band is black = 0, third band is brown = 1 zero, fourth band is silver = 10% tolerance.

$$R = 100 \ \Omega \ \pm 10\%$$

(c) First band is green = 5, second band is blue = 6, third band is green = 5 zeros, fourth band is gold = 5% tolerance.

$$R = 5,600,000 \ \Omega \ \pm 5\%$$

Certain precision resistors with tolerances of 1% or 2% are color coded with five bands. Beginning at the banded end, the first band is the first digit of the resistance value, the second band is the second digit, the third band is the third digit, the fourth band is the multiplier, and the fifth band indicates the tolerance. Table 2–1 applies, except that gold indicates 1% and silver indicates 2%.

Numerical labels are also commonly used on certain types of resistors where the resistance value and tolerance are stamped on the body of the resistor. For example, a common system uses R to designate the decimal point and letters to indicate tolerance as follows:

F = ±1%,   G = ±2%,   J = ±5%,   K = ±10%,   M = ±20%

For values above 100 Ω, three digits are used to indicate resistance value, followed by a fourth digit that specifies the number of zeros. For values less than 100 Ω, R indicates the decimal point.

Some examples are as follows: 6R8M is a 6.8-Ω ±20% resistor; 3301F is a 3300-Ω ±1% resistor; and 2202J is a 22,000-Ω ±5% resistor.

Resistor Reliability Band   The fifth band on some color-coded resistors indicates the resistor's reliability in percent of failures per 1000 hours of use. The fifth-band reliability color code is listed in Table 2–2. For example, a brown fifth band means that if a group of like resistors are operated under standard conditions for 1000 hours, 1% of the resistors in that group will fail.

TABLE 2–2
Fifth-band reliability color code.

Color	Failures (%) during 1000 Hours of Operation
Brown	1.0%
Red	0.1%
Orange	0.01%
Yellow	0.001%

Variable Resistors   Variable resistors are designed so that their resistance values can be changed easily with a manual or an automatic adjustment.

Two basic types of manually adjustable resistors are the *potentiometer* and the *rheostat*. Schematic symbols for these types are shown in Figure 2–20. The potentiometer is a *three-terminal device,* as indicated in Part (a). Terminals 1 and 2 have a fixed resistance between them, which is the total resistance. Terminal 3 is connected to a moving contact (wiper). We can vary the resistance between 3 and 1 or between 3 and 2 by moving the contact up or down.

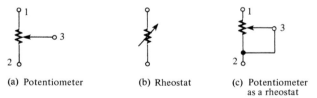

(a) Potentiometer   (b) Rheostat   (c) Potentiometer
                                                       as a rheostat

FIGURE 2–20
Potentiometer and rheostat symbols.

Figure 2–20(b) shows the rheostat as a *two-terminal* variable resistor. Part (c) shows how we can use a potentiometer as a rheostat by connecting terminal 3 to either terminal 1 or terminal 2. Some typical potentiometers are pictured in Figure 2–21.

Potentiometers and rheostats can be classified as *linear* or *tapered,* as shown in Figure 2–22, where a potentiometer with a total resistance of 100 Ω is used as an example. As shown in Part (a), in a linear potentiometer, the resistance between either terminal and the moving contact varies linearly with the position of the moving contact. For example, one-half of a turn results in one-half the total resistance. Three-quarters of a turn results in three-quarters of the total resistance between the moving contact and one terminal, or one-quarter of the total resistance between the other terminal and the moving contact.

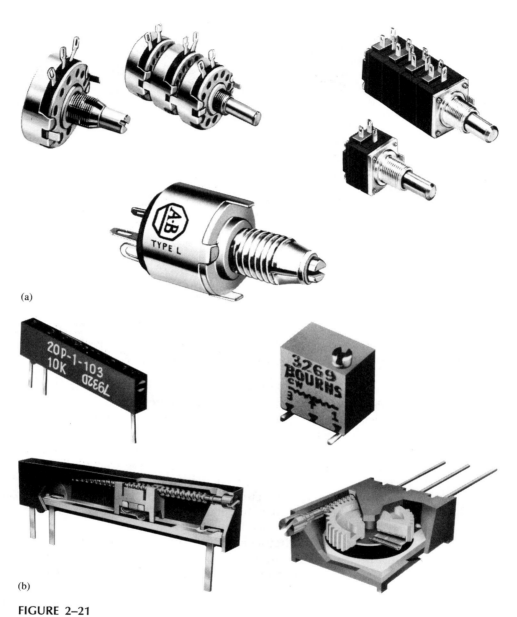

(a)

(b)

FIGURE 2–21
(a) Typical potentiometers (courtesy of Allen-Bradley Co.). (b) Trimmer potentiometers with construction views (courtesy of Bourns, Inc.).

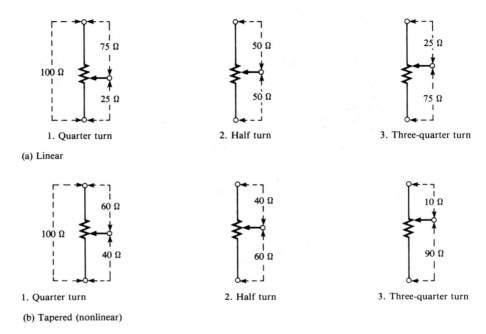

1. Quarter turn   2. Half turn   3. Three-quarter turn

(a) Linear

1. Quarter turn   2. Half turn   3. Three-quarter turn

(b) Tapered (nonlinear)

FIGURE 2–22
Examples of (a) linear and (b) tapered potentiometers.

In the tapered potentiometer, the resistance varies nonlinearly with the position of the moving contact, so that one-half of a turn does not necessarily result in one-half the total resistance. This concept is illustrated in Figure 2–22(b), where the nonlinear values are arbitrary.

The potentiometer is used as a *voltage-control* device because when a fixed voltage is applied across the end terminals, a variable voltage is obtained at the wiper contact with respect to either end terminal. The rheostat is used as a *current-control* device because the current can be changed by changing the wiper position. You will learn the reasons for these functions later.

Thermistors and Photoconductive Cells A thermistor is a type of variable resistor that is temperature-sensitive. Its resistance changes inversely with temperature. That is, it has a *negative temperature coefficient*. When temperature increases, the resistance decreases, and vice versa.

The resistance of a photoconductive cell changes with a change in light intensity. This cell also has a negative temperature coefficient. Symbols for both of these devices are shown in Figure 2–23.

FIGURE 2–23
Resistive devices with sensitivities to temperature and light.

(a) Thermistor

(b) Photoconductive cell

SECTION REVIEW 2–5

1. Define *resistance* and name its unit.
2. What are the two main categories of resistors? Briefly explain the difference between them.
3. In the resistor color code, what does each band represent?
4. Determine the resistance and tolerance for each of the following color codes:
   (a) yellow, violet, red, gold.
   (b) blue, red, orange, silver.
   (c) brown, gray, black, gold.
5. What is the basic difference between a rheostat and a potentiometer?
6. What is a thermistor?

2–6   THE ELECTRIC CIRCUIT

Basically, an electric circuit consists of a *voltage source*, a *load*, and a *path for current* between the source and the load. Figure 2–24 shows an example of a simple electric circuit: a battery connected to a lamp with two conductors (wires). The battery is the voltage source, the lamp is the load on the battery because it draws current from the battery, and the two wires provide the current path from the negative terminal of the battery to the lamp and back to the positive terminal of the battery, as shown in Part (b). The current through the filament of the lamp (which has a resistance), causes it to emit visible light. Current is produced in the battery by chemical action. In many practical cases, one terminal of the battery is connected to a *ground* point. For example, in automobiles, the negative battery terminal is connected to the metal chassis of the car. The chassis is the ground for the automobile electrical system. The concept of circuit ground is covered in detail in a later chapter.

FIGURE 2–24
A simple electric circuit.

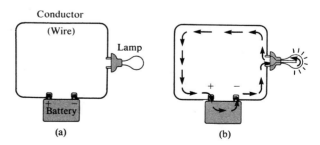

(a)   (b)

THE ELECTRICAL SCHEMATIC

An electric circuit can be represented by a *schematic diagram* using standard symbols for each element, as shown in Figure 2–25 for the simple circuit in Figure 2–24. The purpose of a schematic diagram is to show in an organized manner how the various components in a given circuit are interconnected so that the operation of the circuit can be determined.

FIGURE 2–25
Schematic diagram for the circuit in Figure 2–24.

CLOSED AND OPEN CIRCUITS

The example circuit in Figure 2–24 illustrated a *closed circuit*—that is, a circuit in which there is a complete path for current. When the current path is broken so that there is no current, the circuit is called an *open circuit*.

Switches   Switches are commonly used for controlling the opening or closing of circuits by either mechanical or electronic means. For example, a switch is used to turn a lamp on or off as illustrated in Figure 2–26. Each circuit pictorial is shown with its associated schematic diagram. The type of switch indicated is a *single-pole–single-throw* (SPST) toggle switch.

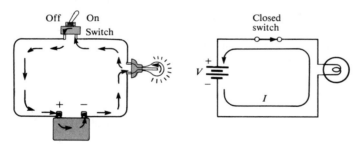

(a) There is current in a *closed* circuit (switch is ON or in the *closed* position).

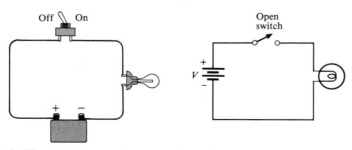

(b) There is no current in an *open* circuit (switch is OFF or in the *open* position).

FIGURE 2–26
Basic closed and open circuits using an SPST switch for control.

Figure 2–27 shows a somewhat more complicated circuit using a *single-pole–double-throw* (SPDT) type of switch to control the current to two different lamps. When one lamp is on, the other is off, and vice versa, as illustrated by the two schematic diagrams which represent each of the switch positions.

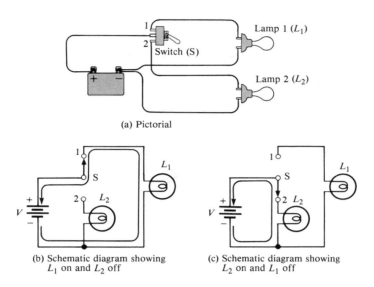

(a) Pictorial

(b) Schematic diagram showing
L_1 on and L_2 off

(c) Schematic diagram showing
L_2 on and L_1 off

FIGURE 2–27
An example of an SPDT switch controlling two lamps.

The term *pole* refers to the movable arm in a switch, and the term *throw* indicates the number of contacts that are affected (either opened or closed) by a single switch action (a single movement of a pole).

In addition to the SPST and the SPDT switches already introduced, several other types are of importance:

☐ *Double-pole–single-throw* (DPST). The DPST switch permits simultaneous opening or closing of two sets of contacts. The symbol is shown in Figure 2–28(a). The dashed line indicates that the contact arms are mechanically linked so that both move with a single switch action.

☐ *Double-pole–double-throw* (DPDT). The DPDT switch provides connection from one set of contacts to either of two other sets. The schematic symbol is shown in Figure 2–28(b).

☐ *Push-button* (PB). In the normally open push-button switch (NOPB), shown in Figure 2–28(c), connection is made between two contacts when the button is depressed, and connection is broken when the button is released. In the normally closed push-button switch (NCPB), shown in Figure 2–28(d), connection between the two contacts is broken when the button is depressed.

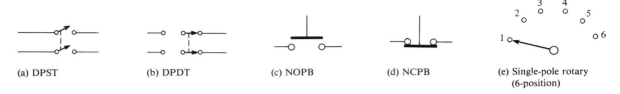

(a) DPST **(b) DPDT** **(c) NOPB** **(d) NCPB** **(e) Single-pole rotary (6-position)**

FIGURE 2–28
Switch symbols.

☐ *Rotary.* In a rotary switch, a knob is turned to make connection between one contact and any one of several others. A symbol for a simple six-position rotary switch is shown in Figure 2–28(e).

Several varieties of switches are shown in Figure 2–29.

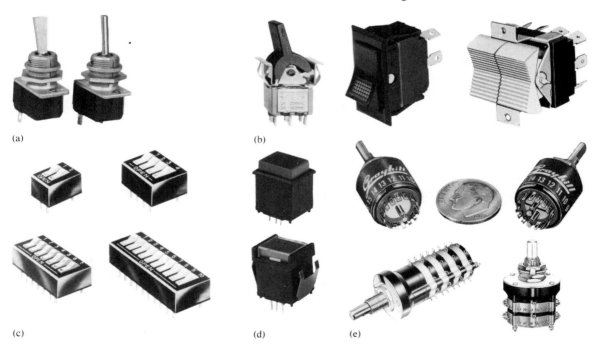

FIGURE 2–29
Switches. (a) Typical toggle-lever switches (courtesy of Eaton Corporation). (b) Rocker switches (courtesy of Eaton Corporation). (c) Rocker DIP (dual-in-line package) switches (courtesy of AMP, Inc.). (d) Push-button switches (courtesy of Eaton Corporation). (e) Rotary-position switches (courtesy of Grayhill, Inc.).

Protective Devices *Fuses* and *circuit breakers* are used to deliberately create an open circuit when the current exceeds a specified number of amperes due to a malfunction or other abnormal condition in a circuit. For example, a 20-A fuse or circuit breaker will open a circuit when the current exceeds 20 A.

The basic difference between a fuse and a circuit breaker is that when a fuse is "blown," it must be replaced, but when a circuit breaker opens, it can be reset and reused repeatedly. The purpose of both of these devices is to protect against damage to a circuit due to excess current or to prevent a hazardous condition created by the overheating of wires and other components when the current is too great. Several typical fuses and circuit breakers, along with their schematic symbols, are shown in Figure 2–30.

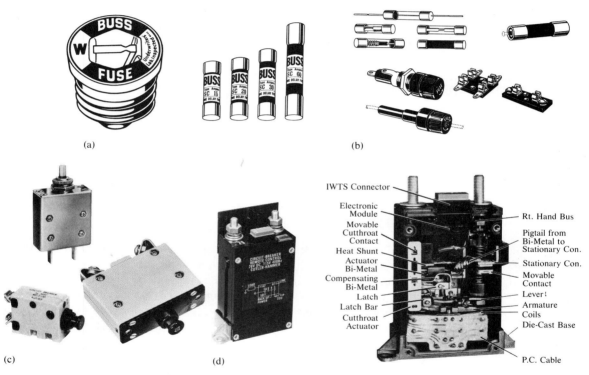

FIGURE 2–30
Fuses and circuit breakers. (a) Power fuses (courtesy of Bussman Manufacturing Corp.). (b) Fuses and fuse holders (courtesy of Bussman Manufacturing Corp.). (c) Circuit breakers (courtesy of Bussman Manufacturing Corp.). (d) Remote control circuit breaker (courtesy of Eaton Corp.).

DIRECTION OF CURRENT

Early in the history of our knowledge of electricity, it was assumed that all current consisted of positive moving charges. Later, of course, the electron was identified as the charge carrier in conductive materials.

Today, there are two accepted conventions for the direction of electrical current. *Electron flow direction,* preferred by many in the fields of electrical and electronics technology, assumes current out of the negative terminal of a voltage source, through the circuit, and into the positive terminal of the source. *Conventional current direction* assumes current out of the positive terminal of a voltage source, through the circuit, and into the negative terminal of the source.

It actually makes little difference which direction of current is assumed as long as you are *consistent.* The outcome of electric circuit analysis is not affected by the direction of current that is assumed for analytical purposes. The direction assumed is largely a matter of preference. Electron flow direction is used throughout this text.

SECTION REVIEW 2–6

1. What are the basic elements of an electric circuit?

2. What is an open circuit?

3. What is a closed circuit?

4. What is the resistance across an open?

2–7   BASIC CIRCUIT MEASUREMENTS

Voltage, current, and resistance measurements are commonly required in electronics work. Special types of instruments are used to measure these basic electrical quantities.

The instrument used to measure voltage is a *voltmeter,* the instrument used to measure current is an *ammeter,* and the instrument used to measure resistance is an *ohmmeter.* Commonly, all three instruments are combined into a single instrument known as a *multimeter,* or VOM (volt-ohm-milliammeter), in which you can choose what specific quantity to measure by selecting the switch setting.

Typical multimeters are shown in Figure 2–31. Part (a) shows an analog meter, that is, with a needle pointer, and Part (b) shows a digital multimeter (DMM), which provides a digital readout of the measured quantity.

METER SYMBOLS

Throughout this book, certain symbols will be used to represent the different meters, as shown in Figure 2–32. You may see any of three types of symbols for voltmeters, ammeters, and ohmmeters, depending on which symbol most effectively conveys the information required. Generally, the pictorial analog symbol is used when relative mea-

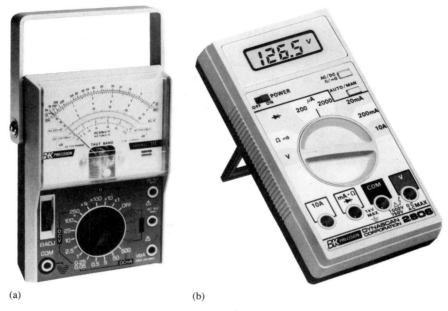

FIGURE 2–31
Typical portable multimeters. (a) Analog and (b) digital (courtesy of B&K Precision).

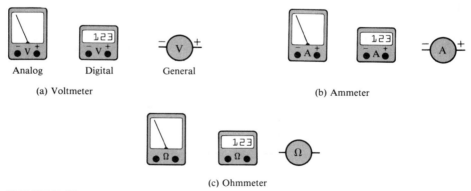

FIGURE 2–32
Meter symbols.

surements or changes in quantities are to be depicted by the position or movement of the pointer. The pictorial digital symbol is used when fixed values are to be indicated in a circuit. The general schematic symbol is used to indicate placement of meters in a circuit when no values or value changes need to be shown.

HOW TO MEASURE CURRENT WITH AN AMMETER

Figure 2–33 illustrates how to measure current with an ammeter. Part (a) shows the simple circuit in which the current through the resistor is to be measured. Connect the ammeter *in the current path* by first opening the circuit, as shown in Part (b). Then insert the meter as shown in Part (c). As you will learn later, such a connection is a *series* connection. The polarity of the meter must be such that the current is in at the negative terminal and out at the positive.

FIGURE 2–33
Example of an ammeter connection.

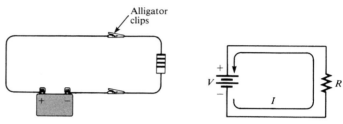

(a) Circuit in which the current is to be measured

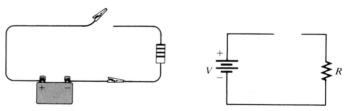

(b) Open the circuit either between the resistor and the positive terminal or between the resistor and the negative terminal of source.

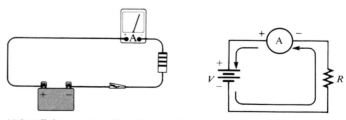

(c) Install the ammeter with polarity as shown (negative to negative–positive to positive).

HOW TO MEASURE VOLTAGE WITH A VOLTMETER

To measure voltage, connect the voltmeter *across the component* for which the voltage is to be found. As you will learn later, such a connection is a *parallel* connection. The negative terminal of the meter must be connected to the negative side of the circuit, and the positive terminal of the meter to the positive side of the circuit. Figure 2–34 shows a voltmeter connected to measure the voltage across the resistor.

FIGURE 2–34
Example of a voltmeter connection.

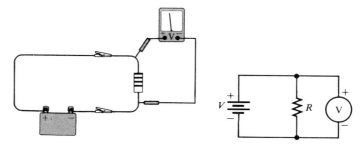

HOW TO MEASURE RESISTANCE WITH AN OHMMETER

To measure resistance, connect the ohmmeter across the resistor. *The resistor must first be removed or disconnected from the circuit.* This procedure is shown in Figure 2–35.

READING ANALOG MULTIMETERS

A typical analog multimeter is shown in Figure 2–36. This particular instrument can be used to measure both direct current (dc) and alternating current (ac) quantities as well as resistance values. It has four selectable functions: dc volts (DC VOLTS), dc milliamperes (DC MA), ac volts (AC VOLTS), and OHMS. Most analog multimeters are similar to this one.

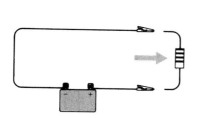

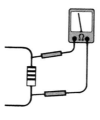

(a) Disconnect the resistor from the circuit to avoid damage to the meter and/or incorrect measurement.

(b) Measure the resistance. (Polarity is not important.)

FIGURE 2–35
Example of using an ohmmeter.

FIGURE 2–36
A typical analog multimeter (courtesy of Triplett Corp.).

Within each function there are several ranges, as indicated by the brackets around the selector switch. For example, the DC VOLTS function has 0.3-V, 3-V, 12-V, 60-V, 300-V, and 600-V ranges. Thus, dc voltages from 0.3 V full-scale to 600 V full-scale can be measured. On the DC MA function, direct currents from 0.06 mA full-scale to 120 mA full-scale can be measured. On the ohm scale, the settings are × 1, × 10, × 100, × 1000, and × 100 000.

The Ohm Scale   Ohms are read on the top scale of the meter. This scale is *nonlinear;* that is, the values represented by each division (large or small) vary as you go across the scale. In Figure 2–36, notice how the scale becomes more compressed as you go from right to left.

To read the actual value in ohms, multiply the number on the scale as indicated by the pointer by the factor selected by the switch. For example, when the switch is set at × 100 and the pointer is at 20, the reading is 20 × 100 = 2000 Ω.

As another example, assume that the switch is at × 10 and the pointer is at the seventh small division between the 1 and 2 marks, indicating 17 Ω (1.7 × 10). Now, if the meter remains connected to the same resistance and the switch setting is changed to × 1, the pointer will move to the second small division between the 15 and 20 marks. This, of course, is also a 17-Ω reading, illustrating that a given resistance value can often be read at more than one switch setting.

The ac-dc Scales   The second, third, and fourth scales from the top, labeled "AC" and "DC," are used in conjunction with the DC VOLTS, DC MA, and AC VOLTS functions. The upper ac-dc scale ends at the 300 mark and is used with the range settings that are multiples of three, such as 0.3, 3, and 300. For example, when the switch is at 3 on the DC VOLTS function, the 300 scale has a full-scale value of 3 V. At the range setting of 300, the full-scale value is 300 V, and so on.

The middle ac-dc scale ends at 60. This scale is used in conjunction with range settings that are multiples of 6, such as 0.06, 60, and 600. For example, when the switch is at 60 on the DC VOLTS function, the full-scale value is 60 V.

The lower ac-dc scale ends at 12 and is used in conjunction with switch settings that are multiples of 12, such as 1.2, 12, and 120.

The remaining scales are for ac current and for decibels, which are discussed later in the book.

EXAMPLE 2–5

In Figure 2–37, determine the quantity that is being measured and its value.

Solution:
(a)   The switch is set on the dc volts (DC-V) function and the 30-V range. The reading taken from the lower dc scale is 18 V.
(b)   The switch is set on the dc ampere (DC-A) function and the 100-μA range. The reading taken from the upper dc scale is 44 μA.
(c)   The switch is set on the ohm (OHMS) function and the × 1K range. The reading taken from the ohm scale (top scale) is 6.7 kΩ.

FIGURE 2-37

(a)

(b)

(c)

DIGITAL MULTIMETERS (DMMs)

DMMs are perhaps the most widely used type of electronic measuring instrument. Generally, DMMs provide more functions, better accuracy, greater ease of reading, and greater reliability than do many analog meters. Analog meters have at least one advantage over DMMs, however: They can track short-term variations and trends in a measured quantity that many DMMs are too slow to respond to. Several typical DMMs are shown in Figure 2-38.

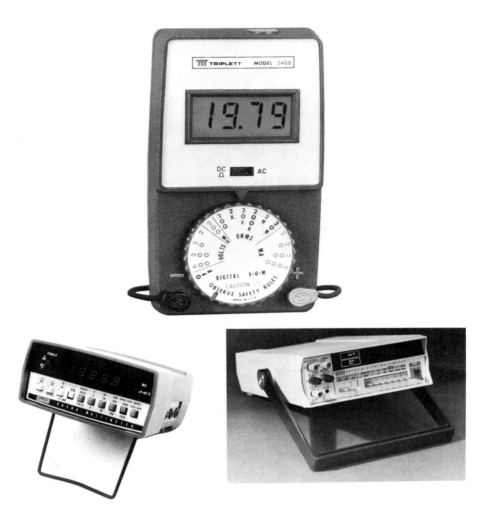

FIGURE 2–38
Typical digital multimeters (DMMs) [(top) courtesy of Triplett Corp. (bottom) courtesy of John Fluke Manufacturing Co.].

DMM Functions   The basic functions found on most DMMs include the following:

- ☐ ohms
- ☐ dc voltage and current
- ☐ ac voltage and current

Some DMMs provide special functions such as transistor or diode tests, power measurement, and decibel measurement for audio amplifier tests.

DMM Displays   DMMs are available with either LCD (liquid-crystal display) or LED (light-emitting diode) readouts. The LCD is the most commonly used readout in battery-powered instruments because it requires only very small amounts of current. A typical battery-powered DMM with an LCD readout operates on a 9-V battery that will last from a few hundred hours to 2000 hours and more. The disadvantages of LCD readouts are that (a) they are difficult or impossible to see in low-light conditions and (b) they are relatively slow to respond to measurement changes. LEDs, on the other hand, can be seen in the dark and respond quickly to changes in measured values. LED displays require much more current than LCDs, and, therefore, battery life is shortened when they are used in portable equipment.

Both LCD and LED DMM displays are in a *seven-segment format*. Each digit in the display consists of seven separate segments, as shown in Figure 2–39(a). Each of the ten decimal digits is formed by activation of appropriate segments, as illustrated in Figure 2–39(b). In addition to the seven segments, there is also a decimal point.

(a)   (b)

FIGURE 2–39
Seven-segment display.

Resolution   The resolution of a meter is the smallest increment of a quantity that the meter can measure. The smaller the increment, the better the resolution. One factor that determines the resolution of a meter is the number of digits in the display.

Because many meters have 3½ digits in their display, we will use this case for illustration. A 3½-digit multimeter has three digit positions that can indicate from 0 through 9, and one digit position that can indicate only a value of 1. This latter digit, called the *half-digit,* is always the most significant digit in the display. For example, suppose that a DMM is reading 0.999 volt, as shown in Figure 2–40(a). If the voltage increases by 0.001 V to 1 V, the display correctly shows 1.000 V, as shown in Part (b). The "1" is the half-digit. Thus, with 3½ digits, a variation of 0.001 V, which is the resolution, can be observed.

Now, suppose that the voltage increases to 1.999 V. This value is indicated on the meter as shown in Part (c) of the figure. If the voltage increases by 0.001 V to 2 V, the half-digit cannot display the "2," so the display shows 2.00. The half-digit is blanked and only three digits are active, as indicated in Part (d). With only three digits active, the resolution is 0.01 V rather than 0.001 V as it is with 3½ active digits. The resolution remains 0.01 V up to 19.99 V. The resolution goes to 0.1 V for readings of 20.0 V to 199.9 V. At 200 V, the resolution goes to 1 V, and so on.

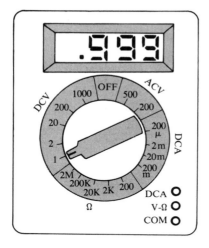

(a) Resolution: 0.001 V

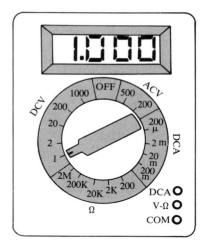

(b) Resolution: 0.001 V

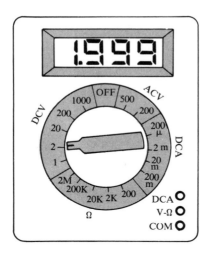

(c) Resolution: 0.001 V

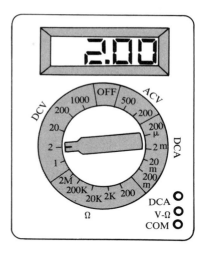

(d) Resolution: 0.01 V

FIGURE 2–40

A 3½-digit DMM illustrates how the resolution changes with the number of digits in use.

The resolution capability of a DMM is also determined by the internal circuitry and the rate at which the measured quantity is *sampled*. DMMs with displays of 4½ through 8½ digits are also available.

Accuracy The accuracy of a DMM is established strictly by its internal circuitry. For typical meters, accuracies range from 0.01% to 0.5%, with some precision laboratory-grade meters going to 0.002%.

SECTION REVIEW 2–7

1. Name the meters for measurement of (a) current, (b) voltage, and (c) resistance.

2. Place two ammeters in the circuit of Figure 2–27 to measure the current through either lamp (be sure to observe the polarities). How can the same measurements be accomplished with only one ammeter?

3. Show how to place a voltmeter to measure the voltage across lamp 2 in Figure 2–27.

4. The multimeter in Figure 2–36 is set on the 3-V range to measure dc voltage. The pointer is at 150 on the upper ac-dc scale. What voltage is being measured?

5. How do you set up the meter to measure 275 V dc, and on what scale do you read the voltage?

6. If you expect to measure a resistance in excess of 20 kΩ, where do you set the switch?

7. List two common types of DMM displays, and discuss the advantages and disadvantages of each.

8. Define *resolution* in a DMM.

SUMMARY

1. An atom is the smallest particle of an element that retains the characteristics of that element.

2. The electron is the basic particle of negative electrical charge.

3. The proton is the basic particle of positive charge.

4. An ion is an atom that has gained or lost an electron and is no longer neutral.

5. When electrons in the outer orbit of an atom (valence electrons) break away, they become free electrons.

6. Free electrons make current possible.

7. Like charges repel each other, and opposite charges attract each other.

8. Voltage must be applied to a circuit to produce current.

9. Resistance limits the current.

10. Basically, an electric circuit consists of a source, a load, and a current path.

11. An open circuit is one in which the current path is broken.

12. A closed circuit is one which has a complete current path.

13. An ammeter is connected in line with the current path.

14. A voltmeter is connected across the current path.

15. An ohmmeter is connected across a resistor (resistor must be removed from circuit).

16. *One coulomb*—the charge of 6.25×10^{18} electrons.

17. *Voltage*—the amount of energy available to move a charge from one point to another.
18. *One volt*—the potential difference (voltage) between two points when one joule of energy is used to move one coulomb from one point to the other.
19. *Current*—the rate of flow of electrons.
20. *One ampere*—the amount of current when one coulomb of charge passes a given point in one second.
21. *Resistance*—the opposition to current.
22. *One ohm*—the resistance when there is one ampere of current in a material with one volt applied across the material.

FORMULAS

$$V = \frac{W}{Q} \qquad (2\text{–}1)$$

$$I = \frac{Q}{t} \qquad (2\text{–}2)$$

$$G = \frac{1}{R} \qquad (2\text{–}3)$$

SELF-TEST

Solutions appear at the end of the book.

1. How many electrons are in a neutral atom with an atomic number of three?
2. In atomic theory, what are *shells?*
3. Discuss how conductors, semiconductors, and insulators basically differ.
4. What is static electricity?
5. When a positively charged material is placed near a negatively charged material, what will happen? Why?
6. What is the charge on a single electron?
7. Describe how a positive ion is created.
8. What is the more common term for *potential difference?* What is its unit?
9. What is the unit of energy?
10. Batteries, power supplies, solar cells, and generators are types of _____.

11. Can voltage exist between two points when there is no current?

12. Explain how current is produced in a material when voltage is applied across the material.

13. Define *one ampere* of current.

14. Name two applications of resistors.

15. Name two types of variable resistors.

16. How do linear and tapered potentiometers differ?

17. A lamp and switch are connected to a voltage source to form a simple circuit. The switch is in a position such that the lamp is off. Is the circuit open or closed?

18. Is a voltage measurement or a current measurement more difficult to make? Why?

19. Fifty coulombs of charge flow past a point in a circuit in 5 seconds. What is the current?

20. If 2 A of current exist in a circuit, how many coulombs pass a given point in 10 seconds?

21. Five hundred joules of energy are used to move 100 C of charge through a resistor. What is the voltage across the resistor?

22. What is the conductance of a 10-Ω resistor?

23. Figure 2–41 shows a color-coded resistor. What are the resistance and the tolerance values?

FIGURE 2–41

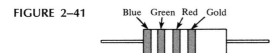

24. The adjustable contact of a linear potentiometer is set at the center of its adjustment. If the total resistance is 1000 Ω, what is the resistance between the end terminal and the contact?

PROBLEMS

Section 2–2

2–1 How many coulombs of charge do 50×10^{31} electrons possess?

2–2 How many electrons does it take to make 80 μC (microcoulombs) of charge?

Section 2–3

2–3 Determine the voltage in each of the following cases:
   (a) 10 J/C   **(b)** 5 J/2 C   **(c)** 100 J/25 C

2–4 Five hundred joules of energy are used to move 100 C of charge through a resistor. What is the voltage across the resistor?

2–5 What is the voltage of a battery that uses 800 J of energy to move 40 C of charge through a resistor?

2–6 How much energy does a 12-V battery use to move 2.5 C through a circuit?

2–7 If a resistor with a current of 2 A through it converts 1000 joules of electrical energy into heat energy in 15 s, what is the voltage across the resistor?

Section 2–4

2–8 Determine the current in each of the following cases:
(a) 75 C in 1 s   **(b)** 10 C in 0.5 s   **(c)** 5 C in 2 s

2–9 Six-tenths coulomb passes a point in 3 s. What is the current in amperes?

2–10 How long does it take 10 C to flow past a point if the current is 5 A?

2–11 How many coulombs pass a point in 0.1 s when the current is 1.5 A?

2–12 574×10^{15} electrons flow through a wire in 250 ms. What is the current in amperes?

Section 2–5

2–13 Determine the resistance value and tolerance for each of the following color codes. The first band is listed first.
(a) Blue, gray, orange, gold
(b) Yellow, violet, red, silver
(c) Orange, white, brown, gold

2–14 Find the minimum and the maximum resistance within the tolerance limits for each resistor in Problem 2–13.

2–15 If you need a 270-Ω resistor, what color bands should you look for?

2–16 The adjustable contact of a linear potentiometer is set at the mechanical center of its adjustment. If the total resistance is 1000 Ω, what is the resistance between each end terminal and the adjustable contact?

2–17 Find the conductance for each of the following resistance values:
(a) 5 Ω **(b)** 25 Ω **(c)** 100 Ω

2–18 Find the resistance corresponding to the following conductances:
(a) 0.1 S   **(b)** 0.5 S   **(c)** 0.02 S

2–19 A 120-V source is to be connected to a 1500-Ω resistive load by two lengths of wire as shown in Figure 2–42. The voltage source is to be located 50 ft from the load. Using the wire table in Appendix A, determine the gage number of the

FIGURE 2–42

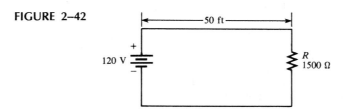

smallest wire that can be used if the *total* resistance of the two lengths of wire is not to exceed 6 Ω.

2–20 Determine the resistance and tolerance of each resistor labeled as follows:

    (a) 4R7J   **(b)** 5602M   **(c)** 1501F

Section 2–6

2–21 Trace the current path in Figure 2–43(a) with the switch in position 2.

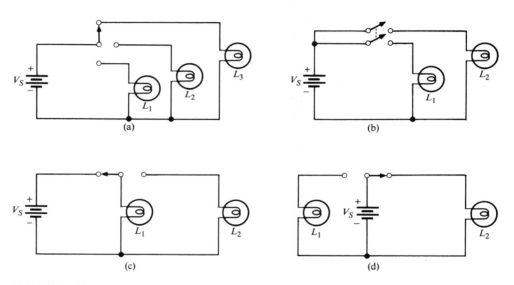

FIGURE 2–43

2–22 With the switch in either position, redraw the circuit in Figure 2–43(d) with a fuse connected to protect the circuit against excessive current.

2–23 There is only one circuit in Figure 2–43 in which it is possible to have all lamps on at the same time. Determine which circuit it is.

2–24 Through which resistor in Figure 2–44 is there always current, regardless of the position of the switches?

FIGURE 2–44

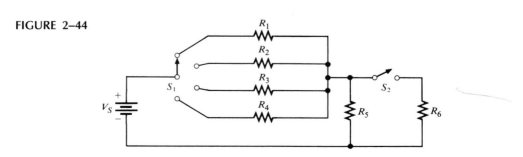

2–25 Devise a switch arrangement whereby two voltage sources (V_{S1} and V_{S2}) can be connected simultaneously to either of two resistors (R_1 and R_2) as follows:

V_{S1} connected to R_1 and V_{S2} connected to R_2 or

V_{S1} connected to R_2 and V_{S2} connected to R_1

2–26 The different sections of a stereo system are represented by the blocks in Figure 2–45. Show how a single switch can be used to connect the phonograph, the CD (compact disc) player, the tape deck, the AM tuner, or the FM tuner to the amplifier by a single knob control. Only one section can be connected to the amplifier at any time.

FIGURE 2–45

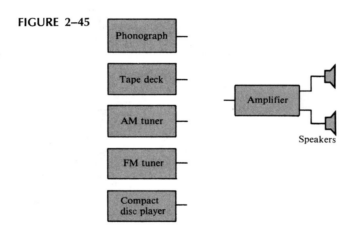

Section 2–7

2–27 Show the placement of an ammeter and a voltmeter to measure the current and the source voltage in Figure 2–46.

2–28 Show how you would measure the resistance of R_2 in Figure 2–46.

2–29 In Figure 2–47, how much voltage does each meter indicate when the switch is in position 1? In position 2?

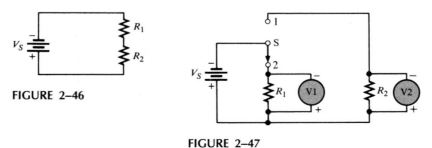

FIGURE 2–46

FIGURE 2–47

2–30 In Figure 2–47, indicate how to connect an ammeter to measure the current from the voltage source regardless of the switch position.

2–31 In Figure 2–44, show the proper placement of ammeters to measure the current through each resistor and the current out of the battery.

2–32 Show the proper placement of voltmeters to measure the voltage across each resistor in Figure 2–44.

2–33 What are the voltage readings in Figures 2–48(a) and 2–48(b)?

FIGURE 2–48

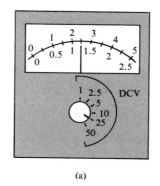   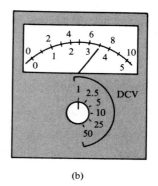

(a)   (b)

2–34 How much resistance is the ohmmeter in Figure 2–49 measuring?

FIGURE 2–49

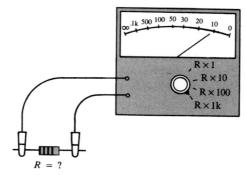

2–35 Determine the resistance indicated by each of the following ohmmeter readings and range settings:
   (a) pointer at 2, range setting at $R \times 100$.
   (b) pointer at 15, range setting at $R \times 10M$.
   (c) pointer at 45, range setting at $R \times 100$.

2–36 What is the maximum resolution of a 4½-digit DMM?

ANSWERS TO SECTION REVIEWS

Section 2–1
1. Electron.
2. The smallest particle of an element that retains the unique characteristics of the element.
3. A positively charged nucleus surrounded by orbiting electrons.   **4.** No.
5. An outer-shell electron that has drifted away from the parent atom.

Section 2–2
1. Q.   **2.** Coulomb, C.   **3.** 1.6×10^{-6} C.

Section 2–3
1. Energy per unit charge.   **2.** 2.4 V.
3. Battery, power supply, solar cell, and generator.

Section 2–4
1. Rate of flow of electrons, ampere (A).   **2.** 6.25×10^{18} **3.** 5 A.

Section 2–5
1. Opposition to current, Ω.
2. Fixed and variable. The value of a fixed resistor cannot be changed, but that of a variable resistor can.
3. *First band:* first digit of resistance value. *Second band:* second digit of resistance value. *Third band:* number of zeros. *Fourth band:* tolerance.
4. (a) 4700 Ω $\pm 5\%$;   **(b)** 62,000 Ω $\pm 10\%$;   **(c)** 18 Ω $\pm 5\%$.
5. A rheostat has two terminals; a potentiometer has three terminals.
6. A thermistor is a temperature-sensitive resistor with a negative coefficient.

Section 2–6
1. Source, load, and current path between source and load.
2. A circuit that has no path for current.
3. A circuit that has a complete path for current.   **4.** Infinite.

Section 2–7
1. (a) Ammeter;   **(b)** voltmeter;   **(c)** ohmmeter.
2. See Figure 2–50.   **3.** See Figure 2–51.

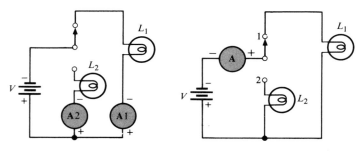

FIGURE 2–50

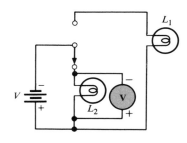

FIGURE 2–51

4. 1.5 V.   **5.** Set the range switch to 300 and read on the upper ac-dc scale.

6. × 1000.

7. The LCD requires little current, but it is difficult to see in low light and is slow to respond.

The LED can be seen in the dark, and it responds quickly. However, it requires much more current than does the LCD.

8. The smallest increment of a quantity that the meter can measure.

THREE

OHM'S LAW

Georg Simon Ohm found that current, voltage, and resistance are related in a specific way. This basic relationship, known as Ohm's law, is one of the most important laws in electrical and electronics work. In this chapter, you will learn Ohm's law and how to use it in solving circuit problems. Also, a computer program is introduced that can be used for solving problems related to Ohm's law.

In this chapter, you will learn:

☐   The definition of Ohm's law.
☐   The formula for Ohm's law.
☐   How to use the Ohm's law formula to determine current, voltage, or resistance.
☐   How to deal with various units of current, voltage, and resistance.
☐   The meaning of the linear relationship between current and voltage.
☐   How to create a flowchart for a problem and write a computer program in BASIC.

3-1   DEFINITION OF OHM'S LAW

Ohm's law describes mathematically how current, voltage, and resistance are related. Georg Ohm determined experimentally that if the voltage across a resistor is increased, the current will also increase, and, likewise, if the voltage is decreased, the current will decrease. For example, if the voltage is doubled, the current will double. If the voltage is halved, the current will also be halved. This relationship is illustrated in Figure 3–1, with meter indications of voltage and current.

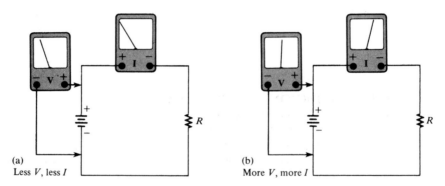

(a)
Less V, less I

(b)
More V, more I

FIGURE 3–1
Effect of changing the voltage with the same resistance in both circuits.

Ohm's law also states that if the voltage is kept constant, *less* resistance results in *more* current, and, also, *more* resistance results in *less* current. For example, if the resistance is halved, the current doubles. If the resistance is doubled, the current is halved. This concept is illustrated by the meter indications in Figure 3–2, where the resistance is varied and the voltage is constant.

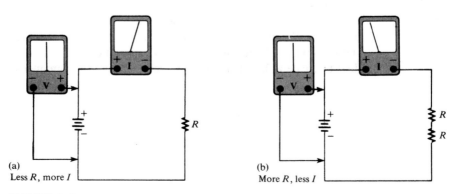

(a)
Less R, more I

(b)
More R, less I

FIGURE 3–2
Effect of changing the resistance with the same voltage in both circuits.

FORMULA FOR CURRENT

Ohm's law can be stated as follows:

$$I = \frac{V}{R} \tag{3-1}$$

This formula describes what was indicated by the circuits of Figures 3–1 and 3–2. For a constant value of R, if the value of V is increased, the value of I increases; if V is decreased, I decreases. Also notice in Equation (3–1) that if V is constant and R is increased, I decreases. Similarly, if V is constant and R is decreased, I increases.

Using Equation (3–1), we can calculate the *current* if the values of voltage and resistance are known.

FORMULA FOR VOLTAGE

Ohm's law can also be stated another way. By multiplying both sides of Equation (3–1) by R, we obtain an *equivalent* form of Ohm's law, as follows:

$$V = IR \tag{3-2}$$

With this equation, we can calculate *voltage* if the current and resistance are known.

FORMULA FOR RESISTANCE

There is a third *equivalent* way to state Ohm's law. By dividing both sides of Equation (3–2) by I, we obtain

$$R = \frac{V}{I} \tag{3-3}$$

This form of Ohm's law is used to determine *resistance* if voltage and current values are known.

Remember, *the three formulas you have learned in this section are all equivalent.* They are simply three different ways of expressing Ohm's law.

SECTION REVIEW 3–1

1. Ohm's law defines how three basic quantities are related. What are these quantities?
2. Write the Ohm's law formula for current.
3. Write the Ohm's law formula for voltage.
4. Write the Ohm's law formula for resistance.
5. If the voltage across a fixed value resistor is tripled, does the current increase or decrease, and by how much?
6. If the voltage across a fixed resistor is cut in half, how much will the current change?

7. There is a fixed voltage across a resistor, and you measure a current of 1 A. If you replace the resistor with one that has twice the resistance value, how much current will you measure?

8. In a circuit the voltage is doubled and the resistance is cut in half. Would you observe any change in the current value?

3–2   CALCULATING CURRENT

In this section you will learn to determine current values when you know the values of voltage and resistance. In these problems, the formula $I = V/R$ is used. In order to get current in *amperes,* you must express the value of V in *volts* and the value of R in *ohms*.

EXAMPLE 3–1

How many amperes of current are in the circuit of Figure 3–3?

FIGURE 3–3

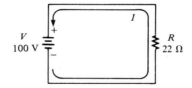

Solution:
Substitute into the formula $I = V/R$, 100 V for V, and 22 Ω for R. Divide 22 Ω into 100 V as follows:

$$I = \frac{V}{R} = \frac{100 \text{ V}}{22 \text{ Ω}} = 4.55 \text{ A}$$

There are 4.55 A of current in this circuit.

EXAMPLE 3–2

If the resistance in Figure 3–3 is changed to 47 Ω, what is the new value of current?

Solution:
We still have 100 V. Substituting 47 Ω into the formula for I gives 2.13 A as follows:

$$I = \frac{V}{R} = \frac{100 \text{ V}}{47 \text{ Ω}} = 2.13 \text{ A}$$

LARGER UNITS OF RESISTANCE

In electronics work, resistance values of thousands of ohms or even millions of ohms are common. As you learned in Chapter 1, large values of resistance are indicated by the metric system prefixes *kilo* (k) and *mega* (M). Thus, thousands of ohms are expressed in kilohms (kΩ), and millions of ohms in megohms (MΩ). The following examples illustrate use of kilohms and megohms when using Ohm's law to calculate current.

**EXAMPLE
3–3**

Calculate the current in Figure 3–4.

FIGURE 3–4

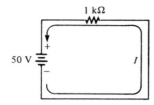

1 kΩ

50 V

I

Solution:
Remember that 1 kΩ is the same as 1×10^3 Ω. Substituting 50 V for V and 1×10^3 Ω for R gives the current in amperes as follows:

$$I = \frac{V}{R} = \frac{50 \text{ V}}{1 \times 10^3 \text{ Ω}} = 50 \times 10^{-3} \text{ A}$$

$$= 0.05 \text{ A}$$

The calculator sequence is

| 5 | 0 | ÷ | 1 | EXP | 3 | = |

In Example 3–3, 50×10^{-3} A can be expressed as 50 milliamperes (50 mA). This fact can be used to advantage when we divide *volts* by *kilohms*. The current will be in *milliamperes,* as Example 3–4 illustrates.

**EXAMPLE
3–4**

How many milliamperes of current are in the circuit of Figure 3–5?

FIGURE 3–5

I

30 V

5.6 kΩ

Solution:
When we divide *volts* by *kilohms,* we get *milliamperes.* In this case, 30 V divided by 5.6 kΩ gives 5.36 mA as follows:

$$I = \frac{V}{R} = \frac{30 \text{ V}}{5.6 \text{ k}\Omega} = 5.36 \text{ mA}$$

If *volts* are applied when resistance values are in *megohms,* the current is in *microamperes* (μA), as Example 3–5 shows. *why ?*

EXAMPLE 3–5

Determine the amount of current in the circuit of Figure 3–6.

FIGURE 3–6

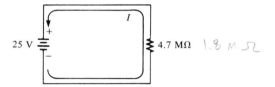

Solution:
Recall that 4.7 MΩ equals 4.7×10^6 Ω. Substituting 25 V for V and 4.7×10^6 Ω for R gives the following result:

$$I = \frac{V}{R} = \frac{25 \text{ V}}{4.7 \times 10^6 \, \Omega} = 5.32 \times 10^{-6} \text{ A}$$

Notice that 5.32×10^{-6} A equals 5.32 microamperes (5.32 μA).

EXAMPLE 3–6

Change the value of R in Figure 3–6 to 1.8 MΩ. What is the new value of current?

Solution:
When we divide *volts* by *megohms,* we get *microamperes.* In this case, 25 V divided by 1.8 MΩ gives 13.9 μA as follows:

$$I = \frac{V}{R} = \frac{25 \text{ V}}{1.8 \text{ M}\Omega} = 13.9 \text{ μA}$$

LARGER UNITS OF VOLTAGE

Small voltages, usually less than 50 volts, are common in semiconductor circuits. Occasionally, however, large voltages are encountered. For example, the high-voltage supply in a television receiver is around 20,000 volts (20 kilovolts, or 20 kV), and transmission voltages generated by the power companies may be as high as 345,000 V (345 kV). The following two examples illustrate use of voltage values in the kilovolt range when calculating current.

EXAMPLE 3–7

How much current is produced by a voltage of 24 kV across a 12-kΩ resistance?

Solution:

Since we are dividing kV by kΩ, the prefixes cancel, and we get amperes:

$$I = \frac{V}{R} = \frac{24 \text{ kV}}{12 \text{ k}\Omega}$$

$$= \frac{24 \times 10^3 \text{ V}}{12 \times 10^3 \text{ k}\Omega} = 2 \text{ A}$$

EXAMPLE 3–8

How much current is there through 100 MΩ when 50 kV are applied?

Solution:

In this case, we divide 50 kV by 100 MΩ to get the current. Using 50×10^3 V for 50 kV and 100×10^6 Ω for 100 MΩ, we obtain the current as follows:

$$I = \frac{V}{R} = \frac{50 \text{ kV}}{100 \text{ M}\Omega} = \frac{50 \times 10^3 \text{ V}}{100 \times 10^6 \text{ }\Omega}$$

$$= 0.5 \times 10^{-3} \text{ A} = 0.5 \text{ mA}$$

Remember that the power of ten in the denominator is subtracted from the power of ten in the numerator. So 50 was divided by 100, giving 0.5, and 6 was subtracted from 3, giving 10^{-3}.

The calculator sequence is

[5] [0] [EXP] [3] [÷] [1] [0] [0] [EXP] [6] [=]

SECTION REVIEW 3–2

In Problems 1–4, calculate I when

1. $V = 10$ V and $R = 5.6$ Ω.
2. $V = 100$ V and $R = 560$ Ω.
3. $V = 5$ V and $R = 2.2$ kΩ.
4. $V = 15$ V and $R = 4.7$ MΩ.
5. If a 4.7-MΩ resistor has 20 kV across it, how much current is produced?
6. How much current will 10 kV across 2.2 kΩ produce?

3–3

CALCULATING VOLTAGE

In this section you will learn to determine voltage values when the current and resistance are known. In these problems, the formula $V = IR$ is used. To obtain voltage in *volts*, you must express the value of I in *amperes* and the value of R in *ohms*.

EXAMPLE 3–9

In the circuit of Figure 3–7, how much voltage is needed to produce 5 A of current?

FIGURE 3–7

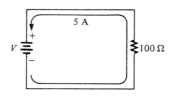

Solution:
Substitute 5 A for I and 100 Ω for R into the formula $V = IR$ as follows:

$$V = IR = (5 \text{ A})(100 \text{ } \Omega) = 500 \text{ V}$$

Thus, 500 V are required to produce 5 A of current through a 100-Ω resistor.

SMALLER UNITS OF CURRENT

The following two examples illustrate use of current values in the milliampere (mA) and microampere (μA) ranges when calculating voltage.

EXAMPLE 3–10

How much voltage will be measured across the resistor in Figure 3–8?

FIGURE 3–8

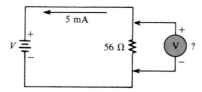

Solution:
Note that 5 mA equals 5×10^{-3} A. Substituting the values for I and R into the formula $V = IR$, we get the following result:

$$V = IR = (5 \text{ mA})(56 \text{ } \Omega)$$
$$= (5 \times 10^{-3} \text{ A})(56 \text{ } \Omega) = 280 \times 10^{-3} \text{ V}$$

Since 280×10^{-3} V equals 280 mV, when *milliamperes* are multiplied by *ohms*, we get *millivolts*.

EXAMPLE 3–11

Suppose that there are 8 μA through a 10-Ω resistor. How much voltage is across the resistor?

Solution:
Note that 8 μA equals 8×10^{-6} A. Substituting the values for I and R into the formula $V = IR$, we get the voltage as follows:

$$V = IR = (8 \ \mu A)(10 \ \Omega)$$
$$= (8 \times 10^{-6} \ A)(10 \ \Omega) = 80 \times 10^{-6} \ V$$

Since 80×10^{-6} V equals 80 μV, when *microamperes* are multiplied by *ohms*, we get *microvolts*.

These examples have demonstrated that when we multiply *milliamperes* and *ohms*, we get *millivolts*. When we multiply *microamperes* and *ohms*, we get *microvolts*.

LARGER UNITS OF RESISTANCE

The following two examples illustrate use of resistance values in the kilohm (kΩ) and megohm (MΩ) ranges when calculating voltage.

EXAMPLE 3–12

The circuit in Figure 3–9 has a current of 10 mA. What is the voltage?

FIGURE 3–9

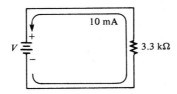

Solution:
Note that 10 mA equals 10×10^{-3} A and that 3.3 kΩ equals 3.3×10^{3} Ω. Substituting these values into the formula $V = IR$, we get

$$V = IR = (10 \ mA)(3.3 \ k\Omega)$$
$$= (10 \times 10^{-3} \ A)(3.3 \times 10^{3} \ \Omega) = 33 \ V$$

Since 10^{-3} and 10^{3} cancel, *milliamperes* cancel *kilohms* when multiplied, and the result is *volts*.

The calculator sequence is

$\boxed{1}\ \boxed{0}\ \boxed{\text{EXP}}\ \boxed{+/-}\ \boxed{3}\ \boxed{\times}\ \boxed{3}\ \boxed{\cdot}\ \boxed{3}\ \boxed{\text{EXP}}\ \boxed{3}\ \boxed{=}$

**EXAMPLE
3–13**

If there are 50 μA through a 4.7-MΩ resistor, what is the voltage?

Solution:
Note that 50 μA equals 50×10^{-6} A and that 4.7 MΩ is 4.7×10^6 Ω. Substituting these values into $V = IR$, we get

$$V = IR = (50\ \mu\text{A})(4.7\ \text{M}\Omega)$$
$$= (50 \times 10^{-6}\ \text{A})(4.7 \times 10^6\ \Omega) = 235\ \text{V}$$

Since 10^{-6} and 10^6 cancel, *microamperes* cancel *megohms* when multiplied, and the result is *volts*.

SECTION REVIEW 3–3

In Problems 1–7, calculate V when

1. $I = 1$ A and $R = 10\ \Omega$.
2. $I = 8$ A and $R = 470\ \Omega$.
3. $I = 3$ mA and $R = 100\ \Omega$.
4. $I = 25\ \mu$A and $R = 56\ \Omega$.
5. $I = 2$ mA and $R = 1.8$ kΩ.
6. $I = 5$ mA and $R = 100$ MΩ.
7. $I = 10\ \mu$A and $R = 2.2$ MΩ.
8. How much voltage is required to produce 100 mA through 4.7 kΩ?
9. What voltage do you need to cause 3 mA of current in a 3.3-kΩ resistance?
10. A battery produces 2 A of current into a 6.8-Ω resistive load. What is the battery voltage?

3–4   CALCULATING RESISTANCE

In this section you will learn to determine resistance values when the current and voltage are known. In these problems, the formula $R = V/I$ is used. To get resistance in *ohms*, you must express the value of I in *amperes* and the value of V in *volts*.

**EXAMPLE
3–14**

In the circuit of Figure 3–10, how much resistance is needed to draw 3.077 A of current from the battery?

FIGURE 3–10

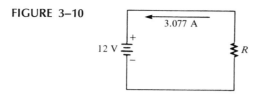

Solution:
Substitute 12 V for V and 3.077 A for I into the formula $R = V/I$:

$$R = \frac{V}{I} = \frac{12 \text{ V}}{3.077 \text{ A}} = 3.9 \ \Omega$$

SMALLER UNITS OF CURRENT

The following two examples illustrate use of current values in the milliampere (mA) and microampere (μA) ranges when calculating resistance.

**EXAMPLE
3–15**

Suppose that the ammeter in Figure 3–11 indicates 4.55 mA of current and the volt-meter reads 150 V. What is the value of R?

FIGURE 3–11

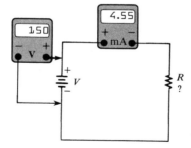

Solution:
Note that 4.55 mA equals 4.55×10^{-3} A. Substituting the voltage and current values into the formula $R = V/I$, we get

$$R = \frac{V}{I} = \frac{150 \text{ V}}{4.55 \text{ mA}} = \frac{150 \text{ V}}{4.55 \times 10^{-3} \text{ A}}$$

$$= 33 \times 10^3 \ \Omega = 33 \text{ k}\Omega$$

Thus, if *volts* are divided by *milliamperes*, the resistance will be in *kilohms*.

**EXAMPLE
3–16**

Suppose that the value of the resistor in Figure 3–11 is changed. If the battery voltage is still 150 V and the ammeter reads 68.2 μA, what is the new resistor value?

Solution:
Note that 68.2 μA equals 68.2 $\times$ 10^{-6} A. Substituting V and I values into the equation for R, we get

$$R = \frac{V}{I} = \frac{150 \text{ V}}{68.2 \ \mu\text{A}} = \frac{150 \text{ V}}{68.2 \times 10^{-6} \text{ A}}$$
$$= 2.2 \times 10^6 \ \Omega = 2.2 \text{ M}\Omega$$

Thus, if *volts* are divided by *microamperes,* the resistance has units of *megohms.*

SECTION REVIEW 3–4

In Problems 1–5, calculate R when
1. $V = 10$ V and $I = 2.13$ A.
2. $V = 270$ V and $I = 10$ A.
3. $V = 20$ kV and $I = 5.13$ A.
4. $V = 15$ V and $I = 2.68$ mA.
5. $V = 5$ V and $I = 2.27 \ \mu$A.
6. You have a resistor across which you measure 25 V, and your ammeter indicates 53.2 mA of current. What is the resistor's value in kilohms? In ohms?

3–5

THE RELATIONSHIP OF CURRENT, VOLTAGE, AND RESISTANCE

THE LINEAR RELATIONSHIP OF CURRENT AND VOLTAGE

Ohm's law brings out a very important relationship between current and voltage: They are *linearly proportional.* You may have already recognized this relationship from our discussions in the previous sections.

When we say that the current and voltage are linearly proportional, we mean that if one is increased or decreased by a certain percentage, the other will increase or decrease by the same percentage, assuming that the resistance is constant in value. For example, if the voltage across a resistor is tripled, the current will triple.

**EXAMPLE
3–17**

Show that if the voltage in the circuit of Figure 3–12 is increased to three times its present value, the current will triple in value.

FIGURE 3–12

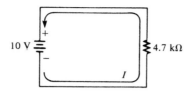

Solution:
With 10 V, the current is

$$I = \frac{V}{R} = \frac{10 \text{ V}}{4.7 \text{ k}\Omega} = 2.127 \text{ mA}$$

If the voltage is increased to 30 V, the current will be

$$I = \frac{V}{R} = \frac{30 \text{ V}}{4.7 \text{ k}\Omega} = 6.383 \text{ mA}$$

The current went from 2.127 mA to 6.383 mA (tripled) when the voltage was tripled to 30 V.

A GRAPH OF CURRENT VERSUS VOLTAGE

Let us take a constant value of resistance, for example, 10 Ω, and calculate the current for several values of voltage ranging from 10 V to 100 V. The current values obtained are shown in Figure 3–13(a). The graph of the *I* values versus the *V* values is shown in

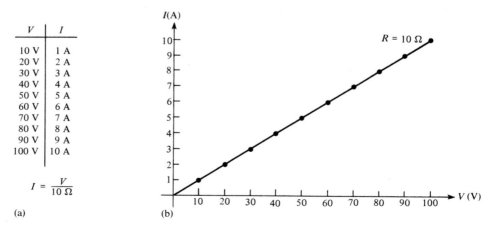

V	I
10 V	1 A
20 V	2 A
30 V	3 A
40 V	4 A
50 V	5 A
60 V	6 A
70 V	7 A
80 V	8 A
90 V	9 A
100 V	10 A

$$I = \frac{V}{10 \text{ }\Omega}$$

(a)

(b)

FIGURE 3–13
Graph of current versus voltage for $R = 10 \text{ }\Omega$.

Figure 3–13(b). Note that it is a straight line graph. This graph tells us that a change in voltage results in a linearly proportional change in current.

No matter what value R is, assuming that R is constant, the graph of I versus V will always be a straight line. Example 3–18 illustrates a use for the linear relationship between voltage and current in a resistive circuit.

EXAMPLE 3–18

Assume that you are measuring the current in a circuit that is operating with 25 V. The ammeter reads 50 mA. Later, you notice that the current has dropped to 40 mA. Assuming that the resistance did not change, you must conclude that the voltage has changed. How much has the voltage changed, and what is its new value?

Solution:
The current has dropped from 50 mA to 40 mA, which is a decrease of 20%. Since the voltage is linearly proportional to the current, the voltage has decreased by the same percentage that the current did. Taking 20% of 25 V, we get

$$\text{Change in voltage} = (0.2)(25 \text{ V}) = 5 \text{ V}$$

Subtracting this change from the original voltage, we get the new voltage as follows:

$$\text{New voltage} = 25 \text{ V} - 5 \text{ V} = 20 \text{ V}$$

Notice that we did not need the resistance value in order to find the new voltage.

CURRENT AND RESISTANCE ARE INVERSELY RELATED

As you have seen, current varies *inversely* with resistance as expressed by Ohm's law, $I = V/R$. When the resistance is reduced, the current goes up; when the resistance is increased, the current goes down. For example, if the source voltage is held constant and the resistance is halved, the current doubles in value; when the resistance is doubled, the current is reduced by half.

Let us take a constant value of voltage, for example, 10 V, and calculate the current for several values of resistance ranging from 10 Ω to 100 Ω. The values obtained are shown in Figure 3–14(a). The graph of the I values versus the R values is shown in Figure 3–14(b).

SECTION REVIEW 3–5

1. What does *linearly proportional* mean?
2. In a circuit, $V = 2$ V and $I = 10$ mA. If V is changed to 1 V, what will I equal?
3. If $I = 3$ A at a certain voltage, what will it be if the voltage is doubled?
4. By how many volts must you increase a 12-V source in order to increase the current in a circuit by 50%?

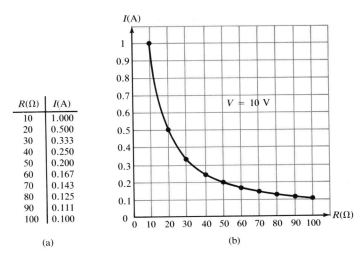

$R(\Omega)$	$I(A)$
10	1.000
20	0.500
30	0.333
40	0.250
50	0.200
60	0.167
70	0.143
80	0.125
90	0.111
100	0.100

(a)

(b)

FIGURE 3–14
Graph of current versus resistance for $V = 10$ V.

3–6

COMPUTER ANALYSIS

As mentioned in Chapter 1, several examples of computer-aided analysis are presented throughout the book. These examples illustrate how computer programs can be written in BASIC to solve typical circuit problems. The program in this chapter is the first of these examples.

This program is based on Ohm's law and computes the unknown value when two known values are provided. For example, if voltage and current values are provided, the resistance is calculated. If voltage and resistance values are given, the current is calculated. If current and resistance values are given, the voltage is calculated. The generalized circuit for which solutions are computed is shown in Figure 3–15(a).

The flowchart in Figure 3–15(b) shows the sequence of operations that the computer follows when the program is executed.

The listing of the program itself appears below. Note that it consists of 34 lines. Lines 10 through 80 display the program options and allow the selection of a desired option. Line 100 is a branching statement that causes the computer to go to the segment of the program required for the selected option. Lines 110 through 180 are for option 1; lines 190 through 260 are for option 2; and lines 270 through 340 are for option 3.

The program must be entered into the computer via the keyboard exactly as it appears and then executed.

```
10   CLS
20   PRINT "THIS PROGRAM PROVIDES THREE OPTIONS:"
30   PRINT
```

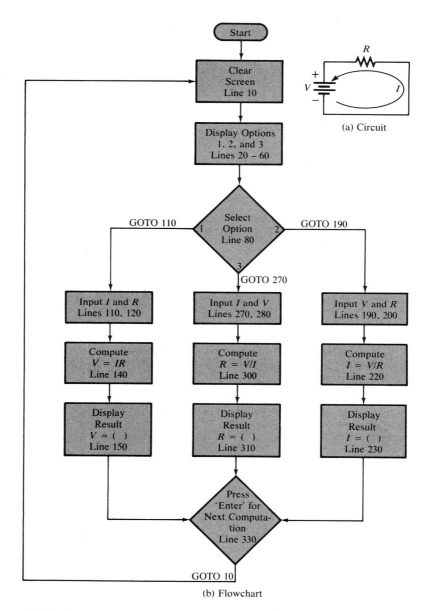

FIGURE 3–15
Program flowchart for the computation of *V*, *R*, and *I* in a simple circuit, as shown in Part (a).

```
40   PRINT "(1) COMPUTE VOLTAGE IF CURRENT AND RESISTANCE ARE
     KNOWN"
50   PRINT "(2) COMPUTE CURRENT IF VOLTAGE AND RESISTANCE ARE
     KNOWN"
60   PRINT "(3) COMPUTE RESISTANCE IF VOLTAGE AND CURRENT ARE
     KNOWN"
70   PRINT:PRINT
80   INPUT "SELECT OPTION 1, 2, OR 3";X
90   CLS
100  ON X GOTO 110, 190, 270
110  INPUT "CURRENT IN AMPS";I
120  INPUT "RESISTANCE IN OHMS";R
130  CLS
140  V=I*R
150  PRINT "V=";V;"VOLTS"
160  PRINT:PRINT:PRINT
170  INPUT "FOR NEXT COMPUTATION PRESS 'ENTER'";Y
180  GOTO 10
190  INPUT "VOLTAGE IN VOLTS";V
200  INPUT "RESISTANCE IN OHMS";R
210  CLS
220  I=V/R
230  PRINT "I=";I;"AMPS"
240  PRINT:PRINT:PRINT
250  INPUT "FOR NEXT COMPUTATION PRESS 'ENTER'";Y
260  GOTO 10
270  INPUT "VOLTAGE IN VOLTS";V
280  INPUT "CURRENT IN AMPS";I
290  CLS
300  R=V/I
310  PRINT "R=";R;"OHMS"
320  PRINT:PRINT:PRINT
330  INPUT "FOR NEXT COMPUTATION PRESS 'ENTER'";Y
340  GOTO 10
```

SECTION REVIEW 3–6

1. What function do the INPUT statements perform?

2. What is the purpose of line 100?

SUMMARY

1. There are three forms of Ohm's law, all of which are equivalent.

2. Use $I = V/R$ when calculating the current.

3. Use $V = IR$ when calculating the voltage.

4. Use $R = V/I$ when calculating the resistance.

5. Voltage and current are linearly proportional.

6. Current and resistance are inversely related.

FORMULAS

$$I = \frac{V}{R} \qquad\qquad \textbf{(3–1)}$$

$$V = IR \qquad\qquad \textbf{(3–2)}$$

$$R = \frac{V}{I} \qquad\qquad \textbf{(3–3)}$$

SELF-TEST

Solutions appear at the end of the book.

1. Explain Ohm's law.
2. In a circuit consisting of a voltage source and a resistor, describe what happens to the current when
   (a) the voltage is tripled.
   (b) the voltage is reduced by 75%.
   (c) the resistance is doubled.
   (d) the resistance is reduced by 35%.
   (e) the voltage is doubled and the resistance is cut in half.
   (f) the voltage is doubled and the resistance is doubled.
3. State the formula used to find I when the values of V and R are known.
4. State the formula used to find V when the values of I and R are known.
5. State the formula used to find R when the values of V and I are known.
6. (a) Volts divided by kΩ gives _____.
   (b) Volts divided by MΩ gives _____.
   (c) Milliamperes times kΩ gives _____.
   (d) Milliamperes times MΩ gives _____.
   (e) Volts divided by mA gives _____.
   (f) Volts divided by μA gives _____.
7. Calculate I for $V = 10$ V and $R = 5.6$ Ω.
8. Calculate I for $V = 75$ V and $R = 10$ kΩ.
9. Calculate V for $I = 2.5$ A and $R = 22$ Ω.
10. Calculate V for $I = 30$ mA and $R = 2.2$ kΩ.
11. Calculate R for $V = 12$ V and $I = 4.44$ A.
12. Calculate R for $V = 9$ V and $I = 109.8$ μA.
13. Determine the current if $V = 50$ kV and $R = 2.7$ kΩ.
14. Determine the current if $V = 15$ mV and $R = 10$ kΩ.

15. Determine the voltage if $I = 4\ \mu A$ and $R = 100\ k\Omega$.

16. Determine the voltage if $I = 100\ mA$ and $R = 1\ M\Omega$.

17. Determine the resistance if $V = 5\ V$ and $I = 1.85\ mA$.

18. If a 1.5-V battery is connected across a 4.7-kΩ resistor, what is the current?

PROBLEMS

Section 3–2

3–1 Determine the current in each case:
(a) $V = 5\ V, R = 1\ \Omega$
(b) $V = 15\ V, R = 10\ \Omega$
(c) $V = 50\ V, R = 100\ \Omega$
(d) $V = 30\ V, R = 15\ k\Omega$
(e) $V = 250\ V, R = 5.6\ M\Omega$

3–2 Determine the current in each case:
(a) $V = 9\ V, R = 2.7\ k\Omega$
(b) $V = 5.5\ V, R = 10\ k\Omega$
(c) $V = 40\ V, R = 68\ k\Omega$
(d) $V = 1\ kV, R = 2.2\ k\Omega$
(e) $V = 66\ kV, R = 10\ M\Omega$

3–3 A 10-Ω resistor is connected across a 12-V battery. What is the current through the resistor?

3–4 A certain resistor has the following color code: orange, orange, red, gold. Determine the maximum and minimum currents you should expect to measure when a 12-V source is connected across the resistor.

3–5 A resistor is connected across the terminals of a dc voltage source that can be varied from 1 V to 10 V. Determine the minimum and maximum current for each of the following resistor values.
(a) Red, red, red
(b) Green, blue, brown
(c) White, red, orange

3–6 The rheostat in Figure 3–16 is used to control the current to a heating element. When the rheostat is adjusted to a value of 8 Ω or less, the heating element can burn out. What is the rated value of the fuse needed to protect the circuit if the voltage across the heating element at the point of maximum current is 100 V?

FIGURE 3–16

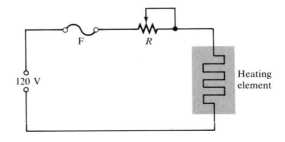

Section 3–3

3–7 Calculate the voltage for each value of I and R:

    (a) $I = 2$ A, $R = 18$ Ω   **(b)** $I = 5$ A, $R = 56$ Ω

    (c) $I = 2.5$ A, $R = 680$ Ω   **(d)** $I = 0.6$ A, $R = 47$ Ω

    (e) $I = 0.1$ A, $R = 560$ Ω

3–8 Calculate the voltage for each value of I and R:

    (a) $I = 1$ mA, $R = 10$ Ω   **(b)** $I = 50$ mA, $R = 33$ Ω

    (c) $I = 3$ A, $R = 5.6$ kΩ   **(d)** $I = 1.6$ mA, $R = 2.2$ kΩ

    (e) $I = 250$ μA, $R = 1$ kΩ   **(f)** $I = 500$ mA, $R = 1.5$ MΩ

    (g) $I = 850$ μA, $R = 10$ MΩ   **(h)** $I = 75$ μA, $R = 47$ Ω

3–9 Three amperes of current are measured through a 27-Ω resistor connected across a voltage source. How much voltage does the source produce?

3–10 Assign a voltage value to each source in the circuits of Figure 3–17 to obtain the indicated amounts of current.

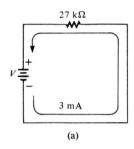 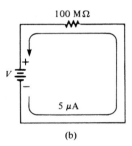 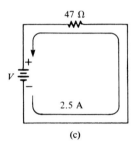

      (a)   (b)   (c)

FIGURE 3–17

3–11 A 6-V source is connected to a 100-Ω resistor by two 12-ft lengths of 18-gage copper wire. Determine the following:

    (a) Current

    (b) Resistor voltage drop

    (c) Voltage drop across each length of wire

Section 3–4

3–12 Calculate the resistance of a rheostat for each value of V and I:

    (a) $V = 10$ V, $I = 2$ A   **(b)** $V = 90$ V, $I = 45$ A

    (c) $V = 50$ V, $I = 5$ A   **(d)** $V = 5.5$ V, $I = 10$ A

    (e) $V = 150$ V, $I = 0.5$ A

3–13 Calculate the resistance of a rheostat for each set of V and I values:

    (a) $V = 10$ kV, $I = 5$ A   **(b)** $V = 7$ V, $I = 2$ mA

    (c) $V = 500$ V, $I = 250$ mA   **(d)** $V = 50$ V, $I = 500$ μA

    (e) $V = 1$ kV, $I = 1$ mA

3–14 Six volts are applied across a resistor. A current of 2 mA is measured. What is the value of the resistor?

3–15 The filament of a light bulb in the circuit of Figure 3–18(a) has a certain amount of resistance, represented by an equivalent resistance in Figure 3–18(b). If the bulb operates with 120 V and 0.8 A of current, what is the resistance of its filament?

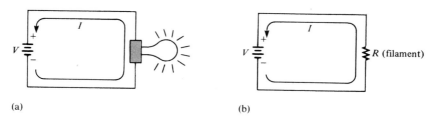

(a)   (b)

FIGURE 3–18

3–16 A certain electrical device has an unknown resistance. You have available a 12-V battery and an ammeter. How would you determine the value of the unknown resistance? Draw the necessary circuit connections.

3–17 By varying the rheostat (variable resistor) in the circuit of Figure 3–19, you can change the amount of current. The setting of the rheostat is such that the current is 750 mA. What is the resistance value of this setting? To adjust the current to 1 A, to what resistance value must you set the rheostat?

FIGURE 3–19

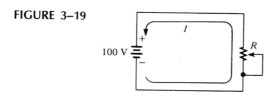

3–18 A 120-V lamp-dimming circuit is controlled by a rheostat and protected from excessive current by a 2-A fuse. To what minimum resistance value can the rheostat be set without blowing the fuse? Assume a lamp resistance of 15 Ω.

Section 3–5

3–19 A variable voltage source is connected to the circuit of Figure 3–20. Start at 0 V and increase the voltage in 10-V steps up to 100 V. Determine the current at each voltage point, and plot a graph of V versus I. Is the graph a straight line? What does the graph indicate?

FIGURE 3–20

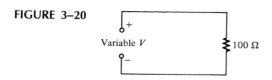

3–20 In a certain circuit, $V = 1$ V and $I = 5$ mA. Determine the current for each of the following voltages in the same circuit:
  (a) $V = 1.5$ V   **(b)** $V = 2$ V   **(c)** $V = 3$ V
  (d) $V = 4$ V   **(e)** $V = 10$ V

3–21 Figure 3–21 is a graph of voltage versus current foi three resistance values. Determine R_1, R_2, and R_3.

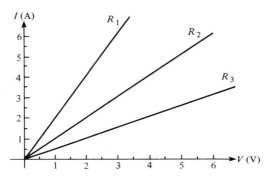

FIGURE 3–21

3–22 Which circuit in Figure 3–22 has the most current? The least current?

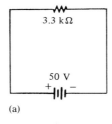

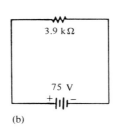

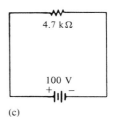

(a)   (b)   (c)

FIGURE 3–22

3–23 You are measuring the current in a circuit that is operated on a 10-V battery. The ammeter reads 50 mA. Later, you notice that the current has dropped to 30 mA. Eliminating the possibility of a resistance change, you must conclude that the voltage has changed. How much has the voltage of the battery changed, and what is its new value?

3–24 If you wish to increase the amount of current in a resistor from 100 mA to 150 mA by changing the 20-V source, by how many volts should you change the source? To what new value should you set it?

3–25 Plot a graph of current versus voltage for voltage values ranging from 10 V to 100 V in 10-V steps for each of the following resistance values:
  (a) $1 \, \Omega$ **(b)** $5 \, \Omega$ **(c)** $20 \, \Omega$ **(d)** $100 \, \Omega$

Section 3–6

3–26 Modify the program in Section 3–6 so that the two known values are displayed along with the final result for each option.

3–27 Write a program to compute values of current for a specified range of voltage values and for a single resistor value. Draw the flowchart.

ANSWERS TO SECTION REVIEWS

Section 3–1
1. Current, voltage, and resistance.   **2.** $I = V/R$.   **3.** $V = IR$.   **4.** $R = V/I$.
5. Increases by three times.   **6.** Reduces to one-half of original value.   **7.** 0.5 A.
8. Yes, it would increase by four times.

Section 3–2
1. 1.79 A.   **2.** 0.179 A.   **3.** 2.27 mA.   **4.** 3.19 μA.   **5.** 4.26 mA.
6. 4.55 A.

Section 3–3
1. 10 V.   **2.** 3760 V.   **3.** 300 mV or 0.3 V.   **4.** 1400 μV or 1.4 mV.
5. 3.6 V.   **6.** 500 kV.   **7.** 22 V.   **8.** 470 V.   **9.** 9.9 V.
10. 13.6 V.

Section 3–4
1. 4.7 Ω.   **2.** 27 Ω.   **3.** 3.9 kΩ.   **4.** 5.6 kΩ.   **5.** 2.2 MΩ.
6. 0.47 kΩ and 470 Ω.

Section 3–5
1. The same percentage change occurs in two quantities.   **2.** 5 mA.   **3.** 6 A.
4. 6 V.

Section 3–6
1. To allow data to be input to the computer.
2. To branch the computer to the selected option.

FOUR

POWER AND ENERGY

Energy is the ability to do work, and power is the rate at which energy is used. Current carries electrical energy through a circuit. As the electrons pass through the resistance of the circuit, they give up their energy when they collide with atoms in the resistive material. The electrical energy given up by the electrons is converted into heat energy. The rate at which the electrical energy is lost is the power in the circuit.

In this chapter, you will learn:

☐   The definition of power and its unit, the watt (W).
☐   How to determine the amount of power in a resistive circuit.
☐   How to choose resistors to handle the power.
☐   The definition of the kilowatthour (kWh).
☐   How energy loss and voltage drop are related.
☐   The characteristics of power supplies and batteries.
☐   What the ampere-hour rating means.
☐   The definition of power supply efficiency.
☐   The basic operation of ohmmeters and how to use them to check resistors.
☐   The use of a simple computer program in analysis work.

4–1   POWER AND ENERGY

Power is the rate at which energy is used. In other words, power, symbolized by *P,* is a certain amount of energy used in a certain length of time, expressed as follows:

$$\text{Power} = \frac{\text{energy}}{\text{time}}$$

$$P = \frac{W}{t} \tag{4–1}$$

Energy is measured in *joules* (J), time is measured in *seconds* (s), and power is measured in *watts* (W). Note that an italic *W* is used to represent energy in the form of work and a roman W is used for watts, the unit of power.

Energy in *joules* divided by time in *seconds* gives power in *watts*. For example, if 50 J of energy are used in 2 s, the power is 50 J/2 s = 25 W.

By definition, *one watt is the amount of power when one joule of energy is consumed in one second.* Thus, the number of joules consumed in one second is always equal to the number of watts. For example, if 75 J are used in 1 s, the power is 75 W. The following example illustrates use of these units.

EXAMPLE 4–1

An amount of energy equal to 100 J is used in 5 s. What is the power in watts?

Solution:

$$P = \frac{\text{energy}}{\text{time}} = \frac{W}{t} = \frac{100 \text{ J}}{5 \text{ s}} = 20 \text{ W}$$

Amounts of power much less than one watt are common in certain areas of electronics. As with small current and voltage values, metric prefixes are used to designate small amounts of power. Thus, *milliwatts* (mW) and *microwatts* (μW) are commonly found in some applications.

In the electrical utilities field, *kilowatts* (kW) and *megawatts* (MW) are common units. Radio and television stations also use large amounts of power to transmit signals.

EXAMPLE 4–2

Express the following powers using appropriate metric prefixes:
(a) 0.045 W   **(b)** 0.000012 W   **(c)** 3500 W   **(d)** 10,000,000 W

Solution:
(a) 0.045 W = 45 mW   **(b)** 0.000012 W = 12 μW
(c) 3500 W = 3.5 kW   **(d)** 10,000,000 W = 10 MW

THE KILOWATTHOUR (kWh) UNIT OF ENERGY

Since power is the rate of energy usage, power utilized over a period of time represents energy consumption. If we multiply *power* and *time,* we have *energy,* symbolized by *W:*

$$\text{Energy} = \text{power} \times \text{time}$$
$$W = Pt \qquad\qquad\qquad\qquad \textbf{(4–2)}$$

Earlier, the *joule* was defined as a unit of energy. However, there is another way of expressing energy. Since power is expressed in *watts* and time in *seconds,* we can use units of energy called the *wattsecond* (Ws), *watthour* (Wh), and *kilowatthour* (kWh).

When you pay your electric bill, you are charged on the basis of the amount of *energy* you use. Because power companies deal in huge amounts of energy, the most practical unit is the *kilowatthour. You use a kilowatthour of energy when you use 1000 watts of power for one hour.* For example, a 100-W light bulb burning for 10 hours uses 1 kWh of energy.

EXAMPLE 4–3

Determine the number of kilowatthours for each of the following energy consumptions:
(a) 1400 W for 1 h **(b)** 2500 W for 2 h **(c)** 100,000 W for 5 h

Solution:
(a) 1400 W = 1.4 kW
Energy = (1.4 kW)(1 h) = 1.4 kWh
(b) 2500 W = 2.5 kW
Energy = (2.5 kW)(2 h) = 5 kWh
(c) 100,000 W = 100 kW
Energy = (100 kW)(5 h) = 500 kWh

SECTION REVIEW 4–1

1. Define *power*.
2. Write the formula for power in terms of energy and time.
3. Define *watt*.
4. Express each of the following values of power in the most appropriate units:
   (a) 68,000 W **(b)** 0.005 W **(c)** 0.000025 W
5. If you use 100 W of power for 10 h, how much energy (in kilowatthours) have you consumed?
6. Convert 2000 Wh to kilowatthours.
7. Convert 360,000 Ws to kilowatthours.

4–2   POWER IN AN ELECTRIC CIRCUIT

When there is current through resistance, the collisions of the electrons give off heat, resulting in a loss of energy as indicated in Figure 4–1. There is always a certain amount of power in an electric circuit, and it is dependent on the amount of *resistance* and on the amount of *current,* expressed as follows:

$$P = I^2R \qquad (4-3)$$

We can produce an *equivalent* expression for power by substituting V for IR (I^2 is $I \times I$).

$$P = I^2R = (I \times I)R = I(IR) = (IR)I$$
$$P = VI \qquad (4-4)$$

We obtain another *equivalent* expression by substituting V/R for I (Ohm's law) as follows:

$$P = VI = V\left(\frac{V}{R}\right)$$
$$P = \frac{V^2}{R} \qquad (4-5)$$

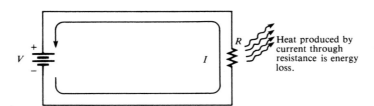

FIGURE 4–1
Power in an electric circuit is seen as heat given off by the resistance.

HOW TO USE THE APPROPRIATE POWER FORMULA

To calculate the power in a resistance, you can use any one of the three power formulas, depending on what information you have. For example, assume that you know the values of current and voltage. In this case you calculate the power with the formula $P = VI$. If you know I and R, use the formula $P = I^2R$. If you know V and R, use the formula $P = V^2/R$.

EXAMPLE 4–4

Calculate the power in each of the three circuits of Figure 4–2.

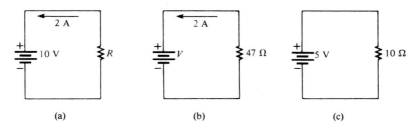

(a)   (b)   (c)

FIGURE 4–2

Solution:
In circuit (a), V and I are known. The power is determined as follows:

$$P = VI = (10 \text{ V})(2 \text{ A}) = 20 \text{ W}$$

In circuit (b), I and R are known. The power is determined as follows:

$$P = I^2R = (2 \text{ A})^2(47 \text{ }\Omega) = 188 \text{ W}$$

In circuit (c), V and R are known. The power is determined as follows:

$$P = \frac{V^2}{R} = \frac{(5 \text{ V})^2}{10 \text{ }\Omega} = 2.5 \text{ W}$$

The calculator sequence for circuit (c) is

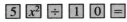

EXAMPLE 4–5

A 100-W light bulb operates on 120 V. How much current does it require?

Solution:
Use the formula $P = VI$ and solve for I as follows: Transpose the terms to get I on the left of the equation:

$$VI = P$$

Divide both sides of the equation by V to get I by itself:

$$\frac{\cancel{V}I}{\cancel{V}} = \frac{P}{V}$$

The V's cancel on the left, leaving

$$I = \frac{P}{V}$$

Substituting 100 W for P and 120 V for V yields

$$I = \frac{P}{V} = \frac{100 \text{ W}}{120 \text{ V}} = 0.833 \text{ A}$$

SECTION REVIEW 4–2

1. If there are 10 V across a resistor and a current of 3 A through it, what is the power?
2. How much power does the source in Figure 4–3 generate? What is the power in the resistor? Are the two values the same? Why?

FIGURE 4–3

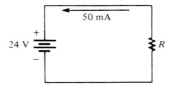

3. If there is a current of 5 A through a 56-Ω resistor, what is the power?
4. How much power is produced by 20 mA through a 4.7-kΩ resistor?
5. Five volts are applied to a 10-Ω resistor. What is the power?
6. How much power does a 2.2-kΩ resistor with 8 V across it produce?
7. What is the resistance of a 75-W bulb that takes 0.5 A?

4–3 RESISTOR POWER RATINGS

As you know, a resistor gives off heat when there is current through it. There is a limit to the amount of heat that a resistor can give off, which is specified by its power rating.

The *power rating* is the maximum amount of power that a resistor can dissipate without being damaged by excessive heat buildup. The power rating is not related to the ohmic value (resistance) but rather is determined mainly by the physical size and shape of the resistor. The larger the *surface area* of a resistor, the more power it can dissipate. *The surface area of a cylindrically shaped resistor is equal to the length (l) times the circumference (c),* as indicated in Figure 4–4.

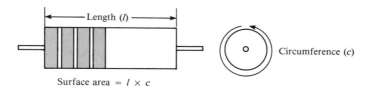

Surface area = $l \times c$

FIGURE 4–4
The power rating of a resistor is directly related to its surface area.

       Carbon-composition resistors are available in standard power ratings from 1/8 W to 2 W, as shown in Figure 4–5. Available power ratings for other types of resistors vary. For example, carbon-film and metal-film resistors have ratings up to 10 W, and wirewound resistors have ratings up to 225 W or greater. Figure 4–6 shows some of these resistors.

FIGURE 4–5
Carbon-composition resistors with standard power ratings of 1/8 W, 1/4 W, 1/2 W, 1 W, and 2 W (courtesy of Allen-Bradley Co.).

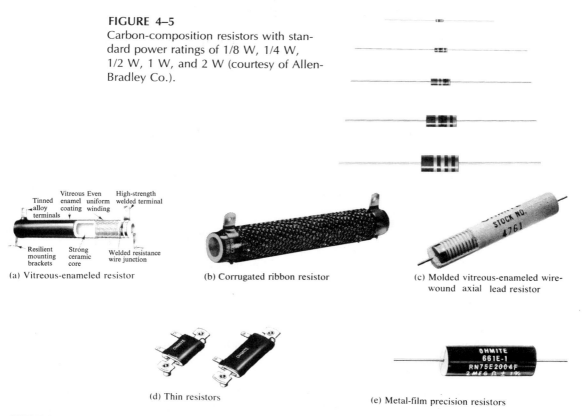

(a) Vitreous-enameled resistor

(b) Corrugated ribbon resistor

(c) Molded vitreous-enameled wire-wound axial lead resistor

(d) Thin resistors

(e) Metal-film precision resistors

FIGURE 4–6
Typical resistors with high power ratings (courtesy of Ohmite Manufacturing Co.).

HOW TO SELECT THE PROPER POWER RATING FOR AN APPLICATION

When a resistor is used in a circuit, its power rating must be greater than the maximum power that it will have to handle. For example, if a carbon-composition resistor is to dissipate 0.75 W in a circuit application, its rating should be at least the next higher standard value which is 1 W. It is common practice to use a rating that is approximately double the actual power when possible.

**EXAMPLE
4–6**

Choose an adequate power rating for each of the carbon-composition resistors in Figure 4–7.

FIGURE 4–7

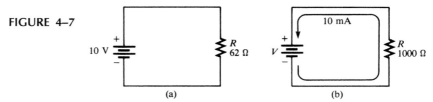

(a)   (b)

Solution:
(a) The actual power is

$$P = \frac{V^2}{R} = \frac{(10\ \text{V})^2}{62\ \Omega} = \frac{100\ \text{V}^2}{62\ \Omega} = 1.6\ \text{W}$$

Select a resistor with a power rating higher than the actual power. In this case, a 2-W resistor should be used.

(b) The actual power is

$$P = I^2R = (10\ \text{mA})^2(1000\ \Omega) = 0.1\ \text{W}$$

A 1/8-W (0.125-W) resistor should be used in this case.

RESISTOR FAILURES

When the power into a resistor is greater than its rating, the resistor will become excessively hot. As a result, either the resistor will burn open or its resistance value will be greatly altered.

A resistor that has been damaged because of overheating can often be detected by the charred or altered appearance of its surface. If there is no visual evidence, a resistor that is suspected of being damaged can be checked with an ohmmeter for an open or increased resistance value.

**EXAMPLE
4–7**

Determine whether the resistor in each circuit of Figure 4–8 has possibly been damaged by overheating.

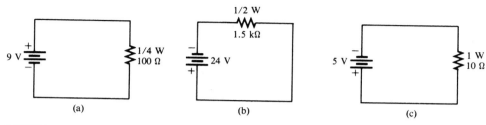

FIGURE 4–8

Solution:
(a)

$$P = \frac{V^2}{R} = \frac{(9 \text{ V})^2}{100 \ \Omega} = 0.81 \text{ W}$$

The rating of the resistor is 1/4 W (0.25 W), which is insufficient to handle the power. The resistor may be burned out.

(b)

$$P = \frac{V^2}{R} = \frac{(24 \text{ V})^2}{1.5 \text{ k}\Omega} = 0.384 \text{ W}$$

The rating of the resistor is 1/2 W (0.5 W), which is sufficient to handle the power.

(c)

$$P = \frac{V^2}{R} = \frac{(5 \text{ V})^2}{10 \ \Omega} = 2.5 \text{ W}$$

The rating of the resistor is 1 W, which is insufficient to handle the power. The resistor may be burned out.

HOW TO CHECK FOR RESISTOR DAMAGE WITH AN OHMMETER

A typical analog multimeter (VOM) and a digital multimeter are shown in Figures 4–9(a) and 4–9(b), respectively. The large switch on the analog meter is called a *range switch*. Notice the resistance (OHMS) settings on both meters. For the analog meter in Part (a), each setting indicates the amount by which the *ohms scale* (top scale) on the meter is to be *multiplied*. For example, if the pointer is at 50 on the ohms scale and the range switch is set at ×100, the resistance being measured is 50 × 100 Ω = 5000 Ω.

(a)   (b)

FIGURE 4–9
Typical multimeters (courtesy of Triplett Corp.).

If the resistor is open, the pointer will stay at full left scale (∞ means infinite) regardless of the range switch setting. For the digital meter, you use the range switch to select the appropriate setting for the value of resistance being measured. You do not have to multiply to get the correct reading because you have a direct digital readout of the resistance value.

The Basic Ohmmeter   A basic ohmmeter, shown in Figure 4–10(a), contains a battery and a variable resistor in series with the meter movement. To measure resistance, connect the leads across the external resistor to be measured, as shown in Part (b). This connection completes the circuit, allowing the internal battery to produce current through the movement, in turn causing a deflection of the pointer proportional to the value of the external resistance being measured.

OHMMETER SCALES

Figure 4–10(c) shows one type of ohmmeter scale. Between zero and infinity (∞), the scale is marked to indicate various resistor values. Because the values decrease from left to right, this scale is called a *back-off* scale.

Let us assume that a certain ohmmeter uses a 50-μA, 1000-Ω movement and has an internal 1.5-V battery. A current of 50 μA produces a full-scale deflection when the test leads are shorted. To have 50 μA, the *total* ohmmeter resistance is

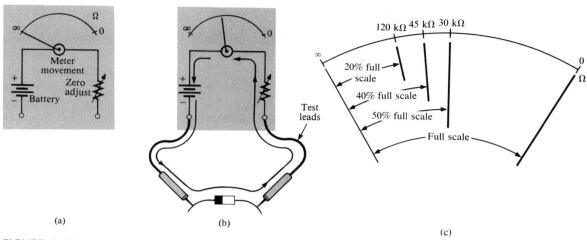

FIGURE 4–10
Basic ohmmeter circuit and simplified ohmmeter scale illustrating nonlinearity with a few
example values.

1.5 V/50 μA = 30 kΩ. Therefore, since the coil resistance is 1 kΩ, the variable zero adjustment resistor must be set at 30 kΩ − 1 kΩ = 29 kΩ.

Suppose that a 120-kΩ resistor is connected to the ohmmeter leads. Combined with the 30-kΩ internal meter resistance, the total R is 150 kΩ. The current is 1.5 V/150 kΩ = 10 μA, which is 20% of the full-scale current and which appears on the scale as shown in Figure 4–10(c). Now, for example, a 45-kΩ resistor connected to the ohmmeter leads results in a current of 1.5 V/75 kΩ = 20 μA, which is 40% of the full-scale current and which is marked on the scale as shown. Additional calculations of this type show that the scale is *nonlinear*. It is more compressed toward the left side than the right side.

The *center scale* point corresponds to an external resistance equal to the internal meter resistance (30 kΩ in this case). The reason is as follows: With 30 kΩ connected to the leads, the current is 1.5 V/60 kΩ = 25 μA, which is half of the full-scale current of 50 μA.

Zero Adjustment When the ohmmeter leads are *open,* as in Figure 4–11(a), the pointer is at full left scale, indicating *infinite* (∞) resistance (open circuit). When the leads are *shorted,* as in Figure 4–11(b), the pointer is at full right scale, indicating *zero* resistance.

The purpose of the variable resistor is to adjust the current so that the pointer is at exactly zero when the leads are shorted. It is used to compensate for changes in the internal battery voltage due to aging.

Because of the ohmmeter's internal battery, any voltage source must be disconnected from a circuit before a resistance measurement is made. If the voltage is left on, there is a risk of damaging the meter.

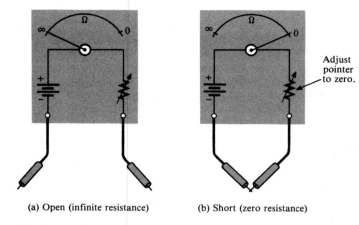

(a) Open (infinite resistance)   (b) Short (zero resistance)

FIGURE 4–11
Zero adjustment.

The following steps should be used in measuring resistance:

1. Disconnect the voltage from the circuit.
2. To prevent other parts of the circuit from altering the measurement, disconnect the resistor to be checked from its circuit.
3. Touch the test leads together, and use the *ohms adjust* knob to set the pointer at 0 Ω (analog meter only).
4. Select the proper range switch setting. On the analog meter, it is best to select a setting that produces a reading on the right half of the scale for better accuracy. On the digital meter, set the switch at the lowest setting that is greater than the value you expect to measure.
5. Connect the meter leads across the resistor. Polarity is not important in resistance measurements. Be careful not to touch the test lead contacts with your fingers because your body resistance will affect the reading.

SECTION REVIEW 4–3

1. Name two important values associated with a resistor.
2. How does the physical size of a resistor determine the amount of power that it can handle?
3. List the standard power ratings of carbon-composition resistors.
4. A resistor must handle 0.3 W. What size carbon resistor should be used to dissipate the energy properly?
5. Why is an internal battery required in an ohmmeter?
6. If the pointer indicates 8 on the ohmmeter scale and the range switch is set at R × 1k, what resistance is being measured?

4-4   ENERGY LOSS AND VOLTAGE DROP IN A RESISTANCE

As you have seen, when there is current through a resistance, energy is dissipated in the form of heat. This heat loss is caused by collisions of the free electrons within the atomic structure of the resistive material. When a collision occurs, heat is given off, and the electron loses some of its acquired energy.

In Figure 4–12, electrons are flowing out of the negative terminal of the battery. They have acquired energy from the battery and are at their highest energy level at the negative side of the circuit. As the electrons move through the resistor, they *lose* energy. The electrons emerging from the upper end of the resistor are at a lower energy level than those entering the lower end. The drop in energy level through the resistor creates a potential difference, or *voltage drop,* across the resistor having the polarity shown in Figure 4–12.

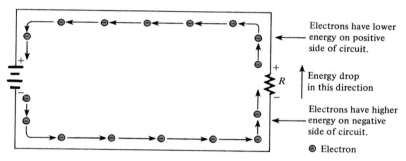

FIGURE 4–12
Illustration of electron flow and voltage drop.

The upper end of the resistor in Figure 4–12 is *less negative* (more positive) than the lower end. When we follow the direction of the current, this change corresponds to a voltage drop from a more negative potential to a less negative potential.

SECTION REVIEW 4–4

1. What is the basic reason for energy loss in a resistor?
2. What is a voltage drop?
3. What is the polarity of a voltage drop in relation to the direction of current?

4–5

POWER SUPPLIES

A *power supply* is a device that provides power to a load. A *load* is any electrical device or circuit that is connected to the output of the power supply and draws current from the supply.

Figure 4–13 shows a block diagram of a power supply with a loading device connected to it. The load can be anything from a light bulb to a computer. The power supply produces a voltage across its two output terminals and provides current through the load, as indicated in the figure. The product IV_{OUT} is the amount of power produced by the supply and consumed by the load. For a given output voltage (V_{OUT}), more current drawn by the load means more power from the supply.

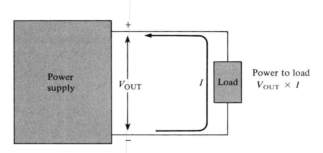

FIGURE 4–13
Block diagram of power supply and load.

Power supplies range from simple batteries to regulated electronic circuits where an accurate output voltage is automatically maintained. A battery is a dc power supply that converts chemical energy into electrical energy. Electronic power supplies normally convert 110 V ac (alternating current) from a wall outlet into a regulated dc (direct current) or ac voltage.

AMPERE-HOUR RATINGS OF BATTERIES

Batteries convert chemical energy into electrical energy. Because of their limited source of chemical energy, batteries have a certain *capacity* which limits the amount of time over which they can produce a given power level. This capacity is measured in *ampere-hours (Ah). The ampere-hour rating determines the length of time that a battery can deliver a certain amount of current to a load at the rated voltage.*

A *rating of one* ampere-hour means that a battery can deliver *one* ampere of current to a load for *one* hour at the rated voltage output. This same battery can deliver two amperes for one-half hour. The more current the battery is required to deliver, the shorter is the life of the battery. In practice, a battery usually is rated for a specified current level and output voltage. For example, a 12-V automobile battery may be rated for 70 Ah at 3.5 A. This means that it can produce 3.5 A for 20 hours.

EXAMPLE 4–8

For how many hours can a battery deliver 2 A if it is rated at 70 Ah?

Solution:
The ampere-hour rating is the current times the hours:

$$70 \text{ Ah} = (2 \text{ A})(x \text{ h})$$

Solving for x hours, we get

$$x = \frac{70 \text{ Ah}}{2 \text{ A}} = 35 \text{ h}$$

POWER SUPPLY EFFICIENCY

An important characteristic of electronic power supplies is efficiency. *Efficiency is the ratio of the output power to the input power:*

$$\text{Efficiency} = \frac{\text{output power}}{\text{input power}}$$

$$\text{Efficiency} = \frac{P_{\text{OUT}}}{P_{\text{IN}}} \tag{4–6}$$

Efficiency is usually expressed as a percentage. For example, if the input power is 100 W and the output power is 50 W, the efficiency is (50 W/100 W) × (100) = 50%.

All power supplies require that power be put into them. For example, an electronic power supply might use the ac power from a wall outlet as its input. Its output may be regulated dc or ac. The output power is *always* less than the input power because some of the total power must be used internally to operate the power supply circuitry. This amount is normally called the *power loss*. The output power is the input power minus the amount of internal power loss:

$$P_{\text{OUT}} = P_{\text{IN}} - P_{\text{LOSS}} \tag{4–7}$$

High efficiency means that little power is lost and there is a higher proportion of output power for a given input power.

EXAMPLE
4–9

A certain power supply unit requires 25 W of input power. It can produce an output power of 20 W. What is its efficiency, and what is the power loss?

Solution:

$$\text{Efficiency} = \left(\frac{P_{OUT}}{P_{IN}}\right)100\% = \left(\frac{20\ W}{25\ W}\right)100\%$$

$$= 80\%$$

$$P_{LOSS} = P_{IN} - P_{OUT} = 25\ W - 20\ W$$

$$= 5\ W$$

SECTION REVIEW 4–5

1. When a loading device draws an increased amount of current from a power supply, does this change represent a greater or a smaller load on the supply?

2. A power supply produces an output voltage of 10 V. If the supply provides 0.5 A to a load, what is the power output?

3. If a battery has an ampere-hour rating of 100 Ah, how long can it provide 5 A to a load?

4. If the battery in Question 3 is a 12-V device, what is its power output for the specified value of current?

5. An electronic power supply used in the lab operates with an input power of 1 W. It can provide an output power of 750 mW. What is its efficiency?

4–6

COMPUTER ANALYSIS

The following program is a simple example of how the computer can be used to quickly perform many calculations and display the results in a convenient format.

The program computes the power for a specified resistance and for each of a specified number of current values within a specified range and displays the results in tabular form. A flowchart is shown in Figure 4–14.

```
10   CLS
20   PRINT "THIS PROGRAM COMPUTES THE POWER FOR EACH OF A
     SPECIFIED"
30   PRINT "NUMBER OF CURRENT VALUES AND A SPECIFIED
     RESISTANCE."
40   PRINT "THE RESULTS ARE DISPLAYED IN TABULAR FORM."
50   PRINT:PRINT:PRINT
60   INPUT "TO CONTINUE PRESS 'ENTER'";X
70   CLS
80   INPUT "RESISTANCE VALUE IN OHMS";R
90   INPUT "MINIMUM CURRENT IN AMPS";I1
100  INPUT "MAXIMUM CURRENT IN AMPS";I2
```

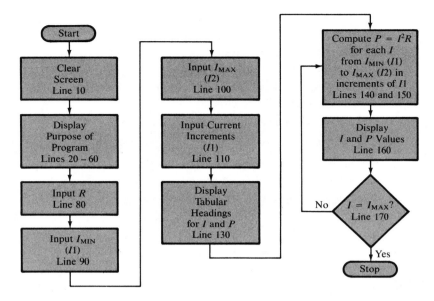

FIGURE 4–14
Program flowchart.

```
110 INPUT "CURRENT INCREMENTS IN AMPS";II
120 CLS
130 PRINT TAB(15),"CURRENT (AMPS)"; TAB(35),"POWER (WATTS)"
140 FOR I=I1 TO I2 STEP II
150 P=I*I*R
160 PRINT TAB(15),I; TAB(35), P
170 NEXT
```

SECTION REVIEW 4–6

1. State the purpose of lines 80 through 110.

2. In which line is the power computation performed?

SUMMARY

1. Power is the rate at which energy is used.

2. One watt equals one joule per second.

3. Watt is the unit of power, joule is a unit of energy, and second is a unit of time.

4. Use $P = VI$ to calculate power when voltage and current are known.

5. Use $P = I^2R$ to calculate power when current and resistance are known.

6. Use $P = V^2/R$ to calculate power when voltage and resistance are known.

7. The power rating of a resistor determines the maximum power that it can handle safely.

8. Resistors with a larger physical size can dissipate more power in the form of heat than smaller ones.

9. A resistor should have a power rating higher than the maximum power that it is expected to handle in the circuit.

10. Power rating is not related to resistance value.

11. A resistor normally opens when it burns out.

12. Energy is equal to power multiplied by time.

13. The kilowatthour is a unit of energy.

14. One kilowatthour is 1000 watts used for one hour.

15. The polarity of a voltage drop is from minus ($-$) to plus ($+$) in the direction of current.

16. A power supply is an energy source used to operate electrical and electronic devices.

17. A battery is one type of power supply that converts chemical energy into electrical energy.

18. An electronic power supply converts commercial energy (ac from the power company) to regulated dc or ac at various voltage levels.

19. The output power of a supply is the output voltage times the load current.

20. A load is a device that draws current from the power supply.

21. The capacity of a battery is measured in ampere-hours (Ah).

22. One ampere-hour equals one ampere used for one hour, or any other combination of amperes and hours that has a product of one.

23. Efficiency of a power supply is equal to the output power divided by the input power. Multiply by 100 to get percentage of efficiency.

24. A power supply with a high efficiency wastes less power than one with a lower efficiency.

FORMULAS

$$P = \frac{W}{t} \tag{4-1}$$

$$W = Pt \tag{4-2}$$

$$P = I^2R \tag{4-3}$$

$$P = VI \tag{4-4}$$

$$P = \frac{V^2}{R} \tag{4-5}$$

$$\text{Efficiency} = \frac{P_{OUT}}{P_{IN}} \qquad\qquad (4\text{--}6)$$

$$P_{OUT} = P_{IN} - P_{LOSS} \qquad\qquad (4\text{--}7)$$

SELF-TEST

Solutions appear at the end of the book.

1. Distinguish between power and energy.

2. **(a)** How many watts are in a kilowatt?
   (b) How many watts are in a megawatt?

3. Show how to use a voltmeter and an ammeter to measure the power in a resistor when it is connected to a voltage source.

4. **(a)** State the formula used to find power when V and I are known.
   (b) State the formula used to find power when I and R are known.
   (c) State the formula used to find power when V and R are known.

5. What power rating would you use for a carbon-composition resistor that will be required to handle 1.1 W?

6. What reading would you expect on an analog ohmmeter when the resistor being measured is open?

7. Two hundred joules of energy are consumed in 10 s. What is the power?

8. If it takes 300 ms to use 10,000 J of energy, what is the power?

9. How many watts are there in 50 kW?

10. How many milliwatts are there in 0.045 W?

11. If a resistive load draws 350 mA from a 10-V battery, how much power is handled by the resistor? How much power is delivered by the battery?

12. If a 1000-Ω resistor is connected across a 50-V source, what is the power?

13. A 5.6-kΩ resistor draws 500 mA from a voltage source. How much power is handled by the resistor?

14. What is the power of a light bulb that operates on 120 V and 2 A of current?

15. If 15 W of power are handled by a resistor, how many joules of energy are used in one minute?

16. If you have used 500 W of power for 24 h, how many kilowatthours have you used?

17. How many watthours represent 75 W used for 10 h?

18. Two 25-V power supplies are sitting on the lab bench. Power supply 1 is providing 100 mA to a load. Power supply 2 is supplying 0.5 A to a load. Which supply has the greater load? Which resistive load has the smaller ohmic value?

19. What is the power output of each power supply in Question 15?

20. A 12-V battery is connected to a 600-Ω load. Under these conditions, it is rated at 50 Ah. How long can it supply current to the load?

21. A given power supply is capable of providing 8 A for 2.5 h. What is its ampere-hour rating?

22. A laboratory power supply requires 250 mW internally. If it takes 2 W of input power, what is the output power?

23. A power supply produces a 0.5-W output with an input of 0.6 W. What is its percentage of efficiency?

PROBLEMS

Section 4–1

4–1 What is the power when energy is consumed at the rate of 350 J/s?

4–2 How many watts are used when 7500 J of energy are consumed in 5 h?

4–3 How many watts does 1000 J in 50 ms equal?

4–4 Convert the following to kilowatts:
  (a) 1000 W   **(b)** 3750 W   **(c)** 160 W
  (d) 50,000 W

4–5 Convert the following to megawatts:
  (a) 1,000,000 W   **(b)** 3×10^6 W   **(c)** 15×10^7 W
  (d) 8700 kW

4–6 Convert the following to milliwatts:
  (a) 1 W   **(b)** 0.4 W   **(c)** 0.002 W
  (d) 0.0125 W

4–7 Convert the following to microwatts:
  (a) 2 W   **(b)** 0.0005 W   **(c)** 0.25 mW
  (d) 0.00667 mW

4–8 Convert the following to watts:
  (a) 1.5 kW   **(b)** 0.5 MW   **(c)** 350 mW
  (d) 9000 μW

4–9 A particular electronic device uses 100 mW of power. If it runs for 24 h, how many joules of energy does it consume?

4–10 A certain appliance uses 300 W. If it is allowed to run continuously for 30 days, how many kilowatthours of energy does it consume?

4–11 At the end of a 31-day period, your utility bill shows that you have used 1500 kWh. What is your average daily power?

4–12 Convert 5×10^6 wattminutes to kWh.

4–13 Convert 6700 wattseconds to kWh.

4–14 For how many seconds must a 5-A current exist in a 47-Ω resistor in order to consume 25 J?

Section 4–2

4–15 If a 75-V source is supplying 2 A to a load, what is the resistance value of the load?

4–16 If a resistor has 5.5 V across it and 3 mA through it, what is the power?

4–17 An electric heater works on 120 V and draws 3 A of current. How much power does it use?

4–18 How much power is produced by 500 mA of current through a 4.7-kΩ resistor?

4–19 Calculate the power handled by a 10-kΩ resistor carrying 100 μA.

4–20 If there are 60 V across a 680-Ω resistor, what is the power?

4–21 A 56-Ω resistor is connected across the terminals of a 1.5-V battery. What is the power dissipation in the resistor?

4–22 If a resistor is to carry 2 A of current and handle 100 W of power, how many ohms must it be? Assume that the voltage can be adjusted to any required value.

4–23 A 12-V source is connected across a 10-Ω resistor.
   (a) How much energy is used in 2 minutes?
   (b) If the resistor is disconnected after 1 minute, is the power greater than, less than, or equal to the power during the 2-minute interval?

Section 4–3

4–24 A 6.8-kΩ resistor has burned out in a circuit. You must replace it with another resistor with the same value. If the resistor carries 10 mA, what should its power rating be? Assume that you have available carbon-composition resistors in all the standard power ratings.

4–25 A certain type of power resistor comes in the following ratings: 3 W, 5 W, 8 W, 12 W, 20 W. Your particular application requires a resistor that can handle approximately 8 W. Which rating would you use? Why?

Section 4–4

4–26 For each circuit in Figure 4–15, assign the proper polarity for the voltage drop across the resistor.

FIGURE 4–15

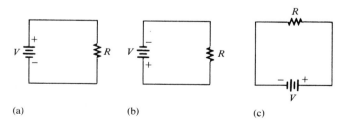

(a)   (b)   (c)

Section 4–5

4–27 A 50-Ω load consumes 1 W of power. What is the output voltage of the power supply?

4–28 A battery can provide 1.5 A of current for 24 h. What is its ampere-hour rating?

4–29 How much continuous current can be drawn from an 80-Ah battery for 10 h?

4–30 If a battery is rated at 650 mAh, how much current will it provide for 48 h?

4–31 If the input power is 500 mW and the output power is 400 mW, how much power is lost? What is the efficiency of this power supply?

4–32 To operate at 85% efficiency, how much output power must a source produce if the input power is 5 W?

4–33 A certain power supply provides a continuous 2 W to a load. It is operating at 60% efficiency. In a 24-h period, how many kilowatthours does it consume?

Section 4–6

4–34 Modify the program in Section 4–6 to compute the powers for a range of voltage values rather than for current values.

4–35 Modify the program in Section 4–6 to compute the resistor voltage for each current value, and display these values in a third column.

4–36 Devise a computer program to compute the energy consumption (kWh) for a home during any specified period of time, given the wattage rating of each appliance (for any number of appliances), and the number of hours per day that each appliance is used. Also, the program should compute the cost of energy during the specified period given the cost per kWh.

ANSWERS TO SECTION REVIEWS

Section 4–1

1. The rate at which energy is used.　**2.** $P = W/t$.

3. The unit of power. One watt is the power when 1 J of energy is used in 1 s.

4. **(a)** 68 kW;　**(b)** 5 mW;　**(c)** 25 μW.

5. 1 kWh.　**6.** 2 kWh.　**7.** 0.1 kWh.

Section 4–2

1. 30 W.

2. 1.2 W, 1.2 W. Yes, because all energy produced by the source is dissipated by the resistance.

3. 1400 W.　**4.** 1.88 W.　**5.** 2.5 W.　**6.** 29 mW.　**7.** 300 Ω.

Section 4–3

1. Resistance value, power rating.　**2.** A larger surface area dissipates more energy.

3. 0.125 W, 0.25 W, 0.5 W, 1 W, 2 W.　**4.** At least 0.5 W.

5. To produce current through the resistance being measured.　**6.** 8 kΩ.

Section 4–4
1. Collisions of the electrons within the atomic structure.
2. Potential difference between two points due to energy loss.
3. Minus to plus.

Section 4–5
1. Greater.   **2.** 5 W.   **3.** 20 h.   **4.** 60 W.   **5.** 75%.

Section 4–6
1. For data input.   **2.** Line 150.

FIVE

SERIES RESISTIVE CIRCUITS

Resistive circuits can be of two basic forms: *series* or *parallel*. In this chapter we discuss series circuits. Parallel circuits are presented in Chapter 6, and combinations of series and parallel are examined in Chapter 7. In this chapter you will see how Ohm's law is used in series circuits, and you will study another very important law, called *Kirchhoff's voltage law*. Also, several important applications of series circuts are given.

In this chapter, you will learn:

- [ ] How to identify a series circuit.
- [ ] How to determine the current in a series circuit.
- [ ] How to determine total resistance.
- [ ] How to apply Ohm's law to find the current, the voltages, and the resistances in a series circuit.
- [ ] How to connect voltage sources in series to achieve a higher voltage.
- [ ] The definition of Kirchhoff's voltage law and how to apply it.
- [ ] What voltage dividers are and how they can be used.
- [ ] How to determine the total power in a series circuit.
- [ ] The meaning of ground in a circuit.
- [ ] How to identify and troubleshoot some of the common problems in series circuits.
- [ ] How to use a simple computer program for circuit analysis.

5–1   RESISTORS IN SERIES

Resistors in series are connected end-to-end or in a "string," as shown in Figure 5–1. Figure 5–1(a) shows two resistors connected in series between point A and point B. Part (b) of the figure shows three in series, and Part (c) shows four in series. Of course, there can be any number of resistors in a series connection.

FIGURE 5–1
Resistors in series.

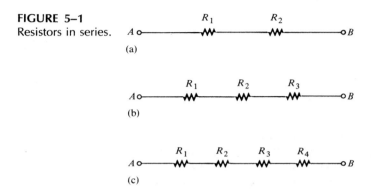

The only way for electrons to get from point B in any of the connections of Figure 5–1 is to go through *each* of the resistors. The following is an important way to identify a series connection:

A series connection provides only one path for current between two points in a circuit so that there is the same current through each series resistor.

IDENTIFYING SERIES CONNECTIONS

In an actual circuit diagram, a series connection may not always be as easy to identify as those in Figure 5–1. For example, Figure 5–2 shows series resistors drawn in other ways. Remember, *if there is only one current path between two points, the resistors between those two points are in series,* no matter how they appear in a diagram.

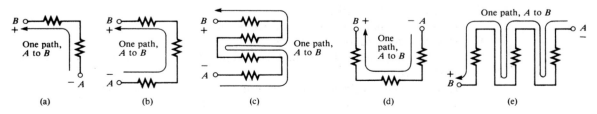

FIGURE 5–2
Some examples of series connections. Notice that the current must be the same at all points.

**EXAMPLE
5–1**

Suppose that there are five resistors positioned on a circuit board as shown in Figure 5–3. Wire them together in series so that, starting from the positive (+) terminal, R_1 is first, R_2 is second, R_3 is third, and so on. Draw a schematic diagram showing this connection.

FIGURE 5–3

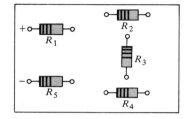

Solution:
The wires are connected as shown in Figure 5–4(a), which is the *assembly diagram*. The *schematic diagram* is shown in Figure 5–4(b). Note that the schematic diagram does not necessarily show the actual physical arrangement of the resistors as does the assembly diagram. The purpose of the *schematic* diagram is to show how components are connected *electrically*. The purpose of the *assembly* diagram is to show how components are arranged *physically*.

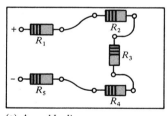

(a) Assembly diagram

(b) Schematic diagram

FIGURE 5–4

**EXAMPLE
5–2**

Describe how the resistors on the printed circuit (PC) board in Figure 5–5 are related electrically.

Solution:
Resistors R_1 through R_7 are in series with each other. This series combination is connected between pins 1 and 2 on the PC board.

　　　　Resistors R_8 through R_{13} are in series with each other. This series combination is connected between pins 3 and 4 on the PC board.

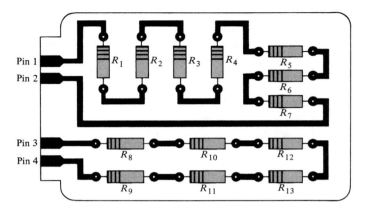

FIGURE 5–5

SECTION REVIEW 5–1

1. How are the resistors connected in a series circuit?

2. How can you identify a series connection?

3. Complete the schematic diagrams for the circuits in each part of Figure 5–6 by connecting the resistors in series in numerical order from terminal A to B.

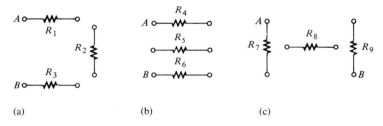

(a)   (b)   (c)

FIGURE 5–6

4. Now connect each *group* of series resistors in Figure 5–6 in series.

5–2

CURRENT IN A SERIES CIRCUIT

There is the same amount of current through all points in a series circuit. Figure 5–7 shows three series resistors connected to a voltage source. *At any point in this circuit, the current entering that point must equal the current leaving that point,* as illustrated by the current directional arrows at points A, B, C, and D.

FIGURE 5–7
Current in a series circuit.

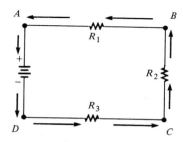

Notice also that the current out of each of the resistors must equal the current in, because there are no other paths through which part of the current can branch off and go somewhere else. Therefore, the current in each section of the circuit is the same as the current in all other sections. The current has only one path going from the nega-tive (−) side of the source to the positive (+) side.

Let us assume that the battery in Figure 5–7 supplies two amperes of current to the series resistance. There are two amperes out of the negative terminal. If we connect ammeters at several points in the circuit as shown in Figure 5–8, *each* meter will read two amperes.

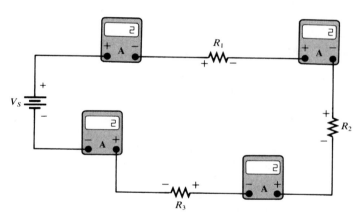

FIGURE 5–8
Current is the same at all points in a series circuit.

SECTION REVIEW 5–2

1. In a series circuit with a 10-Ω and a 4.7-Ω resistor in series, there is 1 A through the 10-Ω resistor. What is the current through the 4.7-Ω resistor?

2. A milliammeter is connected between points A and B in Figure 5–9. It measures 50 mA. If you move the meter and connect it between points C and D, how much current will it indicate? Between E and F?

FIGURE 5–9

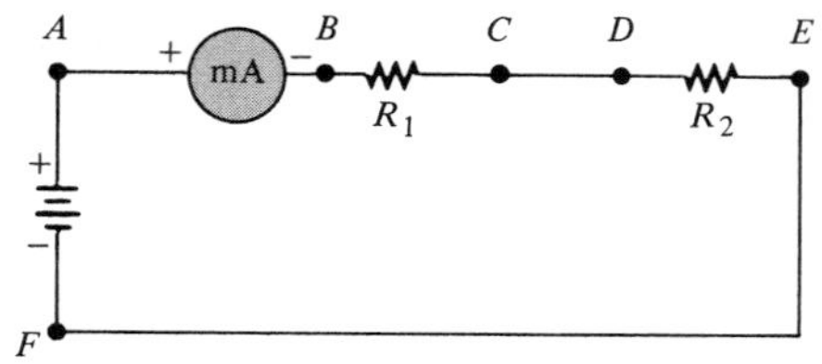

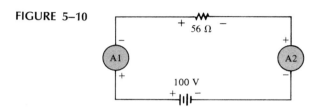

3. In Figure 5–10, how much current does ammeter 1 indicate? How much current does ammeter 2 indicate?

FIGURE 5–10

4. What statement can you make about the amount of current in a series circuit?

5–3

TOTAL SERIES RESISTANCE

The total resistance of a series connection is equal to the sum of the resistances of each individual resistor.

This fact is understandable because each of the resistors in series offers opposition to the current in direct proportion to its resistance. A greater number of resistors connected in series creates *more opposition* to current. More opposition to current implies a higher ohmic value of resistance. Thus, every time a resistor is added in series, the total resistance increases.

SERIES RESISTOR VALUES ADD

Figure 5–11 illustrates how series resistances add to *increase* the total resistance. Figure 5–11(a) has a single 10-Ω resistor. Figure 5–11(b) shows another 10-Ω resistor connected in series with the first one, making a total resistance of 20 Ω. If a third 10-Ω resistor is connected in series with the first two, as shown in Figure 5–11(c), the total resistance becomes 30 Ω.

SERIES RESISTANCE FORMULA

For *any number* of individual resistors connected in series, the total resistance is the sum of each of the individual values:

$$R_T = R_1 + R_2 + R_3 + \cdots + R_n \qquad (5\text{–}1)$$

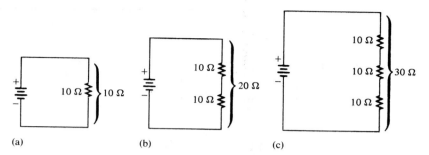

FIGURE 5–11
Total resistance increases with each additional series resistor.

where R_T is the total resistance and R_n is the last resistor in the series string (n can be any positive integer equal to the number of resistors in series). For example, if we have four resistors in series ($n = 4$), the total resistance formula is

$$R_T = R_1 + R_2 + R_3 + R_4$$

If we have six resistors in series ($n = 6$), the total resistance formula is

$$R_T = R_1 + R_2 + R_3 + R_4 + R_5 + R_6$$

To illustrate the calculation of total series resistance, let us take the circuit of Figure 5–12 and determine its R_T. (V_S is the source voltage.)

FIGURE 5–12
Example of five resistors in series.

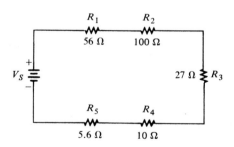

The circuit of Figure 5–12 has five resistors in series. To get the total resistance, we simply add the values as follows:

$$R_T = 56 \ \Omega + 100 \ \Omega + 27 \ \Omega + 10 \ \Omega + 5.6 \ \Omega = 198.6 \ \Omega$$

This equation illustrates an important point: In Figure 5–12, the order in which the resistances are added does not matter; we still get the same total. Also, we can physically change the positions of the resistors in the circuit without affecting the total resistance.

**EXAMPLE
5–3**

Connect the resistors in Figure 5–13 in series, and determine the total resistance, R_T.

FIGURE 5–13

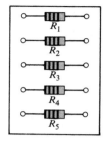

R_1 Orange, orange, black

R_2 Blue, gray, black

R_3 Brown, black, brown

R_4 Yellow, violet, black

R_5 Brown, black, black

Solution:
The resistors are connected as shown in Figure 5–14.

FIGURE 5–14

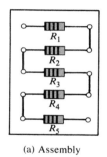

(a) Assembly

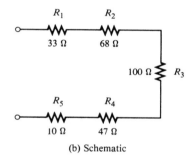

(b) Schematic

We find the total resistance by adding all the values as follows:

$$R_T = 33 \ \Omega + 68 \ \Omega + 100 \ \Omega + 47 \ \Omega + 10 \ \Omega = 258 \ \Omega$$

**EXAMPLE
5–4**

What is the total resistance (R_T) in the circuit of Figure 5–15?

FIGURE 5–15

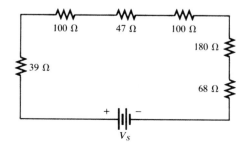

Solution:

Sum all the values as follows:

$$R_T = 39 \ \Omega + 100 \ \Omega + 47 \ \Omega + 100 \ \Omega + 180 \ \Omega + 68 \ \Omega$$
$$= 534 \ \Omega$$

EXAMPLE 5–5

Determine R_4 in the circuit of Figure 5–16.

FIGURE 5–16

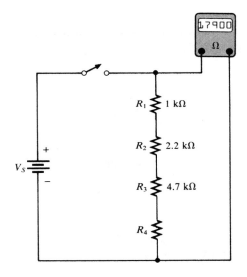

Solution:

From the ohmmeter reading:

$$R_T = 17.9 \ \text{k}\Omega$$
$$R_T = R_1 + R_2 + R_3 + R_4$$

Solving for R_4,

$$R_4 = R_T - (R_1 + R_2 + R_3)$$
$$= 17.9 \ \text{k}\Omega - (1 \ \text{k}\Omega + 2.2 \ \text{k}\Omega + 4.7 \ \text{k}\Omega)$$
$$= 10 \ \text{k}\Omega$$

The calculator sequence is

$\boxed{1}\ \boxed{7}\ \boxed{.}\ \boxed{9}\ \boxed{\text{EXP}}\ \boxed{3}\ \boxed{-}\ \boxed{(}\ \boxed{1}\ \boxed{\text{EXP}}\ \boxed{3}\ \boxed{+}\ \boxed{2}\ \boxed{.}\ \boxed{2}$
$\boxed{\text{EXP}}\ \boxed{3}\ \boxed{+}\ \boxed{4}\ \boxed{.}\ \boxed{7}\ \boxed{\text{EXP}}\ \boxed{3}\ \boxed{)}\ \boxed{=}$

EQUAL-VALUE SERIES RESISTORS

When a circuit has more than one resistor of the *same* value in series, there is a shortcut method to obtain the total resistance: Simply multiply the *resistance* of the resistors having the *same* value by the *number* of equal-value resistors that are in series. This method is essentially the same as adding the values. For example, five 100-Ω resistors in series have an R_T of 5(100 Ω) = 500 Ω. In general, the formula is expressed as

$$R_T = nR \qquad \qquad (5\text{–}2)$$

where n is the number of equal-value resistors and R is the value.

EXAMPLE 5–6

Find the R_T of eight 22-Ω resistors in series.

Solution:
We find R_T by adding the values as follows:

$$R_T = 22\ \Omega + 22\ \Omega + 22\ \Omega + 22\ \Omega + 22\ \Omega + 22\ \Omega + 22\ \Omega + 22\ \Omega$$
$$= 176\ \Omega$$

However, it is much easier to multiply:

$$R_T = 8(22\ \Omega) = 176\ \Omega$$

SECTION REVIEW 5–3

1. The following resistors (one each) are in series: 1 Ω, 2.2 Ω, 3.3 Ω, and 4.7 Ω. What is the total resistance?

2. The following resistors are in series: one 100 Ω, two 56 Ω, four 12 Ω, and one 330 Ω. What is the total resistance?

3. Suppose that you have one resistor each of the following values: 1 kΩ, 2.7 kΩ, 5.6 kΩ, and 560 Ω. To get a total resistance of 13.76 kΩ, you need one more resistor. What should its value be?

4. What is the R_T for twelve 56-Ω resistors in series?

5. What is the R_T for twenty 5.6-kΩ resistors and thirty 8.2-kΩ resistors in series?

5–4   OHM'S LAW IN SERIES CIRCUITS

In this section we use Ohm's law and the basic concepts of series circuits to analyze several circuits.

**EXAMPLE
5–7**

Find the current in the circuit of Figure 5–17.

FIGURE 5–17

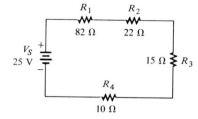

Solution:
The current is determined by the voltage and the *total resistance*. First, we calculate the total resistance as follows:

$$R_T = R_1 + R_2 + R_3 + R_4$$
$$= 82\ \Omega + 22\ \Omega + 15\ \Omega + 10\ \Omega$$
$$= 129\ \Omega$$

Next, using Ohm's law, we calculate the current as follows:

$$I = \frac{V_S}{R_T} = \frac{25\ \text{V}}{129\ \Omega}$$
$$= 0.194\ \text{A}$$

Remember, the *same* current exists at all points in the circuit. Thus, *each* resistor has 0.194 A through it.

**EXAMPLE
5–8**

In the circuit of Figure 5–18, there is 1 mA of current. For this amount of current, what must the source voltage V_S be?

FIGURE 5–18

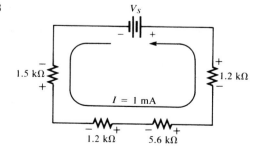

Solution:

In order to calculate V_S, we must determine R_T as follows:

$$R_T = 1.2 \text{ k}\Omega + 5.6 \text{ k}\Omega + 1.2 \text{ k}\Omega + 1.5 \text{ k}\Omega$$

$$= 9.5 \text{ k}\Omega$$

Now we use Ohm's law to get V_S:

$$V_S = IR_T = (1 \text{ mA})(9.5 \text{ k}\Omega)$$

$$= 9.5 \text{ V}$$

**EXAMPLE
5–9**

Calculate the voltage across each resistor in Figure 5–19, and find the value of V_S. To what maximum value can V_S be raised before the fuse blows?

FIGURE 5–19

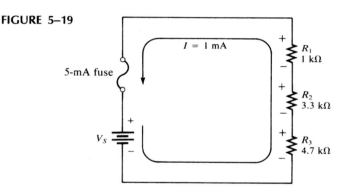

Solution:

By Ohm's law, the voltage across each resistor is equal to its resistance multiplied by the current through it. Using the Ohm's law formula $V = IR$, we determine the voltage across each of the resistors. Keep in mind that there is the same current through each series resistor. The fuse is effectively like a wire and has negligible resistance.

Voltage across R_1:

$$V_1 = IR_1 = (1 \text{ mA})(1 \text{ k}\Omega) = 1 \text{ V}$$

Voltage across R_2:

$$V_2 = IR_2 = (1 \text{ mA})(3.3 \text{ k}\Omega) = 3.3 \text{ V}$$

Voltage across R_3:

$$V_3 = IR_3 = (1 \text{ mA})(4.7 \text{ k}\Omega) = 4.7 \text{ V}$$

The source voltage V_S is equal to the current times the *total resistance:*

$$R_T = 1 \text{ k}\Omega + 3.3 \text{ k}\Omega + 4.7 \text{ k}\Omega$$
$$= 9 \text{ k}\Omega$$
$$V_S = (1 \text{ mA})(9 \text{ k}\Omega)$$
$$= 9 \text{ V}$$

Notice that if you add the voltage drops of the resistors, they total 9 V, which is the same as the source voltage.

The fuse has a rating of 5 mA; thus V_S can be increased to a value where $I = 5$ mA.

$$V_S = IR_T = (5 \text{ mA})(9 \text{ k}\Omega) = 45 \text{ V}$$

**EXAMPLE
5–10**

As you have seen, some resistors are not color coded with bands but have the values stamped on the resistor body. When the circuit board shown in Figure 5–20 was assembled, someone mounted the resistors with the labels turned down, and there is no documentation showing the resistor values. Without using an ohmmeter and without removing the resistors from the board, use Ohm's law to determine the resistance of each one.

FIGURE 5–20

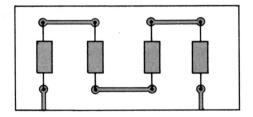

Solution:
The resistors are all in series, so the current is the same through each one. Measure the current by connecting a 12-V source (arbitrary value) and an ammeter as shown in Figure 5–21. Measure the voltage across each resistor by placing the voltmeter across the first resistor. Then repeat this measurement for the other three resistors. For illustration, the values indicated are assumed to be the measured values.

Determine the resistance of each resistor by substituting the measured values of current and voltage into the Ohm's law formula as follows:

$$R_1 = \frac{V_1}{I} = \frac{2.5 \text{ V}}{25 \text{ mA}} = 100 \text{ } \Omega$$

$$R_2 = \frac{V_2}{I} = \frac{3 \text{ V}}{25 \text{ mA}} = 120 \text{ } \Omega$$

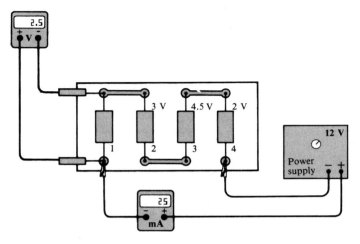

FIGURE 5-21
The voltmeter readings for each resistor are indicated.

$$R_3 = \frac{V_3}{I} = \frac{4.5 \text{ V}}{25 \text{ mA}} = 180 \ \Omega$$

$$R_4 = \frac{V_4}{I} = \frac{2 \text{ V}}{25 \text{ mA}} = 80 \ \Omega$$

The calculator sequence for R_1 is

SECTION REVIEW 5-4

1. A 10-V battery is connected across three 100-Ω resistors in series. What is the current through each resistor?

2. How much voltage is required to produce 5 A through the circuit of Figure 5-22?

FIGURE 5-22

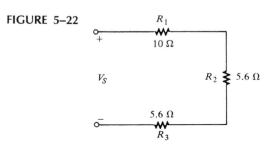

3. How much voltage is dropped across each resistor in Figure 5–22 when the current is 5 A?

4. There are four equal-value resistors connected in series with a 5-V source. A current of 4.63 milliamperes is measured. What is the value of each resistor?

5–5

VOLTAGE SOURCES IN SERIES

A voltage source is an energy source that provides a constant voltage to a load. Batteries and dc power supplies are practical examples.

When two or more voltage sources are in series, the total voltage is equal to the algebraic sum of the individual source voltages. The *algebraic sum* means that the polarities of the sources must be included when the sources are combined in series. Sources with opposite polarities have voltages with opposite signs.

$$V_T = V_{S1} + V_{S2} + \cdots + V_{Sn}$$

When the sources are all in the same direction in terms of their polarities, as in Figure 5–23(a), all of the voltages have the same sign when added, and we get a total of 4.5 V from terminal A to terminal B with A more positive than B:

$$V_{AB} = 1.5 \text{ V} + 1.5 \text{ V} + 1.5 \text{ V} = +4.5 \text{ V}$$

The voltage has a double subscript, AB, to indicate that it is the voltage at point A with respect to point B.

In Figure 5–23(b), the middle source is opposite to the other two; so its voltage has an opposite sign when added to the others. For this case the total voltage from A to B is

$$V_{AB} = +1.5 \text{V} - 1.5 \text{ V} + 1.5 \text{ V} = +1.5 \text{ V}$$

Terminal A is 1.5 V more positive than terminal B.

A familiar example of sources in series is the flashlight. When you put two 1.5-V batteries in your flashlight, they are connected in *series*, giving a total of 3 V.

FIGURE 5–23
Voltage sources in series add algebraically.

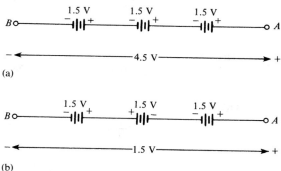

When connecting batteries or other voltage sources in series to increase the total voltage, always connect from the positive (+) terminal of one to the negative (−) terminal of another. Such a connection is illustrated in Figure 5–24.

FIGURE 5–24
Connection of three 6-V batteries to get 18 V.

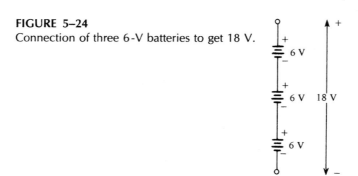

The following two examples illustrate calculation of total source voltage.

EXAMPLE 5–11

What is the total source voltage (V_{ST}) in Figure 5–25?

FIGURE 5–25

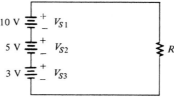

Solution:
The polarity of each source is the same (the sources are connected in the same direction in the circuit). So we sum the three voltages to get the total:

$$V_{ST} = V_{S1} + V_{S2} + V_{S3} = 10 \text{ V} + 5 \text{ V} + 3 \text{ V}$$
$$= 18 \text{ V}$$

The three individual sources can be replaced by a single equivalent source of 18 V with its polarity as shown in Figure 5–26.

FIGURE 5–26

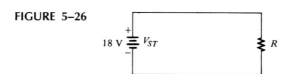

**EXAMPLE
5–12**

Determine V_{ST} in Figure 5–27.

FIGURE 5–27

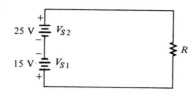

Solution:
These sources are connected in *opposing* directions. If you go clockwise around the circuit, you go from plus to minus through V_{S1}, and minus to plus through V_{S2}. The total voltage is the *difference* of the two source voltages (algebraic sum of oppositely signed values). The total voltage has the same polarity as the larger-value source. Here we will choose V_{S2} to be positive:

$$V_{ST} = V_{S2} - V_{S1} = 25 \text{ V} - 15 \text{ V}$$
$$= 10 \text{ V}$$

The two sources in Figure 5–27 can be replaced by a single 10-V equivalent source with polarity as shown in Figure 5–28.

FIGURE 5–28

SECTION REVIEW 5–5

1. Four 1.5-V flashlight batteries are connected in series plus to minus. What is the total voltage of all four cells?

2. How many 12-V batteries must be connected in series to produce 60 V? Sketch a schematic that shows the battery connections.

3. The resistive circuit in Figure 5–29 is used to bias a transistor amplifier. Show how to connect two 15-V power supplies in order to get 30 V across the two resistors.

FIGURE 5–29

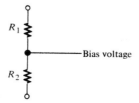

4. Determine the total source voltage in each circuit of Figure 5–30.

5. Sketch the *equivalent single source* circuit for each circuit of Figure 5–30.

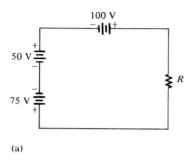

(a)

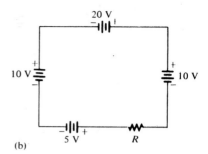

(b)

FIGURE 5–30

5–6 ## KIRCHHOFF'S VOLTAGE LAW

Kirchhoff's voltage law states:

The algebraic sum of all the voltages around a closed path is zero or, in other words, the sum of the voltage drops equals the total source voltage.

For example, in the circuit of Figure 5–31, there are three voltage drops and one voltage source. If we sum all of the voltages around the circuit, we get

$$V_S - V_1 - V_2 - V_3 = 0 \tag{5–3}$$

Notice that the source voltage has a sign opposite to that of the voltage drops. Thus, the algebraic sum equals zero. Equation (5–3) can be written another way by transposing of the voltage drop terms to the right side of the equation:

$$V_S = V_1 + V_2 + V_3 \tag{5–4}$$

FIGURE 5–31
The sum of the voltage drops equals the source voltage, V_S.

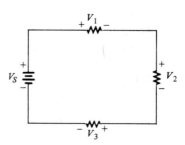

In words, Equation (5–4) says that the source voltage equals the sum of the voltage drops. Both Equations (5–3) and (5–4) are equivalent ways of expressing Kirchhoff's voltage law for the circuit in Figure 5–31.

The previous illustration of Kirchhoff's voltage law was for the special case of three voltage drops and one voltage source. In general, Kirchhoff's voltage law applies to any number of series voltage drops and any number of series voltage sources, as illustrated in Figure 5–32.

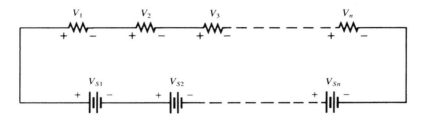

FIGURE 5–32
Sum of the voltage drops equals the total source voltage.

The *general form* of Kirchhoff's voltage law is

$$V_{S1} + V_{S2} + \cdots + V_{Sn} = V_1 + V_2 + \cdots + V_n \tag{5-5}$$

POLARITIES OF VOLTAGE DROPS

Always remember that the total source voltage has a polarity in the circuit *opposite* to that of the voltage drops. As we go around the series circuit of Figure 5–33 in the counterclockwise direction, the voltage drops have polarities as shown. Notice that the negative ($-$) side of each resistor is the one *nearest* the negative terminal of the source as we follow the current path. The positive ($+$) side of each resistor is the one *nearest* the positive terminal of the source as we follow the current path.

FIGURE 5–33
The polarity of the voltage drops is opposite to that of the source voltage.

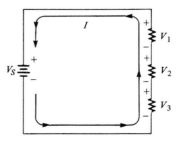

Refering to Figure 5–33, let us examine polarities in a series loop in relation to the current. *The current is out of the negative side of the source* and through the resistors as shown. *The current is into the negative side of each resistor and out of the positive side.*

The following three examples use Kirchhoff's voltage law to solve circuit problems.

**EXAMPLE
5–13**

Determine the applied voltage V_S in Figure 5–34 where the two voltage drops are given.

FIGURE 5–34

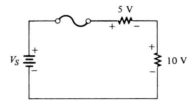

Solution:
By Kirchhoff's voltage law, the source voltage (applied voltage) must equal the sum of the voltage drops. Adding the voltage drops gives us the value of the source voltage:

$$V_S = 5 \text{ V} + 10 \text{ V} = 15 \text{ V}$$

**EXAMPLE
5–14**

Determine the unknown voltage drop, V_3, in Figure 5–35.

FIGURE 5–35

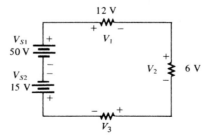

Solution:
By Kirchhoff's voltage law, the algebraic sum of all the voltages around the circuit is zero:

$$V_{S1} - V_{S2} - V_1 - V_2 - V_3 = 0$$

The value of each voltage drop except V_3 is known. Substitute these values into the equation as follows:

$$50 \text{ V} - 15 \text{ V} - 12 \text{ V} - 6 \text{ V} - V_3 = 0 \text{ V}$$

Next combine the known values:

$$17 \text{ V} - V_3 = 0 \text{ V}$$

Transpose 17 V to the right side of the equation, and cancel the minus signs:

$$-V_3 = -17 \text{ V}$$
$$V_3 = 17 \text{ V}$$

The voltage drop across R_3 is 17 V, and its polarity is as shown in Figure 5–35.

EXAMPLE 5–15

Find the value of R_4 in Figure 5–36.

FIGURE 5–36

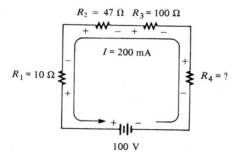

$R_2 = 47 \, \Omega$ $R_3 = 100 \, \Omega$

$I = 200$ mA

$R_1 = 10 \, \Omega$

$R_4 = ?$

100 V

Solution:
In this problem we must use both Ohm's law *and* Kirchhoff's voltage law. Follow this procedure carefully.

First find the voltage drop across each of the *known* resistors. Use Ohm's law:

$$V_1 = IR_1 = (200 \text{ mA})(10 \, \Omega) = 2 \text{ V}$$
$$V_2 = IR_2 = (200 \text{ mA})(47 \, \Omega) = 9.4 \text{ V}$$
$$V_3 = IR_3 = (200 \text{ mA})(100 \, \Omega) = 20 \text{ V}$$

Next, use Kirchhoff's voltage law to find V_4, the voltage drop across the *unknown* resistor:

$$V_S - V_1 - V_2 - V_3 - V_4 = 0 \text{ V}$$

$$100 \text{ V} - 2 \text{ V} - 9.4 \text{ V} - 20 \text{ V} - V_4 = 0 \text{ V}$$

$$68.6 \text{ V} - V_4 = 0 \text{ V}$$

$$V_4 = 68.6 \text{ V}$$

Now that we know V_4, we can use Ohm's law to calculate R_4 as follows:

$$R_4 = \frac{V_4}{I} = \frac{68.6 \text{ V}}{200 \text{ mA}} = 343 \ \Omega$$

This is most likely a 330-Ω resistor because 343 Ω is within a reasonable tolerance range ($+5\%$) of 330 Ω.

SECTION REVIEW 5–6

1. State Kirchhoff's voltage law in two ways.

2. A 50-V source is connected to a series resistive circuit. What is the total of the voltage drops in this circuit?

3. Two equal-value resistors are connected in series across a 10-V battery. What is the voltage drop across each resistor?

4. In a series circuit with a 25-V source, there are three resistors. One voltage drop is 5 V, and the other is 10 V. What is the value of the third voltage drop?

5. The individual voltage drops in a series string are as follows: 1 V, 3 V, 5 V, 8 V, and 7 V. What is the total voltage applied across the series string?

5–7 VOLTAGE DIVIDERS

A series circuit acts as a *voltage divider*. In this section you will see what this term means and why voltage dividers are an important application of series circuits.

To illustrate how a series string of resistors acts as a voltage divider, we will examine Figure 5–37 where there are two resistors in series. As you already know, there are two voltage drops: one across R_1 and one across R_2. We call these voltage drops V_1 and V_2, respectively, as indicated in the diagram.

Since there is the same current through each resistor, the voltage drops are proportional to the values of the resistors. For example, if the value of R_2 is twice that of R_1, then the value of V_2 is twice that of V_1. In other words, the total voltage drop *divides* among the series resistors in amounts directly proportional to the resistance values.

FIGURE 5–37
Two-resistor voltage divider.

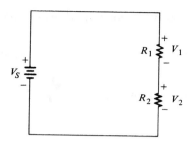

For example, in Figure 5–37, if V_S is 10 V, R_1 is 50 Ω, and R_2 is 100 Ω, then V_1 is one-third the *total* voltage, or 3.33 V, because R_1 is one-third the *total* resistance. Likewise, V_2 is two-thirds V_S, or 6.67 V.

VOLTAGE DIVIDER FORMULA

With a few calculations, a formula for determining how the voltages divide among series resistors can be developed. Let us assume that we have several resistors in series as shown in Figure 5–38. This figure shows five resistors, but there can be any number.

FIGURE 5–38
Five-resistor voltage divider.

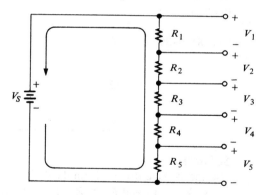

Let us call the voltage drop across any one of the resistors V_x, where x represents the number of a particular resistor (1, 2, 3, and so on). By Ohm's law, the voltage drop across any of the resistors in Figure 5–38 can be written as follows:

$$V_x = IR_x$$

where $x = 1, 2, 3, 4,$ or 5 and R_x is any one of the series resistors.

The current is equal to the source voltage divided by the total resistance. For our example circuit of Figure 5–38, the total resistance is $R_1 + R_2 + R_3 + R_4 + R_5$, and the current is

$$I = \frac{V_S}{R_T}$$

Substituting V_S/R_T for I in the expression for V_x, we get

$$V_x = \left(\frac{V_S}{R_T}\right)R_x$$

By rearranging, we get

$$V_x = \left(\frac{R_x}{R_T}\right)V_S \qquad\qquad (5\text{--}6)$$

Equation (5–6) is the *general voltage divider formula*. It tells us the following:

The voltage drop across any resistor or combination of resistors in a series circuit is equal to the ratio of that resistance value to the total resistance, multiplied by the source voltage.

The following three examples illustrate use of the voltage divider formula.

EXAMPLE 5–16

Determine the voltage across R_1 and the voltage across R_2 in the voltage divider in Figure 5–39.

FIGURE 5–39

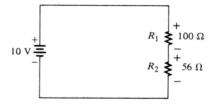

Solution:
Use the voltage divider formula, $V_x = (R_x/R_T)V_S$. In this problem we are looking for V_1; so $V_x = V_1$ and $R_x = R_1$. The total resistance is

$$R_T = R_1 + R_2 = 100\ \Omega + 56\ \Omega = 156\ \Omega$$

R_1 is 100 Ω and V_S is 10 V. Substituting these values into the voltage divider formula, we have

$$V_1 = \left(\frac{R_1}{R_T}\right)V_S = \left(\frac{100\ \Omega}{156\ \Omega}\right)10\ \text{V}$$

$$= 6.41\ \text{V}$$

There are two ways to find the value of V_2 in this problem: Kirchhoff's voltage law or the voltage divider formula.

First, using Kirchhoff's voltage law, we know that $V_S = V_1 + V_2$. By substituting the values for V_S and V_1, we can solve for V_2 as follows:

$$V_2 = 10 \text{ V} - 6.41 \text{ V} = 3.59 \text{ V}$$

Second, using the voltage divider formula, we have

$$V_2 = \left(\frac{R_2}{R_T}\right)V_S = \left(\frac{56 \text{ }\Omega}{156 \text{ }\Omega}\right)10 \text{ V}$$

$$= 3.59 \text{ V}$$

We get the same result either way.

EXAMPLE 5–17

Calculate the voltage drop across each resistor in the voltage divider of Figure 5–40.

FIGURE 5–40

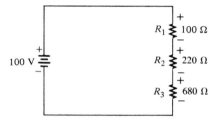

Solution:
Look at the circuit for a moment and consider the following: The total resistance is 1000 Ω. We can examine the circuit and determine that 10% of the total voltage is across R_1 because it is 10% of the total resistance (100 Ω is 10% of 1000 Ω). Likewise, we see that 22% of the total voltage is dropped across R_2 because it is 22% of the total resistance (220 Ω is 22% of 1000 Ω). Finally, R_3 drops 68% of the total voltage because 680 Ω is 68% of 1000 Ω.

Because of the convenient values in this problem, it is easy to figure the voltages mentally. Such is not always the case, but sometimes a little thinking will produce a result more efficiently and eliminate some calculating. This is also a good way to roughly estimate what your results should be so that you will recognize an unreasonable answer as a result of a calculation or measurement error.

Although we have already reasoned through this problem, the calculations will verify our results:

$$V_1 = \left(\frac{R_1}{R_T}\right)V_S = \left(\frac{100\ \Omega}{1000\ \Omega}\right)100\ V = 10\ V$$

$$V_2 = \left(\frac{R_2}{R_T}\right)V_S = \left(\frac{220\ \Omega}{1000\ \Omega}\right)100\ V = 22\ V$$

$$V_3 = \left(\frac{R_3}{R_T}\right)V_S = \left(\frac{680\ \Omega}{1000\ \Omega}\right)100\ V = 68\ V$$

Notice that the sum of the voltage drops is equal to the source voltage, in accordance with Kirchhoff's voltage law. This check is a good way to verify your results.

$$10\ V + 22\ V + 68\ V = 100\ V$$

EXAMPLE 5–18

Determine the voltages between the following points in the voltage divider of Figure 5–41:

(a) A to B (b) A to C (c) B to C (d) B to D (e) C to D

FIGURE 5–41

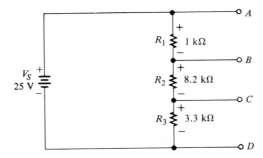

Solution:
First determine R_T:

$$R_T = 1\ k\Omega + 8.2\ k\Omega + 3.3\ k\Omega = 12.5\ k\Omega$$

Now apply the voltage divider formula to obtain each required voltage:

(a) The voltage A to B is the voltage drop across R_1. The calculation is as follows:

$$V_{AB} = \left(\frac{R_1}{R_T}\right)V_S = \left(\frac{1\ k\Omega}{12.5\ k\Omega}\right)25\ V$$

$$= 2\ V$$

(b) The voltage from A to C is the combined voltage drop across both R_1 and R_2. In this case, R_x in the general formula [Equation (5–6)] is $R_1 + R_2$. The calculation is as follows:

$$V_{AC} = \left(\frac{R_1 + R_2}{R_T}\right)V_S = \left(\frac{9.2 \text{ k}\Omega}{12.5 \text{ k}\Omega}\right)25 \text{ V}$$
$$= 18.4 \text{ V}$$

(c) The voltage from B to C is the voltage drop across R_2. The calculation is as follows:

$$V_{BC} = \left(\frac{R_2}{R_T}\right)V_S = \left(\frac{8.2 \text{ k}\Omega}{12.5 \text{ k}\Omega}\right)25 \text{ V}$$
$$= 16.4 \text{ V}$$

(d) The voltage from B to D is the combined voltage drop across both R_2 and R_3. In this case, R_x in the general formula is $R_2 + R_3$. The calculation is as follows:

$$V_{BD} = \left(\frac{R_2 + R_3}{R_T}\right)V_S = \left(\frac{11.5 \text{ k}\Omega}{12.5 \text{ k}\Omega}\right)25 \text{ V}$$
$$= 23 \text{ V}$$

(e) Finally, the voltage from C to D is the voltage drop across R_3. The calculation is as follows:

$$V_{CD} = \left(\frac{R_3}{R_T}\right)V_S = \left(\frac{3.3 \text{ k}\Omega}{12.5 \text{ k}\Omega}\right)25 \text{ V}$$
$$= 6.6 \text{ V}$$

If you connect this voltage divider in the lab, you can verify each of the calculated voltages by connecting a voltmeter between the appropriate points in each case.

THE POTENTIOMETER AS AN ADJUSTABLE VOLTAGE DIVIDER

Recall from Chapter 2 that a potentiometer is a variable resistor with three terminals. A potentiometer connected to a voltage source is shown in Figure 5–42(a). Notice that the two end terminals are labeled 1 and 2. The adjustable terminal or wiper is labeled 3. The potentiometer acts as a voltage divider. We can illustrate this concept better by separating the total resistance into two parts, as shown in Figure 5–42(b). The resistance between terminal 1 and terminal 3 (R_{13}) is one part, and the resistance between terminal 3 and terminal 2 (R_{32}) is the other part. So this potentiometer actually is a two-resistor voltage divider that can be manually adjusted.

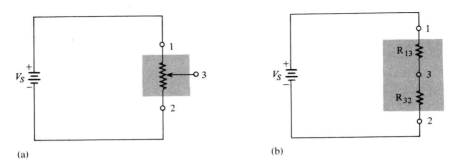

FIGURE 5–42
The potentiometer as a voltage divider.

Figure 5–43 shows what happens when the wiper terminal (3) is moved. In Part (a) of Figure 5–43, the wiper is exactly centered, making the two resistances equal. If we measure the voltage across terminals 3 to 2 as indicated by the voltmeter symbol, we have one-half of the total source voltage. When the wiper is moved up, as in Figure 5–43(b), the resistance between terminals 3 and 2 increases, and the voltage across it increases proportionally. When the wiper is moved down, as in Figure 5–43(c), the resistance between terminals 3 and 2 decreases, and the voltage decreases proportionally.

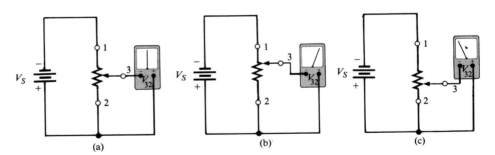

FIGURE 5–43
Adjusting the voltage divider.

APPLICATIONS OF VOLTAGE DIVIDERS

The volume control of radio or TV receivers is a common application of a potentiometer used as a voltage divider. Since the loudness of the sound is dependent on the amount of voltage associated with the audio signal, you can increase or decrease the volume by adjusting the potentiometer, that is, by turning the knob of the volume control on the set. The block diagram in Figure 5–44 shows how a potentiometer can be used for volume control in a typical receiver.

Another application of a voltage divider is illustrated in Figure 5–45, which depicts a potentiometer voltage divider as a fuel-level sensor in an automobile gas tank.

FIGURE 5–44
A voltage divider used for volume control.

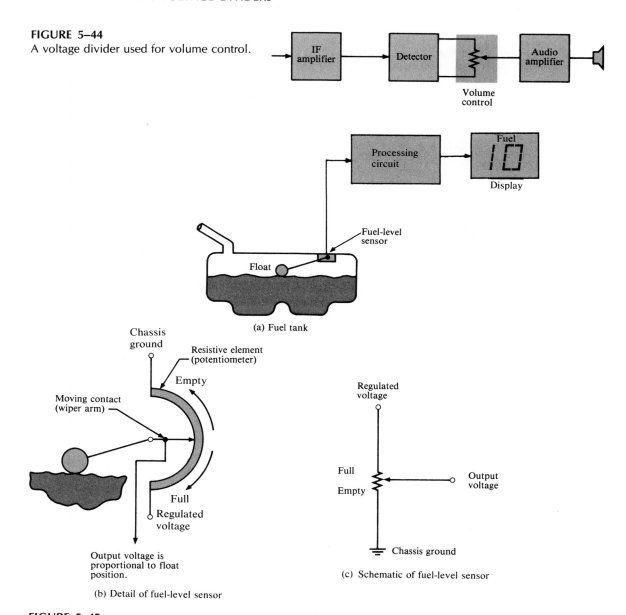

(a) Fuel tank

(b) Detail of fuel-level sensor

(c) Schematic of fuel-level sensor

FIGURE 5–45
A potentiometer voltage divider used as an automotive fuel-level sensor.

As shown in Part (a), the float moves up as the tank is filled and moves down as the tank empties. The float is mechanically linked to the wiper arm of a potentiometer, as shown in Part (b). The output voltage varies proportionally with the position of the wiper arm. As the fuel in the tank decreases, the sensor output voltage also decreases.

The output voltage goes to the indicator circuitry, which controls the fuel gauge or digital readout to show the fuel level. The schematic of this system is shown in Part (c).

Still another application for voltage dividers is in setting the dc operating voltage *(bias)* in transistor amplifiers. Figure 5–46 shows a voltage divider used for this purpose. You will study transistor amplifiers and biasing later, so it is important that you understand the basics of voltage dividers at this point.

These examples are only three out of many possible applications of voltage dividers.

FIGURE 5–46
The voltage divider as a bias circuit for a transistor amplifier.

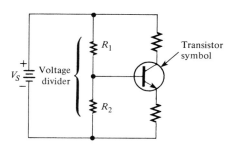

SECTION REVIEW 5–7

1. What is a voltage divider?

2. How many resistors can there be in a series voltage divider circuit?

3. Write the general formula for voltage dividers.

4. If two series resistors of equal value are connected across a 10-V source, how much voltage is there across each resistor?

5. A 47-Ω resistor and a 82-Ω resistor are connected as a voltage divider. The source voltage is 100 V. Sketch the circuit, and determine the voltage across each of the resistors.

6. The circuit of Figure 5–47 is an adjustable voltage divider. If the potentiometer is linear, where would you set the wiper in order to get 5 V from *A* to *B* and 5 V from *B* to *C*?

FIGURE 5–47

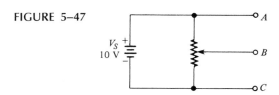

5–8

POWER IN A SERIES CIRCUIT

The total amount of power in a series resistive circuit is equal to the sum of the powers in each resistor in series:

$$P_T = P_1 + P_2 + P_3 + \cdots + P_n \qquad (5-7)$$

where n is the number of resistors in series, P_T is the total power, and P_n is the power in the last resistor in series. In other words, the powers are additive.

The power formulas that you learned in Chapter 4 are, of course, directly applicable to series circuits. Since there is the same current through each resistor in series, the following formulas are used to calculate the total power:

$$P_T = V_S I$$

$$P_T = I^2 R_T$$

$$P_T = \frac{V_S^2}{R_T}$$

where V_S is the total source voltage across the series connection and R_T is the total resistance. Example 5–19 illustrates how to calculate total power in a series circuit.

EXAMPLE 5–19

Determine the total amount of power in the series circuit in Figure 5–48.

FIGURE 5–48

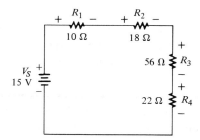

Solution:

We know that the source voltage is 15 V. The total resistance is

$$R_T = 10 \ \Omega + 18 \ \Omega + 56 \ \Omega + 22 \ \Omega = 106 \ \Omega$$

The easiest formula to use is $P_T = V_S^2/R_T$ since we know both V_S and R_T:

$$P_T = \frac{V_S^2}{R_T} = \frac{(15 \text{ V})^2}{106 \ \Omega} = \frac{225 \text{ V}^2}{106 \ \Omega}$$

$$= 2.12 \text{ W}$$

If the power in each resistor is determined separately and all of these powers are added, the same result is obtained. We will rework the problem using this approach. First, find the current as follows:

$$I = \frac{V_S}{R_T} = \frac{15 \text{ V}}{106 \text{ }\Omega}$$

$$= 0.142 \text{ A}$$

Next, calculate the power for each resistor using $P = I^2R$:

$$P_1 = (0.142 \text{ A})^2(10 \text{ }\Omega) = 0.202 \text{ W}$$
$$P_2 = (0.142 \text{ A})^2(18 \text{ }\Omega) = 0.363 \text{ W}$$
$$P_3 = (0.142 \text{ A})^2(56 \text{ }\Omega) = 1.129 \text{ W}$$
$$P_4 = (0.142 \text{ A})^2(22 \text{ }\Omega) = 0.444 \text{ W}$$

Now, add these powers to get the total power:

$$P_T = 0.202 \text{ W} + 0.363 \text{ W} + 1.129 \text{ W} + 0.444 \text{ W}$$

$$= 2.138 \text{ W}$$

This result compares closely to the total power as determined previously by the formula $P_T = V_S^2/R_T$. The small difference is due to rounding.

The amount of power in a resistor is important because the power rating of the resistors must be high enough to handle the expected power in the circuit. The following example illustrates practical considerations relating to power in a series circuit.

EXAMPLE 5–20

Determine if the indicated power rating (1/2 W) of each resistor in Figure 5–49 is sufficient to handle the actual power. If a rating is not adequate, specify the required minimum rating.

FIGURE 5–49

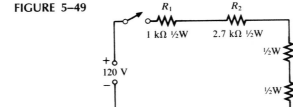

Solution:

$$R_T = R_1 + R_2 + R_3 + R_4 = 1\ k\Omega + 2.7\ k\Omega + 910\ \Omega + 3.3\ k\Omega = 7.91\ k\Omega$$

$$I = \frac{V_S}{R_T} = \frac{120\ V}{7.91\ k\Omega} = 15.17\ mA$$

The power in each resistor is

$$P_1 = I^2R_1 = (15.17\ mA)^2(1\ k\Omega) = 0.230\ W$$

$$P_2 = I^2R_2 = (15.17\ mA)^2(2.7\ k\Omega) = 0.621\ W$$

$$P_3 = I^2R_3 = (15.17\ mA)^2(910\ \Omega) = 0.209\ W$$

$$P_4 = I^2R_4 = (15.17\ mA)^2(3.3\ k\Omega) = 0.759\ W$$

R_2 and R_4 do not have a rating sufficient to handle the actual power, which exceeds 1/2 W in each of these two resistors, and they will burn out if the switch is closed. These resistors must be replaced by 1-W resistors.

SECTION REVIEW 5–8

1. If you know the power in each resistor in a series circuit, how can you find the total power?

2. The resistors in a series circuit dissipate the following powers: 2 W, 5 W, 1 W, and 8 W. What is the total power in the circuit?

3. A circuit has a 100-Ω, a 330-Ω, and a 680-Ω resistor in series. There is a current of 1 A through the circuit. What is the total power?

5–9   CIRCUIT GROUND

Voltage is relative. That is, the voltage at one point in a circuit is always measured relative to another point. For example, if we say that there are +100 V at a certain point in a circuit, we mean that the point is 100 V more positive than some *reference point* in the circuit. This *reference point* in a circuit is usually called *ground*.

The term *ground* derives from the method used in ac power lines, in which one side of the line is neutralized by connecting it to a water pipe or a metal rod driven into the ground. This method of grounding is called *earth ground*.

In most electronic equipment, the metal chassis that houses the assembly or a large conductive area on a printed circuit board is used as the *common* or *reference point*, called the *chassis ground* or *circuit ground*. This ground provides a convenient way of connecting all common points within the circuit back to one side of the battery or other energy source. The chassis or circuit ground does not necessarily have to be connected to earth ground. However, in many cases it is earth grounded in order to prevent a shock hazard due to a potential difference between chassis and earth ground.

In summary, ground is the reference point in electronic circuits. It has a potential of zero volts (0 V) *with respect to all other points in the circuit that are referenced to it,* as illustrated in Figure 5–50. In Part (a), the negative side of the source is grounded, and all voltages indicated are positive with respect to ground. In Part (b), the positive side of the source is ground. The voltages at all other points are therefore negative with respect to ground.

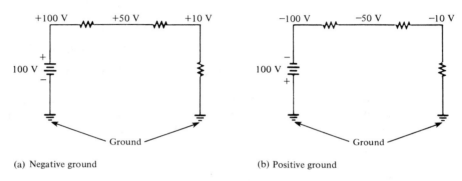

(a) Negative ground   (b) Positive ground

FIGURE 5–50
Example of negative and positive grounds.

MEASUREMENT OF VOLTAGES WITH RESPECT TO GROUND

When voltages are measured in a circuit, one meter lead is connected to the circuit ground, and the other to the point at which the voltage is to be measured. In a negative ground circuit, the negative meter terminal is connected to the circuit ground. The positive terminal of the voltmeter is then connected to the positive voltage point. Measurement of positive voltage is illustrated in Figure 5–51, where the meter reads the voltage at point *A* with respect to ground.

FIGURE 5–51
Measuring a voltage with respect to negative ground.

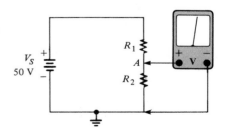

For a circuit with a positive ground, the positive voltmeter lead is connected to ground, and the negative lead is connected to the negative voltage point, as indicated in Figure 5–52. Here the meter reads the voltage at point *A* with respect to ground.

FIGURE 5–52
Measuring a voltage with respect to positive ground.

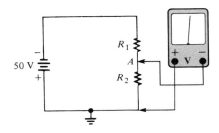

When voltages must be measured at several points in a circuit, the ground lead can be clipped to ground at one point in the circuit and left there. The other lead is then moved from point to point as the voltages are measured. This method is illustrated in Figure 5–53.

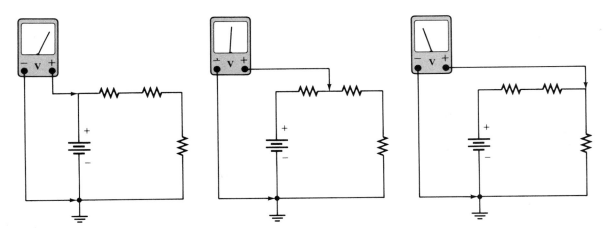

FIGURE 5–53
Measuring voltages at several points in a circuit.

MEASUREMENT OF VOLTAGE ACROSS AN UNGROUNDED RESISTOR

Voltage can normally be measured across a resistor, as shown in Figure 5–54, even though neither side of the resistor is connected to circuit ground. In some cases, when

FIGURE 5–54
Measuring voltage across a resistor.

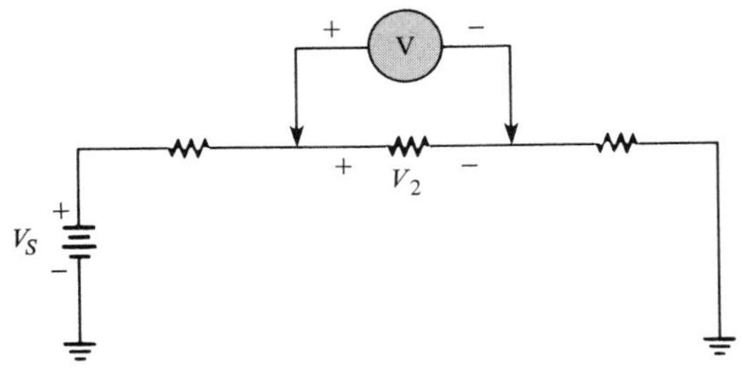

the meter is not isolated from power line ground, the negative lead of the meter will ground one side of the resistor and alter the operation of the circuit. In this situation, another method must be used, as illustrated in Figure 5–55. The voltages on each side of the resistor are measured *with respect to ground*. The *difference* of these two measurements is the voltage drop across the resistor.

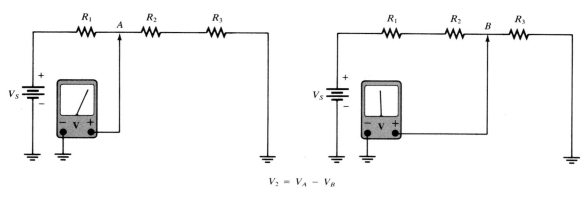

$$V_2 = V_A - V_B$$

FIGURE 5–55
Measuring voltage across a resistor with two separate measurements to ground.

EXAMPLE 5–21
Determine the voltages of each of the indicated points in each circuit of Figure 5–56. Assume that 25 V are dropped across each resistor.

FIGURE 5–56

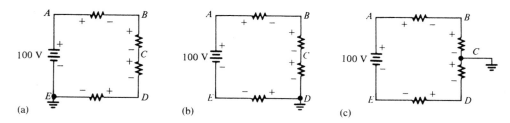

(a)　　　　(b)　　　　(c)

Solution:
In Circuit (a), the voltage polarities are as shown. Point E is ground. Single-letter subscripts denote voltage at a point with respect to ground. The voltages with respect to ground are as follows:

$$V_E = 0 \text{ V}$$
$$V_D = +25 \text{ V}$$
$$V_C = +50 \text{ V}$$
$$V_B = +75 \text{ V}$$
$$V_A = +100 \text{ V}$$

In Circuit (b), the voltage polarities are as shown. Point D is ground. The voltages with respect to ground are as follows:

$$V_E = -25 \text{ V}$$
$$V_D = 0 \text{ V}$$
$$V_C = +25 \text{ V}$$
$$V_B = +50 \text{ V}$$
$$V_A = +75 \text{ V}$$

In Circuit (c), the voltage polarities are as shown. Point C is ground. The voltages with respect to ground are as follows:

$$V_E = -50 \text{ V}$$
$$V_D = -25 \text{ V}$$
$$V_C = 0 \text{ V}$$
$$V_B = +25 \text{ V}$$
$$V_A = +50 \text{ V}$$

SECTION REVIEW 5–9

1. The common point in a circuit is called _____.

2. Most voltages in a circuit are referenced to ground (T or F).

3. The housing or chassis is often used as circuit ground (T or F).

4. What is the symbol for ground?

5. What does *earth ground* mean?

5–10 TROUBLESHOOTING SERIES CIRCUITS

OPEN CIRCUIT

The most common failure in a series circuit is an *open*. For example, when a resistor or a lamp burns out, it creates a break in the current path, as illustrated in Figure 5–57.

An open in a series circuit prevents current.

How to Check for an Open Element Sometimes a visual check will reveal a charred resistor or an open lamp filament. However, it is possible for a resistor to open without showing visible signs of damage. In this situation, a *voltage check* of the series circuit is required. The general procedure is as follows: Measure the voltage across each resistor in series. *The voltage across all of the good resistors will be zero. The voltage across the open resistor will equal the total voltage across the series combination.*

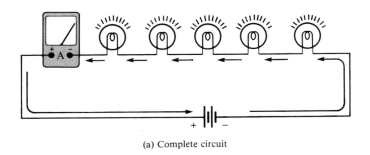

(a) Complete circuit

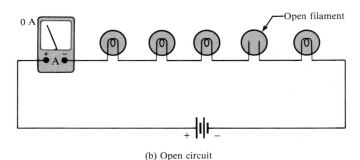

(b) Open circuit

FIGURE 5–57
An open circuit prevents current.

The above condition occurs because an open resistor will prevent current through the series circuit. With no current, there can be no voltage drop across any of the good resistors. Since $IR = 0$, in accordance with *Ohm's law*, the voltage on each side of a good resistor is the same. The total voltage must then appear across the open resistor in accordance with *Kirchhoff's voltage law*, as illustrated in Figure 5–58. To fix the circuit, replace the open resistor.

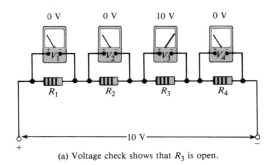

(a) Voltage check shows that R_3 is open.

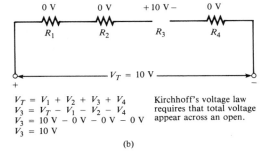

$V_T = V_1 + V_2 + V_3 + V_4$
$V_3 = V_T - V_1 - V_2 - V_4$
$V_3 = 10\text{ V} - 0\text{ V} - 0\text{ V} - 0\text{ V}$
$V_3 = 10\text{ V}$

Kirchhoff's voltage law requires that total voltage appear across an open.

(b)

FIGURE 5–58
Troubleshooting a series circuit for an open element.

SHORT CIRCUIT

Sometimes a short occurs when two conductors touch or a foreign object such as solder or a wire clipping accidentally connects two sections of a circuit together. This situation is particularly common in circuits with a high component density. Several potential causes of short circuits are illustrated on the PC board in Figure 5–59.

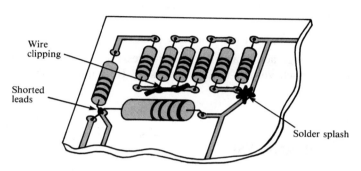

FIGURE 5–59
Examples of shorts on a PC board.

When there is a short, a portion of the series resistance is bypassed (all of the current goes through the short), thus reducing the total resistance as illustrated in Figure 5–60. Notice that the current increases as a result of the short.

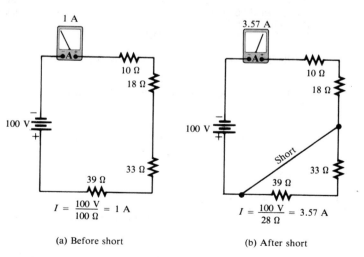

FIGURE 5–60
The effect of a short in a series circuit.

SECTION REVIEW 5–10

1. Define *short circuit.*

2. Define *open circuit.*

3. What happens when a series circuit opens?

4. Name two general ways in which an open circuit can occur in practice. A short circuit.

5. When a resistor fails, it will normally open (T or F).

6. The total voltage across a string of series resistors is 24 V. If one of the resistors is open, how much voltage is there across it? How much is there across each of the good resistors?

5–11　COMPUTER ANALYSIS

The program that follows can be used to compute the voltage across each resistor in a voltage divider with any specified number of resistors and a specified source voltage. The program computes and tabulates the results.

```
10   CLS
20   PRINT "THIS PROGRAM COMPUTES AND TABULATES THE VOLTAGES"
30   PRINT "ACROSS THE RESISTORS IN A SPECIFIED VOLTAGE"
40   PRINT "DIVIDER CIRCUIT."
50   PRINT:PRINT:PRINT
60   PRINT "YOU ARE REQUIRED TO ENTER THE NUMBER OF RESISTORS,"
70   PRINT "THE RESISTOR VALUES, AND THE SOURCE VOLTAGE."
80   PRINT:PRINT:PRINT
90   INPUT "TO CONTINUE PRESS 'ENTER'.";X
100  CLS
110  INPUT "HOW MANY RESISTORS ARE IN THE VOLTAGE DIVIDER";N
120  CLS
130  FOR X=1 TO N
140  PRINT "ENTER THE VALUE OF R";X;"IN OHMS"
150  INPUT R(X)
160  NEXT X
170  CLS
180  INPUT "THE SOURCE VOLTAGE";VS
190  CLS
200  FOR X=1 TO N
210  RT=RT+R(X)
220  NEXT X
230  FOR X=1 TO N
240  V(X)=(R(X)/RT)*VS
250  PRINT "R";X;"=";R(X);"OHMS","V";X;"=";V(X);"VOLTS"
260  NEXT X
```

The flowchart for this program is shown in Figure 5–61.

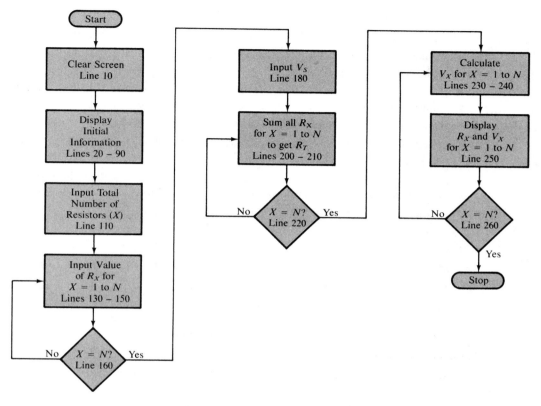

FIGURE 5–61
Program flowchart.

SECTION REVIEW 5–11

1. There are three FOR/NEXT loops in the program. Identify each by line numbers, and explain their purpose.
2. What variable stands for the number of resistors in the divider circuit?

SUMMARY

1. Resistors in series are connected end-to-end.
2. A series connection has only *one* path for current.

3. There is the same amount of current at all points in a series circuit.

4. The total series resistance is the sum of all resistors in the series circuit.

5. The total resistance between any two points in a series circuit is equal to the sum of all resistances connected in series between those two points.

6. If all of the resistors in a series connection are of equal resistance value, the total resistance is the number of resistors multiplied by the value.

7. Voltage sources in series add algebraically.

8. First statement of Kirchhoff's voltage law: The sum of all the voltages around a closed path is zero.

9. Second statement of Kirchhoff's voltage law: The sum of the voltage drops equals the total source voltage. (Items 8 and 9 say the same thing.)

10. The voltage drops in a circuit are always opposite in polarity to the total source voltage.

11. Current is out of the negative side of a source and into the positive side.

12. Current is into the negative side of each resistor and out of the more positive side.

13. A voltage divider is a series arrangement of resistors.

14. A voltage divider is so named because the voltage drop across any resistor in the series circuit is divided down from the total voltage by an amount proportional to that resistance value in relation to the total resistance.

15. A potentiometer can be used as an adjustable voltage divider.

16. The total power in a resistive circuit is the sum of all the individual powers of the resistors making up the series circuit.

17. *Ground* is the common or reference point in a circuit.

18. All voltages in a circuit are referenced to ground unless otherwise specified.

19. Ground is zero volts with respect to all points referenced to it in the circuit.

20. *Negative ground* is the term used when the negative side of the source is grounded.

21. *Positive ground* is the term used when the positive side of the source is grounded.

22. An open circuit is one in which the current is interrupted.

23. A short in a series circuit causes an increase in current.

24. The voltage across an open series element equals the source voltage.

FORMULAS

$$R_T = R_1 + R_2 + R_3 + \cdots + R_n \qquad (5\text{--}1)$$

$$R_T = nR \qquad (5\text{--}2)$$

$$V_S - V_1 - V_2 - V_3 = 0 \qquad (5\text{--}3)$$

$$V_S = V_1 + V_2 + V_3 \qquad (5\text{--}4)$$

$$V_{S1} + V_{S2} + \cdots + V_{Sn} = V_1 + V_2 + \cdots + V_n \qquad (5\text{--}5)$$

$$V_x = \left(\frac{R_x}{R_T}\right)V_S \qquad (5\text{--}6)$$

$$P_T = P_1 + P_2 + P_3 + \cdots + P_n \qquad (5\text{--}7)$$

SELF-TEST

Solutions appear at the end of the book.

1. Sketch a series circuit having four resistors in series with a voltage source.
2. There are five amperes of current into a series string of ten resistors. What is the current out of the sixth resistor? The tenth resistor?
3. Connect each set of resistors in Figure 5–62 in series between points A and B.
4. If you measured the current between each of the points marked in Figure 5–63, how much current would you read in each case? The total current out of the source is 3 A.

(a)

(b)

(c)

FIGURE 5–62

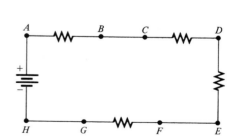

FIGURE 5–63

5. If an 82-Ω resistor and a 470-Ω resistor are connected in series, what is the total resistance?

6. Eight 56-Ω resistors are in series. Determine the total resistance.

7. Suppose that you need a total resistance of 22 kΩ. The resistors that you have available are two 1 kΩ, one 5.6 kΩ, and one 3.3 kΩ. What other single resistor do you need to make the required total?

8. Three 1-kΩ resistors are connected in series, and 5 V are applied across the series circuit. How much current is there through each resistor?

9. Which circuit in Figure 5–64 has more current?

10. Six resistors of equal value are in series with a 12-V source. The current through the circuit is 2 mA. What is the total resistance? Determine the value of each resistor.

11. Determine the total source voltage in the circuit in Figure 5–65.

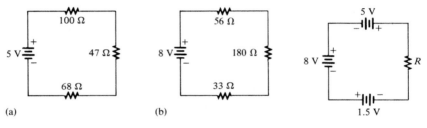

(a)　　　　　　　　　(b)

FIGURE 5–64　　　　　　　　　　　　　　FIGURE 5–65

12. Two 9-V batteries are connected in opposite directions in a series circuit. What is the total source voltage?

13. Show how to connect two 12-V automobile batteries to get 24 V.

14. Suppose you put four 1.5-V batteries in a flashlight, but you accidentally put one of them in backwards. Will the light be brighter or dimmer than it should be? Why?

15. Determine the value and the polarity of the total source voltage in each circuit of Figure 5–66. Sketch the equivalent single-source circuit.

FIGURE 5–66

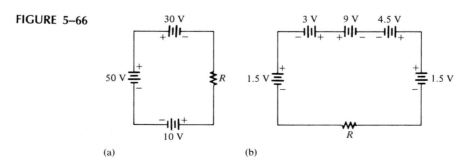

(a)　　　　　　　　　(b)

16. If you measure all the voltage drops and the source voltage in a series circuit and add them together, taking into consideration the polarities, what result will you get? What circuit law does this result verify?

17. There are six resistors in a given series circuit, and each resistor has 5 V dropped across it. What is the source voltage?

18. Five equal-value resistors are in series with a voltage source. The voltage drop across the first resistor is 2 V. What is the value of the source voltage?

19. A series circuit consists of a 4.7-kΩ resistor, a 5.6-kΩ resistor, and a 10-kΩ resistor. Which resistor has the most voltage across it?

20. A circuit with a 50-V source has three resistors in series. The voltage drop of the first resistor is 10 V. The voltage drop of the second resistor is 15 V. What is the voltage drop of the third resistor?

21. Determine the unknown resistance (R_3) in the circuit in Figure 5–67.

FIGURE 5–67

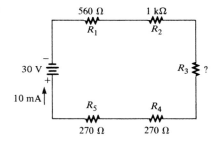

22. The following resistors (one each) are connected in series: 10 Ω, 47 Ω, 100 Ω, and 39 Ω. The total voltage is 20 V. Using the voltage divider formula, calculate the voltage drop across the 47-Ω resistor.

23. Calculate the voltage across each resistor in the voltage divider of Figure 5–68.

FIGURE 5–68

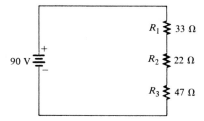

24. You have a 9-V battery available and need 3 V to bias a transistor amplifier for proper operation. You must use two resistors to form a voltage divider that will divide the 9 V down to 3 V. Determine the ideal values needed to achieve this if their total must be 100 kΩ.

25. In a series circuit with three resistors, one handles 2.5 W, one handles 5 W, and one handles 1.2 W. What is the total power in the circuit?

26. Which dissipates more power when connected across a 100-V source, one 100-Ω resistor or two 100-Ω resistors in series?

27. The total power in a given series circuit is 10 W. There are five equal-value resistors in the circuit. How much power does each resistor dissipate?

28. When you connect an ammeter in a given series circuit and turn on the source voltage, the meter reads zero. What should you check for, and how will you know when you have found the problem?

29. You are checking out a series circuit and find that the current is higher than it should be. What type of problem should you look for?

PROBLEMS

Section 5–1

5–1 Connect each set of resistors in Figure 5–69 in series between points A and B.

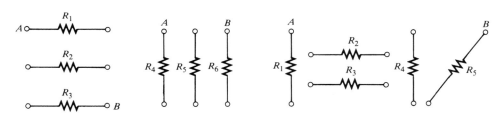

FIGURE 5–69

5–2 Determine which resistors in Figure 5–70 are in series. Show how to interconnect the pins to put all the resistors in series.

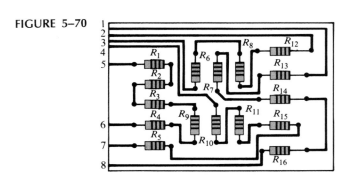

FIGURE 5–70

5–3 On the double-sided PC board in Figure 5–71, identify each group of series resistors. Note that many of the interconnections feed through the board from the top side to the bottom side.

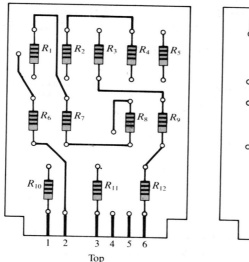
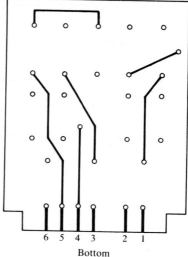

Top   Bottom

FIGURE 5–71

Section 5–2

5–4 What is the current through each resistor in a series circuit if the total voltage is 12 V and the total resistance is 120 Ω?

5–5 The current from the source in Figure 5–72 is 5 mA. How much current does each milliammeter in the circuit indicate?

FIGURE 5–72

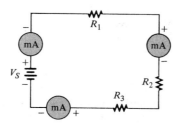

5–6 Show how to connect a voltage source and an ammeter to the PC board in Figure 5–70 to measure the current in R_1. Which other resistor currents are measured by this setup?

5–7 Using 1.5-V batteries, a switch, and three lamps, devise a circuit to apply 4.5 V across either one lamp, two lamps in series, or three lamps in series with a single-control switch. Draw the schematic diagram.

Section 5–3

5–8   The following resistors (one each) are connected in a series circuit: 1 Ω, 2.2 Ω, 5.6 Ω, 12 Ω, and 22 Ω. Determine the total resistance.

5–9   Find the total resistance of each of the following groups of series resistors:
(a) 560 Ω and 1000 Ω   (b) 47 Ω and 56 Ω
(c) 1.5 kΩ, 2.2 kΩ, and 10 kΩ   (d) 1 MΩ, 470 kΩ, 1 kΩ, 2.2 MΩ

5–10   Calculate R_T for each circuit of Figure 5–73.

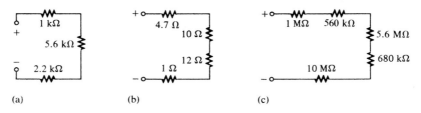

(a)   (b)   (c)

FIGURE 5–73

5–11   What is the total resistance of twelve 5.6-kΩ resistors in series?

5–12   Six 56-Ω resistors, eight 100-Ω resistors, and two 22-Ω resistors are all connected in series. What is the total resistance?

5–13   If the total resistance in Figure 5–74 is 17.4 kΩ, what is the value of R_5?

FIGURE 5–74

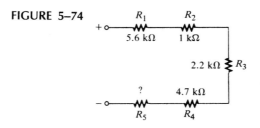

5–14   You have the following resistor values available to you in the lab in unlimited quantities: 10 Ω, 100 Ω, 470 Ω, 560 Ω, 680 Ω, 1 kΩ, 2.2 kΩ, and 5.6 kΩ. All of the other standard values are out of stock. A project that you are working on requires an 18-kΩ resistance. What combinations of the available values would you use in series to achieve this total resistance?

5–15   Find the total resistance if all the circuits in Figure 5–73 are connected in series.

5–16   What is the total resistance from *A* to *B* for each switch position in Figure 5–75?

FIGURE 5-75

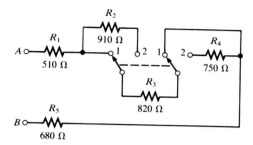

Section 5-4

5-17 What is the current in each circuit of Figure 5-76?

FIGURE 5-76

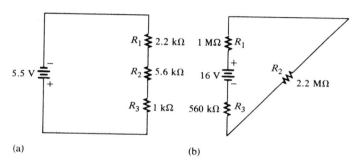

(a)   (b)

5-18 Determine the voltage drop across each resistor in Figure 5-76.

5-19 Three 470-Ω resistors are connected in series with a 500-V source. How much current is in the circuit?

5-20 Four equal-value resistors are in series with a 5-V battery, and 2.23 mA are measured. What is the value of each resistor?

5-21 What is the value of each resistor in Figure 5-77?

5-22 Determine V_{R1}, R_2, and R_3 in Figure 5-78.

FIGURE 5-77   **FIGURE 5-78**

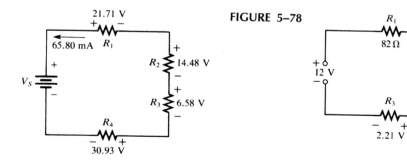

5–23 Determine the current measured by the meter in Figure 5–79 for each switch position.

5–24 Determine the current measured by the meter in Figure 5–80 for each position of the ganged switch.

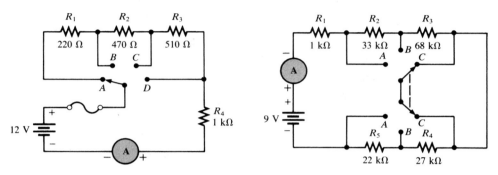

FIGURE 5–79   FIGURE 5–80

Section 5–5

5–25 *Series aiding* is a term sometimes used to describe voltage sources of the *same* polarity in series. If a 5-V and a 9-V source are connected in this manner, what is the total voltage?

5–26 The term *series opposing* means that sources are in series with *opposite* polarities. If a 12-V and a 3-V battery are series opposing, what is the total voltage?

5–27 Determine the total source voltage in each circuit of Figure 5–81.

FIGURE 5–81

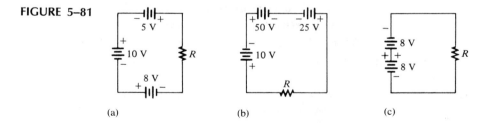

(a)   (b)   (c)

Section 5–6

5–28 The following voltage drops are measured across three resistors in series: 5.5 V, 8.2 V, and 12.3 V. What is the value of the source voltage to which these resistors are connected?

5–29 Five resistors are in series with a 20-V source. The voltage drops across four of the resistors are 1.5 V, 5.5 V, 3 V, and 6 V. How much voltage is dropped across the fifth resistor?

5–30 Determine the unspecified voltage drop(s) in each circuit of Figure 5–82. Show how to connect a voltmeter to measure each unknown voltage drop.

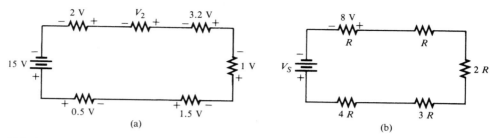

FIGURE 5–82

5–31 In the circuit of Figure 5–83, determine the resistance of R_4.

5–32 Find R_1, R_2, and R_3 in Figure 5–84.

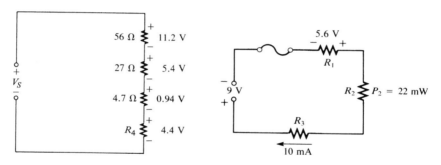

FIGURE 5–83   FIGURE 5–84

5–33 Determine the voltage across R_5 for each position of the switch in Figure 5–85. The current in each position is as follows: A, 3.35 mA; B, 3.73 mA; C, 4.5 mA; D, 6 mA.

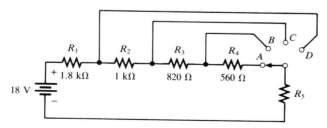

FIGURE 5–85

5–34 Determine the voltage across each resistor in Figure 5–85 if the switch is in position A and the source voltage is changed to 10 V.).

Section 5–7

5–35 The total resistance of a circuit is 560 Ω. What percentage of the total voltage appears across a 27-Ω resistor that makes up part of the total series resistance?

5–36 Determine the voltage between points A and B in each voltage divider of Figure 5–86.

5–37 What is the voltage across each resistor in Figure 5–87? R is the lowest-value resistor, and all others are multiples of that value as indicated.

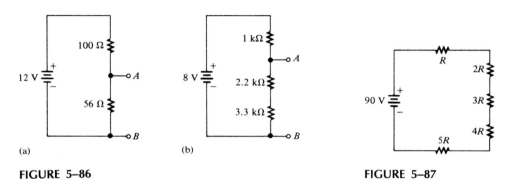

FIGURE 5–86   FIGURE 5–87

5–38 Determine the voltage at each point in Figure 5–88 with respect to the negative side of the battery.

5–39 If there are 10 V across R_1 in Figure 5–89, what is the voltage across each of the other resistors?

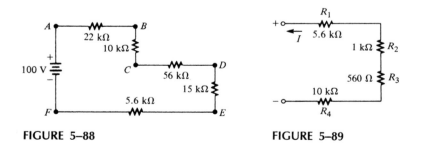

FIGURE 5–88   FIGURE 5–89

5–40 With the table of standard resistor values given in Appendix B, design a voltage divider to provide the following approximate voltages with respect to ground using a 30-V source: 8.18 V, 14.73 V, and 24.55 V. The current drain on the source must be limited to no more than 1 mA. The number of resistors, their ohmic values, and their wattage ratings must be specified. A schematic diagram showing the circuit arrangement and resistor placement must be provided.

5–41 Design a variable voltage divider to provide an output voltage adjustable from a minimum of 10 V to a maximum of 100 V using a 120-V source. The maximum voltage must occur at the maximum resistance setting of the potentiometer, and the minimum voltage must occur at the minimum resistance (zero) setting. The maximum current is to be 10 mA.

Section 5–8

5–42 Five series resistors each handle 50 mW. What is the total power?

5–43 What is the total power in the circuit in Figure 5–89? Use the results of Problem 5–39.

5–44 The following 1/4-W resistors are in series: 1.2 kΩ, 2.2 kΩ, 3.9 kΩ, and 5.6 kΩ. What is the maximum voltage that can be applied across the series resistors without exceeding a power rating? Which resistor will burn out first if excessive voltage is applied?

5–45 Find R_T in Figure 5–90.

FIGURE 5–90

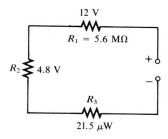

5–46 A certain series circuit consists of a 1/8-W resistor, a 1/4-W resistor, and a 1/2-W resistor. The total resistance is 2400 Ω. If each of the resistors is operating in the circuit at its maximum power dissipation, determine the following:
(a) I (b) V_T (c) The value of each resistor.

Section 5–9

5–47 Determine the voltage at each point with respect to ground in Figure 5–91.

5–48 In Figure 5–92, how would you determine the voltage across R_2 by measuring, without connecting a meter directly across the resistor?

5–49 Determine the voltage at each point with respect to ground in Figure 5–92.

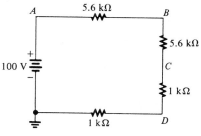

FIGURE 5–91

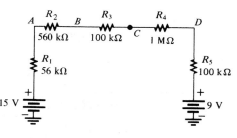

FIGURE 5–92

Section 5–10

5–50 A string of five series resistors is connected across a 12-V battery. Zero volts is measured across all of the resistors except R_2. What is wrong with the circuit? What voltage will be measured across R_2?

5–51 By observing the meters in Figure 5–93, determine the types of failures in the circuits and which components have failed.

FIGURE 5–93

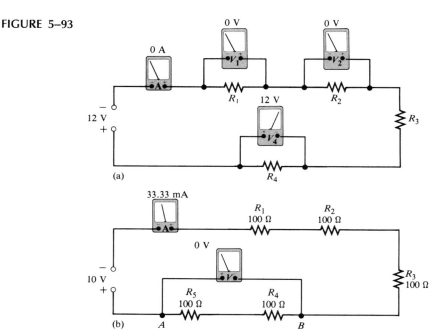

5–52 A circuit has a 2.2 kΩ, 5.6 kΩ, 10 kΩ, 8.2 kΩ, and a 12 kΩ in series. An ohmmeter connected across the circuit reads 28 kΩ. What is wrong?

Section 5–11

5–53 Modify the program and flowchart in Section 5–11 to compute and tabulate the power dissipation in each resistor in addition to the voltages.

5–54 Modify the program and flowchart in Section 5–11 to compute and tabulate the voltages across any specified combination of resistors in a series voltage divider rather than just the voltages across the individual resistors.

ANSWERS TO SECTION REVIEWS

Section 5–1

1. End-to-end.   **2.** There is a single current path.   **3.** See Figure 5–94.

FIGURE 5–94

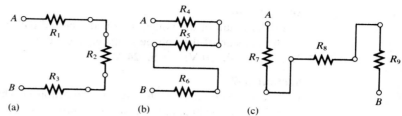

(a) (b) (c)

4. See Figure 5–95.

FIGURE 5–95

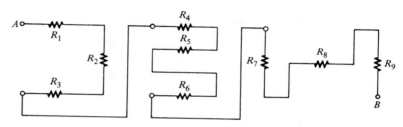

Section 5–2
1. 1 A.   **2.** 50 mA between C and D; 50 mA between E and F.
3. 1.79 A, 1.79 A.   **4.** Current is the same at all points.

Section 5–3
1. 11.2 Ω.   **2.** 590 Ω.   **3.** 3.9 kΩ.   **4.** 672 Ω.   **5.** 358 Ω.

Section 5–4
1. 0.033 A or 33 mA.   **2.** 106 V.   **3.** $V_1 = 50$ V, $V_2 = 28$ V, $V_3 = 28$ V.
4. 270 Ω.

Section 5–5
1. 6 V.   **2.** Five; see Figure 5–96.   **3.** See Figure 5–97.
4. **(a)** 75 V;   **(b)** 15 V.   **5.** See Figure 5–98.

FIGURE 5–96

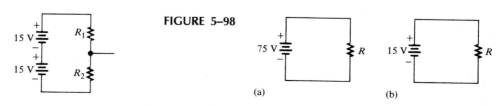

FIGURE 5–97

FIGURE 5–98

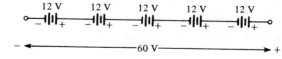

(a) (b)

Section 5–6

1. (a) The sum of the voltages around a closed path is zero;
(b) the sum of the voltage drops equals the total source voltage.   **2.** 50 V.
3. 5 V.   **4.** 10 V.   **5.** 24 V.

Section 5–7

1. Two or more resistors in a series connection in which the voltage taken across any resistor or combination of resistors is proportional to the value of that resistance.
2. Two or more.   **3.** $V_x = (R_x/R_T)V_S$.   **4.** 5 V.
5. 36.4 V across the 47-Ω; 63.6 V across the 82-Ω; see Figure 5–99.
6. At the midpoint.

FIGURE 5–99

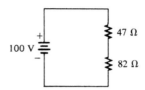

100 V   47 Ω 82 Ω

Section 5–8

1. Add the power in each resistor.   **2.** 16 W.   **3.** 1110 W.

Section 5–9

1. Ground.   **2.** T.   **3.** T.   **4.** See Figure 5–100.
5. A connection to earth through a metal rod or a water pipe.

FIGURE 5–100

Section 5–10

1. A zero resistance path that bypasses a portion of a circuit.
2. A break in the current path.   **3.** Current ceases.
4. An open can be created by a switch or by a component failure. A short can be created by a switch or, unintentionally, by a wire clipping or solder splash.
5. T.   **6.** 24 V; 0 V.

Section 5–11

1. Lines 130–160: Inputs resistor values. Lines 200–220: Calculates total resistance. Lines 230–260: Computes each voltage and prints out results.   **2.** N.

SIX

PARALLEL RESISTIVE CIRCUITS

In Chapter 5, you learned about series circuits. In this chapter, you will study circuits with parallel resistors.

In this chapter, you will learn:

☐ How to identify a parallel circuit.
☐ How to determine total resistance in a parallel circuit.
☐ How to apply Ohm's law to parallel circuits.
☐ How to use Kirchhoff's current law.
☐ How to analyze and apply current dividers.
☐ How to determine power in a parallel circuit.
☐ How to recognize the effects of open paths.
☐ How to troubleshoot parallel circuits.
☐ How to use a simple computer program for circuit analysis.

6–1   RESISTORS IN PARALLEL

When two or more components are connected across the same voltage source, they are in parallel. A parallel circuit provides more than one path for current.

Each parallel path is called a *branch*. Two resistors connected in parallel are shown in Figure 6–1(a).

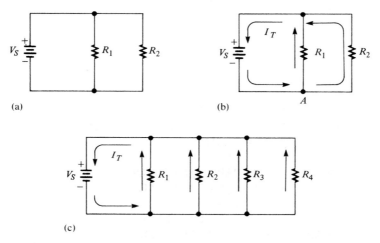

(a)

(b)

(c)

FIGURE 6–1
Resistors in parallel.

In Figure 6–1(b), the current out of the source divides when it gets to point A. Part of it goes through R_1 and part through R_2. If additional resistors are connected in parallel with the first two, more current paths are provided, as shown in Figure 6–1(c).

IDENTIFYING PARALLEL CONNECTIONS

In Figure 6–1, the resistors obviously are connected in parallel. Often, in actual circuit diagrams, the parallel relationship is not as clear. It is important that you learn to recognize parallel connections regardless of how they may be drawn.

A rule for identifying parallel circuits is as follows: *If there is more than one current path (branch) between two points, and if the voltage between those two points appears across each of the branches, then there is a parallel circuit between those two points.* Figure 6–2 shows parallel resistors drawn in different ways between two points labeled A and B. Notice that in each case, the current "travels" two paths going from

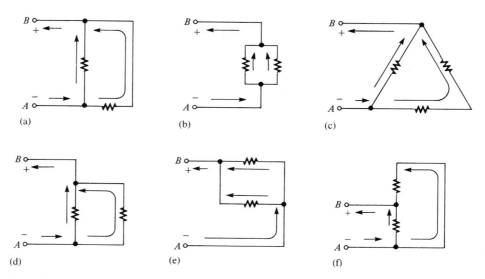

FIGURE 6-2
Examples of circuits with two parallel paths.

A to *B*. Notice that in each case, the current "travels" two paths going from *A* to *B*, and the voltage across each branch is the same. Although these figures show only two parallel paths, there can be any number of resistors in parallel.

EXAMPLE
6-1

Suppose that there are five resistors positioned on a circuit board as shown in Figure 6-3. Wire them together in parallel between the positive (+) and the negative (−) terminals. Draw a schematic diagram showing this connection.

FIGURE 6-3

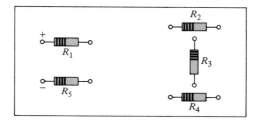

Solution:
Wires are connected as shown in the assembly diagram of Figure 6-4(a). The schematic diagram is shown in Figure 6-4(b). Again, note that the schematic does not necessarily have to show the actual physical arrangement of the resistors. The purpose of the schematic is to show how components are connected electrically.

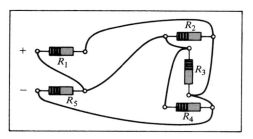

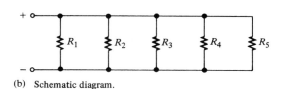

(a) Assembly diagram. Compare this to the same resistor arrangement connected in series in Figure 5–4(a).

(b) Schematic diagram.

FIGURE 6–4

EXAMPLE 6–2

Describe how the resistors on the PC board in Figure 6–5 are related electrically.

FIGURE 6–5

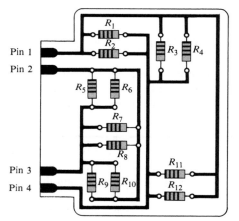

Solution:
Resistors R_1 through R_4 and R_{11} and R_{12} are all in parallel. This parallel combination is connected to pins 1 and 4.

Resistors R_5 through R_{10} are all in parallel. This combination is connected to pins 2 and 3.

SECTION REVIEW 6–1

1. How are the resistors connected in a parallel circuit?

2. How do you identify a parallel connection?

3. Complete the schematic diagrams for the circuits in each part of Figure 6–6 by connecting the resistors in parallel between points *A* and *B*.

FIGURE 6–6

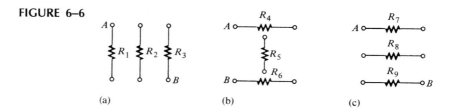

(a)   (b)   (c)

4. Now connect each *group* of parallel resistors in Figure 6–6 in parallel with each other.

6–2   VOLTAGE DROP IN A PARALLEL CIRCUIT

As mentioned, each path in a parallel circuit is sometimes called a *branch*. *The voltage across any branch of a parallel combination is equal to the voltage across each of the other branches in parallel.*

To illustrate voltage drop in a parallel circuit, let us examine Figure 6–7(a). Points *A, B, C,* and *D* along the top of the parallel circuit are *electrically the same point* because the voltage is the same along this line. You can think of all of these points as being connected by a single wire to the positive terminal of the battery. The points *E, F, G,* and *H* along the bottom of the circuit are all at a potential equal to the negative terminal of the source. Thus, each voltage across each parallel resistor is the same, and each is equal to the source voltage.

Figure 6–7(b) is the same circuit as in Part (a), drawn in a slightly different way. Here the tops of the resistors are connected to a single point, which is the positive battery terminal. The bottoms of the resistors are all connected to the same point, which is the negative battery terminal. The resistors are still all in parallel across the source.

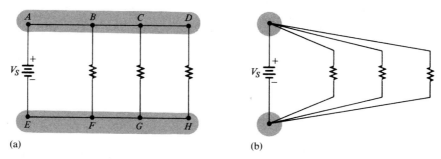

(a)   (b)

FIGURE 6–7
Voltage across parallel branches is the same.

EXAMPLE
6–3

Determine the voltage across each resistor in Figure 6–8.

FIGURE 6–8

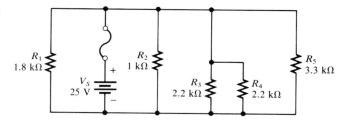

Solution:
The five resistors are in parallel; so the voltage drop across each one is equal to the applied source voltage:

$$V_S = V_1 = V_2 = V_3 = V_4 = V_5$$

$$= 25 \text{ V}$$

SECTION REVIEW 6–2

1. A 10-Ω and a 22-Ω resistor are connected in parallel with a 5-V source. What is the voltage across each of the resistors?

2. A voltmeter is connected across R_1 in Figure 6–9. It measures 118 V. If you move the meter and connect it across R_2, how much voltage will it indicate? What is the source voltage?

3. In Figure 6–10, how much voltage does voltmeter 1 indicate? Voltmeter 2?

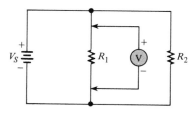

FIGURE 6–9

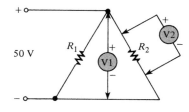

FIGURE 6–10

4. How are voltages across each branch of a parallel circuit related?

6–3

KIRCHHOFF'S CURRENT LAW

Kirchhoff's current law states:

> **The sum of the currents into a junction is equal to the sum of the currents out of that junction.**

A *junction* is any point in a circuit where two or more circuit paths come together. In a parallel circuit, a junction is where the parallel branches connect together. Another way to state Kirchhoff's current law is to say:

> **The total current into a junction is equal to the total current out of that junction.**

For example, in the circuit of Figure 6–11, point A is one junction and point B is another. Let us start at the positive terminal of the source and follow the current. The total current I_T goes from the negative side of the source and *into* the junction at point A. At this point, the current splits up among the three branches as indicated. Each of the three branch currents (I_1, I_2, and I_3) goes *out* of junction A. Kirchhoff's current law says that the total current into junction A is equal to the total current out of junction A; that is,

$$I_T = I_1 + I_2 + I_3$$

Now, following the currents in Figure 6–11 through the three branches, you see that they come back together at point B. Currents I_1, I_2, and I_3 go into junction B, and I_T comes out. Kirchhoff's current law formula at this junction is therefore the same as at junction A:

$$I_T = I_1 + I_2 + I_3$$

FIGURE 6–11
Total current into junction A equals sum of currents out of junction A. Sum of currents into junction B equals total current out of junction B.

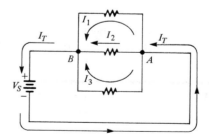

GENERAL FORMULA FOR KIRCHHOFF'S CURRENT LAW

The previous discussion was a specific case to illustrate Kirchhoff's current law. Now let us look at the general case. Figure 6–12 shows a *generalized* circuit junction where

FIGURE 6–12
Generalized circuit junction illustrating
Kirchhoff's current law.

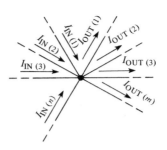

a number of branches are connected to a point in the circuit. Currents $I_{IN(1)}$ through $I_{IN(n)}$ enter the junction (n can be any number). Currents $I_{OUT(1)}$ through $I_{OUT(m)}$ leave the junction (m can be any number, but not necessarily equal to n). By Kirchhoff's current law, the sum of the currents into a junction must equal the sum of the currents out of the junction. With reference to Figure 6–12, the general formula for Kirchhoff's current law is

$$I_{IN(1)} + I_{IN(2)} + \cdots + I_{IN(n)} = I_{OUT(1)} + I_{OUT(2)} + \cdots + I_{OUT(m)} \quad \textbf{(6–1)}$$

If all the terms on the right side of Equation (6–1) are brought over to the left side, their signs change to negative, and a zero is left on the right side as follows:

$$I_{IN(1)} + I_{IN(2)} + \cdots + I_{IN(n)} - I_{OUT(1)} - I_{OUT(2)} - \cdots - I_{OUT(m)} = 0$$

Notice that in this case currents entering the junction are positive and those leaving are negative. Kirchhoff's current law is sometimes stated in this way:

The algebraic sum of all the currents entering and leaving a junction is equal to zero.

This statement is just another, equivalent way of stating what we have just discussed. The following three examples illustrate use of Kirchhoff's current law.

EXAMPLE 6–4

The branch currents in the circuit of Figure 6–13 are known. Determine the total current entering junction A and the total current leaving junction B.

FIGURE 6–13

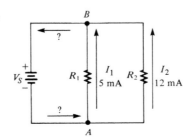

Solution:
The total current leaving junction A is the sum of the two branch currents. So the total current entering A is

$$I_T = I_1 + I_2 = 5 \text{ mA} + 12 \text{ mA} = 17 \text{ mA}$$

The total current entering point B is the sum of the two branch currents. So the total current leaving B is

$$I_T = I_1 + I_2 = 5 \text{ mA} + 12 \text{ mA} = 17 \text{ mA}$$

The calculator sequence is

EXAMPLE 6–5

Determine the current through R_2 in Figure 6–14.

FIGURE 6–14

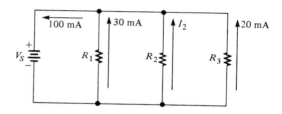

Solution:
The total current leaving the top junction of the three branches is known. Two of the branch currents are known. The current equation at this junction is

$$I_T = I_1 + I_2 + I_3$$

Solving for I_2, we get

$$I_2 = I_T - I_1 - I_3 = 100 \text{ mA} - 30 \text{ mA} - 20 \text{ mA}$$
$$= 50 \text{ mA}$$

EXAMPLE 6–6

Use Kirchhoff's current law to find the current measured by ammeters A_1 and A_2 in Figure 6–15.

FIGURE 6–15

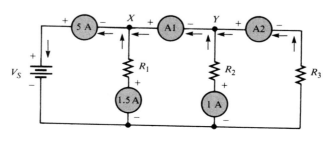

Solution:
The total current leaving junction X is 5 A. Two currents enter junction X: 1.5 A through resistor R_1 and the current through A_1. Kirchhoff's current law at junction X is

$$5 \text{ A} = 1.5 \text{ A} + I_{A1}$$

Solving for I_{A1} yields

$$I_{A1} = 5 \text{ A} - 1.5 \text{ A} = 3.5 \text{ A}$$

The total current leaving junction Y is $I_{A1} = 3.5$ A. Two currents enter junction Y: 1 A through resistor R_2 and the current through A_2 and R_3. Kirchhoff's current law applied at junction Y gives

$$3.5 \text{ A} = 1 \text{ A} + I_{A2}$$

Solving for I_{A2} yields

$$I_{A2} = 3.5 \text{ A} - 1 \text{ A} = 2.5 \text{ A}$$

SECTION REVIEW 6–3

1. State Kirchhoff's current law in two ways.
2. A total current of 2.5 A enters the junction of three parallel branches. What is the sum of all three branch currents?
3. In Figure 6–16, 100 mA and 300 mA enter the junction. What is the amount of current out of the junction?
4. Determine I_1 in the circuit of Figure 6–17.

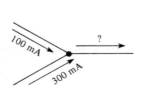

FIGURE 6–16

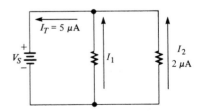

FIGURE 6–17

5. Two branch currents enter a junction, and two branch currents leave the same junction. One of the currents entering the junction is 1 A, and one of the currents leaving the junction is 3 A. The total current entering and leaving the junction is 8 A. Determine the value of the unknown current entering the junction and the value of the unknown current leaving the junction.

6–4

TOTAL PARALLEL RESISTANCE

When resistors are connected in parallel, the total resistance of the circuit decreases. The total resistance of a parallel combination is always less than the value of the smallest resistor.

For example, if a 10-Ω resistor and a 100-Ω resistor are connected in parallel, the total resistance is less than 10 Ω. The exact value must calculated, and you will learn how to do so later in this section.

HOW THE NUMBER OF CURRENT PATHS AFFECTS RESISTANCE

As you know, when resistors are connected in parallel, the current has more than one path. The number of current paths is equal to the number of parallel branches.

For example, in Figure 6–18(a), there is only one current path since it is a series circuit. There is a certain amount of current, I_1, through R_1. If resistor R_2 is connected in parallel with R_1, as shown in Figure 6–18(b), there is an additional amount of current, I_2, through R_2. The total current coming from the source has increased with the addition of the parallel branch. An increase in the *total current* from the source means that the total resistance has decreased, in accordance with Ohm's law. Additional resistors connected in parallel will further reduce the resistance and increase the total current.

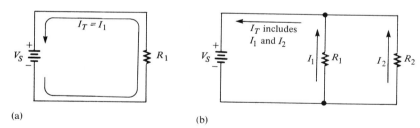

(a)   (b)

FIGURE 6–18
Addition of resistors in parallel reduces total resistance and increases total current.

FORMULA FOR TOTAL PARALLEL RESISTANCE

The circuit in Figure 6–19 shows a general case of n resistors in parallel (n can be any number). From Kirchhoff's current law, the current equation is

$$I_T = I_1 + I_2 + I_3 + \cdots + I_n$$

Since V_S is the voltage across each of the parallel resistors, by Ohm's law, $I_1 = V_S/R_1$, $I_2 = V_S/R_2$, and so on. By substituting into the current equation, we get

$$\frac{V_S}{R_T} = \frac{V_S}{R_1} + \frac{V_S}{R_2} + \frac{V_S}{R_3} + \cdots + \frac{V_S}{R_n}$$

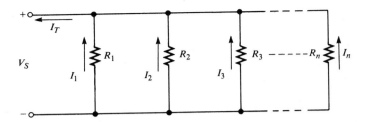

FIGURE 6–19
Circuit with *n* resistors in parallel.

We factor V_S out of the right side of the equation and cancel it with V_S on the left side, leaving only the resistance terms:

$$\frac{1}{R_T} = \frac{1}{R_1} + \frac{1}{R_2} + \frac{1}{R_3} + \cdots + \frac{1}{R_n} \qquad (6\text{--}2)$$

We get another useful form of Equation (6–2) by taking the reciprocal of (that is, by inverting) both sides of the equation:

$$R_T = \frac{1}{\left(\dfrac{1}{R_1}\right) + \left(\dfrac{1}{R_2}\right) + \left(\dfrac{1}{R_3}\right) + \cdots + \left(\dfrac{1}{R_n}\right)} \qquad (6\text{--}3)$$

Equation (6–3) shows that to find the total parallel resistance, add all the $1/R$ terms and then take the reciprocal of the sum. Example 6–7 shows how to use this formula in a specific case, and Example 6–8 shows a simple method of using the calculator in determining parallel resistances. Recall that the reciprocal of resistance ($1/R$) is called *conductance* and is symbolized by G. The unit of conductance is the siemen (S). We will, however, continue to use $1/R$.

EXAMPLE 6–7

Calculate the total parallel resistance between points A and B of the circuit in Figure 6–20.

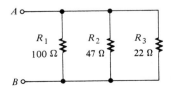

FIGURE 6–20

Solution:
First, find the reciprocal of each of the three resistors as follows:

$$\frac{1}{R_1} = \frac{1}{100 \ \Omega} = 0.01 \text{ S}$$

$$\frac{1}{R_2} = \frac{1}{47 \ \Omega} = 0.0213 \text{ S}$$

$$\frac{1}{R_3} = \frac{1}{22 \ \Omega} = 0.0455 \text{ S}$$

Next, calculate R_T by adding $1/R_1$, $1/R_2$, and $1/R_3$ and taking the reciprocal of the sum as follows:

$$R_T = \frac{1}{0.01 \text{ S} + 0.0213 \text{ S} + 0.0455 \text{ S}} = \frac{1}{0.076 \text{ S}} = 13.02 \ \Omega$$

For a quick accuracy check, notice that the value of R_T (13.02 Ω) is smaller than the smallest value in parallel, which is R_3 (22 Ω), as it should be.

CALCULATOR SOLUTION

The parallel resistance formula is easily solved on the electronic calculator. The general procedure is to enter the value of R_1 and then take its reciprocal by pressing the 2nd F 1/x keys. Next press the + key; then enter the value of R_2 and take its reciprocal. Repeat this procedure until all of the resistor values have been entered and the reciprocal of each has been added. The final step is to press the 2nd F 1/x keys to convert $1/R_T$ to R_T. The total parallel resistance is now on the display. This calculator procedure is illustrated in Example 6–8.

EXAMPLE 6–8

Show the steps required for a calculator solution of Example 6–7.

Solution:
Step 1: Enter 100. Display shows 100.
Step 2: Press 2nd F 1/x. Display shows 0.01.
Step 3: Press + key. Display shows 0.01.
Step 4: Enter 47. Display shows 47.
Step 5: Press 2nd F 1/x. Display shows 0.021276595.
Step 6: Press + key. Display shows 0.031276595.
Step 7: Enter 22. Display shows 22.
Step 8: Press 2nd F 1/x. Display shows 0.045454545.
Step 9: Press = key. Display shows 0.076731141.
Step 10: Press 2nd F 1/x. Display shows 13.03251828.
The number displayed in Step 10 is the total *resistance* in ohms.

TWO RESISTORS IN PARALLEL

Equation (6–3) is a general formula for finding the total resistance for any number of resistors in parallel. It is often useful to consider only two resistors in parallel because this setup occurs commonly in practice. Also, any number of resistors in parallel can be broken down into *pairs* as an alternate way to find the R_T. Based on Equation (6–3), the formula for two resistors in parallel is

$$R_T = \frac{1}{\left(\dfrac{1}{R_1}\right) + \left(\dfrac{1}{R_2}\right)}$$

Combining the terms in the denominator, we get

$$R_T = \frac{1}{\left(\dfrac{R_1 + R_2}{R_1 R_2}\right)}$$

This equation can be rewritten as follows:

$$R_T = \frac{R_1 R_2}{R_1 + R_2} \tag{6-4}$$

Equation (6–4) states:

The total resistance for two resistors in parallel is equal to the product of the two resistances divided by the sum of the two resistances.

This equation is sometimes referred to as the "product over the sum" formula. Example 6–9 illustrates how to use it.

EXAMPLE 6–9

Calculate the total resistance between the positive and negative terminals of the source of the circuit in Figure 6–21.

FIGURE 6–21

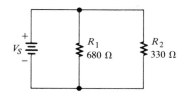

Solution:
Use Equation (6–4) as follows:

$$R_T = \frac{R_1 R_2}{R_1 + R_2} = \frac{(680 \; \Omega)(330 \; \Omega)}{680 \; \Omega + 330 \; \Omega}$$

$$= \frac{224{,}400 \; \Omega^2}{1010 \; \Omega} = 222.18 \; \Omega$$

The calculator sequence is

$$\boxed{6}\,\boxed{8}\,\boxed{0}\,\boxed{\times}\,\boxed{3}\,\boxed{3}\,\boxed{0}\,\boxed{\div}\,\boxed{(}\,\boxed{(}\,\boxed{6}\,\boxed{8}\,\boxed{0}\,\boxed{+}\,\boxed{3}\,\boxed{3}\,\boxed{0}\,\boxed{)}\,\boxed{=}$$

RESISTORS OF EQUAL VALUE IN PARALLEL

Another special case of parallel circuits is the parallel connection of several resistors having the same value. There is a shortcut method of calculating R_T when this case occurs.

If several resistors in parallel have the same resistance, they can be assigned the same symbol R. For example, $R_1 = R_2 = R_3 = \cdots = R_n = R$. Starting with Equation (6–3), we can develop a special formula for finding R_T:

$$R_T = \frac{1}{\left(\dfrac{1}{R}\right) + \left(\dfrac{1}{R}\right) + \left(\dfrac{1}{R}\right) + \cdots + \left(\dfrac{1}{R}\right)}$$

Notice that in the denominator, the same term, $1/R$, is added n times (n is the number of equal resistors in parallel). Therefore, the formula can be written as

$$R_T = \frac{1}{n/R}$$

Rewriting, we obtain

$$R_T = \frac{R}{n} \tag{6–5}$$

Equation (6–5) says that when any number of resistors (n), all having the same resistance (R), are connected in parallel, R_T is equal to the resistance divided by the number of resistors in parallel. Example 6–10 shows how to use this formula.

EXAMPLE
6–10

Four 8-Ω speakers are connected in parallel to the output of an amplifier. What is the total resistance across the output of the amplifier?

Solution:
There are four 8-Ω resistors in parallel. Use Equation (6–5) as follows:

$$R_T = \frac{R}{n} = \frac{8\ \Omega}{4} = 2\ \Omega$$

DETERMINING AN UNKNOWN PARALLEL RESISTOR

Sometimes it is necessary to determine the values of resistors that are to be combined to produce a desired total resistance. For example, consider the case where *two* parallel resistors are used to obtain a desired total resistance. One resistor value is arbitrarily chosen, and then the second resistor value is calculated using Equation (6–6). This equation is derived from the formula for two parallel resistors as follows:

$$R_T = \frac{R_x R_A}{R_x + R_A}$$

$$R_T(R_x + R_A) = R_x R_A$$

$$R_T R_x + R_T R_A = R_x R_A$$

$$R_A R_x - R_T R_x = R_T R_A$$

$$R_x(R_A - R_T) = R_T R_A$$

$$R_x = \frac{R_A R_T}{R_A - R_T} \qquad\qquad \textbf{(6–6)}$$

where R_x is the unknown resistor and R_A is the selected value. Example 6–11 illustrates use of this formula.

EXAMPLE
6–11

Suppose that you wished to obtain a resistance as close to 150 Ω as possible by combining two resistors in parallel. There is a 330-Ω resistor available. What other value is needed?

Solution: $\qquad\qquad R_T = 150\ \Omega \quad$ and $\quad R_A = 330\ \Omega$

$$R_x = \frac{R_A R_T}{R_A - R_T} = \frac{(330\ \Omega)(150\ \Omega)}{330\ \Omega - 150\ \Omega}$$

$$= 275\ \Omega$$

The closest standard value is 270 Ω.

NOTATION FOR PARALLEL RESISTORS

Sometimes, for convenience, parallel resistors are designated by two parallel vertical marks. For example, R_1 in parallel with R_2 can be rewritten as $R_1\|R_2$. Also, when several resistors are in parallel with each other, this notation can be used. For example, $R_1\|R_2\|R_3\|R_4\|R_5$ indicates that R_1 through R_5 are all in parallel.

This notation is also used with resistance values. For example, 10 kΩ||5 kΩ means that 10 kΩ are in parallel with 5 kΩ.

SECTION REVIEW 6–4

1. Does the total resistance increase or decrease as more resistors are connected in parallel?

2. The total parallel resistance is always less than _____.

3. From memory, write the general formula for R_T with any number of resistors in parallel.

4. Write the special formula for two resistors in parallel.

5. Write the special formula for any number of equal-value resistors in parallel.

6. Calculate R_T for Figure 6–22.

7. Determine R_T for Figure 6–23.

8. Find R_T for Figure 6–24.

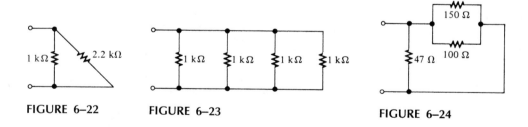

FIGURE 6–22   FIGURE 6–23   FIGURE 6–24

6–5   OHM'S LAW IN PARALLEL CIRCUITS

In this section you will see how Ohm's law can be applied to parallel circuit problems. First you will learn how to find the total current in a parallel circuit, and then you will learn how to determine branch currents using Ohm's law. Let us start with Example 6–12. Then, in Example 6–13, we use Ohm's law to find branch currents.

**EXAMPLE
6–12**

Find the total current produced by the battery in Figure 6–25.

FIGURE 6–25

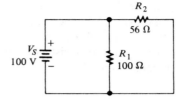

Solution:
The battery "sees" a total parallel resistance which determines the amount of current that it generates. First, we calculate R_T:

$$R_T = \frac{R_1 R_2}{R_1 + R_2} = \frac{(100\ \Omega)(56\ \Omega)}{100\ \Omega + 56\ \Omega}$$

$$= \frac{5600\ \Omega^2}{156\ \Omega} = 35.9\ \Omega$$

The battery voltage is 100 V. Use Ohm's law to find I_T:

$$I_T = \frac{100\ \text{V}}{35.9\ \Omega} = 2.79\ \text{A}$$

**EXAMPLE
6–13**

Determine the current through each resistor in the parallel circuit of Figure 6–26.

FIGURE 6–26

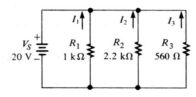

Solution:
The voltage across each resistor (branch) is equal to the source voltage. That is, the voltage across R_1 is 20 V, the voltage across R_2 is 20 V, and the voltage across R_3 is 20 V. The current through each resistor is determined as follows:

$$I_1 = \frac{V_S}{R_1} = \frac{20\ \text{V}}{1\ \text{k}\Omega} = 20\ \text{mA}$$

$$I_2 = \frac{V_S}{R_2} = \frac{20\ \text{V}}{2.2\ \text{k}\Omega} = 9.09\ \text{mA}$$

$$I_3 = \frac{V_S}{R_3} = \frac{20\ \text{V}}{560\ \Omega} = 35.71\ \text{mA}$$

In Example 6–14, we use Ohm's law to determine the unknown voltage across a parallel circuit.

EXAMPLE 6–14

Find the voltage across the parallel circuit in Figure 6–27.

FIGURE 6–27

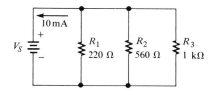

Solution:
We know the total current through the parallel circuit. We need to know the total resistance, and then we can apply Ohm's law to get the voltage. The total resistance is

$$R_T = \cfrac{1}{\left(\cfrac{1}{R_1}\right) + \left(\cfrac{1}{R_2}\right) + \left(\cfrac{1}{R_3}\right)} = \cfrac{1}{0.00455 \text{ S} + 0.00179 \text{ S} + 0.001 \text{ S}}$$

$$= \cfrac{1}{0.00734 \text{ S}} = 136.24 \ \Omega$$

$$V_S = I_T R_T = (10 \text{ mA})(136.24 \ \Omega)$$

$$= 1.3624 \text{ V}$$

EXAMPLE 6–15

The circuit board in Figure 6–28 has three resistors in parallel used for bias modification in an instrumentation amplifier. The values of two of the resistors are known from the color codes, but the third resistor is not clearly marked (maybe the bands are worn off from handling). Determine the value of the unknown resistor without using an ohmmeter.

FIGURE 6–28

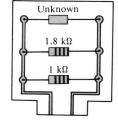

Solution:
If we can determine the total resistance of the three resistors in parallel, then we can use the parallel resistance formula to calculate the unknown resistance. We can use Ohm's law to find the total resistance if voltage and total current are known.

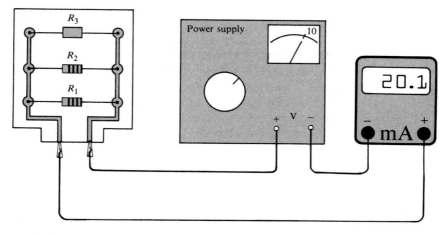

FIGURE 6–29

In Figure 6–29, a 10-V source (arbitrary value) is connected across the resistors, and the total current is measured. Using these measured values, we find that the total resistance is

$$R_T = \frac{V}{I_T} = \frac{10 \text{ V}}{20.1 \text{ mA}} = 498 \text{ }\Omega$$

Use the reciprocal resistance formula to find the unknown resistance as follows:

$$\frac{1}{R_T} = \frac{1}{R_1} + \frac{1}{R_2} + \frac{1}{R_3}$$

$$\frac{1}{R_3} = \frac{1}{R_T} - \frac{1}{R_1} - \frac{1}{R_2} = \frac{1}{498 \text{ }\Omega} - \frac{1}{1 \text{ k}\Omega} - \frac{1}{1.8 \text{ k}\Omega} = 0.000453 \text{ S}$$

$$R_3 = \frac{1}{0.000453 \text{ S}} \cong 2.2 \text{ k}\Omega$$

The calculator sequence for R_3 is

| 4 | 9 | 8 | 2nd F | 1/x | − | 1 | EXP | 3 | 2nd F | 1/x |

| − | 1 | · | 8 | EXP | 3 | 2nd F | 1/x | = | 2nd F | 1/x |

SECTION REVIEW 6–5

1. A 10-V battery is connected across three 68-Ω resistors that are in parallel. What is the total current from the battery?

2. How much voltage is required to produce 2 A of current through the circuit of Figure 6–30?

FIGURE 6–30

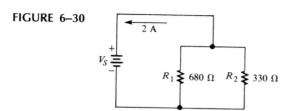

3. How much current is there through each resistor of Figure 6–30?

4. There are four equal-value resistors in parallel with a 12-V source, and 5.85 mA of current from the source. What is the value of each resistor?

5. A 1-kΩ and a 2.2-kΩ resistor are connected in parallel. There is a total of 100 mA through the parallel combination. How much voltage is dropped across the resistors?

6–6

CURRENT SOURCES IN PARALLEL

A current source is a type of energy source that provides a constant value of current to a load even when that load changes in resistance value. A transistor can be used as a current source, and thus current sources are important in circuit analysis. At this point, you are not prepared to study transistor circuits in detail, but you do need to understand how current sources act in circuits. The only practical case that we need to consider is that of current sources connected in parallel.

The general rule to remember is that the total current produced by current sources in parallel is equal to the algebraic sum of the individual current sources. The *algebraic sum* means that you must consider the direction of current when combining the sources in parallel. For example, in Figure 6–31(a), the three current sources in parallel provide current in the same direction (into point A). So the total current into point A is $I_T = 1 \text{ A} + 2 \text{ A} + 2 \text{ A} = 5 \text{ A}$.

In Figure 6–31(b), the 1-A source provides current in a direction opposite to the other two. The total current into point A in this case is $I_T = 2 \text{ A} + 2 \text{ A} - 1 \text{ A} = 3 \text{ A}$.

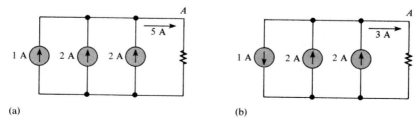

(a)   (b)

FIGURE 6–31

**EXAMPLE
6–16**

Determine the current through R_L in Figure 6–32.

FIGURE 6–32

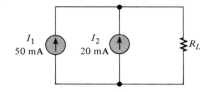

Solution:
The two current sources are in the same direction; so the current through R_L is

$$I_L = I_1 + I_2 = 50 \text{ mA} + 20 \text{ mA} = 70 \text{ mA}$$

SECTION REVIEW 6–6

1. Four 0.5-A current sources are connected in parallel in the same direction. What current will be produced through a load resistor?

2. How many 1-A current sources must be connected in parallel to produce a total current output of 3 A? Sketch a schematic showing the sources connected.

3. In a transistor amplifier circuit, the transistor can be represented by a 10-mA current source, as shown in Figure 6–33. The transistors act in parallel, as in the case of a differential amplifier. How much is the current through the resistor R_E?

FIGURE 6–33

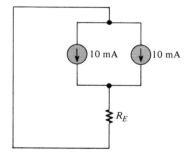

6–7

CURRENT DIVIDERS

A parallel circuit acts as a *current divider*. In this section you will see what this term means and why current dividers are an important application of parallel circuits.

To understand how a parallel connection of resistors acts as a current divider, look at Figure 6–34, where there are two resistors in parallel. As you already know, there is a current through R_1 and a current through R_2. We call these branch currents I_1 and I_2, respectively, as indicated in the diagram.

FIGURE 6–34
Total current divides between two
branches.

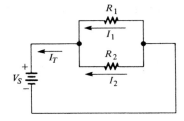

Since the same voltage is across each of the resistors in parallel, the branch currents are inversely proportional to the values of the resistors. For example, if the value of R_2 is twice that of R_1, then the value of I_2 is one-half that of I_1. In other words,

the total current divides among parallel resistors in a manner inversely proportional to the resistance values.

The branches with higher resistance have less current, and the branches with lower resistance have more current, in accordance with Ohm's law.

GENERAL CURRENT DIVIDER FORMULA FOR ANY NUMBER OF BRANCHES

With a few steps, a formula for determining how currents divide among parallel resistors can be developed. Let us assume that we have n resistors in parallel, as shown in Figure 6–35, where n can be any number.

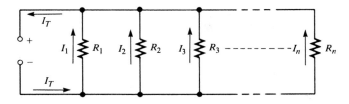

FIGURE 6–35
Generalized parallel circuit with n branches.

Let us call the current through any one of the parallel resistors I_x, where x represents the number of a particular resistor (1, 2, 3, and so on). By Ohm's law, the current through any one of the resistors in Figure 6–35 can be written as follows:

$$I_x = \frac{V_S}{R_x}$$

where $x = 1, 2, 3$, and so on. The source voltage V_S appears across each of the parallel resistors, and R_x represents any one of the parallel resistors. The total source voltage V_S

is equal to the total current times the total parallel resistance:

$$V_S = I_T R_T$$

Substituting $I_T R_T$ for V_S in the expression for I_x, we get

$$I_x = \frac{I_T R_T}{R_x}$$

By rearranging, we get

$$I_x = \left(\frac{R_T}{R_x}\right) I_T \qquad (6\text{--}7)$$

Equation (6–7) is the general current divider formula. It tells us that the current (I_x) through any branch equals the total parallel resistance (R_T), divided by the resistance (R_x) of that branch and multiplied by the total current (I_T) flowing into the parallel junction. This formula applies to a parallel circuit with any number of branches.

Example 6–17 illustrates application of Equation (6–7).

EXAMPLE 6–17

Determine the current through each resistor in the circuit of Figure 6–36.

FIGURE 6–36

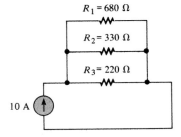

Solution:
First calculate the total parallel resistance:

$$R_T = \frac{1}{\left(\dfrac{1}{R_1}\right) + \left(\dfrac{1}{R_2}\right) + \left(\dfrac{1}{R_3}\right)}$$

$$= \frac{1}{\left(\dfrac{1}{680 \ \Omega}\right) + \left(\dfrac{1}{330 \ \Omega}\right) + \left(\dfrac{1}{220 \ \Omega}\right)} = 110.54 \ \Omega$$

The total current is 10 A. Using Equation (6–7), we calculate each branch current as follows:

$$I_1 = \left(\frac{R_T}{R_1}\right)I_T = \left(\frac{110.54\ \Omega}{680\ \Omega}\right)10\ \text{A} = 1.63\ \text{A}$$

$$I_2 = \left(\frac{R_T}{R_2}\right)I_T = \left(\frac{110.54\ \Omega}{330\ \Omega}\right)10\ \text{A} = 3.35\ \text{A}$$

$$I_3 = \left(\frac{R_T}{R_3}\right)I_T = \left(\frac{110.54\ \Omega}{220\ \Omega}\right)10\ \text{A} = 5.02\ \text{A}$$

CURRENT DIVIDER FORMULA FOR TWO BRANCHES

For the special case of two parallel branches, Equation (6–7) can be modified. The reason for doing so is that two parallel resistors are often found in practical circuits. We start by restating Equation (6–4), the formula for the total resistance of two parallel branches:

$$R_T = \frac{R_1 R_2}{R_1 + R_2}$$

When we have two parallel resistors, as in Figure 6–37, we may want to find the current through either or both of the branches. To do so, we need two special formulas.

FIGURE 6–37

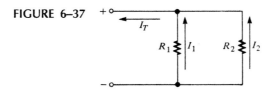

Using Equation (6–7), we write the formulas for I_1 and I_2 as follows:

$$I_1 = \left(\frac{R_T}{R_1}\right)I_T$$

$$I_2 = \left(\frac{R_T}{R_2}\right)I_T$$

Substituting $R_1 R_2/(R_1 + R_2)$ for R_T, we get

$$I_1 = \frac{\left(\dfrac{\cancel{R_1} R_2}{R_1 + R_2}\right)}{\cancel{R_1}}\, I_T$$

$$I_2 = \frac{\left(\dfrac{R_1 \cancel{R_2}}{R_1 + R_2}\right)}{\cancel{R_2}}\, I_T$$

Canceling as shown, we get the final formulas:

$$I_1 = \left(\frac{R_2}{R_1 + R_2}\right)I_T \qquad \textbf{(6–8)}$$

$$I_2 = \left(\frac{R_1}{R_1 + R_2}\right)I_T \qquad \textbf{(6–9)}$$

When there are only *two* resistors in parallel, these equations are a little easier to use than Equation (6–7), because it is not necessary to know R_T.

Note that in Equations (6–8) and (6–9), the current in one of the branches is equal to the *opposite* branch resistance over the *sum* of the two resistors, all times the total current. In all applications of the current divider equations, you must know the total current going into the parallel branches. Example 6–18 illustrates application of these special current divider formulas.

EXAMPLE 6–18

Find I_1 and I_2 in Figure 6–38.

FIGURE 6–38

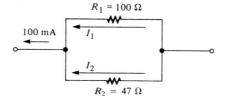

Solution:
Using Equation (6–8), we determine I_1 as follows:

$$I_1 = \left(\frac{R_2}{R_1 + R_2}\right)I_T = \left(\frac{47\ \Omega}{147\ \Omega}\right)100\ \text{mA}$$

$$= 31.97\ \text{mA}$$

Using Equation (6–9), we determine I_2 as follows:

$$I_2 = \left(\frac{R_1}{R_1 + R_2}\right)I_T = \left(\frac{100\ \Omega}{147\ \Omega}\right)100\ \text{mA}$$

$$= 68.03\ \text{mA}$$

SECTION REVIEW 6–7

1. Write the general current divider formula.

2. Write the two special formulas for calculating each branch current for a two-branch circuit.

3. A parallel circuit has the following resistors in parallel: 220 Ω, 100 Ω, 82 Ω, 47 Ω, and 22 Ω. Which resistor has the most current through it? The least current?

4. Determine the current through R_3 in Figure 6–39.

5. Find I_1 and I_2 in the circuit of Figure 6–40.

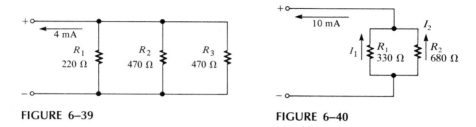

FIGURE 6–39　　　　　　　　　　　　　　　FIGURE 6–40

6–8　POWER IN A PARALLEL CIRCUIT

The total amount of power in a parallel resistive circuit is equal to the sum of the powers in each resistor in parallel. Equation (6–10) states this in a concise way for any number of resistors in parallel:

$$P_T = P_1 + P_2 + P_3 + \cdots + P_n \tag{6–10}$$

where P_T is the total power and P_n is the power in the last resistor in parallel. As you can see, the power losses are additive, just as in the series circuit.

　　　The power formulas in Chapter 4 are directly applicable to parallel circuits. The following formulas are used to calculate the total power P_T:

$$P_T = VI_T$$

$$P_T = I_T^2 R_T$$

$$P_T = \frac{V^2}{R_T}$$

where V is the voltage across the parallel circuit, I_T is the total current into the parallel circuit, and R_T is the total resistance of the parallel circuit. Example 6–19 shows how total power can be calculated in a parallel circuit.

EXAMPLE 6–19

Determine the total amount of power in the parallel circuit in Figure 6–41.

FIGURE 6–41

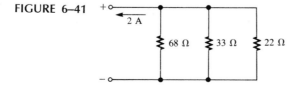

Solution:
We know that the total current is 2 A. The total resistance is

$$R_T = \frac{1}{\left(\dfrac{1}{68 \ \Omega}\right) + \left(\dfrac{1}{33 \ \Omega}\right) + \left(\dfrac{1}{22 \ \Omega}\right)} = 11.05 \ \Omega$$

The easiest formula to use is $P_T = I_T^2 R_T$ since we know both I_T and R_T. Thus,

$$P_T = I_T^2 R_T = (2 \ A)^2 (11.05 \ \Omega) = 44.2 \ W$$

To demonstrate that if the power in each resistor is determined and if all of these values are added together, you get the same result, we will work through another calculation. First, find the voltage across each branch of the circuit:

$$V_S = I_T R_T = (2 \ A)(11.05 \ \Omega) = 22.1 \ V$$

Remember that the voltage across all branches is the same.
Next, calculate the power for each resistor using $P = V^2/R$:

$$P_1 = \frac{(22.1 \ V)^2}{68 \ \Omega} = 7.18 \ W$$

$$P_2 = \frac{(22.1 \ V)^2}{33 \ \Omega} = 14.80 \ W$$

$$P_3 = \frac{(22.1 \ V)^2}{22 \ \Omega} = 22.20 \ W$$

Now, add these powers to get the total power:

$$P_T = 7.18 \ W + 14.80 \ W + 22.20 \ W$$
$$= 44.18 \ W \ \text{(round to 44.2 W)}$$

This calculation shows that the sum of the individual powers is equal to the total power as determined by one of the power formulas.

**EXAMPLE
6–20**

The amplifier in the stereo system of Figure 6–42 drives four speakers as shown. If the maximum voltage to the speakers is 15 V, how much power must the amplifier be able to deliver to the speakers?

Solution:
The speakers are connected in parallel to the amplifier output, so the voltage across each is the same. The maximum power to each speaker is

$$P = \frac{V^2}{R} = \frac{(15 \ V)^2}{8 \ \Omega} = 28.125 \ W$$

FIGURE 6–42

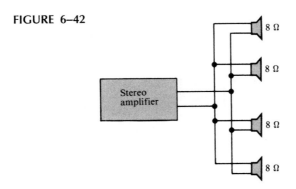

The total power that the amplifier must be capable of delivering to the speaker system is four times the power in an individual speaker because the total power is the sum of the individual powers:

$$P_T = P + P + P + P = 4P = 112.5 \text{ W}$$

SECTION REVIEW 6–8

1. If you know the power in each resistor in a parallel circuit, how can you find the total power?
2. The resistors in a parallel circuit dissipate the following powers: 2.38 W, 5.12 W, 1.09 W, and 8.76 W. What is the total power in the circuit?
3. A circuit has a 1-kΩ, a 2.7-kΩ, and a 3.9-kΩ resistor in parallel. There is a total current of 1 A into the parallel circuit. What is the total power?

6–9 APPLICATIONS OF PARALLEL CIRCUITS

Parallel circuits are found in some form in virtually every electronic device. In many of these applications, the parallel relationship of components may not be obvious until you have covered some advanced topics that will come later. Therefore, in this section, some common and familiar applications of parallel circuits are discussed.

AUTOMOTIVE

One advantage of a parallel circuit over a series circuit is that when one branch opens, the other branches are not affected. For example, Figure 6–43 shows a simplified diagram of an automobile lighting system. When one headlight on your car goes out, it does not cause the other lights to go out, because they are all in parallel.

Notice that the brake lights are switched on independently of the head and tail lights. They come on only when the driver closes the brake light switch by depressing the brake pedal.

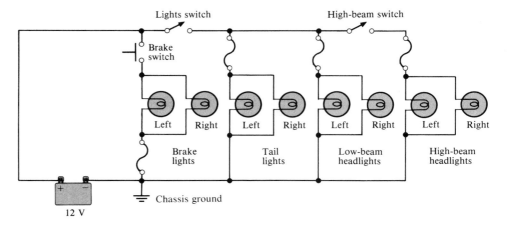

FIGURE 6–43
Simplified diagram of the exterior light system of an automobile.

When the *lights* switch is closed, both low-beam headlights and both taillights are on. The high-beam headlights are on only when both the *lights* switch and the *high-beam* switch are closed. If any one of the lights burns out (opens), there is still current in each of the other lights.

RESIDENTIAL

Another common use of parallel circuits is in residential electrical systems. All the lights and appliances in a home are wired in parallel. Figure 6–44(a) shows a typical room wiring arrangement with two switch-controlled lights and three wall outlets in parallel.

Figure 6–44(b) shows a simplified parallel arrangement of four heating elements on an electric range. The four-position switches in each branch allow the user to control the amount of current through the heating elements by selecting the appropriate limiting resistor. The lowest resistor value (H setting) allows the highest amount of current for maximum heat. The highest resistor value (L setting) allows the least amount of current for minimum heat; M designates the medium settings.

AMMETERS

Another example in which parallel circuits are used is the familiar analog (needle-type) ammeter or milliammeter. Parallel circuits are an important part of the operation of the ammeter because they allow the user to select various ranges in order to measure many different current values.

The mechanism in an ammeter that causes the pointer to move in proportion to the current is called the *meter movement,* which is based on a magnetic principle that you will learn later. Right now, all you need to know is that a meter movement has a certain resistance and a maximum current. This maximum current, called the *full-scale deflection* current, causes the pointer to go all the way to the end of the scale. For

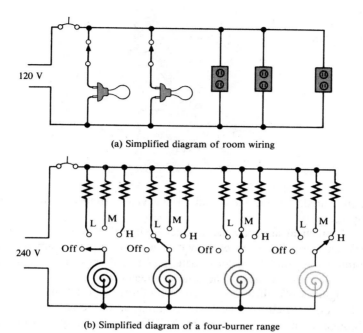

(a) Simplified diagram of room wiring

(b) Simplified diagram of a four-burner range

FIGURE 6–44
Examples of parallel circuits in residential wiring and appliances.

example, a certain meter movement has a 50-Ω resistance and a full-scale deflection current of 1 mA. A meter with this particular movement can measure currents of 1 mA or less. Currents greater than 1 mA will cause the pointer to "peg" (or stop) at full scale. Figure 6–45 illustrates a 1-mA meter.

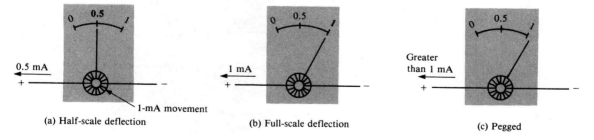

(a) Half-scale deflection   (b) Full-scale deflection   (c) Pegged

FIGURE 6–45
A 1-mA meter.

Figure 6–46 shows a simple ammeter with a resistor in parallel with the meter movement; this resistor is called a *shunt* resistor. Its purpose is to bypass any current in excess of 1 mA around the meter movement. The figure specifically shows 9 mA through the shunt resistor and 1 mA through the meter movement. Thus, up to 10 mA

FIGURE 6–46
A 10-mA meter.

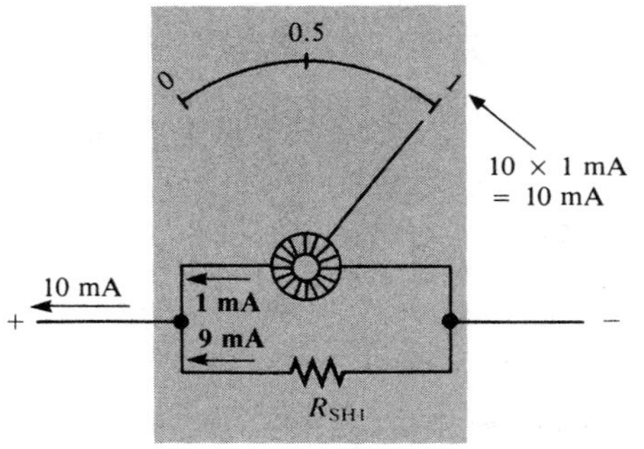

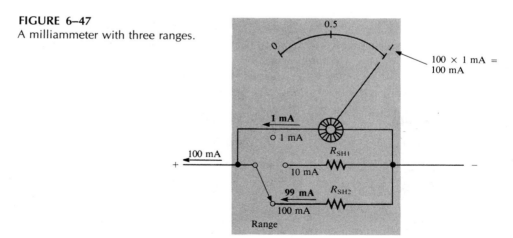

can be measured. To find the actual current value, simply multiply the reading on the scale by 10.

A practical ammeter has a range switch that permits the selection of several full-scale current settings. In each switch position, a certain amount of current is by-passed through a parallel resistor as determined by the resistance value. In our example, the current through the movement is never greater than 1 mA.

Figure 6–47 illustrates a meter with three ranges: 1 mA, 10 mA, and 100 mA. When the range switch is in the 1-mA position, all of the current coming into the meter goes through the meter movement. In the 10-mA setting, up to 9 mA goes through R_1 and up to 1 mA through the movement. In the 100-mA setting, up to 99 mA goes through R_2, and the movement can still have only 1 mA for full-scale.

FIGURE 6–47
A milliammeter with three ranges.

For example, in Figure 6–47, if 50 mA of current are being measured, the needle points at the 0.5 mark on the scale; we must multiply 0.5 by 100 to find the current value. In this situation, 0.5 mA goes through the movement (half-scale deflection), and 49.5 mA go through R_2.

Effect of the Ammeter on a Circuit As you know, an ammeter is connected in *series* to measure the current in a circuit. Ideally, the meter should not alter the current that it is intended to measure. In practice, however, the meter unavoidably has some effect on the circuit, because its *internal resistance* is connected in series with the circuit resistance. However, in most cases, the meter's internal resistance is so small compared to the circuit resistance that it can be neglected.

For example, if the meter has a 50-Ω movement (R_M) and a 100-μA full-scale current (I_M), the voltage dropped across the movement is

$$V_M = I_M R_M = (100\ \mu A)(50\ \Omega) = 5\ mV$$

The shunt resistance (R_{SH}) for the 10-mA range, for example, is

$$R_{SH} = \frac{V_M}{I_{SH}} = \frac{5\ mV}{9.9\ mA} = 0.505\ \Omega$$

As you can see, the total resistance of the ammeter on the 10-mA range is the resistance of the movement in parallel with the shunt resistance:

$$R_M \| R_{SH} = 50\ \Omega \| 0.505\ \Omega = 0.5\ \Omega$$

**EXAMPLE
6–21**

How much does an ammeter with a 100-μA, 50-Ω movement affect the current in the circuit of Figure 6–48?

FIGURE 6–48

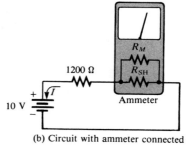

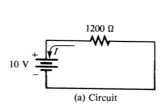

(a) Circuit   (b) Circuit with ammeter connected

Solution:
The true current in the circuit is

$$I = \frac{10\ V}{1200\ \Omega} = 8.3333\ mA$$

The meter is set on the 10-mA range in order to measure this particular amount of current. It was found that the meter's resistance on the 10-mA range is 0.5 Ω. When the meter is connected in the circuit, its resistance is in series with the 1200-Ω resistor. Thus, there is a total of 1200.5 Ω.

The current actually measured by the ammeter is

$$I = \frac{10 \text{ V}}{1200.5 \text{ }\Omega} = 8.3299 \text{ mA}$$

This current differs from the true circuit current by only 0.04%. Therefore, the meter does not significantly alter the current value, a situation which, of course, is necessary because the measuring instrument should not change the quantity that is to be measured accurately.

SECTION REVIEW 6–9

1. For the ammeter in Figure 6–47, what is the maximum resistance that the meter will have when connected in a circuit? What is the maximum current that can be measured at the setting?

2. Do the shunt resistors have resistance values considerably less than or more than that of the meter movement? Why?

6–10

TROUBLESHOOTING PARALLEL CIRCUITS

OPEN BRANCHES

Recall that an open circuit is one in which the current path is interrupted and no current flows. In this section we examine what happens when a branch of a parallel circuit opens.

If a switch is connected in a branch of a parallel circuit, as shown in Figure 6–49, an open or a closed path can be made by the switch. When the switch is closed, as in Figure 6–49(a), R_1 and R_2 are in parallel. The total resistance is 50 Ω (two 100-Ω resistors in parallel). There is current through both resistors. If the switch is opened, as in Figure 6–49(b), R_1 is effectively removed from the circuit, and the total resistance is 100 Ω. There is current now only through R_2.

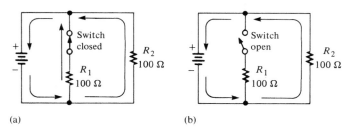

(a)   (b)

FIGURE 6–49
When switch opens, total current decreases and current through R_2 remains unchanged.

In general,

when an open circuit occurs in a parallel branch, the total resistance increases, the total current decreases, and current continues through the remaining parallel paths.

The decrease in total current equals the amount of current that was previously in the open branch. The other branch currents remain the same.

Consider the lamp circuit in Figure 6–50. There are four bulbs in parallel with a 120-V source. In Part (a), there is current through each bulb. Now suppose that one of the bulbs burns out, creating an open path as shown in Figure 6–50(b). This light will go out because there is no current through the open path. Notice, however, that current continues through all the other parallel bulbs, and they continue to glow. The open branch does not change the voltage across the parallel branches; it remains at 120 V.

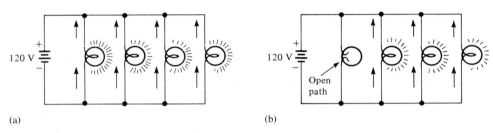

(a)   (b)

FIGURE 6–50
When lamp opens, total current decreases and other branch currents remain unchanged.

You can see that a parallel circuit has an advantage over a series connection in lighting systems because if one or more of the parallel bulbs burn out, the others will stay on. In a *series* circuit, when one bulb goes out, all of the others go out also because the current path is *completely* interrupted.

When a resistor in a parallel circuit opens, the open resistor cannot be located by measurement of the voltage across the branches, because the same voltage exists across all the branches. Thus, there is no way to tell which resistor is open by simply measuring voltage. The good resistors will always have the same voltage as the open one, as illustrated in Figure 6–51 (note that the middle resistor is open).

If a visual inspection does not reveal the open resistor, it must be located by current measurements. In practice, measuring current is more difficult than measuring voltage because you must insert the ammeter in *series* to measure the current. Thus, a

FIGURE 6–51
Parallel branches (open or not) have the same voltage.

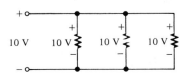

wire or a PC connection must be cut or disconnected, or one end of a component must be lifted off the circuit board, in order to connect the ammeter in series. This procedure, of course, is not required when voltage measurements are made, because the meter leads are simply connected *across* a component.

WHERE TO MEASURE THE CURRENT

In a parallel circuit, the total current should be measured. When a parallel resistor opens, I_T is less than its normal value. Once I_T and the voltage across the branches are known, a few calculations will determine the open resistor when all the resistors are of different ohmic values.

Consider the two-branch circuit in Figure 6–52(a). If one of the resistors opens, the total current will equal the current in the good resistor. Ohm's law quickly tells us what the current in each resistor should be:

$$I_1 = \frac{50 \text{ V}}{560 \text{ }\Omega} = 0.089 \text{ A}$$

$$I_2 = \frac{50 \text{ V}}{100 \text{ }\Omega} = 0.5 \text{ A}$$

If R_2 is open, the total current is 0.089 A, as indicated in Figure 6–52(b). If R_1 is open, the total current is 0.5 A, as indicated in Figure 6–52(c).

This procedure can be extended to any number of branches having unequal resistances. If the parallel resistances are all equal, the current in each branch must be checked until a branch is found with no current. This is the open resistor.

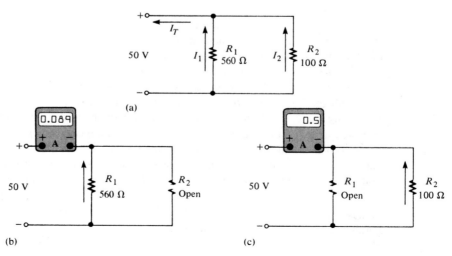

(a)

(b)   (c)

FIGURE 6–52
Finding an open path by current measurement.

**EXAMPLE
6–22**

In Figure 6–53, there is a total current of 31.09 mA, and the voltage across the parallel branches is 20 V. Is there an open resistor, and, if so, which one is it?

FIGURE 6–53

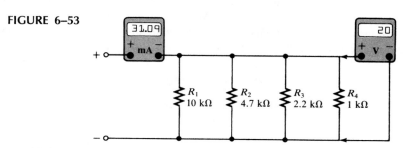

Solution:
Calculate the current in each branch:

$$I_1 = \frac{V}{R_1} = \frac{20 \text{ V}}{10 \text{ k}\Omega} = 2 \text{ mA}$$

$$I_2 = \frac{V}{R_2} = \frac{20 \text{ V}}{4.7 \text{ k}\Omega} = 4.26 \text{ mA}$$

$$I_3 = \frac{V}{R_3} = \frac{20 \text{ V}}{2.2 \text{ k}\Omega} = 9.09 \text{ mA}$$

$$I_4 = \frac{V}{R_4} = \frac{20 \text{ V}}{1 \text{ k}\Omega} = 20 \text{ mA}$$

The total current *should* be

$$I_T = I_1 + I_2 + I_3 + I_4 = 2 \text{ mA} + 4.26 \text{ mA} + 9.09 \text{ mA} + 20 \text{ mA}$$
$$= 35.35 \text{ mA}$$

The actual *measured* current is 31.09 mA, as stated, which is 4.26 mA less than normal, indicating that the branch carrying 4.26 mA is open. Thus, R_2 must be open.

SECTION REVIEW 6–10

1. If a parallel branch opens, what changes can be detected in the circuit's voltage and the currents, assuming that the parallel circuit is across a constant voltage source?

2. What happens to the total resistance if one branch opens?

3. If several light bulbs are connected in parallel and one of the bulbs opens (burns out), will the others continue to glow?

4. There is one ampere of current in each branch of a parallel circuit. If one branch opens, what is the current in each of the remaining branches?

6–11   COMPUTER ANALYSIS

The program listed in this section can be used to compute the currents in each branch of any specified parallel circuit. Also, the total resistance and total current are provided in the display. The inputs required are the number of resistors, the value of the resistors, and the value of the source voltage.

```
10   CLS
20   PRINT "THIS PROGRAM COMPUTES AND TABULATES THE TOTAL
     RESISTANCE,"
30   PRINT "TOTAL CURRENT, AND THE CURRENTS THROUGH EACH OF THE"
40   PRINT "INDIVIDUAL RESISTORS IN A SPECIFIED PARALLEL
     CIRCUIT."
50   PRINT:PRINT:PRINT
60   PRINT "YOU ARE REQUIRED TO ENTER THE NUMBER OF
     RESISTORS,"
70   PRINT "THE RESISTOR VALUES, AND THE SOURCE VOLTAGE."
80   PRINT:PRINT:PRINT
90   INPUT "TO CONTINUE PRESS 'ENTER'.";X:CLS
100  INPUT "HOW MANY RESISTORS ARE IN THE PARALLEL
     CIRCUIT";N:CLS
110  FOR X=1 TO N
120  PRINT "ENTER THE VALUE OF R";X;"IN OHMS"
130  INPUT R(X)
140  NEXT X:CLS
150  INPUT "THE SOURCE VOLTAGE IN VOLTS";VS:CLS
160  RT=R(1)
170  FOR X=2 TO N
180  RT=1/(1/RT+1/R(X))
190  NEXT X
200  IT=VS/RT
210  PRINT "TOTAL RESISTANCE = ";RT;"OHMS"
220  PRINT:PRINT "TOTAL CURRENT = ";IT;"AMPS":PRINT
230  FOR X=1 TO N
240  I(X)=(RT/R(X))*IT
250  PRINT "R";X;"=";R(X);"OHMS","T";X;"=";I(X);"AMPS"
260  NEXT X
```

SECTION REVIEW 6–11

1. Identify by line numbers the series of instructions that permit entering the resistor values.

2. Explain the purpose of line 160.

SUMMARY

1. Resistors in parallel are connected across the same points.

2. A parallel combination has more than one path for current.

3. The number of current paths equals the number of resistors in parallel.

4. The total parallel resistance is less than the lowest-value resistor.

5. The voltages across all branches of a parallel circuit are the same.

6. Current sources in parallel add algebraically.

7. One way to state Kirchhoff's current law: The algebraic sum of all the currents at a junction is zero.

8. Another way to state Kirchhoff's current law: The sum of the currents into a junction (total current in) equals the sum of the currents out of the junction (total current out).

9. A parallel circuit is a current divider, so called because the total current entering the parallel junction divides up into each of the branches.

10. If all of the branches of a parallel circuit have equal resistance, the currents through all of the branches are equal.

11. The total power in a parallel resistive circuit is the sum of all of the individual powers of the resistors making up the parallel circuit.

12. The total power for a parallel circuit can be calculated with the power formulas using values of total current, total resistance, or total voltage.

13. If one of the branches of a parallel circuit opens, the total resistance increases, and therefore the total current decreases.

14. If a branch of a parallel circuit opens, there is still current through the remaining branches.

FORMULAS

$$I_{\text{IN}(1)} + I_{\text{IN}(2)} + \cdots + I_{\text{IN}(n)} = I_{\text{OUT}(1)} + I_{\text{OUT}(2)} + \cdots + I_{\text{OUT}(m)} \qquad \textbf{(6–1)}$$

$$\frac{1}{R_T} = \frac{1}{R_1} + \frac{1}{R_2} + \frac{1}{R_3} + \cdots + \frac{1}{R_n} \qquad \textbf{(6–2)}$$

$$R_T = \frac{1}{\left(\dfrac{1}{R_1}\right) + \left(\dfrac{1}{R_2}\right) + \left(\dfrac{1}{R_3}\right) + \cdots + \left(\dfrac{1}{R_n}\right)} \qquad \textbf{(6–3)}$$

$$R_T = \frac{R_1 R_2}{R_1 + R_2} \qquad \textbf{(6–4)}$$

$$R_T = \frac{R}{n} \qquad \textbf{(6–5)}$$

$$R_x = \frac{R_A R_T}{R_A - R_T} \qquad \textbf{(6–6)}$$

$$I_x = \left(\frac{R_T}{R_x}\right) I_T \qquad \textbf{(6–7)}$$

$$I_1 = \left(\frac{R_2}{R_1 + R_2}\right)I_T \tag{6-8}$$

$$I_2 = \left(\frac{R_1}{R_1 + R_2}\right)I_T \tag{6-9}$$

$$P_T = P_1 + P_2 + P_3 + \cdots + P_n \tag{6-10}$$

SELF-TEST

Solutions appear at the end of the book.

1. Sketch a parallel circuit having four resistors and a voltage source.
2. There are five amperes of current into a parallel circuit having two branches of equal resistance. How much current is there through each of the branches?
3. Connect each set of resistors in Figure 6–54 in parallel between points A and B.
4. If you measured the voltage across each of the points marked in Figure 6–55, how much voltage would you read in each case?

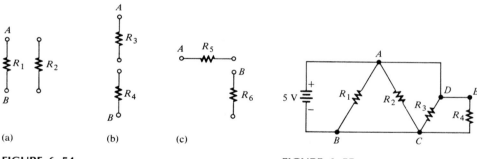

FIGURE 6–54   FIGURE 6–55

5. If an 82-Ω resistor and a 150-Ω resistor are connected in parallel, what is the total resistance?
6. The following resistors are in parallel: 1000 Ω, 820 Ω, 560 Ω, 220 Ω, and 100 Ω. What is the total resistance?
7. Eight 56-Ω resistors are connected in parallel. Determine the total resistance.
8. Suppose that you need a total resistance of 110 Ω. The only resistors that are immediately available are one 220-Ω and several 330-Ω resistors. How would you connect these resistors to get a total of 110 Ω?
9. Three 680-Ω resistors are connected in parallel, and 5 V are applied across the parallel circuit. How much current is out of the source?
10. Which circuit of Figure 6–56 has more total current?

FIGURE 6–56

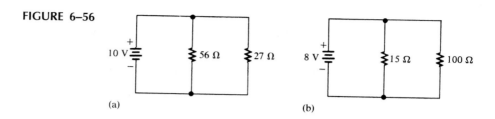

(a)   (b)

11. Six resistors of equal value are in parallel, with 12 V across them. The total current is 3 mA. Determine the value of each resistor. What is the total resistance?

12. Determine the current through R_L in the multiple-current-source circuit of Figure 6–57.

FIGURE 6–57

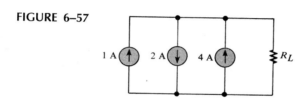

13. Two 1-A current sources are connected in opposite directions in parallel. What is the total current produced?

14. Each branch in a five-branch parallel circuit has 25 mA of current through it. What is the total current into the parallel circuit?

15. The total current into the junction of three parallel resistors is 0.5 A. One branch has a current of 0.1 A, and another branch has a current of 0.2 A. What is the current through the third branch?

16. Determine the unknown resistances in the circuit of Figure 6–58.

17. The following resistors are connected in parallel: 2.2 kΩ, 6.8 kΩ, 3.3 kΩ, and 1 kΩ. The total current is 1 A. Using the general current divider formula, calculate the current through each of the resistors.

18. Determine the current through each resistor in Figure 6–59.

19. Determine the total power in the circuit of Figure 6–59.

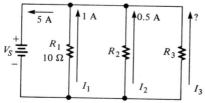

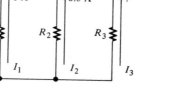

FIGURE 6–58   **FIGURE 6–59**

20. There are two resistors in parallel across a 10-V source. One is 220 Ω, and the other is 560 Ω. You measure a total current of 17.86 mA. What is wrong with the circuit?

21. What resistance value in parallel with 100 Ω produces a total resistance of approximately 45 Ω?

PROBLEMS

Section 6–1

6–1 Connect the resistors in Figure 6–60(a) in parallel across the battery.

6–2 Determine whether or not all the resistors in Figure 6–60(b) are connected in parallel on the printed circuit (PC) board.

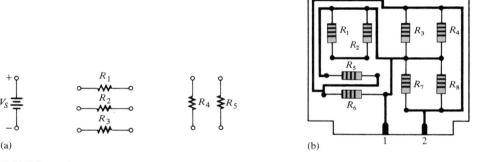

(a)

(b)

FIGURE 6–60

6–3 Identify which groups of resistors are in parallel on the double-sided PC board in Figure 6–61.

Section 6–2

6–4 What is the voltage across and the current through each parallel resistor if the total voltage is 12 V and the total resistance is 550 Ω? There are four resistors, all of equal value.

6–5 The source voltage in Figure 6–62 is 100 V. How much voltage does each of the meters read?

6–6 What is the voltage across each resistor in Figure 6–63 for each switch position?

Section 6–3

6–7 The following currents are measured in the same direction in a three-branch parallel circuit: 250 mA, 300 mA, and 800 mA. What is the value of the current into the junction of these three branches?

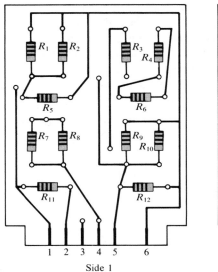
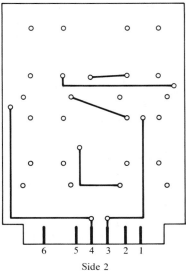

Side 1

Side 2

FIGURE 6–61

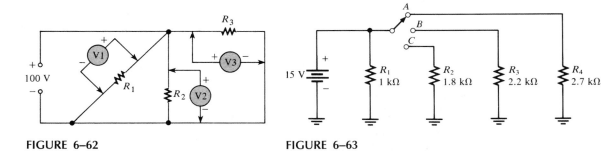

FIGURE 6–62

FIGURE 6–63

6–8 There are five hundred milliamperes (500 mA) into five parallel resistors. The currents through four of the resistors are 50 mA, 150 mA, 25 mA, and 100 mA. How much current is there through the fifth resistor?

6–9 In the circuit of Figure 6–64, determine the resistances R_2, R_3, and R_4.

FIGURE 6–64

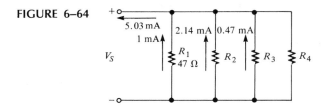

6–10 The electrical circuit in a room has a ceiling lamp that draws 1.25 A and four wall outlets. Two table lamps that each draw 0.833 A are plugged into two outlets, and a TV set that draws 1 A is connected to the third outlet. When all of these items are in use, how much current is in the main line serving the room? If the main line is protected by a 5-A circuit breaker, how much current can be drawn from the fourth outlet? Draw a schematic of this wiring.

6–11 The total resistance of a parallel circuit is 25 Ω. How much is the current through a 220-Ω resistor that makes up part of the parallel circuit if the total current is 100 mA?

Section 6–4

6–12 The following resistors are connected in parallel: 1 MΩ, 2.2 MΩ, 5.6 MΩ, 12 MΩ, and 22 MΩ. Determine the total resistance.

6–13 Find the total resistance for each following group of parallel resistors:
(a) 560 Ω and 1000 Ω.   (b) 47 Ω and 56 Ω.
(c) 1.5 kΩ, 2.2 kΩ, 10 kΩ.   (d) 1 MΩ, 470 kΩ, 1 kΩ, 2.7 MΩ.

6–14 Calculate R_T for each circuit in Figure 6–65.

FIGURE 6–65

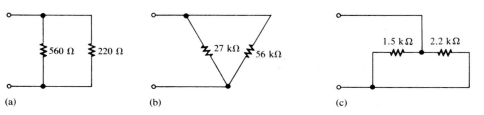

(a)   (b)   (c)

6–15 What is the total resistance of twelve 6.8-kΩ resistors in parallel?

6–16 Five 47-Ω, ten 100-Ω, and two 10-Ω resistors are all connected in parallel. What is the total resistance?

6–17 Find the total resistance if all resistors in Problem 6–13 are connected in parallel.

6–18 If the total resistance in Figure 6–66 is 389.2 Ω, what is the value of R_2?

6–19 What is the total resistance between point A and ground in Figure 6–67 for the following conditions?
(a) S_1 and S_2 open   (b) S_1 closed, S_2 open
(c) S_1 open, S_2 closed   (d) S_1 and S_2 closed

FIGURE 6–66

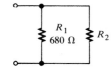

FIGURE 6–67

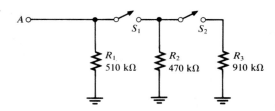

Section 6–5

6–20 What is the total current in each circuit of Figure 6–68?

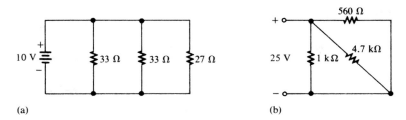

(a)   (b)

FIGURE 6–68

6–21 Three 33-Ω resistors are connected in parallel with a 110-V source. How much current is there from the source?

6–22 Four equal-value resistors are connected in parallel. Five volts are applied across the parallel circuit, and 1.11 mA are measured from the source. What is the value of each resistor?

6–23 Christmas tree lights are usually connected in parallel. If a set of lights is connected to a 110-V source and the filament of each bulb has a resistance of 2.2 kΩ, how much current is there through each bulb? Why is it better to have these bulbs in parallel rather than in series?

6–24 Find the values of the unspecified labeled quantities in each circuit of Figure 6–69.

FIGURE 6–69

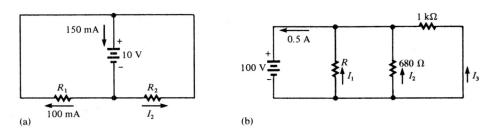

(a)   (b)

6–25 To what minimum value can the 100-Ω rheostat in Figure 6–70 be adjusted before the 0.5-A fuse blows?

FIGURE 6–70

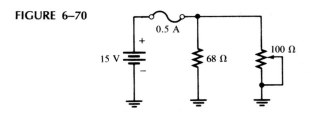

6–26 Determine the total current from the source and the current through each resistor for each switch position in Figure 6–71.

FIGURE 6–71

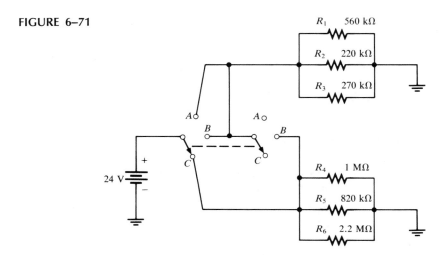

6–27 Find the values of the unspecified quantities in Figure 6–72.

FIGURE 6–72

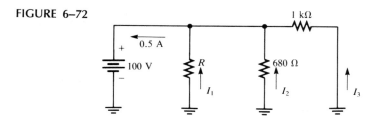

Section 6–6

6–28 Determine the current through R_L in each circuit in Figure 6–73.

FIGURE 6–73

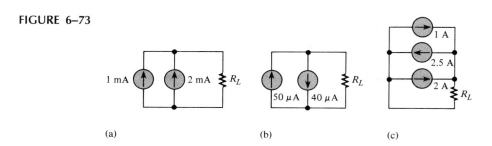

(a)   (b)   (c)

6–29 Find the current through the resistor for each position of the ganged switch in Figure 6–74.

FIGURE 6–74

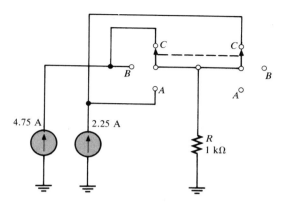

Section 6–7

6–30 How much branch current should each meter in Figure 6–75 indicate?

FIGURE 6–75

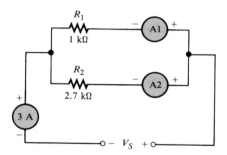

6–31 Determine the current in each branch of the current dividers of Figure 6–76.

FIGURE 6–76

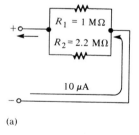

(a)

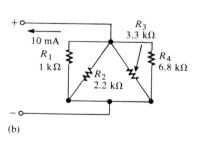

(b)

6–32 What is the current through each resistor in Figure 6–77? R is the lowest-value resistor, and all others are multiples of that value as indicated.

FIGURE 6–77

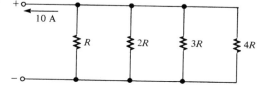

6–33 Determine all of the resistor values in Figure 6–78. $R_T = 773\ \Omega$.

FIGURE 6–78

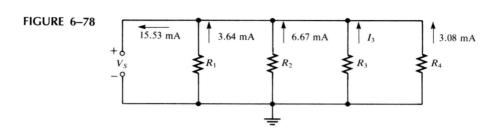

6–34 **(a)** Determine the required value of the shunt resistor R_{SH1} in the ammeter of Figure 6–46 if the resistance of the meter movement is 50 Ω.
(b) Find the required value for R_{SH2} in the meter circuit of Figure 6–47 ($R_M = 50\ \Omega$).

6–35 Add a fourth range position to the meter in Figure 6–47 so that currents up to 1 A (1000 mA) can be measured. Specify the value of the additional shunt resistor. Assume a 1-mA, 50-Ω meter movement.

Section 6–8

6–36 Five parallel resistors each handle 40 mW. What is the total power?

6–37 Determine the total power in each circuit of Figure 6–76.

6–38 Six light bulbs are connected in parallel across 110 V. Each bulb is rated at 75 W. How much current is there through each bulb, and what is the total current?

6–39 Find the values of the unspecified quantities in Figure 6–79.

FIGURE 6–79

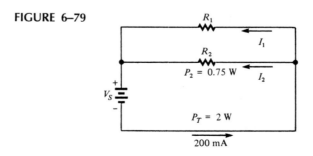

6–40 A certain parallel network consists of only 1/2-W resistors. The total resistance is 1 kΩ, and the total current is 50 mA. If each resistor is operating at one-half its maximum power level, determine the following:
(a) The number of resistors   (b) The value of each resistor
(c) The current in each branch   (d) The applied voltage

Section 6–10

6–41 If one of the bulbs burns out in Problem 6–40, how much current will there be through each of the remaining bulbs? What will the total current be?

6–42 In Figure 6–80, the current and voltage measurements are indicated. Has a resistor opened, and, if so, which one?

FIGURE 6–80

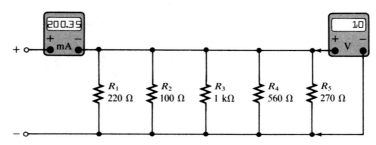

6–43 What is wrong with the circuit in Figure 6–81?

FIGURE 6–81

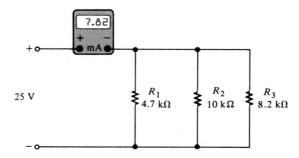

6–44 A parallel circuit consisting of a 4.7 kΩ and two 10-kΩ resistors is connected across a 25-V source. An ammeter in series with the source reads 5 mA. What is wrong with the circuit?

6–45 Develop a test procedure to check the circuit in Figure 6–82 to make sure that there are no open components. You must do this test without removing any component from the board. List the procedure in a detailed step-by-step format.

FIGURE 6–82

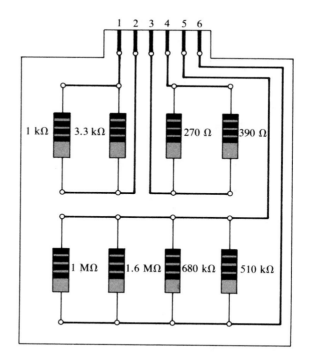

Section 6–11

6–46 Modify the program in Section 6–11 to include computation and display of the power dissipation in each branch and the total power.

6–47 Write a program to compute the value of an unknown resistor in an n-resistor parallel network when the desired total resistance and the values of the other $n - 1$ resistors are specified.

6–48 Develop a computer program to determine which resistor in an n-resistor parallel network is open (if any) based on inputs of each resistor value and the total measured current.

6–49 Develop a flowchart for the program in Section 6–11.

ANSWERS TO SECTION REVIEWS

Section 6–1
1. Between the same two points.
2. More than one current path between two given points.   3. See Figure 6–83.
4. See Figure 6–84.

FIGURE 6–83

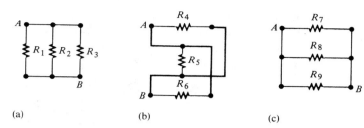

(a)   (b)   (c)

FIGURE 6–84

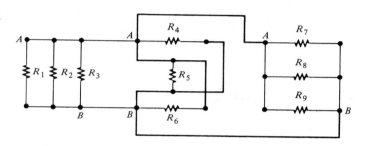

Section 6–2
1. 5 V.   **2.** 118 V, 118 V.   **3.** 50 V and 50 V, respectively.
4. Voltage is the same across all branches.

Section 6–3
1. (a) The algebraic sum of all the currents at a junction is zero; **(b)** the sum of the currents entering a junction equals the sum of the currents leaving that junction.   **2.** 2.5 A.   **3.** 400 mA.   **4.** 3 μA.   **5.** 7 A entering, 5 A leaving.

Section 6–4
1. Decrease.   **2.** The smallest resistance value.   **3.** The equation is as follows:

$$R_T = \frac{1}{\left(\dfrac{1}{R_1}\right) + \left(\dfrac{1}{R_2}\right) + \cdots + \left(\dfrac{1}{R_n}\right)}$$

4. $R_T = \dfrac{R_1 R_2}{R_1 + R_2}$.
5. $R_T = R/n$.   **6.** 687.5 Ω.   **7.** 250 Ω.   **8.** 26.36 Ω.

Section 6–5
1. 0.441 A.   **2.** 444.4 V.
3. 0.654 A through the 680 Ω; 1.35 A through the 330 Ω.   **4.** 8.2 kΩ.
5. 68.75 V.

Section 6–6
1. 2 A.   **2.** Three. See Figure 6–85.   **3.** 20 mA.

FIGURE 6–85

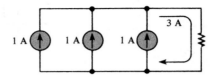

Section 6–7
1. The equation is as follows:

$$I_x = \left(\frac{R_T}{R_x}\right)I_T$$

2. The equations are as follows:

$$I_1 = \left(\frac{R_2}{R_1 + R_2}\right)I_T$$

$$I_2 = \left(\frac{R_1}{R_1 + R_2}\right)I_T$$

3. The 22 Ω has most current; the 220 Ω has least current.   **4.** 0.967 mA.
5. $I_1 = 6.73$ mA, $I_2 = 3.27$ mA.

Section 6–8
1. Add the power of each resistor.   **2.** 17.35 W.   **3.** 614.71 W.

Section 6–9
1. 50 Ω; 1 mA.
2. Yes, because the shunt resistors must allow currents much greater than the current through the meter movement.

Section 6–10
1. There is no change in voltage; the total current decreases.   **2.** It increases.
3. Yes   **4.** 1 A

Section 6–11
1. Lines 120–140.   **2.** Line 160 initializes RT to a value equal to R1.

SEVEN

SERIES-PARALLEL CIRCUITS

Various combinations of both series and parallel resistors are often used in electronic circuits. In this chapter, examples of such series-parallel arrangements are examined and analyzed. An important circuit called the *Wheatstone bridge* is introduced. Troubleshooting series-parallel circuits for shorts and opens is also covered.

In this chapter, you will learn:

☐ How to identify series and parallel parts of a series-parallel circuit and how to recognize the relationships of all the resistors.

☐ How to determine the total resistance of a series-parallel circuit.

☐ How to determine the currents and voltages in a series-parallel circuit.

☐ How resistive loads affect voltage divider circuits.

☐ Under what conditions a voltmeter can affect the value of the voltage being measured.

☐ What a Wheatstone bridge circuit is and how it can be used.

☐ How to analyze ladder networks.

☐ How to use troubleshooting techniques to identify opens and shorts in a series-parallel circuit.

☐ How a computer program can be used for the analysis of ladder networks.

7–1   IDENTIFICATION OF SERIES-PARALLEL RELATIONSHIPS

A series-parallel circuit consists of combinations of both series and parallel current paths. It is important to be able to identify how the components in a circuit are arranged in terms of their series and parallel relationships.

Figure 7–1(a) shows a simple series-parallel combination of resistors. Notice that the resistance from point A to point B is R_1. The resistance from point B to point C is R_2 and R_3 in parallel ($R_2\|R_3$). The resistance from point A to point C is R_1 in series with the parallel combination of R_2 and R_3, as indicated in Figure 7–1(b).

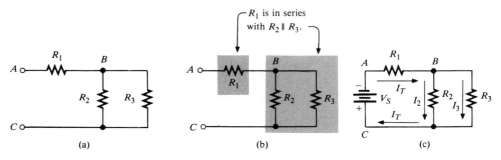

FIGURE 7–1
A simple series-parallel circuit.

When the circuit of Figure 7–1(a) is connected to a voltage source as shown in Part (c), the *total* current goes through R_1 and divides at point B into the two parallel paths. These two branch currents then recombine, and the total current goes into the positive source terminal as shown.

Now, to illustrate series-parallel relationships, we will increase the complexity of the circuit in Figure 7–1(a) step-by-step. In Figure 7–2(a), another resistor (R_4) is connected in series with R_1. The resistance between points A and B is now $R_1 + R_4$, and this combination is in series with the parallel combination of R_2 and R_3, as illustrated in Part (b).

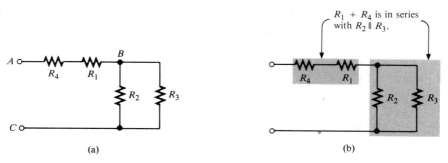

FIGURE 7–2
R_4 is added to the circuit in series with R_1.

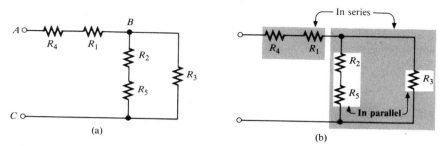

FIGURE 7–3

R_5 is added to the circuit in series with R_2.

In Figure 7–3(a), R_5 is connected in series with R_2. The series combination of R_2 and R_5 is in parallel with R_3. This entire series-parallel combination is in series with the $R_1 + R_4$ combination, as illustrated in Part (b).

In Figure 7–4(a), R_6 is connected in parallel with the series combination of R_1 and R_4. The series-parallel combination of R_1, R_4, and R_6 is in series with the series-parallel combination of R_2, R_3, and R_5, as indicated in Part (b).

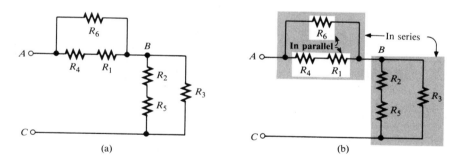

FIGURE 7–4

R_6 is added to the circuit in parallel with the series combination of R_1 and R_4.

EXAMPLE 7–1

Identify the series-parallel relationships in Figure 7–5.

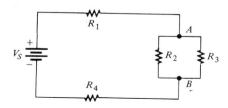

FIGURE 7–5

Solution:
Starting at the positive terminal of the source, follow the current paths. All of the current produced by the source must go through R_1, which is in series with the rest of the circuit.

   The total current takes two paths when it gets to point A. Part of it goes through R_2, and part of it through R_3. R_2 and R_3 are in parallel with each other, and this parallel combination is in series with R_1.

   At point B, the currents through R_2 and R_3 come together again. Thus, the total current goes through R_4. R_4 is in series with R_1 and the parallel combination of R_2 and R_3. The currents are shown in Figure 7–6, where I_T is the total current.

   In summary, R_1 and R_4 are in series with the parallel combination of R_2 and R_3.

FIGURE 7–6

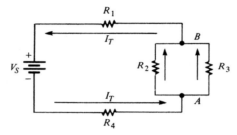

EXAMPLE 7–2

Identify the series-parallel relationships in Figure 7–7.

FIGURE 7–7

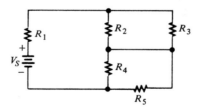

Solution:
Sometimes it is easier to see a particular circuit arrangement if it is drawn in a different way. In this case, the circuit schematic is redrawn in Figure 7–8, which better illustrates the series-parallel relationships. Now you can see that R_2 and R_3 are in parallel with each other and also that R_4 and R_5 are in parallel with each other. Both parallel combinations are in series with each other and with R_1.

FIGURE 7–8

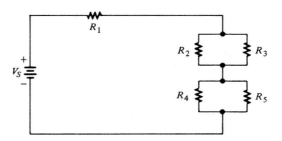

EXAMPLE 7–3

Describe the series-parallel combination between points A and D in Figure 7–9.

FIGURE 7–9

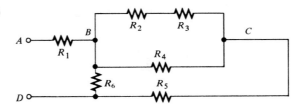

Solution:
Between points B and C, there are two parallel paths. The lower path consists of R_4, and the upper path consists of a *series* combination of R_2 and R_3. This parallel combination is in series with R_5. The R_2, R_3, R_4, R_5 combination is in parallel with R_6. R_1 is in series with this entire combination.

EXAMPLE 7–4

Describe the total resistance between each pair of points in Figure 7–10.

FIGURE 7–10

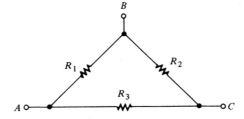

Solution:
1. From point A to B: R_1 is in *parallel* with the *series* combination of R_2 and R_3.
2. From point A to C: R_3 is in *parallel* with the *series* combination of R_1 and R_2.
3. From point B to C: R_2 is in *parallel* with the *series* combination of R_1 and R_3.

DETERMINING RELATIONSHIPS ON A PC BOARD

Usually, the physical arrangement of components on a PC board bears no resemblance to the actual circuit relationships. By tracing out the circuit on the PC board and rearranging the components on paper into a recognizable form, you can determine the series-parallel relationships. An example will illustrate.

EXAMPLE 7–5

Determine the relationships of the resistors on the PC board in Figure 7–11.

FIGURE 7–11

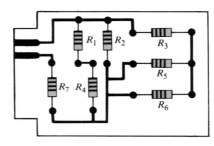

Solution:

In Figure 7–12(a), the schematic is drawn in the same arrangement as that of the resistors on the board. In Part (b), the resistors are reoriented so that the series-parallel relationships are obvious. R_1 and R_4 are in series; $R_1 + R_4$ is in parallel with R_2; R_5 and R_6 are in parallel and this combination is in series with R_3. The R_3, R_5, and R_6 series-parallel combination is in parallel with both R_2 and the $R_1 + R_4$ combination. This entire series-parallel combination is in series with R_7, as Figure 7–12(c) illustrates.

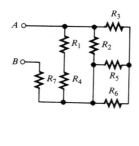

(a)

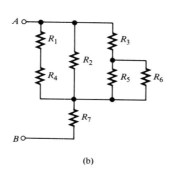

(b)

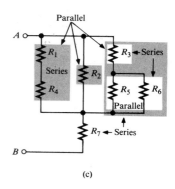

(c)

FIGURE 7–12

SECTION REVIEW 7–1

1. Define *series-parallel resistive circuit.*
2. A certain series-parallel circuit is described as follows: R_1 and R_2 are in parallel. This parallel combination is in series with another parallel combination of R_3 and R_4. Sketch the circuit.
3. In the circuit of Figure 7–13, describe the series-parallel relationships of the resistors.
4. Which resistors are in parallel in Figure 7–14?

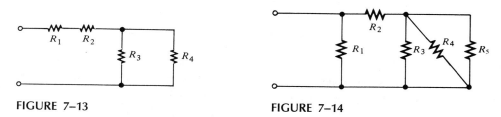

FIGURE 7–13   FIGURE 7–14

5. Describe the parallel arrangements in Figure 7–15.
6. Are the parallel combinations in Figure 7–15 in series?

FIGURE 7–15

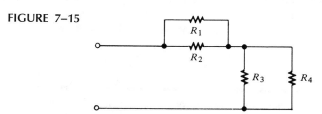

7–2

ANALYSIS OF SERIES-PARALLEL CIRCUITS

Several quantities are important when you have a circuit that is a series-parallel configuration of resistors. In this section you will learn how to determine total resistance, total current, branch currents, and the voltage across any portion of a circuit.

TOTAL RESISTANCE

In Chapter 5, you learned how to determine total series resistance. In Chapter 6, you learned how to determine total parallel resistance.

To find the total resistance R_T of a series-parallel combination, simply define the series and parallel relationships, and then perform the calculations that you have previously learned. The following two examples illustrate this general approach.

EXAMPLE 7–6

Determine R_T of the circuit in Figure 7–16 between points A and B.

FIGURE 7–16

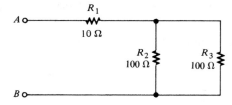

Solution:
First calculate the equivalent parallel resistance of R_2 and R_3. Since R_2 and R_3 are equal in value, we can use Equation (6–5):

$$R_{EQ} = \frac{R}{n} = \frac{100\ \Omega}{2} = 50\ \Omega$$

Notice that we used the term R_{EQ} here to designate the total resistance of a *portion* of a circuit in order to distinguish it from the total resistance R_T of the *complete* circuit.
Now, since R_1 is in series with R_{EQ}, their values are added as follows:

$$R_T = R_1 + R_{EQ} = 10\ \Omega + 50\ \Omega = 60\ \Omega$$

EXAMPLE 7–7

Find the total resistance between the positive and negative terminals of the battery in Figure 7–17.

FIGURE 7–17

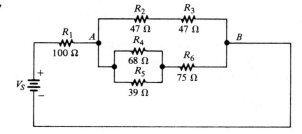

Solution:
In the *upper branch*, R_2 is in series with R_3. We will call the series combination R_{EQ1}. It is equal to $R_2 + R_3$:

$$R_{EQ1} = R_2 + R_3 = 47\ \Omega + 47\ \Omega = 94\ \Omega$$

In the *lower branch*, R_4 and R_5 are in parallel with each other. We will call this parallel combination R_{EQ2}. It is calculated as follows:

$$R_{EQ2} = \frac{R_4 R_5}{R_4 + R_5} = \frac{(68 \ \Omega)(39 \ \Omega)}{68 \ \Omega + 39 \ \Omega} = 24.79 \ \Omega$$

Also in the *lower branch*, the parallel combination of R_4 and R_5 is in series with R_6. This series-parallel combination is designated R_{EQ3} and is calculated as follows:

$$R_{EQ3} = R_6 + R_{EQ2} = 75 \ \Omega + 24.79 \ \Omega = 99.79 \ \Omega$$

Figure 7–18 shows the original circuit in a simplified *equivalent* form.

FIGURE 7–18

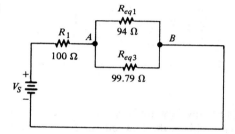

Now we can find the equivalent resistance between points A and B. It is R_{EQ1} in parallel with R_{EQ3}. The equivalent resistance is calculated as follows:

$$R_{AB} = \frac{1}{(1/94 \ \Omega) + (1/99.79 \ \Omega)} = 48.40 \ \Omega$$

Finally, the total resistance is R_1 in series with R_{AB}:

$$R_T = R_1 + R_{AB} = 100 \ \Omega + 48.40 \ \Omega = 148.40 \ \Omega$$

TOTAL CURRENT

Once the total resistance and the source voltage are known, we can find total current in a circuit by applying Ohm's law. Total current is the total source voltage divided by the total resistance:

$$I_T = \frac{V_S}{R_T}$$

For example, let us find the total current in the circuit of Example 7–7 (Figure 7–17). Assume that the source voltage is 30 V. The calculation is as follows:

$$I_T = \frac{V_S}{R_T} = \frac{30 \ \text{V}}{148.40 \ \Omega} = 0.202 \ \text{A}$$

BRANCH CURRENTS

Using the *current divider formula, Kirchhoff's current law,* or *Ohm's law,* or combinations of these, we can find the current in any branch of a series-parallel circuit. In some cases it may take repeated application of the formula to find a given current. Working through some examples aids in understanding the procedure.

**EXAMPLE
7–8**

Find the current through R_2 and the current through R_3 in Figure 7–19.

FIGURE 7–19

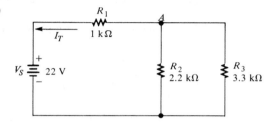

Solution:
First we need to know how much current there is out of the junction (point A) of the parallel branches. This is the total circuit current. To find I_T, we need to know R_T:

$$R_T = R_1 + \frac{R_2 R_3}{R_2 + R_3} = 1 \text{ k}\Omega + \frac{(2.2 \text{ k}\Omega)(3.3 \text{ k}\Omega)}{2.2 \text{ k}\Omega + 3.3 \text{ k}\Omega}$$

$$= 1 \text{ k}\Omega + 1.32 \text{ k}\Omega = 2.32 \text{ k}\Omega$$

$$I_T = \frac{V_S}{R_T} = \frac{22 \text{ V}}{2.32 \text{ k}\Omega}$$

$$= 9.48 \text{ mA}$$

Using the current divider rule for two branches as given in Chapter 6, we find the current through R_2 as follows:

$$I_2 = \left(\frac{R_3}{R_2 + R_3} \right) I_T = \left(\frac{3.3 \text{ k}\Omega}{5.5 \text{ k}\Omega} \right) 9.48 \text{ mA}$$

$$= 5.69 \text{ mA}$$

Now we can use Kirchhoff's current law to find the current through R_3 as follows:

$$I_T = I_2 + I_3$$

$$I_3 = I_T - I_2 = 9.48 \text{ mA} - 5.69 \text{ mA}$$

$$= 3.79 \text{ mA}$$

EXAMPLE 7–9

Determine the current through R_4 in Figure 7–20 if $V_S = 50$ V.

FIGURE 7–20

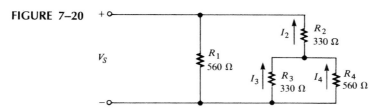

Solution:

First the current (I_2) out of the junction of R_3 and R_4 must be found. Once we know this current, we can use the current divider formula to find I_4.

Notice that there are two main branches in the circuit. The left-most branch consists of only R_1. The right-most branch has R_2 in series with the parallel combination of R_3 and R_4. The voltage across both of these main branches is the same and equal to 50 V. We can find the current (I_2) out of the junction of R_3 and R_4 by calculating the equivalent resistance (R_{EQ}) of the right-most main branch and then applying Ohm's law, because this current is the total current through this main branch. Thus,

$$R_{EQ} = R_2 + \frac{R_3 R_4}{R_3 + R_4}$$

$$= 330\ \Omega + \frac{(560\ \Omega)(330\ \Omega)}{890\ \Omega} = 537.6\ \Omega$$

$$I_2 = \frac{V_S}{R_{EQ}} = \frac{50\ \text{V}}{537.6\ \Omega}$$

$$= 0.093\ \text{A}$$

Using the current divider formula, we calculate I_4 as follows:

$$I_4 = \left(\frac{R_3}{R_3 + R_4}\right) I_2 = \left(\frac{330\ \Omega}{890\ \Omega}\right) 0.093\ \text{A}$$

$$= 0.035\ \text{A}$$

You can find the current in R_3 by applying Kirchhoff's law ($I_2 = I_3 + I_4$) and subtracting I_4 from I_2 ($I_3 = I_2 - I_4$).

You can find the current in R_1 by using Ohm's law ($I_1 = V_S/R_1$).

The calculator sequence for I_4 is

| 5 | 6 | 0 | × | 3 | 3 | 0 | ÷ | (| (| 5 | 6 | 0 | + | 3 | 3 | 0 |) | = | + | 3 | 3 | 0 | = |

| x→M | 5 | 0 | ÷ | RM | = | x→M | 3 | 3 | 0 | ÷ | (| 5 | 6 | 0 | + |

| 3 | 3 | 0 |) | = | × | RM | = |

VOLTAGE DROPS

It is often necessary to find the voltages across certain parts of a series-parallel circuit. We can find these voltages by using the voltage divider formula given in Chapter 5, Kirchhoff's voltage law, Ohm's law, or combinations of each. The following three examples illustrate use of the formulas.

EXAMPLE 7–10

Determine the voltage drop from A to B in Figure 7–21, and then find the voltage across R_1 (V_1).

FIGURE 7–21

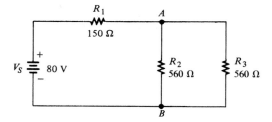

Solution:

Note that R_2 and R_3 are in parallel in this circuit. Since they are equal in value, their equivalent resistance is

$$R_{AB} = \frac{560 \ \Omega}{2} = 280 \ \Omega$$

As shown in the equivalent circuit (Figure 7–22), R_1 is in series with R_{AB}. The total circuit resistance as seen from the source is

$$R_T = R_1 + R_{AB} = 150 \ \Omega + 280 \ \Omega = 430 \ \Omega$$

FIGURE 7–22

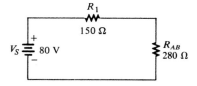

Now we can use the voltage divider formula to find the voltage across the parallel combination of Figure 7–21 (between points A and B):

$$V_{AB} = \left(\frac{R_{AB}}{R_T}\right)V_S = \left(\frac{280 \ \Omega}{430 \ \Omega}\right)80 \text{ V}$$

$$= 52.1 \text{ V}$$

We can now use Kirchhoff's voltage law to find V_1 as follows:

$$V_S = V_1 + V_{AB}$$
$$V_1 = V_S - V_{AB} = 80 \text{ V} - 52.1 \text{ V}$$
$$= 27.9 \text{ V}$$

**EXAMPLE
7–11**

Determine the voltages across each resistor in the circuit of Figure 7–23.

FIGURE 7–23

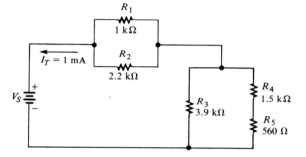

Solution:
The source voltage is not given, but the total current is known. Since R_1 and R_2 are in parallel, they each have the same voltage. The current through R_1 is

$$I_1 = \left(\frac{R_2}{R_1 + R_2}\right)I_T = \left(\frac{2.2 \text{ k}\Omega}{3.2 \text{ k}\Omega}\right)1 \text{ mA}$$
$$= 0.688 \text{ mA}$$

The voltages are

$$V_1 = I_1R_1 = (0.688 \text{ mA})(1 \text{ k}\Omega)$$
$$= 0.688 \text{ V}$$
$$V_2 = V_1 = 0.688 \text{ V}$$

The current through R_3 is found by applying the current divider formula as follows:

$$I_3 = \left[\frac{R_4 + R_5}{R_3 + (R_4 + R_5)}\right]I_T = \left(\frac{2.06 \text{ k}\Omega}{5.96 \text{ k}\Omega}\right)1 \text{ mA}$$
$$= 0.346 \text{ mA}$$

The voltage across R_3 is

$$V_3 = I_3R_3 = (0.346 \text{ mA})(3.9 \text{ k}\Omega)$$
$$= 1.349 \text{ V}$$

The currents through R_4 and R_5 are the same because these resistors are in series:

$$I_4 = I_5 = I_T - I_3$$
$$= 1 \text{ mA} - 0.346 \text{ mA} = 0.654 \text{ mA}$$

The voltages across R_4 and R_5 are

$$V_4 = I_4R_4 = (0.654 \text{ mA})(1.5 \text{ k}\Omega)$$
$$= 0.981 \text{ V}$$
$$V_5 = I_5R_5 = (0.654 \text{ mA})(560 \text{ }\Omega)$$
$$= 0.366 \text{ V}$$

The small difference in V_3 and the sum of V_4 and V_5 is a result of rounding to three places.

**EXAMPLE
7–12**

Determine the voltage drop across each resistor in Figure 7–24.

FIGURE 7–24

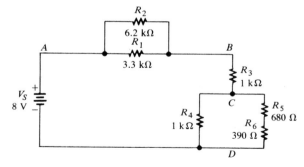

Solution:
Because we know the total voltage, we can solve this problem using the voltage divider formula. First reduce each parallel combination to an equivalent resistance. Since R_1 and R_2 are in parallel between points A and B, we combine their values:

$$R_{AB} = \frac{R_1R_2}{R_1 + R_2} = \frac{(3.3 \text{ k}\Omega)(6.2 \text{ k}\Omega)}{9.5 \text{ k}\Omega}$$
$$= 2.15 \text{ k}\Omega$$

Since R_4 is in parallel with the series combination of R_5 and R_6 between points C and D, we combine these values to obtain

$$R_{CD} = \frac{(R_4)(R_5 + R_6)}{R_4 + R_5 + R_6} = \frac{(1 \text{ k}\Omega)(1.07 \text{ k}\Omega)}{2.07 \text{ k}\Omega}$$

$$= 517 \ \Omega$$

The equivalent circuit is drawn in Figure 7–25.

FIGURE 7–25

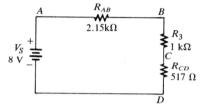

Applying the voltage divider formula to solve for the voltages, we get the following results:

$$V_{AB} = \left(\frac{R_{AB}}{R_T}\right)V_S = \left(\frac{2.15 \text{ k}\Omega}{3.67 \text{ k}\Omega}\right)8 \text{ V}$$

$$= 4.69 \text{ V}$$

$$V_{CD} = \left(\frac{R_{CD}}{R_T}\right)V_S = \left(\frac{517 \ \Omega}{3.67 \text{ k}\Omega}\right)8 \text{ V}$$

$$= 1.13 \text{ V}$$

$$V_3 = \left(\frac{R_3}{R_T}\right)V_S = \left(\frac{1 \text{ k}\Omega}{3.67 \text{ k}\Omega}\right)8 \text{ V}$$

$$= 2.18 \text{ V}$$

V_{AB} equals the voltage across both R_1 and R_2:

$$V_1 = V_2 = V_{AB} = 4.69 \text{ V}$$

V_{CD} is the voltage across R_4 and across the series combination of R_5 and R_6:

$$V_4 = V_{CD} = 1.13 \text{ V}$$

Now apply the voltage divider formula to the series combination of R_5 and R_6 to get V_5 and V_6:

$$V_5 = \left(\frac{R_5}{R_5 + R_6}\right)V_{CD} = \left(\frac{680 \ \Omega}{1070 \ \Omega}\right)1.13 \text{ V}$$

$$= 0.718 \text{ V}$$

$$V_6 = \left(\frac{R_6}{R_5 + R_6}\right)V_{EQ2} = \left(\frac{390\ \Omega}{1070\ \Omega}\right)1.13\ V$$

$$= 0.412\ V$$

As you have seen in this section, the analysis of series-parallel circuits can be approached in many ways, depending on what information you need and what circuit values you know. The examples in this section do not represent an exhaustive coverage. They are meant only to give you an idea of how to approach series-parallel circuit analysis.

If you know Ohm's law, Kirchhoff's laws, the voltage divider formula, and the current divider formula, and if you know how to apply these laws, you can solve most resistive circuit analysis problems. The ability to recognize series and parallel combinations is, of course, essential.

SECTION REVIEW 7–2

1. List the circuit laws and formulas that may be necessary in the analysis of a series-parallel circuit.

2. Find the total resistance between A and B in the circuit of Figure 7–26.

3. Find I_3 in Figure 7–26.

4. Find V_2 in Figure 7–26.

5. Determine R_T in Figure 7–27 as "seen" by the source. What is I_T?

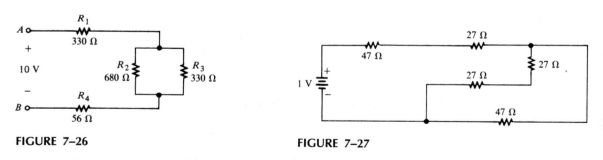

FIGURE 7–26   FIGURE 7–27

7–3   **VOLTAGE DIVIDERS WITH RESISTIVE LOADS**

Voltage dividers were introduced in Chapter 5. In this section we will discuss the effects of resistive loads on voltage dividers. For example, the voltage divider in Figure 7–28(a) produces an output voltage (V_{OUT}) of 5 V because the two resistors are of equal value. This voltage is the *unloaded output voltage*. When a load resistor, R_L, is con-

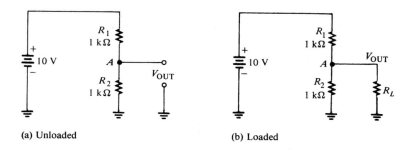

FIGURE 7–28
A voltage divider with both unloaded and loaded outputs.

nected from the output to ground as shown in Figure 7–28(b), the output voltage is reduced by an amount that depends on the value of R_L. The load resistor is in parallel with R_2, reducing the resistance from point A to ground and, as a result, also reducing the voltage across the parallel combination. This is one effect of *loading* a voltage divider. Another effect of a loaded condition is that more current is drawn from the source because the total resistance of the circuit is reduced.

The larger R_L is compared to R_2, the less the output voltage is reduced from its unloaded value, as illustrated in Figure 7–29. The reason is that when two resistors are connected in parallel and one of the resistors is much greater than the other, the total resistance is close to the value of the smaller resistance.

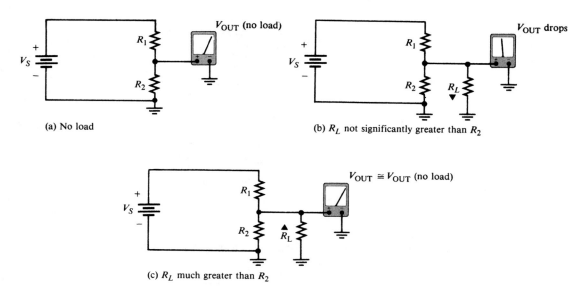

FIGURE 7–29
The effect of a load resistor.

EXAMPLE 7–13

Determine both the unloaded and the loaded output voltages of the voltage divider in Figure 7–30 for the following two values of load resistance:
(a) $R_L = 10 \text{ k}\Omega$; (b) $R_L = 100 \text{ k}\Omega$.

FIGURE 7–30

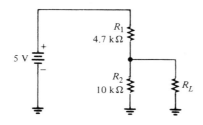

Solution:

(a) The *unloaded* output voltage is

$$V_{OUT} = \left(\frac{10 \text{ k}\Omega}{14.7 \text{ k}\Omega}\right)5 \text{ V} = 3.4 \text{ V}$$

With the 10-kΩ load resistor connected, R_L is in parallel with R_2, which gives 5 kΩ, as shown by the equivalent circuit in Figure 7–31(a). The *loaded* output voltage is

$$V_{OUT} = \left(\frac{5 \text{ k}\Omega}{9.7 \text{ k}\Omega}\right)5 \text{ V} = 2.6 \text{ V}$$

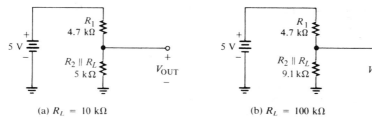

(a) $R_L = 10 \text{ k}\Omega$ (b) $R_L = 100 \text{ k}\Omega$

FIGURE 7–31

(b) With the 100-kΩ load, the resistance from output to ground is

$$\frac{R_2 R_L}{R_2 + R_L} = \frac{(10 \text{ k}\Omega)(100 \text{ k}\Omega)}{110 \text{ k}\Omega} = 9.1 \text{ k}\Omega$$

as shown in Figure 7–31(b). The *loaded* output voltage is

$$V_{OUT} = \left(\frac{9.1 \text{ k}\Omega}{13.8 \text{ k}\Omega}\right)5 \text{ V} = 3.3 \text{ V}$$

Notice that with the larger value of R_L, the output is reduced from its unloaded value by much less than it is with the smaller R_L. This problem illustrates the loading effect of R_L on the voltage divider.

Voltage dividers are sometimes useful in obtaining various voltages from a power supply. For example, suppose that we wish to derive 12 V and 6 V from a 24-V supply. To do so requires a voltage divider with two *taps*, as shown in Figure 7–32. In this example, R_1 must equal $R_2 + R_3$, and R_2 must equal R_3. The actual values of the resistors are set by the amount of current that is to be drawn from the source under unloaded conditions. This current, called the *bleeder current*, represents a continuous drain on the source. With these ideas in mind, in Example 7–14 we design a voltage divider to meet certain specified requirements.

FIGURE 7–32
Voltage divider with two output taps.

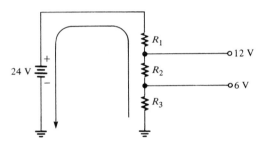

EXAMPLE 7–14

A power supply requires 12 V and 6 V from a 24-V battery. The unloaded current drain on this battery is not to exceed 1 mA. Determine the values of the resistors. Also determine the output voltage at the 12-V tap when both outputs are loaded with 100 kΩ each.

Solution:
A circuit as shown in Figure 7–32 is required. In order to have an unloaded current of 1 mA, the total resistance must be as follows:

$$R_T = \frac{V_S}{I} = \frac{24 \text{ V}}{1 \text{ mA}} = 24 \text{ k}\Omega$$

To get exactly 12 V, R_1 must equal $R_2 + R_3 = 12$ kΩ. To get exactly 6 V, R_2 must equal $R_3 = 6$ kΩ.

The closest standard value to 6 kΩ is 6.04 k in a 1% resistor and 6.2 kΩ in a 5% resistor. Because it is more commonly available in labs, we will use 6.2 kΩ for R_2 and R_3, although this will result in small differences from the desired unloaded output voltages as follows:

$$V_{12V} = \left(\frac{12.4 \text{ k}\Omega}{24.4 \text{ k}\Omega}\right)24 \text{ V} = 12.2 \text{ V}$$

$$V_{6V} = \left(\frac{6.2 \text{ k}\Omega}{24.4 \text{ k}\Omega}\right)24 \text{ V} = 6.1 \text{ V}$$

Now if the 100-kΩ loads are connected across the outputs as shown in Figure 7–33, the loaded output voltages are as determined below.

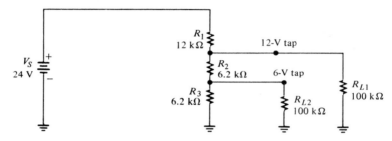

FIGURE 7–33

The equivalent resistance from the 12-V tap to ground is the 100-kΩ load resistor R_{L1} in parallel with the combination of R_2 in series with the parallel combination of R_3 and R_{L2}. This solution is as follows.

R_3 in parallel with R_{L2}:

$$R_{EQ1} = \frac{R_3 R_{L2}}{R_3 + R_{L2}} = \frac{(6.2 \text{ k}\Omega)(100 \text{ k}\Omega)}{106.2 \text{ k}\Omega} = 5.84 \text{ k}\Omega$$

R_2 in series with R_{EQ1}:

$$R_2 + R_{EQ1} = R_{EQ2} = 6.2 \text{ k}\Omega + 5.84 \text{ k}\Omega = 12.04 \text{ k}\Omega$$

R_{L1} in parallel with R_{EQ2}:

$$R_{EQ3} = \frac{R_{L1} R_{EQ2}}{R_{L1} + R_{EQ2}} = \frac{(100 \text{ k}\Omega)(12.04 \text{ k}\Omega)}{112.04 \text{ k}\Omega} = 10.75 \text{ k}\Omega$$

R_{EQ3} is the equivalent resistance from the 12-V tap to ground. The equivalent circuit from the 12-V tap to ground is shown in Figure 7–34. Using this equivalent

FIGURE 7–34

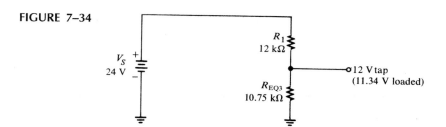

circuit, we calculate the *loaded* voltage at the 12-V tap by using the voltage divider formula as follows:

$$V_{12V} = \left(\frac{R_{EQ3}}{R_T}\right)V_S = \left(\frac{10.75 \text{ k}\Omega}{22.75 \text{ k}\Omega}\right)24 \text{ V} = 11.34 \text{ V}$$

As you can see, the output voltage at the 12-V tap decreases slightly from its unloaded value when the 100-kΩ loads are connected. Smaller values of load resistance would result in a greater decrease in the output voltage.

LOADING EFFECT OF A VOLTMETER

As you know, a voltmeter is always connected in parallel with the circuit component across which the voltage is to be measured. Thus, it is much easier to measure voltage than current, because you must break a circuit to insert an ammeter in series. You simply connect a voltmeter across the circuit without disrupting the circuit or breaking a connection.

Since some current is required through the voltmeter to operate the movement, the voltmeter has some effect on the circuit to which it is connected regardless of the type of circuit. This effect is called *loading*. However, as long as the meter resistance is much greater than the resistance of the circuit across which it is connected, the loading effect is negligible. This characteristic is necessary because we do not want the measuring instrument to change the voltage that it is measuring.

Assume that the meter movement has a full-scale coil current of 50 μA and a coil resistance of 1 kΩ. This current goes through the multiplier resistors since they are in series with the coil. For example, on the 10-V range, the total multiplier resistance is

$$R_X = \frac{10 \text{ V} - (50 \text{ }\mu\text{A})(1 \text{ k}\Omega)}{50 \text{ }\mu\text{A}} = \frac{9.95 \text{ V}}{50 \text{ }\mu\text{A}} = 199 \text{ k}\Omega$$

The internal resistance of the voltmeter is the multiplier resistance in series with the coil resistance:

$$R_X + R_M = 199 \text{ k}\Omega + 1 \text{ k}\Omega = 200 \text{ k}\Omega$$

The internal resistance of a voltmeter is determined by its ohms per volt (Ω/V) rating and by the voltage range setting. The meter just discussed has an Ω/V rating of 20 kΩ/V, so that on the 10-V range as indicated, the internal resistance is 200 kΩ. If the 5-V range were selected, for example, the internal resistance is 100 kΩ.

**EXAMPLE
7–15**

How much does a voltmeter with a 50-μA, 1000-Ω movement affect the voltage being measured in Figure 7–35? The 10-V range setting is used.

FIGURE 7–35

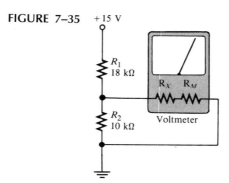

Solution:
The true voltage across R_2 is

$$V_{R2} = \left(\frac{10 \text{ k}\Omega}{28 \text{ k}\Omega}\right) 15 \text{ V} = 5.357 \text{ V}$$

The meter's resistance on the 10-V range is 200 kΩ. When the meter is connected across the circuit, its resistance is in *parallel* with R_2 of the circuit, giving a total of

$$\frac{R_2(R_X + R_M)}{R_2 + R_X + R_M} = \frac{(10 \text{ k}\Omega)(200 \text{ k}\Omega)}{210 \text{ k}\Omega} = 9.52 \text{ k}\Omega$$

The voltage actually measured by the voltmeter is

$$V_{R2} = \left(\frac{9.52 \text{ k}\Omega}{27.52 \text{ k}\Omega}\right) 15 \text{ V} = 5.189 \text{ V}$$

This differs from the true voltage across R_2 by [(5.357 V − 5.189 V)/5.357 V]100 = 3.1%. Therefore, the meter does not significantly alter the voltage value because its internal resistance is much greater than the circuit resistance across which it is connected.

SECTION REVIEW 7–3

1. A load resistor is connected to an output tap on a voltage divider. What effect does the load resistor have on the output voltage at this tap?

2. A larger-value load resistor will cause the output voltage to change less than a smaller-value one will (T or F).

3. For the voltage divider in Figure 7–36, determine the unloaded output voltage. Also determine the output voltage with a 10-kΩ load resistor connected across the output.

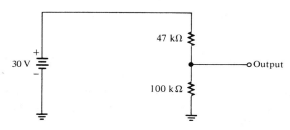

FIGURE 7–36

7–4   LADDER NETWORKS

A ladder network is a special type of series-parallel circuit. One form is commonly used to scale down voltages to certain weighted values for digital-to-analog conversion. You will study this process in later courses. In this section, we examine basic resistive ladder networks of limited complexity, beginning with the one shown in Figure 7–37.

One approach to the analysis of a ladder network is to simplify it one step at a time, *starting at the side farthest from the source*. In this way the current in any branch or the voltage at any point can be determined, as illustrated in Example 7–16.

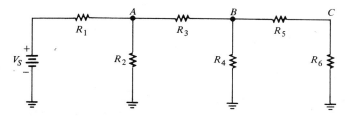

FIGURE 7–37
Basic three-step ladder circuit.

**EXAMPLE
7–16**

Determine each branch current and the voltage at each point in the ladder circuit of Figure 7–38.

FIGURE 7–38

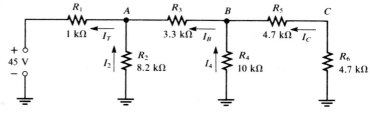

Solution:
To find the branch currents, we must know the total current (I_T). To obtain I_T, we must find the total resistance "seen" by the source.

We determine R_T in a step-by-step process, starting at the right of the circuit diagram. First notice that R_5 and R_6 are in series across R_4. So the resistance from point B to ground is as follows:

$$R_B = \frac{R_4(R_5 + R_6)}{R_4 + (R_5 + R_6)} = \frac{(10 \text{ k}\Omega)(9.4 \text{ k}\Omega)}{19.4 \text{ k}\Omega}$$

$$= 4.9 \text{ k}\Omega$$

Using R_B (the resistance from point B to ground), the equivalent circuit is shown in Figure 7–39.

FIGURE 7–39

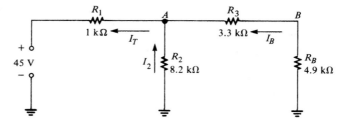

Next, the resistance from point A to ground (R_A) is R_2 in parallel with the series combination of R_3 and R_B. It is calculated as follows:

$$R_A = \frac{R_2(R_3 + R_B)}{R_2 + (R_3 + R_B)} = \frac{(8.2 \text{ k}\Omega)(8.2 \text{ k}\Omega)}{16.4 \text{ k}\Omega}$$

$$= 4.1 \text{ k}\Omega$$

Using R_A, the equivalent circuit of Figure 7–39 is further simplified to the circuit in Figure 7–40.

FIGURE 7–40

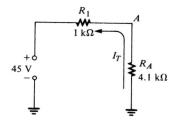

Finally, the total resistance "seen" by the source is R_1 in series with R_A:

$$R_T = R_1 + R_A = 1 \text{ k}\Omega + 4.1 \text{ k}\Omega = 5.1 \text{ k}\Omega$$

The total circuit current is

$$I_T = \frac{V_S}{R_T} = \frac{45 \text{ V}}{5.1 \text{ k}\Omega} = 8.82 \text{ mA}$$

As indicated in Figure 7–39, I_T divides between R_2 (I_2) and the branch containing $R_3 + R_B$ (I_B). Since the branch resistances are equal in this particular example, half the total current is through R_2 and half into point B. So I_2 is 4.41 mA and I_B is 4.41 mA.

If the branch resistances are not equal, the current divider formula is used. As indicated in Figure 7–38, I_B is divided between R_4 and the branch containing $R_5 + R_6$:

$$I_4 = \frac{R_5 + R_6}{R_4 + (R_5 + R_6)} I_B = \left(\frac{9.4 \text{ k}\Omega}{19.4 \text{ k}\Omega}\right)4.41 \text{ mA} = 2.14 \text{ mA}$$

$$I_5 = I_6 = I_B - I_4 = 4.41 \text{ mA} - 2.14 \text{ mA} = 2.27 \text{ mA}$$

To determine V_A, V_B, and V_C, we apply Ohm's law as follows:

$$V_A = I_2R_2 = (4.41 \text{ mA})(8.2 \text{ k}\Omega)$$

$$= 36.2 \text{ V}$$

$$V_B = I_4R_4 = (2.14 \text{ mA})(10 \text{ k}\Omega)$$

$$= 21.4 \text{ V}$$

$$V_C = I_6R_6 = (2.27 \text{ mA})(4.7 \text{ k}\Omega)$$

$$= 10.67 \text{ V}$$

THE R/2R LADDER NETWORK

A basic R/2R ladder circuit is shown in Figure 7–41. As you can see, the name comes from the relationship of the resistor values (one set of resistors has twice the value of

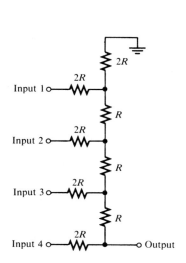

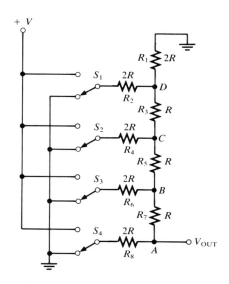

FIGURE 7–41
A basic four-step $R/2R$ ladder network.

FIGURE 7–42
$R/2R$ ladder with switch inputs to simulate a two-level (digital) code.

the others). This type of ladder circuit is used in applications where digital codes are converted to speech, music, or other types of analog signals as found, for example, in the area of digital recording and reproduction. This application area is called *digital-to-analog* (D/A) conversion.

We will now examine general operation of a basic $R/2R$ ladder using the four-step circuit in Figure 7–42. In a later course in digital fundamentals you will learn specifically how this type of circuit is used in D/A conversion.

The switches are used in this illustration to simulate the digital (two-level) inputs. One switch position is connected to ground (0 V) and the other position is connected to a positive voltage (V). The analysis is as follows: Start by assuming that switch S_4 in Figure 7–42 is at the V position and the others are at ground so that the inputs are as shown in Figure 7–43(a).

The total resistance from point A to ground is found by first combining R_1 and R_2 in parallel from point D to ground to simplify the circuit as shown in Part (b).

$$R_1\|R_2 = \frac{2R}{2} = R$$

$R_1\|R_2$ is in series with R_3 from point C to ground as illustrated in Part (c).

$$R_1\|R_2 + R_3 = R + R = 2R$$

Next, the above combination is in parallel with R_4 from point C to ground as shown in Part (d).

$$(R_1\|R_2 + R_3)\|R_4 = 2R\|2R = \frac{2R}{2} = R$$

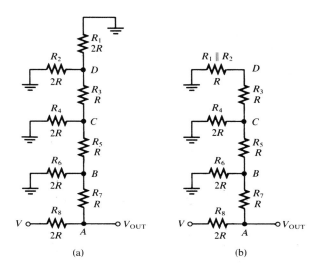

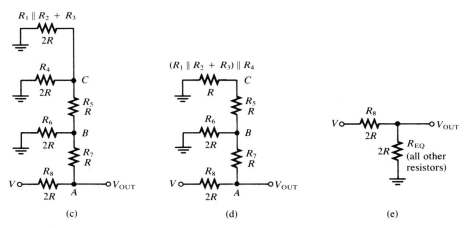

FIGURE 7–43
Simplification of R/2R ladder for analysis.

Continuing this simplification process results in the circuit in Part (e) in which the output voltage can be expressed as

$$V_{OUT} = \left(\frac{2R}{4R}\right)V = \frac{V}{2}$$

A similar analysis, except with switch S_3 in Figure 7–42 connected to V and the other switches connected to ground, results in the simplified circuit shown in Figure 7–44. The analysis for this case is as follows: The resistance from point B to ground is

$$R_B = (R_7 + R_8)\|2R = 3R\|2R = \frac{6R}{5}$$

FIGURE 7–44

Simplified ladder with V input at S_3 in Figure 7–42.

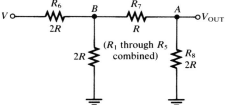

Using the voltage divider formula,

$$V_B = \left(\frac{R_B}{R_6 + R_B}\right)V = \left(\frac{6R/5}{2R + 6R/5}\right)V = \left(\frac{6R/5}{10R/5 + 6R/5}\right)V$$

$$= \left(\frac{6R/5}{16R/5}\right)V = \left(\frac{6R}{16R}\right)V = \frac{3V}{8}$$

The output voltage is therefore

$$V_{OUT} = \left(\frac{R_8}{R_7 + R_8}\right)V_B = \left(\frac{2R}{3R}\right)\left(\frac{3V}{8}\right) = \frac{V}{4}$$

Notice that the output voltage in this case ($V/4$) is one half the output voltage ($V/2$) for the case where V is connected at switch S_4.

A similar analysis for each of the remaining switch inputs in Figure 7–42 results in output voltages as follows: For S_2 connected to V and the rest connected to ground

$$V_{OUT} = \frac{V}{8}$$

For S_1 connected to V and the rest to ground,

$$V_{OUT} = \frac{V}{16}$$

These particular relationships among the output voltages for the various levels of inputs are very important in the application of $R/2R$ ladder networks to digital-to-analog conversion.

SECTION REVIEW 7–4

1. Sketch a basic four-step ladder network.

2. Determine the total circuit resistance presented to the source by the ladder network of Figure 7–45:

3. What is the total current in Figure 7–45?

4. What is the current through the 2.2-kΩ resistor in Figure 7–45?

5. What is the voltage at point A with respect to ground in Figure 7–45?

FIGURE 7–45

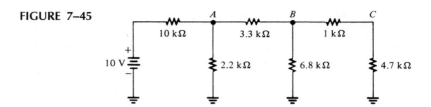

7–5 THE WHEATSTONE BRIDGE

The *bridge circuit* is widely used in measurement devices and other applications that you will learn later. For now, we will consider the *balanced bridge,* which can be used to measure unknown resistance values. This circuit, shown in Figure 7–46(a), is known as a *Wheatstone bridge.* Figure 7–46(b) is the same circuit electrically, but it is drawn in a different way.

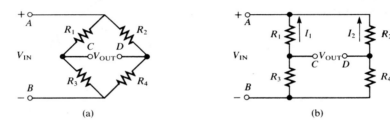

(a)   (b)

FIGURE 7–46
Wheatstone bridge.

A bridge is said to be *balanced* when the voltage across the output terminals C and D is *zero;* that is, $V_{AC} = V_{AD}$. If V_{AC} equals V_{AD}, then $I_1R_1 = I_2R_2$, since one side of both R_1 and R_2 is connected to point A. Also, $I_1R_3 = I_2R_4$, since one side of both R_3 and R_4 connects to point B. Because of these equalities, we can write the ratios of the voltages as follows:

$$\frac{I_1R_1}{I_1R_3} = \frac{I_2R_2}{I_2R_4}$$

The currents cancel to give

$$\frac{R_1}{R_3} = \frac{R_2}{R_4}$$

Solving for R_1, we get

$$R_1 = R_3\left(\frac{R_2}{R_4}\right) \tag{7–1}$$

How can this formula be used to determine an unknown resistance? First, let us make R_3 a *variable* resistor and call it R_V. Also, we set the ratio R_2/R_4 to a known value. If R_V is adjusted until the bridge is balanced, the product of R_V and the ratio R_2/R_4 is equal to R_1, which is our unknown resistor (R_{UNK}). Equation (7–1) is restated in Equation (7–2), using the new subscripts:

$$R_{UNK} = R_V\left(\frac{R_2}{R_4}\right) \tag{7–2}$$

The bridge is balanced when the voltage across the output terminals equals zero ($V_{AC} = V_{AD}$). A *galvanometer* (a meter that measures small currents in either direction and is zero at center scale) is connected between the output terminals. Then R_V is adjusted until the galvanometer shows zero current ($V_{AC} = V_{AD}$), indicating a balanced condition. The setting of R_V multiplied by the ratio R_2/R_4 gives the value of R_{UNK}. Figure 7–47 shows this arrangement. For example, if $R_2/R_4 = \frac{1}{10}$ and $R_V = 680\ \Omega$, then $R_{UNK} = (680\ \Omega)(\frac{1}{10}) = 68\ \Omega$.

FIGURE 7–47
Balanced Wheatstone bridge.

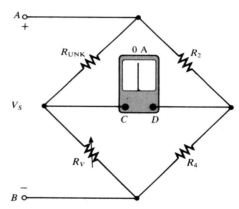

EXAMPLE 7–17

What is R_{UNK} under the balanced bridge conditions shown in Figure 7–48?

FIGURE 7–48

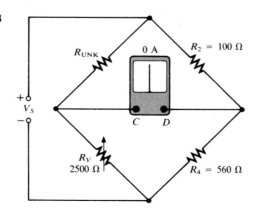

Solution:

$$R_{\text{UNK}} = R_V\left(\frac{R_2}{R_4}\right) = 2500 \ \Omega\left(\frac{100 \ \Omega}{560 \ \Omega}\right)$$

$$= 446.43 \ \Omega$$

A BRIDGE APPLICATION

Bridge circuits are used for many measurements other than that of unknown resistance. One application of the Wheatstone bridge is in accurate temperature measurement. A temperature-sensitive element such as a thermistor is connected in a Wheatstone bridge as shown in Figure 7–49. An amplifier is connected across the output from *A* to *B* in order to increase the output voltage from the bridge to a usable value. The bridge is calibrated so that it is balanced at a specified reference temperature. As the temperature changes, the resistance of the sensing element changes proportionately, and the bridge becomes unbalanced. As a result, V_{AB} changes and is amplified (increased) and converted to a form for direct temperature readout on a gauge or a digital-type display.

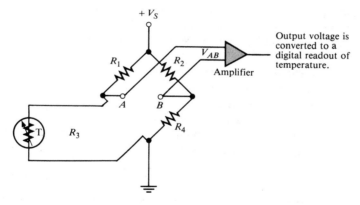

FIGURE 7–49
A simplified circuit for temperature measurement.

SECTION REVIEW 7–5

1. Sketch a basic Wheatstone bridge circuit.
2. Under what condition is the bridge balanced?
3. What formula is used to determine the value of the unknown resistance when the bridge is balanced?
4. What is the unknown resistance for the values shown in Figure 7–50?

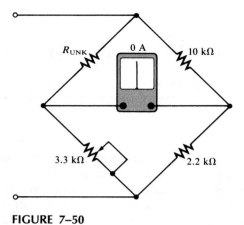

FIGURE 7–50

7–6

TROUBLESHOOTING

Troubleshooting is the process of identifying and locating a failure or problem in a circuit. Some troubleshooting techniques have already been discussed in relation to both series circuits and parallel circuits. Now we will extend these methods to the series-parallel networks.

Opens and shorts are typical problems that occur in electric circuits. As mentioned before, if a resistor burns out, it will normally produce an open circuit. Bad solder connections, broken wires, and poor contacts can also be causes of open paths. Pieces of foreign material, such as solder splashes, broken insulation on wires, and so on, can often lead to shorts in a circuit. A short is a zero resistance path between two points. The following three examples illustrate troubleshooting in series-parallel resistive circuits.

EXAMPLE 7–18

From the indicated voltmeter reading, determine if there is a fault in Figure 7–51. If there is a fault, identify it as either a short or an open.

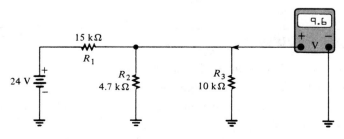

FIGURE 7–51

Solution:

First determine what the voltmeter *should* be indicating. Since R_2 and R_3 are in parallel, their equivalent resistance is

$$R_{EQ} = \frac{R_2 R_3}{R_2 + R_3} = \frac{(4.7 \text{ k}\Omega)(10 \text{ k}\Omega)}{14.7 \text{ k}\Omega}$$
$$= 3.2 \text{ k}\Omega$$

The voltage across the equivalent parallel combination is determined by the voltage divider formula as follows:

$$V_{R_{EQ}} = \left(\frac{R_{EQ}}{R_1 + R_{EQ}}\right) V_S = \left(\frac{3.2 \text{ k}\Omega}{18.2 \text{ k}\Omega}\right) 24 \text{ V}$$
$$= 4.22 \text{ V}$$

Thus, 4.22 V is the voltage reading that you *should* get on the meter. But the meter reads 9.6 V instead. This value is incorrect, and, because it is higher than it should be, R_2 or R_3 is probably open. Why? Because if either of these two resistors is open, the resistance across which the meter is connected is larger than expected. A higher resistance will drop a higher voltage in this circuit, which is, in effect, a voltage divider.

Let us start by assuming that R_2 is open. If it is, the voltage across R_3 is as follows:

$$V_{R3} = \left(\frac{R_3}{R_1 + R_3}\right) V_S = \left(\frac{10 \text{ k}\Omega}{25 \text{ k}\Omega}\right) 24 \text{ V}$$
$$= 9.6 \text{ V}$$

This calculation shows that R_2 is open. Replace R_2 with a new resistor.

EXAMPLE 7–19

Suppose that you measure 24 V with the voltmeter in Figure 7–52. Determine if there is a fault, and, if there is, isolate it.

FIGURE 7–52

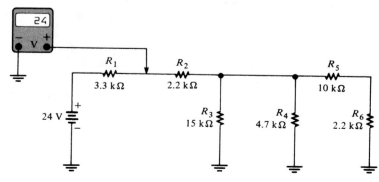

Solution:
There is no voltage drop across R_1 because both sides of the resistor are at $+24$ V. Either there is no current through R_1 from the source, which tells us that R_2 is open in the circuit, or R_1 is shorted.

If R_1 were open, the meter would not read 24 V. The most logical failure is in R_2. If it is open, then there is no current from the source. To verify this, measure across R_2 with the voltmeter as shown in Figure 7–53. If R_2 is open, the meter will indicate 24 V. The right side of R_2 will be at zero volts because there no is current through any of the other resistors to cause a voltage drop.

FIGURE 7–53

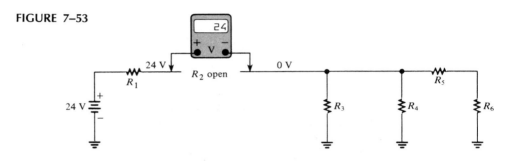

**EXAMPLE
7–20**

The two voltmeters in Figure 7–54 indicate the voltages shown. Determine if there are any opens or shorts in the circuit and, if so, where they are located.

FIGURE 7–54

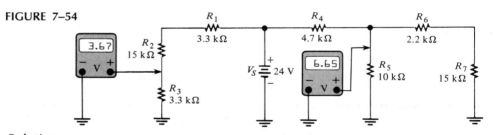

Solution:
First let us see if the voltmeter readings are correct. R_1, R_2, and R_3 act as a voltage divider on the left side of the source. The voltage across R_3 is calculated as follows:

$$V_{R3} = \left(\frac{R_3}{R_1 + R_2 + R_3}\right) V_S$$

$$= \left(\frac{3.3 \text{ k}\Omega}{21.6 \text{ k}\Omega}\right) 24 \text{ V} = 3.67 \text{ V}$$

The voltmeter A reading (V_A) is correct.

Now let us see if the voltmeter B reading (V_B) is correct. The part of the circuit to the right of the source also acts as a voltage divider. The series-parallel combination of R_5, R_6, and R_7 is in series with R_4. The equivalent resistance of the R_5, R_6, and R_7 combination is calculated as follows:

$$R_{EQ} = \frac{(R_6 + R_7)R_5}{R_5 + R_6 + R_7}$$

$$= \frac{(17.2 \text{ k}\Omega)(10 \text{ k}\Omega)}{27.2 \text{ k}\Omega} = 6.3 \text{ k}\Omega$$

where R_5 is in parallel with R_6 and R_7 in series. R_{EQ} and R_4 form a voltage divider. Voltmeter B is measuring the voltage across R_{EQ}. Is it correct? We check as follows:

$$V_{R_{EQ}} = \left(\frac{R_{EQ}}{R_4 + R_{EQ}}\right)V_S$$

$$= \left(\frac{6.3 \text{ k}\Omega}{11 \text{ k}\Omega}\right)24 \text{ V} = 13.75 \text{ V}$$

Thus, the actual measured voltage at this point is incorrect. Some further thought will help to isolate the problem.

We know that R_4 is not open, because if it were, the meter would read 0 V. If there were a short across it, the meter would read 24 V. Since the actual voltage is much less than it should be, R_{EQ} must be less than the calculated value. The most likely problem is a short across R_7. If there is a short from the top of R_7 to ground, R_6 is effectively in parallel with R_5. In this case, R_{EQ} is

$$R_{EQ} = \frac{R_5 R_6}{R_5 + R_6}$$

$$= \frac{(2.2 \text{ k}\Omega)(10 \text{ k}\Omega)}{12.2 \text{ k}\Omega} = 1.8 \text{ k}\Omega$$

Then V_{EQ} is

$$V_{EQ} = \left(\frac{1.8 \text{ k}\Omega}{6.5 \text{ k}\Omega}\right)24 \text{ V} = 6.65 \text{ V}$$

This value for V_{EQ} agrees with the voltmeter B reading. So there is a short across R_7. If this were an actual circuit, you would try to find the physical cause of the short.

SECTION REVIEW 7–6

1. Name two types of common circuit faults.

2. In Figure 7–55, one of the resistors in the circuit is open. Based on the meter reading, determine which is the bad resistor.

FIGURE 7–55

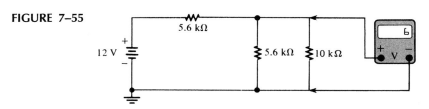

3. For the following faults in Figure 7–56, what voltage would be measured at point *A?*
   (a) No faults   **(b)** R_1 open   **(c)** Short across R_5 **(d)** R_3 and R_4 open
   (e) R_2 open

FIGURE 7–56

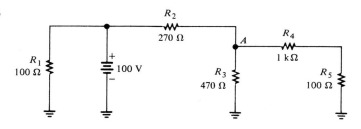

7–7   COMPUTER ANALYSIS

The program listed in this section provides for the analysis of a three-step ladder network similar to that in Example 7–16. All unknown voltages and branch currents are computed and displayed when the program is run.

```
10  CLS
20  PRINT "THIS PROGRAM ANALYZES A BASIC THREE-STEP LADDER
    NETWORK"
30  PRINT "SUCH AS THE ONE SHOWN IN FIGURE 7-37."
40  PRINT:PRINT "YOU PROVIDE THE INPUT VOLTAGE AND RESISTOR
    VALUES AND"
50  PRINT "THE COMPUTER WILL CALCULATE THE VOLTAGE AT EACH
    POINT"
60  PRINT "AND THE CURRENT IN EACH BRANCH."
70  FOR T=1 TO 5000:NEXT:CLS
80  INPUT "WHAT IS THE INPUT VOLTAGE IN VOLTS";VIN
90  PRINT:PRINT
100 FOR X=1 TO 6
110 PRINT "VALUE OF R";X;"IN OHMS"
```

```
120 INPUT R(X)
130 NEXT
140 CLS
150 RB=(R(4)*(R(5)+R(6)))/(R(4)+R(5)+R(6))
160 RA=(R(2)*(R(3)+RB))/(R(2)+R(3)+RB)
170 RT=RA+R(1)
180 IT=VIN/RT
190 VA=IT*RA
200 I(1)=IT
210 I(2)=VA/R(2)
220 I(3)=IT-I(2)
230 VB=I(3)*RB
240 I(4)=VB/R(4)
250 I(5)=I(3)-I(4)
260 I(6)=I(5)
270 VC=I(6)*R(6)
280 PRINT "VA=";VA;"VOLTS","VB=";VB;"VOLTS","VC=";VC;"VOLTS"
290 PRINT
300 FOR Y=1 TO 6
310 PRINT "I";Y;"=";I(Y);"AMPS"
320 NEXT
```

A flowchart for this program is shown in Figure 7–57.

FIGURE 7–57

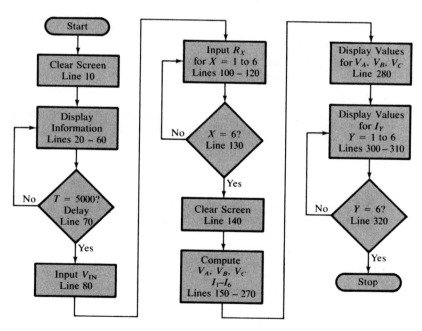

SECTION REVIEW 7–7

1. What is the purpose of line 70?
2. Which sequence of lines displays the results of the computations?

SUMMARY

1. A series-parallel circuit is a combination of both series paths and parallel paths.
2. To determine total resistance in a series-parallel circuit, identify the series and parallel relationships, and then apply the formulas for series resistance and parallel resistance from Chapters 5 and 6.
3. To find the total current, divide the total voltage by the total resistance.
4. To determine branch currents, apply the current divider formula, Kirchhoff's current law, or Ohm's law. Consider each circuit problem individually to determine the most appropriate method.
5. To determine voltage drops across any portion of a series-parallel circuit, use the voltage divider formula, Kirchhoff's voltage law, or Ohm's law. Consider each circuit problem individually to determine the most appropriate method.
6. When a load resistor is connected across a voltage divider output, the output voltage decreases.
7. The load resistor should be large compared to the resistance across which it is connected, in order that the loading effect may be minimized. A *10-times* value is sometimes used as a rule of thumb, but the value depends on the accuracy required for the output voltage.
8. To find total resistance of a ladder network, start at the point farthest from the source and reduce the resistance in steps.
9. A Wheatstone bridge can be used to measure an unknown resistance.
10. A bridge is *balanced* when the output voltage is *zero*. The balanced condition produces zero current through a load connected across the output terminals of the bridge.
11. *Troubleshooting* is the process of identifying and locating a fault in a circuit.
12. *Open circuits* and *short circuits* are typical circuit faults.
13. Resistors normally open when they burn out.

FORMULAS

$$R_1 = R_3 \left(\frac{R_2}{R_4} \right) \tag{7-1}$$

$$R_{\text{UNK}} = R_V \left(\frac{R_2}{R_4} \right) \tag{7-2}$$

SELF-TEST

Solutions appear at the end of the book.

1. Draw the schematic diagram for a series-parallel circuit that is described as follows: R_1 and R_2 are in series with each other, and this series combination is in parallel with a series combination of R_3, R_4, and R_5.

2. How do you find the total resistance for the circuit described in Question 1?

3. If all the resistors in Question 1 are the same value, through which resistors is there the most current when voltage is applied?

4. Two 1-kΩ resistors are in series, and this series combination is in parallel with a 2.2-kΩ resistor. The voltage across one of the 1-kΩ resistors is 6 V. How much voltage is across the 2.2-kΩ resistor?

5. A 330-Ω resistor is in series with the parallel combination of four 1-kΩ resistors. A 100-V source is connected to the circuit. Which resistor has the most current through it? Which one has the most voltage across it?

6. In Question 5, what percentage of the total current goes through each of the 1-kΩ resistors?

7. In a certain circuit, the voltage at point A is 5 V with respect to ground. The voltage at point B is 8 V with respect to ground. The voltage at point C is 12 V with respect to ground. Determine the voltages at points B and C with respect to point A.

8. The output of a certain voltage divider is 9 V with no load. When a load is connected, will the voltage increase or decrease?

9. A certain voltage divider consists of two 10-kΩ resistors in series. Which will have more effect on the output voltage, a 1-MΩ load or a 100-kΩ load?

10. Would you expect more or less current to be drawn from the source when a load resistor is connected to the output of a voltage divider?

11. What is the output voltage of a balanced Wheatstone bridge?

12. What is a galvanometer?

13. You are measuring the voltage at a given point in a circuit that has very high resistance values. The measured value is somewhat lower than it should be. What could be causing this discrepancy?

14. Identify the series-parallel relationships in Figure 7–58.

FIGURE 7–58

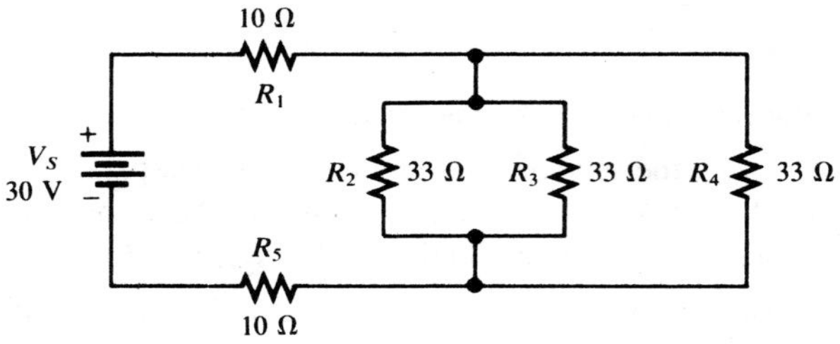

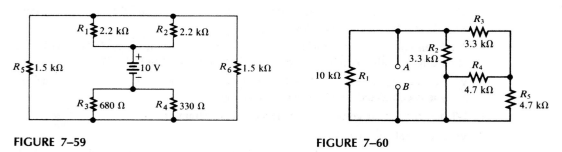

15. For the circuit of Figure 7–58, determine the following:
  (a) Total resistance as "seen" by the source
  (b) Total current drawn from the source
  (c) Current through R_3
  (d) Voltage across R_4

16. For the circuit in Figure 7–59, determine the following:
  (a) Total resistance   **(b)** Total current
  (c) Current through R_1 **(d)** Voltage across R_6

17. In Figure 7–60, find the following:
  (a) Total resistance between terminals A and B
  (b) Total current drawn from a 6-V source connected from A to B
  (c) Current through R_5
  (d) Voltage across R_2

FIGURE 7–59

FIGURE 7–60

18. Determine the voltages with respect to ground for each point in Figure 7–61.

FIGURE 7–61

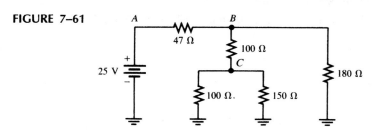

19. If R_2 in Figure 7–62 opens, what voltages will be read at points A, B, and C?

FIGURE 7–62

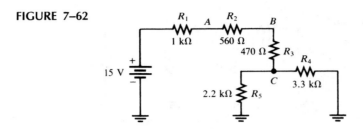

20. Check the meter readings in Figure 7–63 and locate any fault that may exist.

21. Determine the unloaded output voltage in Figure 7–64. If a 220-kΩ load is connected, what is the loaded output voltage?

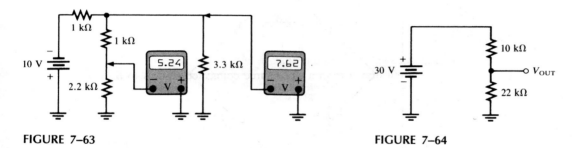

FIGURE 7–63   FIGURE 7–64

22. In Figure 7–65, determine the voltage at point A when the switch is open. Also determine the voltage at point A when the switch is closed.

23. For the ladder network in Figure 7–66, determine the following:
   (a) Total resistance   **(b)** Total current   **(c)** Current through R_3
   (d) Current through R_4 **(e)** Voltage at point A **(f)** Voltage at point B

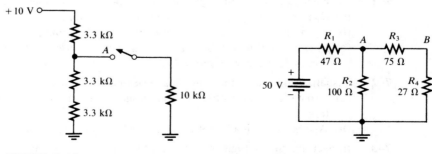

FIGURE 7–65   FIGURE 7–66

24. Calculate V_A, V_B, and V_C for the ladder in Figure 7–67.

FIGURE 7–67

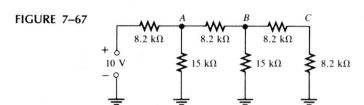

25. In the bridge circuit of Figure 7–68, what is the value of the unknown resistance when the other values are as shown?

FIGURE 7–68

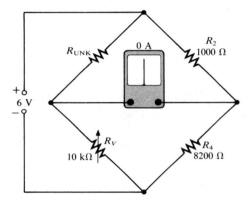

PROBLEMS

Section 7–1

7–1 Visualize and sketch the following series-parallel combinations:
   (a) R_1 in series with the parallel combination of R_2 and R_3
   (b) R_1 in parallel with the series combination of R_2 and R_3
   (c) R_1 in parallel with a branch containing R_2 in series with a parallel combination of four other resistors

7–2 Visualize and sketch the following series-parallel circuits:
   (a) A parallel combination of three branches, each containing two series resistors
   (b) A series combination of three parallel circuits, each containing two resistors

7–3 In each circuit of Figure 7–69, identify the series and parallel relationships of the resistors viewed from the source.

FIGURE 7–69

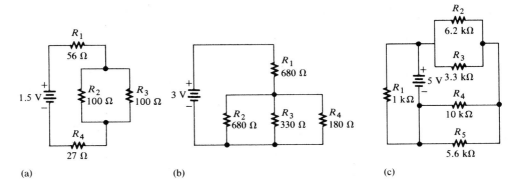

(a)   (b)   (c)

7–4 For each circuit in Figure 7–70, identify the series and parallel relationships of the resistors viewed from the source.

FIGURE 7–70

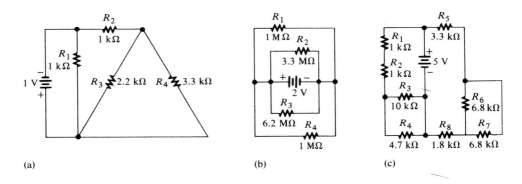

(a)   (b)   (c)

7–5 Draw the schematic of the PC board layout in Figure 7–71 and identify the series-parallel relationships. Which resistors, if any, can be removed with no effect on R_T?

FIGURE 7–71

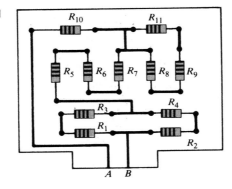

7–6 Develop a schematic for the double-sided PC board in Figure 7–72.

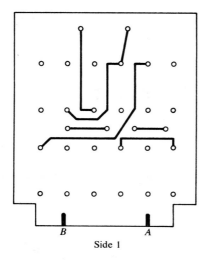
Side 1

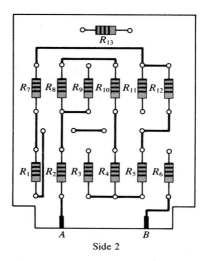
Side 2

FIGURE 7–72

7–7 Lay out a PC board for the circuit in Figure 7–70(c). The battery is to be connected external to the board.

Section 7–2

7–8 A certain circuit is composed of two parallel resistors. The total resistance is 667 Ω. One of the resistors is 1 kΩ. What is the other resistor?

7–9 For each circuit in Figure 7–69, determine the total resistance presented to the source.

7–10 Repeat Problem 7–9 for each circuit in Figure 7–70.

7–11 Determine the current through each resistor in Figure 7–69; then calculate each voltage drop.

7–12 Determine the current through each resistor in Figure 7–70; then calculate each voltage drop.

7–13 Find R_T for all combinations of switch positions in Figure 7–73.

7–14 Determine the resistance between A and B in Figure 7–72 with R_1 removed for the resistor values listed below:
10 kΩ: R_6, R_{12}
4.7 kΩ: R_2, R_3, R_{13}
22 kΩ: R_4, R_9, R_{10}, R_{11}
9.1 kΩ: R_5, R_8
6.8 kΩ: R_7

7–15 Determine the voltage at each point with respect to ground in Figure 7–74.

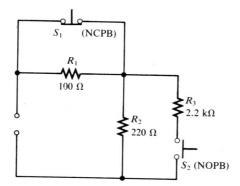

FIGURE 7–73

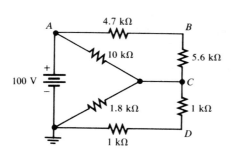

FIGURE 7–74

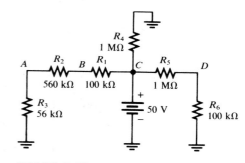

FIGURE 7–75

7–16 Determine the voltage at each point with respect to ground in Figure 7–75.

7–17 In Figure 7–75, how would you determine the voltage across R_2 by measuring without connecting a meter directly across the resistor?

7–18 Determine the voltage, V_{AB}, in Figure 7–76.

7–19 Find the value of R_2 in Figure 7–77.

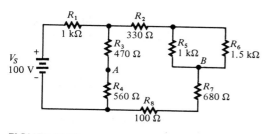

FIGURE 7–76

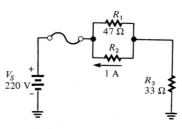

FIGURE 7–77

7–20 Find the resistance between point A and each of the other points (R_{AB}, R_{AC}, R_{AD}, etc.) in Figure 7–78.

FIGURE 7-78

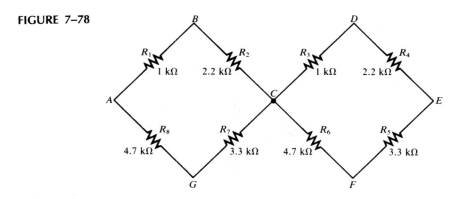

7-21 Find the resistance between each set of points (*AB*, *BC*, and *CD*) in Figure 7-79.

7-22 Determine the value of each resistor in Figure 7-80.

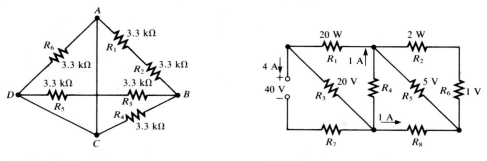

FIGURE 7-79   FIGURE 7-80

Section 7-3

7-23 A voltage divider consists of two 56-kΩ resistors and a 15-V source. Calculate the unloaded output voltage. What will the output voltage be if a load resistor of 1 MΩ is connected to the output?

7-24 A 12-V battery output is divided down to obtain two output voltages. Three 3.3-kΩ resistors are used to provide the two taps. Determine the output voltages. If a 10-kΩ load is connected to the higher of the two outputs, what will its loaded value be?

7-25 Which will cause a smaller decrease in output voltage for a given voltage divider, a 10-kΩ load or a 47-kΩ load?

7-26 In Figure 7-81, determine the continuous current drain on the battery with no load across the output terminals. With a 10-kΩ load, what is the battery current?

7-27 Determine the resistance values for a voltage divider that must meet the following specifications: The current drain under unloaded condition is not to exceed

FIGURE 7–81

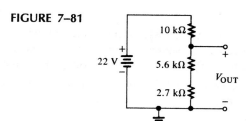

5 mA. The source voltage is to be 10 V. A 5-V output and a 2.5-V output are required. Sketch the circuit. Determine the effect on the output voltages if a 1-kΩ load is connected to each tap.

7–28 The voltage divider in Figure 7–82 has a switched load. Determine the voltage at each tap (V_1, V_2, and V_3) for each position of the switch.

7–29 Figure 7–83 shows a dc *biasing* arrangement for a field-effect transistor amplifier. Biasing is a common method for setting up certain dc voltage levels required for proper amplifier operation. Although you are probably not familiar with transistor amplifiers at this point, the dc voltages and currents in the circuit can be determined using methods that you already know.
(a) Find V_G and V_S **(b)** Determine I_1, I_2, I_D, and I_S
(c) Find V_{DS} and V_{DG}

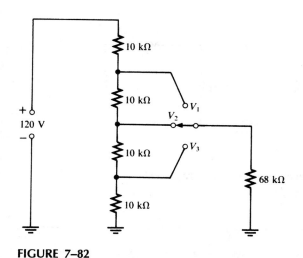

FIGURE 7–82

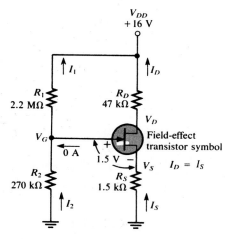

FIGURE 7–83

7–30 Design a voltage divider to provide a 6-V output with no load and a minimum of 5.5 V across a 1-kΩ load. The source voltage is 24 V, and the unloaded current drain is not to exceed 100 mA.

Section 7–4

7–31 For the circuit shown in Figure 7–84, calculate the following:
    (a) Total resistance across the source   **(b)** Total current from the source
    (c) Current through the 910-Ω resistor   **(d)** Voltage from point A to point B

7–32 Determine the total resistance and the voltage at points A, B, and C in the ladder network of Figure 7–85.

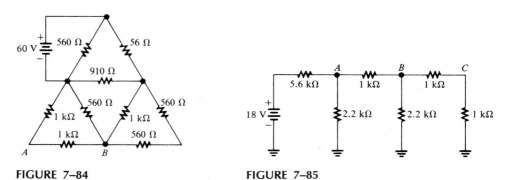

FIGURE 7–84   FIGURE 7–85

7–33 Determine the total resistance between terminals A and B of the ladder network in Figure 7–86. Also calculate the current in each branch with 10 V between A and B.

FIGURE 7–86

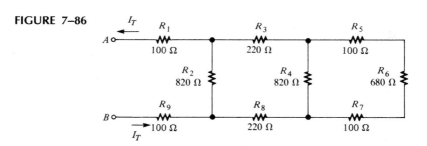

7–34 What is the voltage across each resistor in Figure 7–86?

7–35 Find I_T and V_{OUT} in Figure 7–87.

FIGURE 7–87

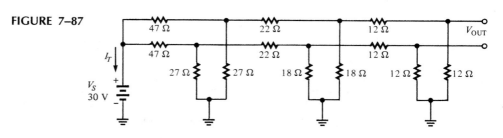

FIGURE 7–88

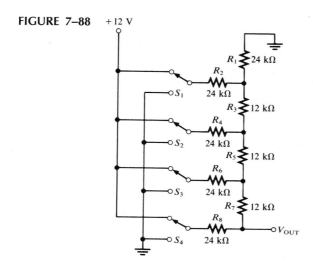

7–36 Determine V_{OUT} for the $R/2R$ ladder network in Figure 7–88 for the following conditions:
(a) Switch S_2 connected to +12 V and the rest to ground.
(b) Switch S_1 connected to +12 V and the rest to ground.

7–37 Repeat Problem 7–36 for the following conditions:
(a) S_3 and S_4 to +12 V, S_1 and S_2 to ground.
(b) S_3 and S_1 to +12 V, S_2 and S_4 to ground.
(c) All switches to +12 V.

Section 7–5

7–38 A resistor of unknown value is connected to a Wheatstone bridge circuit. The bridge parameters are set as follows: $R_V = 18$ kΩ and $R_2/R_4 = 0.02$. What is R_{UNK}?

7–39 A bridge network is shown in Figure 7–89. To what value must R_V be set in order to balance the bridge?

FIGURE 7–89

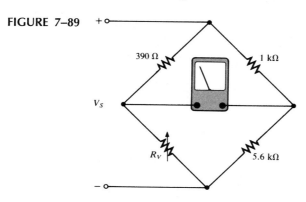

7–40 The temperature-sensitive bridge circuit in Figure 7–49 is used to detect when the temperature in a chemical manufacturing process reaches 100°C. The resistance of the thermistor drops from 5 kΩ at a nominal 20°C to 100 Ω at 100°C. If $R_1 = 1$ kΩ and $R_2 = 2.2$ kΩ, to what value must R_4 be set to produce a balanced bridge when the temperature reaches 100°C?

Section 7–6

7–41 Is the voltmeter reading in Figure 7–90 correct?

7–42 Are the meter readings in Figure 7–91 correct?

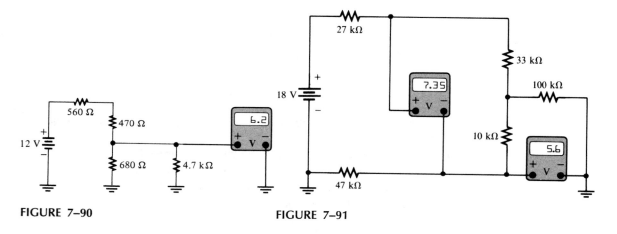

FIGURE 7–90 **FIGURE 7–91**

7–43 There is one fault in Figure 7–92. Based on the meter indications, determine what the fault is.

FIGURE 7–92

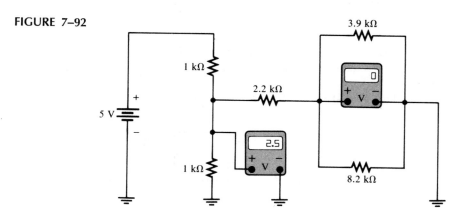

7–44 Look at the meters in Figure 7–93 and determine if there is a fault in the circuit. If there is a fault, identify it.

7–45 Check the meter readings in Figure 7–94 and locate any fault that may exist.

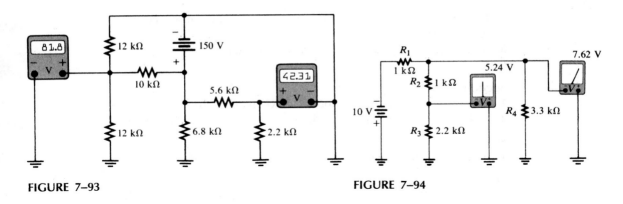

FIGURE 7–93   **FIGURE 7–94**

7–46 A voltmeter connected across R_4 in Figure 7–95 reads 9.29 V. Is the voltmeter reading correct? If not, what is the problem?

7–47 If R_2 in Figure 7–95 opens, what voltages will be read at points A, B, and C?

FIGURE 7–95

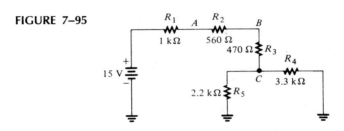

Section 7–7

7–48 Modify the program in Section 7–7 to compute and display the power dissipation in each resistor. Revise the flowchart accordingly.

7–49 Modify the program in Section 7–7 to analyze a basic four-step ladder circuit rather than the three-step circuit. Revise the flowchart accordingly.

ANSWERS TO SECTION REVIEWS

Section 7–1

1. A circuit consisting of both series and parallel connections.

2. See Figure 7–96.

FIGURE 7–96

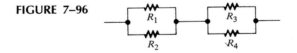

3. R_1 and R_2 are in series with the parallel combination of R_3 and R_4.
4. All resistors.
5. R_1 and R_2 are in parallel; R_3 and R_4 are in parallel.
6. Yes.

Section 7–2
1. Voltage and current divider formulas, Kirchhoff's laws, and Ohm's law.
2. 608.2 Ω.   **3.** 0.011 A.   **4.** 3.63 V.   **5.** 99.13 Ω, 0.0101 A.

Section 7–3
1. It decreases the output voltage.   **2.** T.   **3.** 20.41 V, 4.86 V.

Section 7–4
1. See Figure 7–97.   **2.** 11.64 kΩ.   **3.** 0.859 mA.   **4.** 0.639 mA.
5. 1.41 V.

FIGURE 7–97

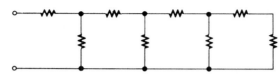

Section 7–5
1. See Figure 7–98.   **2.** $V_A = V_B$.   **3.** $R_V(R_2/R_4)$.   **4.** 15 kΩ.

FIGURE 7–98

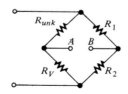

Section 7–6
1. Opens and shorts.   **2.** The 10-kΩ resistor.
3. **(a)** 54.95 V; **(b)** 54.95 V; **(c)** 54.22 V; **(d)** 100 V; **(e)** 0 V.

Section 7–7
1. It provides a delay to keep the initial message on the screen for a fixed time and then clears the screen.   **2.** Lines 280–320.

EIGHT

CIRCUIT THEOREMS
AND CONVERSIONS

There are many ways to simplify circuits in order to ease analysis or, in some cases, to make the analysis a manageable task that would otherwise be almost impossible.

The purpose of the theorems and conversions presented in this chapter is not to change the nature of a given dc circuit but to change the circuit to an *equivalent* form that is easier to analyze.

In this chapter, you will learn:

☐ The characteristics of voltage sources.
☐ The characteristics of current sources.
☐ How to convert voltage sources to current sources and vice versa.
☐ The superposition theorem and how to apply it to circuit analysis.
☐ Thevenin's theorem and how to use it to reduce any dc circuit to a simple equivalent form in order to simplify the analysis.
☐ How to determine the values for the Thevenin circuit by measurement.
☐ Norton's theorem and its application to circuit analysis.
☐ How Millman's theorem is used to reduce parallel voltage sources to a single equivalent voltage source.

☐ How to determine when maximum power is being transferred from a source to a load.
☐ How to make conversions between delta-wye and wye-delta networks.
☐ How to use a computer program for the analysis of delta-wye networks.

8–1   THE VOLTAGE SOURCE

Figure 8–1(a) is the familiar symbol for an ideal dc voltage source. The voltage across its terminals A and B remains fixed regardless of the value of load resistance that may be connected across its output. Figure 8–1(b) shows a load resistor R_L connected. All of the source voltage, V_S, is dropped across R_L. R_L can be changed to any value except zero, and the voltage will remain fixed. The ideal voltage source has an internal resistance of *zero*.

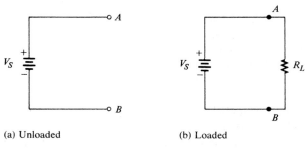

(a) Unloaded   (b) Loaded

FIGURE 8–1
Ideal dc voltage source.

In reality, no voltage source is ideal. That is, all have some inherent internal resistance as a result of their physical and/or chemical makeup, which can be represented by a resistor in series with an ideal source, as shown in Figure 8–2(a). R_S is the internal source resistance and V_S is the source voltage. With no load, the output voltage (voltage from A to B) is V_S. This voltage is sometimes called the *open circuit voltage*.

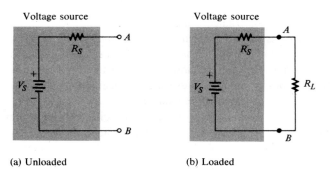

(a) Unloaded   (b) Loaded

FIGURE 8–2
Practical voltage source.

LOADING OF THE VOLTAGE SOURCE

When a load resistor is connected across the output terminals, as shown in Figure 8–2(b), all of the source voltage does not appear across R_L. Some of the voltage is dropped across R_S because of the current through R_S to the load, R_L.

If R_S is very small compared to R_L, the source approaches ideal, because almost all of the source voltage, V_S, appears across the larger resistance R_L. Very little voltage is dropped across the internal resistance, R_S. If R_L changes, most of the source voltage remains across the output as long as R_L is much larger than R_S. As a result, very little change occurs in the output voltage. The larger R_L is compared to R_S, the less change there is in the output voltage. As a rule, R_L should be at least ten times R_S ($R_L \geq 10R_S$).

Example 8–1 illustrates the effect of changes in R_L on the output voltage when R_L is much greater than R_S. Example 8–2 shows the effect of smaller load resistances.

EXAMPLE 8–1

Calculate the voltage output of the source in Figure 8–3 for the following values of R_L: 100 Ω, 560 Ω, and 1 kΩ.

FIGURE 8–3

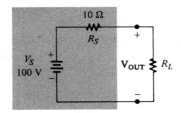

Solution:
For $R_L = 100$ Ω:

$$V_{\text{OUT}} = \left(\frac{R_L}{R_S + R_L}\right)V_S = \left(\frac{100\ \Omega}{110\ \Omega}\right)100\ \text{V}$$

$$= 90.91\ \text{V}$$

For $R_L = 560$ Ω:

$$V_{\text{OUT}} = \left(\frac{560\ \Omega}{570\ \Omega}\right)100\ \text{V}$$

$$= 98.25\ \text{V}$$

For $R_L = 1$ kΩ:

$$V_{\text{OUT}} = \left(\frac{1000\ \Omega}{1010\ \Omega}\right)100\ \text{V}$$

$$= 99.01\ \text{V}$$

Notice that the output voltage is within 10% of the source voltage, V_S, for all three values of R_L, because R_L is at least ten times R_S.

EXAMPLE 8–2

Determine V_{OUT} for $R_L = 10 \ \Omega$ and $R_L = 1 \ \Omega$ in Figure 8–3.

Solution:

For $R_L = 10 \ \Omega$:

$$V_{OUT} = \left(\frac{R_L}{R_S + R_L}\right)V_S = \left(\frac{10 \ \Omega}{20 \ \Omega}\right)100 \ V$$

$$= 50 \ V$$

For $R_L = 1 \ \Omega$:

$$V_{OUT} = \left(\frac{1 \ \Omega}{11 \ \Omega}\right)100 \ V$$

$$= 9.09 \ V$$

Notice in Example 8–2 that the output voltage decreases significantly as R_L is made smaller compared to R_S. This example illustrates the requirement that R_L must be much larger than R_S in order to maintain the output voltage near its open circuit value.

SECTION REVIEW 8–1

1. What is the symbol for the ideal voltage source?

2. Sketch a practical voltage source.

3. What is the internal resistance of the ideal voltage source?

4. What effect does the load have on the output voltage of the practical voltage source?

8–2

THE CURRENT SOURCE

Figure 8–4(a) shows a symbol for the ideal current source. The arrow indicates the direction of current, and I_S is the value of the source current. An ideal current source produces a *fixed* or constant value of current through a load, regardless of the value of the load. This concept is illustrated in Figure 8–4(b), where a load resistor is connected to the current source between terminals A and B. The ideal current source has an infinitely large internal resistance.

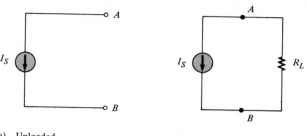

(a) Unloaded   (b) Loaded

FIGURE 8–4
Ideal current source.

Transistors act basically as current sources, and for this reason knowledge of the current source concept is important. You will find that the equivalent model of a transistor does contain a current source.

Although the ideal current source can be used in most analysis work, no actual device is ideal. A practical current source representation is shown in Figure 8–5. Here the internal resistance appears in parallel with the ideal current source.

FIGURE 8–5
Practical current source with load.

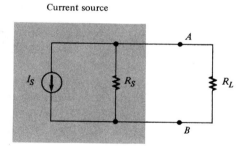

Current source

If the internal source resistance, R_S, is much *larger* than a load resistor, the practical source approaches ideal. The reason is illustrated in the practical current source shown in Figure 8–5. Part of the current I_S is through R_S, and part through R_L. R_S and R_L act as a current divider. If R_S is much larger than R_L, most of the current will be through R_L, and very little through R_S. As long as R_L remains much smaller than R_S, the current through it will stay almost constant, no matter how much R_L changes.

If we have a constant current source, we normally assume that R_S is so much larger than the load that R_S can be neglected. This simplifies the source to ideal, making the analysis easier.

Example 8–3 illustrates the effect of changes in R_L on the load current when R_L is much smaller than R_S. Generally, R_L should be at least ten times smaller $(10R_L \leq R_S)$.

**EXAMPLE
8–3**

Calculate the load current in Figure 8–6 for the following values of R_L: 100 Ω, 560 Ω, and 1 kΩ.

FIGURE 8–6

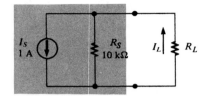

Solution:
For R_L = 100 Ω:

$$I_L = \left(\frac{R_S}{R_S + R_L}\right)I_S = \left(\frac{10 \text{ k}\Omega}{10.1 \text{ k}\Omega}\right)1 \text{ A}$$

$$= 0.99 \text{ A}$$

For R_L = 560 Ω:

$$I_L = \left(\frac{10 \text{ k}\Omega}{10.56 \text{ k}\Omega}\right)1 \text{ A}$$

$$= 0.95 \text{ A}$$

For R_L = 1 kΩ:

$$I_L = \left(\frac{10 \text{ k}\Omega}{11 \text{ k}\Omega}\right)1 \text{ A}$$

$$= 0.91 \text{ A}$$

Notice that the load current I_L is within 10% of the source current for each value of R_L.

SECTION REVIEW 8–2

1. What is the symbol for an ideal current source?

2. Sketch the practical current source.

3. What is the internal resistance of the ideal current source?

4. What effect does the load have on the load current of the practical current source?

8–3   SOURCE CONVERSIONS

In circuit analysis, it is sometimes useful to convert a voltage source to an *equivalent* current source, or vice versa.

CONVERTING A VOLTAGE SOURCE INTO A CURRENT SOURCE

The source voltage, V_S, divided by the source resistance, R_S, gives the value of the equivalent source current:

$$I_S = \frac{V_S}{R_S}$$

The value of R_S is the same for both sources. As illustrated in Figure 8–7, the directional arrow for the current points from plus to minus. The equivalent current source is the source in parallel with R_S.

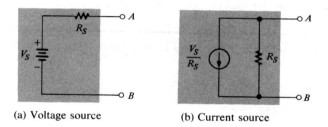

(a) Voltage source   (b) Current source

FIGURE 8–7
Conversion of voltage source to equivalent current source.

Equivalency of two sources means that for any given load resistance connected to the two sources, the same load voltage and current are produced by both sources. This concept is called *terminal equivalency*.

We can show that the voltage source and the current source in Figure 8–7 are equivalent by connecting a load resistor to each, as shown in Figure 8–8, and then calculating the load current as follows: For the voltage source,

$$I_L = \frac{V_S}{R_S + R_L}$$

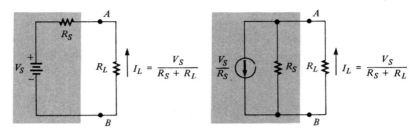

(a) Loaded voltage source   (b) Loaded current source

FIGURE 8–8
Equivalent sources with loads.

For the current source,

$$I_L = \left(\frac{R_S}{R_S + R_L}\right)\frac{V_S}{R_S} = \frac{V_S}{R_S + R_L}$$

As you see, both expressions for I_L are the same. These equations prove that the sources are equivalent as far as the load or terminals AB are concerned.

**EXAMPLE
8–4**

Convert the voltage source in Figure 8–9 to an equivalent current source.

FIGURE 8–9

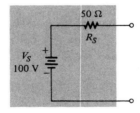

Solution:

$$I_S = \frac{V_S}{R_S} = \frac{100 \text{ V}}{50 \text{ }\Omega} = 2 \text{ A}$$

$$R_S = 50 \text{ }\Omega$$

The equivalent current source is shown in Figure 8–10.

FIGURE 8–10

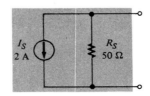

CONVERTING A CURRENT SOURCE INTO A VOLTAGE SOURCE

The source current, I_S, multiplied by the source resistance, R_S, gives the value of the equivalent source voltage:

$$V_S = I_S R_S$$

Again, R_S remains the same. The polarity of the voltage source is plus to minus in the direction of the current. The equivalent voltage source is the voltage in series with R_S, as illustrated in Figure 8–11.

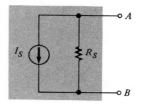 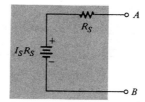

(a) Current source   (b) Voltage source

FIGURE 8–11
Conversion of current source to equivalent voltage source.

EXAMPLE 8–5

Convert the current source in Figure 8–12 to an equivalent voltage source.

FIGURE 8–12

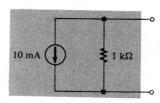

Solution:

$$V_S = I_S R_S = (10 \text{ mA})(1 \text{ k}\Omega)$$
$$= 10 \text{ V}$$
$$R_S = 1 \text{ k}\Omega$$

The equivalent voltage source is shown in Figure 8–13.

FIGURE 8–13

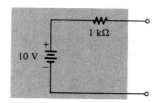

SECTION REVIEW 8–3

1. Write the formula for converting a voltage source to a current source.

2. Write the formula for converting a current source to a voltage source.

3. Convert the voltage source in Figure 8–14 to an equivalent current source.

4. Convert the current source in Figure 8–15 to an equivalent voltage source.

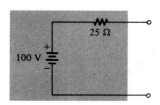

FIGURE 8–14

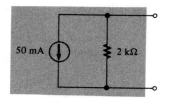

FIGURE 8–15

8–4   ## THE SUPERPOSITION THEOREM

Some circuits require more than one voltage source. For example, certain types of amplifiers require both a positive and a negative voltage source for proper operation.

The superposition method is a way to determine currents and voltages in a circuit that has multiple sources by taking *one source at a time*. The other sources are replaced by their internal resistances. Recall that the *ideal* voltage source has a *zero* internal resistance. In this section, all voltage sources will be treated as ideal in order to simplify the coverage.

A general statement of the superposition theorem is as follows:

The current in any given branch of a multiple-source linear circuit can be found by determining the currents in that particular branch produced by each source acting alone, with all other sources replaced by their internal resistances. The total current in the branch is the algebraic sum of the individual source currents in that branch.

The steps in applying the superposition method are as follows:

1. Take one voltage source at a time and replace each of the other voltage sources with a short (a short represents zero resistance).

2. Determine the current or voltage that you need just as if there were only *one* source in the circuit.

3. Take the next source in the circuit and repeat Steps 1 and 2 for each source.

4. To find the actual current or voltage, add or subtract the currents or voltages due to each individual source. If the currents are in the same direction or the voltages of the same polarity, add them. If the currents are in opposite directions or the voltages of opposite polarities, subtract them with the direction of the resulting current or voltage the same as the larger of the subtracted quantities.

An example of the approach to superposition is demonstrated in Figure 8–16 for a series-parallel circuit with two voltage sources. Study the steps in this figure. The following four examples clarify this procedure.

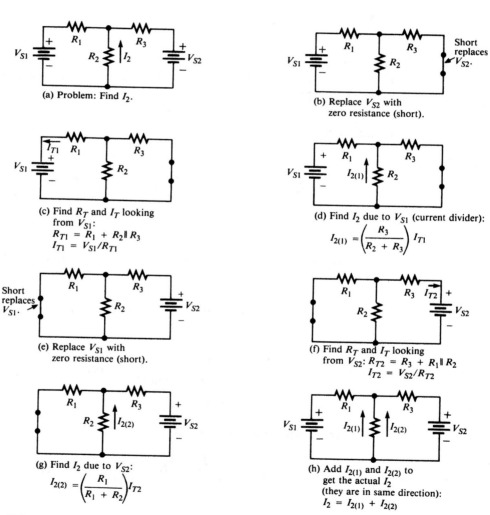

(a) Problem: Find I_2.

(b) Replace V_{S2} with zero resistance (short).

(c) Find R_T and I_T looking from V_{S1}:
$R_{T1} = R_1 + R_2 \| R_3$
$I_{T1} = V_{S1}/R_{T1}$

(d) Find I_2 due to V_{S1} (current divider):
$$I_{2(1)} = \left(\frac{R_3}{R_2 + R_3}\right) I_{T1}$$

(e) Replace V_{S1} with zero resistance (short).

(f) Find R_T and I_T looking from V_{S2}: $R_{T2} = R_3 + R_1 \| R_2$
$I_{T2} = V_{S2}/R_{T2}$

(g) Find I_2 due to V_{S2}:
$$I_{2(2)} = \left(\frac{R_1}{R_1 + R_2}\right) I_{T2}$$

(h) Add $I_{2(1)}$ and $I_{2(2)}$ to get the actual I_2 (they are in same direction):
$I_2 = I_{2(1)} + I_{2(2)}$

FIGURE 8–16
Demonstration of the superposition method.

EXAMPLE
8–6

Find the current in R_2 of Figure 8–17 by using the superposition theorem.

FIGURE 8–17

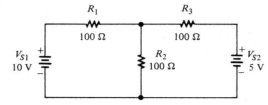

Solution:
First, by replacing V_{S2} with a short, find the current in R_2 due to voltage source V_{S1}, as shown in Figure 8–18.

FIGURE 8–18

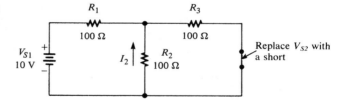

To find I_2, we can use the current divider formula from Chapter 6. Looking from V_{S1},

$$R_T = R_1 + \frac{R_3}{2} = 100\ \Omega + 50\ \Omega$$

$$= 150\ \Omega$$

$$I_T = \frac{V_{S1}}{R_T} = \frac{10\ \text{V}}{150\ \Omega} = 0.0667\ \text{A}$$

$$= 66.7\ \text{mA}$$

The current in R_2 due to V_{S1} is

$$I_2 = \left(\frac{R_3}{R_2 + R_3}\right)I_T = \left(\frac{100\ \Omega}{200\ \Omega}\right)66.7\ \text{mA}$$

$$= 33.3\ \text{mA}$$

Note that this current is *upward* through R_2.

Next find the current in R_2 due to voltage source V_{S2} by replacing V_{S1} with a short, as shown in Figure 8–19.

FIGURE 8–19

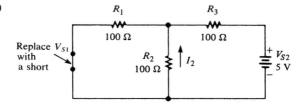

Looking from V_{S2},

$$R_T = R_3 + \frac{R_1}{2} = 100 \ \Omega + 50 \ \Omega$$

$$= 150 \ \Omega$$

$$I_T = \frac{V_{S2}}{R_T} = \frac{5 \ V}{150 \ \Omega} = 0.0333 \ A$$

$$= 33.3 \ mA$$

The current in R_2 due to V_{S2} is

$$I_2 = \left(\frac{R_1}{R_1 + R_2}\right)I_T = \left(\frac{100 \ \Omega}{200 \ \Omega}\right)33.3 \ mA$$

$$= 16.7 \ mA$$

Note that this current is *upward* through R_2.

Both component currents are *upward* through R_2, so they have the same algebraic sign. Therefore, we add the values to get the total current through R_2:

$$I_2 \ (total) = I_2 \ (due \ to \ V_{S1}) + I_2 \ (due \ to \ V_{S2})$$

$$= 33.3 \ mA + 16.7 \ mA = 50 \ mA$$

EXAMPLE 8–7

Find the current through R_2 in the circuit of Figure 8–20.

FIGURE 8–20

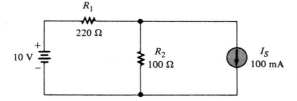

Solution:

First find the current in R_2 due to V_S by replacing I_S with an *open*, as shown in Figure 8–21. Notice that all of the current produced by V_S goes through R_2.

FIGURE 8–21

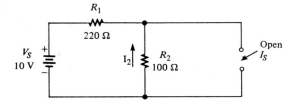

Looking from V_S,

$$R_T = R_1 + R_2 = 320 \ \Omega$$

The current through R_2 due to V_S is

$$I_2 = \frac{V_S}{R_T} = \frac{10 \ \text{V}}{320 \ \Omega} = 0.031 \ \text{A}$$

$$= 31 \ \text{mA}$$

Note that this current is *upward* through R_2.

Next find the current through R_2 due to I_S by replacing V_S with a *short*, as shown in Figure 8–22.

FIGURE 8–22

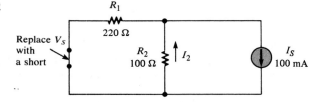

Using the current divider formula, we get the current through R_2 due to I_S as follows:

$$I_2 = \left(\frac{R_1}{R_1 + R_2} \right) I_S = \left(\frac{220 \ \Omega}{320 \ \Omega} \right) 100 \ \text{mA}$$

$$= 69 \ \text{mA}$$

Note that this current is also *upward* through R_2.

Both currents are in the same direction through R_2, so we add them to get the total:

$$I_2 \ (\text{total}) = I_2 \ (\text{due to } V_S) + I_2 \ (\text{due to } I_S)$$

$$= 31 \ \text{mA} + 69 \ \text{mA} = 100 \ \text{mA}$$

EXAMPLE 8–8

Find the current through the 100-Ω resistor in Figure 8–23.

FIGURE 8–23

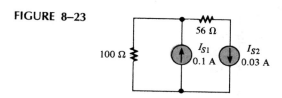

Solution:
First find the current through the 100-Ω resistor due to current source I_{S1} by replacing source I_{S2} with an *open*, as shown in Figure 8–24. As you can see, the entire 0.1 A from the current source I_{S1} is *downward* through the 100-Ω resistor.

FIGURE 8–24

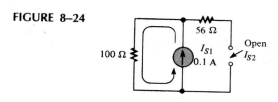

Next find the current through the 100-Ω resistor due to source I_{S2} by replacing source I_{S1} with an *open*, as indicated in Figure 8–25. Notice that all of the 0.03 A from source I_{S2} is *upward* through the 100-Ω resistor.

FIGURE 8–25

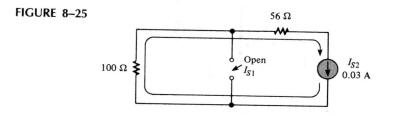

To get the total current through the 100-Ω resistor, we subtract the smaller current from the larger because they are in opposite directions. The resulting total current flows in the direction of the larger current from source I_{S1}:

$$I_{100\Omega} \text{ (total)} = I_{100\Omega} \text{ (due to } I_{S1}) - I_{100\Omega} \text{ (due to } I_{S2})$$

$$= 0.1 \text{ A} - 0.03 \text{ A} = 0.07 \text{ A}$$

The resulting current is downward through the resistor.

EXAMPLE 8–9

Find the total current through R_3 in Figure 8–26.

FIGURE 8–26

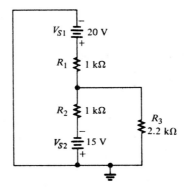

Solution:
First find the current through R_3 due to source V_{S1} by replacing source V_{S2} with a short, as shown in Figure 8–27.

FIGURE 8–27

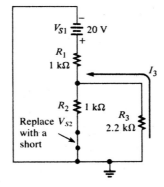

Looking from V_{S1},

$$R_T = R_1 + \frac{R_2 R_3}{R_2 + R_3}$$

$$= 1 \text{ k}\Omega + \frac{(1 \text{ k}\Omega)(2.2 \text{ k}\Omega)}{3.2 \text{ k}\Omega} = 1.69 \text{ k}\Omega$$

$$I_T = \frac{V_{S1}}{R_T} = \frac{20 \text{ V}}{1.69 \text{ k}\Omega}$$

$$= 11.83 \text{ mA}$$

Now apply the current divider formula to get the current through R_3 due to source V_{S1} as follows:

$$I_3 = \left(\frac{R_2}{R_2 + R_3}\right)I_T = \left(\frac{1\ k\Omega}{3.2\ k\Omega}\right)12\ mA$$

$$= 3.75\ mA$$

Notice that this current is *upward* through R_3.

Next find I_3 due to source V_{S2} by replacing source V_{S1} with a short, as shown in Figure 8–28.

FIGURE 8–28

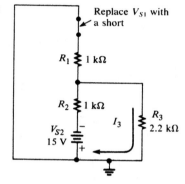

Looking from V_{S2},

$$R_T = R_2 + \frac{R_1 R_3}{R_1 + R_3}$$

$$= 1\ k\Omega + \frac{(2.2\ k\Omega)(1\ k\Omega)}{3.2\ k\Omega} = 1.69\ k\Omega$$

$$I_T = \frac{V_{S2}}{R_T} = \frac{15\ V}{1.69\ k\Omega}$$

$$= 8.88\ mA$$

Now apply the current divider formula to find the current through R_3 due to source V_{S2} as follows:

$$I_3 = \left(\frac{R_1}{R_1 + R_3}\right)I_T = \left(\frac{1\ k\Omega}{3.2\ k\Omega}\right)9\ mA$$

$$= 2.81\ mA$$

Notice that this current is *downward* through R_3.

Calculation of the total current through R_3 is as follows:

$$I_3 \text{ (total)} = I_3 \text{ (due to } V_{S1}) - I_3 \text{ (due to } V_{S2})$$

$$= 3.75 \text{ mA} - 2.81 \text{ mA} = 0.94 \text{ mA}$$

This current flows *upward* through R_3.

SECTION REVIEW 8–4

1. State the superposition theorem.
2. Why is the superposition theorem useful for analysis of multiple-source linear circuits?
3. Why is a voltage source shorted and a current source opened when the superposition theorem is applied?
4. Using the superposition theorem, find the current through R_1 in Figure 8–29.

FIGURE 8–29

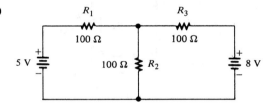

5. If, as a result of applying the superposition theorem, two currents are in opposing directions through a branch of a circuit, in what direction is the net current?

8–5 THEVENIN'S THEOREM

Thevenin's theorem gives us a method for simplifying a circuit to a standard equivalent form. The Thevenin equivalent form of any resistive circuit consists of an *equivalent voltage source* (V_{TH}) and an *equivalent resistance* (R_{TH}), arranged as shown in Figure 8–30. The values of the equivalent voltage and resistance depend on the values in the original circuit. Any resistive circuit can be simplified regardless of its complexity.

FIGURE 8–30
The general form of a Thevenin equivalent circuit is simply a nonideal voltage source. Any resistive circuit can be reduced to this form.

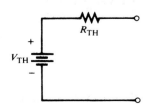

THEVENIN'S EQUIVALENT VOLTAGE (V_{TH}) AND EQUIVALENT RESISTANCE (R_{TH})

As you have seen, the equivalent voltage, V_{TH}, is one part of the complete Thevenin equivalent circuit. The other part is R_{TH}. V_{TH} *is defined to be the open circuit voltage between two points in a circuit.* Any component connected between these two points effectively "sees" V_{TH} in series with R_{TH}. As defined by Thevenin's theorem, R_{TH} *is the total resistance appearing between two terminals in a given circuit with all sources replaced by their internal resistances.*

THE MEANING OF EQUIVALENCY IN THEVENIN'S THEOREM

Although a Thevenin equivalent circuit is not the same as its original circuit, it *acts* the same in terms of the output voltage and current. For example, as shown in Figure 8–31, place a resistive circuit of any complexity in a box with only the output terminals exposed. Then place the Thevenin equivalent of that circuit in an *identical* box with, again, only the output terminals exposed. Connect identical load resistors across the output terminals of each box. Next connect a voltmeter and an ammeter to measure the voltage and current for each load as shown in the figure. The measured values will be identical (neglecting tolerance variations), and you will not be able to determine which box contains the original circuit and which contains the Thevenin equivalent. That is,

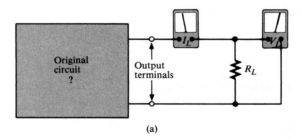

(a)

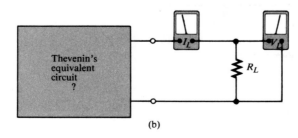

(b)

FIGURE 8–31
Which box contains the original circuit and which contains the Thevenin equivalent circuit? You cannot tell by observing the meters.

in terms of your observations, both circuits are the same. This condition is sometimes known as *terminal equivalency,* because both circuits look the same from the ''view-point'' of the two output terminals.

HOW TO FIND THE THEVENIN EQUIVALENT OF A CIRCUIT

To find the Thevenin equivalent of any circuit, determine the equivalent voltage, V_{TH}, and the equivalent resistance, R_{TH}. For example, in Figure 8–32, the Thevenin equivalent for the circuit between points A and B is found as follows:

In Part (a), the voltage across the designated points A and B is the Thevenin equivalent voltage. In this particular circuit, the voltage from A to B is the same as the

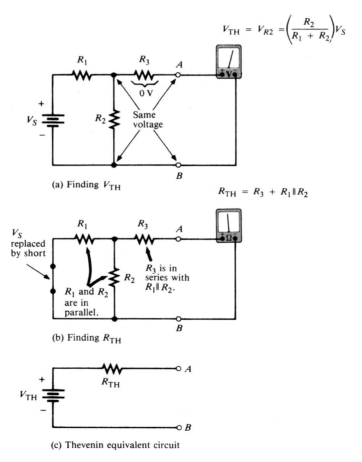

$$V_{TH} = V_{R2} = \left(\frac{R_2}{R_1 + R_2}\right)V_S$$

(a) Finding V_{TH}

$$R_{TH} = R_3 + R_1 \| R_2$$

(b) Finding R_{TH}

(c) Thevenin equivalent circuit

FIGURE 8–32
Example of the simplification of a circuit by Thevenin's theorem.

voltage across R_2 because there is no current through R_3 and, therefore, no voltage drop across it. V_{TH} is expressed as follows for this particular example:

$$V_{TH} = \left(\frac{R_2}{R_1 + R_2}\right)V_S$$

In Part (b), the resistance between points A and B with the source replaced by a short (zero internal resistance) is the Thevenin equivalent resistance. In this particular circuit, the resistance from A to B is R_3 in series with the parallel combination of R_1 and R_2. Therefore, R_{TH} is expressed as follows:

$$R_{TH} = R_3 + \frac{R_1 R_2}{R_1 + R_2}$$

The Thevenin equivalent circuit is shown in Part (c).

EXAMPLE 8–10

Find the Thevenin equivalent between the output terminals of the circuit in Figure 8–33.

FIGURE 8–33

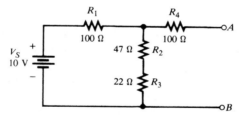

Solution:
V_{TH} equals the voltage across $R_2 + R_3$ as shown in Figure 8–34(a). Use the voltage divider principle to find V_{TH}:

$$V_{TH} = \left(\frac{R_2 + R_3}{R_1 + R_2 + R_3}\right)V_S = \left(\frac{69\ \Omega}{169\ \Omega}\right)10\ V = 4.08\ V$$

To find R_{TH}, first replace the source with a short to simulate a zero internal resistance. Then R_1 appears in parallel with $R_2 + R_3$, and R_4 is in series with the series-parallel combination of R_1, R_2, and R_3 as indicated in Figure 8–34(b):

$$R_{TH} = R_4 + \frac{R_1(R_2 + R_3)}{R_1 + R_2 + R_3} = 100\ \Omega + \frac{(100\ \Omega)(69\ \Omega)}{169\ \Omega} = 140.8\ \Omega$$

The resulting Thevenin equivalent circuit is shown in Part (c).

FIGURE 8–34

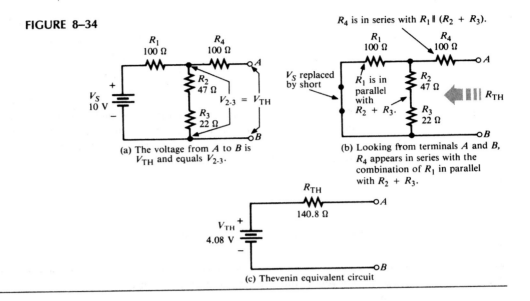

(a) The voltage from A to B is V_{TH} and equals V_{2-3}.

R_4 is in series with $R_1 \| (R_2 + R_3)$.

(b) Looking from terminals A and B, R_4 appears in series with the combination of R_1 in parallel with $R_2 + R_3$.

(c) Thevenin equivalent circuit

HOW THE THEVENIN EQUIVALENCY DEPENDS ON THE VIEWPOINT

The Thevenin equivalent for any circuit depends on the location of the two points from between which the circuit is "viewed." In Figure 8–33, we viewed the circuit from between the two points labeled A and B. Any given circuit can have more than one Thevenin equivalent, depending on how the viewpoints are designated. For example, if you view the circuit in Figure 8–35 from between points A and C, you obtain a completely different result than if you viewed it from between points A and B or from between points B and C.

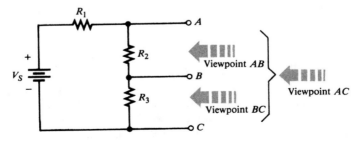

FIGURE 8–35
Thevenin's equivalent depends on viewpoint.

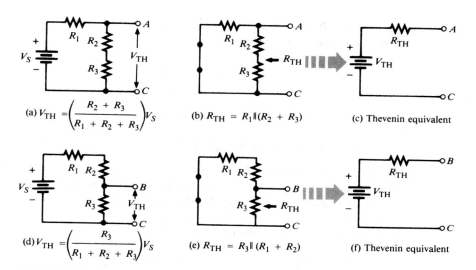

FIGURE 8–36
Example of circuit Thevenized from two viewpoints, resulting in two different equivalent circuits. (The V_{TH} and R_{TH} values are different.)

In Figure 8–36(a), when viewed from points A and C, V_{TH} is the voltage across $R_2 + R_3$ and can be expressed using the voltage divider formula as follows:

$$V_{TH} = \left(\frac{R_2 + R_3}{R_1 + R_2 + R_3}\right)V_S$$

Also, as shown in Part (b), the resistance between points A and C is $R_2 + R_3$ in parallel with R_1 (the source is replaced by a short) and can be expressed as follows:

$$R_{TH} = \frac{R_1(R_2 + R_3)}{R_1 + R_2 + R_3}$$

The resulting Thevenin equivalent circuit is shown in Part (c).
When viewed from points B and C as indicated in Figure 8–36(d), V_{TH} is the voltage across R_3 and can be expressed as follows:

$$V_{TH} = \left(\frac{R_3}{R_1 + R_2 + R_3}\right)V_S$$

As shown in Part (e), the resistance between points B and C is R_3 in parallel with the series combination of R_1 and R_2:

$$R_{TH} = \frac{R_3(R_1 + R_2)}{R_1 + R_2 + R_3}$$

The resulting Thevenin equivalent is shown in Part (f).

THEVENIZING A PORTION OF A CIRCUIT

In many cases it helps to Thevenize only a portion of a circuit. For example, when we need to know the equivalent circuit as viewed by one particular resistor in the circuit, we remove that resistor and apply Thevenin's theorem to the remaining part of the circuit as viewed from the points between which that resistor was connected. Figure 8–37 illustrates the Thevenizing of part of a circuit.

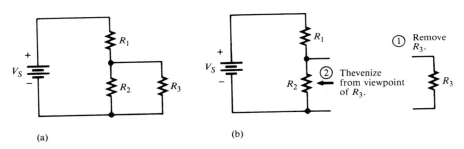

(a) (b)

FIGURE 8–37

Example of Thevenizing a portion of a circuit. In this case, the circuit is Thevenized from the viewpoint of R_3.

Using this type of approach, you can easily find the voltage and current for a specified resistor for any number of resistor values using only Ohm's law. This method eliminates the necessity of reanalyzing the original circuit for each different resistance value.

THEVENIZING A BRIDGE CIRCUIT

The usefulness of Thevenin's theorem is perhaps best illustrated when it is applied to a Wheatstone bridge circuit. For example, when a load resistor is connected to the output terminals of a Wheatstone bridge, as shown in Figure 8–38, the circuit is very difficult to analyze because it is not a straightforward series-parallel arrangement. If you doubt that this analysis is difficult, try to identify which resistors are in parallel and which are in series.

Using Thevenin's theorem, we can simplify the bridge circuit to an equivalent circuit viewed from the load resistor as shown step-by-step in Figure 8–39. Study carefully the steps in this figure. Once the equivalent circuit for the bridge is found, the voltage and current for any value of load resistor can easily be determined.

FIGURE 8–38
Wheatstone bridge with load resistor is
not a series-parallel circuit.

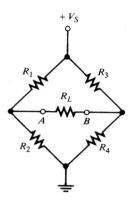

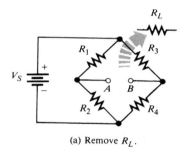

(a) Remove R_L.

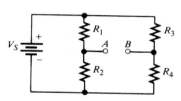

(b) Redraw and find V_{TH}.

(c) $V_{TH} = V_A - V_B = \left(\dfrac{R_2}{R_1 + R_2}\right) V_S - \left(\dfrac{R_4}{R_3 + R_4}\right) V_S$

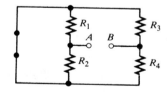

(d) Replace V_S with a short.
Note: The colored lines
are the same electrical
point as the colored lines
in Part (e).

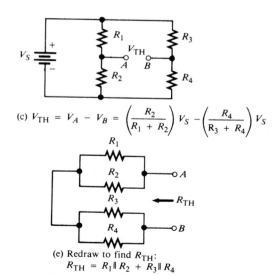

(e) Redraw to find R_{TH}:
$R_{TH} = R_1 \| R_2 + R_3 \| R_4$

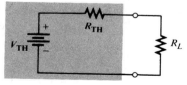

(f) Thevenin's equivalent
with R_L reconnected

FIGURE 8–39
Simplifying a Wheatstone bridge with Thevenin's theorem.

EXAMPLE
8–11

Determine the voltage and current for the load resistor, R_L, in the bridge circuit of Figure 8–40.

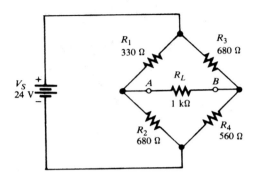

FIGURE 8–40

Solution:
Step 1: Remove R_L.
Step 2: To Thevenize the bridge as viewed from between points A and B, as was shown in Figure 8–39, first determine V_{TH}.

$$V_{TH} = V_A - V_B = \left(\frac{R_2}{R_1 + R_2}\right)V_S - \left(\frac{R_4}{R_3 + R_4}\right)V_S .$$

$$= \left(\frac{680\ \Omega}{1010\ \Omega}\right)24\ V - \left(\frac{560\ \Omega}{1240\ \Omega}\right)24\ V$$

$$= 16.16\ V - 10.84\ V = 5.32\ V$$

Step 3: Determine R_{TH}.

$$R_{TH} = \frac{R_1 R_2}{R_1 + R_2} + \frac{R_3 R_4}{R_3 + R_4} = \frac{(330\ \Omega)(680\ \Omega)}{1010\ \Omega} + \frac{(680\ \Omega)(560\ \Omega)}{1240\ \Omega}$$

$$= 222.18\ \Omega + 307.1\ \Omega = 529.28\ \Omega$$

Step 4: Place V_{TH} and R_{TH} in series to form the Thevenin equivalent circuit.
Step 5: Connect the load resistor from points A to B of the equivalent circuit as illustrated in Figure 8–41, and determine the load voltage and current.

$$V_{RL} = \left(\frac{R_L}{R_L + R_{TH}}\right)V_{TH} = \left(\frac{1\ k\Omega}{1.52928\ k\Omega}\right)5.32\ V = 3.48\ V$$

$$I_{RL} = \frac{V_{RL}}{R_L} = \frac{3.48\ V}{1\ k\Omega} = 3.48\ mA$$

FIGURE 8–41 Thevenin's equivalent for the Wheatstone bridge

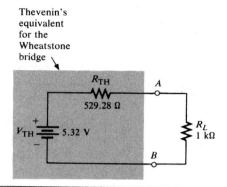

SUMMARY OF THEVENIN'S THEOREM

Remember, the Thevenin equivalent circuit is *always* of the series form regardless of the original circuit that it replaces. The significance of Thevenin's theorem is that the equivalent circuit can replace the original circuit as far as any external load is concerned. Any load resistor connected between the terminals of a Thevenin equivalent circuit will have the same current through it and the same voltage across it as if it were connected to the terminals of the original circuit.

   A summary of steps for applying Thevenin's theorem is as follows:

1. Open the two terminals (remove any load) between which you want to find the Thevenin equivalent circuit.
2. Determine the voltage (V_{TH}) across the two open terminals.
3. Determine the resistance (R_{TH}) between the two terminals with all voltage sources shorted and all current sources opened.
4. Connect V_{TH} and R_{TH} in series to produce the complete Thevenin equivalent for the original circuit.
5. Place the load resistor removed in Step 1 across the terminals of the Thevenin equivalent circuit. The load current can now be calculated using only Ohm's law, and it has the same value as the load current in the original circuit.

DETERMINING V_{TH} AND R_{TH} BY MEASUREMENT

Thevenin's theorem is largely an analytical tool that is applied theoretically in order to simplify circuit analysis. However, in many cases, Thevenin's equivalent can be found for an actual physical circuit by the following general measurement methods. These steps are illustrated in Figure 8–42.

1. Remove any load from the output terminals of the circuit.
2. Measure the open terminal voltage. The voltmeter used must have an internal resistance much greater (at least 10 times greater) than the R_{TH} of the circuit. (V_{TH} is the open terminal voltage.)

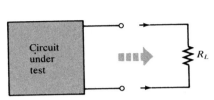

Step 1: Open the terminals (remove load).

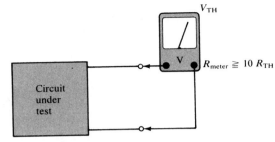

Step 2: Measure V_{TH}.

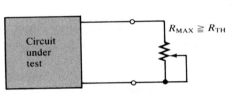

Step 3: Connect variable load resistance across the terminals.

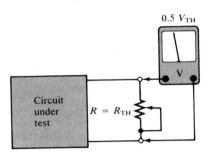

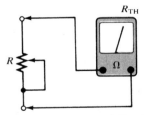

Step 4: Adjust R until $V_1 = 10.5\ V_{TH}$. When $V = 0.5\ V_{TH}$,
$R = R_{TH}$. Remove R and measure.

FIGURE 8–42
Determination of Thevenin's equivalent by measurement.

> **3.** Connect a variable resistor (rheostat) across the output terminals. Its maximum value must be greater than R_{TH}.
>
> **4.** Adjust the rheostat and measure the terminal voltage. When the terminal voltage equals $0.5V_{TH}$, the resistance of the rheostat is equal to R_{TH}. It should be disconnected from the terminals and measured with an ohmmeter.

This procedure for determining R_{TH} differs from the theoretical procedure because it is impractical to short voltage sources or open current sources in an actual circuit.

Also, when measuring R_{TH}, be certain that the circuit is capable of providing the required current to the variable resistor load and that the variable resistor can handle the required power. These considerations may make the procedure impractical in some cases.

SECTION REVIEW 8–5

1. What are the two components of a Thevenin equivalent circuit?
2. Draw the general form of a Thevenin equivalent circuit.
3. How is V_{TH} determined?
4. How is R_{TH} determined?
5. For the original circuit in Figure 8–43, draw the Thevenin equivalent circuit as viewed by R_L.

FIGURE 8–43

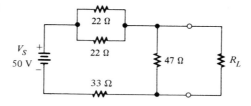

8–6

NORTON'S THEOREM

Like Thevenin's theorem, Norton's theorem provides a method of reducing a more complex circuit to a simpler form. The basic difference is that Norton's theorem gives *an equivalent current source in parallel with an equivalent resistance*. The form of Norton's equivalent circuit is shown in Figure 8–44. Regardless of how complex the original circuit is, it can always be reduced to this equivalent form. The equivalent current source is designated I_N, and the equivalent resistance, R_N.

To apply Norton's theorem, you must know how to find the two quantities I_N and R_N. Once you know them for a given circuit, simply connect them in parallel to get the complete Norton circuit.

FIGURE 8–44
Form of Norton's equivalent circuit.

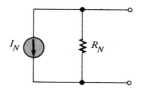

NORTON'S EQUIVALENT CURRENT (I_N)

As stated, I_N is one part of the complete Norton equivalent circuit; R_N is the other part. I_N is defined to be the *short circuit current* between two points in a circuit. Any component connected between these two points effectively "sees" a current source of value I_N in parallel with R_N.

To illustrate, suppose that a resistive circuit of some kind has a resistor connected between two points in the circuit, as shown in Figure 8–45(a). We wish to find the Norton circuit that is equivalent to the one shown as "seen" by R_L. To find I_N, calculate the current between points A and B with these two points *shorted*, as shown in Figure 8–45(b). Example 8–12 demonstrates how to find I_N.

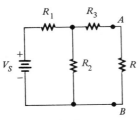
(a) Original circuit

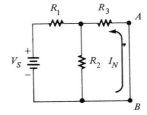
(b) Short the terminals to get I_N.

FIGURE 8–45
Determining the Norton equivalent current, I_N.

EXAMPLE 8–12

Determine I_N for the circuit within the shaded area in Figure 8–46(a).

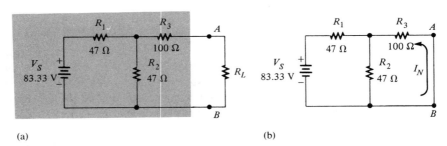

(a)   (b)

FIGURE 8–46

Solution:
Short terminals A and B as shown in Figure 8–46(b). I_N is the current through the short and is calculated as follows: First, the total resistance seen by the voltage source is

$$R_T = R_1 + \frac{R_2 R_3}{R_2 + R_3}$$

$$= 47\ \Omega + \frac{(47\ \Omega)(100\ \Omega)}{147\ \Omega} = 78.97\ \Omega$$

The total current from the source is

$$I_T = \frac{V_S}{R_T} = \frac{83.33 \text{ V}}{78.97 \text{ }\Omega}$$

$$= 1.06 \text{ A}$$

Now apply the current divider formula to find I_N (the current through the short):

$$I_N = \left(\frac{R_2}{R_2 + R_3}\right)I_T = \left(\frac{47 \text{ }\Omega}{147 \text{ }\Omega}\right)1.06 \text{ A}$$

$$= 0.339 \text{ A}$$

This is the value for the equivalent Norton current source.

NORTON'S EQUIVALENT RESISTANCE (R_N)

We define R_N in the same way as R_{TH}: it is the total resistance appearing between two terminals in a given circuit with all sources replaced by their internal resistances. Example 8–13 demonstrates how to find R_N.

EXAMPLE 8–13

Find R_N for the circuit within the shaded area of Figure 8–46 (see Example 8–12).

Solution:
First reduce V_S to zero by shorting it, as shown in Figure 8–47.

FIGURE 8–47

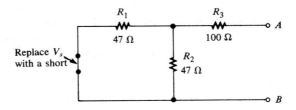

Looking in at terminals A and B, we see that the parallel combination of R_1 and R_2 is in series with R_3. Thus,

$$R_N = R_3 + \frac{R_1}{2} = 100 \text{ }\Omega + \frac{47 \text{ }\Omega}{2}$$

$$= 123.5 \text{ }\Omega$$

The last two examples have shown how to find the two equivalent components of a Norton equivalent circuit, I_N and R_N. Keep in mind that these values can be found for any linear circuit. Once these are known, they must be connected in *parallel* to form the Norton equivalent circuit, as illustrated in Example 8–14.

**EXAMPLE
8–14**

Draw the complete Norton circuit for the original circuit in Figure 8–46 (Example 8–12).

Solution:
We found in Examples 8–12 and 8–13 that $I_N = 0.339$ A and $R_N = 123.5$ Ω. The Norton equivalent circuit is shown in Figure 8–48.

FIGURE 8–48

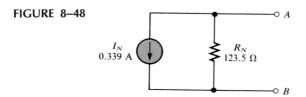

SUMMARY OF NORTON'S THEOREM

Any load resistor connected between the terminals of a Norton equivalent circuit will have the same current through it and the same voltage across it as if it were connected to the terminals of the original circuit. A summary of steps for theoretically applying Norton's theorem is as follows:

1. Short the two terminals between which you want to find the Norton equivalent circuit.
2. Determine the current (I_N) through the shorted terminals.
3. Determine the resistance (R_N) between the two terminals (opened) with all voltage sources shorted and all current sources opened ($R_N = R_{TH}$).
4. Connect I_N and R_N in parallel to produce the complete Norton equivalent for the original circuit.

Norton's equivalent circuit can also be derived from Thevenin's equivalent circuit by use of the source conversion method discussed in Section 8–3.

SECTION REVIEW 8–6

1. State Norton's theorem.
2. What are the two components of a Norton equivalent circuit?
3. Draw the general form of a Norton equivalent circuit.
4. How is I_N determined?
5. How is R_N determined?
6. Find the Norton circuit as seen by R_L in Figure 8–49.

FIGURE 8–49

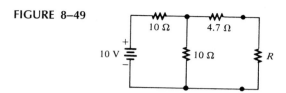

8–7

MILLMAN'S THEOREM

Millman's theorem allows us to reduce *any number of parallel voltage sources to a single equivalent voltage source*. It simplifies finding the voltage across or current through a load. Millman's theorem gives the same results as Thevenin's theorem for the special case of parallel voltage sources. A conversion by Millman's theorem is illustrated in Figure 8–50.

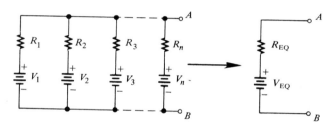

FIGURE 8–50
Reduction of parallel voltage sources to a single equivalent voltage source.

MILLMAN'S EQUIVALENT VOLTAGE (V_{EQ}) AND EQUIVALENT RESISTANCE (R_{EQ})

Millman's theorem gives us a formula for calculating the equivalent voltage, V_{EQ}. To find V_{EQ}, convert each of the parallel voltage sources into current sources, as shown in Figure 8–51.

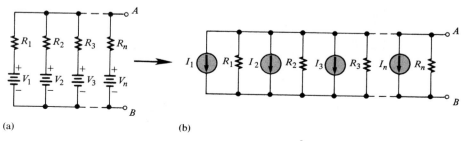

(a)   (b)

FIGURE 8–51
Parallel voltage sources converted to current sources.

In Figure 8–51(b), the total current from the parallel current sources is

$$I_T = I_1 + I_2 + I_3 + \cdots + I_n$$

The total conductance between terminals A and B is

$$G_T = G_1 + G_2 + G_3 + \cdots + G_n$$

where $G_T = 1/R_T$, $G_1 = 1/R_1$, and so on. Remember, the current sources are effectively open. Therefore, by Millman's theorem, the *equivalent resistance* is the total resistance R_T.

$$R_{EQ} = \frac{1}{G_T} = \frac{1}{(1/R_1) + (1/R_2) + (1/R_3) + \cdots + (1/R_n)} \qquad (8\text{–}1)$$

By Millman's theorem, the *equivalent voltage* is $I_T R_{EQ}$, where I_T is expressed as follows:

$$I_T = I_1 + I_2 + I_3 + \cdots + I_n$$

$$= \frac{V_1}{R_1} + \frac{V_2}{R_2} + \frac{V_3}{R_3} + \cdots + \frac{V_n}{R_n}$$

The following is the formula for the equivalent voltage:

$$V_{EQ} = \frac{(V_1/R_1) + (V_2/R_2) + (V_3/R_3) + \cdots + (V_n/R_n)}{(1/R_1) + (1/R_2) + (1/R_3) + \cdots + (1/R_n)} \qquad (8\text{–}2)$$

Equations (8–1) and (8–2) are the two Millman formulas. The equivalent voltage source has a polarity such that the total current through a load will be in the same direction as in the original circuit.

**EXAMPLE
8–15**

Use Millman's theorem to find the voltage across R_L and the current through R_L in Figure 8–52.

FIGURE 8–52

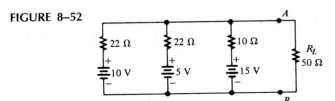

Solution:

Apply Millman's theorem as follows:

$$R_{EQ} = \cfrac{1}{(1/R_1) + (1/R_2) + (1/R_3)}$$

$$= \cfrac{1}{(1/22\ \Omega) + (1/22\ \Omega) + (1/10\ \Omega)}$$

$$= \frac{1}{0.19} = 5.24\ \Omega$$

$$V_{EQ} = \cfrac{(V_1/R_1) + (V_2/R_2) + (V_3/R_3)}{(1/R_1) + (1/R_2) + (1/R_3)}$$

$$= \cfrac{(10\ \text{V}/22\ \Omega) + (5\ \text{V}/22\ \Omega) + (15\ \text{V}/10\ \Omega)}{(1/22\ \Omega) + (1/22\ \Omega) + (1/10\ \Omega)}$$

$$= \frac{2.18\ \text{A}}{0.19\ \text{S}} = 11.47\ \text{V}$$

The single equivalent voltage source is shown in Figure 8–53.

FIGURE 8–53

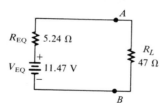

Now we calculate I_L and V_L for the load resistor.

$$I_L = \frac{V_{EQ}}{R_{EQ} + R_L} = \frac{11.47\ \text{V}}{52.24\ \Omega}$$

$$= 0.220\ \text{A}$$

$$V_L = I_L R_L = (0.220\ \text{A})(47\ \Omega)$$

$$= 10.34\ \text{V}$$

SECTION REVIEW 8–7

1. To what type of circuit does Millman's theorem apply?

2. Write the Millman theorem formula for R_{EQ}.

3. Write the Millman theorem formula for V_{EQ}.

4. Find the load current and the load voltage in Figure 8–54.

FIGURE 8–54

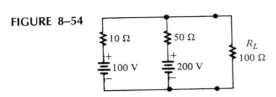

8–8   MAXIMUM POWER TRANSFER THEOREM

The maximum power transfer theorem states:

When a circuit is connected to a load, maximum power is delivered to the load when the load resistance is equal to the output resistance of the circuit.

The output or source resistance R_S of a circuit is the equivalent resistance as viewed from the output terminals using Thevenin's theorem. An equivalent circuit with its output resistance and load is shown in Figure 8–55. When $R_L = R_S$, the maximum power possible is transferred from the voltage source to R_L.

FIGURE 8–55
Maximum power is transferred to the load when $R_L = R_S$.

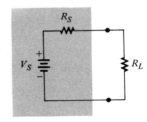

Practical applications of this theorem include audio systems such as stereo, radio, public address, and so on. In these systems the resistance (impedance) of the speaker is the load. The circuit that drives the speaker is a power amplifier. The systems are typically optimized for maximum power to the speakers. Thus, the resistance (impedance) of the speaker must equal the source resistance (impedance) of the amplifier.

Example 8–16 shows that maximum power occurs at the *matched* condition (that is, when $R_L = R_S$).

EXAMPLE 8–16

The source in Figure 8–56 has an output resistance of 75 Ω. Determine the power in each of the following values of load resistance:

(a) 25 Ω (b) 50 Ω

(c) 75 Ω (d) 100 Ω

(e) 125 Ω

Draw a graph showing the load power versus the load resistance.

FIGURE 8–56

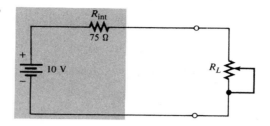

Solution:
We will use Ohm's law ($I = V/R$) and the power formula ($P = I^2R$) to find the load power for each value of load resistance.

(a) $R_L = 25 \; \Omega$:

$$I = \frac{V_S}{R_S + R_L} = \frac{10 \text{ V}}{75 \; \Omega + 25 \; \Omega} = 0.1 \text{ A}$$

$$P_L = I^2R_L = (0.1 \text{ A})^2(25 \; \Omega) = 0.25 \text{ W}$$

(b) $R_L = 50 \; \Omega$:

$$I = \frac{V_S}{R_S + R_L} = \frac{10 \text{ V}}{125 \; \Omega} = 0.08 \text{ A}$$

$$P_L = I^2R_L = (0.08 \text{ A})^2(50 \; \Omega) = 0.32 \text{ W}$$

(c) $R_L = 75 \; \Omega$:

$$I = \frac{V_S}{R_S + R_L} = \frac{10 \text{ V}}{150 \; \Omega} = 0.067 \text{ A}$$

$$P_L = I^2R_L = (0.067 \text{A})^2(75 \; \Omega) = 0.337 \text{ W}$$

(d) $R_L = 100 \; \Omega$:

$$I = \frac{V_S}{R_S + R_L} = \frac{10 \text{ V}}{175 \; \Omega} = 0.057 \text{ A}$$

$$P_L = I^2R_L = (0.057 \text{ A})^2(100 \; \Omega) = 0.33 \text{ W}$$

(e) $R_L = 125 \; \Omega$:

$$I = \frac{V_S}{R_S + R_L} = \frac{10 \text{ V}}{200 \; \Omega} = 0.05 \text{ A}$$

$$P_L = I^2R_L = (0.05 \text{ A})^2(125 \; \Omega) = 0.31 \text{ W}$$

Notice that the load power is greatest when $R_L = 75 \; \Omega$, which is the same as the source resistance. When the load resistance is less than or greater than this value, the power drops off, as the curve in Figure 8–57 graphically illustrates.

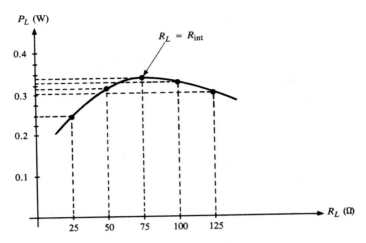

FIGURE 8–57
Curve showing that the load power is maximum when $R_L = R_{int}$.

The calculator sequence for **(a)** is

$$\boxed{1}\ \boxed{0}\ \boxed{\div}\ \boxed{(}\ \boxed{7}\ \boxed{5}\ \boxed{+}\ \boxed{2}\ \boxed{5}\ \boxed{)}\ \boxed{=}\ \boxed{x^2}\ \boxed{\times}\ \boxed{2}\ \boxed{5}\ \boxed{=}$$

SECTION REVIEW 8–8

1. State the maximum power transfer theorem.
2. When is maximum power delivered from a source to a load?
3. A given circuit has a source resistance of 50 Ω. What will be the value of the load to which the maximum power is delivered?

8–9

DELTA-WYE (Δ-Y) AND WYE-DELTA (Y-Δ) NETWORK CONVERSIONS

A resistive delta (Δ) network has the form shown in Figure 8–58(a). A wye (Y) network is shown in Figure 8–58(b). Notice that letter subscripts are used to designate resistors

FIGURE 8–58
Delta and wye networks.

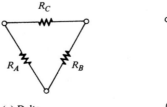

(a) Delta

(b) Wye

in the delta network and that numerical subscripts are used to designate resistors in the wye.

Conversion between these two forms of circuits is sometimes helpful in areas such as bridge analysis and three-phase power systems. In this section, the conversion formulas and rules for remembering them are given.

Δ-TO-Y CONVERSION

It is convenient to think of the wye positioned within the delta, as shown in Figure 8–59. To convert from delta to wye, we need R_1, R_2, and R_3 in terms of R_A, R_B, and R_C. The conversion rule is as follows: *Each resistor in the wye is equal to the product of the resistors in two adjacent delta branches, divided by the sum of all three delta resistors.*

FIGURE 8–59
"Y within Δ" aid for conversion formulas.

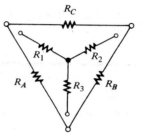

In Figure 8–59, R_A and R_C are "adjacent" to R_1:

$$R_1 = \frac{R_A R_C}{R_A + R_B + R_C} \tag{8-3}$$

Also, R_B and R_C are "adjacent" to R_2:

$$R_2 = \frac{R_B R_C}{R_A + R_B + R_C} \tag{8-4}$$

and R_A and R_B are "adjacent" to R_3:

$$R_3 = \frac{R_A R_B}{R_A + R_B + R_C} \tag{8-5}$$

Y-TO-Δ CONVERSION

To convert from wye to delta, we need R_A, R_B, and R_C in terms of R_1, R_2, and R_3. The conversion rule is as follows: *Each resistor in the delta is equal to the sum of all possible products of wye resistors taken two at a time, divided by the opposite wye resistor.*

In Figure 8–59, R_2 is "opposite" to R_A:

$$R_A = \frac{R_1R_2 + R_1R_3 + R_2R_3}{R_2} \tag{8–6}$$

Also, R_1 is "opposite" to R_B:

$$R_B = \frac{R_1R_2 + R_1R_3 + R_2R_3}{R_1} \tag{8–7}$$

and R_3 is "opposite" to R_C:

$$R_C = \frac{R_1R_2 + R_1R_3 + R_2R_3}{R_3} \tag{8–8}$$

The following two examples illustrate conversion between these two forms of circuits.

EXAMPLE 8–17

Convert the delta network in Figure 8–60 to a wye network.

FIGURE 8–60

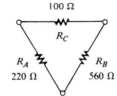

100 Ω
R_C
R_A
220 Ω
R_B
560 Ω

Solution:
Use Equations (8–3), (8–4), and (8–5):

$$R_1 = \frac{R_AR_C}{R_A + R_B + R_C} = \frac{(220\ \Omega)(100\ \Omega)}{220\ \Omega + 560\ \Omega + 100\ \Omega}$$

$$= 25\ \Omega$$

$$R_2 = \frac{R_BR_C}{R_A + R_B + R_C} = \frac{(560\ \Omega)(100\ \Omega)}{880\ \Omega}$$

$$= 63.64\ \Omega$$

$$R_3 = \frac{R_AR_B}{R_A + R_B + R_C} = \frac{(220\ \Omega)(560\ \Omega)}{880\ \Omega}$$

$$= 140\ \Omega$$

The resulting wye network is shown in Figure 8–61.

FIGURE 8–61

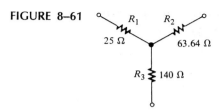

EXAMPLE 8–18

Convert the wye network in Figure 8–62 to a delta network.

FIGURE 8–62

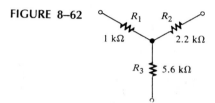

Solution:
Use Equations (8–6), (8–7), and (8–8):

$$R_A = \frac{R_1 R_2 + R_1 R_3 + R_2 R_3}{R_2}$$

$$= \frac{(1 \text{ k}\Omega)(2.2 \text{ k}\Omega) + (1 \text{ k}\Omega)(5.6 \text{ k}\Omega) + (2.2 \text{ k}\Omega)(5.6 \text{ k}\Omega)}{2.2 \text{ k}\Omega} = 9.15 \text{ k}\Omega$$

$$R_B = \frac{R_1 R_2 + R_1 R_3 + R_2 R_3}{R_1}$$

$$= \frac{(1 \text{ k}\Omega)(2.2 \text{ k}\Omega) + (1 \text{ k}\Omega)(5.6 \text{ k}\Omega) + (2.2 \text{ k}\Omega)(5.6 \text{ k}\Omega)}{1 \text{ k}\Omega} = 20.12 \text{ k}\Omega$$

$$R_C = \frac{R_1 R_2 + R_1 R_3 + R_2 R_3}{R_3}$$

$$= \frac{(1 \text{ k}\Omega)(2.2 \text{ k}\Omega) + (1 \text{ k}\Omega)(5.6 \text{ k}\Omega) + (2.2 \text{ k}\Omega)(5.6 \text{ k}\Omega)}{5.6 \text{ k}\Omega} = 3.59 \text{ k}\Omega$$

The resulting delta network is shown in Figure 8–63.

FIGURE 8–63

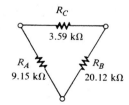

APPLICATION OF Δ-Y CONVERSION TO THE SIMPLIFICATION OF A BRIDGE CIRCUIT

You have already seen how Thevenin's theorem can be used to simplify a bridge circuit. Now we will see how Δ-Y conversion can be used for conversion of a bridge circuit to a series-parallel form for easier analysis.

Figure 8–64 illustrates how the delta (Δ) formed by R_A, R_B, and R_C can be converted to a wye (Y), thus creating an equivalent series-parallel circuit. Equations (8–3), (8–4), and (8–5) are used in this conversion.

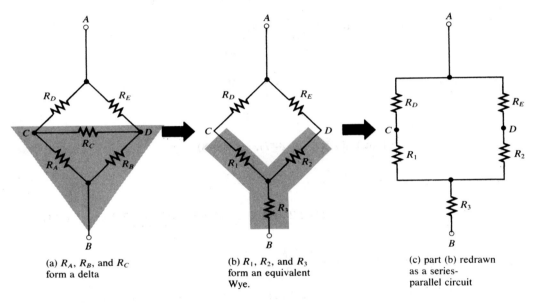

(a) R_A, R_B, and R_C form a delta

(b) R_1, R_2, and R_3 form an equivalent Wye.

(c) part (b) redrawn as a series-parallel circuit

FIGURE 8–64
Conversion of a bridge circuit to a series-parallel configuration.

In a bridge circuit, the load is connected across points C and D. In Figure 8–64(a), R_C represents the load resistor. When voltage is applied across points A and B, the voltage from C to D (V_{CD}) can be determined using the equivalent series-parallel circuit in Figure 8–64(c) as follows.

The total resistance from point A to point B is

$$R_T = \frac{(R_1 + R_D)(R_2 + R_E)}{(R_1 + R_D) + (R_2 + R_E)} + R_3$$

Then

$$I_T = \frac{V_{AB}}{R_T}$$

The current through the left branch is

$$I_{AC} = \left(\frac{R_T}{R_1 + R_D}\right)I_T$$

The current through the right branch is

$$I_{AD} = \left(\frac{R_T}{R_2 + R_E}\right)I_T$$

The voltage at point C with respect to point A is

$$V_{CA} = V_A - I_{AC}R_D$$

The voltage at point D with respect to point A is

$$V_{DA} = V_A - I_{AD}R_E$$

The voltage from point C to point D is

$$V_{CD} = V_{CA} - V_{DA} = (V_A - I_{AC}R_D) - (V_A - I_{AD}R_E)$$
$$= I_{AD}R_E - I_{AC}R_D$$

V_{CD} is the voltage across the load (R_C) in the bridge circuit of Figure 8–64(a). The current through the load can be found by Ohm's law:

$$I_C = \frac{V_{CD}}{R_C}$$

**EXAMPLE
8–19**

Determine the load voltage and the load current in the bridge circuit in Figure 8–65. Notice that the resistors are labeled for convenient conversion using the previous formulas. R_C is the load resistor.

FIGURE 8–65

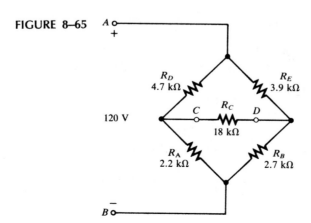

Solution:

First convert the delta formed by R_A, R_B, and R_C to a wye:

$$R_1 = \frac{R_A R_C}{R_A + R_B + R_C} = \frac{(2.2 \text{ k}\Omega)(18 \text{ k}\Omega)}{2.2 \text{ k}\Omega + 2.7 \text{ k}\Omega + 18 \text{ k}\Omega} = 1.73 \text{ k}\Omega$$

$$R_2 = \frac{R_B R_C}{R_A + R_B + R_C} = \frac{(2.7 \text{ k}\Omega)(18 \text{ k}\Omega)}{22.9 \text{ k}\Omega} = 2.12 \text{ k}\Omega$$

$$R_3 = \frac{R_A R_B}{R_A + R_B + R_C} = \frac{(2.2 \text{ k}\Omega)(2.7 \text{ k}\Omega)}{22.9 \text{ k}\Omega} = 259 \text{ }\Omega$$

The resulting equivalent series-parallel circuit is shown in Figure 8–66.

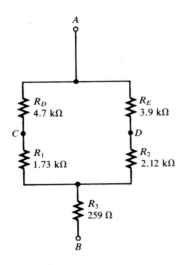

FIGURE 8–66

Now determine R_T and the branch currents in Figure 8–66.

$$R_T = \frac{(R_1 + R_D)(R_2 + R_E)}{(R_1 + R_D) + (R_2 + R_E)} + R_3 = \frac{(6.43 \text{ k}\Omega)(6.02 \text{ k}\Omega)}{6.43 \text{ k}\Omega + 6.02 \text{ k}\Omega} + 259 \text{ }\Omega = 3.37 \text{ k}\Omega$$

$$I_T = \frac{V_{AB}}{R_T} = \frac{120 \text{ V}}{3.37 \text{ k}\Omega} = 35.61 \text{ mA}$$

$$I_{AC} = \left(\frac{R_T}{R_1 + R_D}\right)I_T = \left(\frac{3.37 \text{ k}\Omega}{1.73 \text{ k}\Omega + 4.7 \text{ k}\Omega}\right)35.61 \text{ mA} = 18.66 \text{ mA}$$

$$I_{AD} = \left(\frac{R_T}{R_2 + R_E}\right)I_T = \left(\frac{3.37 \text{ k}\Omega}{2.12 \text{ k}\Omega + 3.9 \text{ k}\Omega}\right)35.61 \text{ mA} = 19.94 \text{ mA}$$

The voltage across the load is

$$V_{CD} = I_{AD}R_E - I_{AC}R_D$$
$$= (19.94 \text{ mA})(3.9 \text{ k}\Omega) - (18.66 \text{ mA})(4.7 \text{ k}\Omega)$$
$$= 77.77 \text{ V} - 87.70 \text{ V} = -9.93 \text{ V}$$

The load current is

$$I_C = \frac{V_{CD}}{R_C} = \frac{-9.93 \text{ V}}{18 \text{ k}\Omega} = -0.552 \text{ mA}$$

SECTION REVIEW 8–9

1. Sketch a delta network.

2. Sketch a wye network.

3. Write the formulas for delta-to-wye conversion.

4. Write the formulas for wye-to-delta conversion.

8–10   COMPUTER ANALYSIS

Delta-wye conversion is used in this section as an example for computer analysis. The program listed below converts a specified delta network into a wye network. The resistor labeling conforms to that in Figure 8–58.

```
10   CLS
20   PRINT "THIS PROGRAM CONVERTS A SPECIFIED DELTA NETWORK TO
     THE"
30   PRINT "CORRESPONDING WYE NETWORK. REFER TO FIGURE 8-58
     FOR"
40   PRINT "THE APPROPRIATE RESISTOR LABELS."
50   PRINT:PRINT:PRINT
60   INPUT "TO CONTINUE PRESS 'ENTER'";X
70   CLS
80   PRINT "PLEASE PROVIDE THE DELTA RESISTOR VALUES WHEN
     PROMPTED."
90   PRINT:PRINT
100  INPUT "RA IN OHMS";RA
110  INPUT "RB IN OHMS";RB
120  INPUT "RC IN OHMS";RC
130  CLS
140  RT=RA+RB+RC
150  R1=RA*RC/RT
160  R2=RB*RC/RT
170  R3=RA*RB/RT
180  PRINT "THE VALUES FOR THE WYE NETWORK ARE AS FOLLOWS:"
190  PRINT:PRINT "R1=";R1;"OHMS"
200  PRINT "R2=";R2;"OHMS"
210  PRINT "R3=";R3;"OHMS"
```

SECTION REVIEW 8–10

1. Identify the calculation statements.
2. What is the purpose of line 90?

SUMMARY

1. An ideal voltage source has zero internal resistance. It provides a constant voltage across its terminals regardless of the load resistance.
2. A practical voltage source has a nonzero internal resistance. Its terminal voltage is essentially constant when $R_L \geq 10R_S$ (rule of thumb).
3. An ideal current source has infinite internal resistance. It provides a constant current regardless of the load resistance.
4. A practical current source has a finite internal resistance. Its current is essentially constant when $10R_L \leq R_S$.
5. The superposition theorem is useful for multiple-source circuits.
6. Thevenin's theorem provides for the reduction of any linear resistive circuit to an equivalent form consisting of an equivalent voltage source in series with an equivalent resistance.
7. The term *equivalency,* as used in Thevenin's and Norton's theorems, means that when a given load resistance is connected to the equivalent circuit, it will have the same voltage across it and the same current through it as when it was connected to the original circuit.
8. Norton's theorem provides for the reduction of any linear resistive circuit to an equivalent form consisting of an equivalent current source in parallel with an equivalent resistance.
9. Millman's theorem provides for the reduction of parallel voltage sources to a single equivalent voltage source consisting of an equivalent voltage and an equivalent series resistance.
10. Maximum power is transferred to a load from a source when the load resistance equals the source resistance.

FORMULAS

$$R_{EQ} = \frac{1}{G_T} = \frac{1}{(1/R_1) + (1/R_2) + (1/R_3) + \cdots + (1/R_n)} \tag{8-1}$$

$$V_{EQ} = \frac{(V_1/R_1) + (V_2/R_2) + (V_3/R_3) + \cdots + (V_n/R_n)}{(1/R_1) + (1/R_2) + (1/R_3) + \cdots + (1/R_n)} \tag{8-2}$$

Δ-to-Y Conversions:

$$R_1 = \frac{R_A R_C}{R_A + R_B + R_C} \tag{8-3}$$

$$R_2 = \frac{R_B R_C}{R_A + R_B + R_C} \tag{8-4}$$

$$R_3 = \frac{R_A R_B}{R_A + R_B + R_C} \tag{8-5}$$

Y-to-Δ Conversions:

$$R_A = \frac{R_1 R_2 + R_1 R_3 + R_2 R_3}{R_2} \tag{8-6}$$

$$R_B = \frac{R_1 R_2 + R_1 R_3 + R_2 R_3}{R_1} \tag{8-7}$$

$$R_C = \frac{R_1 R_2 + R_1 R_3 + R_2 R_3}{R_3} \tag{8-8}$$

SELF-TEST

Solutions appear at the end of the book.

1. A voltage source has the values $V_S = 25$ V and $R_S = 5$ Ω. What are the values for the equivalent current source?

2. Convert the voltage source in Figure 8–67 to an equivalent current source.

3. Convert the current source in Figure 8–68 to an equivalent voltage source.

4. Explain the basic steps used to analyze a circuit with two or more voltage sources.

5. In a two-source circuit, one source acting alone produces 10 mA through a given branch. The other source acting alone produces 8 mA in the opposite direction through the same branch. What is the actual current through the branch?

6. In Figure 8–69, use the superposition theorem to find the total current in R_3.

7. In Figure 8–69, what is the total current through R_2?

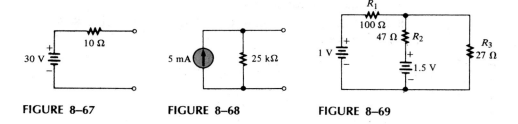

FIGURE 8–67   FIGURE 8–68   FIGURE 8–69

8. The Thevenin values for a certain circuit are found to be 8.5 V and 280 Ω. Draw the Thevenin equivalent circuit.

9. For Figure 8–70, determine the Thevenin equivalent circuit as seen by R.

10. Using Thevenin's theorem, find the current in R_L for the circuit in Figure 8–71.

11. Reduce the circuit in Figure 8–71 to its Norton equivalent as seen by R_L.

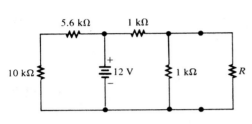

FIGURE 8–70

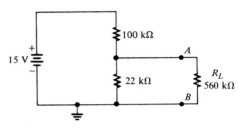

FIGURE 8–71

12. Use Millman's theorem to simplify the circuit in Figure 8–72 to a single voltage source.

13. What value of R_L in Figure 8–72 is required for maximum power transfer?

14. **(a)** Convert the delta network in Figure 8–73(a) to a wye.
(b) Convert the wye network in Figure 8–73(b) to a delta.

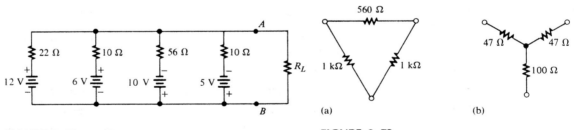

FIGURE 8–72

FIGURE 8–73

PROBLEMS

Sections 8–1 through 8–3

8–1 A voltage source has the values $V_S = 300$ V and $R_S = 50\ \Omega$. Convert it to an equivalent current source.

8–2 Convert the practical voltage sources in Figure 8–74 to equivalent current sources.

FIGURE 8–74

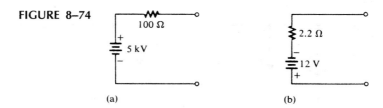

(a)   (b)

8–3 A current source has an I_S of 600 mA and an R_S of 1.2 kΩ. Convert it to an equivalent voltage source.

8–4 Convert the practical current sources in Figure 8–75 to equivalent voltage sources.

FIGURE 8–75

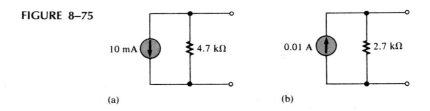

(a)   (b)

Section 8–4

8–5 Using the superposition method, calculate the current in the right-most branch of Figure 8–76.

8–6 Use the superposition theorem to find the current in and the voltage across the R_2 branch of Figure 8–76.

8–7 Using the superposition theorem, solve for the current through R_3 in Figure 8–77.

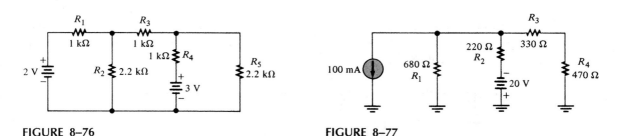

FIGURE 8–76   FIGURE 8–77

8–8 Using the superposition theorem, find the load current in each circuit of Figure 8–78.

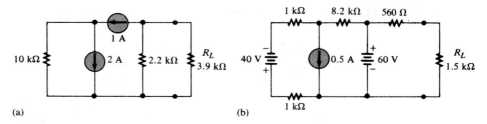

(a)   (b)

FIGURE 8–78

8–9   Determine the voltage from point A to point B in Figure 8–79.

8–10   The switches in Figure 8–80 are closed in sequence, S_1 first. Find the current through R_L after each switch closure.

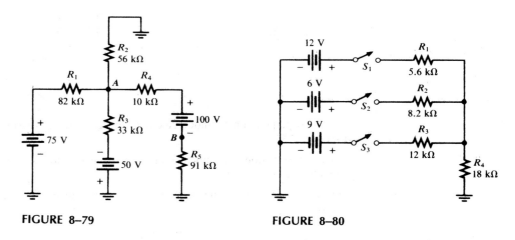

FIGURE 8–79   **FIGURE 8–80**

8–11   Figure 8–81 shows two ladder networks. Determine the current drain on each of the batteries when terminals A are connected (A to A) and terminals B are connected (B to B).

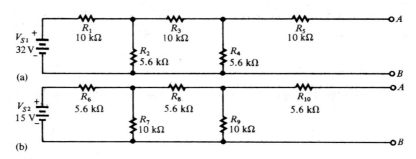

(a)

(b)

FIGURE 8–81

Section 8–5

8–12 For each circuit in Figure 8–82, determine the Thevenin equivalent as seen by R_L.

FIGURE 8–82

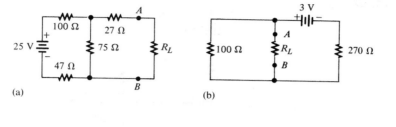

(a)

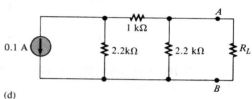

(b)

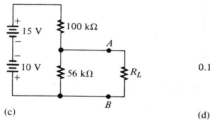

(c)

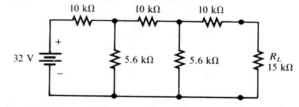

(d)

8–13 Using Thevenin's theorem, determine the current through the load R_L in Figure 8–83.

FIGURE 8–83

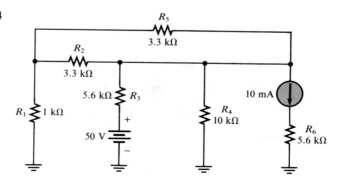

8–14 Using Thevenin's theorem, find the voltage across R_4 in Figure 8–84.

FIGURE 8–84

8–15 Find the Thevenin equivalent for the circuit external to the amplifier in Figure 8–85.

FIGURE 8–85

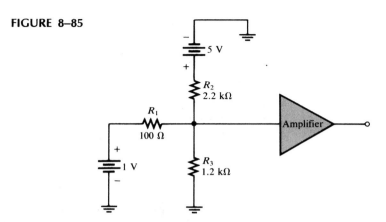

8–16 Determine the current out of point A when R_8 is 1 kΩ, 5 kΩ, and 10 kΩ in Figure 8–86.

FIGURE 8–86

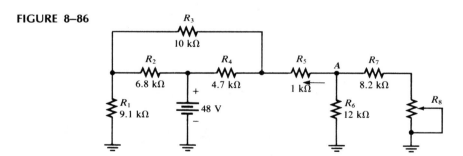

8–17 Find the current in the load resistor in the bridge circuit of Figure 8–87.

FIGURE 8–87

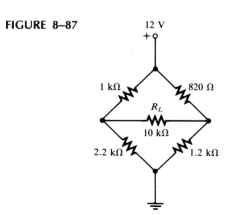

8–18 Determine the Thevenin equivalent looking from terminals *AB* for the circuit in Figure 8–88.

FIGURE 8–88

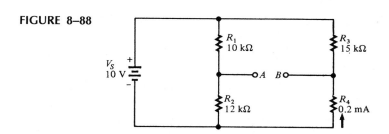

Section 8–6

8–19 For each circuit in Figure 8–82, determine the Norton equivalent as seen by R_L.

8–20 Using Norton's theorem, find the current through the load resistor R_L in Figure 8–83.

8–21 Using Norton's theorem, find the voltage across R_5 in Figure 8–84.

8–22 Using Norton's theorem, find the current through R_1 in Figure 8–86 when $R_8 = 8\ \text{k}\Omega$.

8–23 Determine the Norton equivalent circuit for the bridge in Figure 8–87 with R_L removed.

8–24 Reduce the circuit between terminals *A* and *B* in Figure 8–89 to its Norton equivalent.

FIGURE 8–89

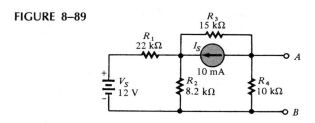

Section 8–7

8–25 Apply Millman's theorem to the circuit of Figure 8–90.

FIGURE 8–90

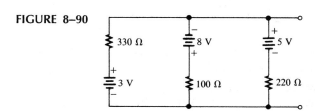

8–26 Use Millman's theorem and source conversions to reduce the circuit in Figure 8–91 to a single voltage source.

FIGURE 8–91

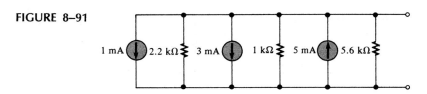

8–27 Use Millman's theorem to find the current in R_L for each case in which at least two switches are closed in Figure 8–80.

Section 8–8

8–28 For each circuit in Figure 8–92, maximum power is to be transferred to the load R_L. Determine the appropriate value for R_L in each case.

FIGURE 8–92

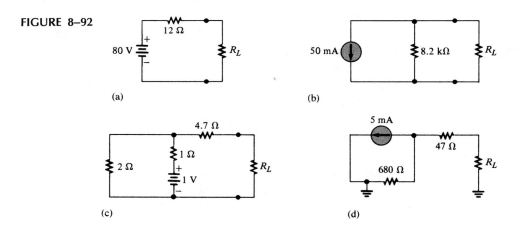

8–29 Determine the value of R_L for maximum power in Figure 8–93.

FIGURE 8–93

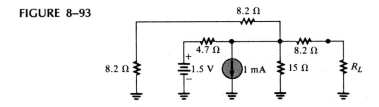

8–30 How much power is delivered to the load when R_L is 10% higher than its value for maximum power in Figure 8–93?

8–31 What are the values of R_4 and R_{TH} when maximum power is transferred from the Thevenized source to the ladder network in Figure 8–94?

FIGURE 8–94

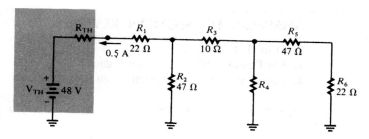

Section 8–9

8–32 In Figure 8–95, convert each delta network to a wye network.

FIGURE 8–95

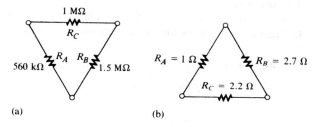

(a) (b)

8–33 In Figure 8–96, convert each wye network to a delta network.

8–34 Find all currents in the circuit of Figure 8–97.

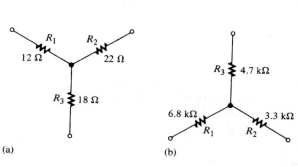

(a) (b)

FIGURE 8–96

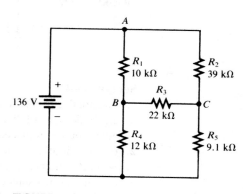

FIGURE 8–97

Section 8–10

8–35 Change the program in Section 8–10 so that the resistor values can be entered in kilohms and displayed in kilohms.

8–36 Develop a flowchart for the program in Section 8–10.

8–37 Write a program to convert a specified wye network to a corresponding delta network.

ANSWERS TO SECTION REVIEWS

Section 8–1
1. See Figure 8–98.   **2.** See Figure 8–99.   **3.** Zero ohms.
4. Output voltage varies directly with load resistance.

FIGURE 8–98

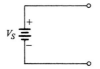

FIGURE 8–99

Section 8–2
1. See Figure 8–100.   **2.** See Figure 8–101.   **3.** Infinite.
4. Load current varies inversely with load resistance.

FIGURE 8–100

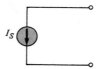

FIGURE 8–101

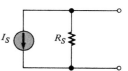

Section 8–3
1. $I_S = V_S/R_S$.   **2.** $V_S = I_SR_S$.   **3.** See Figure 8–102.   **4.** See Figure 8–103.

FIGURE 8–102

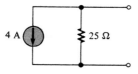

FIGURE 8–103

Section 8–4
1. The total current in any branch of a multiple-source linear circuit is equal to the algebraic sum of the currents due to the individual sources acting alone, with the other sources replaced by their internal resistances.
2. Because it allows each source to be treated independently.
3. A short simulates the internal resistance of an ideal voltage source; an open simulates the internal resistance of an ideal current source.
4. 6.67 mA.   **5.** In the direction of the larger current.

Section 8–5
1. V_{TH} and R_{TH}.   **2.** See Figure 8–104.
3. V_{TH} is the open circuit voltage between two terminals in a circuit.
4. R_{TH} is the resistance as viewed from two terminals in a circuit, with all sources replaced by their internal resistances.
5. See Figure 8–105.

FIGURE 8–104

FIGURE 8–105

Section 8–6

1. Any linear network can be replaced by an equivalent circuit consisting of an equivalent current source and an equivalent parallel resistance.
2. I_N and R_N.   **3.** See Figure 8–106.

FIGURE 8–106

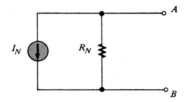

4. I_N is the short circuit current between two terminals in a circuit.
5. R_N is the resistance as viewed from the two open terminals in a circuit.
6. See Figure 8–107.

FIGURE 8–107

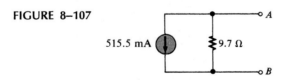

Section 8–7

1. Parallel voltage sources.   **2.** The equation is as follows:

$$R_{EQ} = \frac{1}{(1/R_1) + (1/R_2) + (1/R_3) + \cdots + (1/R_n)}$$

3. The equation is as follows:

$$V_{EQ} = \frac{(V_1/R_1) + (V_2/R_2) + (V_3/R_3) + \cdots + (V_n/R_n)}{(1/R_1) + (1/R_2) + (1/R_3) + \cdots + (1/R_n)}$$

4. $I_L = 1.08$ A; $V_L = 108$ V.

Section 8–8

1. Maximum power is transferred from a source to a load when the load resistance is equal to the source resistance.
2. When $R_L = R_S$.   **3.** 50 Ω.

Section 8–9
1. See Figure 8–108.   **2.** See Figure 8–109.

FIGURE 8–108 R_C

FIGURE 8–109

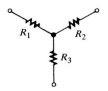

3. The equations are as follows:

$$R_1 = \frac{R_A R_C}{R_A + R_B + R_C}$$

$$R_2 = \frac{R_B R_C}{R_A + R_B + R_C}$$

$$R_3 = \frac{R_A R_B}{R_A + R_B + R_C}$$

4. The equations are as follows:

$$R_A = \frac{R_1 R_2 + R_1 R_3 + R_2 R_3}{R_2}$$

$$R_B = \frac{R_1 R_2 + R_1 R_3 + R_2 R_3}{R_1}$$

$$R_C = \frac{R_1 R_2 + R_1 R_3 + R_2 R_3}{R_3}$$

Section 8–10
1. Lines 140–170.   **2.** Double space.

NINE

BRANCH, MESH, AND NODE ANALYSIS

In this chapter, three circuit analysis methods are discussed. These methods are based on Ohm's law and Kirchhoff's voltage and current laws. The methods presented are particularly useful in the analysis of circuits with two or more voltage or current sources and can be used alone or in conjunction with the techniques covered in the previous chapters. Each of these methods can be applied to a given circuit analysis problem. With experience, you will learn which method is best for a particular problem, or you may develop a preference for one of them.

In this chapter, you will learn:

☐   How to identify loops and nodes in a circuit.
☐   How to develop a set of branch current equations and use them to solve for an unknown quantity.
☐   How to use determinants in the solution of simultaneous equations.
☐   How to develop a set of mesh (loop) equations for a given circuit and use them to solve for an unknown current.
☐   How to develop a set of node equations for a given circuit and use them to solve for an unknown voltage.
☐   How to use a computer program to find an unknown circuit voltage using the node voltage method.

9–1   BRANCH CURRENT METHOD

In the branch current method, we use Kirchhoff's voltage law and Kirchhoff's current law to solve for the current in each branch of a circuit. Once we know the currents, we can find the voltages.

LOOPS AND NODES

Figure 9–1 shows a circuit with two voltage sources and three branch currents. It will be used as the basic model throughout the chapter to illustrate each of the circuit analysis methods. In this circuit, there are two *closed loops*, as indicated by the arrows. A loop is a complete current path within a circuit. Also, there are four *nodes* in this circuit, as indicated by the letters *A*, *B*, *C*, and *D*. A node is a junction where two or more current paths come together.

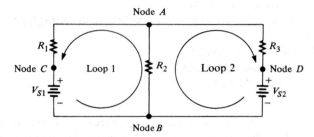

FIGURE 9–1
Basic multiple-source circuit showing loops and nodes.

The following are the general steps used in applying the *branch current method*. These steps are demonstrated with aid of Figure 9–2.

1. Assign a current in each circuit branch in an *arbitrary* direction.
2. Show the polarities of the resistor voltages according to the assigned branch current directions.
3. Apply Kirchhoff's voltage law around each closed loop (sum of voltages is equal to zero).
4. Apply Kirchhoff's current law at the *minimum* number of nodes so that *all* branch currents are included (sum of currents at a node equals zero).
5. Solve the equations resulting from Steps 3 and 4 for the branch current values.

First, the branch currents I_1, I_2, and I_3 are assigned in the direction shown. Do not worry about the *actual* current directions at this point.

Second, the polarities of the voltage drops across R_1, R_2, and R_3 are indicated in the figure according to the current directions.

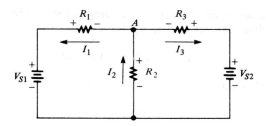

FIGURE 9–2
Circuit for demonstrating branch current analysis.

Third, Kirchhoff's voltage law applied to the two loops gives the following equations:

$$\text{Equation 1:} \quad R_1I_1 + R_2I_2 - V_{S1} = 0 \quad \text{for loop 1}$$

$$\text{Equation 2:} \quad R_2I_2 + R_3I_3 - V_{S2} = 0 \quad \text{for loop 2}$$

Fourth, Kirchhoff's current law is applied to node A, including all branch currents as follows:

$$\text{Equation 3:} \quad I_1 - I_2 + I_3 = 0$$

The negative sign indicates that I_2 is into the junction.

Fifth and last, the three equations must be solved for the three unknown currents, I_1, I_2, and I_3. The three equations in the above steps are called *simultaneous equations* and can be solved in two ways: by *substitution* or by *determinants*. Example 9–1 shows how to solve equations by substitution. In the next section, we study the use of determinants and apply them to the later methods.

EXAMPLE 9–1

Use the branch current method to find each branch current in Figure 9–3.

FIGURE 9–3

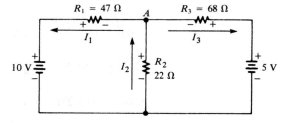

Solution:

Step 1: Assign branch currents as shown in Figure 9–3. Keep in mind that you can assume any current direction at this point, and the final solution will have a negative sign if the *actual* current is opposite to the assigned current.

Step 2:   Mark the polarities of the resistor voltage drops as shown in the figure.

Step 3:   Kirchhoff's voltage law around the *left* loop gives

$$47I_1 + 22I_2 - 10 = 0$$

Around the *right* loop we get

$$22I_2 + 68I_3 - 5 = 0$$

Step 4:   At node A, the current equation is

$$I_1 - I_2 + I_3 = 0$$

Step 5:   The equations are solved by substitution as follows. First find I_1 in terms of I_2 and I_3:

$$I_1 = I_2 - I_3$$

Now substitute $I_2 - I_3$ for I_1 in the left loop equation:

$$47(I_2 - I_3) + 22I_2 = 10$$
$$47I_2 - 47I_3 + 22I_2 = 10$$
$$69I_2 - 47I_3 = 10$$

Next, take the right loop equation and solve for I_2 in terms of I_3:

$$22I_2 = 5 - 68I_3$$
$$I_2 = \frac{5 - 68I_3}{22}$$

Substituting this expression for I_2 into $69I_2 - 47I_3 = 10$, we get the following:

$$69\left(\frac{5 - 68I_3}{22}\right) - 47I_3 = 10$$

$$\frac{345 - 4692I_3}{22} - 47I_3 = 10$$

$$15.68 - 213.27I_3 - 47I_3 = 10$$

$$-260.27I_3 = -5.68$$

$$I_3 = \frac{5.68}{260.27}$$

$$= 0.0218 \text{ A}$$

Now, substitute this value of I_3 into the right loop equation:

$$22I_2 + 68(0.0218) = 5$$

Solve for I_2:

$$I_2 = \frac{5 - 68(0.0218)}{22}$$

$$= \frac{3.52}{22} = 0.16 \text{ A}$$

Substituting I_2 and I_3 values into the current equation at node A, we obtain

$$I_1 - 0.16 + 0.0218 = 0$$

$$I_1 = 0.16 - 0.0218$$

$$= 0.138 \text{ A}$$

SECTION REVIEW 9–1

1. What basic circuit laws are used in the branch current method?

2. When assigning branch currents, you should be careful of the directions (T or F).

3. What is a loop?

4. What is a node?

9–2

THE DETERMINANT METHOD FOR SOLVING SIMULTANEOUS EQUATIONS

When there are several unknown quantities to be found, such as the three currents in the last example, *you must have a number of equations equal to the number of unknowns*. In this section you will learn how to solve for two and three unknowns using a systematic method known as *determinants*. This method is an alternate to the substitution method, which we used in the previous section. Refer to your algebra text for coverage of the theory of determinants.

SOLVING TWO SIMULTANEOUS EQUATIONS FOR TWO UNKNOWNS

To illustrate the method of *second-order determinants,* we will assume two loop equations as follows:

$$10I_1 + 5I_2 = 15$$

$$2I_1 + 4I_2 = 8$$

We want to find the value of I_1 and I_2. To do so, we form a *determinant* with the coefficients of the unknown currents. A *coefficient* is the number associated with an unknown. For example, 10 is the coefficient for I_1 in the first equation.

The first column in the determinant consists of the coefficients of I_1, and the second column consists of the coefficients of I_2. The resulting determinant appears as follows:

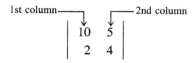

$$\begin{vmatrix} 10 & 5 \\ 2 & 4 \end{vmatrix}$$

This is called the *characteristic determinant* for the set of equations.

Next, we form another determinant and use it in conjunction with the characteristic determinant to solve for I_1. We form this determinant for our example by replacing the coefficients of I_1 in the characteristic determinant with the *constants* on the right side of the equations. Doing this, we get the following determinant:

$$\begin{vmatrix} 15 & 5 \\ 8 & 4 \end{vmatrix}$$ Replace coefficients of I_1 with constants from right sides of equations.

We can now solve for I_1 by *evaluating* both determinants and then dividing by the characteristic determinant. To evaluate the determinants, we *cross-multiply* and subtract the resulting products. An evaluation of the characteristic determinant in this example is illustrated in the following steps:

1. Multiply the first number in the left column by the second number in the right column.

$$\begin{vmatrix} 10 & 5 \\ 2 & 4 \end{vmatrix} = 10 \times 4 = 40$$

2. Multiply the second number in the left column by the first number in the right column and *subtract* from the product in Step 1. This result is the *value* of the determinant (30 in this case).

$$\begin{vmatrix} 10 & 5 \\ 2 & 4 \end{vmatrix} = 40 - (2 \times 5) = 40 - 10 = 30$$

3. Next, the same procedure is repeated for the other determinant that was set up for I_1.

$$\begin{vmatrix} 15 & 5 \\ 8 & 4 \end{vmatrix} = 15 \times 4 = 60$$

$$\begin{vmatrix} 15 & 5 \\ 8 & 4 \end{vmatrix} = 60 - (8 \times 5) = 60 - 40 = 20$$

The value of this determinant is 20. Now we can solve for I_1 by dividing the I_1 determinant by the characteristic determinant as follows:

$$I_1 = \frac{\begin{vmatrix} 15 & 5 \\ 8 & 4 \end{vmatrix}}{\begin{vmatrix} 10 & 5 \\ 2 & 4 \end{vmatrix}} = \frac{20}{30} = 0.667 \text{ A}$$

To find I_2, we form another determinant by substituting the *constants* on the right side of the equations for the coefficients of I_2:

$$\begin{vmatrix} 10 & 15 \\ 2 & 8 \end{vmatrix}$$

Replace coefficients of I_2 with constants from right sides of equations.

We solve for I_2 by dividing this determinant by the characteristic determinant already evaluated:

$$I_2 = \frac{\begin{vmatrix} 10 & 15 \\ 2 & 8 \end{vmatrix}}{30} = \frac{(10 \times 8) - (2 \times 15)}{30}$$

$$= \frac{80 - 30}{30} = \frac{50}{30} = 1.67 \text{ A}$$

EXAMPLE 9–2

Solve the following set of equations for the unknown currents:

$$2I_1 - 5I_2 = 10$$

$$6I_1 + 10I_2 = 20$$

Solution:

The characteristic determinant is

$$\begin{vmatrix} 2 & -5 \\ 6 & 10 \end{vmatrix} = (2)(10) - (-5)(6)$$

$$= 20 - (-30) = 20 + 30 = 50$$

Solving for I_1 yields

$$I_1 = \frac{\begin{vmatrix} 10 & -5 \\ 20 & 10 \end{vmatrix}}{50} = \frac{(10)(10) - (-5)(20)}{50}$$

$$= \frac{100 - (-100)}{50} = \frac{200}{50} = 4 \text{ A}$$

Solving for I_2 yields

$$I_1 = \frac{\begin{vmatrix} 2 & 10 \\ 6 & 20 \end{vmatrix}}{50} = \frac{(2)(20) - (6)(10)}{50}$$

$$= \frac{40 - 60}{50} = -0.4 \text{ A}$$

In a circuit problem, a result with a negative sign indicates that the direction of actual current is opposite to the assigned direction.

Note that multiplication can be expressed either by the multiplication sign such as 2×10 or by parentheses such as $(2)(10)$.

SOLVING THREE SIMULTANEOUS EQUATIONS FOR THREE UNKNOWNS

Third-order determinants can be evaluated in two ways: the *expansion method* and the *cofactor method*. First, we will illustrate the expansion method (which is good only for third order) using the three following equations:

$$1I_1 + 3I_2 - 2I_3 = 7$$

$$0I_1 + 4I_2 + 1I_3 = 8$$

$$-5I_1 + 1I_2 + 6I_3 = 9$$

The three-column characteristic determinant for this set of equations is formed in a similar way to that used earlier for the second-order determinant. The first column consists of the coefficients of I_1, the second column consists of the coefficients of I_2, and the third column consists of the coefficients of I_3, as shown below.

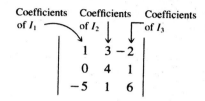

The Expansion Method This third-order determinant is evaluated by the expansion method as shown in the following steps.

1. Rewrite the first two columns immediately to the right of the determinant.

$$\begin{vmatrix} 1 & 3 & -2 \\ 0 & 4 & 1 \\ -5 & 1 & 6 \end{vmatrix} \begin{matrix} 1 & 3 \\ 0 & 4 \\ -5 & 1 \end{matrix}$$

2. Identify the three downward diagonal groups of three coefficients each.

$$\begin{vmatrix} 1 & 3 & -2 \\ 0 & 4 & 1 \\ -5 & 1 & 6 \end{vmatrix} \begin{matrix} 1 & 3 \\ 0 & 4 \\ -5 & 1 \end{matrix}$$

3. Multiply the numbers in each diagonal and add the products.

$$\begin{vmatrix} 1 & 3 & -2 \\ 0 & 4 & 1 \\ -5 & 1 & 6 \end{vmatrix} \begin{matrix} 1 & 3 \\ 0 & 4 \\ -5 & 1 \end{matrix}$$

$$(1)(4)(6) + (3)(1)(-5) + (-2)(0)(1) = 24 + (-15) + (0) = 9$$

4. Repeat Steps 2 and 3 for the three upward diagonal groups of three coefficients.

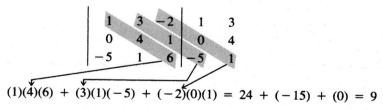

$$(-5)(4)(-2) + (1)(1)(1) + (6)(0)(3) = 40 + 1 + 0 = 41$$

5. Subtract the result in Step 4 from the result in Step 3 to get the value of the determinant.

$$9 - 41 = -32$$

To solve for I_1 in the given set of three equations, a determinant is formed by substituting the constants on the right of the equations for the coefficients of I_1 in the characteristic determinant.

$$\begin{vmatrix} 7 & 3 & -2 \\ 8 & 4 & 1 \\ 9 & 1 & 6 \end{vmatrix}$$

This determinant is evaluated using the method described in the previous steps.

$$\begin{vmatrix} 7 & 3 & -2 \\ 8 & 4 & 1 \\ 9 & 1 & 6 \end{vmatrix}\begin{matrix} 7 & 3 \\ 8 & 4 \\ 9 & 1 \end{matrix}$$

$$= [(7)(4)(6) + (3)(1)(9) + (-2)(8)(1)] - [(9)(4)(-2) + (1)(1)(7) + (6)(8)(3)]$$

$$= (168 + 27 - 16) - (-72 + 7 + 144)$$

$$= 179 - 79 = 100$$

I_1 is found by dividing this determinant by the characteristic determinant.

$$I_1 = \frac{\begin{vmatrix} 7 & 3 & -2 \\ 8 & 4 & 1 \\ 9 & 1 & 6 \end{vmatrix}}{\begin{vmatrix} 1 & 3 & -2 \\ 0 & 4 & 1 \\ -5 & 1 & 6 \end{vmatrix}} = \frac{100}{-32} = -3.125$$

I_2 and I_3 are found in a similar way.

EXAMPLE 9–3

Determine the value of I_2 from the following set of equations:

$$2I_1 + 0.5I_2 + 1I_3 = 0$$

$$0.75I_1 + 0I_2 + 2I_3 = 1.5$$

$$3I_1 + 0.2I_2 + 0I_3 = -1$$

Solution:

The characteristic determinant is evaluated as follows:

$$\begin{vmatrix} 2 & 0.5 & 1 \\ 0.75 & 0 & 2 \\ 3 & 0.2 & 0 \end{vmatrix} \begin{matrix} 2 & 0.5 \\ 0.75 & 0 \\ 3 & 0.2 \end{matrix}$$

$$= [(2)(0)(0) + (0.5)(2)(3) + (1)(0.75)(0.2)]$$
$$- [(3)(0)(1) + (0.2)(2)(2) + (0)(0.75)(0.5)]$$
$$= (0 + 3 + 0.15) - (0 + 0.8 + 0) = 3.15 - 0.8 = 2.35$$

The determinant for I_2 is evaluated as follows:

$$\begin{vmatrix} 2 & 0 & 1 \\ 0.75 & 1.5 & 2 \\ 3 & -1 & 0 \end{vmatrix} \begin{matrix} 2 & 0 \\ 0.75 & 1.5 \\ 3 & -1 \end{matrix}$$

$$= [(2)(1.5)(0) + (0)(2)(3) + (1)(0.75)(-1)]$$
$$- [(3)(1.5)(1) + (-1)(2)(2) + (0)(0.75)(0)]$$
$$= [0 + 0 + (-0.75)] - [4.5 + (-4) + 0]$$
$$= -0.75 - 0.5 = -1.25$$

Finally,

$$I_2 = \frac{-1.25}{2.35} = -0.53 \text{ A}$$

The Cofactor Method   Unlike the expansion method, the cofactor method can be used to evaluate determinants with higher orders than three and is therefore more versatile. We will use a third-order determinant to illustrate the method, keeping in mind that fourth-, fifth-, and higher-order determinants can be evaluated in a similar way. The following specific determinant is used to demonstrate the cofactor method on a step-by-step basis.

$$\begin{vmatrix} 1 & 3 & -2 \\ 4 & 0 & -1 \\ 5 & 1.5 & 6 \end{vmatrix}$$

1. Select any one column or row in the determinant. Each number in the selected column or row is used as a *multiplying factor*. For illustration, we will use the first column.

$$\begin{vmatrix} 1 & 3 & -2 \\ 4 & 0 & -1 \\ 5 & 1.5 & 6 \end{vmatrix}$$

2. Determine the *cofactor* for each number in the selected column (or row). The cofactor for a given number is the determinant formed by all numbers that are *not* in the same column or row as the given number. This is illustrated as follows.

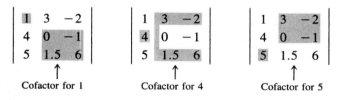

| Cofactor for 1 | Cofactor for 4 | Cofactor for 5 |

3. Assign the proper sign to each multiplying factor according to the following format (note the alternating pattern).

$$\begin{vmatrix} + & - & + \\ - & + & - \\ + & - & + \end{vmatrix}$$

4. Sum all of the products of each multiplying factor and its associated cofactor using the appropriate sign.

$$1 \times \begin{vmatrix} 0 & -1 \\ 1.5 & 6 \end{vmatrix} - 4 \times \begin{vmatrix} 3 & -2 \\ 1.5 & 6 \end{vmatrix} + 5 \times \begin{vmatrix} 3 & -2 \\ 0 & -1 \end{vmatrix}$$

$$= 1[(0)(6) - (1.5)(-1)] - 4[(3)(6) + (1.5)(-2)] + 5[(3)(-1) + (0)(-2)]$$

$$= 1(1.5) - 4(15) + 5(-3)$$

$$= 1.5 - 60 - 15 = -73.5$$

EXAMPLE 9–4

Repeat Example 9–3 using the cofactor method to find I_2. The equations are repeated below.

$$2I_1 + 0.5I_2 + 1I_3 = 0$$

$$0.75I_1 + 0I_2 + 2I_3 = 1.5$$

$$3I_1 + 0.2I_2 + 0I_3 = -1$$

Solution:

Evaluate the characteristic determinant as follows:

$$\begin{vmatrix} 2 & 0.5 & 1 \\ 0.75 & 0 & 2 \\ 3 & 0.2 & 0 \end{vmatrix} = 2\begin{vmatrix} 0 & 2 \\ 0.2 & 0 \end{vmatrix} - 0.75\begin{vmatrix} 0.5 & 1 \\ 0.2 & 0 \end{vmatrix} + 3\begin{vmatrix} 0.5 & 1 \\ 0 & 2 \end{vmatrix}$$

$$= 2[(0)(0) - (0.2)(2)] - 0.75[(0.5)(0) - (0.2)(1)] + 3[(0.5)(2) - (0)(1)]$$

$$= 2(-0.4) - 0.75(-0.2) + 3(1)$$

$$= -0.8 + 0.15 + 3$$

$$= 2.35$$

The determinant for I_2 is

$$\begin{vmatrix} 2 & 0 & 1 \\ 0.75 & 1.5 & 2 \\ 3 & -1 & 0 \end{vmatrix} = 2\begin{vmatrix} 1.5 & 2 \\ -1 & 0 \end{vmatrix} - 0.75\begin{vmatrix} 0 & 1 \\ -1 & 0 \end{vmatrix} + 3\begin{vmatrix} 0 & 1 \\ 1.5 & 2 \end{vmatrix}$$

$$= 2[(1.5)(0) - (-1)(2)] - 0.75[(0)(0) - (-1)(1)] + 3[(0)(2) - (1.5)(1)]$$

$$= 2(2) - 0.75(1) + 3(-1.5)$$

$$= 4 - 0.75 - 4.5$$

$$= -1.25$$

Then

$$I_2 = \frac{-1.25}{2.35} = -0.53 \text{ A}$$

SECTION REVIEW 9–2

1. Evaluate the following determinants:

(a) $\begin{vmatrix} 0 & -1 \\ 4 & 8 \end{vmatrix}$ **(b)** $\begin{vmatrix} 0.25 & 0.33 \\ -0.5 & 1 \end{vmatrix}$ **(c)** $\begin{vmatrix} 1 & 3 & 7 \\ 2 & -1 & 7 \\ -4 & 0 & -2 \end{vmatrix}$

2. Set up the characteristic determinant for the following set of simultaneous equations.

$$2I_1 + 3I_2 = 0$$
$$5I_1 + 4I_2 = 1$$

3. Find I_2 in Question 2 above.

9–3

MESH CURRENT METHOD

In the mesh current method, we will work with *loop currents* rather than branch currents. As you perhaps realize, a branch current is the *actual* current through a branch. An ammeter in that branch will measure the value of the branch current. Loop currents are mathematical quantities that are used to make circuit analysis somewhat easier than it is with the branch current method. Keep this in mind as we proceed through this section. The term *mesh* comes from the fact that a multiple-loop circuit, when drawn out, resembles a wire mesh.

A *systematic* method of mesh analysis is listed in the following steps and is illustrated in Figure 9–4, which is the same circuit configuration used in the branch current section. It demonstrates the basic principles well.

1. Assign a current in the *counterclockwise* (CCW) direction around each closed loop. This may not be the actual current direction, but it does not matter. The number of current assignments must be sufficient to include current through all components in the circuit. No *redundant* current assignments should be made. The direction does not have to be counterclockwise, but we will use CCW for consistency.
2. Indicate the voltage drop polarities in each loop based on the *assigned* current directions.
3. Apply Kirchhoff's voltage law around each closed loop. When more than one loop current passes through a component, include its voltage drop.
4. Using substitution or determinants, solve the resulting equations for the loop currents.

First, the loop currents I_1 and I_2 are assigned in the CCW direction as shown in the figure. A loop current could be assigned around the outer perimeter of the circuit,

FIGURE 9–4
Circuit for mesh analysis.

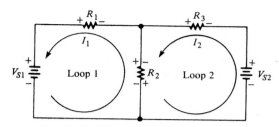

but this information would be redundant since I_1 and I_2 already pass through all of the components.

Second, the polarities of the voltage drops across R_1, R_2, and R_3 are shown based on the loop current directions. Notice that I_1 and I_2 are in opposite directions through R_2 because R_2 is common to both loops. Therefore, two different voltage polarities are indicated. In reality, R_2 currents cannot be separated into two parts, but remember that the loop currents are basically mathematical quantities used for analysis purposes. The polarities of the voltage sources are fixed and are not affected by the current assignments.

Third, Kirchhoff's voltage law applied to the two loops results in the following two equations:

$$R_1I_1 + R_2(I_1 - I_2) = V_{S1} \quad \text{for loop 1}$$

$$R_3I_2 + R_2(I_2 - I_1) = -V_{S2} \quad \text{for loop 2}$$

Fourth, the like terms in the equations are combined and rearranged for convenient solution. The equations are rearranged into the following form. Once the loop currents are evaluated, all of the branch currents can be determined.

$$(R_1 + R_2)I_1 - R_2I_2 = V_{S1} \quad \text{for loop 1}$$

$$-R_2I_1 + (R_2 + R_3)I_2 = -V_{S2} \quad \text{for loop 2}$$

Notice that only *two* equations are required for the same circuit that required *three* equations in the branch current method. The last two equations follow a certain form which can be used as a *format* to make mesh analysis easier. Referring to these last two equations, notice that for loop 1, the total resistance in the loop, $R_1 + R_2$, is multiplied by I_1 (its loop current). Also in the loop 1 equation, the resistance common to both loops, R_2, is multiplied by the other loop current, I_2, and subtracted from the first term. The same general form is seen in the loop 2 equation. From these observations, a set of rules can be stated for applying the same format repeatedly to each loop:

1. Sum the resistances around the loop, and multiply by the loop current.
2. Subtract the common resistance(s) times the adjacent loop current(s).
3. Set the terms in Steps 1 and 2 equal to the total source voltage in the loop. The sign of the source voltage is positive if the assigned loop current is *into* its positive terminal. The sign is negative if the loop current is *out of* its positive terminal.

Example 9–5 illustrates the application of these rules to the mesh current analysis of a circuit.

EXAMPLE
9–5

Using the mesh current method, find the branch currents in Figure 9–5, which is the same circuit as in Example 9–1.

FIGURE 9–5

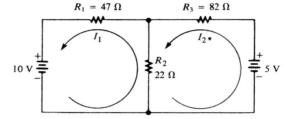

Solution:
The loop currents are assigned as shown. The format rules are followed for setting up the two equations.

$$(47 + 22)I_1 - 22I_2 = 10 \quad \text{for loop 1}$$

$$69I_1 - 22I_2 = 10$$

$$-22I_1 + (22 + 82)I_2 = -5 \quad \text{for loop 2}$$

$$-22I_1 + 104I_2 = -5$$

Using determinants to find I_1, we obtain

$$I_1 = \frac{\begin{vmatrix} 10 & -22 \\ -5 & 104 \end{vmatrix}}{\begin{vmatrix} 69 & -22 \\ -22 & 104 \end{vmatrix}} = \frac{(10)(104) - (-5)(-22)}{(69)(104) - (-22)(-22)}$$

$$= \frac{1040 - 110}{7176 - 484} = 0.139 \text{ A}$$

Solving for I_2 yields

$$I_2 = \frac{\begin{vmatrix} 69 & 10 \\ -22 & -5 \end{vmatrix}}{6692} = \frac{(69)(-5) - (-22)(10)}{6692}$$

$$= \frac{-345 - (-220)}{6692} = -0.019 \text{ A}$$

The negative sign on I_2 means that its direction must be reversed.

Now we find the actual *branch* currents. Since I_1 is the *only* current through R_1, it is also the branch current I_{R1}.

$$I_{R1} = I_1 = 0.139 \text{ A}$$

Since I_2 is the *only* current through R_3, it is also the branch current I_{R3}:

$$I_{R3} = I_2 = -0.019 \text{ A} \quad \text{(opposite direction of that originally assigned to } I_2)$$

Both loop currents I_1 and I_2 go through R_2 in the *same* direction. Remember, the negative I_2 value told us to reverse its assigned direction.

$$I_{R2} = I_1 - I_2 = 0.139 \text{ A} - (-0.019 \text{ A})$$
$$= 0.158 \text{ A}$$

Keep in mind that once we know the branch currents, we can find the voltages by using Ohm's law.

CIRCUITS WITH MORE THAN TWO LOOPS

The mesh method also can be systematically applied to circuits with any number of loops. Of course, the more loops there are, the more difficult is the solution. However, the basic rules still apply. For example, for a three-loop circuit, three simultaneous equations are required. Example 9–6 illustrates the analysis of a three-loop circuit.

EXAMPLE 9–6

Find I_3 in Figure 9–6.

FIGURE 9–6

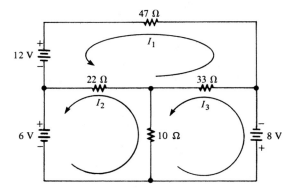

Solution:

Assign three CCW loop currents as shown in the figure. Then use the format rules to write the loop equations. A concise statement of these rules is as follows:

(Sum of resistors in loop) times (loop current) minus (each common resistor) times (associated adjacent loop current) equals (source voltage in the loop). The polarity of a voltage source is positive when the assigned mesh current enters the positive terminal.

$$102I_1 - 22I_2 - 33I_3 = 12 \quad \text{for loop 1}$$
$$-22I_1 + 32I_2 - 10I_3 = 6 \quad \text{for loop 2}$$
$$-33I_1 - 10I_2 + 43I_3 - 8 \quad \text{for loop 3}$$

These three equations can be solved for the currents by substitution or, more easily, with *third-order determinants*. I_3 is found using determinants as follows. The characteristic determinant is

$$\begin{vmatrix} 102 & -22 & -33 \\ -22 & 32 & -10 \\ -33 & -10 & 43 \end{vmatrix} = 102 \begin{vmatrix} 32 & -10 \\ -10 & 43 \end{vmatrix} - (-22) \begin{vmatrix} -22 & -33 \\ -10 & 43 \end{vmatrix} + (-33) \begin{vmatrix} -22 & -33 \\ 32 & -10 \end{vmatrix}$$

$$= 102[(32)(43) - (-10)(-10)] - (-22)[(-22)(43) - (-10)(-33)]$$
$$+ (-33)[(-22)(-10) - (32)(-33)]$$
$$= 102(1276) + 22(-1276) - 33(1276)$$
$$= 130,152 - 28,072 - 42,108$$
$$= 59,972$$

The I_3 determinant is

$$\begin{vmatrix} 102 & -22 & 12 \\ -22 & 32 & 6 \\ -33 & -10 & 8 \end{vmatrix} = 102 \begin{vmatrix} 32 & 6 \\ -10 & 8 \end{vmatrix} - (-22) \begin{vmatrix} -22 & 12 \\ -10 & 8 \end{vmatrix} + (-33) \begin{vmatrix} -22 & 12 \\ 32 & 6 \end{vmatrix}$$

$$= 102[(32)(8) - (-10)(6)] + 22[(-22)(8) - (-10)(12)]$$
$$- 33[(-22)(6) - (32)(12)]$$
$$= 102(316) + 22(-56) - 33(-516)$$
$$= 32,232 - 1232 + 17,028$$
$$= 48,028$$

I_3 is determined by dividing the value of the I_3 determinant by the value of the characteristic determinant.

$$I_3 = \frac{48{,}028}{59{,}972} = 0.806 \text{ A}$$

The other loop currents are found similarly.

SECTION REVIEW 9–3

1. Do the loop currents necessarily represent the actual currents in the branches?
2. When you solve for a loop current and get a negative value, what does it mean?
3. What circuit law is used in the mesh current method?

9–4   NODE VOLTAGE METHOD

Another alternate method of analysis of multiple-source circuits is called the *node voltage method*. It is based on finding the voltages at each node in the circuit using *Kirchhoff's current law*. Remember that a *node* is the junction of two or more current paths.

The general steps for this method are as follows:

1. Determine the number of nodes.
2. Select one node as a *reference*. All voltages will be relative to the reference node. Assign voltage designations to each node where the voltage is unknown.
3. Assign currents at each node where the voltage is unknown, except at the reference node. The directions are arbitrary.
4. Apply Kirchhoff's current law to each node where currents are assigned.
5. Express the current equations in terms of voltages, and solve the equations for the unknown node voltages.

We will use Figure 9–7 to illustrate the general approach to node voltage analysis. First, establish the nodes. In this case there are *four*, as indicated in the figure. Second, let us use node *B* as reference. Think of it as circuit ground. Node voltages *C* and *D* are already known to be the source voltages. The voltage at *node A* is the only unknown in this case. It is designated as V_A. Third, arbitrarily assign the currents at node *A* as indicated in the figure. Fourth, the Kirchhoff current equation at node *A* is

$$I_1 - I_2 + I_3 = 0$$

FIGURE 9–7
Circuit for node voltage analysis.

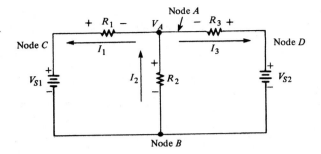

Fifth, express the currents in terms of circuit voltages using Ohm's law as follows:

$$I_1 = \frac{V_1}{R_1} = \frac{V_{S1} - V_A}{R_1}$$

$$I_2 = \frac{V_2}{R_2} = \frac{V_A}{R_2}$$

$$I_3 = \frac{V_3}{R_3} = \frac{V_{S2} - V_A}{R_3}$$

Substituting these into the current equation, we get

$$\frac{V_{S1} - V_A}{R_1} - \frac{V_A}{R_2} + \frac{V_{S2} - V_A}{R_3} = 0$$

The only unknown is V_A; so we can solve the single equation by combining and rearranging terms. Once the voltage is known, all branch currents can be calculated. Example 9–7 illustrates this method further.

EXAMPLE 9–7

Find the node voltages in Figure 9–8.

FIGURE 9–8

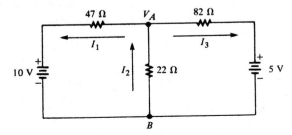

Solution:
The reference node is chosen at B. The unknown node voltage is V_A, as indicated in the figure. This is the only unknown voltage. Currents are assigned at node A as

shown. The current equation is

$$I_1 - I_2 + I_3 = 0$$

Substitution for currents using Ohm's law gives the equation in terms of voltages:

$$\frac{10 - V_A}{47} - \frac{V_A}{22} + \frac{5 - V_A}{82} = 0$$

Solving for V_A yields

$$\frac{10}{47} - \frac{V_A}{47} - \frac{V_A}{22} + \frac{5}{82} - \frac{V_A}{82} = 0$$

$$-\frac{V_A}{47} - \frac{V_A}{22} - \frac{V_A}{82} = -\frac{10}{47} - \frac{5}{82}$$

$$\frac{1804V_A + 3854V_A + 1034V_A}{84,788} = \frac{820 + 235}{3854}$$

$$\frac{6692V_A}{84,788} = \frac{1055}{3854}$$

$$V_A = \frac{(1055)(84,788)}{(6692)(3854)}$$

$$= 3.47 \text{ V}$$

Using the same basic procedure, we can solve circuits with more than one unknown node voltage. Example 9–8 illustrates this calculation for two unknown node voltages.

EXAMPLE 9–8

Using the node analysis method, solve for V_1 and V_2 in the circuit of Figure 9–9.

FIGURE 9–9

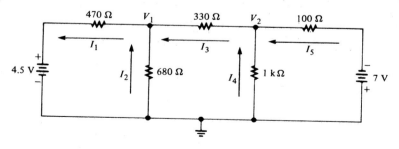

Solution:
First, the branch currents are assigned as shown in the diagram. Next, Kirchhoff's current law is applied at each node. At node 1,

$$I_1 - I_2 - I_3 = 0$$

Using Ohm's law substitution for the currents, we get

$$\left(\frac{4.5 - V_1}{470}\right) - \left(\frac{V_1}{680}\right) - \left(\frac{V_1 - V_2}{330}\right) = 0$$

$$\frac{4.5}{470} - \frac{V_1}{470} - \frac{V_1}{680} - \frac{V_1}{330} + \frac{V_2}{330} = 0$$

$$\left(\frac{1}{470} + \frac{1}{680} + \frac{1}{330}\right)V_1 - \left(\frac{1}{330}\right)V_2 = \frac{4.5}{470}$$

Now, using the $\boxed{1/x}$ key of the calculator, evaluate the coefficients and constant. The resulting equation for node 1 is

$$0.00663V_1 - 0.00303V_2 = 0.00957$$

At node 2,

$$I_3 - I_4 - I_5 = 0$$

Again using Ohm's law substitution, we get

$$\left(\frac{V_1 - V_2}{330}\right) - \left(\frac{V_2}{1000}\right) - \left(\frac{V_2 - (-7)}{100}\right) = 0$$

$$\frac{V_1}{330} - \frac{V_2}{330} - \frac{V_2}{1000} - \frac{V_2}{100} - \frac{7}{100} = 0$$

$$\left(\frac{1}{330}\right)V_1 - \left(\frac{1}{330} + \frac{1}{1000} + \frac{1}{100}\right)V_2 = \frac{7}{100}$$

Evaluating the coefficients and constant, we obtain the equation for node 2:

$$0.00303V_1 - 0.01403V_2 = 0.07$$

Now, these two node equations must be solved for V_1 and V_2. Using determinants, we get the following solutions:

$$V_1 = \frac{\begin{vmatrix} 0.00957 & -0.00303 \\ 0.07 & -0.01403 \end{vmatrix}}{\begin{vmatrix} 0.00663 & -0.00303 \\ 0.00303 & -0.01403 \end{vmatrix}}$$

$$= \frac{(0.00957)(-0.01403) - (0.07)(-0.00303)}{(0.00663)(-0.01403) - (0.00303)(-0.00303)}$$

$$= -0.928 \text{ V}$$

$$V_2 = \frac{\begin{vmatrix} 0.00663 & 0.00957 \\ 0.00303 & 0.07 \end{vmatrix}}{\begin{vmatrix} 0.00663 & -0.00303 \\ 0.00303 & -0.01403 \end{vmatrix}}$$

$$= \frac{(0.00663)(0.07) - (0.00303)(0.00957)}{(0.00663)(-0.01403) - (0.00303)(-0.00303)}$$

$$= -5.19 \text{ V}$$

SECTION REVIEW 9–4

1. What circuit law is the basis for the node voltage method?

2. What is the reference node?

9–5

COMPUTER ANALYSIS

The program listed below provides for the analysis of a circuit like that in Figure 9–7 for one unknown node voltage.

```
10   CLS
20   PRINT "THIS PROGRAM COMPUTES THE UNKNOWN NODE VOLTAGE"
30   PRINT "(NODE A) FOR A CIRCUIT OF THE GENERAL FORM OF"
40   PRINT "FIGURE 9-7."
50   FOR T=1 TO 3500:NEXT:CLS
60   INPUT "VS1 IN VOLTS";V1
70   INPUT "VS2 IN VOLTS";V2
80   INPUT "R1 IN OHMS";R1
90   INPUT "R2 IN OHMS";R2
100  INPUT "R3 IN OHMS";R3
110  CLS
120  VA=((V1/R1)+(V2/R3))/((1/R1)+(1/R2)+(1/R3))
130  PRINT "THE VOLTAGE AT NODE A IS";VA;"VOLTS"
```

SECTION REVIEW 9–5

1. What is the purpose of line 50?

2. If there were a fourth resistor in the circuit in series with R_3 in Figure 9–7, how would you modify the program?

SUMMARY

1. A loop is a closed current path in a circuit.
2. A node is the junction of two or more current paths.
3. The branch current method is based on Kirchhoff's voltage law and Kirchhoff's current law.
4. A branch current is an *actual* current in a branch.
5. Simultaneous equations can be solved by substitution or by determinants.
6. The number of equations must be equal to the number of unknowns.
7. Second-order determinants are evaluated by adding the signed cross-products.
8. Third-order determinants are evaluated by the expansion method or by the cofactor method.
9. The mesh current method is based on Kirchhoff's voltage law.
10. A loop current is not necessarily the actual current in a branch.
11. The node voltage method is based on Kirchhoff's current law.

SELF-TEST

Solutions appear at the end of the book.

1. How many loops and nodes are there in the circuit of Figure 9–10?
2. Use the branch current method to find the currents in Figure 9–11.

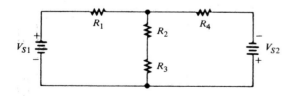

FIGURE 9–10

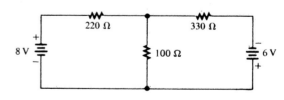

FIGURE 9–11

3. Solve the following two equations by substitution:

$$2I_1 + 5I_2 = 15$$
$$6I_1 - 8I_2 = 10$$

4. Using determinants, solve the following two equations for I_1:

$$15I_1 - 12I_2 = 25$$
$$9I_1 + 3I_2 = 18$$

5. What is the limitation of the expansion method for evaluating determinants?

6. What order of determinant is required for four simultaneous equations, and what method must be used for evaluation?

7. Use the mesh current method to find the current through R_3 in Figure 9–12.

8. Use the node voltage method to find the voltage at node A in Figure 9–13.

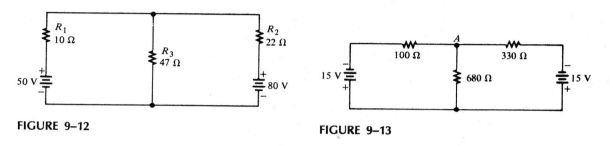

FIGURE 9–12   FIGURE 9–13

PROBLEMS

Section 9–1

9–1   Identify all *possible* loops in Figure 9–14.

FIGURE 9–14

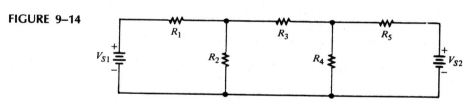

9–2   Identify all nodes in Figure 9–14. Which ones have a *known* voltage?

9–3   Write the Kirchhoff current equation for the current assignment shown at node A in Figure 9–15.

FIGURE 9–15

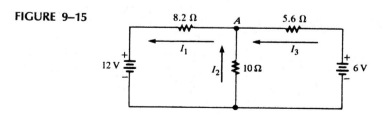

9–4   Solve for each of the branch currents in Figure 9–15.

9–5   Find the voltage drop across each resistor in Figure 9–15 and indicate its actual polarity.

9–6 Using the substitution method, solve the following set of equations for I_1 and I_2:

$$100I_1 + 50I_2 = 30$$
$$75I_1 + 90I_2 = 15$$

9–7 Using the substitution method, solve the following set of three equations for all currents:

$$5I_1 - 2I_2 + 8I_3 = 1$$
$$2I_1 + 4I_2 - 12I_3 = 5$$
$$10I_1 + 6I_2 + 9I_3 = 0$$

9–8 Find the current through each resistor in Figure 9–16.

9–9 In Figure 9–16, determine the voltage across the current source (points A to B).

FIGURE 9–16

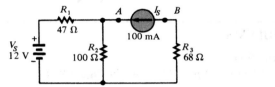

Section 9–2

9–10 Evaluate each determinant:

(a) $\begin{vmatrix} 4 & 6 \\ 2 & 3 \end{vmatrix}$ **(b)** $\begin{vmatrix} 9 & -1 \\ 0 & 5 \end{vmatrix}$ **(c)** $\begin{vmatrix} 12 & 15 \\ -2 & -1 \end{vmatrix}$ **(d)** $\begin{vmatrix} 100 & 50 \\ 30 & -20 \end{vmatrix}$

9–11 Using determinants, solve the following set of equations for both currents:

$$-I_1 + 2I_2 = 4$$
$$7I_1 + 3I_2 = 6$$

9–12 Evaluate each of the determinants using the expansion method:

(a) $\begin{vmatrix} 1 & 0 & -2 \\ 5 & 4 & 1 \\ 2 & 10 & 0 \end{vmatrix}$ **(b)** $\begin{vmatrix} 0.5 & 1 & -0.8 \\ 0.1 & 1.2 & 1.5 \\ -0.1 & -0.3 & 5 \end{vmatrix}$

9–13 Evaluate each of the determinants using the cofactor method:

(a) $\begin{vmatrix} 25 & 0 & -20 \\ 10 & 12 & 5 \\ -8 & 30 & -16 \end{vmatrix}$ (b) $\begin{vmatrix} 1.08 & 1.75 & 0.55 \\ 0 & 2.12 & -0.98 \\ 1 & 3.49 & -1.05 \end{vmatrix}$

9–14 Find I_1 and I_3 in Example 9–3.

9–15 Solve for I_1, I_2, and I_3 in the following set of equations:

$$2I_1 - 6I_2 + 10I_3 = 9$$
$$3I_1 + 7I_2 - 8I_3 = 3$$
$$10I_1 + 5I_2 - 12I_3 = 0$$

9–16 Find V_1, V_2, V_3, and V_4 from the following set of equations:

$$16V_1 + 10V_2 - 8V_3 - 3V_4 = 15$$
$$2V_1 + 0V_2 + 5V_3 + 2V_4 = 0$$
$$-7V_1 - 12V_2 + 0V_3 + 0V_4 = 9$$
$$-1V_1 + 20V_2 - 18V_3 + 0V_4 = 10$$

Section 9–3

9–17 Using the mesh current method, find the loop currents in Figure 9–17.

FIGURE 9–17

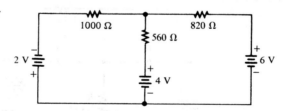

9–18 Find the branch currents in Figure 9–17.

9–19 Determine the voltages and their proper polarities for each resistor in Figure 9–17.

9–20 Write the loop equations for the circuit in Figure 9–18.

9–21 Solve for the loop currents in Figure 9–18.

9–22 Find the current through each resistor in Figure 9–18.

FIGURE 9–18

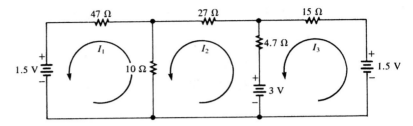

9–23 Determine the voltage across the open bridge terminals, *AB*, in Figure 9–19.

FIGURE 9–19

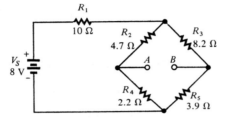

9–24 When a 10-Ω resistor is connected from point *A* to point *B* in Figure 9–19, what is the current through it?

9–25 Find the current through R_1 in Figure 9–20.

FIGURE 9–20

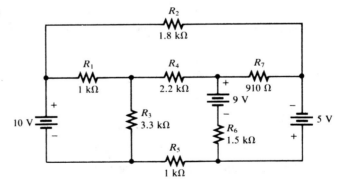

Section 9–4

9–26 In Figure 9–21, use the node voltage method to find the voltage at point *A* with respect to point *B*.

FIGURE 9-21

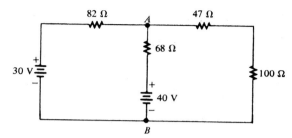

9-27 What are the branch current values in Figure 9-21? Show the actual direction of current in each branch.

9-28 Write the node voltage equations for Figure 9-18.

9-29 Use node analysis to determine the voltage at points *A* and *B* with respect to ground in Figure 9-22.

FIGURE 9-22

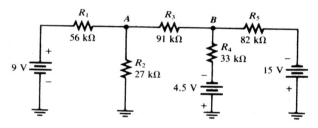

9-30 Find the voltage at points *A*, *B*, and *C* in Figure 9-23.

FIGURE 9-23

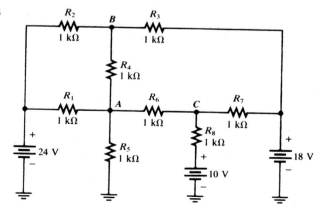

9–31 Use node analysis, mesh analysis, or any other procedure to find all currents and the voltages at each node in Figure 9–24.

FIGURE 9–24

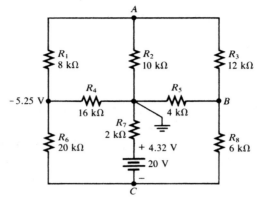

Section 9–5

9–32 Develop a flowchart for the program in Section 9–5.

9–33 Write a program to evaluate any third-order determinant.

ANSWERS TO SECTION REVIEWS

Section 9–1

1. Kirchhoff's voltage law and Kirchhoff's current law.

2. False, but write the equations so that they are consistent with your assigned directions.

3. A closed path within a circuit.   **4.** A junction of two or more current paths.

Section 9–2

1. (a) 4;   (b) 0.415;   (c) −98.

2. $\begin{vmatrix} 2 & 3 \\ 5 & 4 \end{vmatrix}$

3. −0.286 A

Section 9–3

1. No.   **2.** The direction should be reversed.   **3.** Kirchhoff's voltage law.

Section 9–4

1. Kirchhoff's current law.

2. The junction to which all circuit voltages are referenced.

Section 9–5

1. Delay before clearing screen.

2. 95   INPUT "R4 IN OHMS";R4
    120 VA=((V1/R1)+(V2/(R3+R4)))/((1/R1)+(1/R2)+(1/(R3+R4)))

TEN

MAGNETISM AND ELECTROMAGNETISM

This chapter is somewhat of a departure from the previous nine chapters because two new concepts are introduced: magnetism and electromagnetism. The operation of many types of electrical devices is based partially on magnetic or electromagnetic principles.

In this chapter, you will learn:

☐   The principles of magnetic fields.
☐   How materials are magnetized.
☐   How a magnetic switch operates.
☐   How magnetic switches can be applied in alarm systems.
☐   The principles of electromagnetism.
☐   The important characteristics of electromagnetic fields.
☐   How electromagnetism is used to record on magnetic tape.
☐   The principles of operation of solenoids, relays, speakers, and dc generators.
☐   How voltage is created by a conductor moving through a magnetic field.
☐   The principles of Faraday's law and Lenz's law.

10–1   THE MAGNETIC FIELD

A permanent magnet, such as the bar magnet shown in Figure 10–1, has a magnetic field surrounding it. The magnetic field consists of *lines of force* that radiate from the north pole (N) to the south pole (S) and back to the north pole through the magnetic material. For clarity, only a few lines of force are shown in the figure. Imagine, however, that many lines surround the magnet in three dimensions.

FIGURE 10–1
Magnetic lines of force around a bar magnet.

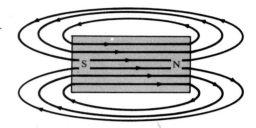

ATTRACTION AND REPULSION OF MAGNETIC POLES

When *unlike* poles of two permanent magnets are placed close together, an attractive force is produced by the magnetic fields, as indicated in Figure 10–2(a). When two *like* poles are brought close together, they repel each other, as shown in Part (b).

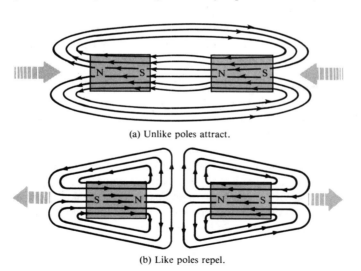

(a) Unlike poles attract.

(b) Like poles repel.

FIGURE 10–2
Magnetic attraction and repulsion.

ALTERING A MAGNETIC FIELD

When a nonmagnetic material, such as paper, glass, wood, or plastic, is placed in a magnetic field, the lines of force are unaltered, as shown in Figure 10–3(a). However, when a *magnetic* material such as iron is placed in the magnetic field, the lines of force tend to change course and pass through the iron rather than through the surrounding air. They do so because the iron provides a magnetic path that is more easily established than that of air. Figure 10–3(b) illustrates this principle.

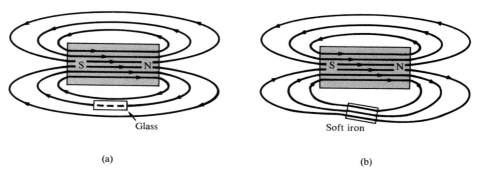

(a)   (b)

FIGURE 10–3
Effect of (a) nonmagnetic and (b) magnetic materials on a magnetic field.

MAGNETIC FLUX (φ)

The group of force lines going from the north pole to the south pole of a magnet is called the *magnetic flux,* symbolized by φ (the lowercase Greek letter phi). The number of lines of force in a magnetic field determines the value of the flux. The more lines of force, the greater the flux and the stronger the magnetic field.

The unit of magnetic flux is the *weber* (Wb). One weber equals 10^8 lines. In most practical situations, the weber is a very large unit; thus, the microweber (μWb) is more common. One microweber equals 100 lines of magnetic flux.

MAGNETIC FLUX DENSITY (B)

The *flux density* is the amount of flux per unit area in the magnetic field. Its symbol is B, and its unit is the tesla (T). One tesla equals one weber per square meter (Wb/m²). The following formula expresses the flux density:

$$B = \frac{\phi}{A}$$

(10–1)

where φ is the flux and A is the cross-sectional area of the magnetic field.

EXAMPLE
10–1

Find the flux density in a magnetic field in which the flux in 0.1 square meter is 800 μWb.

Solution:

$$B = \frac{\phi}{A} = \frac{800\ \mu\text{Wb}}{0.1\ \text{m}^2} = 8000 \times 10^{-6}\ \text{T}$$

EXAMPLE
10–2

If the flux density in a certain magnetic material is 2.3 T and the area of the material is 0.38 in^2, what is the flux through the material?

Solution:
There are 39.37 inches in one meter; therefore

$$A = 0.38\ \text{in}^2[1\ \text{m}^2/(39.37\ \text{in})^2] = 245.16 \times 10^{-6}\ \text{m}^2$$
$$\phi = BA = (2.3\ \text{T})(245.16 \times 10^{-6}\ \text{m}^2) = 5.64 \times 10^{-4}\ \text{Wb}$$

The Gauss Although the T is the SI unit for flux density, another unit called the *gauss,* from the CGS system, is sometimes used (10^4 gauss = 1 T). In fact, the instrument used to measure flux density is the gaussmeter.

HOW MATERIALS BECOME MAGNETIZED

Ferromagnetic materials such as iron, nickel, and cobalt become magnetized when placed in the magnetic field of a magnet. We have all seen a permanent magnet pick up paper clips, nails, iron filings, and so on. In these cases, the object becomes magnetized (that is, it actually becomes a magnet itself) under the influence of the permanent magnetic field and becomes attracted to the magnet. When removed from the magnetic field, the object tends to lose its magnetism.

Ferromagnetic materials have minute *magnetic domains* created within their atomic structure. These domains can be viewed as very small bar magnets with north and south poles. When the material is not exposed to an external magnetic field, the magnetic domains are randomly oriented, as shown in Figure 10–4(a). When the material is placed in a magnetic field, the domains align themselves as shown in Part (b). Thus, the object itself effectively becomes a magnet.

AN APPLICATION

Permanent magnets have almost endless applications, one of which is presented here as an illustration. Figure 10–5 shows a typical magnetically operated, normally closed (NC) switch. When the magnet is near the switch mechanism, the metallic arm is held in its NC position. When the magnet is moved away, the spring pulls the arm up, breaking the contact as shown in Figure 10–6.

Switches of this type are commonly used in perimeter alarm systems to detect entry into a building through windows or doors. As Figure 10–7 shows, several openings can be protected by magnetic switches wired to a common transmitter. When any one of the switches opens, the transmitter is activated and sends a signal to a central receiver and alarm unit.

(a) The magnetic domains (N S) are randomly oriented in the unmagnetized material.

N   S

(b) The magnetic domains become aligned when the material is magnetized.

FIGURE 10–4
Magnetic domains in (a) an unmagnetized and (b) a magnetized material.

FIGURE 10–5
Magnet and switch set (courtesy of Tandy Corp.).

FIGURE 10–6
Operation of a magnetic switch.

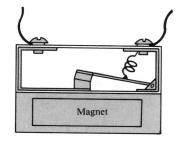

(a) Contact is closed.

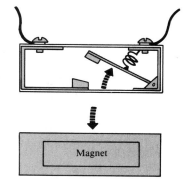

(b) Contact opens.

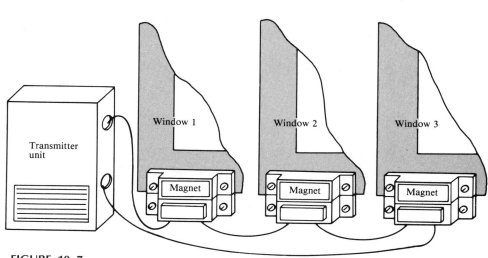

FIGURE 10–7
Connection of a typical perimeter alarm system.

SECTION REVIEW 10–1

1. When the north poles of two magnets are placed close together, do they repel or attract each other?

2. What is magnetic flux?

3. What is the flux density when the flux is 4.5×10^{-6} Wb and the area is 0.5×10^{-2} m^2?

10–2   ELECTROMAGNETISM

Current produces a magnetic field, called an *electromagnetic field,* around a conductor, as illustrated in Figure 10–8. The invisible lines of force of the magnetic field form a concentric circular pattern around the conductor and are continuous along its length.

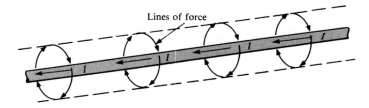

FIGURE 10–8
Magnetic field around a current-carrying conductor.

Although the magnetic field cannot be seen, it is capable of producing visible effects. For example, if a current-carrying wire is inserted through a sheet of paper in a perpendicular direction, iron filings placed on the surface of the paper arrange themselves along the magnetic lines of force in concentric rings, as illustrated in Figure 10–9(a). Part (b) of the figure illustrates that the north pole of a compass placed in the

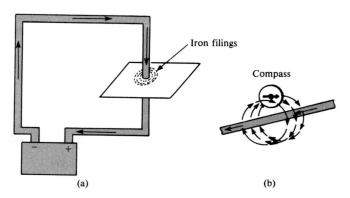

FIGURE 10–9
Visible effects of an electromagnetic field.

electromagnetic field will point in the direction of the lines of force. The field is stronger closer to the conductor and becomes weaker with increasing distance from the conductor.

DIRECTION OF THE LINES OF FORCE

The direction of the magnetic lines of force surrounding the conductor is indicated in Figure 10–10. When the direction of current is right to left, as in Part (a), the magnetic lines are in a *clockwise* direction. When current is left to right, as in Part (b), the lines are in a *counterclockwise* direction.

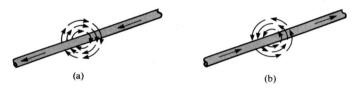

(a)   (b)

FIGURE 10–10
Magnetic lines of force around a current-carrying conductor.

LEFT-HAND RULE

An aid to remembering the direction of the lines of force is illustrated in Figure 10–11. Imagine that you are grasping the conductor with your left hand, with your thumb pointing in the direction of current. Your fingers indicate the direction of the magnetic lines of force.

FIGURE 10–11
Illustration of left-hand rule.

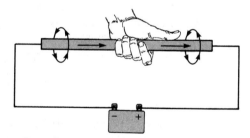

ELECTROMAGNETIC PROPERTIES

Several important properties relating to electromagnetic fields are now presented.

Permeability (μ)   The ease with which a magnetic field can be established in a given material is measured by the *permeability* of that material. The higher the permeability, the more easily a magnetic field can be established.

The symbol of permeability is μ, and its value varies depending on the type of material. The permeability of a vacuum (μ_0) is $4\pi \times 10^{-7}$ weber/ampere-turn·meter

(Wb/At·m) and is used as a reference. Ferromagnetic materials typically have permeabilities hundreds of times larger than that of a vacuum, indicating that a magnetic field can be set up with relative ease in these materials. Ferromagnetic materials include iron, steel, nickel, colbalt, and their alloys.

The *relative permeability* (μ_r) of a material is the ratio of its absolute permeability to the permeability of a vacuum.

$$\mu_r = \frac{\mu}{\mu_0} \tag{10-2}$$

Reluctance ($\mathcal{R}$)   Reluctance ($\mathcal{R}$) is the opposition to the establishment of a magnetic field in a material. The value of reluctance is directly proportional to the length (l) of the magnetic path, and inversely proportional to the permeability (μ) and to the cross-sectional area (A) of the material as expressed by the following equation.

$$\mathcal{R} = \frac{l}{\mu A} \tag{10-3}$$

Reluctance in magnetic circuits is analogous to resistance in electric circuits. The unit of reluctance can be derived using l in meters, A (area) in square meters, and μ in Wb/At·m as follows.

$$\mathcal{R} = \frac{l}{\mu A} = \frac{\text{m}}{(\text{Wb/At·m})(\text{m}^2)} = \frac{\text{At}}{\text{Wb}}$$

At/Wb is ampere-turns/weber.

**EXAMPLE
10–3**

What is the reluctance of a material that has a length of 0.05 m, a cross-sectional area of 0.012 m², and a permeability of 3500 × 10⁻⁶ Wb/At·m?

Solution:
$$\mathcal{R} = \frac{l}{\mu A} = \frac{0.05 \text{ m}}{(3500 \times 10^{-6} \text{ Wb/At·m})(0.012 \text{ m}^2)}$$
$$= 1190.48 \text{ At/Wb}$$

Magnetomotive Force (mmf)   As you have learned, current in a conductor produces a magnetic field. The force that produces the magnetic field is called the *magnetomotive force* (mmf). The unit of mmf, the ampere-turn (At), is established on the basis of the current in a single loop (turn) of wire. The formula for mmf is as follows:

$$F_m = NI \tag{10-4}$$

where F_m is the magnetomotive force, N is the number of turns of wire, and I is the current.

Figure 10–12 illustrates that a number of turns of wire carrying a current around a magnetic material create a force that sets up flux lines through the magnetic path. The amount of flux depends on the magnitude of the mmf and on the reluctance of the material, as expressed by the following equation:

$$\phi = \frac{F_m}{\mathcal{R}}$$

(10–5)

This is known as the *Ohm's law for magnetic circuits* because the flux (ϕ) is analogous to current, the mmf (F_m) is analogous to voltage, and the reluctance ($\mathcal{R}$) is analogous to resistance.

FIGURE 10–12
A basic magnetic circuit.

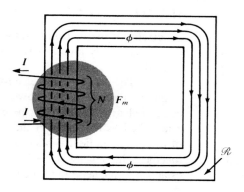

EXAMPLE 10–4

How much flux is established in the magnetic path of Figure 10–13 if the reluctance of the material is 0.28×10^5 At/Wb?

FIGURE 10–13

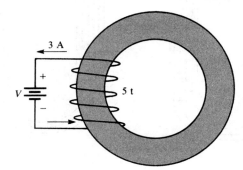

Solution:

$$\phi = \frac{F_m}{\mathcal{R}} = \frac{NI}{\mathcal{R}} = \frac{(5)(3A)}{0.28 \times 10^5 \text{ At/Wb}} = 5.36 \times 10^{-4} \text{ Wb}$$

EXAMPLE
10–5

There are two amperes of current through a wire with 5 turns.

(a) What is the mmf?

(b) What is the reluctance of the circuit if the flux is 250 μWb?

Solution:

(a) $N = 5$ and $I = 2$ A

$$F_m = NI = (5)(2 \text{ A}) = 10 \text{ At}$$

(b) $\mathcal{R} = \dfrac{F_m}{\phi} = \dfrac{10 \text{ At}}{250 \text{ }\mu\text{Wb}} = 0.04 \times 10^6 \text{ At/Wb}$

THE ELECTROMAGNET

An electromagnet is based on the properties that you have just learned. A basic electromagnet is simply a coil of wire wound around a core material that can be easily magnetized.

The shape of the electromagnet can be designed for various applications. For example, Figure 10–14 shows a U-shaped magnetic core. When the coil of wire is connected to a battery and there is current, as shown in Part (a), a magnetic field is established as indicated. If the current is reversed, as shown in Part (b), the direction of the magnetic field is also reversed. The closer the north and south poles are brought together, the smaller the air gap between them becomes, and the easier it becomes to establish a magnetic field, because the reluctance is lessened.

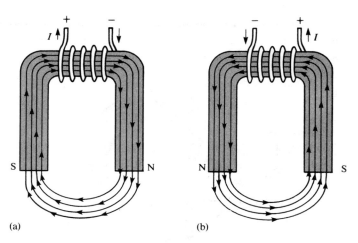

(a)   (b)

FIGURE 10–14
Reversing the current in the coil causes the electromagnetic field to reverse.

A good example of one application of an electromagnet is the process of recording on magnetic tape. In this situation, the *recording head* is an electromagnet with a narrow air gap, as shown in Figure 10–15. Current sets up a magnetic field across the

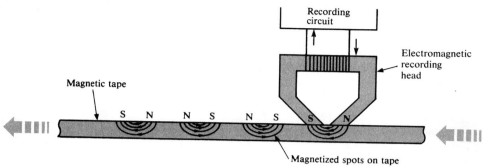

FIGURE 10–15
An electromagnetic recording head recording information on magnetic tape by magnetizing the tape as it passes by.

air gap, and as the recording head passes over the magnetic tape, the tape is permanently magnetized. In digital recording, for example, the tape is magnetized in one direction for a binary 1 and in the other direction for a binary 0, as illustrated in the figure. This magnetization in different directions is accomplished by reversing the coil current in the recording head.

ELECTROMAGNETIC DEVICES

The recording head is one example of an electromagnetic device. Several others are now presented.

The Solenoid Generally, the solenoid is a type of electromagnet that has a *movable* iron core whose movement depends on both an electromagnetic field and a mechanical spring force. The basic structure is shown in Figure 10–16(a). When there is current through the coil, the electromagnetic field magnetizes the core so that, effectively, there are two magnetic fields. The repulsive force of like poles and the attractive force of unlike poles cause the core to move outward against the spring tension, as shown in Figure 10–16(b). When the coil current stops, the core is pulled back in by the spring.

 This mechanical movement in a solenoid is used for many applications, such as opening and closing valves, automobile door locks, and so on.

The Relay Relays differ from solenoids in that the electromagnetic action is used to open or close electrical contacts rather than to provide mechanical movement. Figure 10–17 shows the basic operation of a relay with one normally open (NO) contact and one normally closed (NC) contact (single-pole–double-throw). When there is no coil current, the armature is held against the upper contact by the spring, thus providing continuity from terminal 1 to terminal 2, as shown in Part (a) of the figure. When energized with coil current, the armature is pulled down by the attractive force of the electromagnetic field and makes connection with the lower contact to provide continuity from terminal 1 to terminal 3, as shown in Part (b).

 A typical relay and its schematic symbol are shown in Figure 10–18.

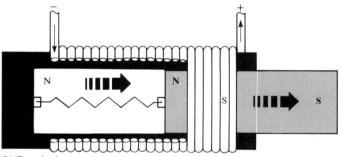

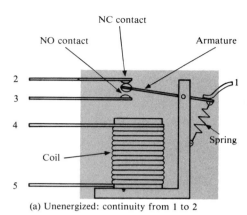

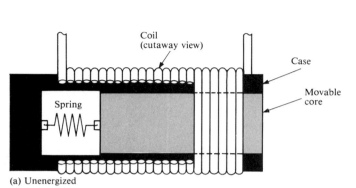

(a) Unenergized

(b) Energized

FIGURE 10–16
A basic solenoid structure with a cutaway view through the windings.

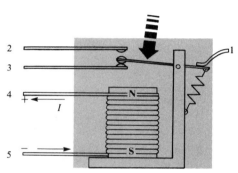

(a) Unenergized: continuity from 1 to 2

(b) Energized: continuity from 1 to 3

FIGURE 10–17
Basic structure of a single-pole–double-throw relay.

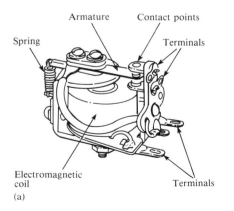

(a)

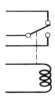

(b)

FIGURE 10–18
A typical relay.

The Speaker Permanent-magnet speakers are commonly used in stereos, radios, and TV, and their operation is based on the principle of electromagnetism. A typical speaker is constructed with a permanent magnet and an electromagnet, as shown in Figure 10–19(a). The cone of the speaker consists of a paper-like diaphragm to which is attached a hollow cylinder with a coil around it, forming an electromagnet. One of the poles of the permanent magnet is positioned within the cylindrical coil. When there is current through the coil in one direction, the interaction of the permanent magnetic field with the electromagnetic field causes the cylinder to move to the right, as indicated in Figure 10–19(b). Current through the coil in the other direction causes the cylinder to move to the left, as shown in Part (c).

The movement of the coil cylinder causes the flexible diaphragm also to move in or out, depending on the direction of the coil current. The amount of coil current determines the intensity of the magnetic field, which controls the amount that the diaphragm moves.

FIGURE 10–19
Basic speaker operation.

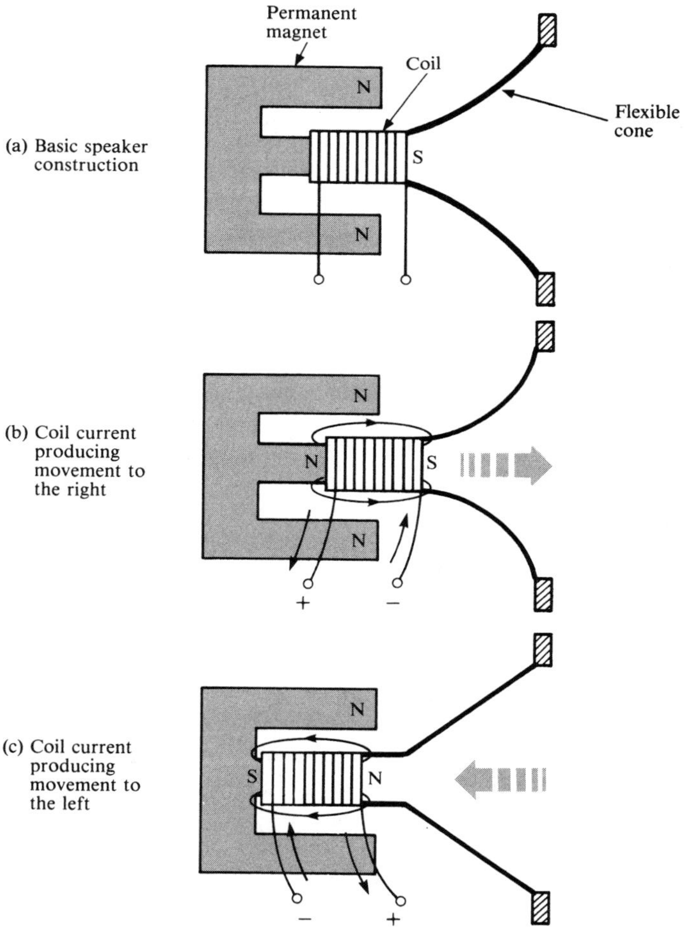

(a) Basic speaker construction

(b) Coil current producing movement to the right

(c) Coil current producing movement to the left

As shown in Figure 10–20, when an audio signal (voice or music) is applied to the coil, the current varies in both direction and amount. In response, the diaphragm will *vibrate* in and out by varying amounts and at varying rates. Vibration in the diaphragm causes the air that is in contact with it to vibrate in the same manner. These air vibrations move through the air as sound waves.

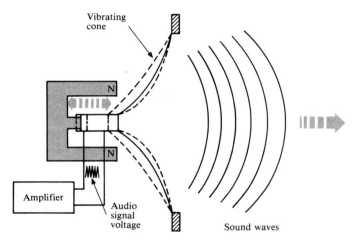

FIGURE 10–20
The speaker converts audio signal voltages into sound waves.

ANALOG METERS

The analog ammeter and voltmeter provide good examples of the application and operation of electromagnetic devices interfaced with additional circuitry used for control purposes. In the case of these meters, the amount of current through the meter coil, and thus the pointer deflection, is determined by series or parallel resistive circuits.

The d'Arsonval Movement The d'Arsonval movement is the most common type used in analog multimeters. In this type of meter movement, the pointer is deflected in proportion to the amount of current through a coil. Figure 10–21(a) shows a basic d'Arsonval meter movement. It consists of a coil of wire wound on a bearing-mounted assembly that is placed between the poles of a permanent magnet. A pointer is attached to the moving assembly. With no current through the coil, a spring mechanism keeps the pointer at its left-most (zero) position. When there is current through the coil, electromagnetic forces act on the coil, causing a rotation to the right. The amount of rotation depends on the amount of current. Figure 10–21(b) shows a construction view of the parts of a typical movement.

Figure 10–22 illustrates how the interaction of magnetic fields produces rotation of the coil assembly. The current is inward at the "cross" and outward at the "dot" in the single winding shown. The inward current produces a counterclockwise

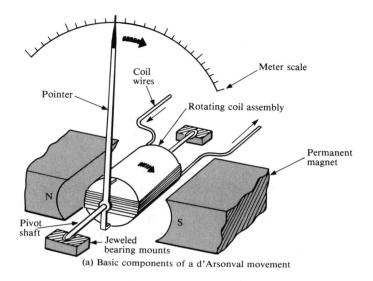

(a) Basic components of a d'Arsonval movement

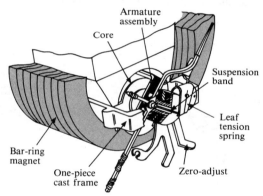

(b) Construction view of a typical d'Arsonval movement.

FIGURE 10–21
The d'Arsonval meter movement.

FIGURE 10–22
When the electromagnetic field interacts with the permanent magnetic field, forces are exerted on the rotating coil assembly, causing it to move clockwise and thus deflecting the pointer.

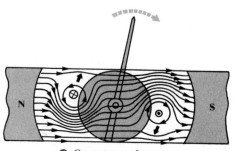

⊕ Current away from you
⊙ Current toward you

electromagnetic field that reinforces the permanent magnetic field below it. The result is an upward force on the left side of the coil as shown. A downward force is developed on the right side of the coil, where the current is outward. These forces produce a clockwise rotation of the coil assembly.

In addition to the d'Arsonval movement, there are two other types of movements that are used in analog meters: the *iron-vane* and the *electrodynamometer,* which are described briefly in the following paragraphs.

Iron-Vane Movement This type of movement consists basically of two iron bars (vanes) that are located side-by-side within a coil, as shown in Figure 10–23(a). One vane is stationary and the other movable. The electromagnetic field produced by the coil current magnetizes the vanes, making them like bar magnets with the same orientation of their north and south poles. The like poles repel each other, and the movable vane is pushed away from the stationary vane. A pointer attached to the movable vane is thus deflected an amount proportional to the coil current.

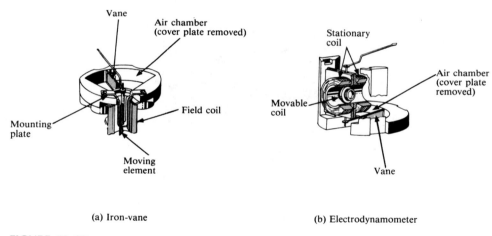

(a) Iron-vane

(b) Electrodynamometer

FIGURE 10–23
Construction views of iron-vane and electrodynamometer movements (courtesy of Triplett Corp.).

The iron-vane movement is very rugged and dependable, but it has the drawback of having a very nonlinear scale. That is, the measurement increments are very crowded on the low end of the scale and are more spread out on the upper end. The most common use for this type of movement is in ac (alternating current) ammeters.

Electrodynamometer Movement This type of movement differs from the d'Arsonval in that it uses an electromagnet rather than a permanent magnet. As shown in Figure 10–23(b), it also has a moving coil to which the pointer is attached, as in the d'Arsonval. The electrodynamometer movement can be used to measure both dc and ac quantities without additional internal circuitry, whereas the d'Arsonval movement requires extra circuitry for measuring ac. The most common use for this movement is in watt-meters for measuring power.

Current Sensitivity and Resistance of the Meter Movement The *current sensitivity* of a meter movement is the amount of current required to deflect the pointer full scale (all the way to its right-most position). For example, a 1-mA sensitivity means that when there is 1 mA through the meter coil, the needle is at its maximum deflection. If there is 0.5 mA through the coil, the needle is at the halfway point of its full deflection.

The movement *resistance* is simply the resistance of the coil of wire used in the movement.

The Ammeter A typical d'Arsonval movement might have a current sensitivity of 1 mA and a resistance of 50 Ω. In order to measure more than 1 mA, additional circuitry must be used with the basic meter movement, as was discussed in Chapter 6. As a basic review, Figure 10–24(a) shows a simple ammeter with a *shunt* (parallel) resistor (R_{SH}) across the movement. The purpose of the shunt resistor is to bypass current in excess of 1 mA around the meter movement. For example, let us assume that this meter must measure currents of up to 10 mA. Thus, for *full-scale* deflection, the movement must carry 1 mA, and the shunt resistor must carry 9 mA, as indicated in Figure 10–24(b).

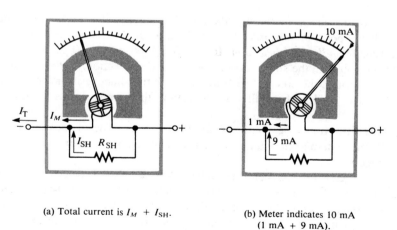

(a) Total current is $I_M + I_{SH}$.

(b) Meter indicates 10 mA (1 mA + 9 mA).

FIGURE 10–24
Shunt resistance (R_{SH}) in an ammeter.

The simple ammeter in Figure 10–24 has only one range. As you saw, it can measure currents from 0 to 10 mA and no higher. Most practical ammeters have several ranges. Each range in a multiple-range ammeter has its own shunt resistance which is selected with a multiple-position switch. For example, Figure 10–25 shows a four-range ammeter with a 0.1-mA (100-μA) meter movement. When the switch is in the 0.1-mA position, all of the current being measured is through the coil. In the other positions,

FIGURE 10-25

Example of a multiple-range ammeter.

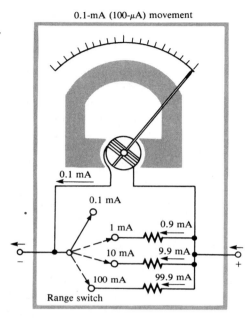

0.1-mA (100-μA) movement

0.1 mA

0.1 mA

1 mA   0.9 mA

10 mA   9.9 mA

100 mA   99.9 mA

Range switch

some of the current is through the coil, but most of it is through the shunt resistor. In the 1-mA position, 100 μA (0.1 mA) is through the coil and 0.9 mA through the shunt resistor for *full-scale deflection*. In the 10-mA position, 100 μA is through the coil and 9.9 mA through the shunt resistor for full-scale deflection. In the 100-mA position, 100 μA is through the coil and 99.9 mA through the shunt resistor for full-scale deflection. Of course, for less than full-scale deflection, the current values are less.

The Voltmeter The voltmeter utilizes the same type of movement as the ammeter. Different external circuitry is added so that the movement will function to measure voltage in a circuit.

As you have seen, the voltage drop across the meter coil is dependent on the current and the coil resistance. For example, a 50-μA, 1000-Ω movement has a full-scale voltage drop of (50 μA)(1000 Ω) = 50 mV. To use the meter to indicate voltages greater than 50 mV, we must add a *series resistance* to drop any additional voltage beyond that which the movement requires for full-scale deflection. This resistance is called the *multiplier resistance* and is designated R_X.

A basic voltmeter is shown in Figure 10-26(a) with a single multiplier resistor for one range. For this meter to measure 1 V full scale, the multiplier resistor (R_X) must drop 0.95 V because the coil drops only 50 mV (0.95 V + 50 mV = 1 V), as shown in Figure 10-26(b).

The simple voltmeter in Figure 10-26 has only one range. It can measure voltages from 0 to 1 V and no higher. Most practical voltmeters have several ranges, each of which has its own multiplier resistor which can be selected by a switch. For example, Figure 10-27 shows a four-range voltmeter with a 50-μA, 1000-Ω meter

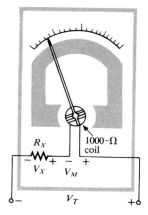

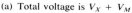

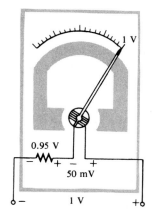

(a) Total voltage is $V_X + V_M$

(b) Meter indicates 1 V full scale (0.95 V + 50 mV).

FIGURE 10-26
Multiplier resistance (R_X) in a voltmeter.

movement. When the switch is in the 50-mV position, all of the voltage is dropped across the coil. In the other positions, some of the total voltage is dropped across the coil, but most is dropped across the multiplier resistor. In the 1-V position, 50 mV is dropped across the coil and 0.95 V across R_{X1} for *full-scale deflection*. In the 10-V position, 50 mV is dropped across the coil and 9.95 V across $R_{X1} + R_{X2}$ for full-scale deflection. In the 100-V position, 50 mV is dropped across the coil and 99.95 V across $R_{X1} + R_{X2} + R_{X3}$ for full-scale deflection. Of course, for less than full-scale deflection, the voltage values are less.

FIGURE 10-27
Example of a multiple-range voltmeter.

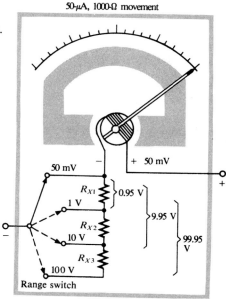

SECTION REVIEW 10–2

1. Explain the difference between magnetism and electromagnetism.

2. What happens to the magnetic field in an electromagnet when the current through the coil is reversed?

10–3

MAGNETIC HYSTERESIS

When a *magnetizing force* is applied to a material, the flux density in the material changes in a certain manner that we will now examine.

MAGNETIZING FORCE (H)

The magnetizing force in a material is defined to be the magnetomotive force (F_m) per unit length (l) of the material, as expressed by the following equation. The unit of magnetizing force (H) is ampere-turns per meter (At/m).

$$H = \frac{F_m}{l} \tag{10–6}$$

where $F_m = NI$. Note that the magnetizing force depends on the number of turns (N) of the coil of wire, the current (I) through the coil, and the length (l) of the material. It does not depend on the type of material.

Since $\phi = F_m/\mathcal{R}$, as F_m increases, the flux increases. Also, the magnetizing force (H) increases. Recall that the flux density (B) is the flux per unit cross-sectional area ($B = \phi/A$), so B is also proportional to H. The curve showing how these two quantities (B and H) are related is called the *B-H* curve or the hysteresis curve. The parameters that influence both B and H are illustrated in Figure 10–28.

THE HYSTERESIS CURVE

H can be readily increased or decreased by varying the current through the coil of wire, and it can be reversed by reversing the voltage polarity across the coil.

We start by assuming a magnetic core is unmagnetized so that $B = 0$. As the magnetizing force H is increased from zero, the flux density B increases proportionally as indicated by the curve in Figure 10–29(a). When H reaches a certain value H_1, the value of B begins to level off and, as H continues to increase, B reaches a *saturation* value when H reaches a value H_{sat} as illustrated in Figure 10–29(b). Once saturation is reached, a further increase in H will not increase B.

Now, if H is decreased to zero, B will fall back along a different path to a *residual value* (B_R) as shown in Figure 10–29(c). This indicates that the material continues to be magnetized even with the magnetizing force removed ($H = 0$). The ability of a material to maintain a magnetized state without the presence of a magnetizing force is called *retentivity*.

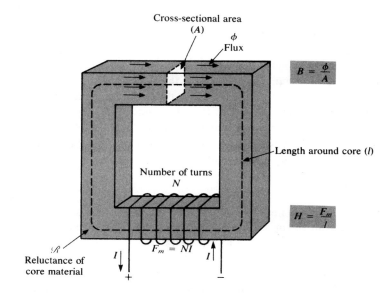

FIGURE 10–28
Parameters that determine the magnetizing force (H) and the flux density (B).

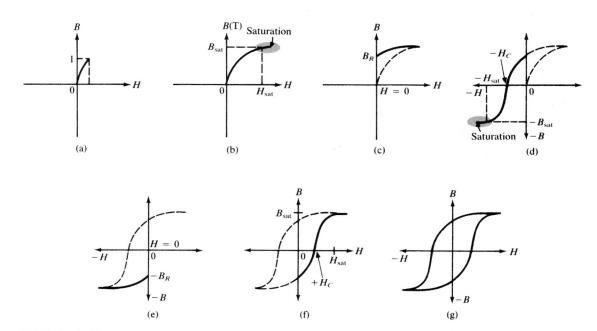

FIGURE 10–29
Development of a magnetic hysteresis curve.

Reversal of the magnetizing force is represented by negative values of H on the curve and is achieved by reversing the current in the coil of wire. An increase in H in the negative direction causes saturation to occur at a value $-H_{sat}$ where the flux density is at its maximum negative value as indicated in Figure 10–29(d). When the magnetizing force is removed ($H = 0$), the flux density goes to its negative *residual* value ($-B_R$) as shown in Part (e). From the $-B_R$ value, the flux density follows the curve indicated in Part (f) back to its maximum positive value when the magnetizing force equals H_{sat} in the positive direction. The complete *B-H* curve is shown in Part (g) and is called the *hysteresis curve*. The magnetizing force required to make the flux density zero is called the *coercive force, H_C*.

RETENTIVITY

The retentivity of a material is indicated by the ratio of B_R to B_{sat}. Materials with a low retentivity do not retain a magnetic field very well while those with high retentivities exhibit values of B_R very close to the saturation value of B. Depending on the application, retentivity in a magnetic material can be an advantage or a disadvantage. In permanent magnets and memory cores, for example, high retentivity is required. In ac motors, retentivity is undesirable because the residual magnetic field must be overcome each time the current reverses, thus wasting energy.

SECTION REVIEW 10–3

1. For a given wirewound core, how does an increase in current through the coil affect the flux density?
2. Define *retentivity*.

10–4 ELECTROMAGNETIC INDUCTION

When a conductor is moved through a magnetic field, a voltage is produced across the conductor. This principle is known as *electromagnetic induction*, and the resulting voltage is an *induced voltage*.

The principle of electromagnetic induction is widely applied in electrical circuits, as you will learn in this chapter and later in the study of transformers. The operation of electrical motors and generators is also based on this principle.

RELATIVE MOTION

When a wire is moved across a magnetic field, there is a relative motion between the wire and the magnetic field. Likewise, when a magnetic field is moved past a stationary wire, there is also relative motion. In either case, there is an *induced voltage* in the wire as a result of this motion, as Figure 10–30 indicates.

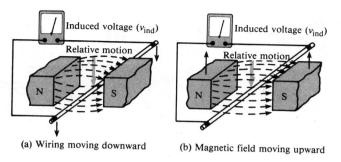

(a) Wiring moving downward   (b) Magnetic field moving upward

FIGURE 10–30
Relative motion between a wire and a magnetic field.

The amount of the induced voltage depends on the *rate* at which the wire and the magnetic field move with respect to each other: The faster the relative speed, the greater the induced voltage.

POLARITY OF THE INDUCED VOLTAGE

If the conductor in Figure 10–30 is moved first one way and then another in the magnetic field, a reversal of the polarity of the induced voltage will be observed. As the wire is moved downward, a voltage is induced with the polarity indicated in Figure 10–31(a). As the wire is moved upward, the polarity is as indicated in Part (b) of the figure. The lowercase v stands for instantaneous voltage.

INDUCED CURRENT

When a load resistor is connected to the wire in Figure 10–31, the voltage induced by the relative motion in the magnetic field will cause a current in the load, as shown in Figure 10–32. This current is called the *induced current*.

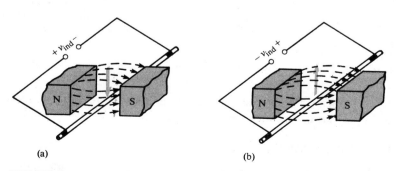

(a)   (b)

FIGURE 10–31
Polarity of induced voltage depends on direction of motion.

FIGURE 10–32
Induced current in a load as the wire
moves through the magnetic field.

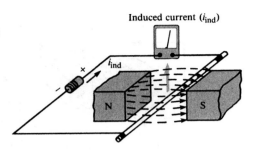

The principle of producing a voltage and a current in a load by moving a
conductor across a magnetic field is the basis for electrical generators. The concept of
a conductor existing in a moving magnetic field is the basis for inductance in an electrical circuit. The lowercase i stands for instantaneous current.

FORCES ON A CURRENT-CARRYING CONDUCTOR IN A MAGNETIC FIELD (MOTOR ACTION)

Figure 10–33(a) shows current inward through a wire in a magnetic field. The electromagnetic field set up by the current interacts with the permanent magnetic field; as a
result, the lines of force *above* the wire tend to cancel, because the permanent lines of
force are opposite in direction to the electromagnetic lines of force. Therefore, the flux
density above is reduced, and the magnetic field is weakened. The flux density below
the conductor is increased, and the magnetic field is strengthened because all the lines
of force are in the same direction. An upward force on the conductor results, and the
conductor tends to move toward the weaker magnetic field.

(a) Upward force: Weak field
above, strong field below.

(b) Downward force: Strong field
above, weak field below.

FIGURE 10–33
Forces on a current-carrying conductor in a magnetic field (motor action).

Figure 10–33(b) shows the current outward, resulting in a force on the conductor in the downward direction. This principle is the basis for electrical motors.

FARADAY'S LAW

Michael Faraday discovered the principle of electromagnetic induction in 1831. He
found that moving a magnet through a coil of wire induced a voltage across the coil,

and that when a complete path was provided, the induced voltage caused an induced current, as you have seen. Faraday's observations are as follows:

1. The amount of voltage induced in a coil is directly proportional to the rate of change of the magnetic field with respect to the coil ($d\phi/dt$).
2. The amount of voltage induced in a coil is directly proportional to the number of turns of wire in the coil (N).

Part 1 of Faraday's law is demonstrated in Figure 10–34, where a bar magnet is moved through a coil, thus creating a changing magnetic field. In Part (a) of the figure, the magnet is moved at a certain rate, and a certain induced voltage is produced as indicated. In Part (b), the magnet is moved at a faster rate through the coil, creating a greater induced voltage.

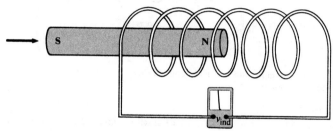

(a) As magnet moves to the right, magnetic field is changing with respect to coil, and a voltage is induced.

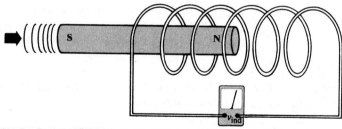

(b) As magnet moves more rapidly to the right, magnetic field is changing more rapidly with respect to coil, and a greater voltage is induced.

FIGURE 10–34
A demonstration of the first part of Faraday's law: The amount of induced voltage is directly proportional to the rate of change of the magnetic field with respect to the coil.

Part 2 of Faraday's law is demonstrated in Figure 10–35. In Part (a), the magnet is moved through the coil and a voltage is induced as shown. In Part (b), the magnet is moved at the same speed through a coil that has a greater number of turns. The greater number of turns creates a greater induced voltage.

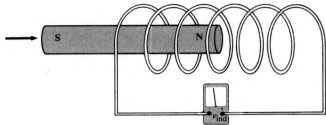

(a) Magnet moves through coil and induces a voltage.

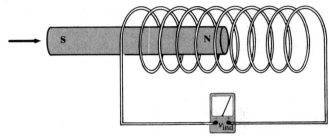

(b) Magnet moves at same rate through a coil with more turns (loops) and induces a greater voltage.

FIGURE 10–35
A demonstration of the second part of Faraday's law: The amount of induced voltage is directly proportional to the number of turns in the coil.

Faraday's law is expressed in equation form as follows:

$$v_{ind} = N\left(\frac{d\phi}{dt}\right) \qquad \textbf{(10–7)}$$

This formula states that the induced voltage across a coil (v_{ind}) is equal to the number of turns (N) times the rate of flux change ($d\phi/dt$).

EXAMPLE 10–6

Apply Faraday's law to find the induced voltage across a coil with 100 turns that is located in a magnetic field that is changing at a rate of 5 Wb/s.

Solution:
$$v_{ind} = N\left(\frac{d\phi}{dt}\right) = (100 \text{ t})(5 \text{ Wb/s}) = 500 \text{ V}$$

LENZ'S LAW

You have learned that a changing magnetic field induces a voltage in a coil that is directly proportional to the rate of change of the magnetic field and the number of turns in the coil. Lenz's law defines the polarity or direction of the induced voltage.

When the amount of current through a coil changes, an induced voltage is created as a result of the changing magnetic field, and the direction of the induced voltage is such that it always opposes the change in current.

SECTION REVIEW 10–4

1. What is the induced voltage across a stationary conductor in a stationary magnetic field?

2. When the speed at which a conductor is moved through a magnetic field is increased, does the induced voltage increase, decrease, or remain the same?

3. When there is current through a conductor in a magnetic field, what happens?

10–5 APPLICATIONS OF ELECTROMAGNETIC INDUCTION

In this section, two interesting applications of electromagnetic induction are discussed: an automotive crankshaft position sensor and a dc generator. Although there are many varied applications, these two are representative.

AUTOMOTIVE CRANKSHAFT POSITION SENSOR

An interesting automotive application is a type of engine sensor that detects the crankshaft position directly using electromagnetic induction. The electronic engine controller in many automobiles uses the position of the crankshaft to set ignition timing and, sometimes, to adjust the fuel control system. Figure 10–36 shows the basic concept. A

FIGURE 10–36
A crankshaft position sensor that produces a voltage when a tab passes through the air gap of the magnet.

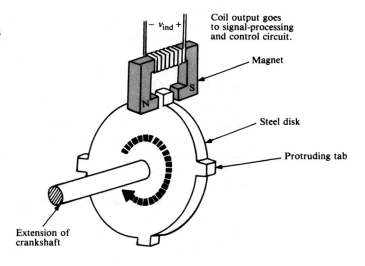

steel disk is attached to the engine's crankshaft by an extension rod; the protruding tabs on the disk represent specific crankshaft positions.

As illustrated in Figure 10–36, as the disk rotates with the crankshaft, the tabs periodically pass through the air gap of the permanent magnet. Since steel has a much lower reluctance than does air (a magnetic field can be established in steel much more easily than in air), the magnetic flux suddenly increases as a tab comes into the air gap, causing a voltage to be induced across the coil. This process is illustrated in Figure 10–37. The electronic engine control circuit uses the induced voltage as an indicator of the crankshaft position.

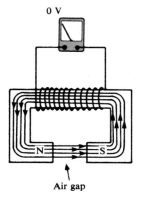

(a) There is no changing magnetic field, so there is no induced voltage.

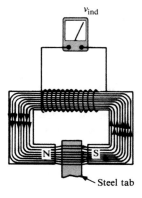

(b) Insertion of the steel tap reduces the reluctance of the air gap, causing the magnetic flux to increase and thus inducing a voltage.

FIGURE 10–37
As the tab passes through the air gap of the magnet, the coil senses a change in the magnetic field, and a voltage is induced.

A dc GENERATOR

Figure 10–38 shows a simplified dc generator consisting of a single loop of wire in a permanent magnetic field. Notice that each end of the loop is connected to a *split-ring* arrangement. This conductive metal ring is called a *commutator*. As the loop is rotated in the magnetic field, the split commutator ring also rotates. Each half of the split ring rubs against the fixed contacts, called *brushes,* and connects the loop to an external circuit.

As the loop rotates through the magnetic field, it "cuts" through the flux lines at varying angles, as illustrated in Figure 10–39. At position *A* in its rotation, the loop of wire is effectively moving *parallel* with the magnetic field. Therefore, at this instant, the rate at which it is cutting through the magnetic flux lines is *zero*.

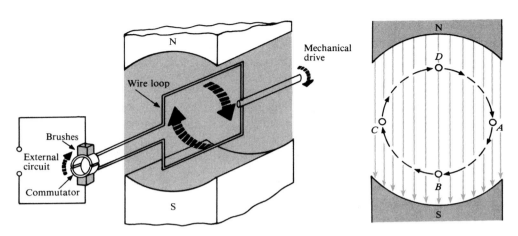

FIGURE 10–38
A basic dc generator.

FIGURE 10–39
End view of loop cutting through the magnetic field.

As the loop moves from position A to position B, it cuts through the flux lines at an increasing rate. At position B, it is moving effectively *perpendicular* to the magnetic field and thus is cutting through a maximum number of lines. As the loop rotates from position B to position C, the rate at which it cuts the flux lines decreases to minimum (zero) at C. From position C to position D, the rate at which the loop cuts the flux lines increases to a maximum at D and then back to a minimum again at A.

As you previously learned, when a wire moves through a magnetic field, a voltage is induced, and by Faraday's law, the amount of induced voltage is proportional to the number of loops (turns) in the wire and the rate at which it is moving with respect to the magnetic field. Now you know that the *angle* at which the wire moves with respect to the magnetic flux lines determines the amount of induced voltage, because the rate at which the wire cuts through the flux lines depends on the angle of motion.

Figure 10–40 illustrates how a voltage is induced in the external circuit as the single loop rotates in the magnetic field. Assume that the loop is in its instantaneous horizontal position, so the induced voltage is zero. As the loop continues in its rotation, the induced voltage builds up to a maximum at position B, as shown in Part (a) of the figure. Then, as the loop continues from B to C, the voltage decreases to zero at C, as shown in Part (b).

During the second half of the revolution, shown in Parts (c) and (d), the brushes switch to opposite commutator sections, so the polarity of the voltage remains the same across the output. Thus, as the loop rotates from position C to position D and then back to A, the voltage increases from zero at C to a maximum at D and back to zero at A.

Figure 10–41 shows how the induced voltage varies as the loop goes through several rotations (three in this case). This voltage is a dc voltage because its polarities do not change. However, the voltage is pulsating between zero and its maximum value.

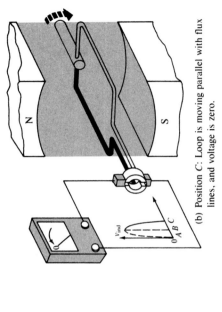

(a) Position *B*: Loop is moving perpendicular to flux lines, and voltage is maximum.

(b) Position *C*: Loop is moving parallel with flux lines, and voltage is zero.

(c) Position *D*: Loop is moving perpendicular to flux lines, and voltage is maximum.

(d) Position *A*: Loop is moving parallel with flux lines, and voltage is zero.

FIGURE 10–40

Operation of a basic dc generator.

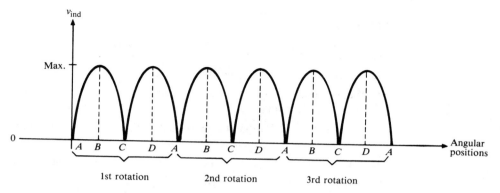

FIGURE 10–41
Induced voltage over three rotations of the loop.

When more loops are added, the voltages induced across each loop are combined across the output. Since the voltages are offset from each other, they do not reach their maximum or zero values at the same time. A smoother dc voltage results, as shown in Figure 10–42 for two loops. The variations can be further smoothed out by *filters* to achieve a nearly constant dc voltage. (Filters are discussed in a later chapter.)

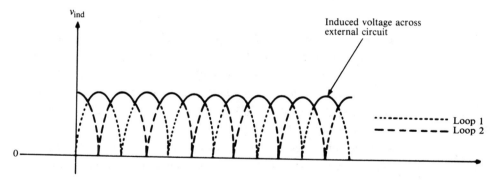

FIGURE 10–42
The induced voltage for a two-loop generator. There is much less variation in the induced voltage.

SECTION REVIEW 10–5

1. If the steel disk in the crankshaft position sensor has stopped, with a tab in the magnet's air gap, what is the induced voltage?

2. What happens to the induced voltage if the loop in the basic dc generator suddenly begins rotating at a faster speed?

SUMMARY

1. Unlike magnetic poles attract each other, and like poles repel each other.

2. Materials that can be magnetized are called *ferromagnetic*.

3. When there is current through a conductor, it produces an electromagnetic field around the conductor.

4. You can use the left-hand rule to establish the direction of the electromagnetic lines of force around a conductor.

5. An electromagnet is basically a coil of wire around a magnetic core.

6. When a conductor moves within a magnetic field, or when a magnetic field moves relative to a conductor, a voltage is induced across the conductor.

7. The faster the relative motion between a conductor and a magnetic field, the greater is the induced voltage.

8. *Weber* (Wb)—the unit of magnetic flux.

9. *Tesla* (T)—the unit of flux density.

10. *Ampere-turn* (At)—the unit of magnetomotive force (mmf).

11. *Magnetic flux*—the lines of force between the north pole and the south pole of a magnet.

12. *Flux density*—the number of lines of force per unit area in a magnetic field.

13. *Electromagnetic field*—a magnetic field created by current in a conductor.

14. *Magnetomotive force* (mmf)—the force that produces an electromagnetic field; it is the product of current and number of turns, and its unit is ampere-turns.

15. *Reluctance*—the opposition to the establishment of a magnetic field.

16. *Permeability*—the measure of the ease with which a magnetic field can be established in a material.

17. *Magnetizing force*—the amount of mmf per unit length of magnetic material.

18. *Retentivity*—the ability of a material to remain magnetized when the magnetizing force is removed.

19. *Hysteresis*—the B versus H characteristic of a material.

20. *Relay*—an electromagnetic device in which electrical contacts are controlled by current in a coil.

21. *Electromagnetic induction*—the process of producing a voltage by the relative motion between a magnetic field and a conductor.

22. *Faraday's law*—the voltage induced across a coil equals the number of turns in the coil times the rate of change of the magnetic flux.

23. *Lenz's law*—the polarity of the induced voltage is in a direction to oppose a change in the current.

FORMULAS

$$B = \frac{\phi}{A} \qquad\qquad (10\text{–}1)$$

$$\mu_r = \frac{\mu}{\mu_0} \qquad\qquad (10\text{–}2)$$

$$\mathcal{R} = \frac{l}{\mu A} \qquad\qquad (10\text{–}3)$$

$$F_m = NI \qquad\qquad (10\text{–}4)$$

$$\phi = \frac{F_m}{\mathcal{R}} \qquad\qquad (10\text{–}5)$$

$$H = \frac{F_m}{l} \qquad\qquad (10\text{–}6)$$

$$v_{\text{ind}} = N\left(\frac{d\phi}{dt}\right) \qquad\qquad (10\text{–}7)$$

SELF-TEST

Solutions appear at the end of the book.

1. If the south poles of two bar magnets are brought close together, will they be attracted or repelled?

2. What is a magnetic field?

3. Match each quantity in (a), (b), and (c) with its appropriate unit in (d), (e), and (f):
   (a) magnetic flux **(b)** magnetomotive force **(c)** flux density
   (d) tesla **(e)** ampere-turns **(f)** weber

4. What is *reluctance,* and to what is it analogous in an electric circuit?

5. Explain basically how information is recorded on magnetic tape.

6. What is the difference between solenoids and relays?

7. What is the basic requirement to produce voltage by electromagnetic induction?

8. If there is current through a wire placed in a magnetic field, a force is exerted on the wire (T or F).

9. A coil is exposed to a changing magnetic field. If the number of turns in the coil is increased, does the induced voltage increase or decrease?

10. In the crankshaft position sensor in Section 10–5, what causes the induced voltage in the coil?

PROBLEMS

Section 10–1

10–1 The cross-sectional area of a magnetic field is increased, but the flux remains the same. Does the flux density increase or decrease?

10–2 In a certain magnetic field the cross-sectional area is 0.5 square meter and the flux is 1500 μWb. What is the flux density?

10–3 What is the flux in a magnetic material when the flux density is 2500×10^{-6} T and the cross-sectional area is 150 cm^2?

Section 10–2

10–4 What happens to the compass needle in Figure 10–9 when the current through the conductor is reversed?

10–5 What is the relative permeability of a ferromagnetic material whose absolute permeability is 750×10^{-6} Wb/At·m?

10–6 Determine the reluctance of a material with a length of 0.28 m and a cross-sectional area of 0.08 m^2 if the absolute permeability is 150×10^{-7} Wb/At·m.

10–7 What is the magnetomotive force in a 50-turn coil of wire when there are 3 A through it?

Section 10–3

10–8 What is the magnetizing force in Problem 10–7 if the length of the core is 0.2 meter?

10–9 How can the flux density in Figure 10–43 be changed without altering the physical characteristics of the core?

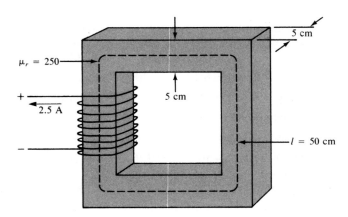

FIGURE 10–43

10–10 In Figure 10–43, determine the following:
   (a) H **(b)** ϕ **(c)** B

10-11 Determine from the hysteresis curves in Figure 10-44 which material has the most retentivity.

FIGURE 10-44

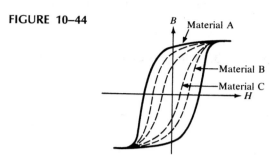

Section 10-4

10-12 According to Faraday's law, what happens to the induced voltage across a given coil if the rate of change of magnetic flux doubles?

10-13 The voltage induced across a certain coil is 100 mV. A 100-Ω resistor is connected to the coil terminals. What is the induced current?

10-14 A magnetic field is changing at a rate of 3500×10^{-3} Wb/s. How much voltage is induced across a 50-turn coil that is placed in the magnetic field?

10-15 How does Lenz's law complement Faraday's law?

Section 10-5

10-16 In Figure 10-36, why is there no induced voltage when the steel disk is not rotating?

10-17 Explain the purpose of the commutator and brushes in Figure 10-38.

10-18 A basic one-loop dc generator is rotated at 60 revolutions per second. How many times each second does the dc output voltage peak (reach a maximum)?

10-19 Assume that another loop, 90 degrees from the first loop, is added to the dc generator in Problem 10-18. Make a graph of voltage versus time to show how the output voltage appears. Let the maximum voltage be 10 V.

ANSWERS TO SECTION REVIEWS

Section 10-1

1. Repel.   **2.** The group of lines of force that make up a magnetic field.
3. 900 T.

Section 10-2

1. Electromagnetism is produced by current through a conductor. An electromagnetic field exists only when there is current. A magnetic field exists independently of current.

2. The direction of the magnetic field also reverses.

Section 10–3
1. An increase in current increases the flux density.
2. The ability of a material to remain magnetized after removal of the magnetizing force.

Section 10–4
1. Zero.   2. Increase.   3. A force is exerted on the conductor.

Section 10–5
1. Zero.   2. Increases.

ELEVEN

INTRODUCTION TO ALTERNATING CURRENT AND VOLTAGE

This chapter provides an introduction to ac circuit analysis in which time-varying electrical signals, particularly the *sine wave,* are presented. An electrical signal, for our purposes, is a voltage or a current that changes in some consistent manner with time. In other words, the voltage or current fluctuates according to a certain pattern called a *waveform.*

Particular emphasis is given to the sine wave because of its basic importance in ac circuit analysis. Other types of waveforms are also introduced, including pulse, triangular, and sawtooth. The use of the oscilloscope for displaying and measuring waveforms is discussed.

In this chapter, you will learn:

☐　How to identify a sinusoidal waveform and measure its characteristics.
☐　How to relate the frequency and period of a periodic waveform.
☐　How to relate the peak, peak-to-peak, and rms values of a sine wave.
☐　How to use Ohm's law in ac resistive circuits.
☐　How to identify points on a sine wave in angular units.
☐　How to relate degrees and radians when working in angular units.
☐　How to measure the relative phase angle of a sine wave.
☐　How to mathematically analyze a sinusoidal waveform.

☐ How to identify the characteristics of pulse, triangular, and sawtooth waveforms.
☐ The basic operation of oscilloscopes and how to use these instruments for waveform measurements.

11–1   THE SINE WAVE

The sine wave is a very common type of alternating current (ac) and alternating voltage. It is also referred to as a sinusoidal wave or, simply, sinusoid. The electrical service provided by the power companies is in the form of sinusoidal voltage and current. In addition, other types of waveforms are composites of many individual sine waves called *harmonics,* as you will see later.

Figure 11–1 shows the general shape of a sine wave, which can be either current or voltage. Notice how the voltage (or current) varies with time. Starting at zero, it *increases* to a positive maximum (peak), returns to zero, and then increases to a *negative* maximum (peak) before returning again to zero.

FIGURE 11–1
Sine wave.

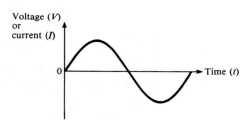

POLARITY OF A SINE WAVE

As you have seen, a sine wave changes *polarity* at its zero value; that is, it *alternates* between positive and negative values. When a sine wave voltage is applied to a resistive circuit, as in Figure 11–2, an alternating sine wave current results. When the voltage changes polarity, the current correspondingly changes direction as indicated.

During the *positive alternation* of the applied voltage V_s, the current is in the direction shown in Figure 11–2(a). During a *negative alternation* of the applied voltage, the current is in the opposite direction, as shown in Part (b). The combined positive and negative alternations make up one *cycle* of a sine wave.

PERIOD OF A SINE WAVE

As you have seen, a sine wave varies with time in a definable manner. Time is designated by t. The time required for a sine wave to complete one full cycle is called the

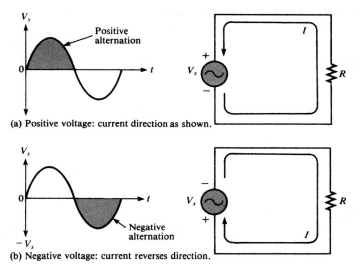

(a) Positive voltage: current direction as shown.

(b) Negative voltage: current reverses direction.

FIGURE 11–2
Alternating current and voltage.

period (*T*), as indicated in Figure 11–3(a). Typically, a sine wave continues to repeat itself in identical cycles, as shown in Part (b). Since all cycles of a repetitive sine wave are the same, the period is always a fixed value for a given sine wave.

The period of a sine wave does not necessarily have to be measured between the zero crossings at the beginning and end of a cycle. It can be measured from any point in a given cycle to the *corresponding* point in the next cycle.

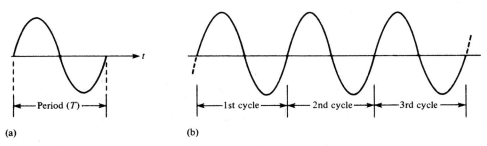

(a)

(b)

FIGURE 11–3
The period of a sine wave is the same for each cycle.

**EXAMPLE
11–1**

What is the period of the sine wave in Figure 11–4?

FIGURE 11–4

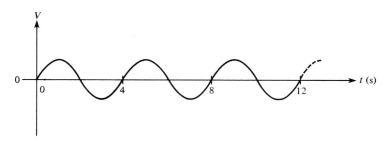

Solution:
As you can see, it takes four seconds (4 s) to complete each cycle. Therefore, the period is 4 s.

$$T = 4 \text{ s}$$

**EXAMPLE
11–2**

Show three possible ways to measure the period of the sine wave in Figure 11–5. How many cycles are shown?

FIGURE 11–5

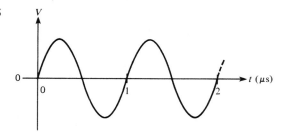

Solution:
Method 1: The period can be measured from one zero crossing to the corresponding zero crossing in the next cycle.
Method 2: The period can be measured from the positive peak in one cycle to the positive peak in the next cycle.
Method 3: The period can be measured from the negative peak in one cycle to the negative peak in the next cycle.

These measurements are indicated in Figure 11–6, where two cycles of the sine wave are shown. Keep in mind that you obtain the same value for the period no matter which points on the waveform you use.

FIGURE 11–6
Measurement of the period.

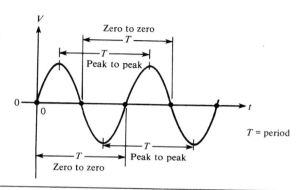

FREQUENCY OF A SINE WAVE

Frequency is the number of cycles that a sine wave completes in one second. The more cycles completed in one second, the higher the frequency.

Figure 11–7 shows two sine waves. The sine wave in Part (a) completes two full cycles in one second. The one in Part (b) completes four cycles in one second. Therefore, the sine wave in Part (b) has twice the frequency of the one in Part (a).

Frequency is measured in units of *hertz*, abbreviated Hz. One hertz is equivalent to one cycle per second; 60 Hz is 60 cycles per second; and so on. The symbol for frequency is f.

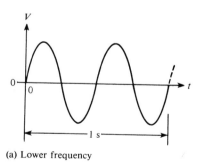

(a) Lower frequency

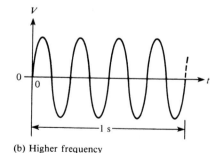

(b) Higher frequency

FIGURE 11–7
Illustration of frequency.

RELATIONSHIP OF FREQUENCY AND PERIOD

The relationship between frequency and period is very important. The formulas for this relationship are as follows:

$$f = \frac{1}{T} \tag{11-1}$$

$$T = \frac{1}{f} \tag{11-2}$$

There is a *reciprocal* relationship between f and T. Knowing one, you can calculate the other with the $\boxed{1/x}$ key on your calculator.

This relationship makes sense because a sine wave with a longer period goes through fewer cycles in one second than one with a shorter period.

EXAMPLE 11-3

Which sine wave in Figure 11-8 has the higher frequency? Determine the period and the frequency of both waveforms.

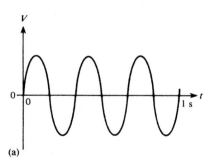
(a)

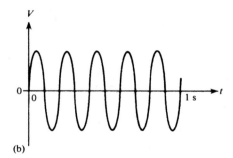
(b)

FIGURE 11-8

Solution:
In Part (a) of the figure, three cycles take 1 s. Therefore, one cycle takes 0.333 s (one-third second), and this is the period.

$$T = 0.333 \text{ s}$$

$$f = \frac{1}{T} = \frac{1}{0.333 \text{ s}} = 3 \text{ Hz}$$

In Part (b) of the figure, five cycles take 1 s. Therefore, one cycle takes 0.2 s (one-fifth second), and this is the period.

$$T = 0.2 \text{ s}$$

$$f = \frac{1}{T} = \frac{1}{0.2 \text{ s}} = 5 \text{ Hz}$$

EXAMPLE 11–4

The period of a certain sine wave is 10 milliseconds (10 ms). What is the frequency?

Solution:
Using Equation (11–1), we obtain the following result:

$$f = \frac{1}{T} = \frac{1}{10 \text{ ms}} = \frac{1}{10 \times 10^{-3} \text{ s}}$$
$$= 0.1 \times 10^3 \text{ Hz} = 100 \text{ Hz}$$

EXAMPLE 11–5

The frequency of a sine wave is 60 Hz. What is the period?

Solution:
Using Equation (11–2), the period is determined as follows:

$$T = \frac{1}{f} = \frac{1}{60 \text{ Hz}} = 0.01667 \text{ s} = 16.67 \text{ ms}$$

SECTION REVIEW 11–1

1. Describe one cycle of a sine wave.
2. At what point does a sine wave change polarity?
3. How many maximum points does a sine wave have during one cycle?
4. How is the period of a sine wave measured?
5. Define *frequency,* and state its unit.
6. Determine f when $T = 5 \text{ } \mu\text{s}$.
7. Determine T when $f = 120 \text{ Hz}$.

11–2   VOLTAGE AND CURRENT VALUES OF A SINE WAVE

There are several ways to express the value of a sine wave in terms of its voltage or its current magnitude. These are *instantaneous, peak, peak-to-peak, rms,* and *average* values.

INSTANTANEOUS VALUE

Figure 11–9 illustrates that at any point in time on a sine wave, the voltage (or current) has an *instantaneous* value. This instantaneous value is different at different points along the curve. Instantaneous values are positive during the positive alternation and negative during the negative alternation. Instantaneous values of voltage and current are symbolized by lowercase v and i, respectively, as shown in Part (a). The curve is shown for

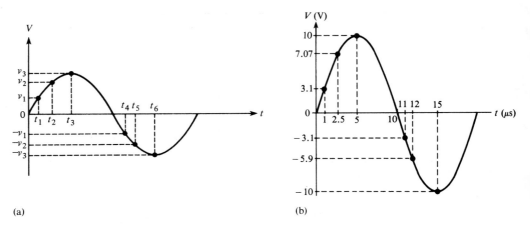

(a)   (b)

FIGURE 11–9
Instantaneous values.

voltage only, but it applies equally for current when the v's are replaced with i's. An example is shown in Part (b) where the instantaneous voltage is 3.1 V at 1 μs, 7.07 V at 2.5 μs, 10 V at 5 μs, 0 V at 10 μs, -3.1 V at 11 μs, and so on.

PEAK VALUE

The peak value of a sine wave is the value of voltage (or current) at the positive or the negative maximum (peaks) with respect to zero. Since the peaks are equal in magnitude, a sine wave is characterized by a single peak value. This is illustrated in Figure 11–10. For a given sine wave, the peak value is constant and is represented by V_p or I_p.

FIGURE 11–10
Peak values.

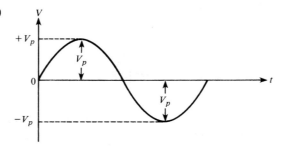

PEAK-TO-PEAK VALUE

The peak-to-peak value of a sine wave, as shown in Figure 11–11, is the voltage or current from the positive peak to the negative peak. Of course, it is always twice the peak value as expressed in the following equations. Peak-to-peak values are represented by the symbols V_{pp} or I_{pp}.

$$V_{pp} = 2V_p \qquad \qquad \textbf{(11–3)}$$

$$I_{pp} = 2I_p \qquad \qquad \textbf{(11–4)}$$

FIGURE 11–11
Peak-to-peak value.

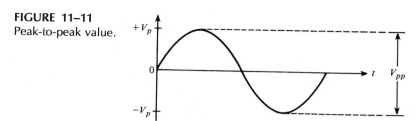

rms VALUE

The term *rms* stands for *root mean square*. It refers to the mathematical process by which this value is derived (see Appendix D for the derivation). The rms value is also referred to as the *effective value*. Most ac voltmeters display rms voltage. The 120 volts at your wall outlet is an rms value.

The rms value of a sine wave is actually a measure of the *heating effect* of the sine wave. For example, when a resistor is connected across an ac (sine wave) voltage source, as shown in Figure 11–12(a), a certain amount of heat is generated by the power in the resistor. Figure 11–12(b) shows the *same* resistor connected across a dc voltage source. The value of the dc voltage can be adjusted so that the resistor gives off the same amount of heat as it does when connected to the ac source.

The rms value of a sine wave is equal to the dc voltage that produces the same amount of heat as the sinusoidal voltage.

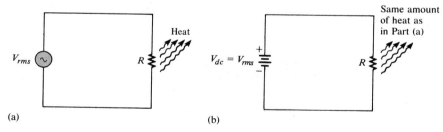

(a)   (b)

FIGURE 11–12
The meaning of the rms value of a sine wave.

The peak value of a sine wave can be converted to the corresponding rms value using the following relationships for either voltage or current:

$$V_{\text{rms}} = \sqrt{0.5}\, V_p \cong 0.707 V_p \tag{11–5}$$

$$I_{\text{rms}} = \sqrt{0.5}\, I_p \cong 0.707 I_p \tag{11–6}$$

Using these formulas, we can also determine the peak value knowing the rms value as follows:

$$V_p = \frac{V_{\text{rms}}}{0.707} = \left(\frac{1}{0.707}\right) V_{\text{rms}}$$

$$V_p = \sqrt{2}\, V_{\text{rms}} \cong 1.414 V_{\text{rms}} \tag{11–7}$$

Similarly,

$$I_p = \sqrt{2}\, I_{\text{rms}} \cong 1.414 I_{\text{rms}} \qquad \text{(11–8)}$$

To get the peak-to-peak value, simply double the peak value:

$$V_{pp} = 2.828 V_{\text{rms}} \qquad \text{(11–9)}$$

and

$$I_{pp} = 2.828 I_{\text{rms}} \qquad \text{(11–10)}$$

AVERAGE VALUE

The average value of a sine wave taken over *one complete cycle* is always zero, because the positive values (above the zero crossing) offset the negative values (below the zero crossing).

EXAMPLE 11–6

Determine V_p, V_{pp}, V_{rms}, and V_{avg} for the sine wave in Figure 11–13.

FIGURE 11–13

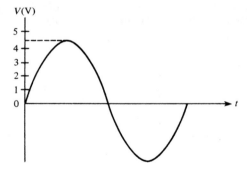

Solution:
$V_p = 4.5$ V as taken directly from the graph. From this, the other values are calculated.

$$V_{pp} = 2V_p = 2(4.5 \text{ V}) = 9 \text{ V}$$

$$V_{\text{rms}} = 0.707 V_p = 0.707(4.5 \text{ V}) = 3.182 \text{ V}$$

$$V_{\text{avg}} = 0 \text{ V}$$

OHM'S LAW IN ac ANALYSIS OF RESISTIVE CIRCUITS

When a sinusoidal voltage is applied to a resistive circuit, as shown in Figure 11–14, a sinusoidal current is generated. *Ohm's law can be used in resistive ac circuits just as in*

FIGURE 11–14
A sine wave voltage produces a sine wave
current.

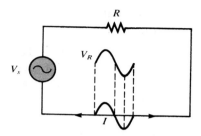

dc circuits. The sine wave values must be consistent; for example, when voltage is
expressed in rms, the current must also be rms. Equations (11–11) through (11–14)
express Ohm's law in terms of the sine wave values of instantaneous, peak, peak-to-
peak, and rms:

$$v = iR \tag{11–11}$$

$$V_p = I_p R \tag{11–12}$$

$$V_{pp} = I_{pp} R \tag{11–13}$$

$$V_{rms} = I_{rms} R \tag{11–14}$$

**EXAMPLE
11–7**

Determine the rms voltage across each resistor and the rms current in Figure 11–15.
The source voltage is given as an rms value.

FIGURE 11–15

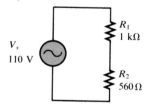

Solution:
The total resistance of the circuit is

$$R_T = R_1 + R_2 = 1 \text{ k}\Omega + 560 \ \Omega = 1.56 \text{ k}\Omega$$

Use Ohm's law to find the rms current:

$$I_{rms} = \frac{V_{s(rms)}}{R_T} = \frac{110 \text{ V}}{1.56 \text{ k}\Omega} = 70.5 \text{ mA}$$

The rms voltage drop across each resistor is

$$V_{1(rms)} = I_{rms} R_1 = (70.5 \text{ mA})(1 \text{ k}\Omega) = 70.5 \text{ V}$$

$$V_{2(rms)} = I_{rms} R_2 = (70.5 \text{ mA})(560 \ \Omega) = 39.5 \text{ V}$$

KIRCHHOFF'S LAWS IN ac ANALYSIS OF RESISTIVE CIRCUITS

Kirchhoff's voltage and current laws apply to ac circuits as well as to dc circuits. Figure 11–16 illustrates Kirchhoff's voltage law in a resistive circuit that has a sine wave voltage source. As indicated, the source voltage is the sum of all the voltage drops across the resistors, just as in a dc circuit.

FIGURE 11–16
Illustration of Kirchhoff's voltage law in a resistive ac circuit.

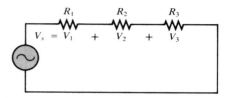

EXAMPLE 11–8

(a) Find the unknown peak voltage drop in Figure 11–17(a). All values are given in rms.

(b) Find the total rms current in Figure 11–17(b). All values are given in rms.

FIGURE 11–17

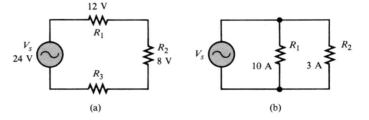

(a)   (b)

Solution:

(a) Use Kirchhoff's voltage law to find V_3:

$$V_s = V_1 + V_2 + V_3$$

Solve for V_3:

$$V_{3(rms)} = V_{s(rms)} - V_{1(rms)} - V_{2(rms)}$$
$$= 24 \text{ V} - 12 \text{ V} - 8 \text{ V} = 4 \text{ V}$$

Convert rms to peak:

$$V_{3(p)} = 1.414V_{3(rms)} = 1.414(4 \text{ V}) = 5.66 \text{ V}$$

(b) Use Kirchhoff's current law to find I_T:

$$I_{T(rms)} = I_{1(rms)} + I_{2(rms)}$$
$$= 10 \text{ A} + 3 \text{ A} = 13 \text{ A}$$

SECTION REVIEW 11–2

1. Determine V_{pp} when (a) $V_p = 1$ V, (b) $V_{rms} = 1.414$ V.
2. Determine V_{rms} when (a) $V_p = 2.5$ V, (b) $V_{pp} = 10$ V.
3. A sinusoidal voltage with an rms value of 5 V is applied to a circuit with a resistance of 10 Ω. What is the rms value of the current? The peak value of the current?

11–3   GENERATION OF SINE WAVE VOLTAGES

There are two basic methods of generating sine wave voltages: *electromagnetic* and *electronic*. Sine waves are produced electromagnetically by ac generators and electronically by oscillator circuits. In this section you will see how they are produced electromagnetically.

AN ac GENERATOR

Figure 11–18 shows a basic ac generator consisting of a single loop of wire in a permanent magnetic field. Notice that each end of the loop is connected to a separate solid conductive ring called a *slip ring*. As the loop rotates in the magnetic field, the slip rings also rotate and rub against the brushes which connect the loop to an external load. Compare this generator to the basic dc generator in Chapter 10, and note the difference in the ring and brush arrangements.

FIGURE 11–18
A basic ac generator.

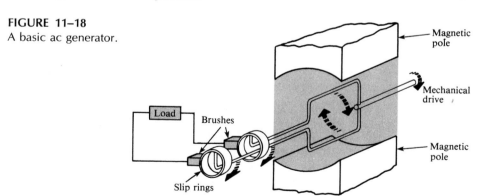

As you learned in Chapter 10, when a conductor moves through a magnetic field, a voltage is induced as the conductor cuts through the magnetic field. Figure 11–19 illustrates how a sine wave voltage is produced by the basic ac generator as the loop rotates. An oscilloscope is used to display the voltage waveform.

To begin, Figure 11–19(a) shows the loop rotating through the first quarter of a revolution. It goes from an instantaneous horizontal position, where the induced voltage is zero, to an instantaneous vertical position, where the induced voltage is maximum. At the horizontal position, the loop is instantaneously moving parallel with the

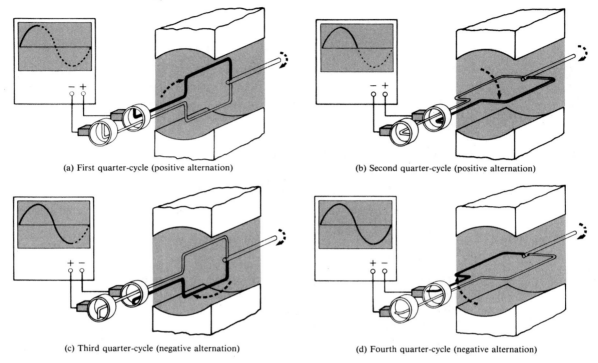

(a) First quarter-cycle (positive alternation)

(b) Second quarter-cycle (positive alternation)

(c) Third quarter-cycle (negative alternation)

(d) Fourth quarter-cycle (negative alternation)

FIGURE 11–19
One revolution of the loop generates one cycle of the sine wave voltage.

flux lines; thus, no lines are being cut and the voltage is zero. As the loop rotates through the first quarter-cycle, it cuts through the flux lines at an increasing rate until it is instantaneously moving perpendicular to the flux lines at the vertical position and cutting through them at a maximum rate. Thus, the induced voltage increases from zero to a peak during the quarter-cycle. As shown on the display, this part of the rotation produces the first quarter of the sine wave cycle as the voltage builds up from zero to its positive maximum. Part (b) of the figure shows the loop completing the first half of a revolution. During this part of the rotation, the voltage decreases from its positive maximum back to zero as the rate at which the loop cuts through the flux lines decreases.

During the second half of the revolution, illustrated in Figures 11–19(c) and 11–19(d), the loop is cutting through the magnetic field in the opposite direction, so the voltage produced has a polarity opposite to that produced during the first half of the revolution. After one complete revolution of the loop, one full cycle of the sine wave voltage has been produced. As the loop continues to rotate, repetitive cycles of the sine wave are generated.

FACTORS THAT AFFECT FREQUENCY IN ac GENERATORS

You have seen that one revolution of the conductor through the magnetic field in the basic ac generator (also called an *alternator*) produces one cycle of induced sinusoidal voltage. It is obvious that the rate at which the conductor is rotated determines the time for completion of one cycle. For example, if the conductor completes 60 revolutions in one second, the period of the resulting sine wave is 1/60 second, corresponding to a frequency of 60 Hz. Thus, the faster the conductor rotates, the higher the resulting frequency of the induced voltage, as illustrated in Figure 11–20.

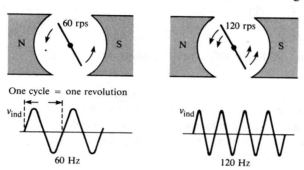

FIGURE 11–20
Frequency is directly proportional to the rate of rotation of the loop.

Another way of achieving a higher frequency is to increase the number of magnetic poles. In the previous discussion, two magnetic poles were used to illustrate the ac generator principle. During one revolution, the conductor passes under a north pole and a south pole, thus producing one cycle of a sine wave. When four magnetic poles are used instead of two, as shown in Figure 11–21, one cycle is generated during *one-half* a revolution. This doubles the frequency for the same rate of rotation.

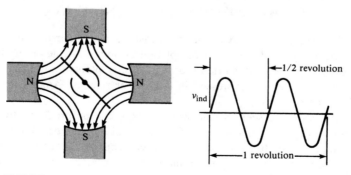

FIGURE 11–21
Four poles achieve a higher frequency than two for the same rps.

An expression for frequency in terms of the number of poles and the number of revolutions per second (rps) is as follows:

$$f = \text{(number of pole pairs)(rps)} \qquad (11\text{–}15)$$

**EXAMPLE
11–9**

A four-pole generator has a rotation speed of 100 rps. Determine the frequency of the output voltage.

Solution: $\quad f = \text{(number of pole pairs)(rps)} = 2(100) = 200 \text{ Hz}$

FACTORS THAT AFFECT VOLTAGE AMPLITUDE IN ac GENERATORS

Recall from Chapter 10 that the voltage induced in a conductor depends on the number of turns (N) and the rate of change with respect to the magnetic field. Therefore, when the speed of rotation of the conductor is increased, not only the frequency of the induced voltage increases—so also does the amplitude. Since the frequency value normally is fixed, the most practical method of increasing the amount of induced voltage is to increase the number of loops.

SECTION REVIEW 11–3

1. What two basic laws govern the electromagnetic generation of sine waves?

2. How are the speed of rotation and the frequency in an ac generator related?

11–4   ANGULAR RELATIONSHIPS OF A SINE WAVE

As you have seen, sine waves can be measured along the horizontal axis on a time basis; however, since the time for completion of one full cycle or any portion of a cycle is frequency-dependent, it is often useful to specify points on the sine wave in terms of an *angular measurement* expressed in degrees or radians. Angular measurement is independent of frequency.

As you have seen, a sine wave voltage can be produced by an ac generator. As the rotor of the ac generator goes through a full 360 degrees of rotation, the resulting voltage output is one full cycle of a sine wave. Thus the angular measurement of a sine wave can be related to the angular rotation of a generator, as shown in Figure 11–22.

ANGULAR MEASUREMENT

A *radian* (rad) is defined as the angular distance along the circumference of a circle equal to the radius of the circle. One radian is equivalent to 57.3 degrees, as illustrated in Figure 11–23.

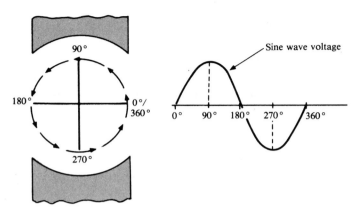

FIGURE 11–22
Relationship of a sine wave to the rotational motion in an ac generator.

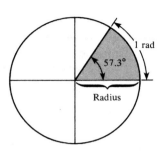

FIGURE 11–23
Angular measurement showing relationship of radian to degrees.

In a 360-degree revolution, there are 2π radians.

π is the ratio of the circumference of any circle to its diameter and has a constant value of 3.1416.

Most calculators have a $\boxed{\pi}$ key so that the actual numerical value does not have to be entered.

Table 11–1 lists several values of degrees and the corresponding radian values. These angular measurements are illustrated in Figure 11–24.

TABLE 11–1

Degrees	Radians (rad)
0	0
45	$\pi/4$
90	$\pi/2$
135	$3\pi/4$
180	π
225	$5\pi/4$
270	$3\pi/2$
315	$7\pi/4$
360	2π

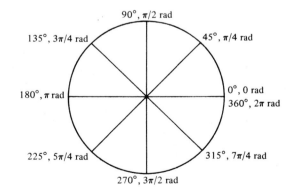

FIGURE 11–24
Angular measurements.

RADIAN/DEGREE CONVERSION

Radians can be converted to degrees using Equation (11–16).

$$\text{rad} = \left(\frac{\pi \text{ rad}}{180°}\right) \times \text{degrees} \qquad \textbf{(11–16)}$$

Similarly, degrees can be converted to radians with Equation (11–17).

$$\text{degrees} = \left(\frac{180°}{\pi \text{ rad}}\right) \times \text{rad} \tag{11–17}$$

**EXAMPLE
11–10**

(a) Convert 60 degrees to radians.
(b) Convert $\pi/6$ radians to degrees.

Solution:

(a) $\text{Rad} = \left(\dfrac{\pi \text{ rad}}{180°}\right)60° = \dfrac{\pi}{3} \text{ rad} = 1.047 \text{ rad}$

(b) $\text{Degrees} = \left(\dfrac{180°}{\pi \text{ rad}}\right)\left(\dfrac{\pi}{6} \text{ rad}\right) = 30°$

SINE WAVE ANGLES

The angular measurement of a sine wave is based on 360 degrees or 2π radians for a complete cycle. A half-cycle is 180 degrees or π radians; a quarter-cycle is 90 degrees or $\pi/2$ radians; and so on. Figure 11–25(a) shows angles in degrees for a full cycle of a sine wave; Part (b) shows the same points in radians.

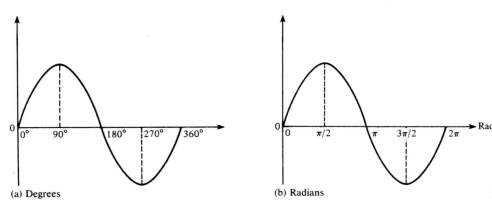

(a) Degrees

(b) Radians

FIGURE 11–25
Sine wave angles.

PHASE

The phase of a sine wave is an angular measurement that specifies the position of that sine wave *relative* to a reference. Figure 11–26 shows one cycle of a sine wave to be used as the *reference*. Note that the first positive-going crossing of the horizontal axis

FIGURE 11–26
Phase reference.

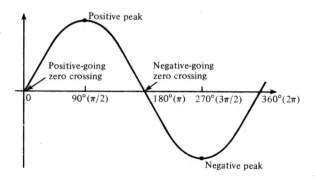

(zero crossing) is at 0 degrees (0 rad), and the positive peak is at 90 degrees ($\pi/2$ rad). The negative-going zero crossing is at 180 degrees (π rad), and the negative peak is at 270 degrees ($3\pi/2$ rad). The cycle is completed at 360 degrees (2π rad). When the sine wave is shifted left or right with respect to this reference, there is a *phase shift*.

Figure 11–27 illustrates phase shifts of a sine wave. In Part (a) of the figure, sine wave *B* is shifted to the right by 90 degrees ($\pi/2$ rad). Thus, there is a *phase angle* of 90 degrees between sine wave *A* and sine wave *B*. In terms of time, the positive peak of sine wave *B* occurs *later* than the positive peak of sine wave *A,* because time increases to the right along the horizontal axis. In this case, sine wave *B* is said to *lag* sine wave *A* by 90 degrees or $\pi/2$ radians. Stated another way, sine wave *A leads* sine wave *B* by 90 degrees.

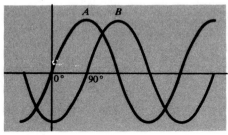

(a) *A* leads *B* by 90°, or *B* lags *A* by 90°.

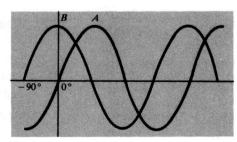

(b) *B* leads *A* by 90°, or *A* lags *B* by 90°.

FIGURE 11–27
Illustration of a phase shift.

In Figure 11–27(b), sine wave *B* is shown shifted left by 90 degrees; so, again, there is a phase angle of 90 degrees between sine wave *A* and sine wave *B*. In this case, the positive peak of sine wave *B* occurs earlier in time than that of sine wave *A;* therefore, sine wave *B* is said to *lead* by 90 degrees. A sine wave shifted 90 degrees to the left with respect to the reference is called a *cosine* wave.

EXAMPLE 11–11

What are the phase angles between the two sine waves in Figure 11–28(a) and (b)?

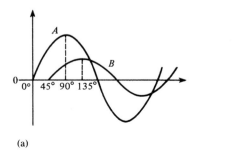

(a)

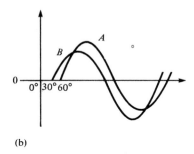

(b)

FIGURE 11–28

Solution:
In Part (a) of the figure, the phase angle is 45 degrees. Sine wave *A* leads sine wave *B* by 45 degrees.

In Part (b), the phase angle is 30 degrees. Sine wave *A* lags sine wave *B* by 30 degrees.

SECTION REVIEW 11–4

1. When the positive-going zero crossing of a sine wave occurs at 0 degrees, at what angle does each of the following points occur?
 (a) Positive peak   **(b)** Negative-going zero crossing
 (c) Negative peak   **(d)** End of first complete cycle

2. A half-cycle is completed in _____ degrees or _____ radians.

3. A full cycle is completed in _____ degrees or _____ radians.

4. Determine the phase angle between the two sine waves in Figure 11–29.

FIGURE 11–29

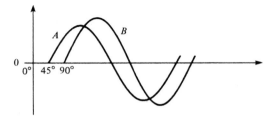

11–5

THE SINE WAVE EQUATION

As you have seen, a sine wave can be graphically represented by voltage (current) values on the vertical axis and by angular measurement (degrees or radians) along the horizontal axis. A generalized graph of one cycle of a sine wave is shown in Figure 11–30. The *amplitude, A,* is the maximum value of the voltage or current on the vertical axis, and angular values run along the horizontal axis.

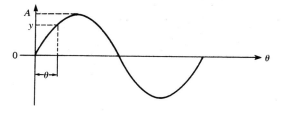

FIGURE 11–30
One cycle of a sine wave showing amplitude and phase angle associated with an instantaneous value (y) on the curve.

A sine wave curve follows a specific mathematical formula. The *general* expression for the sine wave curve in Figure 11–30 is

$$y = A \sin \theta \qquad \qquad \textbf{(11–18)}$$

This expression states that any point on the sine wave, represented by an instantaneous value y, is equal to the maximum value times the sine (sin) of the angle θ at that point. For example, a certain *voltage* sine wave has a peak value of 10 V. The instantaneous voltage at a point 60 degrees along the horizontal axis can be calculated as follows, where $y = v$ and $A = V_p$:

$$v = V_p \sin \theta = 10 \sin 60° = 10(0.866) = 8.66 \text{ V}$$

Figure 11–31 shows this particular instantaneous value on the curve. You can find the sine of any angle on your calculator by first entering the value of the angle and then pressing the $\boxed{\sin}$ key.

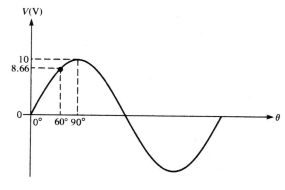

FIGURE 11–31
Illustration of the instantaneous value at $\theta = 60°$.

EXPRESSIONS FOR SHIFTED SINE WAVES

When a sine wave is shifted to the right of the reference by a certain angle, ϕ, as illustrated in Figure 11–32(a), the general expression is

$$y = A \sin(\theta - \phi) \tag{11–19}$$

When a sine wave is shifted to the left of the reference by a certain angle, ϕ, as shown in Figure 11–32(b), the general expression is

$$y = A \sin(\theta + \phi) \tag{11–20}$$

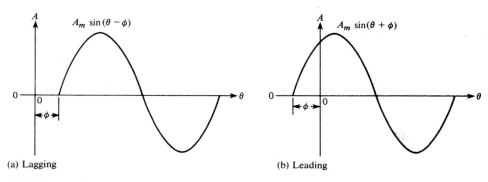

(a) Lagging   (b) Leading

FIGURE 11–32
Shifted sine waves.

EXAMPLE 11–12

Determine the instantaneous value at the 90-degree reference point on the horizontal axis for each sine wave voltage in Figure 11–33.

FIGURE 11–33

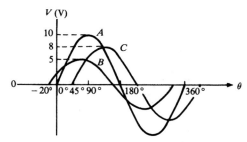

Solution:
Sine wave A is the reference. Sine wave B is shifted left 20 degrees with respect to A, so it leads. Sine wave C is shifted right 45 degrees with respect to A, so it lags.

$$v_A = V_p\sin \theta$$
$$= 10 \sin 90° = 10(1) = 10 \text{ V}$$
$$v_B = V_p\sin(\theta + \phi_B)$$
$$= 5 \sin(90° + 20°) = 5 \sin 110° = 5(0.9397) = 4.7 \text{ V}$$
$$v_C = V_p\sin(\theta - \phi_C)$$
$$= 8 \sin(90° - 45°) = 8 \sin 45° = 8(0.7071) = 5.66 \text{ V}$$

SECTION REVIEW 11–5

1. Calculate the instantaneous value at 120 degrees for the sine wave in Figure 11–31.
2. Determine the instantaneous value at the 45-degree point of a sine wave shifted 10 degrees to the left of the zero reference (V_p = 10 V).
3. Find the instantaneous value at the 90-degree point of a sine wave shifted 25 degrees to the right from the zero reference (V_p = 5 V).

11–6 NONSINUSOIDAL WAVEFORMS

Sine waves are very important in electronics, but they are by no means the only type of ac or time-varying waveform. Two other categories are discussed next; these are the *pulse waveform* and the *triangular waveform*.

PULSE WAVEFORMS

Basically, a pulse can be described as a very rapid transition from one voltage or current level (baseline) to another, and then, after an interval of time, a very rapid transition back to the original baseline level. The transitions in level are called *steps*. An *ideal* pulse consists of two opposite-going steps of equal amplitude. Figure 11–34(a) shows an ideal positive-going pulse consisting of two equal but opposite instantaneous steps separated by an interval of time called the *pulse width*. Part (b) of the figure shows an ideal negative-going pulse. The height of the pulse measured from the baseline is its voltage (or current) amplitude.

FIGURE 11–34
Ideal pulses.

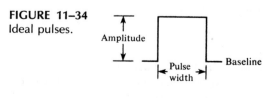

(a) Positive-going pulse

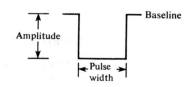

(b) Negative-going pulse

In many applications, analysis is simplified by treating all pulses as ideal (composed of instantaneous steps and perfectly rectangular in shape). Actual pulses, however, are never ideal. All pulses possess certain characteristics that cause them to be different from the ideal.

In practice, pulses cannot change from one level to another instantaneously. Time is always required for a transition (step), as illustrated in Figure 11–35(a). As you can see, there is an interval of time during which the pulse is rising from its lower value to its higher value. This interval is called the *rise time, t_r.*

The accepted definition of rise time is the time required for the pulse to go from 10% of its full amplitude to 90% of its full amplitude.

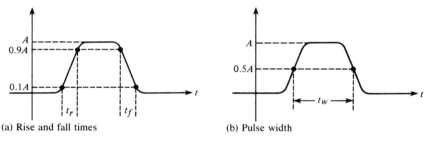

(a) Rise and fall times   (b) Pulse width

FIGURE 11–35
Nonideal pulse.

The interval of time during which the pulse is falling from its higher value to its lower value is called the *fall time, t_f.*

The accepted definition of fall time is the time required for the pulse to go from 90% of its full amplitude to 10% of its full amplitude.

Pulse width also requires a precise definition for the nonideal pulse, because the rising and falling edges are not vertical.

A widely accepted definition of pulse width (t_w) is the time between the point on the rising edge, where the value is 50% of full amplitude, to the point on the falling edge, where the value is 50% of full amplitude.

Pulse width is shown in Figure 11–35(b).

REPETITIVE PULSES

Any waveform that repeats itself at fixed intervals is *periodic*. Some examples of periodic pulse waveforms are shown in Figure 11–36. Notice that in each case, the pulses repeat at regular intervals. The rate at which the pulses repeat is the pulse repetition rate (PRR) or pulse repetition frequency (PRF), which is the *fundamental frequency* of

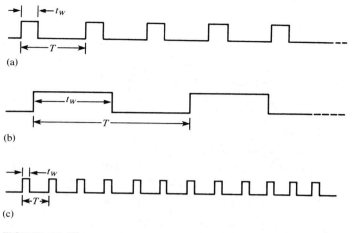

FIGURE 11–36
Repetitive pulse waveforms.

the waveform. The frequency can be expressed in *hertz* or in *pulses per second*. The time from one pulse to the corresponding point on the next pulse is the period T. The relationship between frequency and period is the same as with the sine wave.

$$\text{PRF} = \frac{1}{T} \tag{11–21}$$

$$T = \frac{1}{\text{PRF}} \tag{11–22}$$

A very important characteristic of pulse waveforms is the *duty cycle*.

The duty cycle is defined as the ratio of the pulse width (t_W) to the period T and is usually expressed as a percentage:

$$\text{percent duty cycle} = \left(\frac{t_W}{T}\right)100\% \tag{11–23}$$

EXAMPLE 11–13

Determine the period, PRF, and duty cycle for the pulse waveform in Figure 11–37.

FIGURE 11–37

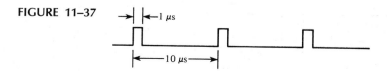

Solution:

$$T = 10 \ \mu s$$

$$\text{PRF} = \frac{1}{T} = \frac{1}{10 \ \mu s} = 0.1 \ \text{MHz}$$

$$\text{percent duty cycle} = \left(\frac{1 \ \mu s}{10 \ \mu s}\right)100\% = 10\%$$

SQUARE WAVES

A square wave is a pulse waveform with a duty cycle of 50%. Thus, the pulse width is equal to one-half of the period. A square wave is shown in Figure 11–38.

FIGURE 11–38
Square wave.

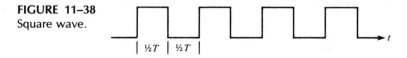

THE AVERAGE VALUE OF A PULSE WAVEFORM

The average value (V_{avg}) of a pulse waveform is equal to its baseline value plus its duty cycle times its amplitude. The lower level of the waveform is taken as the baseline. The formula is as follows:

$$V_{avg} = \text{baseline} + (\text{duty cycle})(\text{amplitude}) \qquad \textbf{(11–24)}$$

The following example illustrates the calculation of the average value.

EXAMPLE 11–14

Determine the average value of each of the waveforms in Figure 11–39.

FIGURE 11–39

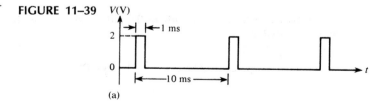

(a)

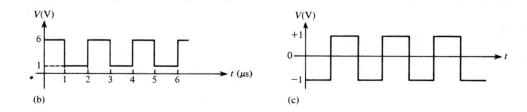

(b)

(c)

Solution:

(a) Here the baseline is at 0 V, the amplitude is 2 V, and the duty cycle is 10%. The average value is as follows:

$$V_{avg} = \text{baseline} + (\text{duty cycle})(\text{amplitude})$$

$$= 0 \text{ V} + (0.1)(2 \text{ V}) = 0.2 \text{ V}$$

(b) This waveform has a baseline of +1 V, an amplitude of 5 V, and a duty cycle of 50%. The average value is

$$V_{avg} = \text{baseline} + (\text{duty cycle})(\text{amplitude})$$

$$= 1 \text{ V} + (0.5)(5 \text{ V}) = 1 \text{ V} + 2.5 \text{ V} = 3.5 \text{ V}$$

(c) This is a square wave with a baseline of −1 V and an amplitude of 2 V. The average value is

$$V_{avg} = \text{baseline} + (\text{duty cycle})(\text{amplitude})$$

$$= -1 \text{ V} + (0.5)(2 \text{ V}) = -1 \text{ V} + 1 \text{ V} = 0 \text{ V}$$

This is an *alternating* square wave, and, as with an alternating sine wave, it has an average of zero.

TRIANGULAR AND SAWTOOTH WAVEFORMS

Triangular and sawtooth waveforms are formed by voltage or current *ramps*. A ramp is a *linear* increase or decrease in the voltage or current. Figure 11–40 shows both positive- and negative-going ramps. In Part (a) of the figure, the ramp has a positive slope; in Part (b), the ramp has a negative slope. The *slope* of a voltage ramp is $\pm \Delta v/\Delta t$ and is expressed in units of V/s. The slope of a current ramp is $\pm \Delta i/\Delta t$ and is expressed in units of A/s.

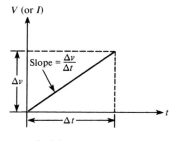

(a)   Positive ramp

(b) Negative ramp

FIGURE 11–40
Ramps.

EXAMPLE 11–15

What are the slopes of the voltage ramps in Figure 11–41?

FIGURE 11–41

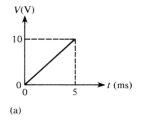

(a)

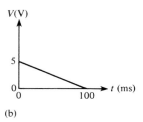

(b)

Solution:

(a) The voltage increases from 0 V to + 10 V in 5 ms. Thus, $\Delta v = 10$ V and $t = 5$ ms. The slope is

$$\frac{\Delta v}{\Delta t} = \frac{10 \text{ V}}{5 \text{ ms}} = 2 \text{ V/ms} = 2 \text{ kV/s}$$

(b) The voltage decreases from + 5 V to 0 V in 100 ms. Thus, $\Delta v = -5$ V and $t = 100$ ms. The slope is

$$\frac{\Delta v}{\Delta t} = \frac{-5 \text{ V}}{100 \text{ ms}} = -0.05 \text{ V/ms} = -50 \text{ V/s}$$

Triangular Waveforms   Figure 11–42 shows that a triangular waveform is composed of positive- and negative-going ramps having equal slopes. The period of this waveform is measured from one peak to the next corresponding peak, as illustrated. This particular triangular waveform is alternating and has an average value of zero.

Figure 11–43 depicts a triangular waveform with a nonzero average value. The frequency for triangular waves is determined in the same way as for sine waves, that is, $f = 1/T$.

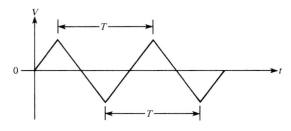

FIGURE 11–42
Alternating triangular waveform.

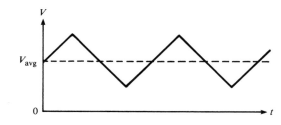

FIGURE 11–43
Triangular waveform with a nonzero average value.

Sawtooth Waveforms   The sawtooth wave is actually a special case of the triangular wave consisting of two ramps, one of much longer duration than the other. Sawtooth waveforms are commonly used in many electronic systems. For example, the electron beam that sweeps across the screen of your TV receiver, creating the picture, is controlled by sawtooth voltages and currents. One sawtooth wave produces the horizontal beam movement, and the other produces the vertical beam movement. The sawtooth is sometimes called a *sweep*.

Figure 11–44 is an example of a sawtooth wave. Notice that it consists of a positive-going ramp of relatively long duration, followed by a negative-going ramp of relatively short duration.

FIGURE 11–44
Sawtooth waveform.

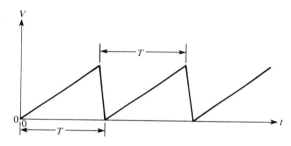

HARMONICS

A repetitive nonsinusoidal waveform is composed of a *fundamental frequency* and *harmonic frequencies*. The fundamental frequency is the repetition rate of the waveform, and the harmonics are higher frequency sine waves that are multiples of the fundamental.

Odd Harmonics   *Odd harmonics* are frequencies that are *odd multiples* of the fundamental frequency of a waveform. For example, a 1-kHz square wave consists of a fundamental of 1 kHz and odd harmonics of 3 kHz, 5kHz, 7 kHz, and so on. The 3-kHz frequency in this case is called the *third harmonic*, the 5-kHz frequency is the *fifth harmonic*, and so on.

Even Harmonics   *Even harmonics* are frequencies that are *even multiples* of the fundamental frequency. For example, if a certain wave has a fundamental of 200 Hz, the second harmonic is 400 Hz, the fourth harmonic is 800 Hz, the sixth harmonic is 1200 Hz, and so on. These are even harmonics.

Composite Waveform   Any variation from a pure sine wave produces harmonics. A nonsinusoidal wave is a composite of the fundamental and the harmonics. Some types of waveforms have only odd harmonics, some have only even harmonics, and some contain both. The shape of the wave is determined by its harmonic content. Generally, only the fundamental and the first few harmonics are of significant importance in determining the wave shape.

A *square wave* is an example of a waveform that consists of a fundamental and only odd harmonics. When the instantaneous values of the fundamental and each odd harmonic are added algebraically at each point, the resulting curve will have the shape of a square wave, as illustrated in Figure 11–45. In Part (a) of the figure, the fundamental and the third harmonic produce a wave shape that begins to resemble a square wave. In Part (b), the fundamental, third, and fifth harmonics produce a closer resemblance. When the seventh harmonic is included, as in Part (c), the resulting wave shape becomes even more like a square wave. As more harmonics are included, a square wave is approached.

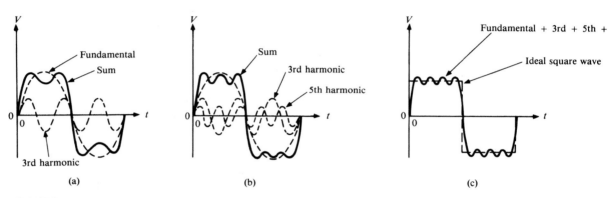

FIGURE 11–45
Odd harmonics produce a square wave.

SECTION REVIEW 11–6

1. Define the following parameters:
   (a) rise time   (b) fall time   (c) pulse width
2. In a certain repetitive pulse waveform, the pulses occur once every millisecond. What is the PRF of this waveform?
3. Determine the duty cycle, amplitude, and average value of the waveform in Figure 11–46(a).
4. What is the period of the triangular wave in Figure 11–46(b)?
5. What is the frequency of the sawtooth wave in Figure 11–46(c)?
6. Define *fundamental frequency*.
7. What is the second harmonic of a fundamental frequency of 1 kHz?
8. What is the fundamental frequency of a square wave having a period of 10 μs?

FIGURE 11–46

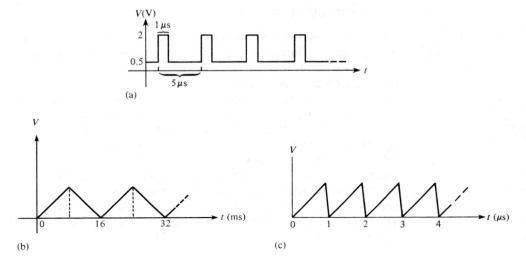

(a)

(b)

(c)

11–7   DISPLAY AND MEASUREMENT OF WAVEFORMS ON THE OSCILLOSCOPE

The oscilloscope, or *scope* for short, is one of the most widely used and versatile test instruments. It displays on a screen the actual shape of a voltage that is changing with time so that various measurements can be made.

Figure 11–47 shows two typical oscilloscopes. The one in Part (a) is a simpler and less expensive model. The one in Part (b) has better performance characteristics and plug-in modules that provide a variety of specialized functions.

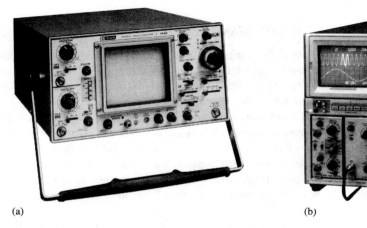

(a)   (b)

FIGURE 11–47
Oscilloscopes [(a) courtesy of B&K Precision. (b) courtesy of Tektronix, Inc.].

CATHODE-RAY TUBE (CRT)

The oscilloscope is built around the cathode-ray tube (CRT), which is the device that displays the waveforms. The screen of the scope is the front of the CRT.

The CRT is a vacuum tube device containing an *electron gun* that emits a narrow, focused *beam* of electrons. A phosphorescent coating on the face of the tube forms the *screen*. The beam is electronically focused and accelerated so that it strikes the screen, causing light to be emitted at the point of impact.

Figure 11–48 shows the basic construction of a CRT. The *electron gun* assembly contains a *heater, a cathode, a control grid,* and *accelerating* and *focusing grids*. The heater carries current that indirectly heats the cathode. The heated cathode emits electrons. The amount of voltage on the control grid determines the flow of electrons and thus the *intensity* of the beam. The electrons are accelerated by the accelerating grid and are focused by the focusing grid into a narrow beam that converges at the screen. The beam is further accelerated to a high speed after it leaves the electron gun by a high voltage on the *anode* surfaces of the CRT.

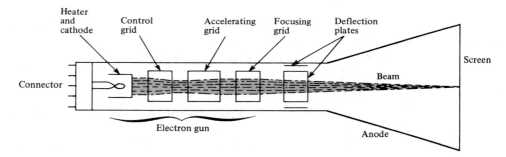

FIGURE 11–48
Basic construction of a CRT.

DEFLECTION OF THE BEAM

The purpose of the *deflection plates* in the CRT is to produce a "bending" or deflection of the electron beam. This deflection allows the position of the point of impact on the screen to be varied. There are two sets of deflection plates: one set for *vertical deflection,* and the other set for *horizontal deflection*.

Figure 11–49 shows a front view of the CRT's deflection plates. One plate from each set normally is grounded as shown. If there is *no voltage* on the other plates, as in Figure 11–49(a), the beam is not deflected and hits the center of the screen. If a *positive* voltage is on the vertical plate, the beam is attracted upward, as indicated in Part (b) of the figure. Remember that opposite charges attract. If a *negative* voltage is applied, the beam is deflected downward because like charges repel, as shown in Part (c).

Likewise, a positive or a negative voltage on the horizontal plate deflects the beam right or left, respectively, as shown in Figure 11–49(d) and (e). The amount of deflection is proportional to the amount of voltage on the plates.

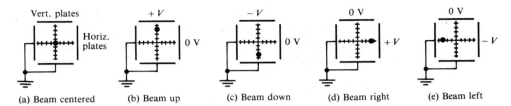

FIGURE 11–49
Deflection of an electron beam in a CRT.

SWEEPING THE BEAM HORIZONTALLY

In normal oscilloscope operation, the beam is horizontally deflected from left to right across the screen at a certain rate. This *sweeping action* produces a horizontal line or *trace* across the screen, as shown in Figure 11–50.

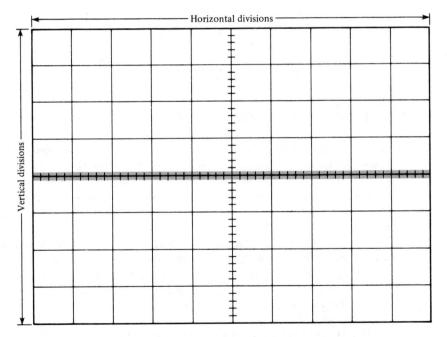

FIGURE 11–50
Scope screen with horizontal trace in color (8 cm × 10 cm).

The *rate* at which the beam is swept across the screen establishes a *time base*. The scope screen is divided into horizontal and vertical divisions, as shown in Figure 11–50. For a given time base, each horizontal division represents a fixed interval of time. For example, if the beam takes 1 second for a full left-to-right sweep, then each division represents 0.1 second. All scopes have provisions for selecting various sweep rates, which will be discussed later.

The actual sweeping of the beam is accomplished by application of a *sawtooth voltage* across the horizontal plates. The basic idea is illustrated in Figure 11–51. When the sawtooth is at its maximum negative peak, the beam is deflected to its left-most screen position. This deflection is due to maximum repulsion from the right deflection plate.

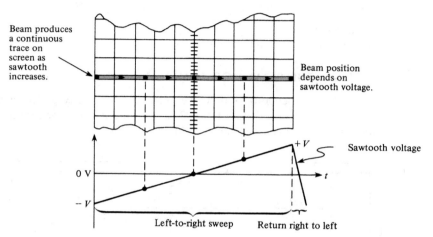

FIGURE 11–51
Sweeping the beam across the screen with a sawtooth voltage.

As the sawtooth voltage *increases,* the beam moves toward the center of the screen. When the sawtooth voltage is *zero,* the beam is at the center of the screen, because there is no repulsion or attraction from the plate. As the voltage increases *positively,* the plate attracts the beam, causing it to move toward the right side of the screen. At the positive peak of the sawtooth, the beam is at its right-most screen position.

The rate at which the sawtooth goes from negative to positive is determined by its frequency. This rate in turn establishes the *sweep rate* of the beam. When the sawtooth makes the abrupt change from positive back to negative, the beam is rapidly returned to the left side of the screen, ready for another sweep. During this "flyback" time, the beam is *blanked* out and thus does not produce a trace on the screen.

HOW A VOLTAGE PATTERN IS PRODUCED

The main purpose of the scope is to display the shape (waveform) of a voltage under test. To do so, we apply the voltage under test across the *vertical plates* through a vertical amplifier circuit. As you have seen, a voltage across the vertical plates causes a vertical deflection of the beam. A negative voltage causes the beam to go below the center of the screen, and a positive voltage makes it go above center.

Assume, for example, that a varying dc voltage (such as the output of a dc generator) is applied across the vertical plates. As a result, the beam will move up and down on the screen. The amount that the beam goes above center depends on the maximum value of the voltage.

At the same time that the beam is being deflected vertically, it is also sweeping horizontally, causing the voltage variation to be traced out across the screen, as illustrated in Figure 11–52. All scopes provide for the calibrated adjustment of the vertical deflection, so each vertical division represents a known amount of voltage.

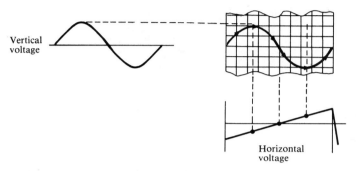

FIGURE 11–52
Vertical and horizontal deflection combined to produce a pattern on the screen.

OSCILLOSCOPE CONTROLS

A wide variety of oscilloscopes are available, ranging from relatively simple instruments with limited capabilities to much more sophisticated models that provide a variety of optional functions and precision measurements. Regardless of their complexity, however, all scopes have certain operational features in common. In this section, we examine the most basic front panel controls. Each control and its basic function are described. Figure 11–53 shows a simplified representative front panel of an oscilloscope.

Screen   In the upper portion of Figure 11–53 is the CRT screen. There are 8 vertical divisions and 10 horizontal divisions indicated with grid lines or graticules. A standard screen size is 8 cm × 10 cm. The screen is coated with phosphor which emits light when struck by the electron beam.

Power Switch and Light   This switch turns the power to the scope on and off. The light indicates when the power is on.

Intensity   The intensity control knob varies the brightness of the trace on the screen. Caution should be used so that the intensity is not left too high for an extended period of time, especially when the beam forms a motionless dot on the screen. Damage to the screen can result from excessive intensity.

Focus   This control focuses the beam so that it converges to a tiny point at the screen. An out-of-focus condition results in a fuzzy trace.

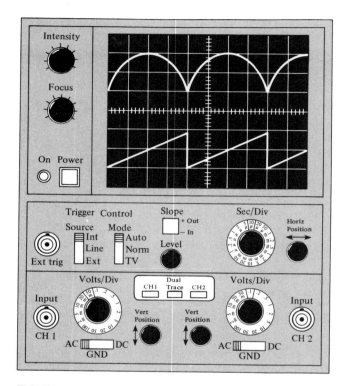

FIGURE 11–53
Representative front panel of a dual-trace oscilloscope.

Horizontal Position This control knob adjusts the neutral horizontal position of the beam. It is used to reposition horizontally a waveform display for more convenient viewing or measurement.

Seconds/Division This selector switch sets the horizontal sweep rate. It is the *time base control*. The switch selects the time interval that is to be represented by each horizontal division in seconds, milliseconds, or microseconds. The setting in Figure 11–53 is at 10 μs. Thus, each of the ten horizontal divisions represents 10 μs, so there are 100 μs from the extreme left of the screen to the extreme right.

Trigger Control The trigger controls allow the beam to be triggered from various selected sources. The triggering of the beam causes it to begin its sweep across the screen. It can be triggered from an internally generated signal derived from an input signal, or from the line voltage, or from an externally applied trigger signal. The *modes* of triggering are auto, normal, and TV. In the auto mode, sweep occurs in the absence of an adequate trigger signal. In the normal mode, a trigger signal must be present for the sweep to occur. The TV mode provides triggering on the TV field or TV line signals. The *slope* switch allows the triggering to occur on either the positive-going slope or the negative-going slope of the trigger waveform. The *level* control selects the voltage level on the trigger signal at which the triggering occurs.

Basically, the trigger controls provide for synchronization of the horizontal sweep waveform and the input signal waveform. As a result, the display of the input signal is stable on the screen, rather than appearing to drift across the screen.

Volts/Division The example scope in Figure 11–53 is a *dual-trace* type, which allows two waveforms to be displayed simultaneously. Some scopes have only single-trace capability. Notice that there are two identical volts/div selectors. There is a set of controls for each of the two *input channels*. Only one is described here, but the same applies to the other as well.

The volts/div selector switch sets the number of volts to be represented by each division on the *vertical* scale. For example, the upper waveform is applied to channel 1 and covers two vertical divisions from maximum to minimum. The volts/div switch for channel 1 is set at 50 mV, which means that each vertical division represents 50 mV. Therefore, the peak-to-peak value of the sine wave is (2 div)(50 mV/div) = 100 mV. If a lower setting were selected, the displayed wave would cover more vertical divisions. If a higher setting were selected, the displayed wave would cover fewer vertical divisions.

Notice that there is a set of three switches for selecting channel 1 (CH 1), channel 2 (CH 2), or dual trace. Either input signal can be displayed separately, or both can be displayed as illustrated.

Vertical Position The two vertical position controls move the traces up or down for easier measurement or observation.

AC-GND-DC Switch This switch, located below the volts/div control, allows the input signal to be ac coupled, dc coupled, or grounded. The ac coupling eliminates any dc component on the input signal. The dc coupling permits dc values to be displayed. The ground position allows a 0-V reference to be established on the screen.

Input The signals to be displayed are connected into the channel 1 and channel 2 input connectors. This connection is normally done with a special *probe* that minimizes the loading effect of the scope's input resistance on the circuit being measured.

AN EXAMPLE OF WAVEFORM MEASUREMENT

Figure 11–54 shows two different waveforms displayed on the screen of an oscilloscope. In Part (a), a sine wave is shown with the volts per division (volts/div) and the time per division (sec/div) controls appropriately set to measure the specific waveform values. As a rule of thumb, when measuring both amplitude and period, use the lowest settings that result in at least one full cycle and the full amplitude to be displayed.

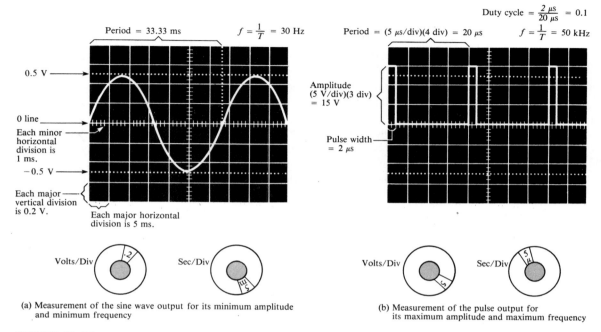

(a) Measurement of the sine wave output for its minimum amplitude and minimum frequency

(b) Measurement of the pulse output for its maximum amplitude and maximum frequency

FIGURE 11–54

SECTION REVIEW 11–7

1. What does *CRT* stand for?

2. On an oscilloscope, voltage is measured (horizontally, vertically) on the screen and time is measured (horizontally, vertical).

3. What can an oscilloscope do that a multimeter cannot?

11–8 COMPUTER ANALYSIS

The program in this section computes the values of a sinusoidal voltage, given the peak voltage and the period. In addition, the instantaneous value is calculated for any specified phase angle. A flowchart is given in Figure 11–55.

```
10   CLS
20   PRINT "THIS PROGRAM COMPUTES ALL SINE WAVE VOLTAGE VALUES
     WHEN THE"
30   PRINT "PEAK VALUE AND PERIOD ARE PROVIDED. THE
     INSTANTANEOUS"
40   PRINT "VALUE FOR A SPECIFIED PHASE ANGLE IS ALSO
     COMPUTED."
50   PRINT:PRINT:PRINT
60   INPUT "TO CONTINUE PRESS 'ENTER'";X
70   CLS
```

```
80    INPUT "THE PEAK VALUE IN VOLTS";VM
90    INPUT "THE PERIOD IN SECONDS";T
100   INPUT "INSTANTANEOUS PHASE ANGLE IN DEGREES";THETA
110   CLS
120   VPP=2*VM
130   VRMS=SQR(0.5)*VM
140   F=1/T
150   V=VM*SIN(THETA/57.29577786)
160   PRINT "VP  =";VM;"V"
170   PRINT "PHASE ANGLE =";THETA;"DEGREES"
180   PRINT "T =";T;"SECONDS"
190   PRINT "VPP =";VPP;"V"
200   PRINT "VRMS =";VRMS;"V"
210   PRINT "F =";F;"HZ"
220   PRINT "INSTANTANEOUS VOLTAGE AT";THETA;"DEGREES =";V;"V"
```

FIGURE 11–55

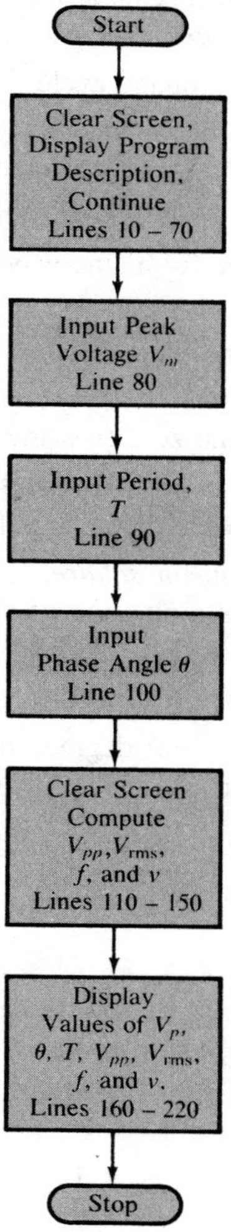

SECTION REVIEW 11–8

1. Add a line to the program to compute the average value of an alternating sine wave.

SUMMARY

1. The sine wave is a time-varying periodic waveform. Voltage or current can vary sinusoidally.

2. The sine wave is a common form of ac (alternating current). Alternating current reverses direction in response to changes in the voltage polarity.

3. A periodic waveform is one that repeats at fixed intervals. The time required for a sine wave to repeat is called the *period*.

4. The period is the time for one complete cycle.

5. Frequency is the rate of change of the sine wave in cycles per second. The unit of frequency is the *hertz*. One hertz is one cycle per second.

6. The frequency is the reciprocal of the period, and vice versa.

7. The instantaneous rate of change for a sine wave is maximum at the zero crossings and minimum at the peaks.

8. The instantaneous value of a sine wave is its value at any point in time.

9. The peak value of a sine wave is its maximum positive value or its maximum negative value measured from the zero crossing.

10. The peak-to-peak value of a sine wave is measured from the positive peak to the negative peak and is equal to twice the peak value.

11. The term *rms* stands for *root mean square*. The rms value of a sine wave, also known as the *effective* value, is a measure of the heating effect of a sine wave. It is 0.707 times the peak value.

12. The average value of a sine wave over a full cycle is zero.

13. A full cycle of a sine wave is 360 degrees or 2π radians. A half-cycle is 180 degrees or π radians. A quarter-cycle is 90 degrees or $\pi/2$ radians.

14. A phase angle is the difference in degrees or radians between two sine waves.

FORMULAS

$$f = \frac{1}{T}$$

(11–1)

$$T = \frac{1}{f}$$

(11–2)

$$V_{pp} = 2V_p \tag{11-3}$$

$$I_{pp} = 2I_p \tag{11-4}$$

$$V_{rms} = \sqrt{0.5}\, V_p \cong 0.707V_p \tag{11-5}$$

$$I_{rms} = \sqrt{0.5}\, I_p \cong 0.707I_p \tag{11-6}$$

$$V_p = \sqrt{2}\, V_{rms} \cong 1.414V_{rms} \tag{11-7}$$

$$I_p = \sqrt{2}\, I_{rms} \cong 1.414I_{rms} \tag{11-8}$$

$$V_{pp} = 2.828V_{rms} \tag{11-9}$$

$$I_{pp} = 2.828I_{rms} \tag{11-10}$$

$$v = iR \tag{11-11}$$

$$V_p = I_pR \tag{11-12}$$

$$V_{pp} = I_{pp}R \tag{11-13}$$

$$V_{rms} = I_{rms}R \tag{11-14}$$

$$f = \text{(number of pole pairs)(rps)} \tag{11-15}$$

$$\text{rad} = \left(\frac{\pi\ \text{rad}}{180°}\right) \text{degrees} \tag{11-16}$$

$$\text{degrees} = \left(\frac{180°}{\pi\ \text{rad}}\right) \text{rad} \tag{11-17}$$

$$y = A \sin \theta \tag{11-18}$$

$$y = A \sin(\theta - \phi) \tag{11-19}$$

$$y = A \sin(\theta + \phi) \tag{11-20}$$

$$\text{PRF} = \frac{1}{T} \tag{11-21}$$

$$T = \frac{1}{\text{PRF}} \tag{11-22}$$

$$\text{percent duty cycle} = \left(\frac{t_W}{T}\right)100\% \tag{11-23}$$

$$V_{avg} = \text{baseline} + \text{(duty cycle)(amplitude)} \tag{11-24}$$

SELF-TEST

Solutions appear at the end of the book.

1. How does alternating current (ac) differ from direct current (dc)?

2. How many times does a sine wave reach a peak value during each cycle?

3. Define *cycle* in relation to a sine wave.

4. What is the difference between a periodic and a nonperiodic waveform?

5. One sine wave has a frequency of 12 kHz, and another has a frequency of 20 kHz. Which sine wave is changing at a faster rate?

6. One sine wave has a period of 2 ms, and another has a period of 5 ms. Which sine wave is changing at a faster rate?

7. How many cycles does a sine wave go through in 10 seconds when its frequency is 60 Hz?

8. A certain sine wave has a period of 200 ms. What is its frequency?

9. A certain sine wave has a frequency of 25 kHz. What is its period?

10. A given sine wave takes 5 μs to complete one cycle. Determine its frequency.

11. For the sine wave in Figure 11–56, determine the peak, peak-to-peak, rms, and average values.

FIGURE 11–56

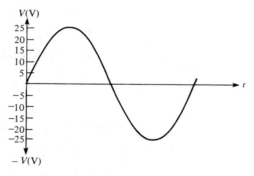

12. The rms value for a given sine wave is 115 V. What is its peak value?

13. A 10-V rms sine wave is applied to a 100-Ω resistive circuit. Determine the rms value of the current.

14. How many degrees are in a half-cycle of a sine wave?

15. How many degrees are there in two radians?

16. One sine wave has a positive-going zero crossing at 10 degrees, and another sine wave has a positive-going zero crossing at 45 degrees. What is the phase angle between the two waveforms?

17. What is the instantaneous value of a sine wave current at 32 degrees from its positive-going zero crossing if its peak value is 15 A?

18. The rms current through a 10-kΩ resistor is 5 mA. What is the rms voltage drop across the resistor?

19. Two series resistors are connected to an ac source. There are 6.5 V rms across one resistor and 3.2 V rms across the other. What is the rms source voltage?

20. One 10-kHz pulse waveform consists of pulses that are 10 μs wide, and another consists of pulses that are 50 μs wide. Which 10-kHz waveform has a higher duty cycle?

21. What is the duty cycle of a square wave?

22. Explain why a sawtooth is a special case of a triangular waveform.

PROBLEMS

Section 11–1

11–1 Calculate the frequency for each of the following values of period:
(a) 1 s   (b) 0.2 s   (c) 50 ms
(d) 1 ms   (e) 500 μs   (f) 10 μs

11–2 Calculate the period for each of the following values of frequency:
(a) 1 Hz   (b) 60 Hz   (c) 500 Hz
(d) 1 kHz   (e) 200 kHz   (f) 5 MHz

11–3 A sine wave goes through 5 cycles in 10 μs. What is its period?

11–4 A sine wave has a frequency of 50 kHz. How many cycles does it complete in 10 ms?

Section 11–2

11–5 A sine wave has a peak value of 12 V. Determine the following values:
(a) rms   (b) peak-to-peak   (c) average

11–6 A sinusoidal current has an rms value of 5 mA. Determine the following values:
(a) peak   (b) average   (c) peak-to-peak

11–7 A sinusoidal voltage is applied to the resistive circuit in Figure 11–57. Determine the following:
(a) I_{rms} (b) I_{avg} (c) I_p (d) I_{pp} (e) i at the positive peak

11–8 Determine the rms voltage across R_3 in Figure 11–58.

FIGURE 11–57

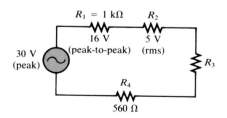

FIGURE 11–58

11–9 Figure 11–59 shows a sinusoidal voltage source in series with a dc source. Effectively, the two voltages are superimposed. Determine the power dissipation in the load resistor.

FIGURE 11–59

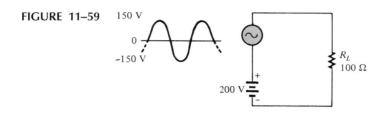

Section 11–3

11–10 The conductive loop on the rotor of a simple two-pole, single-phase generator rotates at a rate of 250 rps. What is the frequency of the induced output voltage?

11–11 A certain four-pole generator has a speed of rotation of 3600 rpm. What is the frequency of the voltage produced by this generator?

11–12 At what speed of rotation must a four-pole generator be operated to produce a 400-Hz sine wave voltage?

Section 11–4

11–13 Convert the following angular values from degrees to radians:
(a) 30°   **(b)** 45°   **(c)** 78°
(d) 135°   **(e)** 200°   **(f)** 300°

11–14 Convert the following angular values from radians to degrees:
(a) $\pi/8$ **(b)** $\pi/3$ **(c)** $\pi/2$
(d) $3\pi/5$ **(e)** $6\pi/5$ **(f)** 1.8π

11–15 Sine wave A has a positive-going zero crossing at 30 degrees. Sine wave B has a positive-going zero crossing at 45 degrees. Determine the phase angle between the two signals. Which signal leads?

11–16 One sine wave has a positive peak at 75 degrees, and another has a positive peak at 100 degrees. How much is each sine wave shifted in phase from the 0-degree reference? What is the phase angle between them?

11–17 Make a sketch of two sine waves as follows: Sine wave A is the reference, and sine wave B lags A by 90 degrees. Both have equal amplitudes.

Section 11–5

11–18 A certain sine wave has a positive-going zero crossing at 0 degrees and an rms value of 20 V. Calculate its instantaneous value at each of the following angles:
(a) 15°   **(b)** 33°   **(c)** 50°
(d) 110°   **(e)** 70°   **(f)** 145°
(g) 250°   **(h)** 325°

11–19 For a particular 0 degree reference sinusoidal current, the peak value is 100 mA. Determine the instantaneous value at each of the following points:
(a) 35° (b) 95° (c) 190°
(d) 215° (e) 275° (f) 360°

11–20 For a 0-degree reference sine wave with an rms value of 6.37 V, determine its instantaneous value at each of the following points:
(a) $\pi/8$ radians (b) $\pi/4$ radians (c) $\pi/2$ radians
(d) $3\pi/4$ radians (e) π radians (f) $3\pi/2$ radians
(g) 2π radians

11–21 Sine wave A lags sine wave B by 30 degrees. Both have peak values of 15 V. Sine wave A is the reference with a positive-going crossing at 0 degrees. Determine the instantaneous value of sine wave B at 30 degrees, 45 degrees, 90 degrees, 180 degrees, 200 degrees, and 300 degrees.

11–22 Repeat Problem 11–21 for the case when sine wave A *leads* sine wave B by 30 degrees.

11–23 A certain sine wave has a frequency of 2.2 kHz and an rms value of 25 V. Assuming a given cycle begins (zero crossing) at $t = 0$ s, what is the change in voltage from 0.12 ms to 0.2 ms?

Section 11–6

11–24 From the graph in Figure 11–60, determine the approximate values of t_r, t_f, t_w, and amplitude.

FIGURE 11–60

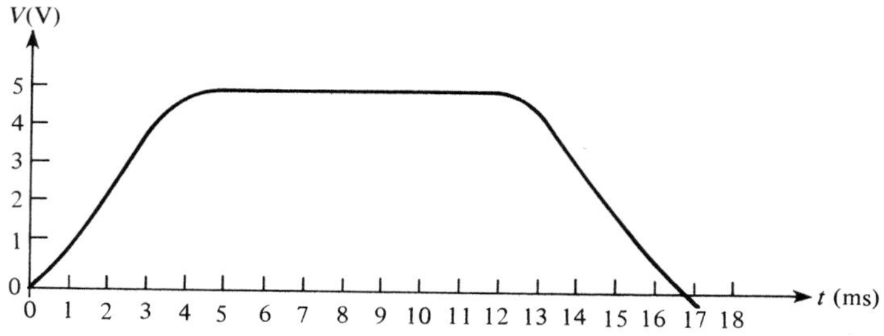

$V(V)$

```
5 ┤        ╭──────────╮
4 ┤       ╱            ╲
3 ┤      ╱              ╲
2 ┤     ╱                ╲
1 ┤    ╱                  ╲
0 ┼──┬──┬──┬──┬──┬──┬──┬──┬──┬──┬──┬──┬──┬──┬──┬──┬──┬──┬→ t (ms)
   0  1  2  3  4  5  6  7  8  9 10 11 12 13 14 15 16 17 18
```

11–25 The repetition frequency of a pulse waveform is 2 kHz, and the pulse width is 1 μs. What is the percent duty cycle?

11–26 Calculate the average value of the pulse waveform in Figure 11–61.

FIGURE 11–61

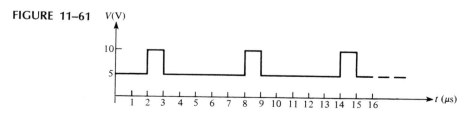

$V(V)$

11–27 Determine the duty cycle for each waveform in Figure 11–62.

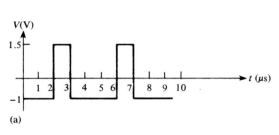

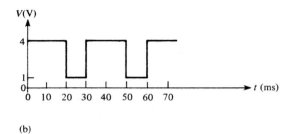

(a)

(b)

FIGURE 11–62

11–28 Find the average value of each pulse waveform in Figure 11–62.

11–29 What is the frequency of each waveform in Figure 11–62?

11–30 What is the frequency of each sawtooth waveform in Figure 11–63?

FIGURE 11–63

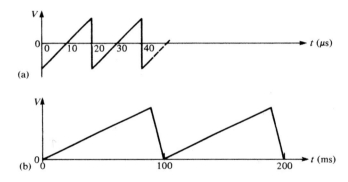

11–31 A nonsinusoidal waveform called a *stairstep* is shown in Figure 11–64. Determine its average value.

FIGURE 11–64

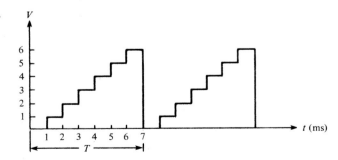

11–32 A square wave has a period of 40 μs. List the first six odd harmonics.

11–33 What is the fundamental frequency of the square wave mentioned in Problem 11–32?

Section 11–8

11–34 Modify the program in Section 11–7 to compute sinusoidal current values instead of voltage.

11–35 Write a program to convert degrees to radians.

11–36 Develop a program to convert complex numbers in rectangular form into polar form.

ANSWERS TO SECTION REVIEWS

Section 11–1
1. From the zero crossing through a positive peak, then through zero to a negative peak and back to the zero crossing.
2. At the zero crossings.   **3.** 2.
4. From one zero crossing to the next *corresponding* zero crossing, or from one peak to the next *corresponding* peak.
5. The number of cycles completed in one second; hertz.   **6.** 200 kHz.
7. 8.33 ms.

Section 11–2
1. (a) 2 V; **(b)** 4 V.   **2. (a)** 1.77 V; **(b)** 3.54 V.   **3.** 0.5 A; 0.707 A.

Section 11–3
1. Faraday's law and Lenz's law.   **2.** Directly.

Section 11–4
1. (a) 90°; **(b)** 180°; **(c)** 270°; **(d)** 360°.   **2.** 180; π.   **3.** 360; 2π.   **4.** 45°.

Section 11–5
1. 8.66 V.   **2.** 8.19 V.   **3.** 4.53 V.

Section 11–6
1. (a) The time interval from 10% to 90% of the rising pulse edge; **(b)** The time interval from 90% to 10% of the falling pulse edge; **(c)** The time interval from 50% of the leading pulse edge to 50% of the trailing pulse edge.
2. 1 kHz.   **3.** 20%, 1.5 V, 0.8 V.   **4.** 16 ms.   **5.** 1 MHz.
6. The repetition rate of the waveform.   **7.** 2 kHz.   **8.** 100 kHz.

Section 11–7
1. Cathode-ray tube.   **2.** Vertically, horizontally.
3. It can display time-varying quantities.

Section 11–8
1. DPR 135 VAVG = 0

TWELVE

PHASORS AND COMPLEX NUMBERS

In this chapter, two important tools for the analysis of ac circuits are introduced. These are *phasors* and *complex numbers*. You will see how phasors are a convenient, graphic means of representing sine wave voltages and currents in terms of their magnitude and phase angle. In later chapters you will see how phasors can also represent other ac circuit quantities.

The complex number system is a means for expressing phasor quantities and for performing mathematical operations with those quantities.

In this chapter, you will learn:

☐ What a phasor is and how it is used to represent sine waves.
☐ How to create a phasor diagram.
☐ How the formula for a sine wave is derived from the phasor diagram.
☐ How to express the sine wave in terms of the phasor's angular velocity.
☐ What the complex plane is.
☐ How to represent a point on the complex plane.
☐ The meaning of real and imaginary numbers.
☐ How a phasor is expressed as a complex number.
☐ How to express a complex number in either rectangular or polar form.
☐ How to convert from one form of complex number to the other.
☐ How to do mathematical operations with complex numbers.
☐ How complex numbers are useful in ac circuits.

12–1   INTRODUCTION TO PHASORS

Phasors provide a graphic means for representing quantities that have both magnitude and direction (angle). Phasors are especially useful for representing sine waves in terms of their amplitude and phase angle and also for analysis of reactive circuits studied in later chapters.

Examples of phasors are shown in Figure 12–1. The *length* of the phasor "arrow" represents the magnitude of a quantity. The angle, θ (relative to 0 degrees), represents the angular position, as shown in Part (a). The specific phasor example in Part (b) has a magnitude of 2 and a phase angle of 45 degrees. The phasor in Part (c) has a magnitude of 3 and a phase angle of 180 degrees. The phasor in Part (d) has a magnitude of 1 and a phase angle of -45 degrees (or $+315$ degrees).

FIGURE 12–1
Examples of phasors.

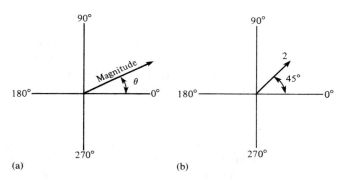

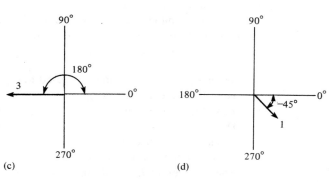

PHASOR REPRESENTATION OF A SINE WAVE

A full cycle of a sine wave can be represented by rotation of a phasor through 360 degrees.

The instantaneous value of the sine wave at any point is equal to the vertical distance from the tip of the phasor to the horizontal axis.

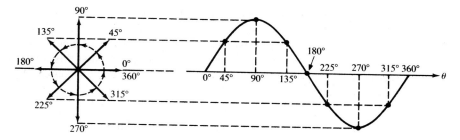

FIGURE 12–2
Sine wave represented by rotational phasor motion.

Figure 12–2 shows how the phasor "traces out" the sine wave as it goes from 0 to 360 degrees. You can relate this concept to the rotation in an ac generator (refer back to Chapter 11).

Notice in Figure 12–2 that the length of the phasor is equal to the *peak* value of the sine wave (observe the 90-degree and the 270-degree points). The angle of the phasor measured from 0 degrees is the corresponding angular point on the sine wave.

PHASORS AND THE SINE WAVE FORMULA

Let's examine a phasor representation at one specific angle. Figure 12–3 shows a voltage phasor at an angular position of 45 degrees and the corresponding point on the sine wave. The instantaneous value of the sine wave at this point is related to both the position and the length of the phasor. As previously mentioned, the vertical distance from the phasor tip down to the horizontal axis represents the instantaneous value of the sine wave at that point.

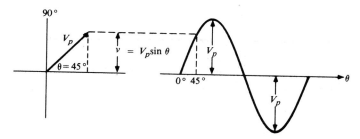

FIGURE 12–3
Right triangle derivation of sine wave formula.

Notice that when a vertical line is drawn from the phasor tip down to the horizontal axis, a *right triangle* is formed as shown in the figure. The length of the phasor is the *hypotenuse* of the triangle, and the vertical projection is the *opposite side*.

From trigonometry, *the opposite side of a right triangle is equal to the hypotenuse times the sine of the angle θ*. In this case, the length of the phasor is the *peak* value of the sine wave voltage, V_p. Thus, the opposite side of the triangle, which is the instantaneous value, can be expressed as $v = V_p\sin\theta$. Recall that this formula is the one stated earlier for calculating instantaneous sine wave values. Of course, this also applies to a sine wave of current.

POSITIVE AND NEGATIVE PHASOR ANGLES

The position of a phasor at any instant can be expressed as a *positive* angle, as you have seen, or as an *equivalent negative angle*. Positive angles are measured counterclockwise from 0 degrees. Negative angles are measured clockwise from 0 degrees. For a given positive angle θ, the corresponding negative angle is $\theta - 360°$, as illustrated in Figure 12–4(a). In Part (b), a specific example is shown. The angle of the phasor in this case can be expressed as $+225$ degrees or -135 degrees.

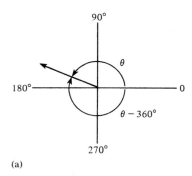
(a)

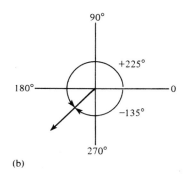
(b)

FIGURE 12–4
Positive and negative phasor angles.

EXAMPLE 12–1

For each phasor in Figure 12–5, determine the instantaneous sine wave value. Also express each positive angle shown as an equivalent negative angle. The length of each phasor represents the peak value of the sine wave.

Solution:
(a) $\theta = 0°$
$v = 10 \sin 0° = 10(0) = 0$ V
(b) $\theta = 30° = -330°$
$v = 10 \sin 30° = 10(0.5) = 5$ V
(c) $\theta = 90° = -270°$
$v = 10 \sin 90° = 10(1) = 10$ V
(d) $\theta = 135° = -225°$
$v = 10 \sin 135° = 10(0.707) = 7.07$ V

FIGURE 12–5

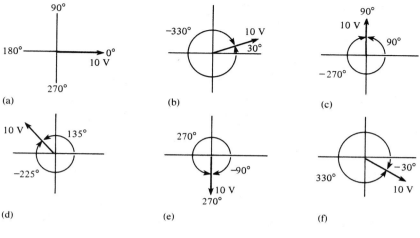

(a)

(b)

(c)

(d)

(e)

(f)

(e) $\theta = 270° = -90°$
 $v = 10 \sin 270° = 10(-1) = -10 \text{ V}$
(f) $\theta = 330° = -30°$
 $v = 10 \sin 330° = 10(-0.5) = -5 \text{ V}$

The equivalent negative angles are shown in Figure 12–5.

PHASOR DIAGRAMS

A phasor diagram can be used to show the *relative relationship* of two or more sine waves of the same frequency. A phasor in a *fixed* position represents a *complete* sine wave, because once the phase angle between two or more sine waves of the same frequency is established, it remains *constant throughout the cycles*. For example, the two sine waves in Figure 12–6(a) can be represented by a phasor diagram, as shown in Part (b). As you can see, sine wave *A* leads sine wave *A* by 30 degrees.

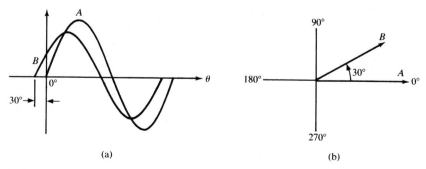

(a)

(b)

FIGURE 12–6
Example of a phasor diagram.

**EXAMPLE
12–2**

Use a phasor diagram to represent the sine waves in Figure 12–7.

FIGURE 12–7

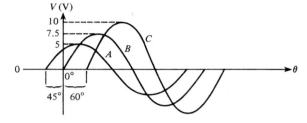

Solution:
The phasor diagram representing the sine waves is shown in Figure 12–8. In this case, the length of each phasor represents the peak value of the sine wave.

FIGURE 12–8

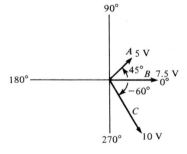

ANGULAR VELOCITY OF A PHASOR

As you have seen, one cycle of a sine wave is traced out when a phasor is rotated through 360 degrees. The faster it is rotated, the faster the sine wave cycle is traced out. Thus, the *period* and *frequency* are related to the *velocity of rotation* of the phasor. The velocity of rotation is called the *angular velocity* and is designated ω (the Greek letter omega).

When a phasor rotates through 360 degrees or 2π radians, one complete cycle is traced out. Therefore, the time required for the phasor to go through 2π radians is the *period* of the sine wave. Because the phasor rotates through 2π radians in a time equal to the period T, the angular velocity can be expressed as

$$\omega = \frac{2\pi}{T}$$

Since $f = 1/T$,

$$\omega = 2\pi f \tag{12-1}$$

When a phasor is rotated at a velocity ω, then ωt is the *angle through which the phasor has passed at any instant*. Therefore, the following relationship can be stated:

$$\theta = \omega t \tag{12-2}$$

With this relationship between angle and time, the equation for the instantaneous value of sine wave voltage can be written as

$$v = V_p \sin \omega t \tag{12-3}$$

The instantaneous value can be calculated at any point in time along the sine wave curve if the frequency and peak value are known. The unit of ωt is the radian.

EXAMPLE 12–3

What is the value of a sine wave voltage at 3 μs from the positive-going zero crossing when $V_p = 10$ V and $f = 50$ kHz?

Solution:

$$v = V_p \sin \omega t$$
$$= 10 \sin[2\pi(50 \times 10^3 \text{ rad/s})(3 \times 10^{-6} \text{ s})]$$
$$= 8.09 \text{ V}$$

SECTION REVIEW 12–1

1. What is a *phasor?*
2. What is the angular velocity of a phasor representing a sine wave with a frequency of 1500 Hz?
3. A certain phasor has an angular velocity of 628 rad/s. To what frequency does this correspond?
4. Sketch a phasor diagram to represent the two sine waves in Figure 12–9. Use peak values.

FIGURE 12–9

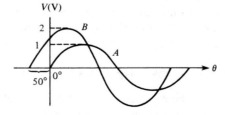

12–2   THE COMPLEX NUMBER SYSTEM

Complex numbers allow mathematical operations with phasor quantities and are very useful in the analysis of ac circuits. With the complex number system, we can add, subtract, multiply, and divide quantities that have both magnitude and angle, such as sine waves and other ac circuit quantities.

POSITIVE AND NEGATIVE NUMBERS

Positive numbers can be represented by points to the right of the origin on the horizontal axis of a graph, and negative numbers can be represented by points to the left of the origin, as illustrated in Figure 12–10(a). Also, positive numbers can be represented by points on the vertical axis above the origin, and negative numbers can be represented by points below the origin, as shown in Figure 12–10(b).

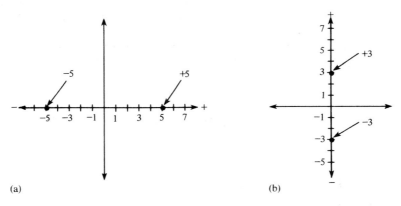

(a)   (b)

FIGURE 12–10
Graphic representation of positive and negative numbers.

THE COMPLEX PLANE

To distinguish between values on the horizontal axis and values on the vertical axis, a *complex plane* is used. In the complex plane, the horizontal axis is called the *real axis,* and the vertical axis is called the *imaginary axis,* as shown in Figure 12–11.

In electrical circuit work, a $\pm j$ prefix is used to designate numbers that lie on the imaginary axis in order to distinguish them from numbers lying on the real axis. This prefix is known as the *j operator.* In mathematics, an *i* is used instead of a *j,* but in electric circuits, the *i* can be confused with instantaneous current, so *j* is used.

ANGULAR POSITION ON THE COMPLEX PLANE

Angular positions can be represented on the complex plane, as shown in Figure 12–12. The positive real axis represents 0 degrees. Proceeding counterclockwise, the $+j$ axis represents 90 degrees, the negative real axis represents 180 degrees, the $-j$ axis is the

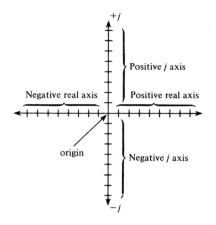

FIGURE 12–11
The complex plane.

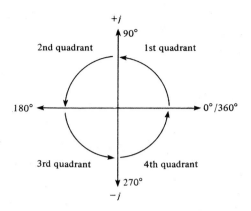

FIGURE 12–12
Angles on the complex plane.

270-degree point, and, after a full rotation of 360 degrees, we are back to the positive real axis. Notice that the plane is sectioned into four *quadrants*.

REPRESENTING A POINT ON THE COMPLEX PLANE

A point located on the complex plane can be classified as real, imaginary ($\pm j$), or a combination of the two. For example, a point located 4 units from the origin on the positive real axis is the *positive real number* $+4$, as shown in Figure 12–13(a). A point 2 units from the origin on the negative real axis is the *negative real number*, -2, as shown in Part (b). A point on the $+j$ axis 6 units from the origin, as in Part (c), is designated $+j6$. Finally, a point 5 units along the $-j$ axis is designated $-j5$, as in Part (d).

When a point lies not on any axis but somewhere in one of the four quadrants, it is a *complex number* and can be defined by its *coordinates*. For example, in Figure 12–14, the point located in the first quadrant has a real value of $+4$ and a j value of $+j4$. The point located in the second quadrant has coordinates -3 and $+j2$ and is expressed -3, $+j2$. The point located in the third quadrant has coordinates -3 and $-j5$. The point located in the fourth quadrant has coordinates of $+6$ and $-j4$.

VALUE OF j

If we multiply the positive real value of $+2$ by j, the result is $+j2$. This multiplication has effectively moved the $+2$ through a 90-degree angle to the $+j$ axis. Similarly, multiplying $+2$ by $-j$ rotates it -90 degrees to the $-j$ axis.

Mathematically, the j operator has a value of $\sqrt{-1}$. If $+j2$ is multiplied by j, we get

$$j^2 2 = (\sqrt{-1})(\sqrt{-1})(2) = (-1)(2) = -2$$

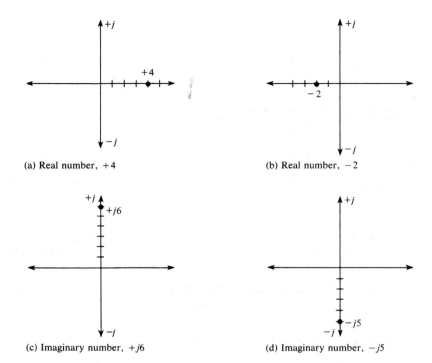

(a) Real number, +4

(b) Real number, −2

(c) Imaginary number, +j6

(d) Imaginary number, −j5

FIGURE 12–13
Real and imaginary (j) numbers on the complex plane.

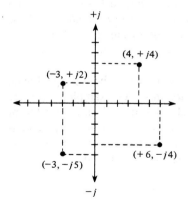

FIGURE 12–14
Coordinate points on the complex plane.

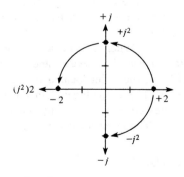

FIGURE 12–15
Effect of the j operator on location of a number on the complex plane.

This calculation effectively places the value on the negative real axis. Therefore, multiplying a positive real number by j^2 converts it to a negative real number, which, in effect, is a rotation of 180 degrees on the complex plane. These operations are illustrated in Figure 12–15.

EXAMPLE 12–4

(a) Locate the following points on the complex plane: 7, $j5$; 5, $-j2$; -3.5, $j1$; and -5.5, $-j6.5$.

(b) Determine the coordinates for each point in Figure 12–16.

FIGURE 12–16

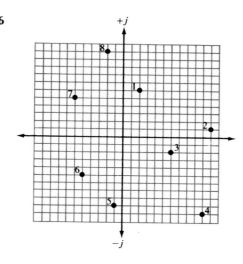

Solution:

(a) See Figure 12–17.

(b)

1: 2, $j6$	2: 11, $j1$	3: 6, $-j2$	4: 10, $-j10$
5: -1, $-j9$	6: -5, $-j5$	7: -6, $j5$	8: -2, $j11$

FIGURE 12–17

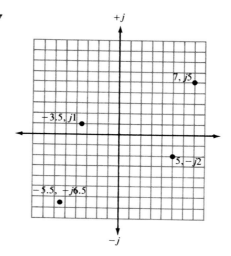

SECTION REVIEW 12–2

1. Locate the following points on the complex plane:
   (a) $+3$ (b) -4 (c) $+j1$
2. What is the angular difference between the following numbers:
   (a) $+4$ and $+j4$ (b) $+j6$ and -6 (c) $+j2$ and $-j2$

12–3

RECTANGULAR AND POLAR FORMS OF COMPLEX NUMBERS

There are two forms of complex numbers that are used to represent phasor quantities: the *rectangular form* and the *polar form*. Each has certain advantages when used in circuit analysis, depending on the particular application.

As you know, a phasor quantity contains both *magnitude* and *phase*. In this text, italic letters such as V and I are used to represent magnitude only, and boldface letters such as **V** and **I** are used to represent complete phasor quantities. Other circuit quantities that can be expressed in phasor form will be introduced later.

RECTANGULAR FORM

A phasor quantity is represented in *rectangular form* by the algebraic sum of the *real value* of the coordinate and the *j value* of the coordinate. An "arrow" drawn from the origin to the coordinate point in the complex plane is used to represent *graphically* the phasor quantity. Examples are $1 + j2$, $5 - j3$, $-4 + j4$, and $-2 - j6$, which are shown on the complex plane in Figure 12–18. As you can see, the rectangular coordinates describe the phasor in terms of its values projected onto the real axis and the *j* axis.

FIGURE 12–18
Examples of phasors specified by rectangular coordinates.

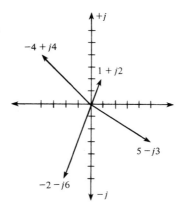

POLAR FORM

Phasor quantities can also be expressed in *polar form,* which consists of the phasor magnitude and the angular position relative to the positive real axis. Examples are $2\angle 45°$, $5\angle 120°$, and $8\angle -30°$. The first number is the *magnitude,* and the symbol $\angle$ precedes the value of the *angle.* Figure 12–19 shows these phasors on the complex plane. The length of the phasor, of course, represents the magnitude of the quantity. Keep in mind that for every phasor expressed in polar form, there is also an equivalent expression in rectangular form.

FIGURE 12–19
Examples of phasors specified by polar values.

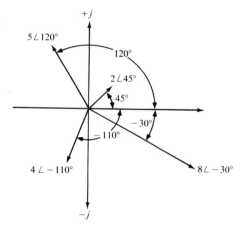

CONVERSION FROM RECTANGULAR TO POLAR FORM

Most scientific calculators have provisions for conversion between rectangular and polar forms. However, we discuss the basic conversion method here so that you will understand the mathematical procedure.

A phasor can exist in any of the four quadrants of the complex plane, as indicated in Figure 12–20. The phase angle θ in each case is measured relative to the positive real axis (0°) as shown.

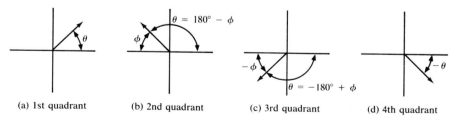

| (a) 1st quadrant | (b) 2nd quadrant | (c) 3rd quadrant | (d) 4th quadrant |

FIGURE 12–20
All possible phasor quadrant locations. θ is the angle relative to the positive real axis in each case.

A phasor can be visualized as forming a *right triangle* in the complex plane, as indicated in Figure 12–21, for each quadrant location. The horizontal side of the triangle is the real value, A, and the vertical side is the j value, B. The hypotenuse of the triangle is the length of the phasor, C, representing the magnitude, and can be expressed as

$$C = \sqrt{A^2 + B^2} \tag{12–4}$$

using the Pythagorean theorem.

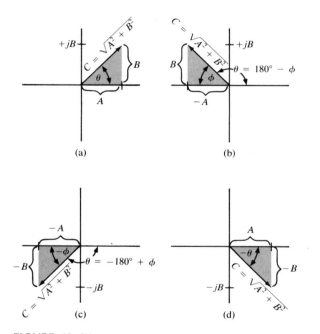

FIGURE 12–21
Right angle relationships in the complex plane.

Next, the angle, θ, indicated in Figure 12–21(a) and (d) is expressed as an *inverse tangent* function.

$$\theta = \tan^{-1}\left(\frac{\pm B}{A}\right) \tag{12–5}$$

The angle θ indicated in Figure 12–21(b) and (c) is

$$\theta = \pm 180° \mp \tan^{-1}\left(\frac{B}{A}\right) \tag{12–6}$$

In each case the appropriate signs must be used in the calculation. Note that $\tan^{-1}$ is INV TAN on some calculators. The general formula for converting from rectangular to polar is as follows:

$$\pm A \pm jB = C\angle \pm \theta \tag{12–7}$$

The following example illustrates the conversion procedure.

EXAMPLE 12–5

Convert the following complex numbers from rectangular form to polar form:
(a) $8 + j6$ (b) $10 - j5$
(c) $-12 - j18$ (d) $-7 + j10$

Solution:
(a) The magnitude of the phasor represented by $8 + j6$ is

$$C = \sqrt{8^2 + 6^2} = \sqrt{100} = 10$$

The angle is

$$\theta = \tan^{-1}\left(\frac{6}{8}\right) = 36.87°$$

Since the phasor is in the first quadrant, θ is the angle relative to the positive real axis. The complete polar expression for this phasor is

$$\mathbf{C} = 10\angle 36.87°$$

(b) The magnitude of the phasor represented by $10 - j5$ is

$$C = \sqrt{10^2 + (-5)^2} = \sqrt{125} = 11.18$$

The angle is

$$\theta = \tan^{-1}\left(\frac{-5}{10}\right) = -26.57°$$

Since the phasor is in the fourth quadrant, θ is the angle relative to the positive real axis. The complete polar expression for this phasor is

$$\mathbf{C} = 11.18\angle -26.57°$$

(c) The magnitude of the phasor represented by $-12 - j18$ is

$$C = \sqrt{(-12)^2 + (-18)^2} = \sqrt{468} = 21.63$$

The angle is

$$\theta = \tan^{-1}\left(\frac{-18}{-12}\right) = -56.31°$$

Since the phasor is in the third quadrant, θ is the angle relative to the *negative* real axis.

    The angle relative to the positive real axis which is the standard reference is

$$\phi = -(180° - \theta) = -(180° - 56.31°) = -123.69°$$

The complete polar expression for this phasor is

$$\mathbf{C} = 21.63\angle - 123.69°$$

(d) The magnitude of the phasor represented by $-7 + j10$ is

$$C = \sqrt{(-7)^2 + 10^2} = \sqrt{149} = 12.21$$

The angle is

$$\theta = \tan^{-1}\left(\frac{10}{-7}\right) = -55°$$

Since the phasor is in the second quadrant, θ is the angle relative to the *negative* real axis. The angle relative to the positive real axis is

$$\phi = 180° - \theta = 180° - 55° = 125°$$

The complete polar expression for this phasor is

$$\mathbf{C} = 12.21\angle 125°$$

    The calculator sequences are

(a) $\boxed{8}$ $\boxed{a}$ $\boxed{6}$ $\boxed{b}$ $\boxed{\text{2nd F}}$ $\boxed{\rightarrow r\theta}$ press $\boxed{b}$ for angle
(b) $\boxed{1}$ $\boxed{0}$ $\boxed{a}$ $\boxed{5}$ $\boxed{+/-}$ $\boxed{b}$ $\boxed{\text{2nd F}}$ $\boxed{\rightarrow r\theta}$ press $\boxed{b}$ for angle
(c) $\boxed{1}$ $\boxed{2}$ $\boxed{+/-}$ $\boxed{a}$ $\boxed{1}$ $\boxed{8}$ $\boxed{+/-}$ $\boxed{b}$ $\boxed{\text{2nd F}}$ $\boxed{\rightarrow r\theta}$ press $\boxed{b}$ for angle
(d) $\boxed{7}$ $\boxed{+/-}$ $\boxed{a}$ $\boxed{1}$ $\boxed{0}$ $\boxed{b}$ $\boxed{\text{2nd F}}$ $\boxed{\rightarrow r\theta}$ press $\boxed{b}$ for angle

CONVERSION FROM POLAR TO RECTANGULAR FORM

The polar form gives the magnitude and angle of a phasor quantity, as indicated in Figure 12–22. To get the rectangular form, sides A and B of the triangle must be found, using the rules from trigonometry stated below:

$$A = C \cos(\pm\theta) \tag{12–8}$$

$$B = C \sin(\pm\theta) \tag{12–9}$$

FIGURE 12–22
Polar components of a phasor.

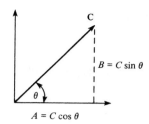

The general polar-to-rectangular conversion formula is as follows:

$$C\angle \pm\theta = C\cos(\pm\theta) + jC\sin(\pm\theta) = \pm A \pm jB \qquad \textbf{(12–10)}$$

The following example demonstrates this conversion.

**EXAMPLE
12–6**

Convert the following polar quantities to rectangular form:
(a) $10\angle 30°$ **(b)** $200\angle -45°$ **(c)** $4\angle 135°$

Solution:
(a) The real part of the phasor represented by $10\angle 30°$ is

$$A = 10\cos 30° = 10(0.866) = 8.66$$

The j part of this phasor is

$$jB = j10\sin 30° = j10(0.5) = j5$$

The complete rectangular expression is

$$A + jB = 8.66 + j5$$

(b) The real part of the phasor represented by $200\angle -45°$ is

$$A = 200\cos(-45°) = 200(0.707) = 141.4$$

The j part is

$$jB = j200\sin(-45°) = j200(-0.707) = -j141.4$$

The complete rectangular expression is

$$A + jB = 141.4 - j141.4$$

(c) The real part of the phasor represented by $4\angle 135°$ is

$$A = 4\cos 135° = 4(-0.707) = -2.828$$

The j part is

$$jB = j4 \sin 135° = 4(0.707) = 2.828$$

The complete rectangular expression is

$$A + jB = -2.828 + j2.828$$

The calculator sequences are

(a) [1] [0] [a] [3] [0] [b] [2nd F] [→xy] press [b] for j part
(b) [2] [0] [0] [a] [4] [5] [+/−] [b] [2nd F] [→xy] press [b] for j part
(c) [4] [a] [1] [3] [5] [b] [2nd F] [→xy] press [b] for j part

SECTION REVIEW 12–3

1. Name the two parts of a complex number in rectangular form.
2. Name the two parts of a complex number in polar form.
3. Convert $2 + j2$ to polar form. In which quadrant does this phasor lie?
4. Convert $5\angle -45°$ to rectangular form. In which quadrant does this phasor lie?

12–4 MATHEMATICAL OPERATIONS WITH COMPLEX NUMBERS

ADDITION

Complex numbers must be in rectangular form in order to add them. The rule is:

Add the real parts of each complex number to get the real part of the sum. Then add the j parts of each complex number to get the j part of the sum.

EXAMPLE 12–7

Add the following sets of complex numbers:
(a) $8 + j5$ and $2 + j1$
(b) $20 - j10$ and $12 + j6$

Solution:
(a) $(8 + j5) + (2 + j1) = (8 + 2) + j(5 + 1) = 10 + j6$
(b) $(20 - j10) + (12 + j6) = (20 + 12) + j(-10 + 6)$
$$= 32 + j(-4) = 32 - j4$$

SUBTRACTION

As in addition, the numbers must be in rectangular form to be subtracted. The rule is:

Subtract the real parts of the numbers to get the real part of the difference, and subtract the *j* parts of the numbers to get the *j* part of the difference.

EXAMPLE 12–8

(a) Subtract $1 + j2$ from $3 + j4$.
(b) Subtract $10 - j8$ from $15 + j15$.

Solution:
(a) $(3 + j4) - (1 + j2) = (3 - 1) + j(4 - 2) = 2 + j2$
(b) $(15 + j15) - (10 - j8) = (15 - 10) + j[15 - (-8)] = 5 + j23$

MULTIPLICATION

Multiplication of two complex numbers in rectangular form is accomplished by multiplying, in turn, each term in one number by both terms in the other number and then combining the resulting real terms and the resulting *j* terms (recall that $j \times j = -1$). As an example,

$$(5 + j3)(2 - j4) = 10 + j6 - j20 + 12 = 22 - j14$$

Multiplication of two complex numbers is most easily performed with both numbers in polar form. The rule is:

Multiply the magnitudes, and add the angles algebraically.

EXAMPLE 12–9

Perform the following multiplications:
(a) $10\angle 45°$ times $5\angle 20°$.
(b) $2\angle 60°$ times $4\angle -30°$

Solution:
(a) $(10\angle 45°)(5\angle 20°) = (10)(5)\angle(45° + 20°) = 50\angle 65°$
(b) $(2\angle 60°)(4\angle -30°) = (2)(4)\angle[60° + (-30°)] = 8\angle 30°$

DIVISION

Division of two complex numbers in rectangular form is accomplished by multiplying both the numerator and the denominator by the complex conjugate of the denominator and then combining terms and simplifying. The complex conjugate of a number is found by changing the sign of the j term. As an example,

$$\frac{10 + j5}{2 + j4} = \frac{(10 + j5)(2 - j4)}{(2 + j4)(2 - j4)} = \frac{20 - j30 + 20}{4 + 16}$$

$$= \frac{40 - j30}{20} = 2 - j1.5$$

Like multiplication, division is done most easily when the numbers are in polar form. The rule is:

> **Divide the magnitude of the numerator by the magnitude of the denominator to get the magnitude of the quotient, and subtract the denominator angle from the numerator angle to get the angle of the quotient.**

EXAMPLE 12–10

Perform the following divisions:
(a) Divide $100\angle50°$ by $25\angle20°$.
(b) Divide $15\angle10°$ by $3\angle-30°$.

Solution:

(a) $\dfrac{100\angle50°}{25\angle20°} = \left(\dfrac{100}{25}\right)\angle(50° - 20°) = 4\angle30°$

(b) $\dfrac{15\angle10°}{3\angle-30°} = \left(\dfrac{15}{3}\right)\angle[10° - (-30°)] = 5\angle40°$

APPLICATION OF COMPLEX NUMBERS TO SINE WAVES

Since sine waves can be represented by phasors, they can be described in terms of complex numbers in rectangular or polar form. For example, four series sine wave voltage sources are shown in Figure 12–23. The sine waves are graphed in Figure

FIGURE 12–23
Superimposed sine wave sources applied to a load.

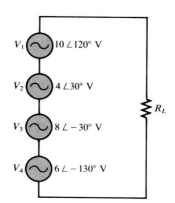

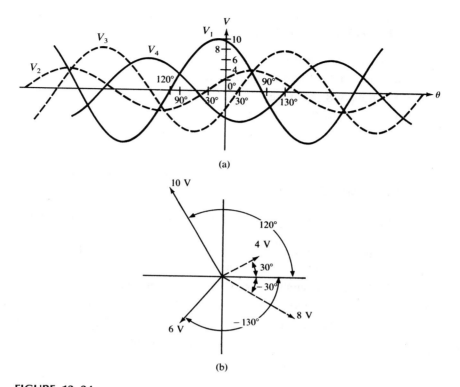

FIGURE 12–24
A phasor diagram representing four out-of-phase sine waves.

12–24(a), and the phasor representation is shown in Part (b). The total voltage across the load in Figure 12–23 can be determined by adding each phasor in complex form as follows:

$$\mathbf{V}_T = \mathbf{V}_1 + \mathbf{V}_2 + \mathbf{V}_3 + \mathbf{V}_4 = 10\angle 120° \text{ V} + 4\angle 30° \text{ V} + 8\angle -30° \text{ V}$$
$$+ 6\angle -130° \text{ V}$$
$$= (-5 \text{ V} + j8.66 \text{ V}) + (3.46 \text{ V} + j2 \text{ V}) + (6.93 \text{ V} - j4\text{V})$$
$$+ (-3.86 \text{ V} - j4.60)$$
$$= 1.53 \text{ V} + j2.06 \text{ V} = 2.57\angle 56.83° \text{ V}$$

The total voltage has a peak value of 2.57 V and a phase angle of 56.83°.

As you have seen, sine waves can be represented in complex form and can be added, subtracted, multiplied, and divided using the rules that we have discussed. Also, as you will learn later, other electrical quantities, such as capacitive and inductive reactances, impedance, and power, can be described in complex form to ease many circuit analysis problems.

SECTION REVIEW 12–4

1. Add $1 + j2$ and $3 - j1$.
2. Subtract $12 + j18$ from $15 + j25$.
3. Multiply $8\angle45°$ times $2\angle65°$.
4. Divide $30\angle75°$ by $6\angle60°$.

SUMMARY

1. A phasor represents a time-varying quantity in terms of both magnitude and direction.
2. The angular position of a phasor represents the angle of the sine wave, and the length of a phasor represents the amplitude.
3. A complex number represents a phasor quantity.
4. Complex numbers can be added, subtracted, multiplied, and divided.
5. The rectangular form of a complex number consists of a real part and a j part.
6. The polar form of a complex number consists of a magnitude and an angle.

FORMULAS

$$\omega = 2\pi f \tag{12-1}$$

$$\theta = \omega t \tag{12-2}$$

$$v = V_p\sin \omega t \tag{12-3}$$

$$C = \sqrt{A^2 + B^2} \tag{12-4}$$

$$\theta = \tan^{-1}\left(\frac{\pm B}{A}\right) \tag{12-5}$$

$$\theta = \pm180° \mp \tan^{-1}\left(\frac{B}{A}\right) \tag{12-6}$$

$$\pm A \pm jB = C\angle\pm\theta \tag{12-7}$$

$$A = C \cos(\pm\theta) \tag{12-8}$$

$$B = C \sin(\pm\theta) \tag{12-9}$$

$$C\angle\pm\theta = C \cos(\pm\theta) + jC \sin(\pm\theta) = \pm A \pm jB \tag{12-10}$$

SELF-TEST

Solutions appear at the end of the book.

1. What types of quantities can phasors represent?

2. Express each of the following angles as an equivalent negative angle:
   (a) 20°   **(b)** 60°   **(c)** 135°
   (d) 200°   **(e)** 315°   **(f)** 330°

3. If the angular velocity, ω, is 1000 rad/s, what is the frequency?

4. Locate the following points in the complex plane:
   (a) $1 + j2$ **(b)** $3 + j4$ **(c)** $5 - j3$ **(d)** $-2 + j3$ **(e)** $-1 - j2$

5. Sketch the phasors specified by the following polar coordinates:
   (a) $2\angle 45°$ **(b)** $4\angle 0°$ **(c)** $5\angle 30°$ **(d)** $1\angle 90°$ **(e)** $3\angle -90°$

6. Convert the following rectangular numbers into polar form:
   (a) $5 + j5$ **(b)** $12 + j9$ **(c)** $8 - j10$ **(d)** $100 - j50$

7. Convert the following polar numbers into rectangular form:
   (a) $1\angle 45°$ **(b)** $12\angle 60°$ **(c)** $100\angle -80°$ **(d)** $40\angle 125°$

8. Identify the quadrant for each phasor in Question 7.

9. Multiply the following complex quantities:
   (a) $(5\angle 20°)(2\angle 45°)$ **(b)** $(3.5\angle -15°)(6\angle 120°)$

10. Add the following:
    (a) $(8 + j7) + (12 + j4)$ **(b)** $(12 - j10) + (-8 + j5)$

PROBLEMS

Section 12–1

12–1 Draw a phasor diagram to represent the sine waves in Figure 12–25.

12–2 Sketch the sine waves represented by the phasor diagram in Figure 12–26. The phasor lengths represent peak values.

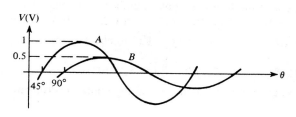

FIGURE 12–25

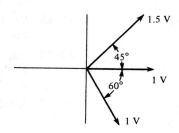

FIGURE 12–26

12–3 Determine the frequency for each angular velocity:
(a) 60 rad/s (b) 360 rad/s (c) 2 rad/s (d) 1256 rad/s

12–4 Determine the value of sine wave A in Figure 12–25 at each of the following times, measured from the positive-going zero crossing. Assume the frequency is 5 kHz.
(a) 30 μs (b) 75 μs (c) 125 μs

12–5 In Figure 12–25, how many microseconds after the zero crossing does sine wave A reach 0.8 V? Assume the frequency is 5 kHz.

Section 12–2
12–6 Locate the following numbers on the complex plane:
(a) +6 (b) −2 (c) +j3 (d) −j8

12–7 Locate the points represented by each of the following coordinates on the complex plane:
(a) 3, j5 (b) −7, j1 (c) −10, −j10

12–8 Determine the coordinates of each point that is located 180 degrees away from each point in Problem 12–7.

12–9 Repeat Problem 12–8 for points 90 degrees away from those in Problem 12–7.

Section 12–3
12–10 Points on the complex plane are described below. Express each point as a complex number in rectangular form:
(a) 3 units to the right of the origin on the real axis, and up 5 units on the j axis.
(b) 2 units to the left of the origin on the real axis, and 1.5 units up on the j axis.
(c) 10 units to the left of the origin on the real axis, and down 14 units on the −j axis.

12–11 What is the value of the hypotenuse of a right triangle whose sides are 10 and 15?

12–12 Convert each of the following rectangular numbers to polar form:
(a) 40 − j40 (b) 50 − j200 (c) 35 − j20 (d) 98 + j45

12–13 Convert each of the following polar numbers to rectangular form:
(a) 1000∠ − 50° (b) 15∠160° (c) 25∠ − 135° (d) 3∠180°

12–14 Express each of the following polar numbers using an equivalent negative angle:
(a) 10∠120° (b) 32∠85° (c) 5∠310°

12–15 Identify the quadrant in which each of the points in Problem 12–12 is located.

12–16 Identify the quadrant in which each point in Problem 12–14 is located.

12–17 Write the polar expressions using positive angles for each phasor in Figure 12–27.

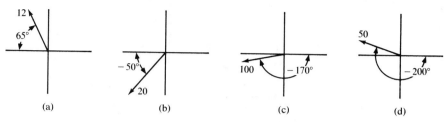

FIGURE 12–27

Section 12–4

12–18 Add the following sets of complex numbers:
    (a) $9 + j3$ and $5 + j8$ **(b)** $3.5 - j4$ and $2.2 + j6$
    (c) $-18 + j23$ and $30 - j15$ **(d)** $12\angle45°$ and $20\angle32°$
    (e) $3.8\angle75°$ and $1 + j1.8$ **(f)** $50 - j39$ and $60\angle-30°$

12–19 Perform the following subtractions:
    (a) $(2.5 + j1.2) - (1.4 + j0.5)$ **(b)** $(-45 - j23) - (36 + j12)$
    (c) $(8 - j4) - 3\angle25°$ **(d)** $48\angle135° - 33\angle-60°$

12–20 Multiply the following numbers:
    (a) $4.5\angle48°$ and $3.2\angle90°$
    (b) $120\angle-220°$ and $95\angle200°$
    (c) $-3\angle150°$ and $4 - j3$
    (d) $67 + j84$ and $102\angle40°$
    (e) $15 - j10$ and $-25 - j30$
    (f) $0.8 + j0.5$ and $1.2 - j1.5$

12–21 Perform the following divisions:
$$\text{(a)} \ \frac{8\angle50°}{2.5\angle39°} \quad \text{(b)} \ \frac{63\angle-91°}{9\angle10°} \quad \text{(c)} \ \frac{28\angle30°}{14 - j12} \quad \text{(d)} \ \frac{40 - j30}{16 + j8}$$

12–22 Perform the following operations:
$$\text{(a)} \ \frac{2.5\angle65° - 1.8\angle-23°}{1.2\angle37°}$$
$$\text{(b)} \ \frac{(100\angle15°)(85 - j150)}{25 + j45}$$
$$\text{(c)} \ \frac{(250\angle90° + 175\angle75°)(50 - j100)}{(125 + j90)(35\angle50°)}$$
$$\text{(d)} \ \frac{(1.5)^2(3.8)}{1.1} + j\left(\frac{8}{4} - j\frac{4}{2}\right)$$

FIGURE 12-28

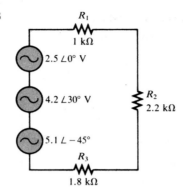

12-23 Three sine wave voltage sources are connected in series as shown in Figure 12-28. Determine the total voltage and current expressed as polar quantities.

12-24 What is the magnitude and phase of the voltages across each resistor in Figure 12-28?

ANSWERS TO SECTION REVIEWS

Section 12-1
1. A graphic representation of the magnitude and angular position of a time-varying quantity.
2. 9425 rad/s.   **3.** 99.95 Hz.   **4.** See Figure 12-29.

FIGURE 12-29

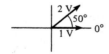

Section 12-2
1. (a) 3 units right of the origin on real axis;   **(b)** 4 units left of the origin on real axis;   **(c)** 1 unit above origin on j axis.
2. (a) 90 degrees;   **(b)** 90 degrees;   **(c)** 180 degrees.

Section 12-3
1. Real part and j part.   **2.** Magnitude and angle.   **3.** $2.828 \angle 45°$, first.
4. $3.54 - j3.54$, fourth.

Section 12-4
1. $4 + j1$.   **2.** $3 + j7$.   **3.** $16 \angle 110°$.   **4.** $5 \angle 15°$.

THIRTEEN

CAPACITORS

The capacitor is an electrical device that can store electrical charge, thereby creating an electric field which in turn stores energy; the measure of that energy-storing ability is called *capacitance*. In this chapter, the basic capacitor is introduced and its characteristics are studied. The physical construction and electric properties of various types of capacitors are discussed.

The basic behavior of capacitors in both dc and ac circuits is studied, and series and parallel combinations are analyzed. Representative applications and methods of testing capacitors also are discussed.

In this chapter, you will learn:

☐ The basic structure of a capacitor.
☐ What a capacitor is and what it does.
☐ The definition of *capacitance* and how it is measured.
☐ How a capacitor stores energy.
☐ The definition of *Coulomb's law* and how it relates to an electric field and the storage of energy.
☐ How a capacitor charges and discharges.
☐ How various physical parameters determine capacitance value.
☐ Several common capacitor classifications according to dielectric material.

☐ What happens when capacitors are connected in series.
☐ What happens when capacitors are connected in parallel.
☐ The meaning of *time constant*.
☐ How the time constant affects the charging and discharging of a capacitor.
☐ Why a capacitor blocks dc.
☐ How a capacitor introduces a phase shift between current and voltage.
☐ The definition of *capacitive reactance* and how to determine its value in a circuit.
☐ Why there is ideally no energy loss in a capacitor.
☐ The significance of reactive power.
☐ Several common capacitor applications.
☐ How to check a capacitor with an ohmmeter.
☐ How an *LC* meter is used to perform various capacitor tests.

13–1 THE BASIC CAPACITOR

BASIC CONSTRUCTION

In its simplest form, a capacitor is an electrical device constructed of two parallel conductive plates separated by an insulating material called the *dielectric*. Connecting leads are attached to the parallel plates. A basic capacitor is shown in Figure 13–1(a), and the schematic symbol is shown in Part (b).

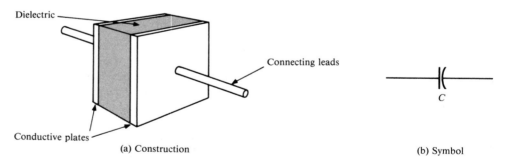

(a) Construction

(b) Symbol

FIGURE 13–1
The basic capacitor.

HOW A CAPACITOR STORES CHARGE

In the neutral state, both plates of a capacitor have an equal number of free electrons, as indicated in Figure 13–2(a). When the capacitor is connected to a voltage source through a resistor, as shown in Part (b), electrons (negative charge) are removed from plate *A,* and an equal number are deposited on plate *B*. As plate *A* loses electrons and plate *B* gains electrons, plate *A* becomes positive with respect to plate *B*. During this charging process, electrons flow only through the connecting leads and the source. No

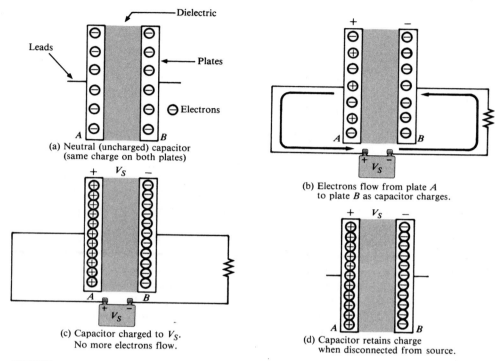

FIGURE 13-2
Illustration of a capacitor storing charge.

electrons flow through the dielectric of the capacitor because it is an insulator. The movement of electrons ceases when the voltage across the capacitor equals the source voltage, as indicated in Part (c). If the capacitor is disconnected from the source, it retains the stored charge for a long period of time (the length of time depends on the type of capacitor) and still has the voltage across it, as shown in Part (d). Actually, the charged capacitor can be considered as a temporary battery.

CAPACITANCE

The amount of charge per unit of voltage that a capacitor can store is its *capacitance,* designated C. That is, capacitance is a measure of a capacitor's ability to store charge. The more charge per unit of voltage that a capacitor can store, the greater its capacitance, as expressed by the following formula:

$$C = \frac{Q}{V} \qquad \qquad \textbf{(13–1)}$$

where C is capacitance, Q is charge, and V is voltage.

By rearranging Equation (13–1), we obtain two other forms as follows:

$$Q = CV \qquad\qquad (13\text{–}2)$$

$$V = \frac{Q}{C} \qquad\qquad (13\text{–}3)$$

THE UNIT OF CAPACITANCE

The *farad* (F) is the basic unit of capacitance. By definition, *one farad* is the amount of capacitance when *one coulomb* of charge is stored with *one volt* across the plates.

Most capacitors that you will use in electronics work have capacitance values in microfarads (μF) and picofarads (pF). A microfarad is one-millionth of a farad (1 μF = 1 × 10^{-6} F), and a picofarad is one-trillionth of a farad (1 pF = 1 × 10^{-12} F). Conversions for farads, microfarads, and picofarads are given in Table 13–1.

EXAMPLE 13–1

(a) A certain capacitor stores 50 microcoulombs (50 μC) with 10 V across its plates. What is its *capacitance* in units of farads?
(b) A 2-μF capacitor has 100 V across its plates. How much charge does it store?
(c) Determine the voltage across a 1-nF (nanofarad) capacitor that is storing 20 microcoulombs (20 μC) of charge.

Solution:

(a) $C = \dfrac{Q}{V} = \dfrac{50 \ \mu C}{10 \ V} = 5 \ \mu F$

(b) $Q = CV = (2 \ \mu F)(100 \ V) = 200 \ \mu C$

(c) $V = \dfrac{Q}{C} = \dfrac{20 \ \mu C}{1 \ nF} = 20 \ kV$

EXAMPLE 13–2

Convert the following values to microfarads:
(a) 0.00001 F   (b) 0.005 F   (c) 1000 pF   (d) 200 pF

Solution:
(a) 0.00001 F × 10^6 = 10 μF   (b) 0.005 F × 10^6 = 5000 μF
(c) 1000 pF × 10^{-6} = 0.001 μF   (d) 200 pF × 10^{-6} = 0.0002 μF

EXAMPLE 13–3

Convert the following values to picofarads:
(a) 0.1 × 10^{-8} F   (b) 0.000025 F   (c) 0.01 μF   (d) 0.005 μF

Solution:
(a) 0.1 × 10^{-8} F × 10^{12} = 1000 pF   (b) 0.000025 F × 10^{12} = 25 × 10^6 pF
(c) 0.01 μF × 10^6 = 10,000 pF   (d) 0.005 μF × 10^6 = 5000 pF

TABLE 13–1
Conversions for farads, microfarads, and picofarads.

To convert from	to	multiply
farads	microfarads	farads by 10^6
farads	picofarads	farads by 10^{12}
microfarads	farads	microfarads by 10^{-6}
microfarads	picofarads	microfarads by 10^6
picofarads	farads	picofarads by 10^{-12}
picofarads	microfarads	picofarads by 10^{-6}

HOW A CAPACITOR STORES ENERGY

A capacitor stores energy in the form of an *electric field* that is established by the opposite charges on the two plates. The electric field is represented by *lines of force* between the positive and negative charges and concentrated within the dielectric, as shown in Figure 13–3.

Coulomb's law states:

A force exists between two charged bodies that is directly proportional to the product of the two charges and inversely proportional to the square of the distance between the bodies.

This relationship is expressed in Equation (13–4):

$$F = \frac{kQ_1Q_2}{d^2} \qquad \textbf{(13–4)}$$

where F is the force in newtons, Q_1 and Q_2 are the charges in coulombs, d is the distance between the charges in meters, and k is a proportionality constant equal to 9×10^9. Figure 13–4(a) illustrates the line of force between a positive and a negative

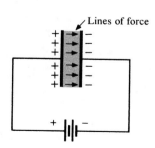

FIGURE 13–3
The electric field stores energy in a capacitor.

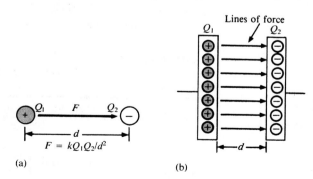

(a)

(b)

FIGURE 13–4
Lines of force are created by opposite charges.

charge; Part (b) shows that many opposite charges on the plates of a capacitor create many lines of force, which form an *electric field* that stores energy within the dielectric.

The greater the forces between the charges on the plates of a capacitor, the more energy is stored. The amount of energy stored therefore is directly proportional to the capacitance, because, from Coulomb's law, the more charge stored, the greater the force.

Also, from Equation (13–2), the amount of charge stored is directly related to the voltage as well as the capacitance. Therefore, the amount of energy stored is also dependent on the square of the voltage across the plates of the capacitor. The formula for the energy stored by a capacitor is as follows:

$$W = \frac{1}{2}CV^2 \tag{13–5}$$

The energy, W, is in joules when C is in farads and V is in volts.

VOLTAGE RATING

Every capacitor has a limit on the amount of voltage that it can withstand across its plates. The *voltage rating* specifies the maximum dc voltage that can be applied without risk of damage to the device. If this maximum voltage, commonly called the *breakdown voltage* or *working voltage,* is exceeded, permanent damage to the capacitor can result.

Both the capacitance and the voltage rating must be taken into consideration before a capacitor is used in a circuit application. The choice of capacitance value is based on particular circuit requirements (and on factors that are studied later). The voltage rating should always be well above the maximum voltage expected in a particular application.

Dielectric Strength   The breakdown voltage of a capacitor is determined by the dielectric strength of the material used. The dielectric strength is expressed in volts/mil (1 mil = 0.001 in). Table 13–2 lists typical values for several materials. Exact values vary depending on the specific composition of the material.

TABLE 13–2
Some common dielectric materials and their dielectric strengths.

Material	Dielectric Strength (volts/mil)
Air	80
Oil	375
Ceramic	1000
Paper (paraffined)	1200
Teflon®	1500
Mica	1500
Glass	2000

The dielectric strength can best be explained by an example. Assume that a certain capacitor has a plate separation of 1 mil and that the dielectric material is ceramic. This particular capacitor can withstand a maximum voltage of 1000 V, because its dielectric strength is 1000 V/mil. If the maximum voltage is exceeded, the dielectric may break down and conduct current, causing permanent damage to the capacitor. Similarly, if the ceramic capacitor has a plate separation of 2 mils, its breakdown voltage is 2000 V.

TEMPERATURE COEFFICIENT

The temperature coefficient indicates the amount and direction of a change in capacitance value with temperature. A *positive* temperature coefficient means that the capacitance increases with an increase in temperature or decreases with a decrease in temperature. A *negative* coefficient means that the capacitance decreases with an increase in temperature or increases with a decrease in temperature.

Temperature coefficients are typically specified in *parts per million per degree Celsius* (ppm/°C). For example, a negative temperature coefficient of 150 ppm/°C for a $1\text{-}\mu\text{F}$ capacitor means that for every degree rise in temperature, the capacitance decreases by 150 pF (there are one million picofarads in one microfarad).

LEAKAGE

No insulating material is perfect. The dielectric of any capacitor will conduct some very small amount of current. Thus, the charge on a capacitor will eventually leak off. Some types of capacitors have higher leakages than others. An equivalent circuit for a nonideal capacitor is shown in Figure 13–5. The parallel resistor R_l represents the extremely high resistance of the dielectric material through which there is leakage current.

FIGURE 13–5
Equivalent circuit for a nonideal capacitor.

PHYSICAL CHARACTERISTICS OF A CAPACITOR

The following parameters are important in establishing the capacitance and the voltage rating of a capacitor.

Plate Area *Capacitance is directly proportional to the physical size of the plates as determined by the plate area, A.* A larger plate area produces a larger capacitance, and vice versa. Figure 13–6(a) shows that the plate area of a parallel plate capacitor is the area of one of the plates. If the plates are moved in relation to each other, as shown in

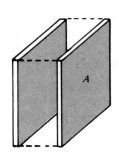 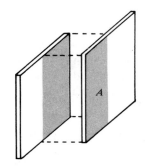

(a) Full plate area:
   more capacitance

(b) Reduced plate area:
   less capacitance

FIGURE 13–6
Capacitance is directly proportional to plate area *(A)*.

Part (b), the *overlapping area* determines the effective plate area. This variation in effective plate area is the basis for a certain type of variable capacitor.

Plate Separation *Capacitance is inversely proportional to the distance between the plates.* The plate separation is designated *d*, as shown in Figure 13–7. A greater separation of the plates produces a smaller capacitance, as illustrated in the figure. The breakdown voltage is directly proportional to the plate separation. The further the plates are separated, the greater the breakdown voltage.

FIGURE 13–7
Capacitance is inversely proportional to the distance between the plates.

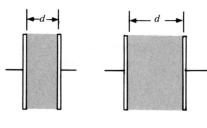

(a) More capacitance   (b) Less capacitance

Dielectric Constant As you know, the insulating material between the plates of a capacitor is called the *dielectric*. Every dielectric material has the ability to concentrate the lines of force of the electric field existing between the oppositely charged plates of a capacitor and thus increase the capacity for energy storage. The measure of a material's ability to establish an electric field is called the *dielectric constant* or *relative permittivity,* symbolized by ϵ_r (the Greek letter epsilon).

Capacitance is directly proportional to the dielectric constant. The dielectric constant of a vacuum is defined as 1, and that of air is very close to 1. These values are used as a reference, and all other materials have values of ϵ_r specified with respect to that of a vacuum or air. For example, a material with $\epsilon_r = 8$ can result in a capacitance eight times greater than that of air with all other factors being equal.

Table 13–3 lists several common dielectric materials and typical dielectric constants for each. Values can vary because they depend on the specific composition of the material.

TABLE 13–3
Some common dielectric materials
and their dielectric constants.

Material	Typical ϵ_r Values
Air (vacuum)	1.0
Teflon®	2.0
Paper (paraffined)	2.5
Oil	4.0
Mica	5.0
Glass	7.5
Ceramic	1200

The dielectric constant (relative permittivity) is dimensionless, because it is a relative measure and is a ratio of the *absolute permittivity*, ϵ, of a material to the *absolute permittivity*, ϵ_0, of a vacuum, as expressed by the following formula:

$$\epsilon_r = \frac{\epsilon}{\epsilon_0} \tag{13–6}$$

The value of ϵ_0 is 8.85×10^{-12} F/m (farads per meter).

FORMULA FOR CAPACITANCE IN TERMS OF PHYSICAL PARAMETERS

You have seen how capacitance is directly related to plate area A and the dielectric constant ϵ_r, and inversely related to plate separation d. An exact formula for calculating the capacitance in terms of these three quantities is as follows:

$$C = \frac{A\epsilon_r(8.85 \times 10^{-12} \text{ F/m})}{d} \tag{13–7}$$

A is in square meters (m^2), d is in meters (m), and C is in farads (F). Recall that 8.85×10^{-12} F/m is the absolute permittivity, ϵ_0, of a vacuum and that $\epsilon_r(8.85 \times 10^{-12})$ F/m is the absolute permittivity of a dielectric, as derived from Equation (13–6).

**EXAMPLE
13–4**

Determine the capacitance of a parallel plate capacitor having a plate area of 0.01 m^2 and a plate separation of 0.02 m. The dielectric is mica which has a dielectric constant of 5.0.

Solution:
Use Equation (13–7):

$$C = \frac{A\epsilon_r(8.85 \times 10^{-12}\ \text{F/m})}{d}$$

$$= \frac{(0.01\ \text{m}^2)(5.0)(8.85 \times 10^{-12}\ \text{F/m})}{0.02\ \text{m}} = 22.13\ \text{pF}$$

The calculator sequence is

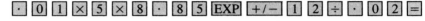

SECTION REVIEW 13–1

1. Define *capacitance*.
2. (a) How many microfarads in a farad?
   (b) How many picofarads in a farad?
   (c) How many picofarads in a microfarad?
3. Convert 0.0015 μF to picofarads. To farads.
4. How much energy in joules is stored by a 0.01-μF capacitor with 15 V across its plates?
5. (a) When the plate area of a capacitor is increased, does the capacitance increase or decrease?
   (b) When the distance between the plates is increased, does the capacitance increase or decrease?
6. The plates of a ceramic capacitor are separated by 10 mils. What is the typical breakdown voltage?
7. A ceramic capacitor has a plate area of 0.2 m^2. The thickness of the dielectric is 0.005 m. What is the capacitance?
8. A capacitor with a value of 2 μF at 25°C has a positive temperature coefficient of 50 ppm/°C. What is the capacitance value when the temperature increases to 125°C?

13–2   TYPES OF CAPACITORS

Capacitors normally are classified according to the type of dielectric material. The most common types of dielectric materials are mica, ceramic, paper/plastic, and electrolytic (aluminum oxide and tantalum oxide). In this section, the characteristics and construction of each of these types of capacitors and the variable capacitors are examined.

MICA CAPACITORS

There are two types of mica capacitors: stacked-foil and silver-mica. The basic construction of the stacked-foil type is shown in Figure 13–8. It consists of alternate layers of metal foil and thin sheets of mica. The metal foil forms the plate, with alternate foil sheets connected together to increase the plate area. More layers are used to increase the plate area, thus increasing the capacitance. The mica/foil stack is encapsulated in an insulating material such as Bakelite®, as shown in Part (b) of the figure. The silver-mica capacitor is formed in a similar way by stacking mica sheets with silver electrode material screened on them.

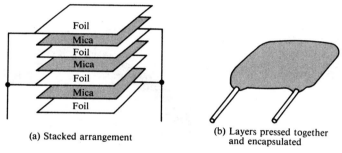

(a) Stacked arrangement

(b) Layers pressed together and encapsulated

FIGURE 13–8
Construction of a typical mica capacitor.

Mica capacitors are available with capacitance values ranging from 1 pF to 0.1 μF and voltage ratings from 100 to 2500 V dc. Temperature coefficients from -20 to $+100$ ppm/°C are common. Mica has a typical dielectric constant of 5.

CERAMIC CAPACITORS

Ceramic dielectrics provide very high dielectric constants (1200 is typical). As a result, comparatively high capacitance values can be achieved in a small physical size. Ceramic capacitors are available in either ceramic disk, as shown in Figure 13–9, or in a multilayer configuration, as shown in Figure 13–10.

Ceramic capacitors typically are available in capacitance values ranging from 1 pF to 2.2 μF with voltage ratings up to 6 kV. A typical temperature coefficient for ceramic capacitors is 200,000 ppm/°C.

PAPER/PLASTIC CAPACITORS

There are several types of plastic-film capacitors and the older paper dielectric capacitors. Polycarbonate, parylene, polyester, polystyrene, polypropylene, mylar, and paper

FIGURE 13–9
A ceramic disk capacitor and its basic construction.

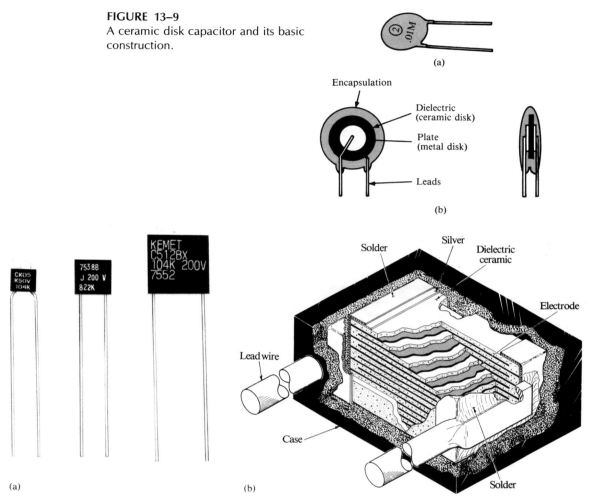

(a)

Encapsulation

Dielectric (ceramic disk)

Plate (metal disk)

Leads

(b)

KEMET
C512BX
104K 200V
7552

CK05
K50V
104K

7538B
J 200 V
822K

Solder

Silver

Dielectric ceramic

Electrode

Lead wire

Case

Solder

(a)   (b)

FIGURE 13–10
Ceramic capacitors. (a) Typical capacitors. (b) Construction view (courtesy of KEMET Electronics Corporation).

are some of the more common dielectric materials used. Some of these types have capacitance values up to 100 μF.

Figure 13–11 shows a common basic construction used in many plastic-film and paper capacitors. A thin strip of plastic-film dielectric is sandwiched between two thin metal strips that act as plates. One lead is connected to the inner plate and one to the outer plate as indicated. The strips are then rolled in a spiral configuration and encapsulated in a molded case. Thus, a large plate area can be packaged in a relatively small physical size, thereby achieving large capacitance values. Figure 13–12 shows a construction view for one type of plastic-film capacitor.

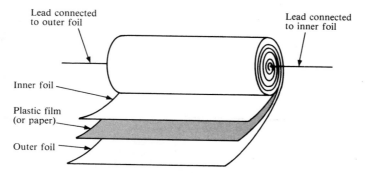

FIGURE 13–11
Basic construction of tubular paper/plastic dielectric capacitors.

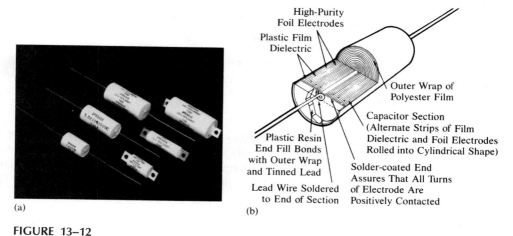

(a)

(b)

FIGURE 13–12
Film capacitors. (a) Typical example (courtesy of Siemens Corp.); (b) Construction view of plastic-film capacitor (courtesy of KEMET Electronics Corporation).

ELECTROLYTIC CAPACITORS

Electrolytic capacitors are *polarized* so that one plate is positive and the other negative. These capacitors are used for high capacitance values up to over 200,000 μF, but they have relatively low breakdown voltages (350 V is a typical maximum) and high amounts of leakage.

Electrolytic capacitors are available in two types: *aluminum* and *tantalum*. The basic construction of an electrolytic capacitor is shown in Figure 13–13(a). The capacitor consists of two strips of either aluminum or tantalum foil separated by a paper or gauze strip saturated with an electrolyte. During manufacturing, an electrochemical reaction is induced which causes an oxide layer (either aluminum oxide or tantalum oxide) to form on the inner surface of the positive plate. This oxide layer acts as the dielectric. Part (b) of Figure 13–13 shows several typical electrolytic capacitors.

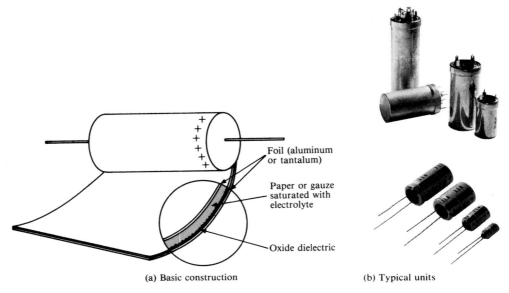

(a) Basic construction

(b) Typical units

FIGURE 13–13
Electrolytic capacitors.

Since an electrolytic capacitor is polarized, *the positive plate must always be connected to the positive side of a circuit.* The positive end is indicated by plus signs or some other obvious marking. Be very careful to make the correct connection and to install the capacitor only in a dc, not ac, circuit.

VARIABLE CAPACITORS

Variable capacitors are used in a circuit when there is a need to adjust the capacitance value either manually or automatically, for example, in radio or TV tuners. The major types of variable or adjustable capacitors are now introduced.

Air Capacitor   Variable capacitors with air dielectrics, such as the one shown in Figure 13–14, are sometimes used as tuning capacitors in applications requiring frequency selection. This type of capacitor is constructed of several plates that mesh together. One set of plates can be moved relative to the other, thus changing the effective plate area and the capacitance. The movable plates are linked together mechanically so that they all move when a shaft is rotated. The schematic symbol for a variable capacitor is shown in Figure 13–15.

Trimmers and Padders   These adjustable capacitors normally have screwdriver adjustments and are used for very fine adjustments in a circuit. Ceramic or mica is a common dielectric in these types of capacitors, and the capacitance usually is changed by adjusting the plate separation. Figure 13–16 shows some typical devices.

FIGURE 13–15
Schematic symbol for a variable capacitor.

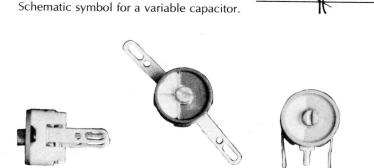

FIGURE 13–14
A typical variable air capacitor.

FIGURE 13–16
Trimmer capacitors (courtesy of Murata Erie, North America, Inc.).

Varactors   The varactor is a semiconductor device that exhibits a capacitance characteristic that is varied by changing the voltage across its terminals. This device usually is covered in detail in a course on electronic devices.

CAPACITOR LABELING

Capacitor values are indicated on the body of the capacitor either by *typographical labels* or by *color codes*. Typographical labels consist of letters and numbers that indicate various parameters such as capacitance, voltage rating, tolerance, and others.

Some capacitors carry no unit designation for capacitance. In these cases, the units are implied by the value indicated. For example, a ceramic capacitor marked .001 or .01 has units of microfarads because picofarad values that small are not available. As another example, a ceramic capacitor labeled 50 or 330 has units of picofarads because microfarad units that large normally are not available in this type.

In some instances, the units are labeled as pF or μF; often the microfarad unit is labeled as MF or MFD. Voltage rating appears on some types of capacitors and is omitted on others. When it is omitted, the voltage rating can be determined from information supplied by the manufacturer. The tolerance of the capacitor is usually labeled as a percentage, such as $\pm 10\%$. The temperature coefficient is indicated by a *parts per million* marking. This type of label consists of a P or an N followed by a number. For example, N750 means a negative temperature coefficient of 750 ppm/°C, and P30 means a positive temperature coefficient of 330 ppm/°C.

SECTION REVIEW 13–2

1. How are capacitors commonly classified?

2. What is the difference between a fixed and a variable capacitor?

3. What type of capacitor is polarized?

4. What precautions must be taken when installing a polarized capacitor in a circuit?

13–3   SERIES CAPACITORS

TOTAL CAPACITANCE

When capacitors are connected in series, the *effective* plate separation increases, and *the total capacitance is less than that of the smallest capacitor.* The reason is as follows: Consider the generalized circuit in Figure 13–17(a), which has n capacitors in series with a voltage source and a switch. When the switch is closed, the capacitors charge as current is established through the circuit. Since this is a series circuit, the current must be the same at all points, as illustrated. Since current is the rate of flow of charge, *the amount of charge stored by each capacitor is equal to the total charge,* expressed as follows:

$$Q_T = Q_1 = Q_2 = Q_3 = \cdots = Q_n \qquad (13\text{–}8)$$

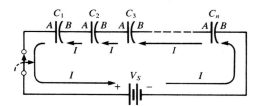 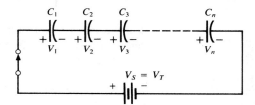

(a) Charging current is same for each capacitor, $I = Q/t$

(b) All capacitors store same amount of charge and $V = Q/C$.

FIGURE 13–17
Series capacitive circuit.

Next, according to *Kirchhoff's voltage law,* the *sum* of the voltages across the charged capacitors must equal the total voltage, V_T, as shown in Figure 13–17(b). This is expressed in equation form as

$$V_T = V_1 + V_2 + V_3 + \cdots + V_n$$

From Equation (13–3), $V = Q/C$. Substituting this relationship into each term of the voltage equation, the following result is obtained:

$$\frac{Q_T}{C_T} = \frac{Q_1}{C_1} + \frac{Q_2}{C_2} + \frac{Q_3}{C_3} + \cdots + \frac{Q_n}{C_n} \qquad (13\text{–}9)$$

Since the charges on all the capacitors are equal, the Q terms can be factored from Equation (13–9) and canceled, resulting in

$$\frac{1}{C_T} = \frac{1}{C_1} + \frac{1}{C_2} + \frac{1}{C_3} + \cdots + \frac{1}{C_n}$$

Taking the reciprocal of both sides, the total series capacitance is

$$C_T = \cfrac{1}{\cfrac{1}{C_1} + \cfrac{1}{C_2} + \cfrac{1}{C_3} + \cdots + \cfrac{1}{C_n}} \qquad \textbf{(13–10)}$$

SPECIAL CASE OF TWO CAPACITORS IN SERIES

When only two capacitors are in series, a special form of Equation (13–9) can be used.

$$\frac{1}{C_T} = \frac{1}{C_1} + \frac{1}{C_2} = \frac{C_1 + C_2}{C_1 C_2}$$

Taking the reciprocal of both sides we get

$$C_T = \frac{C_1 C_2}{C_1 + C_2} \qquad \textbf{(13–11)}$$

CAPACITORS OF EQUAL VALUE IN SERIES

This special case is another in which a formula can be developed from Equation (13–10). When all n values are the same and equal to C, we get

$$\frac{1}{C_T} = \frac{1}{C} + \frac{1}{C} + \frac{1}{C} + \cdots + \frac{1}{C}$$

Adding all the terms on the right, we get

$$\frac{1}{C_T} = \frac{n}{C}$$

Taking the reciprocal of both sides:

$$C_T = \frac{C}{n} \qquad \textbf{(13–12)}$$

The capacitance value of the equal capacitors divided by the number of equal series capacitors gives the total capacitance. Notice that the total series capacitance is calculated in the same manner as *total parallel resistance*.

The total series capacitance is always less than the smallest capacitance.

EXAMPLE 13–5

Determine the total capacitance in Figure 13–18.

FIGURE 13–18

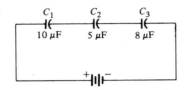

C_1 C_2 C_3
10 μF 5 μF 8 μF

Solution:

$$\frac{1}{C_T} = \frac{1}{C_1} + \frac{1}{C_2} + \frac{1}{C_3}$$

$$= \frac{1}{10\ \mu F} + \frac{1}{5\ \mu F} + \frac{1}{8\ \mu F}$$

Taking the reciprocal of both sides:

$$C_T = \frac{1}{\dfrac{1}{10\ \mu F} + \dfrac{1}{5\ \mu F} + \dfrac{1}{8\ \mu F}}$$

$$= \frac{1}{0.425}\ \mu F$$

$$= 2.35\ \mu F$$

The calculator sequence is

1 0 EXP +/− 6 2nd F 1/x + 5 EXP +/− 6 2nd F 1/x + 8 EXP +/− 6 2nd F 1/x = 2nd F 1/x =

EXAMPLE 13–6

Find C_T in Figure 13–19.

FIGURE 13–19

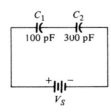

C_1 C_2
100 pF 300 pF

V_S

Solution:

$$C_T = \frac{C_1 C_2}{C_1 + C_2}$$

$$= \frac{(100\ pF)(300\ pF)}{400\ pF}$$

$$= 75\ pF$$

For a calculator solution use the formula

$$C_T = \cfrac{1}{\cfrac{1}{100 \text{ pF}} + \cfrac{1}{300 \text{ pF}}}$$

EXAMPLE 13–7

Determine C_T for the series capacitors in Figure 13–20.

FIGURE 13–20

V_S

C_1 0.02 µF
C_2 0.02 µF
C_3 0.02 µF
C_4 0.02 µF

Solution:

$$C_1 = C_2 = C_3 = C_4 = C$$

$$C_T = \frac{C}{n} = \frac{0.02 \text{ µF}}{4}$$

$$= 0.005 \text{ µF}$$

CAPACITOR VOLTAGES

A series connection of charged capacitors acts as a *voltage divider*. The voltage across each capacitor in series is inversely proportional to its capacitance value, as shown by the formula $V = Q/C$.

The voltage across any capacitor in series can be calculated as follows:

$$V_X = \left(\frac{C_T}{C_X}\right)V_T \qquad (13\text{–}13)$$

where C_X is C_1, or C_2, or C_3, and so on. The derivation is as follows: Since the charge on any capacitor in series is the same as the total charge ($Q_X = Q_T$), and since $Q_X = V_X C_X$ and $Q_T = V_T C_T$, then

$$V_X C_X = V_T C_T$$

Solving for V_X, we get

$$V_X = \frac{C_T V_T}{C_X}$$

The largest capacitor in series will have the smallest voltage, and the smallest capacitor will have the largest voltage.

EXAMPLE 13–8

Find the voltage across each capacitor in Figure 13–21.

FIGURE 13–21

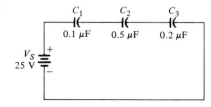

Solution:

$$\frac{1}{C_T} = \frac{1}{C_1} + \frac{1}{C_2} + \frac{1}{C_3}$$

$$= \frac{1}{0.1\ \mu F} + \frac{1}{0.5\ \mu F} + \frac{1}{0.2\ \mu F}$$

$$C_T = \frac{1}{17}\ \mu F = 0.0588\ \mu F$$

$$V_S = V_T = 25\ V$$

$$V_1 = \left(\frac{C_T}{C_1}\right)V_T = \left(\frac{0.0588\ \mu F}{0.1\ \mu F}\right)25\ V = 14.71\ V$$

$$V_2 = \left(\frac{C_T}{C_2}\right)V_T = \left(\frac{0.0588\ \mu F}{0.5\ \mu F}\right)25\ V = 2.94\ V$$

$$V_3 = \left(\frac{C_T}{C_3}\right)V_T = \left(\frac{0.0588\ \mu F}{0.2\ \mu F}\right)25\ V = 7.35\ V$$

SECTION REVIEW 13–3

1. Is the total capacitance of a series connection less than or greater than the value of the smallest capacitor?
2. The following capacitors are in series: 100 pF, 250 pF, and 500 pF. What is the total capacitance?
3. A 0.01-μF and a 0.015-μF capacitor are in series. Determine the total capacitance.
4. Five 100-pF capacitors are connected in series. What is C_T?

5. Determine the voltage across C_1 in Figure 13–22.

FIGURE 13–22

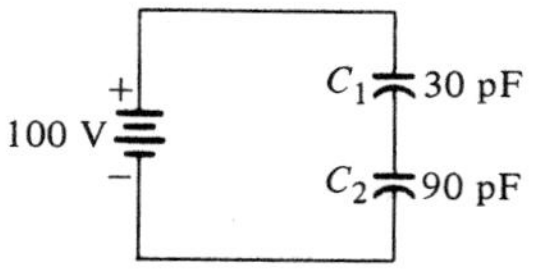

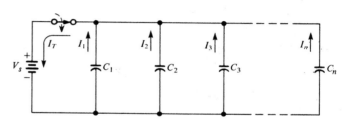

100 V
C_1 30 pF
C_2 90 pF

13–4

PARALLEL CAPACITORS

When capacitors are connected in parallel, the *effective* plate area increases, and *the total capacitance is the sum of the individual capacitances*. To understand this, consider what happens when the switch in Figure 13–23 is closed.

FIGURE 13–23
Capacitors in parallel.

The total charging current from the source divides at the junction of the parallel branches. There is a separate charging current through each branch so that a different charge can be stored by each capacitor. By Kirchhoff's current law, the *sum* of all of the charging currents is equal to the total current. Therefore, the sum of the charges on the capacitors is equal to the total charge. Also, the voltages across all of the parallel branches are equal. These observations are used to develop a formula for total parallel capacitance as follows for the general case of n capacitors in parallel.

$$Q_T = Q_1 + Q_2 + Q_3 + \cdots + Q_n \qquad \textbf{(13–14)}$$

Since $Q = CV$,

$$C_T V_T = C_1 V_1 + C_2 V_2 + C_3 V_3 + \cdots + C_n V_n$$

Since $V_T = V_1 = V_2 = V_3 = \cdots = V_n$, the voltages can be factored and canceled, giving

$$C_T = C_1 + C_2 + C_3 + \cdots + C_n \qquad \textbf{(13–15)}$$

Equation (13–15) is the general formula for total parallel capacitance where n is the number of capacitors. Remember that *capacitors add in parallel*.

For the special case when all of the capacitors have the same value, C, multiply the value by the number of capacitors in parallel:

$$C_T = nC \qquad \qquad \text{(13–16)}$$

Notice that in all cases, the total parallel capacitance is calculated in the same manner as *total series resistance*.

The total parallel capacitance is the sum of all the capacitors in parallel.

EXAMPLE 13–9

What is the total capacitance in Figure 13–24? What is the voltage across each capacitor?

FIGURE 13–24

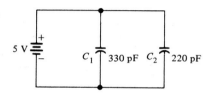

Solution:

$$C_T = C_1 + C_2 = 330 \text{ pF} + 220 \text{ pF}$$
$$= 550 \text{ pF}$$
$$V_1 = V_2 = 5 \text{ V}$$

EXAMPLE 13–10

Determine C_T in Figure 13–25.

FIGURE 13–25

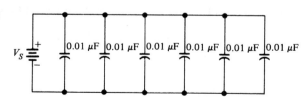

Solution:
There are six equal-valued capacitors in parallel, so $n = 6$.

$$C_T = nC = (6)(0.01 \ \mu\text{F}) = 0.06 \ \mu\text{F}$$

SECTION REVIEW 13–4

1. How is total parallel capacitance determined?

2. In a certain application, you need 0.05 μF. The only values available are 0.01 μF, which are available in large quantities. How can you get the total capacitance that you need?

3. The following capacitors are in parallel: 10 pF, 5 pF, 33 pF, and 50 pF. What is C_T?

13–5 CAPACITORS IN dc CIRCUITS

CHARGING A CAPACITOR

A capacitor charges when it is connected to a dc voltage source, as shown in Figure 13–26. The capacitor in Part (a) of the figure is uncharged; that is, plate A and plate B have equal numbers of free electrons. When the switch is closed, as shown in Part (b), the source moves electrons away from plate A through the circuit to plate B. As plate A loses electrons and plate B gains electrons, plate A becomes positive with respect to plate B. As this charging process continues, the voltage across the plates builds up rapidly until it is equal to the applied voltage, V_S, but opposite in polarity, as shown in Part (c). *When the capacitor is fully charged, there is no current.* A capacitor *blocks* constant dc.

FIGURE 13–26
Charging a capacitor.

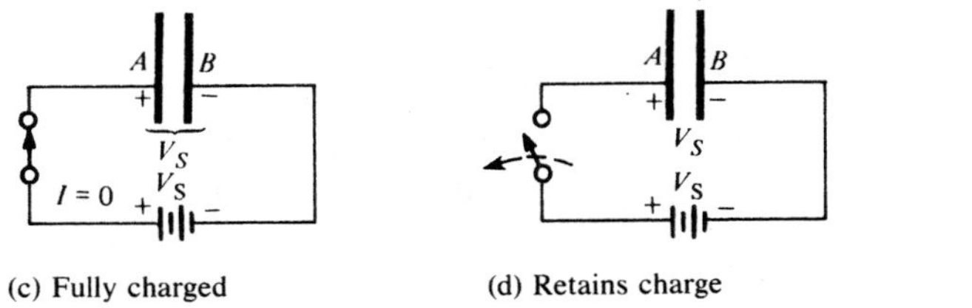

(a) Uncharged

(b) Charging

(c) Fully charged

(d) Retains charge

When the charged capacitor is disconnected from the source, as shown in Figure 13–26(d), it remains charged for long periods of time, depending on its leakage resistance, and can cause severe electrical shock. The charge on an electrolytic capacitor generally leaks off more rapidly than in other types of capacitors.

DISCHARGING A CAPACITOR

When a wire is connected across a charged capacitor, as shown in Figure 13–27, the capacitor will discharge. In this particular case, a very low resistance path (the wire) is connected across the capacitor with a switch. Before the switch is closed, the capacitor is charged to 50 V, as indicated in Part (a). When the switch is closed, as shown in Part (b), the excess electrons on plate B move through the circuit to plate A; as a result of the current through the low resistance of the wire, the energy stored by the capacitor is dissipated in the wire. The charge is neutralized when the numbers of free electrons on both plates are again equal. At this time, the voltage across the capacitor is zero, and the capacitor is completely discharged, as shown in Part (c).

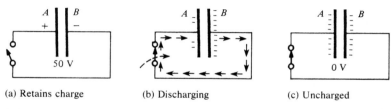

(a) Retains charge   (b) Discharging   (c) Uncharged

FIGURE 13–27
Discharging a capacitor.

CURRENT DURING CHARGING AND DISCHARGING

Notice in Figures 13–26 and 13–27 that the direction of the current during discharge is opposite to that of the charging current. It is important to understand that *there is no current through the dielectric of the capacitor during charging or discharging, because the dielectric is an insulating material*. There is current from one plate to the other only through the external circuit.

THE *RC* TIME CONSTANT

In a practical situation, there cannot be capacitance without some resistance in a circuit. It may simply be the small resistance of a wire, or it may be a designed-in resistance. Because of this, the charging and discharging characteristics of a capacitor must always be considered in light of the associated resistance. The resistance introduces the element of *time* in the charging and discharging of a capacitor.

When a capacitor charges or discharges through a resistance, a certain time is required for the capacitor to charge fully or discharge fully. *The voltage across a capacitor cannot change instantaneously,* because a finite time is required to move charge from one point to another. The rate at which the capacitor charges or discharges is determined by the *time constant* of the circuit.

The time constant of a series *RC* circuit is a time interval that equals the product of the resistance and the capacitance.

The time constant is symbolized by τ (Greek letter tau), and the formula is as follows:

$$\tau = RC \qquad\qquad (13\text{--}17)$$

Recall that $I = Q/t$. The current depends on the amount of charge moved in a given time. When the resistance is increased, the charging current is reduced, thus increasing the charging time of the capacitor. When the capacitance is increased, the amount of charge increases; thus, for the same current, more time is required to charge the capacitor.

EXAMPLE 13–11

A series *RC* circuit has a resistance of 1 MΩ and a capacitance of 5 μF. What is the time constant?

Solution: $\tau = RC = (1 \times 10^6\ \Omega)(5 \times 10^{-6}\ \text{F}) = 5\ \text{s}$

During one time constant interval, the charge on a capacitor changes approximately 63%. Therefore, an uncharged capacitor charges to 63% of its fully charged voltage in one time constant. When discharging, the capacitor voltage drops to approximately 37% (100% − 63%) of its initial value in one time constant, which is a 63% change.

THE CHARGING AND DISCHARGING CURVES

A capacitor charges and discharges following a nonlinear curve, as shown in Figure 13–28. In these graphs, the percentage of full charge is shown at each time-constant interval. This type of curve follows a precise mathematical formula and is called an *exponential curve*. The charging curve is an *increasing exponential,* and the discharging curve is a *decreasing exponential.* As you can see, it takes *five time constants* to approximately reach the final value.

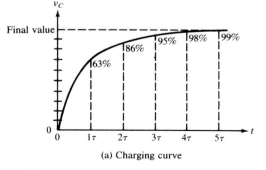

(a) Charging curve

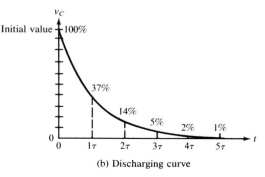

(b) Discharging curve

FIGURE 13–28
Charging and discharging exponential curves for an *RC* circuit.

General Formula The general expressions for either increasing or decreasing exponential curves are given in the following equations for both voltage and current.

$$v = V_F + (V_i - V_F)e^{-t/\tau} \tag{13-18}$$

$$i = I_F + (I_i - I_F)e^{-t/\tau} \tag{13-19}$$

where V_F and I_F are the *final* values, and V_i and I_i are the *initial* values. v and i are the instantaneous values of the capacitor voltage or current at time t, and e is the base of natural logarithms with a value of 2.718. The $\boxed{e^x}$ key or the $\boxed{\text{INV}}$ and $\boxed{\ln x}$ keys on your calculator make it easy to evaluate this exponential term.

The Charging Curve The formula for the special case in which an increasing exponential voltage curve begins at zero ($V_i = 0$) is given in Equation (13-20). It is developed as follows, starting with the general formula.

$$v = V_F + (V_i - V_F)e^{-t/\tau}$$

$$= V_F + (0 - V_F)e^{-t/RC}$$

$$v = V_F(1 - e^{-t/RC}) \tag{13-20}$$

Using Equation (13-20), we can calculate the value of the charging voltage of a capacitor at any instant of time. The same is true for an increasing current.

**EXAMPLE
13–12**

In Figure 13–29, determine the capacitor voltage 50 microseconds (μs) after the switch is closed if the capacitor is initially uncharged. Sketch the charging curve.

FIGURE 13–29

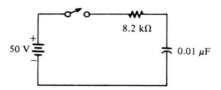

Solution:
The time constant is $RC = (8.2 \text{ k}\Omega)(0.01 \text{ }\mu\text{F}) = 82 \text{ }\mu\text{s}$. The voltage to which the capacitor will fully charge is 50 V (this is V_F). The initial voltage is zero. Notice that 50 μs is less than one time constant; so the capacitor will charge less than 63% of the full voltage in that time.

$$v_C = V_F(1 - e^{-t/RC}) = 50 \text{ V}(1 - e^{-50\mu s/82\mu s})$$

$$= 50 \text{ V}(1 - e^{-0.61}) = 50 \text{ V}(1 - 0.543)$$

$$= 22.85 \text{ V}$$

We determine the value of $e^{-0.61}$ on the calculator by entering -0.61 and then pressing the $\boxed{e^x}$ key (or $\boxed{\text{INV}}$ and then $\boxed{\ln x}$ on some calculators).

The charging curve for the capacitor is shown in Figure 13–30.

FIGURE 13–30

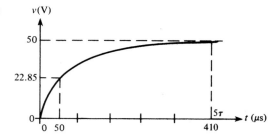

The calculator sequence is

The Discharging Curve   The formula for the special case in which a decreasing exponential voltage curve ends at zero is derived from the general formula as follows:

$$v = V_F + (V_i - V_F)e^{-t/\tau}$$
$$= 0 + (V_i - 0)e^{-t/RC}$$
$$v = V_i e^{-t/RC} \tag{13–21}$$

where V_i is the voltage at the beginning of the discharge. We can use this formula to calculate the discharging voltage at any instant, as Example 13–13 illustrates.

EXAMPLE 13–13

Determine the capacitor voltage in Figure 13–31 at a point in time 6 milliseconds (ms) after the switch is closed. Sketch the discharging curve.

FIGURE 13–31

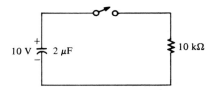

Solution:

The discharge time constant is $RC = (10 \text{ k}\Omega)(2 \text{ }\mu\text{F}) = 20$ ms. The initial capacitor voltage is 10 V. Notice that 6 ms is less than one time constant, so the capacitor will discharge less than 63%. Therefore, it will have a voltage greater than 37% of the initial voltage at 6 ms.

$$v_C = V_i e^{-t/RC} = 10e^{-6ms/20ms}$$

$$= 10e^{-0.3} = 10(0.741)$$

$$= 7.41 \text{ V}$$

Again, the value of $e^{-0.3}$ can be determined with a calculator.

The discharging curve for the capacitor is shown in Figure 13–32.

FIGURE 13–32

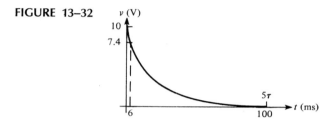

Universal Exponential Curves The universal curves in Figure 13–33 provide a graphic solution of the charge and discharge of capacitors. Example 13–14 illustrates this graphic method.

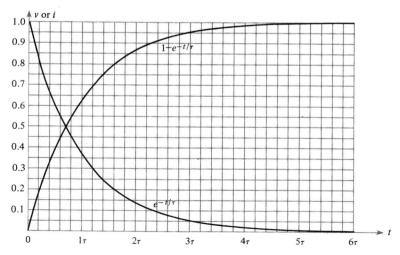

FIGURE 13–33
Universal exponential curves.

**EXAMPLE
13–14**

How long will it take the capacitor in Figure 13–34 to charge to 75 V? What is the capacitor voltage 2 ms after the switch is closed? Use the universal curves in Figure 13–33 to determine the answers.

FIGURE 13–34

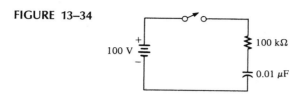

Solution:
The full charge voltage is 100 V, which is the 100% level on the graph. Since 75 V is 75% of the maximum, you can see that this value occurs at 1.4 time constants. One time constant is 1 ms. Therefore, the capacitor voltage reaches 75 V at 1.4 ms after the switch is closed.

     The capacitor is at approximately 87 V in 2 ms. These graphic solutions are shown in Figure 13–35.

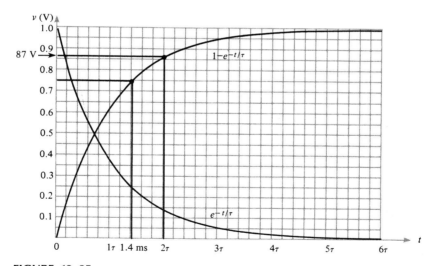

FIGURE 13–35

Time-Constant Percentage Tables The percentages of full charge or discharge at each time-constant interval can be calculated using the exponential formulas, or they can be extracted from the universal graphs. The results are summarized in Tables 13–4 and 13–5.

TABLE 13–4
Percentage of final charge after each charging time-constant interval.

Number of Time Constants	% Final Charge
1	63
2	86
3	95
4	98
5	99 (considered 100%)

TABLE 13–5
Percentage of initial charge after each discharging time-constant interval.

Number of Time Constants	% Initial Charge
1	37
2	14
3	5
4	2
5	1 (considered 0)

SOLVING FOR TIME

Occasionally, it is necessary to determine how long it will take a capacitor to charge or discharge to a specified voltage. Equations (13–19) and (13–20) can be solved for t if v is specified. The natural logarithm (abbreviated 1n) of $e^{-t/RC}$ is the exponent $-t/RC$. Therefore, taking the natural logarithm of both sides of the equation allows us to solve for time. This calculation is done as follows with Equation (13–21):

$$v = V_i e^{-t/RC}$$

$$\frac{v}{V_i} = e^{-t/RC}$$

$$\ln\left(\frac{v}{V_i}\right) = \ln e^{-t/RC}$$

$$\ln\left(\frac{v}{V_i}\right) = \frac{-t}{RC}$$

$$t = -RC \ln\left(\frac{v}{V_i}\right)$$

The same procedure can be used for the increasing exponential formula in Equation (13–20).

EXAMPLE 13–15

In Figure 13–36, how long will it take the capacitor to discharge to 25 V when the switch is closed?

FIGURE 13–36

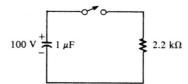

100 V 1 μF 2.2 kΩ

Solution:

$$t = -RC \ln\left(\frac{v}{V_i}\right) = -(2.2 \text{ ms}) \ln\left(\frac{25 \text{ V}}{100 \text{ V}}\right)$$

$$= -(2.2 \text{ ms}) \ln(0.25) = -(2.2 \text{ ms})(-1.39)$$

$$= 3.05 \text{ ms}$$

We can determine ln(0.25) with a calculator by first entering 0.25 and then pressing the $\boxed{\ln x}$ key.

SECTION REVIEW 13–5

1. Determine the time constant when $R = 1.2 \text{ k}\Omega$ and $C = 1000 \text{ pF}$.

2. If the circuit mentioned in Problem 1 is charged with a 5-V source, how long will it take the capacitor to reach full charge? At full charge, what is the capacitor voltage?

3. A certain circuit has a time constant of 1 ms. If it is charged with a 10-V battery, what will the capacitor voltage be at each of the following intervals: 2 ms, 3 ms, 4 ms, and 5 ms?

4. A capacitor is charged to 100 V. If it is discharged through a resistor, what is the capacitor voltage at one time constant?

5. In Figure 13–37, determine the voltage across the capacitor at 2.5 time constants after the switch is closed.

FIGURE 13–37

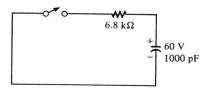

6. In Figure 13–37, how long will it take the capacitor to discharge to 15 V?

13–6

CAPACITORS IN ac CIRCUITS

In order to understand fully the action of capacitors in ac circuits, the concept of the *derivative* must be introduced. *The derivative of a time-varying quantity is the instantaneous rate of change of that quantity.*

Recall that current is the *rate of flow of charge (electrons)*. Therefore, instantaneous current, i, can be expressed as the instantaneous rate of change of charge, q, with respect to time, t:

$$i = \frac{dq}{dt} \tag{13–22}$$

The term dq/dt is the *derivative* of q with respect to time and represents the instantaneous rate of change of q. Also, in terms of instantaneous quantities, $q = Cv$. Therefore, from a basic rule of differential calculus, the derivative of q is $dq/dt = C(dv/dt)$. Since $i = dq/dt$, we get the following relationship:

$$i = C\left(\frac{dv}{dt}\right)$$

(13–23)

This equation says:

> **The instantaneous capacitor current is equal to the capacitance times the instantaneous rate of change of the voltage across the capacitor.**

From this, you can see that the faster the voltage across a capacitor changes, the greater the current.

PHASE RELATIONSHIP OF CURRENT AND VOLTAGE IN A CAPACITOR

Now consider what happens when a sinusoidal voltage is applied across a capacitor, as shown in Figure 13–38. The voltage waveform has a maximum rate of change ($dv/dt = $ max) at the zero crossings and a zero rate of change ($dv/dt = 0$) at the peaks, as indicated in Figure 13–39.

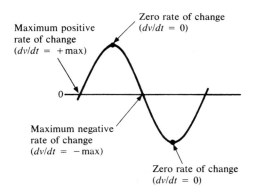

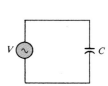

FIGURE 13–38
Sine wave applied to a capacitor.

FIGURE 13–39
The rates of change of a sine wave.

Using Equation (13–23), the phase relationship between the current and the voltage for the capacitor can be established. When $dv/dt = 0$, i is also zero, because $i = C(dv/dt) = C(0) = 0$. When dv/dt is a positive-going maximum, i is a positive maximum; when dv/dt is a negative-going maximum, i is a negative maximum.

A sinusoidal voltage always produces a sinusoidal current in a capacitive circuit. Therefore, the current can be plotted with respect to the voltage by knowing the points on the voltage curve at which the current is zero and those at which it is maximum. This relationship is shown in Figure 13–40(a). Notice that *the current leads the voltage in phase by 90 degrees*. This is always true in a purely capacitive circuit. A phasor diagram of this relationship is shown in Figure 13–40(b).

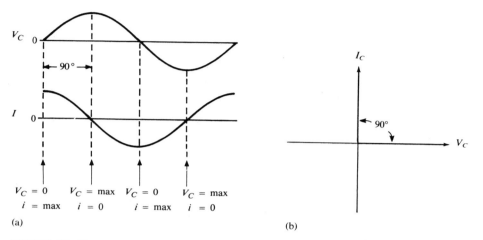

$V_C = 0$ $V_C = \text{max}$ $V_C = 0$ $V_C = \text{max}$
$i = \text{max}$ $i = 0$ $i = \text{max}$ $i = 0$

(a)   (b)

FIGURE 13–40
Current is always leading the capacitor voltage by 90 degrees.

CAPACITIVE REACTANCE, X_C

Capacitive reactance is the opposition to sinusoidal current, expressed in ohms. The symbol for capacitive reactance is X_C.

To develop a formula for X_C, we use the relationship $i = C(dv/dt)$ and the curves in Figure 13–41. The rate of change of voltage is directly related to *frequency*. The faster the voltage changes, the higher the frequency. For example, you can see that in Figure 13–41, the slope of sine wave A at the zero crossings is greater than that of sine wave B. The *slope* of a curve at a point indicates the rate of change at that point.

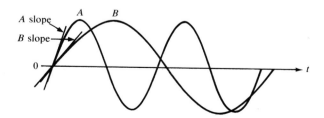

FIGURE 13–41
A higher frequency wave has a greater slope at its zero crossings, corresponding to a higher rate of change.

Sine wave A has a higher frequency than sine wave B, as indicated by a greater maximum rate of change (dv/dt is greater at the zero crossings).

When frequency increases, dv/dt increases, and thus i increases. When frequency decreases, dv/dt decreases, and thus i decreases:

$$\underset{}{\overset{\uparrow}{i}} = C(\overset{\uparrow}{dv}/dt) \qquad \text{and} \qquad \underset{\downarrow}{i} = C(\underset{\downarrow}{dv}/dt)$$

An increase in i means that there is less opposition to current (X_C is less), and a decrease in i means a greater opposition (X_C is greater). Therefore, *X_C is inversely proportional to i and thus inversely proportional to frequency.*

$$X_C \text{ is proportional to } \frac{1}{f}$$

Again, from the relationship $i = C(dv/dt)$, you can see that if dv/dt is constant and C is varied, an increase in C produces an increase in i, and a decrease in C produces a decrease in i.

$$\overset{\uparrow}{i} = \overset{\uparrow}{C}(dv/dt) \qquad \text{and} \qquad \underset{\downarrow}{i} = \underset{\downarrow}{C}(dv/dt)$$

Again, an increase in i means less opposition (X_C is less), and a decrease in i means greater opposition (X_C is greater). Therefore, *X_C is inversely proportional to i and thus inversely proportional to capacitance.*

The capacitive reactance is inversely proportional to both f and C.

$$X_C \text{ is proportional to } \frac{1}{fC}$$

Thus far, we have determined a proportional relationship between X_C and $1/fC$. What is needed is a formula that tells us what X_C is *equal* to so that it can be calculated. This formula is derived in Appendix D and is stated as follows:

$$X_C = \frac{1}{2\pi fC} \tag{13–24}$$

X_C is in *ohms* when f is in *hertz* and C is in *farads*. Notice that 2π appears in the denominator as a constant of proportionality. This term is derived from the relationship of a sine wave to rotational motion as you can see in Appendix D.

EXAMPLE
13–16

A sinusoidal voltage is applied to a capacitor, as shown in Figure 13–42. The frequency of the sine wave is 1 kHz. Determine the capacitive reactance.

FIGURE 13–42

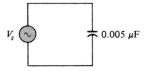

V_s ⏦   ⊥ 0.005 μF

Solution:

$$X_C = \frac{1}{2\pi f C} = \frac{1}{2\pi (1 \times 10^3 \text{ Hz})(0.005 \times 10^{-6} \text{ F})}$$

$$= 31.83 \text{ k}\Omega$$

ANALYSIS OF CAPACITIVE ac CIRCUITS

As you have seen, the current leads the voltage by 90 degrees in capacitive ac circuits. If the applied voltage is assigned a reference phase angle of zero, it can be expressed in polar form as $V_s \angle 0°$. The resulting current can be expressed as $I \angle 90°$ or jI in rectangular form as shown in Figure 13–43.

FIGURE 13–43

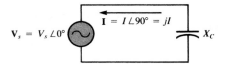

$\mathbf{V}_s = V_s \angle 0°$ ⏦ $\mathbf{I} = I \angle 90° = jI$ X_C

Ohm's law applies to ac circuits containing capacitive reactance with R replaced by $\mathbf{X}_C$ in the Ohm's law formula. The quantities are expressed as complex numbers because of the introduction of phase angles. Applying Ohm's law to the circuit in Figure 13–43 gives the following result:

$$\mathbf{X}_C = \frac{V_s \angle 0°}{I \angle 90°} = \left(\frac{V_s}{I}\right)\angle -90°$$

This shows that $\mathbf{X}_C$ always has a -90-degree angle attached to its magnitude and is written as $X_C \angle -90°$ or $-jX_C$.

**EXAMPLE
13–17**

Determine the rms current in Figure 13–44.

FIGURE 13–44

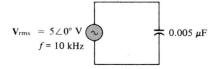

$\mathbf{V}_{rms} = 5\angle 0°$ V
$f = 10$ kHz

0.005 μF

Solution:
The magnitude of X_C is

$$X_C = \frac{1}{2\pi f C} = \frac{1}{2\pi (10 \times 10^3 \text{ Hz})(0.005 \times 10^{-6} \text{ F})}$$
$$= 3.18 \text{ k}\Omega$$

Expressed in polar form, $\mathbf{X}_C$ is

$$\mathbf{X}_C = 3.18\angle -90° \text{ k}\Omega$$

Applying Ohm's law:

$$\mathbf{I} = \frac{\mathbf{V}_{rms}}{\mathbf{X}_C} = \frac{5\angle 0° \text{ V}}{3.18\angle -90° \text{ k}\Omega} = 1.57\angle 90° \text{ mA}$$

Notice that the current expression has a 90-degree phase angle indicating that it leads the voltage by 90 degrees.

POWER IN A CAPACITOR

As discussed earlier in this chapter, a charged capacitor stores energy in the electric field within the dielectric. An ideal capacitor does not dissipate energy; it only stores it. When an ac voltage is applied to a capacitor, energy is stored by the capacitor during a portion of the voltage cycle; then the stored energy is *returned to the source* during another portion of the cycle. *There is no net energy loss.* Figure 13–45 shows the power curve that results from one cycle of capacitor voltage and current.

Instantaneous Power (p) *The product of v and i gives instantaneous power, p.* At points where *v* or *i* is zero, *p* is also zero. When both *v* and *i* are positive, *p* is also positive. When either *v* or *i* is positive and the other negative, *p* is negative. When both *v* and *i* are negative, *p* is positive. As you can see, the power follows a sinusoidal curve. Positive values of power indicate that energy is stored by the capacitor. Negative values of power indicate that energy is returned from the capacitor to the source. Note that the power fluctuates at a frequency twice that of the voltage or current as energy is alternately stored and returned to the source.

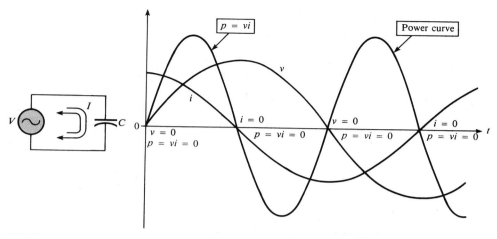

FIGURE 13–45
Power curve.

True Power (P_{true})   Ideally, all of the energy stored by a capacitor during the positive portion of the power cycle is returned to the source during the negative portion. *No net energy is consumed in the capacitor*, so the *true power is zero*. Actually, because of leakage and foil resistance in a practical capacitor, a small percentage of the total power is dissipated.

Reactive Power (P_r)   The rate at which a capacitor stores or returns energy is called its *reactive power, P_r*. The reactive power is a nonzero quantity, because at any instant in time, the capacitor is actually taking energy from the source or returning energy to it. Reactive power does not represent an energy loss. The following formulas apply:

$$P_r = V_{\text{rms}}I_{\text{rms}} \tag{13–25}$$

$$P_r = \frac{V_{\text{rms}}^2}{X_C} \tag{13–26}$$

$$P_r = I_{\text{rms}}^2 X_C \tag{13–27}$$

Notice that these equations are of the same form as those for power in a resistor. The voltage and current are expressed in rms. The unit of reactive power is *volt-amperes reactive* (VAR).

EXAMPLE 13–18

Determine the true power and the reactive power in Figure 13–46.

FIGURE 13–46

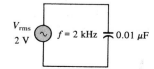

Solution:

The true power, P_{true} is *always zero for a capacitor*. The reactive power is as follows:

$$X_C = \frac{1}{2\pi f C} = \frac{1}{2\pi(2 \times 10^3 \text{ Hz})(0.01 \times 10^{-6} \text{ F})}$$

$$= 7.958 \text{ k}\Omega$$

$$P_r = \frac{V_{\text{rms}}^2}{X_C} = \frac{(2 \text{ V})^2}{7.958 \text{ k}\Omega}$$

$$= 0.503 \times 10^{-3} \text{ VAR}$$

$$= 0.503 \text{ mVAR}$$

SECTION REVIEW 13–6

1. State the phase relationship between current and voltage in a capacitor.

2. Calculate X_C for $f = 5$ kHz and $C = 50$ pF.

3. At what frequency is the reactance of a 0.1-μF capacitor equal to 2 kΩ?

4. Calculate the rms current in Figure 13–47.

FIGURE 13–47

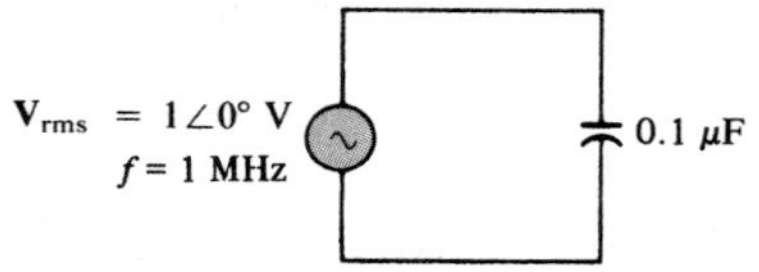

$\mathbf{V}_{\text{rms}} = 1\angle 0° \text{ V}$
$f = 1 \text{ MHz}$

$0.1 \; \mu\text{F}$

5. A 1-μF capacitor is connected to an ac voltage source of 12 V rms. What is the true power?

6. In Question 5, determine reactive power at a frequency of 500 Hz.

13–7 CAPACITOR APPLICATIONS

Capacitors are very widely used in electrical and electronic applications. A few typical applications are discussed here to illustrate the usefulness of this component.

POWER SUPPLY FILTER

A device that converts the 60 Hz sine wave voltage from your wall outlet to a pulsating dc voltage is called a *full-wave rectifier*. The basic concept of a power supply with a full-wave rectifier is shown in Figure 13–48. In order to be useful in powering most systems such as the radio, TV, or computer, the pulsating dc voltage must be converted to a nearly constant level. This conversion is accomplished with a power supply filter as indicated.

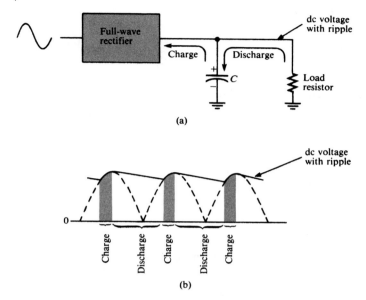

FIGURE 13-48
Basic operation of a capacitor power supply filter.

A basic power supply filter is implemented with a capacitor as shown in Figure 13-48(a). As shown in Part (b), the basic operation is as follows: The capacitor charges as the full-wave voltage increases. When the peak is reached and the full-wave voltage starts to decrease, the capacitor begins to discharge through the load resistance. The value of the capacitance is selected so that the time constant is very long compared to the period of the full-wave voltage. The rectifier circuitry allows current only in a direction to charge the capacitor and prevents discharge current. By the time the full-wave voltage nears its next peak, the capacitor has discharged only a small amount and requires only a small amount of recharging to get it back to the peak. This action results in an almost constant dc voltage as shown. The small fluctuation is caused by the slight discharging and recharging of the capacitor. This fluctuation is called *ripple voltage*.

COUPLING AND BYPASS CAPACITORS

Many applications, such as transistor amplifiers, require that an ac voltage be superimposed on a dc voltage at a certain point in the circuit, while at other points, the ac voltage must be removed without affecting the dc voltage.

The first situation is illustrated in Figure 13-49(a), where a capacitor is used to *couple* an ac voltage from the source to a point on a voltage divider that has a dc voltage. Since a capacitor blocks dc, the ac source is unaffected by the dc level, but the ac signal is passed through and superimposed on the dc level.

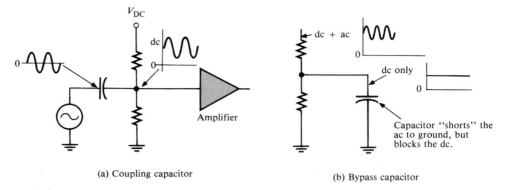

(a) Coupling capacitor

(b) Bypass capacitor

FIGURE 13–49
Coupling and bypass capacitors.

The second situation is illustrated in Figure 13–49(b) where a capacitor is used to *bypass* the ac voltage to ground, leaving only the dc voltage. Details of electronic amplifiers with ac coupling and bypass circuits are covered in a later course.

TUNED CIRCUITS

Capacitors are used in conjunction with other components such as the inductor (to be covered in the next chapter) to provide *frequency selection* in communications systems. These *tuned circuits* allow a narrow band of frequencies to be selected while all other frequencies are rejected. A tuned circuit is one form of *filter*. The tuners in your TV and radio receivers are based on this principle and permit you to select one channel or station out of the many that are available.

Frequency selectivity is based on the fact the reactance of a capacitor depends on the frequency. The basic concept of a tuned circuit is shown in Figure 13–50. This topic will be covered in detail in Chapters 18 and 19.

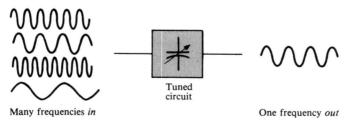

Many frequencies *in*

Tuned circuit

One frequency *out*

FIGURE 13–50
Basic concept of a tuned circuit.

COMPUTER MEMORIES

Many computer memories utilize capacitors as the *storage element* for binary data, which consist of arrangements of only two types of digits, 1s and 0s. A charged capacitor can represent a 1, and a discharged capacitor can represent a 0. Patterns of 1s and 0s can be stored in a memory that consists of an array of capacitors with associated circuitry. You will study this topic later in a computer or digital fundamentals course.

SECTION REVIEW 13–7

1. Explain how pulsating dc voltage is smoothed out by a capacitor in a power supply filter.
2. How can you use a capacitor to remove ac voltage from a given point in a circuit?

13–8   TESTING CAPACITORS

Capacitors are generally very reliable devices. Their useful life can be extended significantly by operating well within the voltage rating and at moderate temperatures.

Failures can be categorized into two areas: *catastrophic* and *degradation*. The catastrophic failures are usually a short circuit caused by dielectric breakdown or an open circuit caused by connection failure. Degradation usually results in a gradual decrease in leakage resistance, hence an increase in leakage current or an increase in equivalent series resistance or dielectric absorption.

OHMMETER CHECK

When there is a suspected problem, the capacitor can be removed from the circuit and checked with an ohmmeter. First, to be sure that the capacitor is discharged, short its leads, as indicated in Figure 13–51(a). Connect the meter, set on a high ohms range

(a) Discharging

(b) Initially: The pointer
jumps to zero.

(c) Charging: The pointer
slowly moves back.

(d) Fully charged

FIGURE 13–51
Checking a capacitor with an ohmmeter. This check shows a good capacitor.

such as R × 1M, to the capacitor, as shown in Part (b), and observe the needle. It should initially indicate near zero ohms. Then it should begin to move toward the high-resistance end of the scale as the capacitor charges from the ohmmeter's battery, as shown in Part (c). When the capacitor is fully charged, the meter will indicate an extremely high resistance as shown in Part (d).

As mentioned, the capacitor charges from the internal battery of the ohmmeter, and the meter responds to the charging current. The larger the capacitance value, the more slowly the capacitor will charge, as indicated by the needle movement. For very small pF values, the meter response may be insufficient to indicate the fast charging action.

If the capacitor is internally shorted, the meter will go to zero and stay there. If it is leaky, the final meter reading will be much less than normal. Most capacitors have a resistance of several hundred megohms. The exception is the electrolytic, which may normally have less than one megohm of leakage resistance. If the capacitor is open, no charging action will be observed, and the meter will indicate an infinite resistance.

TESTING FOR CAPACITANCE VALUE AND OTHER PARAMETERS WITH AN *LC* METER

An *LC* meter such as the one shown in Figure 13–52 can be used to check the value of a capacitor. All capacitors change value over a period of time, some more than others. Ceramic capacitors, for example, often exhibit a 10% to 15% change in value during the first year. Electrolytic capacitors are particularly subject to value change due to drying of the electrolytic solution. In other cases, capacitors may be labeled incorrectly or the wrong value was installed in the circuit. Although a value change represents less

FIGURE 13–52
A typical *LC* meter (courtesy SENCORE).

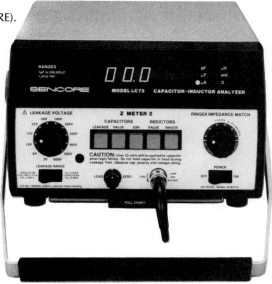

than 25% of defective capacitors, a value check should be made to quickly eliminate this as a source of trouble when troubleshooting a circuit.

Typically values from 1 pF to 200,000 μF can be measured by simply connecting the capacitor, pushing the appropriate button, and reading the value on the display.

Many *LC* meters can also be used to check for *leakage current* in capacitors. In order to check for leakage, a sufficient voltage must be applied across the capacitor to simulate operating conditions. This is automatically done by the test instrument. Over 40% of all defective capacitors have excessive leakage current and electrolytics are particularly susceptible to this problem.

The problem of *dielectric absorption* occurs mostly in electrolytic capacitors when they do not completely discharge during use and retain a residual charge. Approximately 25% of defective capacitors have exhibited this condition.

Another defect sometimes found in capacitors is called *equivalent series resistance*. This problem may be caused by defective lead to plate contacts, resistive leads, or resistive plates and shows up only under ac conditions. This is the least common capacitor defect and occurs in less than 10% of all defects.

SECTION REVIEW 13–8

1. How can a capacitor be discharged after removal from the circuit?

2. Describe how the needle of an ohmmeter responds when a good capacitor is checked.

3. List four common capacitor defects.

13–9 COMPUTER ANALYSIS

This program computes the instantaneous charging voltage for a capacitor after switch closure in a series *RC* circuit connected to a dc source. The program requires that you input the dc source voltage, the resistance value in ohms, the capacitance value in farads, and the number of time intervals for which you want the voltage computed during the full charging interval. The program output is a tabulation of the instantaneous voltages and percent of full charge at each point in time during the charging interval. Note that the exponentiation symbol ([) in Line 190 is the ↑ symbol on the keyboard. A flowchart is shown in Figure 13–53.

```
10   CLS
20   PRINT "THIS PROGRAM COMPUTES AND TABULATES THE
     INSTANTANEOUS"
30   PRINT "CAPACITOR VOLTAGE AND PERCENT OF FULL CHARGE AT"
40   PRINT "NUMBER OF SPECIFIED TIME INTERVALS DURING CHARGING"
50   PRINT "IN A SPECIFIED RC CIRCUIT."
60   PRINT
70   PRINT "ENTER THE DC SOURCE VOLTAGE, RESISTANCE IN OHMS,"
```

FIGURE 13–53

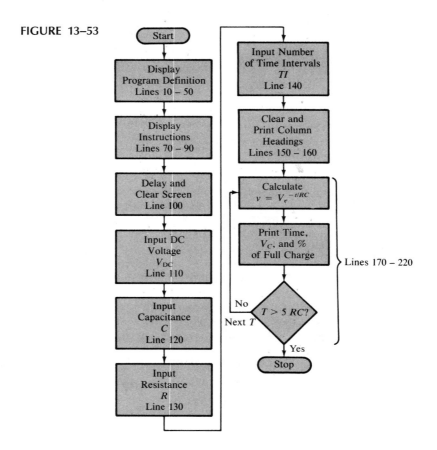

```
80    PRINT "CAPACITANCE IN FARADS, AND DESIRED NUMBER OF TIME"
90    PRINT "INTERVALS."
100   FOR T=0 TO 4000:NEXT:CLS
110   INPUT "DC SOURCE VOLTAGE";VDC
120   INPUT "CAPACITANCE IN FARADS";C
130   INPUT "RESISTANCE IN OHMS";R
140   INPUT "NUMBER OF TIME INTERVALS";TI
150   CLS
160   PRINT "TIME (SEC)","CAPACITOR VOLTAGE","% OF FULL CHARGE"
170   FOR T=0 TO 5*R*C STEP 5*R*C/TI
180   X=T/(R*C)
190   V=VDC*(1-(2.718)[-X)
200   P=(V/VDC)*100
210   PRINT TAB(0) T "S";TAB(20) V "V";TAB(50) P "%"
220   NEXT
```

SECTION REVIEW 13–9

1. Identify by line number each FOR/NEXT loop.
2. State the purpose of each FOR/NEXT loop.

SUMMARY

1. A capacitor is composed of two parallel conducting plates separated by a dielectric insulator.
2. Capacitance is the measure of a capacitor's ability to store electrical charge.
3. Energy is stored by a capacitor in the electric field concentrated in the dielectric.
4. The farad is the unit of capacitance.
5. *One farad* is the amount of capacitance when *one coulomb* of charge is stored with *one volt* across the plates.
6. Capacitance is directly proportional to the plate area and inversely proportional to the plate separation.
7. The dielectric constant is an indication of the ability of a material to establish an electric field.
8. The dielectric strength is one factor that determines the breakdown voltage of a capacitor.
9. The time constant for a series RC circuit is the resistance times the capacitance.
10. In an RC circuit, the voltage and current in a charging or discharging capacitor make a 63% change during each time-constant interval.
11. Five time constants are required for a capacitor to charge fully or to discharge fully.
12. Charging and discharging follow exponential curves.
13. Total series capacitance is less than that of the smallest capacitor in series.
14. Capacitance adds in parallel.
15. Current leads voltage by 90 degrees in a capacitor.
16. Capacitive reactance is the opposition to sinusoidal current and is expressed in ohms.
17. X_C is inversely proportional to frequency and capacitance.
18. The true power in a capacitor is zero; that is, there is no energy loss in an ideal capacitor.

FORMULAS

$$C = \frac{Q}{V} \tag{13–1}$$

$$Q = CV \tag{13–2}$$

$$V = \frac{Q}{C} \tag{13–3}$$

$$F = \frac{kQ_1Q_2}{d^2} \tag{13–4}$$

$$W = \frac{1}{2}CV^2 \tag{13–5}$$

$$\epsilon_r = \frac{\epsilon}{\epsilon_0} \tag{13–6}$$

$$C = \frac{A\epsilon_r(8.85 \times 10^{-12} \text{ F/m})}{d} \tag{13–7}$$

$$Q_T = Q_1 = Q_2 = Q_3 = \cdots = Q_n \tag{13–8}$$

$$\frac{Q_T}{C_T} = \frac{Q_1}{C_1} + \frac{Q_2}{C_2} + \frac{Q_3}{C_3} + \cdots + \frac{Q_n}{C_n} \tag{13–9}$$

$$C_T = \frac{1}{\dfrac{1}{C_1} + \dfrac{1}{C_2} + \dfrac{1}{C_3} + \cdots + \dfrac{1}{C_n}} \tag{13–10}$$

$$C_T = \frac{C_1C_2}{C_1 + C_2} \tag{13–11}$$

$$C_T = \frac{C}{n} \tag{13–12}$$

$$V_X = \left(\frac{C_T}{C_X}\right)V_T \tag{13–13}$$

$$Q_T = Q_1 + Q_2 + Q_3 + \cdots + Q_n \tag{13–14}$$

$$C_T = C_1 + C_2 + C_3 + \cdots + C_n \tag{13–15}$$

$$C_T = nC \tag{13–16}$$

$$\tau = RC \tag{13–17}$$

$$v = V_F + (V_i - V_F)e^{-t/\tau} \tag{13–18}$$

$$i = I_F + (I_i - I_F)e^{-t/\tau} \tag{13–19}$$

$$v = V_F(1 - e^{-t/RC}) \tag{13-20}$$

$$v = V_i e^{-t/RC} \tag{13-21}$$

$$i = \frac{dq}{dt} \tag{13-22}$$

$$i = C\left(\frac{dv}{dt}\right) \tag{13-23}$$

$$X_C = \frac{1}{2\pi fC} \tag{13-24}$$

$$P_r = V_{rms}I_{rms} \tag{13-25}$$

$$P_r = \frac{V_{rms}^2}{X_C} \tag{13-26}$$

$$P_r = I_{rms}^2 X_C \tag{13-27}$$

SELF-TEST

Solutions appear at the end of the book.

1. Indicate true or false for each of the following statements:
   (a) The plates of a capacitor are conductive.
   (b) The dielectric is the insulating material that separates the plates.
   (c) There is constant direct current through a fully charged capacitor.
   (d) A practical capacitor stores charge indefinitely when it is disconnected from the source.

2. Indicate true or false for each of the following statements:
   (a) There is current through the dielectric of a charging capacitor.
   (b) When a capacitor is connected to a dc voltage source, it will charge to the value of the source voltage.
   (c) You can discharge an ideal capacitor by simply disconnecting it from the voltage source.
   (d) When a capacitor is completely discharged, the voltage across its plates is zero.

3. Which is the larger capacitance, 0.01 μF or 0.00001 F?

4. Which is the smaller capacitance, 1000 pF or 0.0001 μF?

5. If the voltage across a given capacitor is increased, does the amount of stored charge increase or decrease?

6. If the voltage across a given capacitor is doubled, how much does the stored energy increase?

7. How can the voltage rating of a capacitor be increased?

8. Select the physical parameter changes that result in an increase in the capacitance value:
   (a) Reduce plate area.   (b) Move plates further apart.
   (c) Move plates closer.   (d) Increase plate area.
   (e) Increase the thickness of the dielectric.
   (f) Decrease the thickness of the dielectric.

9. A 1-μF, a 2.2-μF, and a 0.05-μF capacitor are connected in series. The total capacitance is (greater, less than) _____ μF.

10. Four 0.02-μF capacitors are connected in parallel. What is the total capacitance?

11. An uncharged capacitor and a resistor are placed in series with a switch and are connected to a dc voltage source.
    (a) At the instant the switch is closed, what is the voltage across the capacitor?
    (b) When will the capacitor reach full charge?
    (c) What is the voltage across the capacitor when it is fully charged?
    (d) What is the current in the circuit when the capacitor is fully charged?

12. A sine wave voltage is applied across a capacitor. When the frequency of the voltage is increased, does the current increase or decrease? Why?

13. A capacitor and a resistor are connected in series with a sine wave voltage source. The frequency is set so that the capacitive reactance is equal to the resistance, and an equal amount of voltage is dropped across each component. If the frequency is decreased, which component will have the greater voltage across it?

14. An ohmmeter is connected across a discharged capacitor, and the pointer stabilizes at a low resistance value. What is your opinion of this capacitor?

15. A given capacitor stores 50×10^{-6} coulomb of charge when the voltage across its plates is 5 V. Determine its capacitance.

16. A 0.01-μF capacitor stores 5×10^{-8} C. What is the voltage across its plates?

17. Convert 0.005 μF to picofarads.

18. Convert 2000 pF to microfarads.

19. Convert 1500 μF to farads.

20. A 10-μF capacitor is charged to 100 V. How many joules of energy are stored?

21. Calculate the absolute permittivity, ϵ, for ceramic.

22. Calculate the capacitance for the following physical parameters: $A = 0.009 \text{ m}^2$, $d = 0.0015$ m, and a dielectric of Teflon®.

23. A circuit has $R = 2 \text{ k}\Omega$ in series with $C = 0.05 \ \mu$F. What is the time constant?

24. The capacitor in an RC circuit is charged to 15 V. If it is allowed to discharge for an interval equal to one time constant, what is the voltage across the capacitor?

25. A series RC circuit with a time constant of 10 ms is connected across a 10-V dc source with a switch. If the capacitor is initially uncharged (0 V), what voltage will it reach at 15 ms after switch closure?

26. Determine the total capacitance in each circuit in Figure 13–54.

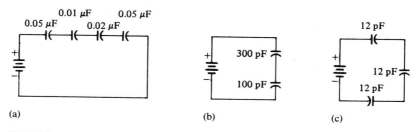

(a)

(b)

(c)

FIGURE 13–54

27. Find the voltage across each capacitor in Figure 13–55.

FIGURE 13–55

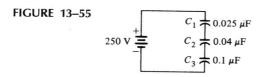

28. Find C_T in each circuit of Figure 13–56.

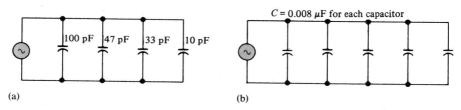

(a)

(b)

FIGURE 13–56

29. A 2-MHz sinusoidal voltage is applied to a 1-μF capacitor. What is the reactance?

30. The frequency of the source in Figure 13–57 can be adjusted. In order to obtain 50 mA of rms current, to what frequency must the source be adjusted?

FIGURE 13–57

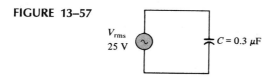

31. Determine the true power and the reactive power when the circuit in Figure 13–57 is adjusted as required in Question 30.

32. An ohmmeter shows 1000 Ω when connected across a capacitor that is suspected to be faulty. What is your conclusion?

PROBLEMS

Section 13–1

13–1   (a) Find the capacitance when $Q = 50 \ \mu C$ and $V = 10$ V.
(b) Find the charge when $C = 0.001 \ \mu F$ and $V = 1$ kV.
(c) Find the voltage when $Q = 2$ mC and $C = 200 \ \mu F$.

13–2   Convert the following values from microfarads to picofarads:
(a) $0.1 \ \mu F$ (b) $0.0025 \ \mu F$ (c) $5 \ \mu F$

13–3   Convert the following values from picofarads to microfarads:
(a) 1000 pF   (b) 3500 pF   (c) 250 pF

13–4   Convert the following values from farads to microfarads:
(a) 0.0000001 F   (b) 0.0022 F   (c) 0.0000000015 F

13–5   Calculate the force of repulsion between two electrons 0.001 m apart.

13–6   What size capacitor is capable of storing 10 mJ of energy with 100 volts across its plates?

13–7   Calculate the absolute permittivity, ϵ, for each of the following materials:
(a) air   (b) oil   (c) glass   (d) Teflon®

13–8   A mica capacitor has a plate area of 0.04 m² and a dielectric thickness of 0.008 m. What is its capacitance?

13–9   An air capacitor has 0.1-m square plates. The plates are separated by 0.01 m. Calculate the capacitance.

13–10   At ambient temperature (25°C), a certain capacitor is specified to be 1000 pF. It has a negative temperature coefficient of 200 ppm/°C. What is its capacitance at 75°C?

13–11   A 0.001-μF capacitor has a positive temperature coefficient of 500 ppm/°C. How much change in capacitance will a 25°C increase in temperature cause?

Section 13–2

13–12   In the construction of a stacked-foil mica capacitor, how is the plate area increased?

13–13   What type of capacitor has the highest dielectric constant, mica or ceramic?

13–14   Show how to connect an electrolytic capacitor across R_2 between points A and B in Figure 13–58.

13–15   Name two types of electrolytic capacitors. How do electrolytics differ from other capacitors?

13–16   Identify the parts of the ceramic disk capacitor shown in the cutaway view of Figure 13–59.

13–17   Determine the value of the ceramic disk capacitors in Figure 13–60.

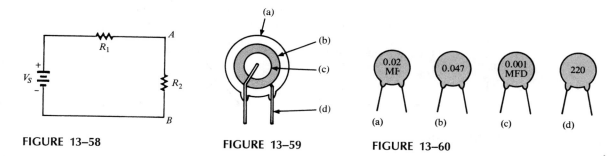

FIGURE 13–58   FIGURE 13–59   FIGURE 13–60

Section 13–3

13–18 Five 1000-pF capacitors are in series. What is the total capacitance?

13–19 Find the total capacitance for each circuit in Figure 13–61.

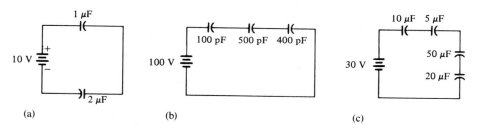

(a)   (b)   (c)

FIGURE 13–61

13–20 For each circuit in Figure 13–61, determine the voltage across each capacitor.

13–21 Two series capacitors (one 1-μF, the other of unknown value) are charged from a 12-V source. The 1-μF capacitor is charged to 8 V, and the other to 4 V. What is the value of the unknown capacitor?

13–22 The total charge stored by the series capacitors in Figure 13–62 is 10 μC. Determine the voltage across each of the capacitors.

FIGURE 13–62

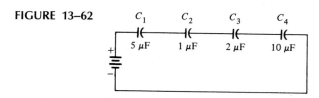

Section 13–4

13–23 Determine C_T for each circuit in Figure 13–63.

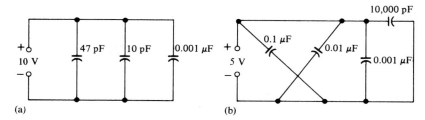

(a) (b)

FIGURE 13–63

13–24 What is the charge on each capacitor in Figure 13–63?

13–25 Determine C_T for each circuit in Figure 13–64.

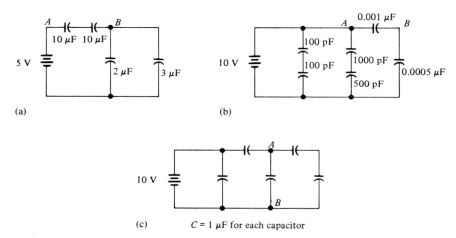

(a) (b)

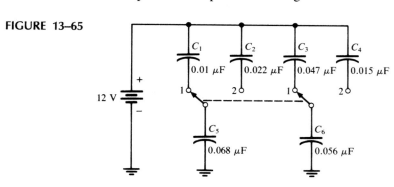

(c) $C = 1 \ \mu F$ for each capacitor

FIGURE 13–64

13–26 What is the voltage between points A and B in each circuit in Figure 13–64?

13–27 How much does the voltage across C_5 and C_6 change when the ganged switch is thrown from position 1 to position 2 in Figure 13–65?

FIGURE 13–65

Section 13–5

13–28 Determine the time constant for each of the following series RC combinations:
  (a) $R = 100 \ \Omega$, $C = 1 \ \mu F$ (b) $R = 10 \ M\Omega$, $C = 50 \ pF$
  (c) $R = 4.7 \ k\Omega$, $C = 0.005 \ \mu F$ (d) $R = 1.5 \ M\Omega$, $C = 0.01 \ \mu F$

13–29 Determine how long it takes the capacitor to reach full charge for each of the following combinations:
  (a) $R = 56 \ \Omega$, $C = 50 \ \mu F$ (b) $R = 3300 \ \Omega$, $C = 0.015 \ \mu F$
  (c) $R = 22 \ k\Omega$, $C = 100 \ pF$ (d) $R = 5.6 \ M\Omega$, $C = 10 \ pF$

13–30 In the circuit of Figure 13–66, the capacitor is initially uncharged. Determine the capacitor voltage at the following times after the switch is closed:
  (a) $10 \ \mu s$ (b) $20 \ \mu s$ (c) $30 \ \mu s$ (d) $40 \ \mu s$ (e) $50 \ \mu s$

13–31 In Figure 13–67, the capacitor is charged to 25 V. When the switch is closed, what is the capacitor voltage after the following times?
  (a) $1.5 \ ms$ (b) $4.5 \ ms$ (c) $6 \ ms$ (d) $7.5 \ ms$

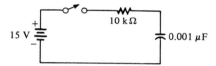

FIGURE 13–66

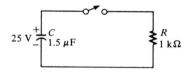

FIGURE 13–67

13–32 Repeat Problem 13–30 for the following time intervals:
  (a) $2 \ \mu s$ (b) $5 \ \mu s$ (c) $15 \ \mu s$

13–33 Repeat Problem 13–31 for the following times:
  (a) $0.5 \ ms$ (b) $1 \ ms$ (c) $2 \ ms$

13–34 Derive the formula for finding the time at any point on an *increasing* exponential voltage curve. Use this formula to find the time at which the voltage in Figure 13–68 reaches 6 V after switch closure.

13–35 How long does it take C to charge to 8 V in Figure 13–66?

13–36 How long does it take C to discharge to 3 V in Figure 13–67?

13–37 Determine the time constant for the circuit in Figure 13–69.

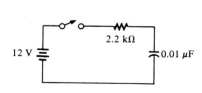

FIGURE 13–68

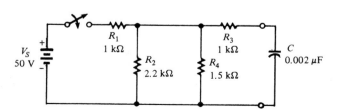

FIGURE 13–69

13–38 In Figure 13–70, the capacitor is initially uncharged. At $t = 10$ μs after the switch is closed, the instantaneous capacitor voltage is 7.2 V. Determine the value of R.

13–39 **(a)** The capacitor in Figure 13–71 is uncharged when the switch is thrown into position 1. The switch remains in position 1 for 10 ms and is then thrown into position 2, where it remains indefinitely. Sketch the complete waveform for the capacitor voltage.

(b) If the switch is thrown back to position 1 after 5 ms in position 2, and then left in position 1, how would the waveform appear?

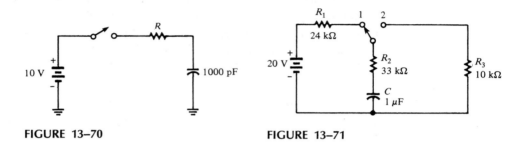

FIGURE 13–70   FIGURE 13–71

Section 13–6

13–40 What is the value of the total capacitive reactance in each circuit in Figure 13–72?

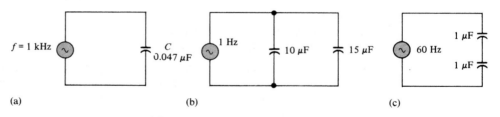

(a)   (b)   (c)

FIGURE 13–72

13–41 In Figure 13–64, each dc voltage source is replaced by a 10-V rms, 2-kHz ac source. Determine the reactance in each case.

13–42 In each circuit of Figure 13–72, what frequency is required to produce an X_C of 100 Ω? An X_C of 1 kΩ?

13–43 A sinusoidal voltage of 20 V rms produces an rms current of 100 mA when connected to a certain capacitor. What is the reactance?

13–44 A 10-kHz voltage is applied to a 0.0047-μF capacitor, and 1 mA of rms current is measured. What is the value of the voltage?

13–45 Determine the true power and the reactive power in Problem 13–44.

13–46 Determine the ac voltage across each capacitor and the current in each branch of the circuit in Figure 13–73. What is the phase angle between the current and the voltage in each case?

FIGURE 13–73

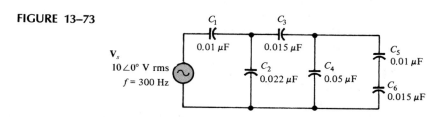

13–47 Find the value of C_1 in Figure 13–74.

FIGURE 13–74

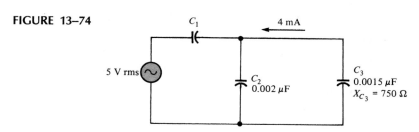

Section 13–7

13–48 If another capacitor is connected in parallel with the existing capacitor in the power supply filter of Figure 13–51, how is the ripple voltage affected?

13–49 Ideally, what should the reactance of a bypass capacitor be in order to eliminate a 10-kHz ac voltage at a given point in an amplifier circuit?

Section 13–8

13–50 Assume that you are checking a capacitor with an ohmmeter, and when you connect the leads across the capacitor, the pointer does not move from its left-end scale position. What is the problem?

13–51 In checking a capacitor with the ohmmeter, you find that the pointer goes all the way to the right end of the scale and stays there. What is the problem?

13–52 If C_4 in Figure 13–73 opened, determine the voltages that would be measured across the other capacitors.

Section 13–9

13–53 Write a computer program in BASIC that will compute and tabulate the capacitive reactance for specified values of R and C over a specified range of frequencies in specified increments.

13–54 Develop a program similar to that in Section 13–9 for capacitor discharging.

ANSWERS TO SECTION REVIEWS

Section 13–1
1. The ability (capacity) to store electrical charge.
2. (a) 10^6;   (b) 10^{12};   (c) 10^6.   3. 1500 pF; 0.0000000015 F.
4. 1.125 μJ.   5. (a) increase;   (b) decrease.   6. 10 kV.
7. 0.425 μF.   8. 2.01 μF.

Section 13–2
1. By the dielectric material.
2. A fixed capacitance cannot be changed; a variable can.   3. Electrolytic.
4. Make sure the voltage rating is sufficient. Connect the positive end to the positive side of the circuit.

Section 13–3
1. Less.   2. 62.5 pF.   3. 0.006 μF.   4. 20 pF.   5. 75 V.

Section 13–4
1. The individual capacitors are added.
2. By using five 0.01-μF capacitors in parallel.   3. 98 pF.

Section 13–5
1. 1.2 μs.   2. 6 μs, 5 V.   3. 8.6 V, 9.5 V, 9.8 V, 9.9 V.   4. 37 V.
5. 4.93 V.   6. 9.43 μs.

Section 13–6
1. Current leads voltage by 90 degrees.   2. 637 kΩ.   3. 796 Hz.
4. 628 mA.   5. 0.   6. 0.453 VAR.

Section 13–7
1. Once the capacitor charges to the peak voltage, it discharges very little before the next peak.
2. By selecting a capacitor that has a reactance of almost zero at the frequency of the ac voltage and connecting that capacitor from that point to ground.

Section 13–8
1. Short its leads.
2. Initially, the needle jumps to zero; than it slowly moves to the high-resistance end of the scale.
3. Shorted, open, leakage, and dielectric absorption.

Section 13–9
1. Line 100; lines 170–220.
2. Line 100 is the time delay loop; lines 170–220 calculate and print capacitor voltage and percent of full charge.

FOURTEEN

INDUCTORS

The principle of electromagnetic induction was introduced in Chapter 10. The inductors studied in this chapter are based on this principle.

Inductance is the property of a coil of wire that opposes a change in current. The basis for inductance is the electromagnetic field that surrounds any conductor when there is current through it. The electrical component designed to have the property of inductance is called an *inductor, coil,* or *choke*. All of these terms refer to essentially the same type of device.

In this chapter, the basic inductor is introduced and its characteristics are studied. Various types of inductors are covered in terms of their physical construction and their electrical properties. The basic behavior of inductors in both dc and ac circuits is studied, and series and parallel combinations are analyzed. Methods of testing inductors are discussed.

In this chapter, you will learn:

☐   What an inductor is and what it does in a circuit.
☐   How an inductor stores energy.
☐   The unit of inductance.
☐   How physical characteristics determine inductance.
☐   Why inductors exhibit resistance and capacitance.

☐ How Lenz's law and Faraday's law apply to inductors.
☐ Various types of inductors.
☐ What happens when inductors are connected in series.
☐ What happens when inductors are connected in parallel.
☐ The meaning of the *time constant* in an inductive circuit.
☐ How an inductor introduces phase shift between current and voltage.
☐ The definition of *inductive reactance* and how to determine its value in a circuit.
☐ The meaning of *reactive power* in an inductive circuit.
☐ Several common applications of inductors.
☐ How to test an inductor.

14–1   THE BASIC INDUCTOR

When a length of wire is formed into a coil, as shown in Figure 14–1(a), it becomes a basic *inductor*. Current through the coil produces a magnetic field, as illustrated in Figure 14–1(b). The magnetic lines of force around each loop (turn) in the coil effectively add to the lines of force around the adjoining loops, forming a strong magnetic field within and around the coil, as shown. The net direction of the total magnetic field creates a north and a south pole, as shown in Part (b).

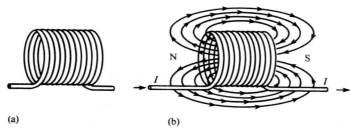

(a)   (b)

FIGURE 14–1
A coil of wire forms an inductor. When current flows through it, a three-dimensional electromagnetic field is created, surrounding the coil in all directions.

To understand the formation of the total magnetic field in a coil, let's discuss the interaction of the magnetic fields around two adjacent loops. The magnetic lines of force around adjacent loops are deflected into an outer path when the loops are brought close together. This effect occurs because the magnetic lines of force are in *opposing* directions between adjacent loops, as illustrated in Figure 14–2(a). The total magnetic field for the two loops is depicted in Part (b) of the figure. For simplicity, only single lines of force are shown. This effect is additive for many closely adjacent loops in a coil; that is, each additional loop adds to the strength of the electromagnetic field.

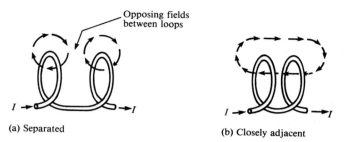

FIGURE 14–2
Interaction of magnetic lines of force in two adjacent loops of a coil.

REVIEW OF FARADAY'S LAW

The principle of Faraday's law was introduced in Chapter 10 and is reviewed here because of its importance in the study of inductors. Michael Faraday discovered the principle of electromagnetic induction in 1831. Faraday found that by moving a magnet through a coil of wire, a voltage was induced across the coil, and that when a complete path was provided, the induced voltage caused an induced current.

> **The amount of induced voltage is directly proportional to the rate of change of the magnetic field with respect to the coil.**

This principle is illustrated in Figure 14–3, where a bar magnet is moved through a coil of wire. An induced voltage is indicated by the voltmeter connected across the coil. The *faster* the magnet is moved, the *greater* is the induced voltage.

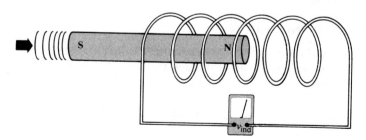

FIGURE 14–3
Induced voltage created by a changing magnetic field.

When a wire is formed into a certain number of loops or turns and is exposed to a *changing magnetic field*, a voltage is induced across the coil. The induced voltage is proportional to the *number of turns* of wire in the coil, N, and to the *rate* at which the magnetic field changes. The rate of change of the magnetic field is designated $d\phi/dt$, where ϕ is the *magnetic flux*. $d\phi/dt$ is expressed in webers/second (Wb/s). Faraday's law expresses this relationship in concise form as follows:

$$v_{\text{ind}} = N\left(\frac{d\phi}{dt}\right) \qquad (14\text{–}1)$$

This formula states that the induced voltage across a coil is equal to the number of turns (loops) times the rate of flux change.

EXAMPLE 14–1

Apply Faraday's law to find the induced voltage across a coil with 500 turns located in a magnetic field that is changing at a rate of 5 Wb/s.

Solution:

$$v_{\text{ind}} = N\left(\frac{d\phi}{dt}\right) = (500 \text{ T})(5 \text{ WB/s}) = 2.5 \text{ kV}$$

SELF-INDUCTANCE

When there is current through an inductor, a magnetic field is established. When the current changes, the magnetic field also changes. An increase in current expands the magnetic field, and a decrease in current reduces it. Therefore, a changing current produces a changing magnetic field around the inductor (coil). In turn, the changing magnetic field induces a voltage across the coil because of a property called *self-inductance*.

Self-inductance is a measure of a coil's ability to establish an induced voltage as a result of a change in its current. Self-inductance is usually referred to as simply *inductance*. Inductance is symbolized by L.

THE UNIT OF INDUCTANCE

The *henry*, symbolized by H, is the basic unit of inductance. By definition, the inductance is *one henry* when current through the coil, changing at the rate of *one ampere per second*, induces *one volt* across the coil. In many practical applications, *millihenries* (mH) and *microhenries* (μH) are the more common units. A common schematic symbol for the inductor is shown in Figure 14–4.

FIGURE 14–4
Symbol for inductor.

L

LENZ'S LAW

Lenz's law was introduced in Chapter 10 and is restated here.

> **When the current through a coil changes and an induced voltage is created as a result of the changing magnetic field, the direction of the induced voltage is such that it always opposes the change in current.**

In Figure 14–5(a), the current is constant and is limited by R_1. There is no induced voltage because the magnetic field is unchanging. In Part (b), the switch suddenly is closed, placing R_2 in parallel with R_1 and thus reducing the resistance. Naturally, the current tries to increase and the magnetic field begins to expand, but the induced voltage opposes this attempted increase in current for an instant.

In Part (c), the induced voltage gradually decreases, allowing the current to increase. In Part (d), the current has reached a constant value as determined by the parallel resistors, and the induced voltage is zero. In Part (e), the switch has been suddenly opened, and, for an instant, the induced voltage prevents any decrease in

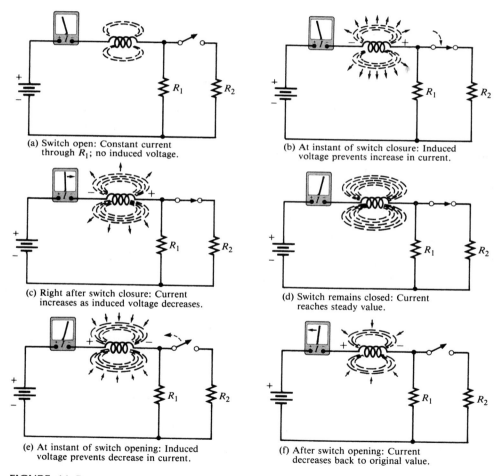

(a) Switch open: Constant current through R_1; no induced voltage.

(b) At instant of switch closure: Induced voltage prevents increase in current.

(c) Right after switch closure: Current increases as induced voltage decreases.

(d) Switch remains closed: Current reaches steady value.

(e) At instant of switch opening: Induced voltage prevents decrease in current.

(f) After switch opening: Current decreases back to original value.

FIGURE 14–5

Demonstration of Lenz's law in an inductive circuit: When the current tries to change suddenly, the electromagnetic field changes and induces a voltage in a direction that opposes that change in current.

current, and arcing results. In Part (f), the induced voltage gradually decreases, allowing the current to decrease back to a value determined by R_1. Notice that the induced voltage has a polarity that opposes any current change. The polarity of the induced voltage is opposite that of the battery voltage for an increase in current and aids the battery voltage for a decrease in current.

THE INDUCED VOLTAGE DEPENDS ON L AND di/dt

L is the symbol for the inductance of a coil, and di/dt is the time rate of change of the current. A change in current causes a change in the magnetic field, which, in turn, induces a voltage across the coil, as you know. The induced voltage is directly proportional to L and di/dt, as stated by the following formula:

$$v_{\text{ind}} = L\left(\frac{di}{dt}\right) \tag{14-2}$$

This equation indicates that the greater the inductance, the greater the induced voltage. Also, the faster the coil current changes (greater di/dt), the greater the induced voltage. Notice the similarity of Equation (14–2) to Equation (13–23): $i = C(dv/dt)$.

EXAMPLE 14–2

Determine the induced voltage across a 1-henry (1-H) inductor when the current is changing at a rate of 2 A/s.

Solution:

$$v_{\text{ind}} = L\left(\frac{di}{dt}\right) = (1 \text{ H})(2 \text{ A/s}) = 2 \text{ V}$$

ENERGY STORAGE

An inductor stores energy in the magnetic field created by the current. The energy stored is expressed as follows:

$$W = \frac{1}{2}LI^2 \tag{14-3}$$

As you can see, the energy stored is proportional to the inductance and the square of the current. When I is in amperes and L is in henries, the energy is in joules.

PHYSICAL CHARACTERISTICS OF INDUCTORS

The following parameters are important in establishing the inductance of a coil: core material, number of turns, core length, and cross-sectional area of the core.

Core Material   As discussed earlier, an inductor is basically a coil of wire. The material around which the coil is formed is called the *core*. Coils are wound on either nonmagnetic or magnetic materials. Examples of nonmagnetic materials are air, wood, copper, plastic, and glass. The permeabilities of these materials are the same as for a vacuum. Examples of magnetic materials are iron, nickel, steel, cobalt, or alloys. These materials have permeabilities that are hundreds or thousands of times greater than that of a vacuum and are classified as *ferromagnetic*. A ferromagnetic core provides a better path for the magnetic lines of force and thus permits a stronger magnetic field.

As you have learned, the permeability (μ) of the core material determines how easily a magnetic field can be established. *The inductance is directly proportional to the permeability of the core material.*

Physical Parameters The number of turns of wire, the length, and the cross-sectional area of the core, as indicated in Figure 14–6, are factors in setting the value of inductance. The inductance is inversely proportional to the length of the core and directly proportional to the cross-sectional area. Also, the inductance is directly related to the number of turns squared. This relationship is as follows:

$$L = \frac{N^2 \mu A}{l} \tag{14–4}$$

where L is the inductance in henries, N is the number of turns, μ is the permeability, A is the cross-sectional area in meters squared, and l is the core length in meters.

FIGURE 14–6
Parameters of an inductor.

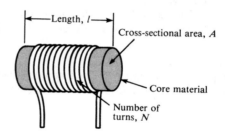

EXAMPLE 14–3

Determine the inductance of the coil in Figure 14–7. The permeability of the core is 0.25×10^{-3}.

FIGURE 14–7

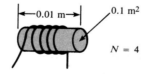

Solution:

$$L = \frac{N^2 \mu A}{l} = \frac{(4)^2(0.25 \times 10^{-3})(0.1)}{0.01} = 40 \text{ mH}$$

The calculator sequence is

4 x^2 × · 2 5 EXP +/− 3 × · 1 ÷ · 0 1 =

WINDING RESISTANCE

When a coil is made with, for example, insulated copper wire, that wire has a certain *resistance* per unit of length. When many turns of wire are used to construct a coil, the total resistance may be significant. This inherent resistance is called the *dc resistance* or the *winding resistance* (R_W). It effectively appears in series with the inductance of the coil, as shown in Figure 14–8. In many applications, the winding resistance can be ignored and the coil considered as an ideal inductor. In other cases, the resistance must be considered.

FIGURE 14–8
Equivalent circuit diagram for a coil and its winding resistance.

WINDING CAPACITANCE

When two conductors are placed side by side, there is always some capacitance between them. Thus, when many turns of wire are placed close together in a coil, a certain amount of *stray* capacitance is a natural side effect. In many applications, this stray capacitance is very small and has no significant effect. In other cases, particularly at high frequencies, it may become quite important.

The equivalent circuit for an inductor with both its winding resistance (R_W) and capacitance (C_W) is shown in Figure 14–9. The capacitance effectively acts in parallel.

FIGURE 14–9
Complete equivalent circuit for a nonideal inductor showing winding capacitance and resistance.

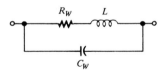

SECTION REVIEW 14–1

1. List the parameters that contribute to the inductance of a coil.
2. The current through a 15-mH inductor is changing at the rate of 500 mA/s. What is the induced voltage?
3. Describe what happens to L when:
   (a) N is increased.
   (b) The core length is increased.
   (c) The cross-sectional area of the core is decreased.
   (d) A ferromagnetic core is replaced by an air core.
4. Explain why inductors inevitably have some winding resistance.

14–2   TYPES OF INDUCTORS

Inductors are made in a variety of shapes and sizes. Basically, they fall into two general categories: *fixed* and *variable*. The standard schematic symbols are shown in Figure 14–10.

FIGURE 14–10
Symbols for fixed and variable inductors.

(a) Fixed   (b) Variable

      Both fixed and variable inductors can be classified according to the type of core material. Three common types are the air core, the iron core, and the ferrite core. Each has a unique symbol, as shown in Figure 14–11.

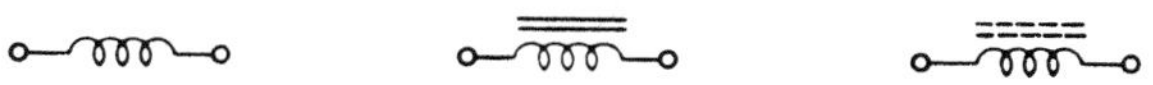

FIGURE 14–11.
Inductor symbols.

(a) Air core   (b) Iron core   (c) Ferrite core

      Adjustable (variable) inductors usually have a screw-type adjustment that moves a sliding core in and out, thus changing the inductance. A wide variety of inductors exist, and some are shown in Figure 14–12.

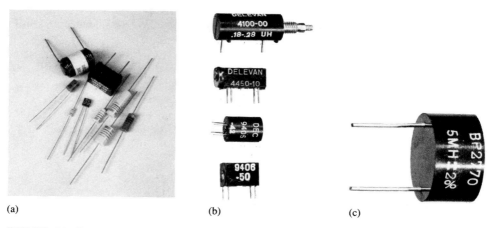

(a)   (b)   (c)

FIGURE 14–12
Typical inductors. (a) Fixed molded inductors; (b) variable coils; (c) toroid inductor [(a) and (b) courtesy of Delevan/American Precision. (c) courtesy of Dale Electronics].

SECTION REVIEW 14–2

1. Name two general categories of inductors.

2. Identify the inductor symbols in Figure 14–13.

FIGURE 14–13

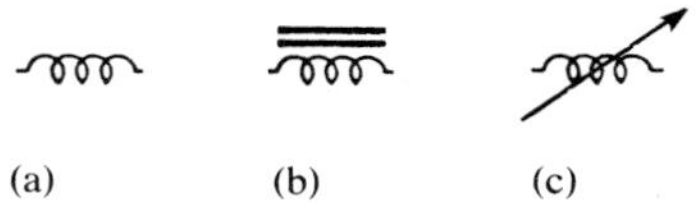

(a)   (b)   (c)

14–3

SERIES INDUCTORS

When inductors are connected in series, as in Figure 14–14, *the total inductance, L_T,* *is the sum of the individual inductances.* The formula for L_T is expressed in the following equation for the general case of n inductors in series:

$$L_T = L_1 + L_2 + L_3 + \cdots + L_n \qquad \text{(14–5)}$$

Notice that inductance in series is similar to resistance in series.

FIGURE 14–14
Inductors in series.

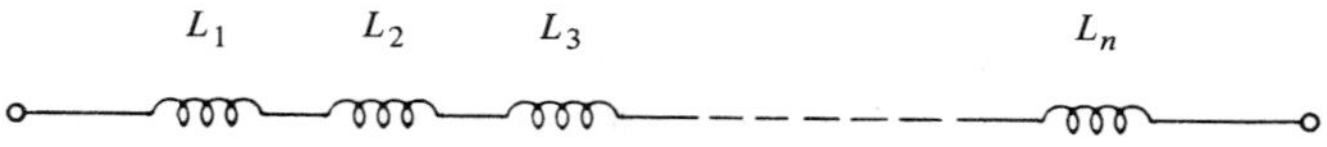

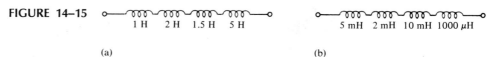

EXAMPLE 14–4

Determine the total inductance for each of the series connections in Figure 14–15.

FIGURE 14–15

1 H   2 H   1.5 H   5 H

5 mH   2 mH   10 mH   1000 μH

(a)   (b)

Solution:
(a) $L_T = 1\ \text{H} + 2\ \text{H} + 1.5\ \text{H} + 5\ \text{H} = 9.5\ \text{H}$
(b) $L_T = 5\ \text{mH} + 2\ \text{mH} + 10\ \text{mH} + 1\ \text{mH} = 18\ \text{mH}$

Note: $1000\ \mu\text{H} = 1\ \text{mH}$

SECTION REVIEW 14–3

1. State the rule for combining inductors in series.

2. What is L_T for a series connection of 100 μH, 500 μH, and 2 mH?

3. Five 100-mH coils are connected in series. What is the total inductance?

14–4   PARALLEL INDUCTORS

When inductors are connected in parallel, as in Figure 14–16, *the total inductance is less than the smallest inductance*. The formula for total inductance in parallel is very similar to that for total parallel resistance or total series capacitance:

$$\frac{1}{L_T} = \frac{1}{L_1} + \frac{1}{L_2} + \frac{1}{L_3} + \cdots + \frac{1}{L_n} \tag{14–6}$$

This general formula states that the reciprocal of the total inductance is equal to the sum of the reciprocals of the individual inductances. L_T can be found by taking the reciprocal of both sides of Equation (14–6).

$$L_T = \frac{1}{\left(\dfrac{1}{L_1}\right) + \left(\dfrac{1}{L_2}\right) + \left(\dfrac{1}{L_3}\right) + \cdots + \left(\dfrac{1}{L_n}\right)} \tag{14–7}$$

FIGURE 14–16
Inductors in parallel.

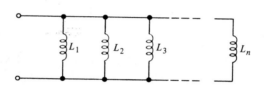

SPECIAL CASE OF TWO PARALLEL INDUCTORS

When only two inductors are in parallel, a special *product over sum* form of Equation (14–6) can be used.

$$L_T = \frac{L_1 L_2}{L_1 + L_2} \tag{14–8}$$

EQUAL-VALUE PARALLEL INDUCTORS

This is another special case in which a short-cut formula can be used. This formula is also derived from the general Equation (14–6) and is stated as follows for *n* equal-value inductors in parallel:

$$L_T = \frac{L}{n} \tag{14–9}$$

EXAMPLE
14–5

Determine L_T in Figure 14–17.

FIGURE 14–17

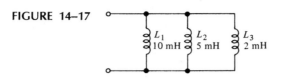

Solution:

$$\frac{1}{L_T} = \frac{1}{L_1} + \frac{1}{L_2} + \frac{1}{L_3}$$

$$= \frac{1}{10 \text{ mH}} + \frac{1}{5 \text{ mH}} + \frac{1}{2 \text{ mH}}$$

$$L_T = \frac{1}{\dfrac{1}{10 \text{ mH}} + \dfrac{1}{5 \text{ mH}} + \dfrac{1}{2 \text{ mH}}}$$

$$= \frac{1}{0.8 \text{ mH}} = 1.25 \text{ mH}$$

The calculator sequence is

EXAMPLE
14–6

Find L_T for both circuits in Figure 14–18.

FIGURE 14–18

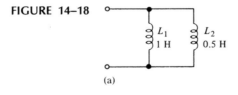 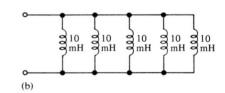

(a)   (b)

Solution:

(a) Using the special formula for two parallel inductors, we obtain

$$L_T = \frac{L_1 L_2}{L_1 + L_2} = \frac{(1 \text{ H})(0.5 \text{ H})}{1.5 \text{ H}} = 0.33 \text{ H}$$

(b) Using the special formula for equal parallel inductors, we obtain

$$L_T = \frac{L}{n} = \frac{10 \text{ mH}}{5} = 2 \text{ mH}$$

SECTION REVIEW 14–4

1. Compare the total inductance in parallel with the smallest-valued individual inductor.
2. The calculation of total parallel inductance is similar to that for parallel resistance (T or F).
3. Determine L_T for each parallel combination:
   (a) 100 mH, 50 mH, and 10 mH   (b) 40 μH and 60 μH
   (c) Ten 1-H coils

14–5   INDUCTORS IN dc CIRCUITS

When there is constant direct current in an inductor, there is no induced voltage. There is, however, a voltage drop due to the winding resistance of the coil. The inductance itself appears as a *short* to dc. Energy is stored in the magnetic field according to the formula previously stated in Equation (14–3), $W = \frac{1}{2}LI^2$. The only energy loss occurs in the winding resistance ($P = I^2R_W$). This condition is illustrated in Figure 14–19.

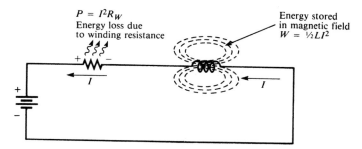

FIGURE 14–19
Energy storage and loss in an inductor in a dc circuit.

TIME CONSTANT

Because the inductor's basic action is to *oppose a change in its current*, it follows that *current cannot change instantaneously in an inductor*. A certain time is required for the current to make a change from one value to another. The rate at which the current changes is determined by the *time constant*. The time constant for a series *RL* circuit is

$$\tau = \frac{L}{R} \tag{14–10}$$

where τ is in seconds when L is in henries and R is in ohms.

EXAMPLE
14–7

A series *RL* circuit has a resistance of 1 kΩ and an inductance of 1 mH. What is the time constant?

Solution:

$$\tau = \frac{L}{R} = \frac{1 \text{ mH}}{1 \text{ k}\Omega} = \frac{1 \times 10^{-3} \text{ H}}{1 \times 10^{3} \text{ }\Omega}$$

$$= 1 \times 10^{-6} \text{ s} = 1 \text{ }\mu\text{s}$$

ENERGIZING CURRENT IN AN INDUCTOR

In a series *RL* circuit, the current will increase to 63% of its full value in one time-constant interval after the switch is closed. This build-up of current is analogous to the build-up of capacitor voltage during the charging in an *RC* circuit; they both follow an exponential curve and reach the approximate percentages of final value as indicated in Table 14–1 and as illustrated in Figure 14–20.

TABLE 14–1
Percentage of final current after each time-constant interval during current build-up.

Number of Time Constants	% Final Value
1	63
2	86
3	95
4	98
5	99 (considered 100%)

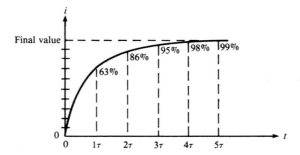

FIGURE 14–20
Energizing current in an inductor.

The change in current over five time-constant intervals is illustrated in Figure 14–21. When the current reaches its final value at approximately 5τ, it ceases to change. At this time, the inductor acts as a short (except for winding resistance) to the constant current. The final value of the current is $V_S/R = 10 \text{ V}/1 \text{ k}\Omega = 10 \text{ mA}$.

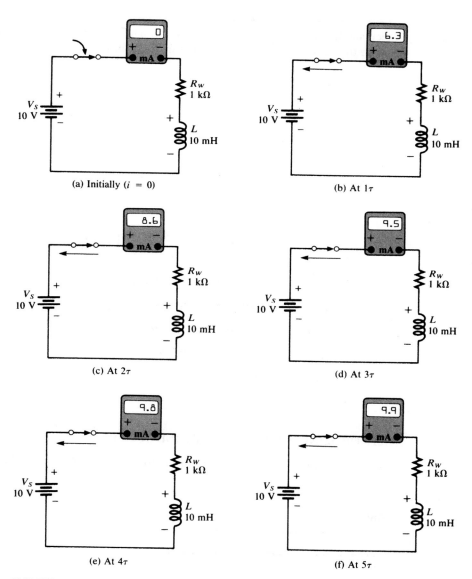

FIGURE 14–21
Current build-up in an inductor.

EXAMPLE 14–8

Calculate the time constant for Figure 14–22. Then determine the current and the time at each time-constant interval, measured from the instant the switch is closed.

FIGURE 14–22

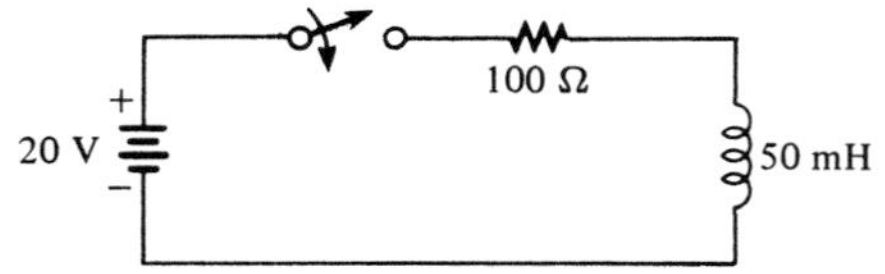

Solution:

$$I_{final} = \frac{V_S}{R} = \frac{20 \text{ V}}{100 \text{ }\Omega}$$

$$= 0.2 \text{ A}$$

$$\tau = \frac{L}{R} = \frac{50 \text{ mH}}{100 \text{ }\Omega} = 0.5 \text{ ms}$$

At $1\tau = 0.5$ ms: $i = 0.63(0.2 \text{ A}) = 0.126 \text{ A}$

At $2\tau = 1$ ms: $i = 0.86(0.2 \text{ A}) = 0.172 \text{ A}$

At $3\tau = 1.5$ ms: $i = 0.95(0.2 \text{ A}) = 0.19 \text{ A}$

At $4\tau = 2$ ms: $i = 0.98(0.2 \text{ A}) = 0.196 \text{ A}$

At $5\tau = 2.5$ ms: $i = 0.99(0.2 \text{ A}) = 0.198 \text{ A}$

$$\cong 0.2 \text{ A}$$

DEENERGIZING CURRENT IN AN INDUCTOR

Current in an inductor decreases exponentially according to the approximate percentage values in Table 14–2.

TABLE 14–2
Percentage of initial current after each time-constant interval while current is decreasing.

Number of Time Constants	% Initial Value
1	37
2	14
3	5
4	2
5	1 (considered 0)

Figure 14–23(a) shows a constant current of 10 mA through the inductor. If switch 2 is closed at the *same* instant that switch 1 is opened, as shown in Part (b), the current decreases to zero in five time constants ($5L/R$) as the inductor deenergizes. This exponential decrease in current is shown in Part (c).

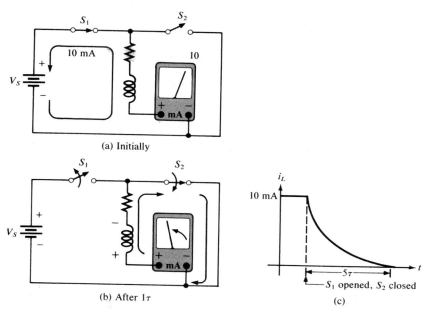

FIGURE 14–23
Deenergizing current in an inductor.

EXAMPLE 14–9

In Figure 14–24, S_1 is opened at the instant that S_2 is closed.
(a) What is the time constant?
(b) What is the initial coil current at the instant of switching?
(c) What is the coil current at 1τ?
Assume steady state current through the coil prior to switch change.

FIGURE 14–24

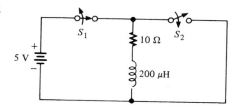

Solution:

(a) $\tau = L/R = 200 \ \mu\text{H}/10 \ \Omega = 20 \ \mu\text{s}.$

(b) Current cannot change instantaneously in an inductor. Therefore, the current at the instant of the switch change is the same as the steady state current.

$$I = \frac{5 \text{ V}}{10 \ \Omega} = 0.5 \text{ A}$$

(c) At 1τ, the current has decreased to 37% of its initial value:

$$i = 0.37(0.5 \text{ A}) = 0.185 \text{ A}$$

INDUCED VOLTAGE IN THE SERIES *RL* CIRCUIT

When current changes in an inductor, a voltage is induced. We now examine what happens to the voltages across the resistor and the coil in a series circuit when a change in current occurs.

Look at the circuit in Figure 14–25(a). When the switch is open, there is no current, and the resistor voltage and the coil voltage are both zero. At the instant the switch is closed, as indicated in Part (b), v_R is zero and v_L is 10 V. The reason for this change is that the induced voltage across the coil is equal and opposite to the applied

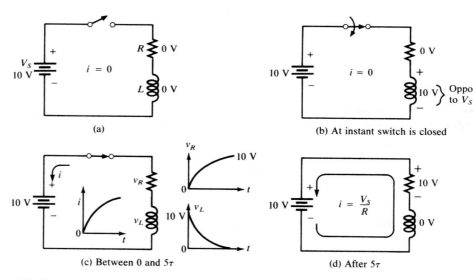

(a)

(b) At instant switch is closed

(c) Between 0 and 5τ

(d) After 5τ

FIGURE 14–25
Voltage in an *RL* circuit as the inductor energizes.

voltage to prevent the current from changing instantaneously. Therefore, at the instant of switch closure, L effectively acts as an *open* with all the applied voltage across it.

During the first five time constants, the current is building up exponentially, and the induced coil voltage is decreasing. The resistor voltage increases with the current, as Part (c) illustrates.

After five time constants have elapsed, the current has reached its final value, V_S/R. At this time, all of the applied voltage is dropped across the resistor and none across the coil. Thus, L effectively acts as a *short* to nonchanging current, as Part (d) illustrates. Keep in mind that the inductor always reacts to a change in current by creating an induced voltage in order to counteract that change.

Now let us examine the case illustrated in Figure 14–26, where the steady state current is switched out, and the inductor discharges through another path. Part (a) shows the steady state condition, and Part (b) illustrates the instant at which the source is removed by opening S_1 and the discharge path is connected with the closure of S_2. There was 1 A through L prior to this. Notice that 10 V are induced in L in the direction to aid the 1 A in an effort to keep it from changing. Then, as shown in Part (c), the current decays exponentially, and so do v_R and v_L. After 5τ, as shown in Part (d), all of the energy stored in the magnetic field of L is dissipated, and all values are zero.

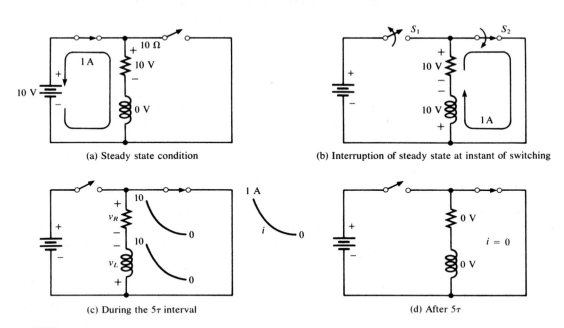

FIGURE 14–26
Voltage in an *RL* circuit as the inductor deenergizes.

EXAMPLE 14–10

(a) In Figure 14–27(a), what is v_L at the instant S_1 is closed? What is v_L after 5τ?

(b) In Figure 14–27(b), what is v_L at the instant S_1 opens and S_2 closes? What is v_L after 5τ?

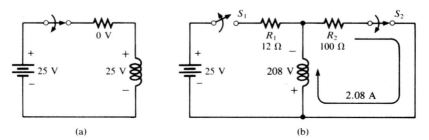

(a)   (b)

FIGURE 14–27

Solution:

(a) At the instant the switch is closed, all of the voltage is across L. Thus, $v_L = 25$ V, with the polarity as shown. After 5τ, L acts as a short, so $v_L = 0$ V.

(b) With S_1 closed and S_2 open, the steady state current is

$$\frac{25 \text{ V}}{12 \ \Omega} = 2.08 \text{ A}$$

When the switches are thrown, an induced voltage is created across L sufficient to keep this 2.08-A current for an instant. In this case, it takes $v_L = IR_2 = (2.08 \text{ A})(100 \ \Omega) = 208$ V. After 5τ, the inductor voltage is zero. These results are indicated in the circuit diagrams.

THE EXPONENTIAL FORMULAS

The formulas for the exponential current and voltage in an *RL* circuit are similar to those used in the last chapter for the *RC* circuit, and the universal exponential curves in Figure 13–33 apply to inductors as well as capacitors. The general formulas for *RL* circuits are stated as follows:

$$v = V_F + (V_i - V_F)e^{-Rt/L} \qquad (14\text{–}11)$$

$$i = I_F + (I_i - I_F)e^{-Rt/L} \qquad (14\text{–}12)$$

where V_F and I_F are the final values; V_i and I_i are the initial values; and v and i are the instantaneous values of the inductor voltage or current at time t.

Increasing Current The formula for the special case in which an increasing exponential current curve begins at zero ($I_i = 0$) is

$$i = I_F(1 - e^{-Rt/L})$$ (14–13)

Using Equation (14–13), we can calculate the value of the increasing inductor current at any instant of time. The same is true for voltage.

EXAMPLE 14–11

In Figure 14–28, determine the inductor current 30 μs after the switch is closed.

FIGURE 14–28

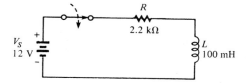

Solution:

The time constant is $L/R = 100$ mH/2.2 k$\Omega = 45.45$ μs. The final current is $V_S/R = 12$ V/2.2 k$\Omega = 5.45$ mA. The initial current is zero. Notice that 30 μs is less than one time constant, so the current will reach less than 63% of its final value in that time.

$$i_L = I_F(1 - e^{-Rt/L}) = 5.45 \text{ mA}(1 - e^{-0.66})$$

$$= 5.45 \text{ mA}(1 - 0.517) = 2.63 \text{ mA}$$

The calculator sequence is

Decreasing Current The formula for the special case in which a decreasing exponential current has a final value of zero is as follows:

$$i = I_i e^{-Rt/L}$$ (14–14)

This formula can be used to calculate the deenergizing current at any instant, as the following example shows.

EXAMPLE 14–12

Determine the inductor current in Figure 14–29 at a point in time 2 ms after the switches are thrown (S_1 opened and S_2 closed).

FIGURE 14–29

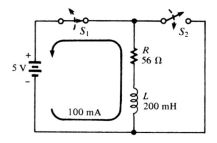

Solution:
The deenergizing time constant is L/R = 200 mH/56 Ω = 3.57 ms. The initial current in the inductor is 89.29 mA. Notice that 2 ms is less than one time constant, so the current will show a less than 63% decrease. Therefore, the current will be greater than 37% of its initial value at 2 ms after the switches are thrown.

$$i = I_i e^{-Rt/L} = (89.29 \text{ mA})e^{-0.56} = 51 \text{ mA}$$

SECTION REVIEW 14–5

1. A 15-mH inductor with a winding resistance of 10 Ω has a constant direct current of 10 mA through it. What is the voltage drop across the inductor?

2. A 20-V dc source is connected to a series RL circuit with a switch. At the instant of switch closure, what are the values of v_R and v_L?

3. In the same circuit, after a time interval equal to 5τ from switch closure, what are v_R and v_L?

4. In a series RL circuit where R = 1 kΩ and L = 500 μH, what is the time constant? Determine the current 0.25 μs after a switch connects 10 V across the circuit.

14–6   INDUCTORS IN ac CIRCUITS

The concept of the derivative was introduced in Chapter 13. The expression for induced voltage in an inductor was stated earlier in Equation (14–2). This formula is $v_{\text{ind}} = L(di/dt)$.

PHASE RELATIONSHIP OF CURRENT AND VOLTAGE IN AN INDUCTOR

From the formula for induced voltage, you can see that the faster the current through an inductor changes, the greater the induced voltage will be. For example, if the rate of change of current is zero, the voltage is zero [$v_{ind} = L(di/dt) = L(0) = 0$ V]. When di/dt is a positive-going maximum, v_{ind} is a positive maximum; when di/dt is a negative-going maximum, v_{ind} is a negative maximum.

A sinusoidal current always induces a sinusoidal voltage in inductive circuits. Therefore, the voltage can be plotted with respect to the current by knowing the points on the current curve at which the voltage is zero and those at which it is maximum. This relationship is shown in Figure 14–30(a). Notice that *the voltage leads the current by 90 degrees.* This is always true in a purely inductive circuit. A phasor diagram of this relationship is shown in Figure 14–30(b).

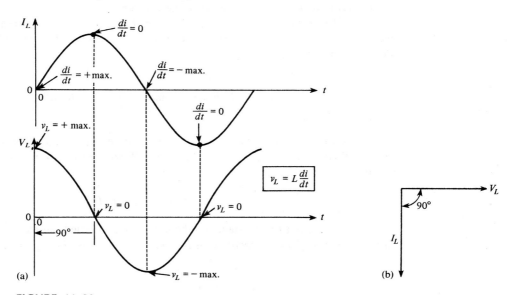

FIGURE 14–30
Phase relation of V_{ind} and I in an inductor current always lags the inductor voltage by 90 degrees.

INDUCTIVE REACTANCE

Inductive reactance is the opposition to sinusoidal current, expressed in ohms. The symbol for inductive reactance is X_L. To develop a formula for X_L, we use the relationship $v_{ind} = (di/dt)$ and the curves in Figure 14–31.

FIGURE 14–31
Slope indicates rate of change. Sine wave A has a greater rate of change at the zero crossing than B, and thus A has a higher frequency.

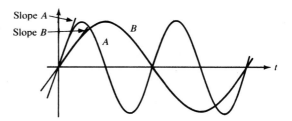

The rate of change of current is directly related to frequency. The faster the current changes, the higher the frequency. For example, you can see that in Figure 14–31, the slope of sine wave A at the zero crossings is greater than that of sine wave B. Recall that the slope of a curve at a point indicates the rate of change at that point. Sine wave A has a higher frequency than sine wave B, as indicated by a greater maximum rate of change (di/dt is greater at the zero crossings).

When frequency increases, di/dt increases, and thus v_{ind} increases. When frequency decreases, di/dt decreases, and thus v_{ind} decreases. The induced voltage is directly dependent on frequency:

$$\overset{\uparrow}{v_{ind}} = L(\overset{\uparrow}{di/dt}) \quad \text{and} \quad \underset{\downarrow}{v_{ind}} = L(\underset{\downarrow}{di/dt})$$

An increase in induced voltage means more opposition (X_L is greater). Therefore, *X_L is directly proportional to induced voltage and thus directly proportional to frequency:*

$$X_L \text{ is proportional to } f$$

Now, if di/dt is constant and the inductance is varied, an increase in L produces an increase in v_{ind}, and a decrease in L produces a decrease *in* v_{ind}, as indicated:

$$\overset{\uparrow}{v_{ind}} = \overset{\uparrow}{L} (di/dt) \quad \text{and} \quad \underset{\downarrow}{v_{ind}} = \underset{\downarrow}{L} (di/dt)$$

Again, an increase in v_{ind} means more opposition (greater X_L). Therefore, *X_L is directly proportional to induced voltage and thus directly proportional to inductance.*
The inductive reactance is directly proportional to both f and L.

$$X_L \text{ is proportional to } fL$$

The complete formula for X_L is as follows:

$$X_L = 2\pi fL \qquad\qquad (14\text{–}15)$$

Notice that 2π appears as a constant factor in the equation. This comes from the relationship of a sine wave to rotational motion, as derived in Appendix D. X_L is in ohms when f is in hertz and L is in henries.

**EXAMPLE
14-13**

A sinusoidal voltage is applied to the circuit in Figure 14–32. The frequency is 1 kHz. Determine the inductive reactance.

FIGURE 14-32

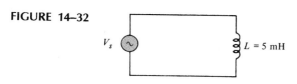

Solution:

$$1 \text{ kHz} = 1 \times 10^3 \text{ Hz}$$
$$5 \text{ mH} = 5 \times 10^{-3} \text{ H}$$
$$X_L = 2\pi fL = 2\pi(1 \times 10^3 \text{ Hz})(5 \times 10^{-3} \text{ H}) = 31.4 \text{ } \Omega$$

ANALYSIS OF INDUCTIVE ac CIRCUITS

As you have seen, the current lags the voltage by 90 degrees in inductive ac circuits. If the applied voltage is assigned a reference phase angle of zero, it can be expressed in polar form as $V_s\angle 0°$. The resulting current can be expressed as $I\angle -90°$ or $-jI$ in rectangular form, as shown in Figure 14–33.

FIGURE 14-33

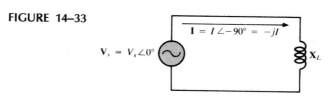

Ohm's law applies to ac circuits with inductive reactance with R replaced by X_L in the Ohm's law formula. The quantities are expressed as complex numbers because of the introduction of phase angles. Applying Ohm's law to the circuit in Figure 14–33 gives the following result:

$$\mathbf{X}_L = \frac{V_s\angle 0°}{I\angle -90°} = \left(\frac{V_s}{I}\right)\angle 90°$$

This shows that $\mathbf{X}_L$ always has a 90 degree angle attached to its magnitude and is written as $X_L\angle 90°$ or jX_L.

EXAMPLE 14–14

Determine the rms current in Figure 14–34.

FIGURE 14–34

$\mathbf{V}_{rms} = 5\angle0°$ V

$f = 10$ kHz

$L = 100$ mH

Solution:

$$10 \text{ kHz} = 10 \times 10^3 \text{ Hz}$$
$$100 \text{ mH} = 100 \times 10^{-3} \text{ H}$$

Calculate the magnitude of X_L:

$$X_L = 2\pi f L = 2\pi(10 \times 10^3 \text{ Hz})(100 \times 10^{-3} \text{ H}) = 6283 \ \Omega$$

Expressed in polar form, X_L is

$$\mathbf{X}_L = 6283\angle90° \ \Omega$$

Applying Ohm's law:

$$\mathbf{I} = \frac{\mathbf{V}_{rms}}{\mathbf{X}_L} = \frac{5\angle0° \text{ V}}{6283\angle90° \ \Omega} = 795.8\angle-90° \ \mu\text{A}$$

POWER IN AN INDUCTOR

As discussed earlier, an inductor stores energy in its magnetic field when there is current through it. An ideal inductor (assuming no winding resistance) does not dissipate energy; it only stores it. When an ac voltage is applied to an inductor, energy is stored by the inductor during a portion of the cycle; then the stored energy is returned to the source during another portion of the cycle. *There is no net energy loss.* Figure 14–35 shows the power curve that results from one cycle of inductor current and voltage.

Instantaneous Power (p) *The product of v and i gives instantaneous power, p.* At points where v or i is zero, p is also zero. When both v and i are positive, p is also positive. When either v or i is positive and the other negative, p is negative. When both v and i are negative, p is positive. As you can see in Figure 14–35, the power follows a sinusoidal curve. Positive values of power indicate that energy is stored by the inductor. Negative values of power indicate that energy is returned from the inductor to the source. Note that the power fluctuates at a frequency twice that of the voltage or current as energy is alternately stored and returned to the source.

True Power (P_{true}) Ideally, all of the energy stored by an inductor during the positive portion of the power cycle is returned to the source during the negative portion. *No net*

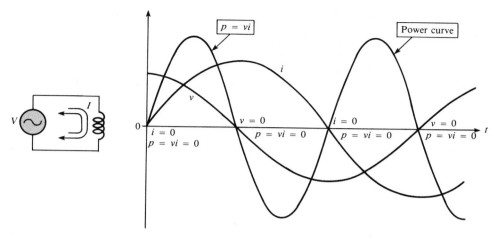

FIGURE 14–35
Power curve.

energy is consumed in the inductance, so the true power is zero. Actually, because of winding resistance in a practical inductor, some power is always dissipated.

$$P_{true} = (I_{rms})^2 R_W \qquad (14\text{–}16)$$

Reactive Power (P_r) The rate at which an inductor stores or returns energy is called its *reactive power, P_r.* The reactive power is a nonzero quantity, because at any instant in time, the inductor is actually taking energy from the source or returning energy to it. Reactive power does not represent an energy loss. The following formulas apply:

$$P_r = V_{rms} I_{rms} \qquad (14\text{–}17)$$

$$P_r = \frac{V_{rms}^2}{X_L} \qquad (14\text{–}18)$$

$$P_r = I_{rms}^2 X_L \qquad (14\text{–}19)$$

**EXAMPLE
14–15**

A 10-V rms signal with a frequency of 1 kHz is applied to a 10-mH coil with a winding resistance of 5 Ω. Determine the reactive power (P_r).

Solution: $\qquad X_L = 2\pi fL = 2\pi(1 \text{ kHz})(10 \text{ mH}) = 62.8 \text{ } \Omega$

$$I = \frac{V_S}{X_L} = \frac{10 \text{ V}}{63 \text{ } \Omega} = 0.159 \text{ A}$$

$$P_r = I^2 X_L = (0.159 \text{ A})^2(62.8 \text{ } \Omega) = 1.59 \text{ VAR}$$

THE QUALITY FACTOR (Q) OF A COIL

The quality factor (Q) is the ratio of the reactive power in the inductor to the true power in the winding resistance of the coil or the resistance in series with the coil. It is a ratio of the power in L to the power in R_W. The quality factor is very important in resonant circuits, which are studied in a later chapter. A formula for Q is developed as follows:

$$Q = \frac{\text{reactive power}}{\text{true power}}$$

$$= \frac{I^2 X_L}{I^2 R_W}$$

In a series circuit, I is the same in L and R; thus, the I^2 terms cancel, leaving

$$Q = \frac{X_L}{R_W} \tag{14-20}$$

When the resistance is just the winding resistance of the coil, the circuit Q and the coil Q are the same. Note that Q is a ratio of like units and, therefore, has no unit itself.

SECTION REVIEW 14-6

1. State the phase relationship between current and voltage in an inductor.

2. Calculate X_L for $f = 5$ kHz and $L = 100$ mH.

3. At what frequency is the reactance of a 50-μH inductor equal to 800 Ω?

4. Calculate the rms current in Figure 14-36.

FIGURE 14-36

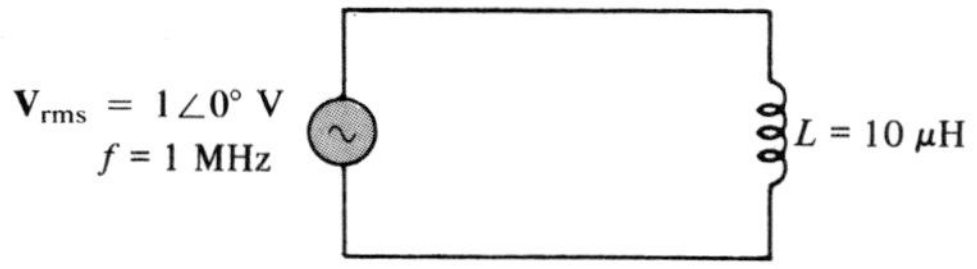

$$V_{rms} = 1 \angle 0° \text{ V}$$
$$f = 1 \text{ MHz}$$

$L = 10 \, \mu\text{H}$

5. An ideal 50-mH inductor is connected to a 12-V rms source. What is the true power? What is the reactive power at a frequency of 1 kHz?

14-7   INDUCTOR APPLICATIONS

Inductors are not as versatile as capacitors and tend to be more limited in their applications. However, there are many practical uses for inductors (coils), such as those discussed in Chapter 10. Recall that relay and solenoid coils, recording and pick-up heads, and sensing elements were introduced as electromagnetic applications of coils. In this section, some additional uses of inductors are presented.

POWER SUPPLY FILTER

In Chapter 13, you saw that a capacitor is used to filter the pulsating dc in a power supply. The final output voltage was a dc voltage with a small amount of *ripple*. In many cases, an inductor is used in the filter, as shown in Figure 14–37(a), to smooth out the ripple voltage. The inductor, placed in series with the load as shown, tends to oppose the current fluctuations caused by the ripple voltage, and thus the voltage developed across the load is more constant, as shown in Figure 14–37(b).

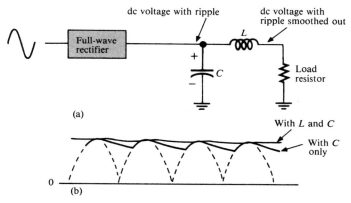

(a)

(b)

FIGURE 14–37
Basic capacitor power supply filter with a series inductor.

rf CHOKE

Certain types of inductors are used in applications where radio frequencies (rf) must be prevented from getting into parts of a system, such as the power supply or the audio section of a receiver. In these situations, an inductor is used as a series filter and "chokes" off any unwanted rf signals that may be picked up on a line. This filtering action is based on the fact that the reactance of a coil increases with frequency. When the frequency of the current is sufficiently high, the reactance of the coil becomes extremely large and essentially blocks the current. A basic illustration of an inductor used as an rf choke is shown in Figure 14–38.

FIGURE 14–38
An inductor used as an rf choke to minimize interfering signals on the power supply line.

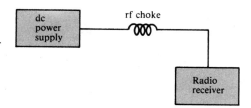

TUNED CIRCUITS

Inductors are used in conjunction with capacitors to provide *frequency selection* in communications systems. These *tuned circuits* allow a narrow band of frequencies to be selected while all other frequencies are rejected. The tuners in your TV and radio receivers are based on this principle and permit you to select one channel or station out of the many that are available.

Frequency selectivity is based on the fact that the reactances of both capacitors and inductors depend on the frequency and on the interaction of these two components when connected in series or parallel. Since the capacitor and the inductor produce opposite phase shifts, their combined opposition to current can be used to obtain a desired response at a selected frequency. Tuned *LC* circuits are covered in detail in a later chapter.

SECTION REVIEW 14–7

1. Explain how the ripple voltage from a power supply filter can be reduced through use of an inductor.

2. How does an inductor connected in series act as an rf choke?

14–8

TESTING INDUCTORS

The most common failure in an inductor is an open. To check for an open, the coil should be removed from the circuit. If there is an open, an ohmmeter check will indicate infinite resistance, as shown in Figure 14–39(a).

(a) Open, reads ∞.

(b) Good, reads R_W.

(c) Shorted windings, reads low R_W or near zero.

FIGURE 14–39
Checking a coil by measuring the resistance.

If the coil is good, the ohmmeter will show the winding resistance. The value of the winding resistance depends on the wire size and length of the coil. It can be anywhere from one ohm to several hundred ohms. Figure 14–39(b) shows a good reading.

Occasionally, when an inductor is overheated with excessive current, the wire insulation will melt, and one or more turns will short together. This must be tested on an *LC* meter because, with one shorted turn (or even several), an ohmmeter check may show the coil to be perfectly good from a resistance standpoint. Single, shorted turns are the most common type of coil failure because the turns are adjacent and can easily short across from poor insulation, voltage breakdown, or simple wear if something is rubbing on them.

SECTION REVIEW 14–8

1. When a coil is checked, a reading of infinity on the ohmmeter indicates a partial short (T or F).

2. An ohmmeter check of a good coil will indicate the value of the inductance (T or F).

14–9   COMPUTER ANALYSIS

The program in this section allows you to compute the inductive reactance as a function of frequency. Inputs required are the inductance, the frequency range, and the increments of frequency.

```
10   CLS
20   PRINT "THIS PROGRAM COMPUTES INDUCTIVE REACTANCE AS A"
30   PRINT "FUNCTION OF FREQUENCY. THE INPUTS REQUIRED ARE"
40   PRINT "INDUCTANCE AND THE FREQUENCY RANGE AND INCREMENTS."
50   FOR T = 1 TO 4000:NEXT
60   CLS
70   INPUT "ENTER L IN HENRIES";L
80   INPUT "LOWEST FREQUENCY IN HERTZ";FL
90   INPUT "HIGHEST FREQUENCY IN HERTZ";FH
100  INPUT "INCREMENTS OF FREQUENCY IN HERTZ";FI
110  CLS
120  PRINT "FREQUENCY (HZ)",,"XL (OHMS)"
130  FOR F=FL TO FH STEP FI
140  XL = 2*3.1416*F*L
150  PRINT F,,XL
160  NEXT
```

SECTION REVIEW 14–9

1. In line 130, explain the purpose of "STEP FI."

2. What is 3.1416 in line 140?

SUMMARY

1. An inductor can also be referred to as a *coil* or a *choke*.

2. Self-inductance is a measure of a coil's ability to establish an induced voltage as a result of a change in its current.

3. An inductor opposes a change in its own current.

4. Faraday's law states that relative motion between a magnetic field and a coil produces a voltage across the coil.

5. The amount of induced voltage is directly proportional to the inductance and to the rate of change in current.

6. Lenz's law states that the polarity of induced voltage is such that the resulting induced current is in a direction that opposes the change in the magnetic field that produced it.

7. Energy is stored by an inductor in its magnetic field.

8. The henry is the unit of inductance.

9. *One henry* is the amount of inductance when current, changing at the rate of *one ampere per second,* induces *one volt* across the inductor.

10. Inductance is directly proportional to the square of the turns, the permeability, and the cross-sectional area of the core. It is inversely proportional to the length of the core.

11. The permeability of a core material is an indication of the ability of the material to establish a magnetic field.

12. The time constant for a series *RL* circuit is the inductance divided by the resistance.

13. In an *RL* circuit, the voltage and current in an energizing or deenergizing inductor make a 63% change during each time-constant interval.

14. Energizing and deenergizing follow exponential curves.

15. Inductors add in series.

16. Total parallel inductance is less than that of the smallest inductor in parallel.

17. Voltage leads current by 90 degrees in an inductor.

18. Inductive reactance is the opposition to sinusoidal current and is expressed in ohms.

19. X_L is directly proportional to frequency and inductance.

20. The true power in an inductor is zero; that is, there is no energy loss in an ideal inductor, only in its winding resistance.

FORMULAS

$$v_{\text{ind}} = N\left(\frac{d\phi}{dt}\right) \tag{14-1}$$

$$v_{\text{ind}} = L\left(\frac{di}{dt}\right) \tag{14-2}$$

$$W = \frac{1}{2}LI^2 \tag{14-3}$$

$$L = \frac{N^2\mu A}{l} \tag{14-4}$$

$$L_T = L_1 + L_2 + L_3 + \cdots + L_n \tag{14-5}$$

$$\frac{1}{L_T} = \frac{1}{L_1} + \frac{1}{L_2} + \frac{1}{L_3} + \cdots + \frac{1}{L_n} \tag{14-6}$$

$$L_T = \frac{1}{\left(\frac{1}{L_1}\right) + \left(\frac{1}{L_2}\right) + \left(\frac{1}{L_3}\right) + \cdots + \left(\frac{1}{L_n}\right)} \tag{14-7}$$

$$L_T = \frac{L_1 L_2}{L_1 + L_2} \tag{14-8}$$

$$L_T = \frac{L}{n} \tag{14-9}$$

$$\tau = \frac{L}{R} \tag{14-10}$$

$$v = V_F + (V_i - V_F)e^{-Rt/L} \tag{14-11}$$

$$i = I_F + (I_i - I_F)e^{-Rt/L} \tag{14-12}$$

$$i = I_F(1 - e^{-Rt/L}) \tag{14-13}$$

$$i = I_i e^{-Rt/L} \tag{14-14}$$

$$X_L = 2\pi f L \tag{14-15}$$

$$P_{\text{true}} = (I_{\text{rms}})^2 R_W \tag{14-16}$$

$$P_r = V_{\text{rms}} I_{\text{rms}} \tag{14-17}$$

$$P_r = \frac{V_{rms}^2}{X_L} \qquad (14\text{--}18)$$

$$P_r = I_{rms}^2 X_L \qquad (14\text{--}19)$$

$$Q = \frac{X_L}{R_W} \qquad (14\text{--}20)$$

SELF-TEST

Solutions appear at the end of the book.

1. Which is the larger inductance, 0.05 μH or 0.000005 H?
2. Which is the smaller inductance, 0.33 mH or 33 μH?
3. If the current through an inductor is increased, does the amount of energy stored in the electromagnetic field increase or decrease?
4. If the current through an inductor is doubled, how much does the stored energy increase?
5. How can the winding resistance of a coil be decreased?
6. Select the changes in physical parameters that result in an increase in the inductance value:
   (a) Decrease the number of turns.
   (b) Change from an air core to an iron core.
   (c) Increase the number of turns.
   (d) Increase the length of the core.
   (e) Decrease the length of the core.
   (f) Use a larger-size wire.
7. An inductor with 50 turns of wire is placed in a magnetic field that is changing at a rate of 25 Wb/s. What is the induced voltage?
8. Convert 1500 μH to millihenries. Convert 20 mH to microhenries.
9. A circuit has $R = 2$ kΩ in series with $L = 10$ mH. What is the time constant?
10. Three 25-μH coils and two 10-μH coils are connected in series. What is the total inductance?
11. The following coils are connected in parallel: two 1-H, two 0.5-H, and one 2-H. What is L_T?
12. An inductor and a resistor are placed in series with a switch and are connected to a dc voltage source.
    (a) At the instant the switch is closed, what is the voltage across the inductor?
    (b) At the instant the switch is closed, what is the current?
    (c) How long will it take the current to reach its maximum value?

(d) What is the voltage across the inductor when the current has reached its maximum value?

(e) What is the voltage across the series resistor at the instant the switch is closed and after the current has reached its maximum value?

13. A sine wave voltage is applied across an inductor. When the frequency of the voltage is increased, does the current increase or decrease? Why?

14. An inductor and a resistor are connected in series with a sine wave voltage source. The frequency is set so that the inductive reactance is equal to the resistance and an equal amount of voltage is dropped across each component. If the frequency is decreased, which component will have the greater voltage across it?

15. A 2-MHz sinusoidal voltage is applied across a 15-μH inductor. What is the frequency of the resulting current? What is the reactance of the inductor?

16. The frequency of the source in Figure 14–40 is adjustable. To get 50 mA of rms current, to what frequency must you adjust the source?

FIGURE 14–40

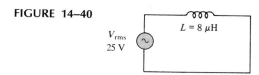

V_{rms}
25 V

$L = 8\ \mu H$

17. What inductance is required for an inductor to produce an X_L of 10 MΩ at 50 MHz?

18. An ohmmeter is connected across an inductor, and the pointer indicates an infinite value. What is your opinion of this inductor?

PROBLEMS

Section 14–1

14–1 Convert the following to millihenries:
(a) 1 H (b) 250 μH (c) 10 μH (d) 0.0005 H

14–2 Convert the following to microhenries:
(a) 300 mH (b) 0.08 H (c) 5 mH (d) 0.00045 mH

14–3 What is the voltage across a coil when $di/dt = 10$ mA/μs and $L = 5\mu$H?

14–4 Fifty volts are induced across a 25-mH coil. At what rate is the current changing?

14–5 The current through a 100-mH coil is changing at a rate of 200 mA/s. How much voltage is induced across the coil?

14–6 How many turns are required to produce 30 mH with a coil wound on a cylindrical core having a cross-sectional area of 10×10^{-5} m² and a length of 0.05 m? The core has a permeability of 1.2×10^{-6}.

14–7   A 12-V battery is connected across a coil with a winding resistance of 12 Ω. How much current is there in the coil?

14–8   How much energy is stored by a 100-mH inductor with a current of 1 A?

Section 14–3

14–9   Five inductors are connected in series. The lowest value is 5 μH. If the value of each inductor is twice that of the preceding one, and if the inductors are connected in order of ascending values, what is the total inductance?

14–10   Suppose that you require a total inductance of 50 mH. You have available a 10-mH coil and a 22-mH coil. How much additional inductance do you need?

14–11   Determine the total inductance in Figure 14–41.

14–12   What is the total inductance between points A and B for each switch position in Figure 14–42?

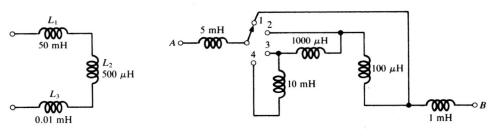

FIGURE 14–41   FIGURE 14–42

Section 14–4

14–13   Determine the total parallel inductance for the following coils in parallel: 75 μH, 50 μH, 25 μH, and 15 μH.

14–14   You have a 12-mH inductor, and it is your smallest value. You need an inductance of 8 mH. What value can you use in parallel with the 12-mH to obtain 8 mH?

14–15   Determine the total inductance of each circuit in Figure 14–43.

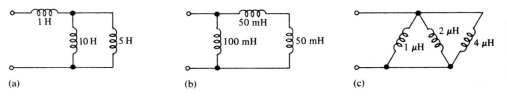

(a)   (b)   (c)

FIGURE 14–43

14–16   Determine the total inductance of each circuit in Figure 14–44.

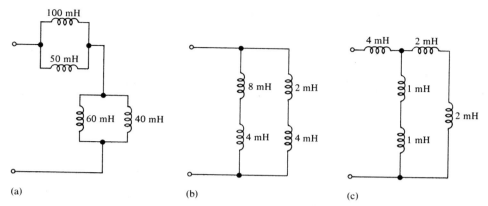

FIGURE 14–44

Section 14–5

14–17 Determine the time constant for each of the following series *RL* combinations:
(a) $R = 100\ \Omega$, $L = 100\ \mu H$ (b) $R = 4.7\ k\Omega$, $L = 10\ mH$
(c) $R = 1.5\ M\Omega$, $L = 3\ H$

14–18 In a series *RL* circuit, determine how long it takes the current to build up to its full value for each of the following:
(a) $R = 55\ \Omega$, $L = 50\ \mu H$ (b) $R = 3300\ \Omega$, $L = 15\ mH$
(c) $R = 22\ k\Omega$, $L = 100\ mH$

14–19 In the circuit of Figure 14–45, there is initially no current. Determine the inductor voltage at the following times after the switch is closed:
(a) 10 μs (b) 20 μs (c) 30 μs (d) 40 μs (e) 50 μs

14–20 In Figure 14–46, there are 100 mA through the coil. When S_1 is opened and S_2 simultaneously closed, find the inductor voltage at the following times:
(a) Initially   (b) 1.5 ms   (c) 4.5 ms   (d) 6 ms

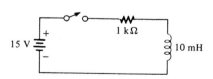

FIGURE 14–45   FIGURE 14–46

14–21 Repeat Problem 14–19 for the following times:
(a) 2 μs (b) 5 μs (c) 15 μs

14–22 Repeat Problem 14–20 for the following times:
(a) 0.5 ms   (b) 1 ms   (c) 2 ms

14–23 In Figure 14–45, at what time after switch closure does the inductor voltage reach 5 V?

14–24 What is the polarity of the induced voltage in Figure 14–47 when the switch is closed? What is the final value of current if $R_W = 10\ \Omega$?

14–25 Determine the time constant for the circuit in Figure 14–48.

FIGURE 14–47

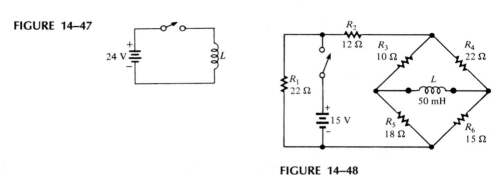

FIGURE 14–48

14–26 Find the inductor current at 10 μs after the switch is thrown from position 1 to position 2 in Figure 14–49. For simplicity, assume that the switch makes contact with position 2 at the same instant it breaks contact with position 1.

14–27 In Figure 14–50, switch 1 is opened and switch 2 is closed at the same instant (t_0). What is the instantaneous voltage across R_2 at t_0?

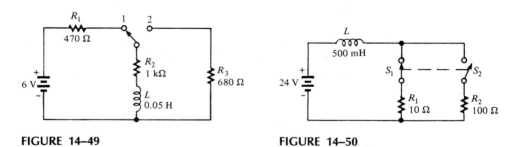

FIGURE 14–49   **FIGURE 14–50**

Section 14–6

14–28 Find the total reactance for each circuit in Figure 14–43 when a voltage with a frequency of 5 kHz is applied across the terminals.

14–29 Find the total reactance for each circuit in Figure 14–44 when a 400-Hz voltage is applied.

14–30 Determine the total rms current in Figure 14–51. What are the currents through L_2 and L_3? Express all currents in polar form.

FIGURE 14–51

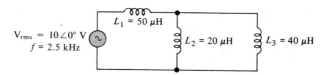

14–31 What frequency will produce 500-mA total rms current in each circuit of Figure 14–44 with an rms input voltage of 10 V?

14–32 Determine the reactive power in Figure 14–51.

14–33 Determine I_{L2} in Figure 14–52.

FIGURE 14–52

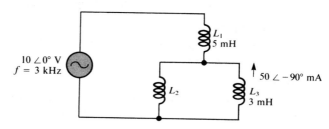

Section 14–8

14–34 A certain coil that is supposed to have a 5-Ω winding resistance is measured with an ohmmeter. The meter indicates 2.8 Ω. What is the problem with the coil?

14–35 What is the indication corresponding to each of the following failures in a coil?
(a) open   **(b)** completely shorted   **(c)** some windings shorted

Section 14–9

14–36 Write a program to compute and display the inductive current at any value of time after a switch is closed in a series *RL* circuit. The values of source voltage, resistance, and inductance are to be specified as input variables.

14–37 Develop a program to determine the energy stored by an inductor and the power dissipated in the winding resistance for specified values of inductance, winding resistance, and direct current.

14–38 Develop a flowchart for the program in Section 14–9.

ANSWERS TO SECTION REVIEWS

Section 14–1

1. Turns, permeability, cross-sectional area, and length.   **2.** 7.5 mV.
3. (a) *L* increases;   **(b)** *L* decreases;   **(c)** *L* decreases;   **(d)** *L* decreases.

4. All wire has some resistance, and since inductors are made from turns of wire, there is always resistance.

Section 14–2
1. Fixed and variable.　　**2.** Air core, iron core, variable.

Section 14–3
1. Inductances are added in series.　　**2.** 2600 μH.　　**3.** 500 mH.

Section 14–4
1. The total parallel inductance is less than that of the smallest individual inductor in parallel.
2. T.　　**3. (a)** 7.69 mH;　　**(b)** 24 μH;　　**(c)** 0.1 H.

Section 14–5
1. 0.1 V.　　**2.** $v_R = 0$ V, $v_L = 20$ V.　　**3.** $v_R = 20$ V, $v_L = 0$ V.
4. 0.5 μs, 3.93 mA.

Section 14–6
1. Voltage leads current by 90 degrees.　　**2.** 3.14 kΩ.　　**3.** 2.55 MHz.
4. $15.9 \angle -90°$ mA.　　**5.** 0, 458.4 mVAR.

Section 14–7
1. The inductor tends to level out the ripple because of its opposition to changes in current.
2. The inductive reactance is very high at radio frequencies and blocks these frequencies.

Section 14–8
1. F.　　**2.** F.

Section 14–9
1. To set the increments of frequency between the lowest and the highest.　　**2.** π.

FIFTEEN

TRANSFORMERS

In Chapter 14, we studied self-inductance. In this chapter, we study the application of another inductive effect called *mutual inductance,* which occurs when two or more coils are in close proximity. Mutual inductance is the basis for transformer action. Because there is no electrical contact between the two magnetically coupled coils that form the basic transformer, the transfer of energy from one coil to the other can be achieved in a situation of complete electrical isolation. There are many uses for the transformer, and a basic understanding of its operation is essential.

In this chapter, you will learn:

☐   The meaning of *mutual inductance.*
☐   How a basic transformer operates.
☐   What the turns ratio is and how it affects the transformer's operation.
☐   How to determine the polarity of the voltages in a transformer.
☐   How a transformer steps up the input voltage.
☐   How a transformer steps down the input voltage.
☐   The meaning of *primary winding* and *secondary winding.*
☐   The effects of connecting a load to the secondary winding.

□ The meaning of *reflected resistance* and how to determine the resistance seen by the primary voltage source.
□ How a transformer can be used for impedance matching.
□ How a transformer can be used as an electrical isolation device.
□ The various types of transformer configurations.
□ How a transformer is constructed.
□ How to interpret transformer ratings.

15–1   MUTUAL INDUCTANCE

When two coils are placed close to each other, as depicted in Figure 15–1, a changing magnetic field produced by current in one coil will cause an induced voltage in the second coil. This induced voltage occurs because the magnetic lines of force (flux) of coil 1 "cut" the winding of coil 2. The two coils are thereby *magnetically linked* or *coupled*. It is important to notice that there is no electrical connection between the coils; therefore they are *electrically isolated*.

FIGURE 15–1
Two magnetically coupled coils.

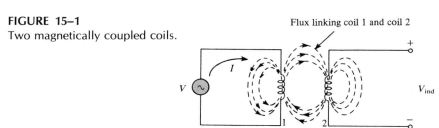

The mutual inductance L_M is a measure of how much voltage is induced in coil 2 as a result of a change in current in coil 1. Mutual inductance is measured in henries (H), just as self-inductance is. A greater L_M means that there is a greater induced voltage in coil 2 for a given change in current in coil 1.

There are several factors that determine L_M: the coefficient of coupling (k), the inductance of coil 1 (L_1), and the inductance of coil 2 (L_2).

COEFFICIENT OF COUPLING

The coefficient of coupling k between two coils is the ratio of the lines of force (flux) produced by coil 1 linking coil 2 (ϕ_{12}) to the total flux produced by coil 1 (ϕ_1):

$$k = \frac{\phi_{12}}{\phi_1} \tag{15–1}$$

For example, if half of the total flux produced by coil 1 links coil 2, then $k = 0.5$. A greater value of k means that more voltage is induced in coil 2 for a certain rate of change of current in coil 1. Note that *k has no units. Recall that the unit of magnetic lines of force (flux) is the weber, abbreviated Wb.*

The coefficient k depends on the physical closeness of the coils and the type of core material on which they are wound. Also, the construction and shape of the cores are factors.

FORMULA FOR MUTUAL INDUCTANCE

The three factors influencing L_M (k, L_1, and L_2) are shown in Figure 15–2. The formula for L_M is

$$L_M = k\sqrt{L_1 L_2} \qquad\qquad (15\text{–}2)$$

FIGURE 15–2
The mutual inductance of two coils.

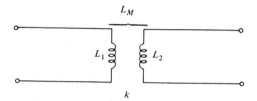

EXAMPLE 15–1

One coil produces a total magnetic flux of 50 μWb, and 20 μWb link coil 2. What is k?

Solution:

$$k = \frac{20\ \mu\text{Wb}}{50\ \mu\text{Wb}} = 0.4$$

EXAMPLE 15–2

Two coils are wound on a single core, and the coefficient of coupling is 0.3. The inductance of coil 1 is 10 μH, and the inductance of coil 2 is 15 μH. What is L_M?

Solution:

$$L_M = k\sqrt{L_1 L_2} = 0.3\sqrt{(10\ \mu\text{H})(15\ \mu\text{H})}$$

$$= 3.67\ \mu\text{H}$$

SECTION REVIEW 15–1

1. Define *mutual inductance*.
2. Two 50-mH coils have $k = 0.9$. What is L_M?
3. If k is increased, what happens to the voltage induced in one coil as a result of a current change in the other coil?

15–2 THE BASIC TRANSFORMER

A *basic transformer* is an electrical device constructed of two coils with mutual inductance. A schematic diagram of a transformer is shown in Figure 15–3(a). As shown, one coil is called the *primary winding,* and the other is called the *secondary winding.*

FIGURE 15–3
The basic transformer.

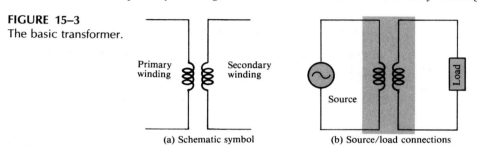

(a) Schematic symbol   (b) Source/load connections

The source voltage is applied to the primary and the load is connected to the secondary, as shown in Figure 15–3(b).

Typical transformers are wound on a common core in several ways. The core can be either an air core, an iron core, or a ferrite core. Figure 15–4(a) and (b) show the primary and secondary coils on a nonferromagnetic cylindrical form as examples of air core transformers. In Part (a), the windings are separated; in Part (b), they overlap for *tighter coupling* and thus a greater mutual inductance (higher k). Figure 15–4(c) illustrates an iron core transformer. Iron and ferrite cores increase the coefficient of coupling. Standard schematic symbols are shown for all types in Part (d), and photos of typical transformers appear in Part (e).

Air-core and ferrite-core transformers generally are used for high-frequency applications and consist of windings on an insulating shell which is hollow (air) or constructed of ferrite, such as depicted in Figure 15–4.

Iron-core transformers generally are used for audio frequency (af) and power applications. These transformers consist of windings on a core constructed from laminated sheets of ferromagnetic material insulated from each other, as shown in the figure. This construction provides an easy path for the magnetic flux and increases the amount of coupling between the windings.

TURNS RATIO

An important parameter of a transformer is its *turns ratio. The turns ratio (n) is the ratio of the number of turns in the secondary winding (N_s) to the number of turns in the primary winding (N_p):*

$$n = \frac{N_s}{N_p} \tag{15–3}$$

In the following sections, you will see how the turns ratio affects the voltages and currents in a transformer.

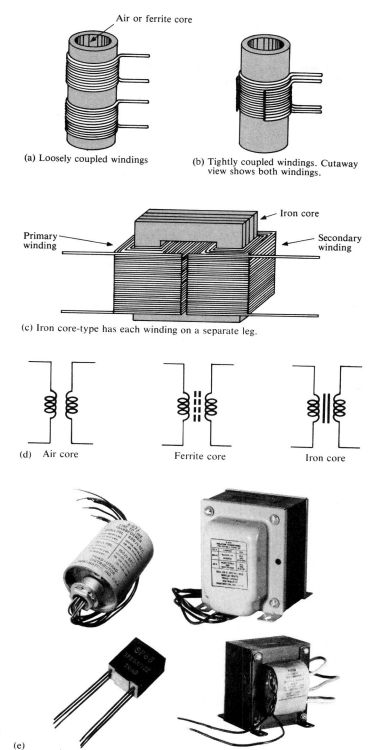

Air or ferrite core

(a) Loosely coupled windings

(b) Tightly coupled windings. Cutaway view shows both windings.

Iron core

Primary winding

Secondary winding

(c) Iron core-type has each winding on a separate leg.

(d)　Air core　　　　Ferrite core　　　　Iron core

(e)

FIGURE 15–4
Basic types of transformers and schematic symbols.

**EXAMPLE
15–3**

A transformer primary has 100 turns, and the secondary has 400 turns. What is the turns ratio?

Solution:

$$N_s = 400 \quad \text{and} \quad N_p = 100$$

$$\text{Turns ratio} = \frac{N_s}{N_p} = \frac{400}{100} = 4$$

DIRECTION OF WINDINGS

Another important transformer parameter is the direction in which the windings are placed around the core. As illustrated in Figure 15–5, the direction of the windings determines the polarity of the secondary voltage with respect to the primary voltage. *Phase dots* are used on the schematic symbols to indicate polarities, as shown in Figure 15–6.

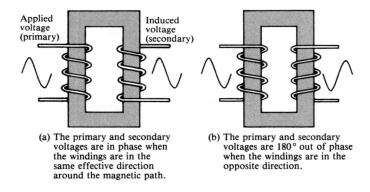

Applied voltage (primary)

Induced voltage (secondary)

(a) The primary and secondary voltages are in phase when the windings are in the same effective direction around the magnetic path.

(b) The primary and secondary voltages are 180° out of phase when the windings are in the opposite direction.

FIGURE 15–5
Relative polarities of the voltages are determined by the direction of the windings.

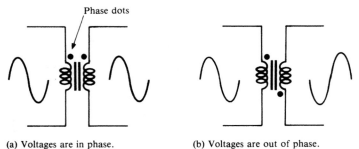

Phase dots

(a) Voltages are in phase.

(b) Voltages are out of phase.

FIGURE 15–6
Phase dots indicate relative polarities of primary and secondary voltages.

SECTION REVIEW 15–2

1. Upon what principle is the operation of a transformer based?
2. Define *turns ratio*.
3. Why are the directions of the windings of a transformer important?
4. A certain transformer has a primary with 500 turns and a secondary with 250 turns. What is the turns ratio?

15–3   STEP-UP TRANSFORMERS

A transformer in which the secondary voltage is greater than the primary voltage is called a step-up transformer. The amount that the voltage is stepped up depends on the turns ratio.

The ratio of secondary voltage to primary voltage is equal to the ratio of the number of secondary turns to the number of primary turns:

$$\frac{V_s}{V_p} = \frac{N_s}{N_p} \qquad (15\text{–}4)$$

where V_s is the secondary voltage and V_p is the primary voltage. From Equation (15–4), we get

$$V_s = \left(\frac{N_s}{N_p}\right) V_p \qquad (15\text{–}5)$$

This equation shows that the secondary voltage is equal to the turns ratio times the primary voltage. This condition assumes that the coefficient of coupling is 1. A good iron core transformer approaches this value.

The turns ratio for a step-up transformer is always *greater* than 1.

EXAMPLE 15–4

The transformer in Figure 15–7 has a 200-turn primary winding and a 600-turn secondary winding. What is the voltage across the secondary?

FIGURE 15–7

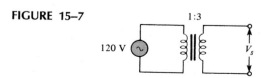

Solution: $N_p = 200$ and $N_s = 600$

$$\text{Turns ratio } = \frac{N_s}{N_p} = \frac{600}{200} = 3$$

$$V_s = 3V_p = 3(120 \text{ V}) = 360 \text{ V}$$

Note that the turns ratio of 3 is indicated on the schematic as 1:3, meaning that there are 3 secondary turns for each primary turn.

SECTION REVIEW 15–3

1. What does a step-up transformer do?

2. If the turns ratio is 5, how much greater is the secondary voltage than the primary voltage?

3. When 240 V ac are applied to a transformer with a turns ratio of 10, what is the secondary voltage?

15–4 STEP-DOWN TRANSFORMERS

A transformer in which the secondary voltage is less than the primary voltage is called a step-down transformer. The amount by which the voltage is stepped down depends on the turns ratio. Equation (15–5) applies also to a step-down transformer; the turns ratio, however, is always *less* than 1.

EXAMPLE 15–5

The transformer in Figure 15–8 has 50 turns in the primary and 10 turns in the secondary. What is the secondary voltage?

FIGURE 15–8

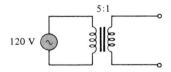

Solution:

$$N_p = 50 \quad \text{and} \quad N_s = 10$$

$$\text{Turns ratio } = \frac{N_s}{N_p} = \frac{10}{50} = 0.2$$

$$V_s = 0.2V_p = 0.2(120 \text{ V}) = 24 \text{ V}$$

SECTION REVIEW 15–4

1. What does a step-down transformer do?
2. A voltage of 120 V ac is applied to the primary of a transformer with a turns ratio of 0.5. What is the secondary voltage?
3. A primary voltage of 120 V ac is reduced to 12 V ac. What is the turns ratio?

15–5 **LOADING THE SECONDARY**

When a load resistor is connected to the secondary, as shown in Figure 15–9, there will be secondary current because of the voltage induced in the secondary coil. It can be shown that the ratio of the primary current I_p to the secondary current I_s is equal to the turns ratio, as expressed in the following equation:

$$\frac{I_p}{I_s} = \frac{N_s}{N_p} \tag{15-6}$$

FIGURE 15–9

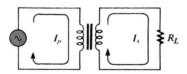

A manipulation of this equation gives Equation (15–7), which shows that I_s is equal to I_p times the *reciprocal* of the turns ratio:

$$I_s = \left(\frac{N_p}{N_s}\right)I_p \tag{15-7}$$

Thus, for a step-up transformer, in which N_s/N_p is greater than 1, the secondary current is less than the primary current. For a step-down transformer, N_s/N_p is less than 1, and I_s is greater than I_p.

EXAMPLE 15–6

The transformers in Figure 15–10(a) and (b) have loaded secondaries. If the primary current is 100 mA in each case, how much is the load current?

Solution:

(a) $I_s = \left(\dfrac{N_p}{N_s}\right)I_p = 0.1(100 \text{ mA}) = 10 \text{ mA}$

FIGURE 15–10

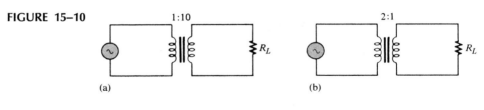

(a)   (b)

(b) $I_s = \left(\dfrac{N_p}{N_s}\right)I_p = 2(100 \text{ mA}) = 200 \text{ mA}$

PRIMARY POWER EQUALS SECONDARY POWER

In an ideal transformer the secondary power is the same as the primary power regardless of the turns ratio, as shown in the following steps:

$$P_p = V_p I_p \quad \text{and} \quad P_s = V_s I_s$$

$$I_s = \left(\dfrac{N_p}{N_s}\right)I_p \quad \text{and} \quad V_s = \left(\dfrac{N_s}{N_p}\right)V_p$$

By substitution, we obtain

$$P_s = \left(\dfrac{\cancel{N_p}}{\cancel{N_s}}\right)\left(\dfrac{\cancel{N_s}}{\cancel{N_p}}\right)V_p I_p$$

Canceling yields

$$P_s = V_p I_p = P_p$$

This result is closely approached in practice because of the very high efficiencies of transformers.

SECTION REVIEW 15–5

1. If the turns ratio of a transformer is 2, is the secondary current greater than or less than the primary current? By how much?

2. A transformer has 100 primary turns and 25 secondary turns, and I_p is 0.5 A. What is the value of I_s?

3. In Problem 2, how much primary current is necessary to produce a secondary load current of 10 A?

15–6   REFLECTED LOAD IN A TRANSFORMER

From the viewpoint of the primary, a load connected across the secondary of a transformer appears to have an ohmic value that is not necessarily equal to the actual ohmic value of the load, but rather depends on the turns ratio. This load that is reflected into the primary is what the source effectively sees, and it determines the amount of primary current. The concept of the *reflected load* is illustrated in Figure 15–11.

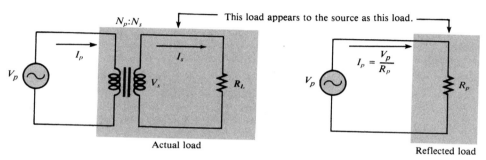

FIGURE 15–11
The load in the secondary of a transformer is reflected into the primary by transformer action. It appears to the source to be a resistance (R_p) with a value determined by the turns ratio and the actual value of the load resistance.

The resistance in the primary circuit of Figure 15–11 is $R_p = V_p/I_p$. The resistance in the secondary circuit is $R_L = V_s/I_s$. From the previous sections, we know that $V_s/V_p = N_s/N_p$ and $I_p/I_s = N_s/N_p$. Using these relationships, we find a formula for R_p in terms of R_L as follows:

$$\frac{R_p}{R_L} = \frac{V_p/I_p}{V_s/I_s} = \left(\frac{V_p}{V_s}\right)\left(\frac{I_s}{I_p}\right) = \left(\frac{N_p}{N_s}\right)\left(\frac{N_p}{N_s}\right)$$
$$= \frac{N_p^2}{N_s^2} = \left(\frac{N_p}{N_s}\right)^2$$

Solving for R_p, we get

$$R_p = \left(\frac{N_p}{N_s}\right)^2 R_L \qquad\qquad (15\text{--}8)$$

Equation (15–8) tells us that *the resistance reflected into the primary is the square of the reciprocal of the turns ratio times the load resistance.*

**EXAMPLE
15–7**

Figure 15–12 shows a source that is transformer-coupled to a load resistor of 100 Ω. The transformer has a turns ratio of 4. What is the reflected resistance seen by the source?

FIGURE 15–12

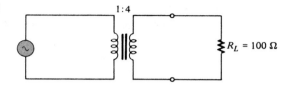

$R_L = 100 \, \Omega$

Solution:
The reflected resistance is determined by Equation (15–8):

$$R_p = \left(\frac{1}{4}\right)^2 R_L = \left(\frac{1}{16}\right)(100 \, \Omega)$$

$$= 6.25 \, \Omega$$

The source sees a resistance of 6.25 Ω just as if it were connected directly, as shown in the equivalent circuit of Figure 15–13.

FIGURE 15–13

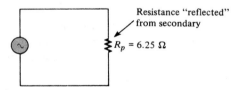

Resistance "reflected" from secondary

$R_p = 6.25 \, \Omega$

The calculator sequence is

4 2nd F 1/x x^2 × 1 0 0 =

On some calculators, $1/x$ is not a secondary function.

**EXAMPLE
15–8**

If a transformer is used in Figure 15–12 having 40 primary turns and 10 secondary turns, what is the reflected resistance?

Solution: Turns ratio = 0.25

$$R_p = \left(\frac{1}{0.25}\right)^2 (100 \, \Omega) = (4)^2 (100 \, \Omega)$$

$$= 1600 \, \Omega$$

This result illustrates the difference that the turns ratio makes.

SECTION REVIEW 15–6

1. Define *reflected resistance*.

2. What transformer characteristic determines the reflected resistance?

3. A given transformer has a turns ratio of 10, and the load is 50 Ω. How much resistance is reflected into the primary?

4. What is the turns ratio required to reflect a 4-Ω load resistance into the primary as 400 Ω?

15–7

MATCHING THE LOAD RESISTANCE TO THE SOURCE RESISTANCE

One application of transformers is in *impedance matching* a load to a source in order to achieve *maximum transfer of power* from the source to the load. Recall that the maximum power transfer theorem was studied in Chapter 8. The term *impedance* will become very familiar to you in the next chapter. Basically, impedance is a general term for the opposition to current, including the effects of both resistance and reactance combined. However, for the time being, we will confine our usage to resistance only.

The concept of impedance matching is illustrated in the basic circuit of Figure 15–14. Part (a) shows an ac voltage source with a series resistance representing its internal resistance. Some internal resistance is inherent in all sources due to their internal circuitry or physical makeup. When the source is connected directly to a load, as shown in Part (b), often the objective is to transfer as much of the power produced by the source to the load as possible. However, a certain amount of the power produced by the source is lost in its internal resistance, and the remaining power goes to the load.

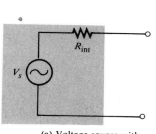

(a) Voltage source with internal resistance

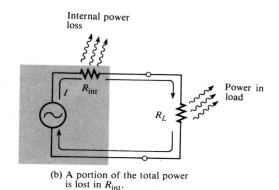

(b) A portion of the total power is lost in R_{int}.

FIGURE 15–14
Power transfer from a nonideal voltage source to a load.

A PRACTICAL APPLICATION

In most practical situations, the internal source resistance of various types of sources is fixed. Also, in many cases, the resistance of a device that acts as a load is fixed and cannot be altered. If you need to connect a given source to a given load, remember that only by chance will their resistances match. In this situation a transformer comes in handy. You can use the reflected-resistance characteristic of a transformer to make the load resistance appear to have the same value as the source resistance, thereby "fooling" the source into "thinking" that there is a match.

Let's take a practical, everyday situation to illustrate. The typical resistance of the input to a TV receiver is 300 Ω. An antenna must be connected to this input by a lead-in cable in order to receive TV signals. In this situation, the antenna and the lead-in act as the source, and the input resistance of the TV receiver is the load, as illustrated in Figure 15–15.

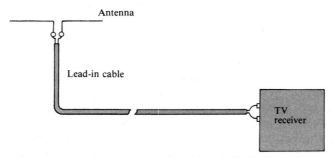

(a) The antenna/lead-in is the source; the TV input is the load.

Source-antenna and lead-in   Load – TV receiver

(b) Circuit equivalent of antenna and TV receiver system

FIGURE 15–15
An antenna directly coupled to a TV receiver.

It is common for an antenna system to have a *characteristic resistance* of 75 Ω. Thus, if the 75-Ω source (antenna and lead-in) is connected directly to the 300-Ω TV input, maximum power will not be delivered to the input to the TV, and you will have poor signal reception. The solution is to use a *matching transformer,* con-

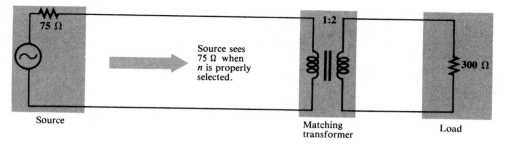

FIGURE 15–16
Example of a load matched to a source by transformer coupling for maximum power transfer.

nected as indicated in Figure 15–16, in order to match the 300-Ω load resistance to the 75-Ω source resistance.

To match the resistances, we must select a proper value of turns ratio (n). We want the 300-Ω load to look like 75 Ω to the source. We solve equation (15–8) for the turns ratio, N_s/N_p, using 300 Ω for R_L and 75 Ω for R_p.

$$R_p = \left(\frac{N_p}{N_s}\right)^2 R_L$$

Transposing terms, we get

$$\left(\frac{N_p}{N_s}\right)^2 = \frac{R_p}{R_L}$$

Taking the square root of both sides, we have

$$\frac{N_p}{N_s} = \sqrt{\frac{R_p}{R_L}}$$

Inverting both sides and solving for n yields

$$\frac{N_s}{N_p} = \sqrt{\frac{R_L}{R_p}} = \sqrt{\frac{300 \ \Omega}{75 \ \Omega}} = \sqrt{4} = 2$$

Therefore, a matching transformer with a turns ratio of 2 must be used in this application.

EXAMPLE
15–9

An amplifier has an 800-Ω internal resistance. In order to provide maximum power to an 8-Ω speaker, what turns ratio must be used in the coupling transformer?

Solution:
The reflected resistance must equal 800 Ω. Thus,

$$\frac{N_s}{N_p} = \sqrt{\frac{R_L}{R_p}} = \sqrt{\frac{8\ \Omega}{800\ \Omega}} = \sqrt{0.01} = 0.1$$

There must be ten primary turns for each secondary turn. The diagram and its equivalent reflected circuit are shown in Figure 15–17.

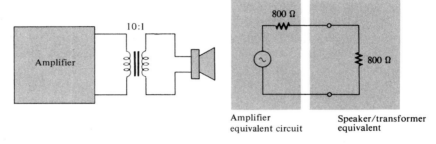

FIGURE 15–17

SECTION REVIEW 15–7

1. What does *impedance matching* mean?
2. What is the advantage of matching the load impedance to the impedance of a source?
3. A transformer has 100 primary turns and 50 secondary turns. What is the reflected resistance with 100 Ω across the secondary?

15–8

THE TRANSFORMER AS AN ISOLATION DEVICE

Transformers are useful in providing *electrical isolation* between the primary and the secondary, since there is no electrical connection between the two windings. As you know, energy is transferred entirely by magnetic coupling.

dc ISOLATION

As illustrated in Figure 15–18, if there is a constant direct current in a transformer primary, nothing happens in the secondary, because a *changing* current in the primary

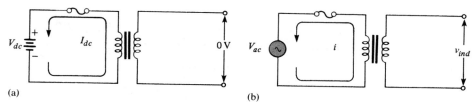

FIGURE 15–18
dc isolation and ac coupling.

is necessary to induce a voltage in the secondary. Therefore, the transformer serves to *isolate* the secondary from any dc in the primary.

In a typical application, a transformer can be used to keep the dc voltage on the output of an amplifier stage from affecting the dc bias of the next amplifier. Only the ac signal is coupled through the transformer from one stage to the next, as Figure 15–19 illustrates.

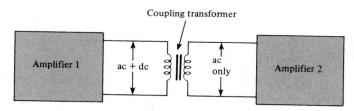

FIGURE 15–19
Amplifier stages with transformer coupling for dc isolation.

POWER LINE ISOLATION

Transformers are often used to isolate the 60-Hz, 120-V ac power line from a piece of electronic equipment, such as a TV set or any test instrument that operates from the 60-Hz ac power.

The reason for using a transformer to couple the 60-Hz ac to the equipment is to prevent a possible shock hazard if the "hot" side (120 V ac) of the power line is connected to the equipment chassis. This condition is possible if the line cord socket can be plugged into the outlet either way. Figure 15–20 illustrates this situation.

A transformer can prevent this hazardous condition, as illustrated in Figure 15–21. With such isolation, there is no way of directly connecting the 120-V ac line to the instrument ground, no matter how the power cord is plugged into the outlet.

Many TV sets, for example, do not have isolation transformers for reasons of economy. When working on the chassis, you should exercise care by using an external isolation transformer or by plugging into the outlet so that chassis ground is not connected to the 120-V side of the outlet.

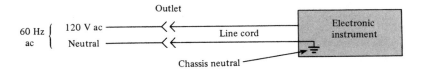

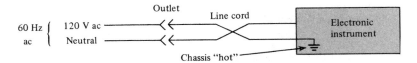

FIGURE 15–20
Instrument powered without transformer isolation.

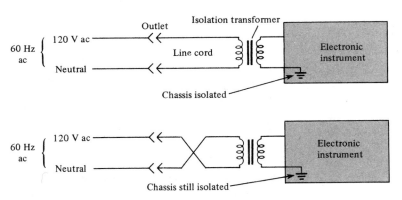

FIGURE 15–21
Power line isolation.

SECTION REVIEW 15–8

1. Name two applications of a transformer as an isolation device.

2. Will a transformer operate with a dc input?

15–9   NONIDEAL TRANSFORMER CHARACTERISTICS

Up to this point, transformer operation has been presented from an ideal point of view, and this approach is valid when you are learning new concepts. However, you should be aware of the nonideal characteristics of practical transformers and how they affect performance.

WINDING RESISTANCE

Both the primary and the secondary windings of a transformer have winding resistance. (You learned about winding resistance in Chapter 14 on inductors.) The winding resistances of a transformer are represented as resistors in series with the windings as shown in Figure 15–22.

FIGURE 15–22
Winding resistance in a nonideal transformer.

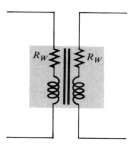

Winding resistance in a transformer results in less voltage across a secondary load. Voltage drops due to the winding resistance effectively subtract from the primary and secondary voltages and result in load voltage that is less than that predicted by the relationships $V_s = (N_s/N_p)V_p$. In most cases the effect is relatively small and is not considered.

LOSSES IN THE CORE

There is always some energy loss in the core material of a nonideal transformer. This loss is seen as a heating of ferrite and iron cores, but it does not occur in air cores. Part of this energy is consumed in the continuous reversal of the magnetic field due to the changing direction of the primary current; this energy loss is called *hysteresis loss*. The rest of the energy loss is caused by *eddy currents* produced when voltage is induced in the core material by the changing magnetic flux, according to Faraday's law. Eddy currents occur in circular patterns in the core resistance, thus causing the energy loss. This loss is greatly reduced by the use of laminated construction of iron cores. The thin layers of ferromagnetic material are insulated from each other to minimize the build-up of eddy currents by confining them to a small area and to keep core losses to a minimum.

MAGNETIC FLUX LEAKAGE

In an ideal transformer, all of the magnetic flux produced by the primary current is assumed to pass through the core to the secondary winding, and vice versa. In an actual transformer, some of the magnetic flux lines break out of the core and pass through the surrounding air back to the other end of the winding, as illustrated in Figure 15–23 for the magnetic field produced by the primary current. Magnetic flux leakage results in a reduced secondary voltage.

FIGURE 15–23
Flux leakage in a nonideal transformer.

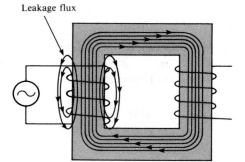

The percentage of magnetic flux that actually reaches the secondary determines the *coefficient of coupling* of the transformer. For example, if nine out of ten flux lines remain inside the core, the coefficient of coupling is 0.90 or 90%. Most iron-core transformers have very high coefficients of coupling (greater than 0.99), while ferrite-core and air-core devices have lower values.

WINDING CAPACITANCE

As you learned in Chapter 14, there is always some stray capacitance between adjacent turns of a winding. These stray capacitances result in an effective capacitance in parallel with each winding of a transformer, as indicated in Figure 15–24.

FIGURE 15–24
Winding capacitance in a nonideal transformer.

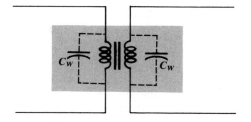

These stray capacitances have very little effect on the transformer's operation at low frequencies, because the reactances (X_C) are very high. However, at higher frequencies, the reactances decrease and begin to produce a bypassing effect across the primary winding and across the secondary load. As a result, there is less of the total primary current through the primary winding, and less of the total secondary current through the load. This effect reduces the load voltage as the frequency goes up.

TRANSFORMER POWER RATING

A transformer is typically rated in volt-amperes (VA), primary/secondary voltage, and operating frequency. For example, a given transformer rating may be specified as

2 kVA, 500/50, 60 Hz. The 2-kVA value is the *apparent power rating*. The 500 and the 50 can be either secondary or primary voltages.

Let's assume, for example, that 50 V is the secondary voltage. In this case the load current is $I_L = P_s/V_s = 2$ kVA/50 V = 40 A. On the other hand, if 500 V is the secondary voltage, then $I_L = P_s/V_s = 2$ kVA/500 V = 4 A. These are the maximum currents that the secondary can handle in either case.

The reason that the power rating is in volt-amperes (VA) rather than in watts (true power) is as follows: If the transformer load is purely capacitive or purely inductive, the true power (watts) delivered to the load is zero. However, the current for $V_s = 500$ V and $X_C = 100$ Ω at 60 Hz, for example, is 5 A. This current exceeds the maximum that the 2-kVA secondary can handle, and the transformer may be damaged. So it is meaningless to specify power in watts.

TRANSFORMER EFFICIENCY

Recall that the secondary power is equal to the primary power in an ideal transformer. Because the nonideal characteristics just discussed result in a power loss in the transformer, the secondary (output) power is always less than the primary (input) power. The *efficiency* (η) of a transformer is a measure of the percentage of the input power that is delivered to the output:

$$\eta = \left(\frac{P_{out}}{P_{in}}\right)100\% \qquad\qquad (15\text{--}9)$$

Most power transformers have efficiencies in excess of 95%.

EXAMPLE 15–10

A certain type of transformer has a primary current of 5 A and a primary voltage of 4800 V. The secondary current is 90 A, and the secondary voltage is 240 V. Determine the efficiency of this transformer.

Solution:
The input power is

$$P_{in} = V_p I_p = (4800 \text{ V})(5 \text{ A}) = 24 \text{ kVA}$$

The output power is

$$P_{out} = V_s I_s = (240 \text{ V})(90 \text{ A}) = 21.6 \text{ kVA}$$

The efficiency is

$$\eta = \left(\frac{P_{out}}{P_{in}}\right)100\% = \left(\frac{21.6 \text{ kVA}}{24 \text{ kVA}}\right)100\% = 90\%$$

SECTION REVIEW 15–9

1. Explain how an actual transformer differs from the ideal model.

2. The coefficient of coupling of a certain transformer is 0.85. What does this mean?

3. A certain transformer has a rating of 10 kVA. If the secondary voltage is 250 V, how much load current can the transformer handle?

15–10 OTHER TYPES OF TRANSFORMERS

There are several important variations of the basic transformer that you have studied so far. They include *tapped transformers*, *multiple-winding transformers*, and *autotransformers*.

TAPPED TRANSFORMERS

A schematic diagram of a transformer with a *center-tapped* secondary winding is shown in Figure 15–25(a). The center tap (CT) is equivalent to two secondary windings with half the total voltage across each.

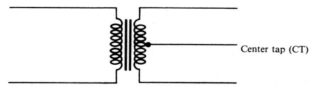

(a) Center-tapped transformer

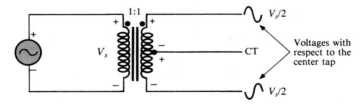

(b) Output voltages with respect to the center tap are 180° out of phase with each other and are one-half the magnitude of the secondary voltage.

FIGURE 15–25
Operation of a center-tapped transformer.

The voltages between either end of the secondary and the center tap are, at any instant, equal in magnitude but opposite in polarity, as illustrated in Figure 15–25(b). Here, for example, at some instant on the sine wave voltage, the polarity across the entire secondary is as shown (top end +, bottom −). At the center tap, the voltage is less positive than the top end but more positive than the bottom end of the secondary. Therefore, measured *with respect to the center tap,* the top end of the secondary is positive, and the bottom end is negative. This center-tapped feature is used in power supply rectifiers in which the ac voltage is converted to dc, as illustrated in Figure 15–26.

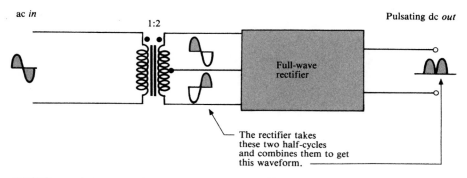

FIGURE 15–26
Application of a center-tapped transformer in ac-to-dc conversion.

Some tapped transformers have taps on the secondary winding at points other than the electrical center. Also, multiple primary and secondary taps are sometimes used in certain applications. Examples of these types of transformers are shown in Figure 15–27.

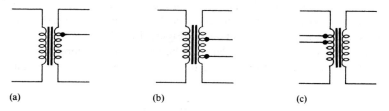

(a)   (b)   (c)

FIGURE 15–27
Tapped transformers.

One example of a transformer with a multiple-tap primary and a center-tapped secondary is the utility-pole transformer used by power companies to step down the high voltage from the power line to 120 V/240 V service for residential and commercial customers, as shown in Figure 15–28. The multiple taps on the primary are used for minor adjustments in the turns ratio in order to overcome line voltages that are slightly too high or too low.

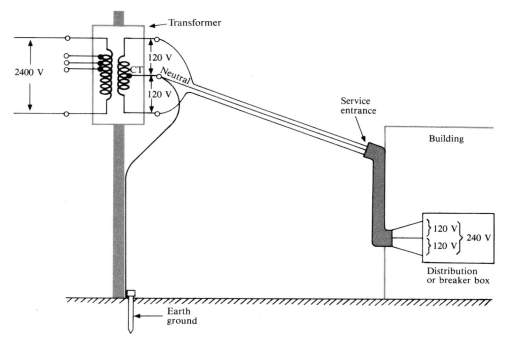

FIGURE 15–28
Utility-pole transformer in a typical power distribution system.

MULTIPLE-WINDING TRANSFORMERS

Some transformers are designed to operate from either 120-V ac or 240-V ac lines. These transformers usually have two primary windings, each of which is designed for 120 V ac. When the two are connected in series, the transformer can be used for 240-V ac operations, as illustrated in Figure 15–29.

More than one secondary can be wound on a common core. Transformers with several secondaries are often used to achieve several voltages by either stepping up or

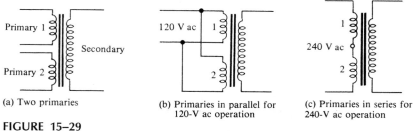

(a) Two primaries

(b) Primaries in parallel for 120-V ac operation

(c) Primaries in series for 240-V ac operation

FIGURE 15–29
Multiple-primary transformer.

stepping down the primary voltage. These types are commonly used in power supply applications in which several voltage levels are required for the operation of an electronic instrument.

A typical schematic of a multiple-secondary transformer is shown in Figure 15–30; this transformer has three secondaries. Sometimes you will find combinations of multiple-primary, multiple-secondary, and tapped transformers all in one unit.

FIGURE 15–30
Multiple-secondary transformer.

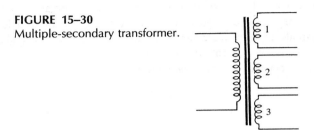

EXAMPLE 15–11

The transformer shown in Figure 15–31 has the numbers of turns (denoted as T) indicated. One of the secondaries is also center-tapped. If 120 V ac are connected to the primary, determine each secondary voltage and the voltages with respect to the center tap (CT) on the middle secondary.

FIGURE 15–31

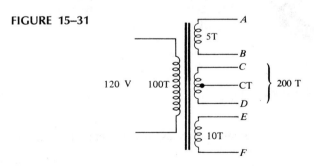

Solution:

$$V_{AB} = \left(\frac{5}{100}\right)120 \text{ V} = 6 \text{ V}$$

$$V_{CD} = \left(\frac{200}{100}\right)120 \text{ V} = 240 \text{ V}$$

$$V_{CTC} = V_{CTD} = \frac{240 \text{ V}}{2} = 120 \text{ V}$$

$$V_{EF} = \left(\frac{10}{100}\right)120 \text{ V} = 12 \text{ V}$$

AUTOTRANSFORMERS

In an autotransformer, *one winding* serves as both the primary and the secondary. The winding is tapped at the proper points to achieve the desired turns ratio for stepping up or stepping down the voltage.

Autotransformers differ from conventional transformers in that there is no electrical isolation between the primary and the secondary because both are on one winding. Autotransformers normally are smaller and lighter than equivalent conventional transformers because they require a much lower kVA rating for a given load. Many autotransformers provide an adjustable tap using a sliding contact mechanism so that the output voltage can be varied (these are often called *variacs*). Figure 15–32 shows a typical autotransformer and several schematic symbols.

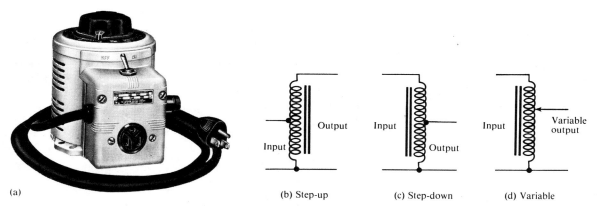

(a)

(b) Step-up (c) Step-down (d) Variable

FIGURE 15–32
The autotransformer (courtesy of Superior Electric Co.).

The following example illustrates why an autotransformer has a kVA requirement that is less than the input or output kVA.

EXAMPLE 15–12

A certain autotransformer is used to change a source voltage of 240 V to a load voltage of 160 V across an 8-Ω load resistance. Determine the input and output power in kVA, and show that the actual kVA requirement is less than this value. Assume that this transformer is ideal.

Solution:
The circuit is shown in Figure 15–33 with the voltages and currents indicated. The current directions have been assigned arbitrarily for convenience.

The load current I_3 is determined as follows:

$$I_3 = \frac{V_3}{R_L} = \frac{160 \text{ V}}{8 \text{ }\Omega} = 20 \text{ A}$$

FIGURE 15–33

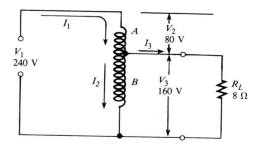

The input power is the total source voltage (V_1) times the total current from the source (I_1):

$$P_{in} = V_1 I_1$$

The output power is the load voltage V_3 times the load current I_3:

$$P_{out} = V_3 I_3$$

For an ideal transformer, $P_{in} = P_{out}$; thus,

$$V_1 I_1 = V_3 I_3$$

Solving for I_1 yields

$$I_1 = \frac{V_3 I_3}{V_1} = \frac{(160 \text{ V})(20 \text{ A})}{240 \text{ V}} = 13.33 \text{ A}$$

Applying Kirchhoff's current law at the tap junction, we get

$$I_1 = I_2 + I_3$$

Solving for I_2 yields

$$I_2 = I_1 - I_3 = 13.33 \text{ A} - 20 \text{ A} = -6.67 \text{ A}$$

The minus sign can be dropped because the current directions are arbitrary. The input and output power are

$$P_{in} = P_{out} = V_3 I_3 = (160 \text{ V})(20 \text{ A}) = 3.2 \text{ kVA}$$

The power in winding A is

$$V_2 I_1 = (80 \text{ V})(13.33 \text{ A}) = 1.07 \text{ kVA}$$

The power in winding B is

$$V_3 I_2 = (160 \text{ V})(6.67 \text{ A}) = 1.07 \text{ kVA}$$

Thus, the power rating required for each winding is less than the power that is delivered to the load.

SECTION REVIEW 15–10

1. A certain transformer has two secondaries. The turns ratio from the primary to the first secondary is 10. The turns ratio from the primary to the other secondary is 0.2. If 240 V ac are applied to the primary, what are the secondary voltages?

2. Name one advantage and one disadvantage of an autotransformer over a conventional transformer.

15–11

TROUBLES IN TRANSFORMERS

The common failures in transformers are opens, shorts, or partial shorts in either the primary or the secondary windings. One cause of such failures is the operation of the device under conditions that exceed its ratings. A few transformer failures and the typical symptoms are discussed in this section.

OPEN PRIMARY WINDING

When there is an open primary winding, there is no primary current and, therefore, no induced voltage or current in the secondary. This condition is illustrated in Figure 15–34(a), and the method of checking with an ohmmeter is shown in Part (b).

FIGURE 15–34
Open primary winding.

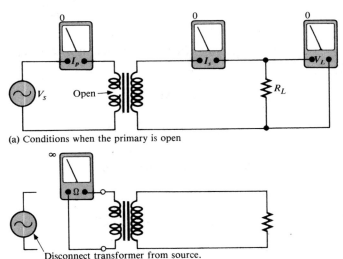

(a) Conditions when the primary is open

(b) Checking the primary with an ohmmeter

OPEN SECONDARY WINDING

When there is an open secondary winding, there is no current in the secondary and, as a result, no voltage across the load. Also, an open secondary causes the primary current to be very small (only a small magnetizing current). In fact, the primary current may be practically zero. This condition is illustrated in Figure 15–35(a), and the ohmmeter check is shown in Part (b).

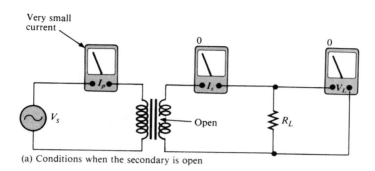

(a) Conditions when the secondary is open

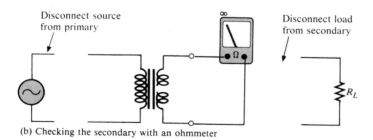

(b) Checking the secondary with an ohmmeter

FIGURE 15–35
Open secondary winding.

SHORTED OR PARTIALLY SHORTED SECONDARY

In this case, there is an excessive primary current because of the low reflected resistance due to the short. Often, this excessive current will burn out the primary and result in an open. The short-circuit current in the secondary causes the load current to be zero (full short) or smaller than normal (partial short), as demonstrated in Figure 15–36(a) and 15–36(b). The ohmmeter check for this condition is shown in Part (c).

FIGURE 15–36
Shorted secondary winding.

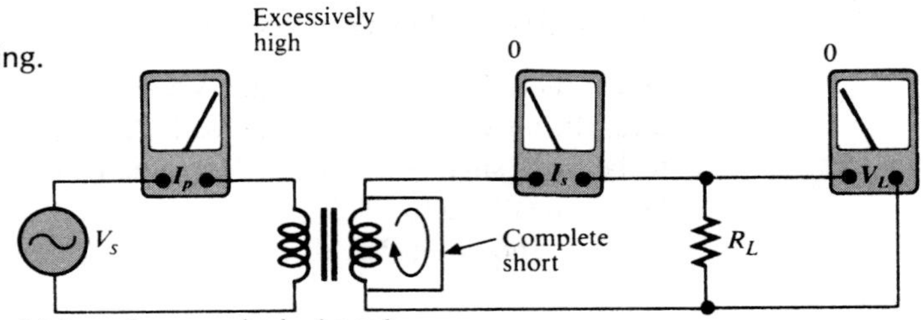

(a) Secondary completely shorted

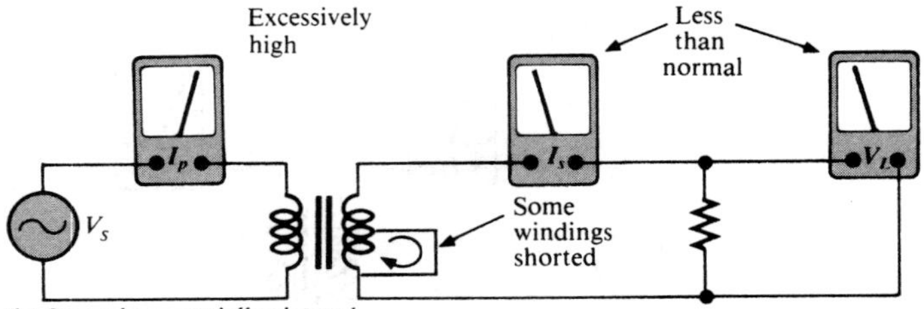

(b) Secondary partially shorted

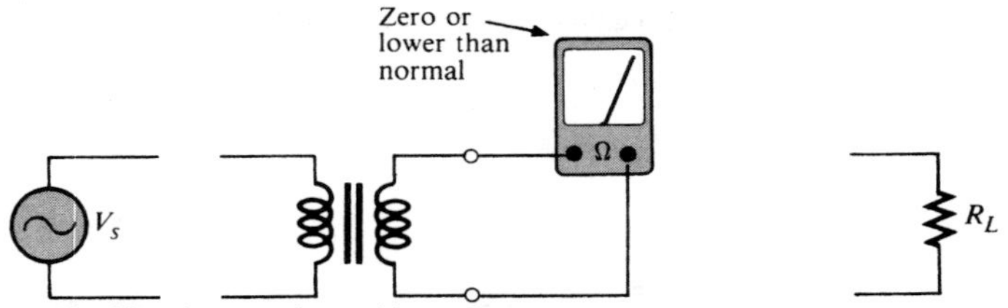

(c) Checking the secondary with an ohmmeter

Normally, when a transformer fails, it is very difficult to repair, and therefore the simplest procedure is to replace it.

SECTION REVIEW 15–11

1. Name two possible failures in a transformer.
2. What is often the cause of transformer failure?

SUMMARY

1. Mutual inductance is the inductance between two coils that are magnetically coupled.
2. A transformer consists of two or more coils that are magnetically coupled on a common core.

3. There is mutual inductance between two magnetically coupled coils.

4. When current in one coil changes, voltage is induced in the other coil.

5. The primary is the winding connected to the source, and the secondary is the winding connected to the load.

6. The number of turns in the primary and the number of turns in the secondary determine the *turns ratio*.

7. The relative polarities of the primary and secondary voltages are determined by the direction of the windings around the core.

8. A step-up transformer has a turns ratio greater than 1.

9. A step-down transformer has a turns ratio less than 1.

10. A transformer cannot increase power.

11. In an ideal transformer, the power from the source (input power) is equal to the power delivered to the load (output power).

12. If the voltage is stepped up, the current is stepped down, and vice versa.

13. A load in the secondary of a transformer appears to the source as a *reflected* load having a value dependent on the reciprocal of the turns ratio squared.

14. A transformer can match a load resistance to a source resistance to achieve maximum power transfer to the load by selecting the proper turns ratio.

15. A transformer does not respond to dc.

16. Energy losses in an actual transformer result from winding resistances, hysteresis loss in the core, eddy currents in the core, and flux leakage.

FORMULAS

$$k = \frac{\phi_{12}}{\phi_1} \qquad (15\text{--}1)$$

$$L_M = k\sqrt{L_1 L_2} \qquad (15\text{--}2)$$

$$n = \frac{N_s}{N_p} \qquad (15\text{--}3)$$

$$\frac{V_s}{V_p} = \frac{N_s}{N_p} \qquad (15\text{--}4)$$

$$V_s = \left(\frac{N_s}{N_p}\right)V_p \qquad (15\text{--}5)$$

$$\frac{I_p}{I_s} = \frac{N_s}{N_p} \qquad (15\text{--}6)$$

$$I_s = \left(\frac{N_p}{N_s}\right)I_p \qquad\qquad (15\text{--}7)$$

$$R_p = \left(\frac{N_p}{N_s}\right)^2 R_L \qquad\qquad (15\text{--}8)$$

$$\eta = \left(\frac{P_{out}}{P_{in}}\right)100\% \qquad\qquad (15\text{--}9)$$

SELF-TEST

Solutions appear at the end of the book.

1. In a given transformer, 90% of the flux produced by the primary coil links the secondary. What is the value of the coefficient of coupling, k?

2. For the transformer in Question 1, the inductance of the primary is 10 μH, and the inductance of the secondary is 5 μH. Determine L_M.

3. What is the turns ratio of a transformer having 12 primary turns and 36 secondary turns? Is it a step-up or a step-down transformer?

4. Determine the polarity of the secondary voltage for each transformer in Figure 15–37.

FIGURE 15–37

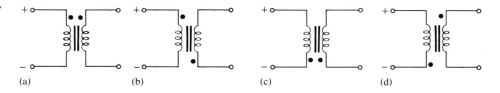

(a)　　　　　(b)　　　　　(c)　　　　　(d)

5. If 120 V ac are connected to the primary of a transformer with 10 primary turns and 15 secondary turns, what is the secondary voltage?

6. With 50 V ac across the primary, what is the secondary voltage if the turns ratio is 0.5?

7. If 240 V ac are applied to a transformer primary, and the secondary voltage is 60 V, what is the turns ratio?

8. If a load is connected to the transformer in Question 7, and if I_p is 0.25 A, what is I_s?

9. What is the reflected resistance in Figure 15–38?

10. If the source resistance in Figure 15–38 is 10 Ω, what must R_L be for maximum power?

11. Determine the voltage for each secondary in Figure 15–39.

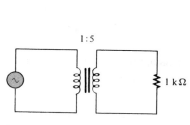

FIGURE 15–38

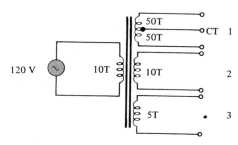

FIGURE 15–39

12. In Figure 15–39, what is the voltage from each end of the upper secondary to the center tap? Are the polarities the same?

13. A 12-V dc battery is connected across the primary of a transformer with a turns ratio of 4. What is the secondary voltage?

14. Why is the output voltage of an actual transformer usually less than what you expect based on the turns ratio?

15. The coefficient of coupling of a certain transformer is 0.95. What does this mean?

PROBLEMS

Section 15–1

15–1 What is the mutual inductance when $k = 0.75$, $L_p = 1$ μH, and $L_s = 4$ μH?

15–2 Determine the coefficient of coupling when $L_M = 1$ μH, $L_p = 8$ μH, and $L_s = 2$ μH.

Section 15–2

15–3 What is the turns ratio of a transformer having 250 primary turns and 1000 secondary turns? What is the turns ratio when the primary has 400 turns and the secondary has 100 turns?

15–4 A certain transformer has 25 turns in its primary. In order to double the voltage, how many turns must be in the secondary?

15–5 For each transformer in Figure 15–40, sketch the secondary voltage showing its relationship to the primary voltage. Also indicate the amplitude.

FIGURE 15–40

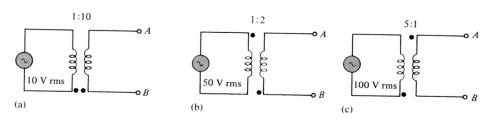

Section 15–3

15–6 To step 240 V ac up to 720 V, what must the turns ratio be?

15–7 The primary of a transformer has 120 V ac across it. What is the secondary voltage if the turns ratio is 5?

15–8 How many primary volts must be applied to a transformer with a turns ratio of 10 to obtain a secondary voltage of 60 V ac?

Section 15–4

15–9 To step 120 V down to 30 V, what must the turns ratio be?

15–10 The primary of a transformer has 1200 V across it. What is the secondary voltage if the turns ratio is 0.2?

15–11 How many primary volts must be applied to a transformer with a turns ratio of 0.1 to obtain a secondary voltage of 6 V ac?

Section 15–5

15–12 Determine I_s in Figure 15–41. What is the value of R_L?

15–13 Determine the following quantities in Figure 15–42:
 (a) Primary current   **(b)** Secondary current
 (c) Secondary voltage   **(d)** Power in the load

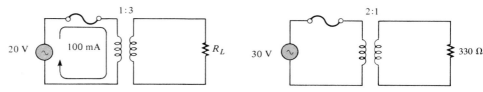

FIGURE 15–41   **FIGURE 15–42**

Section 15–6

15–14 What is the load resistance as seen by the source in Figure 15–43?

15–15 What must the turns ratio be in Figure 15–44, in order to reflect 300 Ω into the primary?

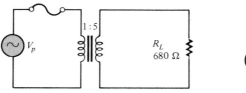
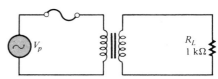

FIGURE 15–43   **FIGURE 15–44**

Section 15–7

15–16 For the circuit in Figure 15–45, find the turns ratio required to deliver maximum power to the 4-Ω speaker.

15–17 In Figure 15–45, what is the maximum power delivered to the speaker?

15–18 Find the appropriate turns ratio for each switch position in Figure 15–46 in order to transfer the maximum power to each load when the source resistance is 10 Ω. Specify the number of turns for the secondary if the primary has 1000 turns.

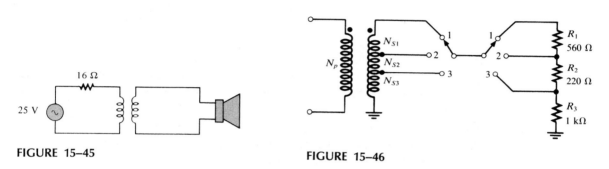

FIGURE 15–45   FIGURE 15–46

Section 15–8

15–19 What is the voltage across the load in each circuit of Figure 15–47?

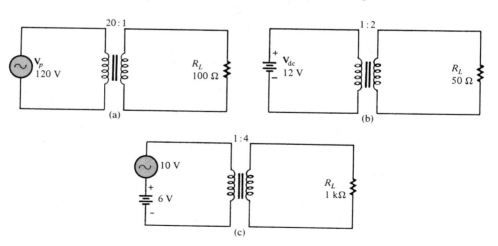

FIGURE 15–47

15–20 Determine the unspecified meter readings in Figure 15–48.

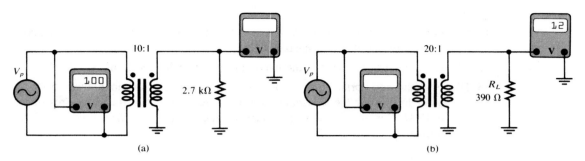

FIGURE 15–48

Section 15–9

15–21 In a certain transformer, the input power to the primary is 100 W. If 5.5 W are lost in the winding resistances, what is the output power to the load, neglecting any other losses?

15–22 What is the efficiency of the transformer in Problem 15–21?

15–23 Determine the coefficient of coupling for a transformer in which 2% of the total flux generated in the primary does not pass through the secondary.

15–24 A certain transformer is rated at 1 kVA. It operates on 60 Hz, 120 V ac. The secondary voltage is 600 V.
   (a) What is the maximum load current?
   (b) What is the smallest R_L that you can drive?
   (c) What is the largest capacitor that can be connected as a load?

15–25 What kVA rating is required for a transformer that must handle a maximum load current of 10 A with a secondary voltage of 2.5 kV?

15–26 A certain transformer is rated at 5 kVA, 2400/120 V, at 60 Hz.
   (a) What is the turns ratio if the 120 V is the secondary voltage?
   (b) What is the current rating of the secondary if 2400 V is the primary voltage?
   (c) What is the current rating of the primary if 2400 V is the primary voltage?

Section 15–10

15–27 Determine each unknown voltage indicated in Figure 15–49.

15–28 Using the indicated secondary voltages in Figure 15–50, determine the turns ratio of the primary to each tapped section.

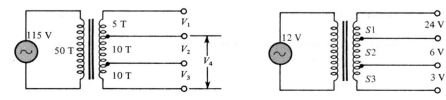

FIGURE 15–49   **FIGURE 15–50**

15–29 Find the secondary voltage for each autotransformer in Figure 15–51.

15–30 In Figure 15–52, each primary can accommodate 120 V ac. How should the primaries be connected for 240-V ac operation? Determine each secondary voltage for 240-V operation.

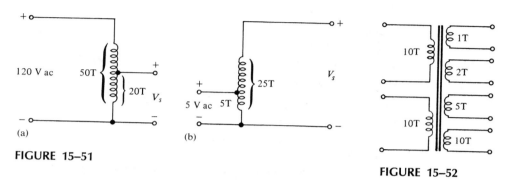

FIGURE 15–51

FIGURE 15–52

15–31 For the loaded, tapped-secondary transformer in Figure 15–53, determine the following:

(a) All load voltages and currents

(b) The impedance looking into the primary

FIGURE 15–53

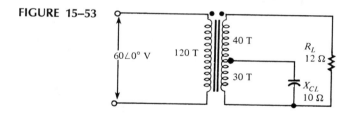

ANSWERS TO SECTION REVIEWS

Section 15–1
1. The inductance between two coils.   **2.** 45 mH.   **3.** It increases.

Section 15–2
1. Mutual inductance.
2. The ratio of secondary turns to primary turns.
3. The directions determine the relative polarities of the voltages.   **4.** 0.5

Section 15–3
1. Increases voltage from primary to secondary.   **2.** Five times greater.
3. 2400 V.

Section 15–4
1. Decreases voltage from primary to secondary.   **2.** 60 V.   **3.** 0.1.

Section 15–5
1. Less; half.   **2.** 2 A.   **3.** 2.5 A.

Section 15–6
1. The resistance in the secondary reflected into the primary.   **2.** The turns ratio.
3. 0.5 Ω.   **4.** 0.1.

Section 15–7
1. Making the load impedance equal the source impedance.
2. Maximum power is delivered to the load.   **3.** 400 Ω.

Section 15–8
1. dc isolation and power line isolation.   **2.** No.

Section 15–9
1. In an actual transformer, energy loss reduces the efficiency. An ideal model has an efficiency of 100%.
2. 85% of the magnetic flux generated in the primary passes through the secondary.
3. 40 A.

Section 15–10
1. 2400 V, 48 V.   **2.** Smaller and lighter for same rating, no electrical isolation.

Section 15–11
1. Open or shorted primary, open or shorted secondary.
2. Operating above rated values.

SIXTEEN

RC CIRCUIT ANALYSIS

An *RC* circuit contains both *resistance* and *capacitance*. It is one of the basic types of reactive circuits that will be studied. In this chapter, basic series and parallel *RC* circuits and their responses to sinusoidal ac voltages are covered. Series-parallel combinations are also analyzed. Power in *RC* circuits is studied, and basic applications are introduced. Troubleshooting and computer analysis are introduced at the end of the chapter. Applications of the *RC* circuit include filters, amplifier coupling, oscillators, and wave-shaping circuits.

In this chapter, you will learn:

☐ How to calculate the impedance of a series *RC* circuit.
☐ How to analyze a series *RC* circuit in terms of phase angle, current, and voltages using Ohm's law and Kirchhoff's voltage law.
☐ How to determine the effects of frequency on a series *RC* circuit.
☐ How to calculate the impedance of a parallel *RC* circuit.
☐ How to use the concepts of conductance, susceptance, and admittance in the analysis of parallel *RC* circuits.
☐ How to analyze parallel *RC* circuits in terms of phase angle, currents, and voltage using Ohm's law and Kirchhoff's current law.

□ How to convert a parallel *RC* circuit to an equivalent series form.
□ How to analyze circuits with combinations of series and parallel elements.
□ How to determine the true power, reactive power, apparent power, and the power factor in *RC* circuits.
□ How to analyze *RC* lead and lag networks and basic *RC* filters.
□ How to troubleshoot *RC* circuits for basic types of component failures.

16–1   SINUSOIDAL RESPONSE OF *RC* CIRCUITS

When a sinusoidal voltage is applied to an *RC* circuit, *each resulting voltage drop and the current in the circuit is also sinusoidal with the same frequency as the applied voltage.* As shown in Figure 16–1, the resistor voltage, the capacitor voltage, and the current are all sine waves with the frequency of the source.

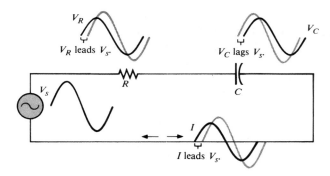

FIGURE 16–1
Illustration of sinusoidal response with general phase relationships of V_R, V_C, and I relative to the source voltage. V_R and I are in phase; V_R leads V_s; V_C lags V_s; and V_R and V_C are 90 degrees out of phase.

Phase shifts are introduced because of the capacitance. As you will learn, the resistor voltage and current lead the source voltage, and the capacitor voltage lags the source voltage. The phase angle between the current and the capacitor voltage is always 90 degrees. These generalized phase relationships are indicated in the figure.

The amplitudes and the phase relationships of the voltages and current depend on the ohmic values of the resistance and the capacitive reactance. When a circuit is purely resistive, the phase angle between the applied (source) voltage and the total current is zero. When a circuit is purely capacitive, the phase angle between the applied voltage and the total current is 90 degrees, with the current leading the voltage. When there is a combination of both resistance and capacitive reactance in a circuit, the phase angle between the applied voltage and the total current is somewhere between zero and 90 degrees, depending on the relative values of the resistance and the reactance.

SIGNAL GENERATORS

When a circuit is hooked up for a laboratory experiment or for troubleshooting, a signal generator similar to those shown in Figure 16–2 is used to provide the source voltage. These instruments, depending on their capability, are classified as *sine wave generators,* which produce only sine waves; *sine/square generators,* which produce both sine waves and square waves; or *function generators,* which produce sine waves, pulse waveforms, and triangular (ramp) waveforms.

(a)

(b)

(c)

FIGURE 16–2

Typical signal (function) generators used in circuit testing and troubleshooting [(a) courtesy of Hewlett-Packard Company. (b) courtesy of Wavetek. (c) courtesy of B&K Precision].

SECTION REVIEW 16–1

1. A 60-Hz sinusoidal voltage is applied to an *RC* circuit. What is the frequency of the capacitor voltage? The current?

2. When the resistance in an *RC* circuit is greater than the capacitive reactance, is the phase angle between the applied voltage and the total current closer to zero or to 90 degrees?

16–2   IMPEDANCE OF A SERIES *RC* CIRCUIT

Impedance is the total opposition to sinusoidal current expressed in units of ohms. In a purely resistive circuit, the impedance is simply equal to the total resistance. In a purely capacitive circuit, the impedance is equal to the total capacitive reactance. The impedance of a series *RC* circuit is determined by both the resistance and the capacitive reactance. These cases are illustrated in Figure 16–3. The magnitude of the impedance is symbolized by *Z*.

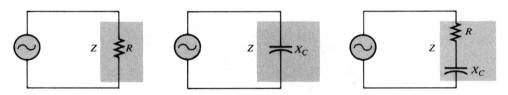

FIGURE 16–3
Three cases of impedance.

Recall from Chapter 13 that capacitive reactance is expressed as a complex number in rectangular form as

$$\mathbf{X}_C = -jX_C \tag{16–1}$$

where boldface $\mathbf{X}_C$ designates a phasor quantity (representing both magnitude and angle) and X_C is just the magnitude.

In the series *RC* circuit of Figure 16–4, the total impedance is the phasor sum of R and $-jX_C$ and is expressed as

$$\mathbf{Z} = R - jX_C \tag{16–2}$$

FIGURE 16–4
Series *RC* circuit.

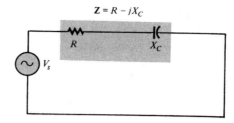

THE IMPEDANCE TRIANGLE

In ac analysis, both R and X_C are treated as phasor quantities, as shown in the phasor diagram of Figure 16–5(a), with X_C appearing at a -90-degree angle with respect to R. This relationship comes from the fact that the capacitor voltage in a series *RC* circuit lags the current, and thus the resistor voltage, by 90 degrees. Since $\mathbf{Z}$ is the phasor sum of R and $-jX_C$, its phasor representation is shown in Figure 16–5(b). A repositioning of the phasors, as shown in Part (c), forms a right triangle. This is called the *impedance triangle*. The length of each phasor represents the *magnitude* in ohms, and the angle θ is the phase angle of the *RC* circuit and represents the phase difference between the applied voltage and the current.

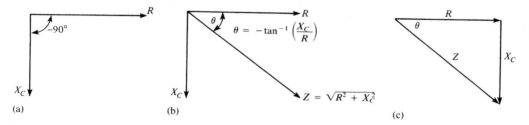

FIGURE 16–5
Development of the impedance triangle for a series *RC* circuit.

From right-angle trigonometry, the magnitude (length) of the impedance can be expressed in terms of the resistance and reactance as

$$Z = \sqrt{R^2 + X_C^2} \tag{16–3}$$

This is the *magnitude of* $\mathbf{Z}$ and is expressed in ohms.

The phase angle, θ, is expressed as

$$\theta = -\tan^{-1}\left(\frac{X_C}{R}\right) \tag{16-4}$$

Tan^{-1} can be found on some calculators by pressing $\boxed{\text{INV}}$, then $\boxed{\text{TAN}}$. Combining the magnitude and the angle, the impedance can be expressed in polar form as

$$\mathbf{Z} = \sqrt{R^2 + X_C^2}\angle -\tan^{-1}\left(\frac{X_C}{R}\right) \tag{16-5}$$

**EXAMPLE
16–1**

For each circuit in Figure 16–6, express the impedance in both rectangular form and polar form.

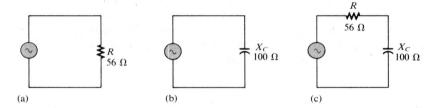

FIGURE 16–6

Solution:

(a) $\mathbf{Z} = R - j0 = R = 56\ \Omega$ in rectangular form ($X_C = 0$)

$\mathbf{Z} = R\angle 0° = 56\angle 0°\ \Omega$ in polar form

The impedance is simply the resistance, and the phase angle is zero because pure resistance does not cause a phase shift between the voltage and current.

(b) $\mathbf{Z} = 0 - jX_C = -j100\ \Omega$ in rectangular form ($R = 0$)

$\mathbf{Z} = X_C\angle -90° = 100\angle -90°\ \Omega$ in polar form

The impedance is simply the capacitive reactance, and the phase angle is -90 degrees because the capacitance causes the current to lead the voltage by 90 degrees.

(c) $\mathbf{Z} = R - jX_C = 56\ \Omega - j100\ \Omega$ in rectangular form

$$\mathbf{Z} = \sqrt{R^2 + X_C^2}\angle -\tan^{-1}\left(\frac{X_C}{R}\right)$$

$$= \sqrt{(56\ \Omega)^2 + (100\ \Omega)^2}\angle -\tan^{-1}\left(\frac{100\ \Omega}{56\ \Omega}\right) = 114.6\angle -60.8°\ \Omega$$

The impedance is the phasor sum of the resistance and the capacitive reactance. The phase angle is fixed by the relative values of X_C and R. Rectangular to polar conversion using a calculator was illustrated in Chapter 12 and can be used in problems like this one to great advantage.

SECTION REVIEW 16–2

1. The impedance of a certain *RC* circuit is 150 Ω − *j*220 Ω. What is the value of the resistance? The capacitive reactance?

2. A series *RC* circuit has a total resistance of 33 kΩ and a capacitive reactance of 50 kΩ. Write the expression for the impedance in rectangular form.

3. For the circuit in Question 2, what is the magnitude of the impedance? What is the phase angle?

16–3   ANALYSIS OF SERIES *RC* CIRCUITS

In the previous section, you learned how to express the impedance of a series *RC* circuit. In this section, Ohm's law and Kirchhoff's voltage law are utilized in the analysis of *RC* circuits.

OHM'S LAW

The application of Ohm's law to series *RC* circuits involves the use of the phasor quantities of **Z**, **V**, and **I**. Keep in mind that the use of boldface letters indicates that both magnitude and angle are included. The three equivalent forms of Ohm's law are as follows:

$$\mathbf{V} = \mathbf{IZ} \tag{16–6}$$

$$\mathbf{I} = \frac{\mathbf{V}}{\mathbf{Z}} \tag{16–7}$$

$$\mathbf{Z} = \frac{\mathbf{V}}{\mathbf{I}} \tag{16–8}$$

From your study of phasor algebra, you should recall that multiplication and division are most easily accomplished with the *polar* forms. Since Ohm's law calculations involve multiplications and divisions, the voltage, current, and impedance should be expressed in polar form, as the next examples show.

EXAMPLE 16–2

If the current in Figure 16–7 is expressed in polar form as $\mathbf{I} = 0.2\angle 0°$ mA, determine the source voltage, and express it in polar form.

FIGURE 16–7

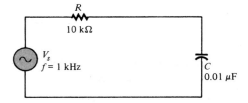

Solution:
The magnitude of the capacitive reactance is

$$X_C = \frac{1}{2\pi fC} = \frac{1}{2\pi (1000 \text{ Hz})(0.01 \text{ } \mu F)} = 15.9 \text{ k}\Omega$$

The impedance is

$$\mathbf{Z} = R - jX_C = 10 \text{ k}\Omega - j15.9 \text{ k}\Omega$$

Converting to polar form:

$$\mathbf{Z} = \sqrt{(10 \text{ k}\Omega)^2 + (15.9 \text{ k}\Omega)^2} \angle -\tan^{-1}\left(\frac{15.9 \text{ k}\Omega}{10 \text{ k}\Omega}\right)$$

$$= 18.78\angle -57.83° \text{ k}\Omega$$

Applying Ohm's law:

$$\mathbf{V}_s = \mathbf{IZ} = (0.2\angle 0° \text{ mA})(18.78\angle -57.83° \text{ k}\Omega) = 3.76\angle -57.83° \text{ V}$$

The magnitude of the source voltage is 3.76 V at an angle of -57.83 degrees with respect to the current; that is, the voltage lags the current by 57.83 degrees, as shown in the phasor diagram of Figure 16–8.

FIGURE 16–8

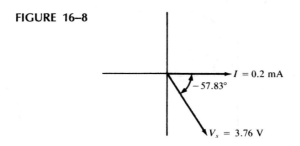

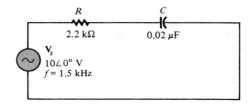

EXAMPLE 16–3

Determine the current in the circuit of Figure 16–9.

FIGURE 16–9

Solution:

$$X_C = \frac{1}{2\pi fC} = \frac{1}{2\pi(1.5 \text{ kHz})(0.02 \ \mu\text{F})} = 5.3 \text{ k}\Omega$$

The total impedance is

$$\mathbf{Z} = R - jX_C = 2.2 \text{ k}\Omega - j5.3 \text{ k}\Omega$$

Converting to polar form:

$$\mathbf{Z} = \sqrt{(2.2 \text{ k}\Omega)^2 + (5.3 \text{ k}\Omega)^2} \angle -\tan^{-1}\left(\frac{5.3 \text{ k}\Omega}{2.2 \text{ k}\Omega}\right)$$

$$= 5.74 \angle -67.46° \text{ k}\Omega$$

Applying Ohm's law:

$$\mathbf{I} = \frac{\mathbf{V}}{\mathbf{Z}} = \frac{10\angle 0° \text{ V}}{5.74 \angle -67.46° \text{ k}\Omega} = 1.74 \angle 67.46° \text{ mA}$$

The magnitude of the current is 1.74 mA. The positive phase angle of 67.46 degrees indicates that the current leads the voltage by that amount, as shown in the phasor diagram of Figure 16–10.

FIGURE 16–10

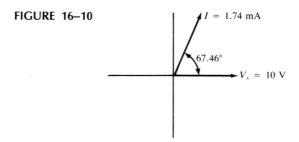

RELATIONSHIPS OF THE CURRENT AND VOLTAGES IN A
SERIES *RC* CIRCUIT

In a series circuit, the current is the same through both the resistor and the capacitor. Thus, *the resistor voltage is in phase with the current, and the capacitor voltage lags the current by 90 degrees.* Therefore, there is a phase difference of 90 degrees between the resistor voltage, V_R, and the capacitor voltage, V_C, as shown in the waveform diagram of Figure 16–11.

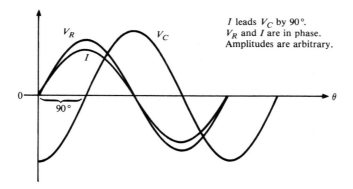

FIGURE 16–11
Phase relation of voltages and current in a series *RC* circuit.

We know from Kirchhoff's voltage law that the sum of the voltage drops must equal the applied voltage. However, since V_R and V_C are not in phase with each other, they must be added as phasor quantities, with V_C lagging V_R by 90 degrees, as shown in Figure 16–12(a). As shown in Part (b), $\mathbf{V}_s$ is the phasor sum of $\mathbf{V}_R$ and $\mathbf{V}_C$, as expressed in the following equation:

$$\mathbf{V}_s = V_R - jV_C \qquad (16\text{–}9)$$

This equation can be expressed in polar form as

$$\mathbf{V}_s = \sqrt{V_R^2 + V_C^2} \angle -\tan^{-1}\left(\frac{V_C}{V_R}\right) \qquad (16\text{–}10)$$

FIGURE 16–12
Voltage phasor diagram for a series *RC*
circuit.

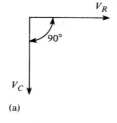

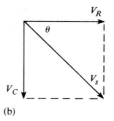

where the magnitude of the source voltage is

$$V_s = \sqrt{V_R^2 + V_C^2} \qquad (16\text{--}11)$$

and the phase angle between the resistor voltage and the source voltage is

$$\theta = -\tan^{-1}\left(\frac{V_C}{V_R}\right) \qquad (16\text{--}12)$$

Since the resistor voltage and the current are in phase, θ also represents the phase angle between the source voltage and the current. Figure 16–13 shows a complete voltage and current phasor diagram representing the waveform diagram of Figure 16–11.

FIGURE 16–13
Voltage and current phasor diagram for
the waveforms in Figure 16–11.

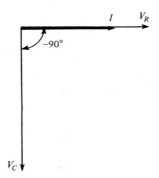

VARIATION OF IMPEDANCE AND PHASE ANGLE WITH FREQUENCY

The impedance triangle is useful in visualizing how the frequency of the applied sinusoidal voltage affects the *RC* circuit. As you know, capacitive reactance varies *inversely* with frequency. Since $Z = \sqrt{R^2 + X_C^2}$, you can see that when X_C increases, the magnitude of the total impedance also increases; and when X_C decreases, the magnitude of the total impedance also decreases. Therefore, *Z is inversely dependent on frequency.*

    The phase angle θ also varies *inversely* with frequency, because $\theta = -\tan^{-1}(X_C/R)$. As X_C increases, so does θ and vice versa.

    Figure 16–14 uses the impedance triangle to illustrate the variations in X_C, Z, and θ as the frequency changes. Of course, R remains constant. The key point is that

FIGURE 16–14
Effect of frequency on impedance and
phase angle.

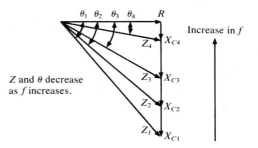

Z and θ decrease
as f increases.

*because X_C varies inversely as the frequency, so also do the magnitude of the total
impedance and the phase angle.* Example 16–4 illustrates this.

**EXAMPLE
16–4**

For the series *RC* circuit in Figure 16–15, determine the magnitude of the total impedance and phase angle for each of the following values of input frequency: 10 kHz,
20 kHz, and 30 kHz.

FIGURE 16–15

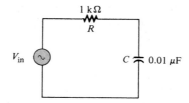

Solution:
For $f = 10$ kHz:

$$X_C = \frac{1}{2\pi fC} = \frac{1}{2\pi (10 \text{ kHz})(0.01 \ \mu\text{F})} = 1592 \ \Omega$$

$$\mathbf{Z} = \sqrt{R^2 + X_C^2} \angle -\tan^{-1}\left(\frac{X_C}{R}\right)$$

$$= \sqrt{(1000 \ \Omega)^2 + (1592 \ \Omega)^2} \angle -\tan^{-1}\left(\frac{1592}{1000}\right)$$

$$= 1880 \angle -57.87° \ \Omega$$

For $f = 20$ kHz:

$$X_C = \frac{1}{2\pi (20 \text{ kHz})(0.01 \ \mu\text{F})} = 796 \ \Omega$$

$$\mathbf{Z} = \sqrt{(1000 \ \Omega)^2 + (796 \ \Omega)^2} \angle -\tan^{-1}\left(\frac{796}{1000}\right)$$

$$= 1278 \angle -38.52° \ \Omega$$

For $f = 30$ kHz:

$$X_C = \frac{1}{2\pi (30 \text{ kHz})(0.01 \text{ }\mu\text{F})} = 531 \text{ }\Omega$$

$$\mathbf{Z} = \sqrt{(1000 \text{ }\Omega)^2 + (531 \text{ }\Omega)^2} \angle -\tan^{-1}\left(\frac{531}{1000}\right)$$

$$= 1132\angle -27.97° \text{ }\Omega$$

Notice that as the frequency increases, X_C, Z, and θ decrease.

SECTION REVIEW 16–3

1. In a certain series RC circuit, $V_R = 4$ V, and $V_C = 6$ V. What is the magnitude of the total voltage?
2. In Question 1, what is the phase angle between the total voltage and the current?
3. What is the phase difference between the capacitor voltage and the resistor voltage in a series RC circuit?
4. When the frequency of the applied voltage in a series RC circuit is increased, what happens to the capacitive reactance? What happens to the magnitude of the total impedance? What happens to the phase angle?

16–4   IMPEDANCE OF A PARALLEL RC CIRCUIT

A basic parallel RC circuit is shown in Figure 16–16. The expression for the total impedance is developed as follows, using the rules of phasor algebra.

$$\mathbf{Z} = \frac{(R\angle 0°)(X_C\angle -90°)}{R - jX_C}$$

FIGURE 16–16
Basic parallel RC circuit.

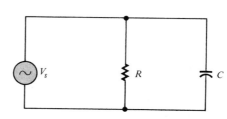

By multiplying the magnitudes, adding the angles in the numerator, and converting the denominator to polar form, we get

$$\mathbf{Z} = \frac{RX_C\angle(0° - 90°)}{\sqrt{R^2 + X_C^2}\angle -\tan^{-1}\left(\dfrac{X_C}{R}\right)}$$

Now, dividing the magnitude expression in the numerator by that in the denominator, and by subtracting the angle in the denominator from that in the numerator, we get

$$\mathbf{Z} = \left(\frac{RX_C}{\sqrt{R^2 + X_C^2}} \right) \angle \left(-90° + \tan^{-1}\left(\frac{X_C}{R}\right) \right) \qquad (16–13)$$

Equation (16–13) is the expression for the total parallel impedance, where the magnitude is

$$Z = \frac{RX_C}{\sqrt{R^2 + X_C^2}} \qquad (16–14)$$

and the phase angle between the applied voltage and the total current is

$$\theta = -90° + \tan^{-1}\left(\frac{X_C}{R}\right) \qquad (16–15)$$

Equivalently, this expression can be written as

$$\theta = \tan^{-1}\left(\frac{R}{X_C}\right)$$

EXAMPLE 16–5

For each circuit in Figure 16–17, determine the magnitude of the total impedance and the phase angle.

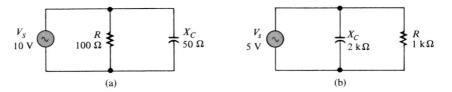

(a)   (b)

FIGURE 16–17

Solution:

(a) $\mathbf{Z} = \left(\dfrac{RX_c}{\sqrt{R^2 + X_C^2}} \right) \angle \left(-90° + \tan^{-1}\left(\dfrac{X_C}{R}\right) \right)$

$= \left[\dfrac{(100\ \Omega)(50\ \Omega)}{\sqrt{(100\ \Omega)^2 + (50\ \Omega)^2}} \right] \angle \left(-90° + \tan^{-1}\left(\dfrac{50}{100}\right) \right)$

$= 44.72 \angle -63.43°\ \Omega$

Thus, $Z = 44.72\ \Omega$, and $\theta = -63.43°$.

(b) $\mathbf{Z} = \left[\dfrac{(1 \ \text{k}\Omega)(2 \ \text{k}\Omega)}{\sqrt{(1 \ \text{k}\Omega)^2 + (2 \ \text{k}\Omega)^2}} \right] \angle \left(-90° + \tan^{-1}\left(\dfrac{2000}{1000} \right) \right)$

$= 894.4 \angle -26.57° \ \Omega$.

Thus, $Z = 894.4 \ \Omega$, and $\theta = -26.57°$.

CONDUCTANCE, SUSCEPTANCE, AND ADMITTANCE

Recall that *conductance* is the reciprocal of resistance and is expressed as

$$\mathbf{G} = \frac{1}{R\angle 0°} = G\angle 0° \tag{16–16}$$

Two new terms are now introduced for use in parallel *RC* circuits. *Capacitive susceptance* (B_C) is the reciprocal of capacitive reactance and is expressed as

$$\mathbf{B}_C = \frac{1}{X_C\angle -90°} = B_C\angle 90° = +jB_C \tag{16–17}$$

Admittance (Y) is the reciprocal of impedance and is expressed as

$$\mathbf{Y} = \frac{1}{Z\angle \pm\theta} = Y\angle \mp\theta \tag{16–18}$$

The unit of each of these terms is, of course, the *siemen*, which is the reciprocal of the ohm.

In working with parallel circuits, it is often easier to use G, B_C, and Y rather than R, X_C, and Z, as we now discuss. In a parallel *RC* circuit, as shown in Figure 16–18, the total admittance is simply the phasor sum of the conductance and the susceptance.

$$\mathbf{Y} = G + jB_C \tag{16–19}$$

FIGURE 16–18
Admittance in a parallel *RC* circuit.

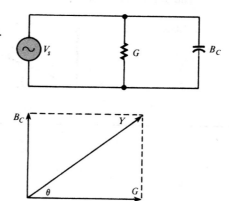

EXAMPLE 16–6

Determine the admittance in Figure 16–19. Sketch the admittance phasor diagram.

FIGURE 16–19

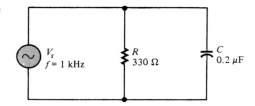

Solution:

$R = 330 \ \Omega$; thus $G = 1/R = 1/330 \ \Omega = 0.003$ S.

$$X_C = \frac{1}{2\pi(1000 \text{ Hz})(0.2 \ \mu\text{F})} = 796 \ \Omega$$

$$B_C = \frac{1}{X_C} = \frac{1}{796 \ \Omega} = 0.00126 \text{ S}$$

$$\mathbf{Y} = G + jB_C = 0.003 \text{ S} + j0.00126 \text{ S}$$

Y can be expressed in polar form as follows:

$$\mathbf{Y} = \sqrt{(0.003 \text{ S})^2 + (0.00126 \text{ S})^2} \angle \tan^{-1}\left(\frac{0.00126}{0.003}\right)$$

$$= 0.00325\angle 22.78° \text{ S}$$

Now, this can be converted to impedance.

$$\mathbf{Z} = \frac{1}{Y} = \frac{1}{(0.00325\angle 22.78° \text{ S})} = 307.69\angle -22.78° \ \Omega$$

The admittance phasor diagram is shown in Figure 16–20.

FIGURE 16–20

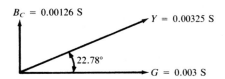

SECTION REVIEW 16–4

1. Define *conductance, capacitive susceptance,* and *admittance.*
2. If $Z = 100 \ \Omega$, what is the value of Y?
3. In a certain parallel *RC* circuit, $R = 47 \ \Omega$ and $X_C = 75 \ \Omega$. Determine **Y.**
4. In Question 3, what is the magnitude of **Y,** and what is the phase angle between the total current and the applied voltage?

16–5

ANALYSIS OF PARALLEL *RC* CIRCUITS

For convenience in the analysis of parallel circuits, the Ohm's law formulas using impedance, previously stated, can be rewritten for admittance using the relation $Y = 1/Z$.

$$V = \frac{I}{Y} \tag{16–20}$$

$$I = VY \tag{16–21}$$

$$Y = \frac{I}{V} \tag{16–22}$$

EXAMPLE 16–7

Determine the total current and phase angle in Figure 16–21.

FIGURE 16–21

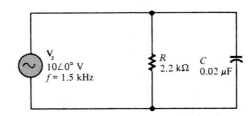

V_s
$10\angle 0° \ V$
$f = 1.5 \ kHz$

R
$2.2 \ k\Omega$

C
$0.02 \ \mu F$

Solution:

$$X_C = \frac{1}{2\pi(1.5 \ kHz)(0.02 \ \mu F)} = 5.3 \ k\Omega$$

The susceptance is

$$B_C = \frac{1}{X_C} = \frac{1}{5.3 \ k\Omega} = 0.189 \ mS$$

The conductance is

$$G = \frac{1}{R} = \frac{1}{2.2 \text{ k}\Omega} = 0.455 \text{ mS}$$

The total admittance is

$$\mathbf{Y} = G + jB_C = 0.455 \text{ mS} + j0.189 \text{ mS}$$

Converting to polar form:

$$\mathbf{Y} = \sqrt{(0.455 \text{ mS})^2 + (0.189 \text{ mS})^2}\angle\tan^{-1}\left(\frac{0.189}{0.455}\right)$$

$$= 0.493\angle22.56° \text{ mS}$$

Applying Ohm's law:

$$\mathbf{I}_T = \mathbf{VY} = (10\angle0° \text{ V})(0.493\angle22.56° \text{ mS}) = 4.93\angle22.56° \text{ mA}$$

The magnitude of the total current is 4.93 mA, and it leads the applied voltage by 22.56 degrees, as the phasor diagram in Figure 16–22 indicates.

FIGURE 16–22

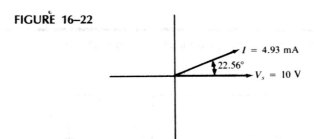

RELATIONSHIPS OF THE CURRENTS AND VOLTAGES IN A
PARALLEL *RC* CIRCUIT

Figure 16–23(a) shows all the currents and voltages in a basic parallel *RC* circuit. As you can see, the applied voltage, V_s, appears across both the resistive and the capacitive branches, so V_s, V_R, and V_C are all in phase and of the same magnitude. The total current, I_T, divides at the junction into the two branch currents, I_R and I_C.

The current through the resistor is in phase with the voltage. The current through the capacitor leads the voltage, and thus the resistive current, by 90 degrees. By Kirchhoff's current law, the total current is the phasor sum of the two branch cur-

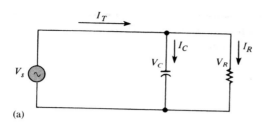 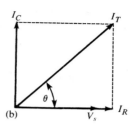

(a)
(b)

FIGURE 16–23
Currents and voltages in a parallel *RC* circuit.

rents, as shown by the phasor diagram in Figure 16–23(b). The total current is expressed as

$$\mathbf{I}_T = I_R + jI_C \qquad (16\text{--}23)$$

This equation can be expressed in polar form as

$$\mathbf{I}_T = \sqrt{I_R^2 + I_C^2} \angle \tan^{-1}\left(\frac{I_C}{I_R}\right) \qquad (16\text{--}24)$$

where the magnitude of the total current is

$$I_T = \sqrt{I_R^2 + I_C^2} \qquad (16\text{--}25)$$

and the phase angle between the resistor current and the total current is

$$\theta = \tan^{-1}\left(\frac{I_C}{I_R}\right) \qquad (16\text{--}26)$$

Since the resistor current and the applied voltage are in phase, θ also represents the phase angle between the total current and the applied voltage. Figure 16–24 shows a complete current and voltage phasor diagram.

FIGURE 16–24
Current and voltage phasor diagram for a parallel *RC* circuit (amplitudes are arbitrary).

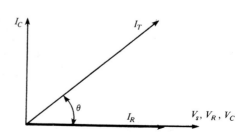

**EXAMPLE
16–8**

Determine the value of each current in Figure 16–25, and describe the phase relationship of each with the applied voltage.

FIGURE 16–25

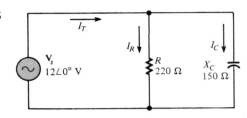

Solution:

$$\mathbf{I}_R = \frac{\mathbf{V}_s}{\mathbf{R}} = \frac{12\angle0° \text{ V}}{220\angle0° \ \Omega} = 54.55\angle0° \text{ mA}$$

$$\mathbf{I}_C = \frac{\mathbf{V}_s}{\mathbf{X}_C} = \frac{12\angle0° \text{ V}}{150\angle-90° \ \Omega} = 80\angle90° \text{ mA}$$

$$\mathbf{I}_T = \mathbf{I}_R + j\mathbf{I}_C = 54.55 \text{ mA} + j80 \text{ mA}$$

Converting $\mathbf{I}_T$ to polar form:

$$\mathbf{I}_T = \sqrt{(54.55 \text{ mA})^2 + (80 \text{ mA})^2}\angle\tan^{-1}\left(\frac{80}{54.55}\right)$$

$$= 96.83\angle55.71° \text{ mA}$$

As the results show, the resistor current is 54.55 mA and is in phase with the voltage. The capacitor current is 80 mA and leads the voltage by 90 degrees. The total current is 96.83 mA and leads the voltage by 55.71 degrees. The phasor diagram in Figure 16–26 illustrates these relationships.

FIGURE 16–26

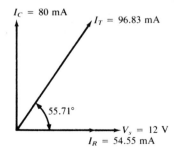

CONVERSION FROM PARALLEL TO SERIES FORM

For every parallel *RC* circuit, there is an *equivalent* series *RC* circuit. Two circuits are equivalent when they both present an equal impedance at their terminals; that is, the magnitude of impedance and the phase angle are identical.

To obtain the equivalent series circuit from a given parallel circuit, express the total impedance of the parallel circuit in *rectangular* form. From this, the values of *R* and X_C are obtained. An example best illustrates this approach.

EXAMPLE 16–9

Convert the parallel circuit in Figure 16–27 to a series form.

FIGURE 16–27

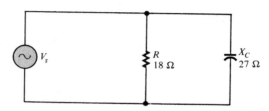

Solution:

First, find the impedance of the parallel circuit.

$$G = \frac{1}{R} = \frac{1}{18\ \Omega} = 0.055\ S$$

$$B_C = \frac{1}{X_C} = \frac{1}{27\ \Omega} = 0.037\ S$$

$$\mathbf{Y} = G + jB_C = 0.055\ S + j0.037\ S$$

Converting to polar form:

$$\mathbf{Y} = \sqrt{(0.055\ S)^2 + (0.037\ S)^2}\angle\tan^{-1}\left(\frac{0.037}{0.055}\right) = 0.066\angle 33.93°\ S$$

The total impedance is

$$\mathbf{Z} = \frac{1}{\mathbf{Y}} = \frac{1}{0.066\angle 33.93°\ S} = 15.15\angle -33.93°\ \Omega$$

Converting to rectangular form:

$$\mathbf{Z} = 15.15\cos(-33.93°) + j15.15\sin(-33.93°)$$
$$= 12.57\ \Omega - j8.46\ \Omega$$

The equivalent series *RC* circuit is a 12.57-Ω resistor in series with a capacitive reactance of 8.46 Ω. This is shown in Figure 16–28.

FIGURE 16–28

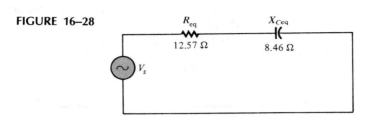

R_{eq} 12.57 Ω
X_{Ceq} 8.46 Ω
V_s

SECTION REVIEW 16–5

1. The admittance of an *RC* circuit is 0.0035 S, and the applied voltage is 6 V. What is the total current?

2. In a certain parallel *RC* circuit, the resistor current is 10 mA, and the capacitor current is 15 mA. Determine the magnitude and phase angle of the total current. This phase angle is measured with respect to what?

3. What is the phase angle between the capacitor current and the applied voltage in a parallel *RC* circuit?

16–6

SERIES-PARALLEL ANALYSIS

In this section, we use the concepts studied in the previous sections to analyze circuits with combinations of both series and parallel *R* and *C* elements. Examples are used to demonstrate the procedures.

**EXAMPLE
16–10**

In the circuit of Figure 16–29, determine the following:

(a) Total impedance   (b) Total current

(c) Phase angle by which I_T leads V_s

FIGURE 16–29

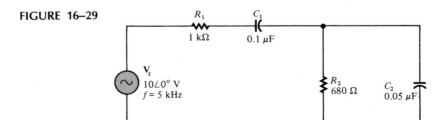

R_1 1 kΩ
C_1 0.1 μF
V_s
10∠0° V
f = 5 kHz
R_2 680 Ω
C_2 0.05 μF

Solution:

(a) First, calculate the magnitudes of capacitive reactance:

$$X_{C1} = \frac{1}{2\pi(5 \text{ kHz})(0.1 \ \mu\text{F})} = 318.3 \ \Omega$$

$$X_{C2} = \frac{1}{2\pi(5 \text{ kHz})(0.05 \ \mu\text{F})} = 636.6 \ \Omega$$

One approach is to find the impedance of the series portion and the impedance of the parallel portion and combine them to get the total impedance.

R_1 and C_1 are in series. R_2 and C_2 are in parallel. The series portion and the parallel portion are in series with each other.

$$\mathbf{Z}_1 = R_1 - jX_{C1} = 1 \text{ k}\Omega - j318.3 \ \Omega$$

$$G_2 = \frac{1}{R_2} = \frac{1}{680 \ \Omega} = 0.00147 \text{ S}$$

$$B_{C2} = \frac{1}{X_{C2}} = \frac{1}{636.6 \ \Omega} = 0.00157 \text{ S}$$

$$\mathbf{Y}_2 = G_2 + jB_{C2} = 0.00147 \text{ S} + j0.00157 \text{ S}$$

Converting to polar form:

$$\mathbf{Y}_2 = \sqrt{(0.00147 \text{ S})^2 + (0.00157 \text{ S})^2} \angle \tan^{-1}\left(\frac{0.00157}{0.00147}\right)$$

$$= 0.00215 \angle 46.88° \text{ S}$$

Then

$$\mathbf{Z}_2 = \frac{1}{\mathbf{Y}_2} = \frac{1}{0.00215 \angle 46.88° \text{ S}} = 465.12 \angle -46.88° \ \Omega$$

Converting to rectangular form:

$$\mathbf{Z}_2 = 465.12 \cos(46.88°) - j465.12 \sin(46.88°)$$

$$= 317.9 \ \Omega - j339.5 \ \Omega$$

Combining $\mathbf{Z}_1$ and $\mathbf{Z}_2$ we get

$$\mathbf{Z}_T = \mathbf{Z}_1 + \mathbf{Z}_2$$

$$= (1000 \ \Omega - j318.3 \ \Omega) + (317.9 \ \Omega - j339.5 \ \Omega)$$

$$= 1317.9 \ \Omega - j657.8 \ \Omega$$

Expressing $\mathbf{Z}_T$ in polar form:

$$\mathbf{Z}_T = \sqrt{(1317.9 \ \Omega)^2 + (657.8 \ \Omega)^2} \angle -\tan^{-1}\left(\frac{657.8}{1317.9}\right)$$

$$= 1472.9 \angle -26.53° \ \Omega$$

(b) The total current can be found with Ohm's law:

$$\mathbf{I}_T = \frac{\mathbf{V}_s}{\mathbf{Z}_T} = \frac{10 \angle 0° \ \text{V}}{1472.9 \angle -26.53° \ \Omega}$$

$$= 6.79 \angle 26.53° \ \text{mA}$$

(c) The total current leads the applied voltage by 26.53 degrees.

**EXAMPLE
16–11**

Determine all currents in Figure 16–30. Sketch a current phasor diagram.

FIGURE 16–30

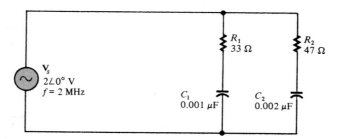

Solution:
First, calculate X_{C1} and X_{C2}:

$$X_{C1} = \frac{1}{2\pi(2 \ \text{MHz})(0.001 \ \mu\text{F})} = 79.58 \ \Omega$$

$$X_{C2} = \frac{1}{2\pi(2 \ \text{MHz})(0.002 \ \mu\text{F})} = 39.79 \ \Omega$$

Now, we determine the impedance of each of the two parallel branches:

$$\mathbf{Z}_1 = R_1 - jX_{C1} = 33 \ \Omega - j79.58 \ \Omega$$

$$\mathbf{Z}_2 = R_2 - jX_{C2} = 47 \ \Omega - j39.79 \ \Omega$$

Convert these to polar form:

$$\mathbf{Z}_1 = \sqrt{(33 \ \Omega)^2 + (79.58 \ \Omega)^2} \angle -\tan^{-1}\left(\frac{79.58}{33}\right) = 86.15 \angle -67.48° \ \Omega$$

$$\mathbf{Z}_2 = \sqrt{(47 \ \Omega)^2 + (39.78 \ \Omega)^2} \angle -\tan^{-1}\left(\frac{39.79}{47}\right) = 61.57 \angle -40.25° \ \Omega$$

Calculate each branch current:

$$\mathbf{I}_1 = \frac{\mathbf{V}_s}{\mathbf{Z}_1} = \frac{2\angle 0° \text{ V}}{86.15\angle -67.48° \ \Omega} = 23.22\angle 67.48° \text{ mA}$$

$$\mathbf{I}_2 = \frac{\mathbf{V}_s}{\mathbf{Z}_2} = \frac{2\angle 0° \text{ V}}{61.57\angle -40.25° \ \Omega} = 32.48\angle 40.25° \text{ mA}$$

To get the total current, we must express each branch current in rectangular form so that they can be added:

$$\mathbf{I}_1 = 23.22 \cos(67.48°) + j23.22 \sin(67.48°)$$

$$= 8.89 \text{ mA} + j21.45 \text{ mA}$$

$$\mathbf{I}_2 = 32.48 \cos(40.25°) + j32.48 \sin(40.25°)$$

$$= 24.79 \text{ mA} + j20.99 \text{ mA}$$

$$\mathbf{I}_T = \mathbf{I}_1 + \mathbf{I}_2$$

$$= (8.89 \text{ mA} + j21.45 \text{ mA}) + (24.79 \text{ mA} + j20.99 \text{ mA})$$

$$= 33.68 \text{ mA} + j42.44 \text{ mA}$$

Converting to polar form:

$$\mathbf{I}_T = \sqrt{(33.68 \text{ mA})^2 + (42.44 \text{ mA})^2}\angle \tan^{-1}\left(\frac{42.44}{33.68}\right)$$

$$= 54.18\angle 51.56° \text{ mA}$$

The current phasor diagram is shown in Figure 16–31.

FIGURE 16–31

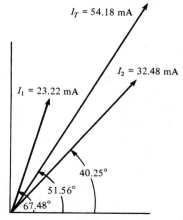

The two previous examples should give you a feel for how to approach the analysis of complex *RC* networks. Further problem work will sharpen your skills.

MEASUREMENT OF Z_T AND θ

Now, let's see how the values of Z_T and θ for the circuit in Example 16–10 can be determined by measurement. First, the total impedance is measured as outlined in the following steps and as illustrated in Figure 16–32 (other ways are also possible):

1. Using a sine wave generator, set the source voltage to a known value (10 V) and the frequency to 5 kHz. It is advisable to check the voltage with an ac voltmeter and the frequency with a frequency counter rather than relying on the marked values on the generator controls.
2. Connect an ac ammeter as shown in the figure, and measure the total current.
3. Calculate the total impedance by using Ohm's law.

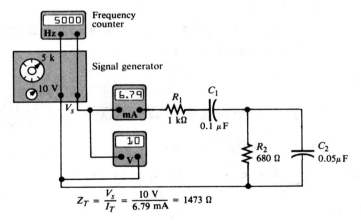

$$Z_T = \frac{V_s}{I_T} = \frac{10\text{ V}}{6.79\text{ mA}} = 1473\ \Omega$$

FIGURE 16–32
Determining Z_T by measurement of V_s and I_T.

Although we could use a phase meter to measure the phase angle, we will use an oscilloscope in this illustration because it is more commonly available. To measure the phase angle, we must have the source voltage and the total current displayed on the screen in the proper time relationship. Two basic types of scope probes are available to measure the quantities with an oscilloscope: the voltage probe and the current probe. Although the current probe is a convenient device, it is often not as readily available as a voltage probe. For this reason, we will confine our phase measurement technique to the use of voltage probes in conjunction with the oscilloscope. A typical oscilloscope voltage probe has two points that are connected to the circuit: the probe tip and the ground lead. Thus, all voltage measurements must be referenced to ground.

Since only voltage probes are to be used, the total current cannot be measured directly. However, for phase measurement, the voltage across R_1 is in phase with the total current and can be used to establish the phase angle. In setting up this circuit, we take the lower side of the source as circuit ground, as shown in Figure 16–33(a).

Before proceeding with the actual phase measurement, note that there is a problem with displaying V_{R1}. If the scope probe is connected across the resistor, as indicated in Figure 16–33(b), the ground lead of the scope will short point B to ground, thus

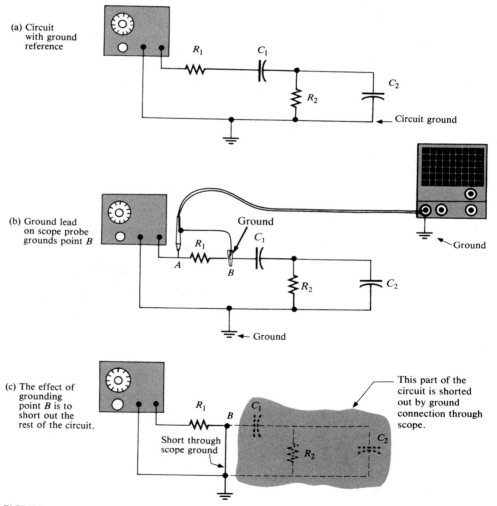

(a) Circuit with ground reference

(b) Ground lead on scope probe grounds point B

(c) The effect of grounding point B is to short out the rest of the circuit.

FIGURE 16–33
Effects of measuring directly across a component when the instrument and the circuit are grounded.

bypassing the rest of the components and effectively removing them from the circuit electrically, as illustrated in Part (c) (assuming that the scope is not isolated from power line ground).

To avoid this problem, we can reposition R_1 in the circuit (when possible) so that one end of it is connected to ground, as shown in Figure 16–34(a). This connection does not alter the circuit electrically, because R_1 still has the same series relationship with the rest of the circuit. Now the scope can be connected across it to display V_{R1}, as indicated in Part (b) of the figure. The other probe is connected across the voltage source to display V_s as indicated. Now channel 1 of the scope has V_s as an input, and channel 2 has V_{R1}. The *trigger source* switch on the scope should be on internal so that each trace on the screen will be triggered by one of the inputs and the other will then be shown in the proper time relationship to it. Since amplitudes are not important, the

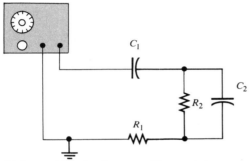

(a) R_1 repositioned so that one end is grounded

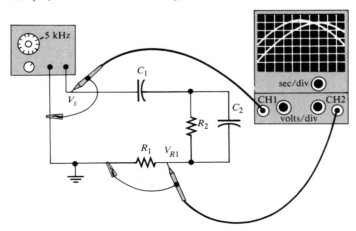

(b) The scope displays a half-cycle of V_{R1} and V_s.
V_{R1} represents the phase of the total current.

FIGURE 16–34
Repositioning R_1 so that a direct voltage measurement can be made with respect to ground.

volts/div settings are arbitrary. The *sec/div* settings should be adjusted so that one half-cycle of the waveforms appears on the screen.

Before connecting the probes to the circuit, we must align the two horizontal lines (traces) so that they appear as a single line across the center of the screen. To do so, ground the probe tips and adjust the *vertical position* knobs to move the traces toward the center line of the screen until they are superimposed. The reason for this procedure is to ensure that both waveforms have the same zero crossing so that an accurate phase measurement can be made.

The resulting oscilloscope display is shown in Figure 16–35. Since there are 180 degrees in one half-cycle, each of the ten horizontal divisions across the screen represents 18 degrees. Thus, the horizontal distance between the corresponding points of the two waveforms is the phase angle in degrees as indicated.

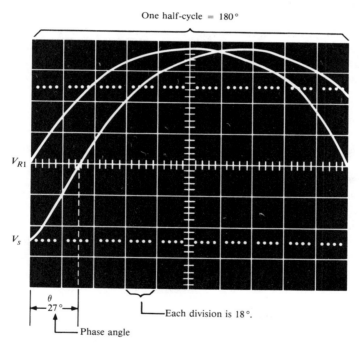

FIGURE 16–35
Measurement of the phase angle on the oscilloscope.

SECTION REVIEW 16–6

1. What is the equivalent series *RC* circuit for the series-parallel circuit in Figure 16–29?

2. What is the total impedance of the circuit in Figure 16–30?

16–7   POWER IN *RC* CIRCUITS

In a purely resistive ac circuit, all of the energy delivered by the source is dissipated in the form of heat by the resistance. In a purely capacitive ac circuit, all of the energy delivered by the source is stored by the capacitor during a portion of the voltage cycle and then returned to the source during another portion of the cycle so that there is no net energy loss. What happens when both resistance and capacitance exist in a circuit? *Some of the energy is alternately stored and returned by the capacitance, and some is dissipated by the resistance.* The amount of energy loss is determined by the relative values of the resistance and the capacitive reactance.

It is reasonable to assume that when the resistance is greater than the reactance, more of the total energy delivered by the source is dissipated by the resistance than is stored by the capacitance. Likewise, when the reactance is greater than the resistance, more of the total energy is stored and returned than is lost.

The power in a resistor, sometimes called *true power* (P_{true}), and the power in a capacitor, called *reactive power* (P_r), were developed in previous chapters and are restated here. The unit of true power is the *watt,* and the unit of reactive power is the VAR (volt-ampere reactive).

$$P_{\text{true}} = I^2R \tag{16–27}$$

$$P_r = I^2X_C \tag{16–28}$$

THE POWER TRIANGLE

The generalized impedance phasor diagram is shown in Figure 16–36(a). A phasor relationship for the powers can also be represented by a similar diagram, because the respective magnitudes of the powers, P_{true} and P_r, differ from R and X_C by a factor of I^2. This is shown in Figure 16–36(b).

The resultant power phasor, I^2Z, represents the *apparent power, P_a.* At any instant in time, P_a is the total power that *appears* to be transferred between the source

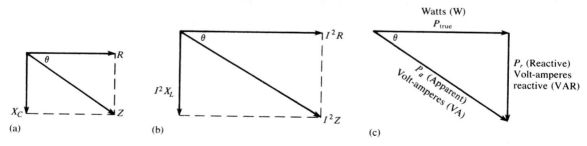

FIGURE 16–36
Development of the power triangle for an *RC* circuit.

and the *RC* circuit. The unit of apparent power is the *volt-ampere*, VA. The expression for apparent power is

$$P_a = I^2 Z \qquad (16\text{–}29)$$

The power phasor diagram in Figure 16–36(b) can be rearranged in the form of a right triangle, as shown in Figure 16–36(c). This is called the *power triangle*. Using the rules of trigonometry, P_{true} can be expressed as

$$P_{true} = P_a \cos \theta \qquad (16\text{–}30)$$

Since P_a equals $I^2 Z$ or VI, the equation for the true power loss in an *RC* circuit can be written as

$$P_{true} = VI \cos \theta \qquad (16\text{–}31)$$

where *V* is the applied voltage and *I* is the total current.

For the case of a purely resistive circuit, $\theta = 0°$ and $\cos 0° = 1$, so P_{true} equals *VI*. For the case of a purely capacitive circuit, $\theta = 90°$ and $\cos 90° = 0$, so P_{true} is zero. As you already know, there is no power loss in an ideal capacitor.

THE POWER FACTOR

The term $\cos \theta$ is called the *power factor* and is stated as follows:

$$PF = \cos \theta \qquad (16\text{–}32)$$

As the phase angle between applied voltage and total current increases, the power factor decreases, indicating an increasingly reactive circuit. The smaller the power factor, the smaller the power dissipation.

The power factor can vary from 0 for a purely reactive circuit to 1 for a purely resistive circuit. In an *RC* circuit, the power factor is referred to as a *leading* power factor because the current leads the voltage.

EXAMPLE 16–12

Determine the power factor and the true power in the circuit of Figure 16–37.

FIGURE 16–37

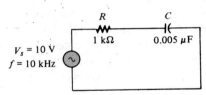

$V_s = 10$ V
$f = 10$ kHz

Solution:

$$X_C = \frac{1}{2\pi fC} = \frac{1}{2\pi (10 \text{ kHz})(0.005 \ \mu\text{F})} = 3183 \ \Omega$$

$$\mathbf{Z} = R - jX_C = 1 \text{ k}\Omega - j3183 \ \Omega$$

$$= \sqrt{(1000 \ \Omega)^2 + (3183 \ \Omega)^2} \angle -\tan^{-1}\left(\frac{3183}{1000}\right) = 3336.4 \angle -72.56°$$

The angle associated with the impedance is θ, the angle between the applied voltage and the total current; therefore

$$PF = \cos \theta = \cos(-72.56°) = 0.2997$$

$$I = \frac{V_s}{Z} = \frac{15 \text{ V}}{3336.4 \ \Omega} = 4.496 \text{ mA}$$

The true power is

$$P_{\text{true}} = V_s I \cos \theta = (15 \text{ V})(4.496 \text{ mA})(0.2997) = 20.21 \text{ mW}$$

THE SIGNIFICANCE OF APPARENT POWER

As mentioned, apparent power is the power that *appears* to be transferred between the source and the load, and it consists of two components: a true power component and a reactive power component.

In all electrical and electronic systems, it is the true power that does the work. The reactive power is simply shuttled back and forth between the source and load. Ideally, in terms of performing useful work, all of the power transferred to the load should be true power and none of it reactive power. However, in most practical situations the load must have some reactance associated with it, and therefore we must deal with both power components.

In the last chapter, the use of apparent power was discussed in relation to transformers. For any reactive load, there are two components of the total current: the resistive component and the reactive component. If we consider only the true power (watts) in a load, we are dealing with only a portion of the total current that the load demands from a source. In order to have a realistic picture of the actual current that a load will draw, we must consider apparent power (in VA).

A source such as an ac generator can provide current to a load up to some maximum value. *If the load draws more than this maximum value, the source can be damaged.* Figure 16–38(a) shows a 120-V generator that can deliver a maximum current of 5 A to a load. Assume that the generator is rated at 600 W and is connected to a purely resistive load of 24 Ω (power factor of 1). The ammeter shows that the current is 5 A, and the wattmeter indicates that the power is 600 W. The generator has no problem under these conditions, although it is operating at maximum current and power.

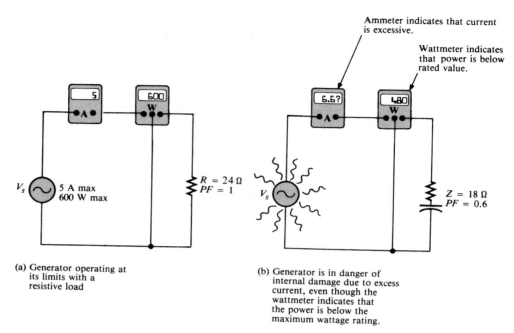

Ammeter indicates that current is excessive.

Wattmeter indicates that power is below rated value.

(a) Generator operating at its limits with a resistive load

(b) Generator is in danger of internal damage due to excess current, even though the wattmeter indicates that the power is below the maximum wattage rating.

FIGURE 16–38

The wattage rating of a source is inappropriate when the load is reactive. The rating should be in VA rather than in watts.

Now, consider what happens if the load is changed to a reactive one with an impedance of 18 Ω and a power factor of 0.6, as indicated in Figure 16–38(b). The current is 120 V/18 Ω = 6.67 A, which *exceeds* the maximum. Even though the wattmeter reads 480 W, which is less than the power rating of the generator, the excessive current probably will cause damage. This example shows that a true power rating can be deceiving and is inappropriate for ac sources. The ac generator should be rated at 600 VA, a rating that manufacturers generally use, rather than 600 W.

**EXAMPLE
16–13**

For the circuit in Figure 16–39, find the true power, the reactive power, and the apparent power. X_C has been determined to be 2 kΩ.

FIGURE 16–39

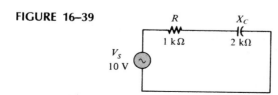

Solution:

We first find the total impedance so that the current can be calculated:

$$\mathbf{Z} = R - jX_C = 1 \text{ k}\Omega - j2 \text{ k}\Omega$$

$$= \sqrt{(1000 \ \Omega)^2 + (2000 \ \Omega)^2} \angle -\tan^{-1}\left(\frac{2000}{1000}\right)$$

$$= 2236 \angle -63.44° \ \Omega$$

$$I = \frac{V_s}{Z} = \frac{10 \text{ V}}{2236 \ \Omega} = 4.47 \text{ mA}$$

The phase angle, θ, is indicated in the polar expression for impedance.

$$\theta = -63.44°$$

The true power is

$$P_{true} = V_s I \cos \theta = (10 \text{ V})(4.47 \text{ mA}) \cos(-63.44°) = 19.99 \text{ mW}$$

(The same result is realized using the formula $P_{true} = I^2 R$.) The reactive power is

$$P_r = I^2 X_C = (4.47 \text{ mA})^2 (2 \text{ k}\Omega) = 39.96 \text{ mVAR}$$

The apparent power is

$$P_a = I^2 Z = (4.47 \text{ mA})^2 (2236 \ \Omega) = 44.68 \text{ mVA}$$

The apparent power is also the phasor sum of P_{true} and P_r.

$$P_a = \sqrt{P_{true}^2 + P_r^2} = 44.68 \text{ mVA}$$

SECTION REVIEW 16–7

1. To which component in an *RC* circuit is the energy loss due?
2. The phase angle, θ, is 45 degrees. What is the power factor?
3. A certain series *RC* circuit has the following parameter values: $R = 330 \ \Omega$, $X_C = 460 \ \Omega$, and $I = 2$ A. Determine the true power, the reactive power, and the apparent power.

16–8   BASIC APPLICATIONS

In this section, some basic applications of *RC* circuits are discussed. These are *phase shift networks, frequency-selective networks (filters),* and *signal coupling*.

THE *RC* LAG NETWORK

The first type of phase shift network that we cover causes the output to lag the input by a specified amount. Figure 16–40(a) shows a series *RC* circuit with the *output voltage taken across the capacitor*. The source voltage is the *input*, V_{in}. As you know, θ, the phase angle between the current and the input voltage, is also the phase angle between the resistor voltage and the input voltage, because V_R and I are in phase with each other.

FIGURE 16–40
RC lag network.

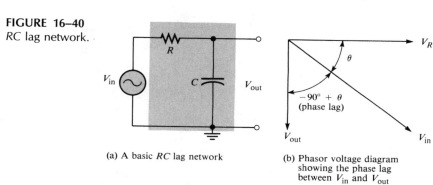

(a) A basic *RC* lag network

(b) Phasor voltage diagram showing the phase lag between V_{in} and V_{out}

Since V_C lags V_R by 90 degrees, the phase angle between the capacitor voltage and the input voltage is the difference between -90 degrees and θ, as shown in Figure 16–40(b). The capacitor voltage is the *output*, and it lags the input, thus creating a basic *lag* network.

When the input and output waveforms of the lag network are displayed on an oscilloscope, a relationship similar to that in Figure 16–41 is observed. The amount of phase difference between the input and the output is dependent on the relative sizes of the capacitive reactance and the resistance, as is the magnitude of the output voltage.

Phase Difference Between Input and Output As already established, θ is the phase angle between I and V_{in}. The angle between V_{out} and V_{in} is designated ϕ (phi) and is developed as follows. The polar expressions for the input voltage and the current are $V_{in}\angle 0°$ and $I\angle\theta$, respectively. The output voltage is

$$\mathbf{V}_{out} = (I\angle\theta)(X_C\angle -90°) = IX_C\angle(-90° + \theta)$$

This shows that the output voltage is at an angle of $-90° + \theta$ with respect to the input voltage. Since $\theta = -\tan^{-1}(X_C/R)$, the angle between the input and output is

$$\phi = -90° + \tan^{-1}\left(\frac{X_C}{R}\right) \tag{16–33}$$

This angle is always negative, indicating that the output voltage *lags* the input voltage, as shown in Figure 16–42.

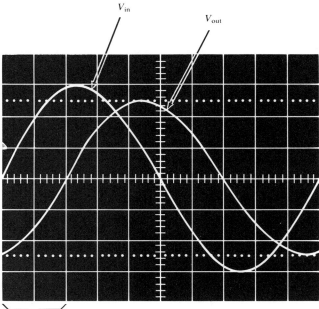

Phase lag
$\phi = -90° + \theta$

FIGURE 16–41
General oscilloscope display of the input and output waveforms of a lag network (V_{out} lags V_{in}).

FIGURE 16–42

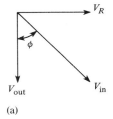

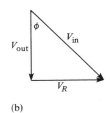

(a)　　　　　　　　(b)

EXAMPLE 16–14

Determine the amount of phase lag from input to output in each lag network in Figure 16–43.

FIGURE 16–43

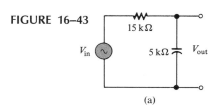

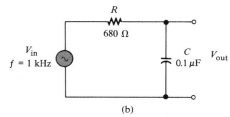

(a)　　　　　　　　(b)

Solution:

(a) $\phi = -90° + \tan^{-1}\left(\dfrac{X_C}{R}\right) = -90° + \tan^{-1}\left(\dfrac{5\ k\Omega}{15\ k\Omega}\right)$

$= -90° + 18.44° = -71.56°$

The output lags the input by 71.56 degrees.

(b) $X_C = \dfrac{1}{2\pi fC} = \dfrac{1}{2\pi\ (1\ \text{kHz})(0.1\ \mu\text{F})} = 1592\ \Omega$

$\phi = -90° + \tan^{-1}\left(\dfrac{X_C}{R}\right) = -90° + \tan^{-1}\left(\dfrac{1592\ \Omega}{680\ \Omega}\right) = -23.13°$

The output lags the input by 23.13 degrees.

Magnitude of the Output Voltage To evaluate the output voltage in terms of its magnitude, visualize the *RC* lag network as a voltage divider. A portion of the total input voltage is dropped across *R* and a portion across *C*. Since the output voltage is V_C, it can be calculated as

$$V_{\text{out}} = \left(\frac{X_C}{\sqrt{R^2 + X_C^2}}\right)V_{\text{in}} \qquad (16\text{–}34)$$

Or it can be calculated using Ohm's law as

$$V_{\text{out}} = IX_C \qquad (16\text{–}35)$$

The total phasor expression for the output voltage of a lag network is

$$\mathbf{V}_{\text{out}} = V_{\text{out}}\angle\phi \qquad (16\text{–}36)$$

EXAMPLE 16–15

For the lag network in Figure 16–43(b), determine the output voltage in phasor form when the input voltage has an rms value of 10 V. Sketch the input and output waveforms showing the proper relationships. ϕ was found in Example 16–14.

Solution:

$$\mathbf{V}_{\text{out}} = \left(\frac{X_C}{\sqrt{R^2 + X_C^2}}\right)V_{\text{in}}\angle\phi$$

$$= \left(\frac{1592\ \Omega}{\sqrt{(680\ \Omega)^2 + (1592\ \Omega)^2}}\right)10\angle -23.13°\ \text{V}$$

$$= 9.2\angle -23.13°\ \text{V rms}$$

The waveforms are shown in Figure 16–44.

FIGURE 16–44

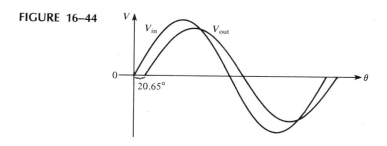

THE *RC* LEAD NETWORK

The second basic type of phase shift network is the *RC* lead network. When the output of a series *RC* circuit is taken across the resistor rather than across the capacitor, as shown in Figure 16–45(a), it becomes a *lead* network. Lead networks cause the phase of the output voltage to *lead* the input by a specified amount.

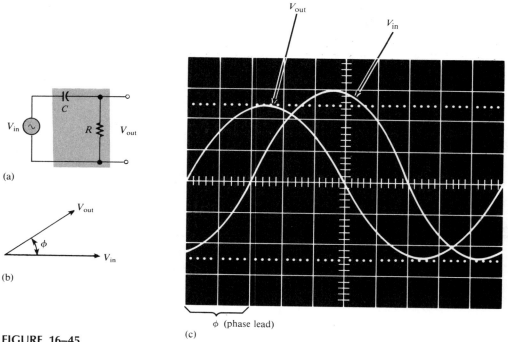

FIGURE 16–45
RC lead network.

Phase Difference Between Input and Output   In a series *RC* circuit, the current leads the input voltage. Also, as you know, the resistor voltage is in phase with the current. Since the output voltage is taken across the resistor, the output *leads* the input, as indicated by the phasor diagram in Figure 16–45(b). A typical oscilloscope display of the waveforms is shown in Part (c).

As in the lag network, the amount of phase difference between the input and output and also the magnitude of the output voltage in the lead network is dependent on the relative values of the resistance and the capacitive reactance. When the input voltage is assigned a reference angle of 0 degrees, the angle of the output voltage is the same as θ (the angle between total current and applied voltage), because the resistor voltage (output) and the current are in phase with each other. Therefore, since $\phi = \theta$ in this case, the expression is

$$\phi = \tan^{-1}\left(\frac{X_C}{R}\right) \tag{16–37}$$

This angle is positive, because the output leads the input. The following example illustrates the computation of phase angles for lead networks.

EXAMPLE 16–16

Calculate the output phase angle for each circuit in Figure 16–46.

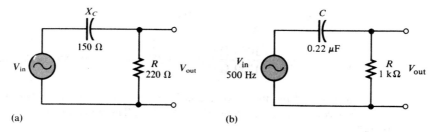

(a)   (b)

FIGURE 16–46

Solution:

(a) $\phi = \tan^{-1}\left(\frac{X_C}{R}\right) = \tan^{-1}\left(\frac{150\ \Omega}{220\ \Omega}\right) = 34.29°$

The output leads the input by 34.29 degrees.

(b) $X_C = \frac{1}{2\pi fC} = \frac{1}{2\pi(500\ \text{Hz})(0.22\ \mu\text{F})} = 1446.86\ \Omega$

$\phi = \tan^{-1}\left(\frac{X_C}{R}\right) = \tan^{-1}\left(\frac{1446.86\ \Omega}{1000\ \Omega}\right) = 55.35°$

The output leads the input by 55.35 degrees.

Magnitude of the Output Voltage   Since the output voltage of an *RC* lead network is taken across the resistor, the magnitude can be calculated using either the voltage divider formula or Ohm's law, stated as

$$V_{out} = \left(\frac{R}{\sqrt{R^2 + X_C^2}}\right)V_{in} \tag{16–38}$$

$$V_{out} = IR \tag{16–39}$$

The expression for the output voltage in phasor form is

$$\mathbf{V}_{out} = V_{out}\angle\phi \tag{16–40}$$

**EXAMPLE
16–17**

The input voltage in Figure 16–46(b) has an rms value of 10 V. Determine the phasor expression for the output voltage. Sketch the waveform relationships for the input and output voltages showing peak values.

Solution:
The phase angle was found to be 55.35 degrees in Example 16–16.

$$\mathbf{V}_{out} = \left(\frac{R}{\sqrt{R^2 + X_C^2}}\right)(V_{in}\angle\phi)$$

$$= \left(\frac{1000\ \Omega}{1759\ \Omega}\right)(10\angle55.35°\ V) = 5.69\angle55.35°\ V\ rms$$

The peak value of the input voltage is 1.414(10 V) = 14.14 V. The peak value of the output voltage is 1.414(5.69 V) = 8.05 V. The waveforms are shown in Figure 16–47.

FIGURE 16–47

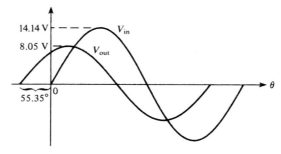

THE *RC* CIRCUIT AS A FILTER

Filters are frequency-selective circuits that permit signals of certain frequencies to pass from the input to the output while blocking all others. That is, all frequencies but the selected ones are *filtered* out. Filters are covered in greater depth in a later chapter, but are introduced here as an application example.

Series *RC* circuits exhibit a frequency-selective characteristic and therefore act as basic filters. There are two types. The first one that we examine, called a *low-pass filter,* is realized by taking the output across the capacitor, just as in a lag network. The second type, called a *high-pass filter,* is implemented by taking the output across the resistor, as in a lead network.

Low-Pass Filter You have already seen what happens to the output magnitude and phase angle in the lag network. In terms of its filtering action, we are interested primarily in the variation of the output magnitude with frequency.

Figure 16–48 shows the filtering action of a series *RC* circuit using specific values for illustration. In Part (a) of the figure, the input is *zero* frequency (dc). Since the capacitor blocks constant direct current, the output voltage equals the full value of the input voltage, because there is no voltage dropped across *R*. Therefore, the circuit passes all of the input voltage to the output (10 V in, 10 V out).

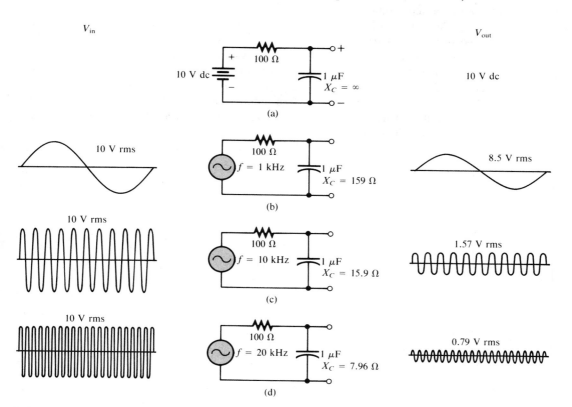

FIGURE 16–48
Low-pass filter action (phase shifts are not indicated).

In Figure 16–48(b), the frequency of the input voltage has been increased to 1 kHz, causing the capacitive reactance to *decrease* to 159 Ω. For an input voltage of 10 V rms, the output voltage is approximately 8.5 V rms, which can be calculated using the voltage divider approach or Ohm's law.

In Figure 16–48(c), the input frequency has been increased to 10 kHz, causing the capacitive reactance to decrease further to 15.9 Ω. For a constant input voltage of 10 V rms, the output voltage is now 1.57 rms.

As the input frequency is increased further, the output voltage continues to decrease and approaches zero as the frequency becomes very high, as shown in Figure 16–48(d). A description of the circuit action is as follows: As the frequency of the input increases, the capacitive reactance decreases. Because the resistance is constant and the capacitive reactance decreases, the voltage across the capacitor (output voltage) also decreases according to the voltage divider principle. The input frequency can be increased until it reaches a value at which the reactance is so small compared to the resistance that the output voltage can be neglected, because it is very small compared to the input voltage. At this value of frequency, the circuit is essentially completely blocking the input signal.

As shown in Figure 16–48, the circuit passes dc (zero frequency) completely. As the frequency of the input increases, less of the input voltage is passed through to the output; that is, the output voltage decreases as the frequency increases. It is apparent that the lower frequencies pass through the circuit much better than the higher frequencies. This *RC* circuit is therefore a very basic form of *low-pass filter*.

Figure 16–49 shows a graph of output voltage magnitude versus frequency for a low-pass filter. This graph, called a *response curve*, indicates that the output decreases as the frequency increases.

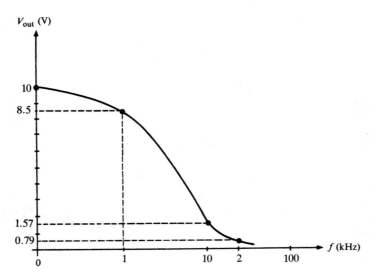

FIGURE 16–49
Frequency response curve for a low-pass filter.

High-Pass Filter Next, refer to Figure 16–50(a), where the output is taken across the resistor, just as in a lead network. When the input voltage is dc (zero frequency), the output is zero volts, because the capacitor blocks direct current; therefore no voltage is developed across R.

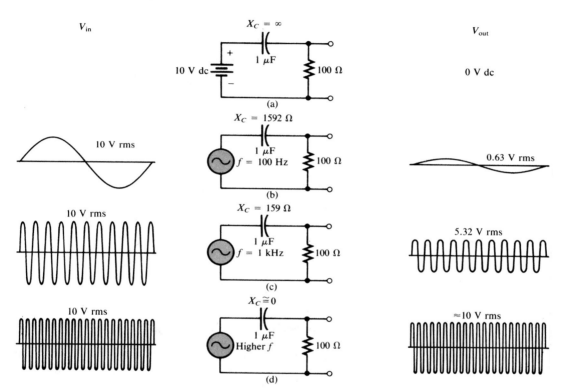

FIGURE 16–50
High-pass filter action (phase shifts are not indicated).

In Figure 16–50(b), the frequency of the input signal has been increased to 100 Hz with an rms value of 10 V. The output voltage is 0.63 V rms. Thus, only a small percentage of the input voltage appears on the output at this frequency.

In Figure 16–50(c), the input frequency is increased further to 1 kHz, causing more voltage to be developed across the resistor because of the further decrease in the capacitive reactance. The output voltage at this frequency is 5.32 V rms. As you can see, the output voltage increases as the frequency increases. A value of frequency is reached at which the reactance is negligible compared to the resistance, and most of the input voltage appears across the resistor, as shown in Figure 16–50(d).

As illustrated, this circuit tends to prevent lower frequencies from appearing on the output but allows higher frequencies to pass through from input to output. Therefore, this RC circuit is a very basic form of *high-pass filter*.

Figure 16–51 shows a plot of output voltage magnitude versus frequency for a high-pass filter. This is a response curve; it shows that the output increases as the frequency increases and then levels off, approaching the value of the input voltage.

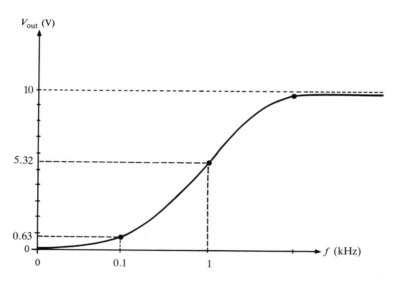

FIGURE 16–51
Frequency response curve for a high-pass filter.

COUPLING AN ac SIGNAL INTO A dc BIAS NETWORK

Figure 16–52 shows an *RC* network that is used to create a dc voltage level with an ac voltage superimposed on it. This type of circuit is commonly found in amplifiers in which the dc voltage is required to *bias* the amplifier to the proper operating point, and the signal voltage to be amplified is coupled through a capacitor and superimposed on the dc level. The capacitor prevents the low internal resistance of the signal source from affecting the dc bias voltage.

FIGURE 16–52
Amplifier bias and signal-coupling circuit.

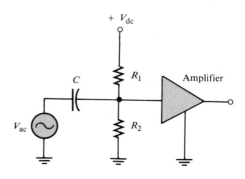

In this type of application, a relatively high value of capacitance is selected so that for the frequencies to be amplified, the reactance is very small compared to the resistance of the bias network. When the reactance is very small (ideally zero), there is practically no phase shift or signal voltage dropped across the capacitor. Therefore, all of the signal voltage passes from the source to the input to the amplifier.

Figure 16–53 illustrates the application of the superposition principle to circuits. In Part (a), we have effectively removed the ac source from the circuit by replacing it with a short to represent its ideal internal resistance. Since C is open to dc, the voltage at point A is determined by the voltage divider action of R_1 and R_2 and the dc voltage source.

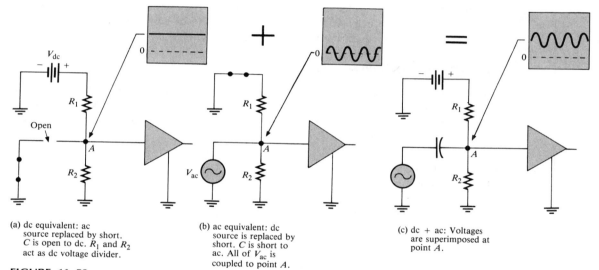

(a) dc equivalent: ac source replaced by short. C is open to dc. R_1 and R_2 act as dc voltage divider.

(b) ac equivalent: dc source is replaced by short. C is short to ac. All of V_{ac} is coupled to point A.

(c) dc + ac: Voltages are superimposed at point A.

FIGURE 16–53
The superposition of dc and ac voltages in an RC bias and coupling circuit.

In Part (b), we have effectively removed the dc source from the circuit by replacing it with a short to represent its ideal internal resistance. Since C appears as a short at the frequency of the ac, the signal voltage is coupled directly to point A and appears across the parallel combination of R_1 and R_2. Part (c) illustrates that the combined effect of the superposition of the dc and the ac voltages results in the signal voltage "riding" on the dc level.

SECTION REVIEW 16–8

1. A certain RC lag network consists of a 4.7-kΩ resistor and a 0.022-μF capacitor. Determine the phase shift between input and output at a frequency of 3 kHz.

2. An *RC* lead network has the same component values as the lag network in Question 1. What is the magnitude of the output voltage at 3 kHz when the input is 10 V rms?

3. When an *RC* circuit is used as a low-pass filter, across which component is the output taken?

4. In a coupling network, is the criterion for effective operation $X_C \ll R$, $X_C = R$, or $X_C \gg R$?

16–9   TROUBLESHOOTING *RC* CIRCUITS

In this section, we consider the effects that typical component failures or degradation have on the response of basic *RC* circuits. The cases to be considered are open resistor, open capacitor, shorted capacitor, and excessive leakage in a capacitor.

EFFECTS OF AN OPEN RESISTOR

It is very easy to see how an open resistor affects the operation of a basic series *RC* circuit, as shown in Figure 16–54. Obviously, there is no path for current, so the capacitor voltage remains at zero; thus, the total voltage, V_s, appears across the open resistor.

FIGURE 16–54
Effect of an open resistor.

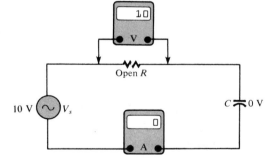

EFFECTS OF AN OPEN CAPACITOR

When the capacitor is open, there is no current; thus, the resistor voltage remains at zero. The total source voltage is across the open capacitor, as shown in Figure 16–55.

FIGURE 16–55
Effect of an open capacitor.

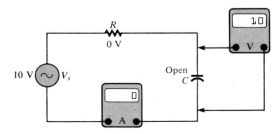

EFFECTS OF A SHORTED CAPACITOR

When a capacitor shorts out, the voltage across it is zero, the current equals V_s/R, and the total voltage appears across the resistor, as shown in Figure 16–56.

FIGURE 16–56
Effect of a shorted capacitor.

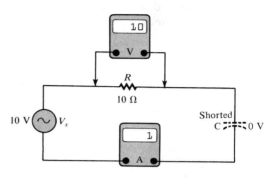

EFFECTS OF A LEAKY CAPACITOR

When a capacitor exhibits a high leakage current, the leakage resistance effectively appears in parallel with the capacitor, as shown in Figure 16–57(a). When the leakage

FIGURE 16–57
Effects of a leaky capacitor.

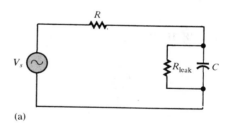

(a)

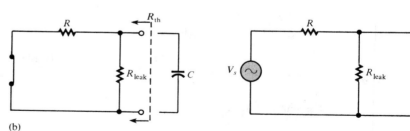

(b)

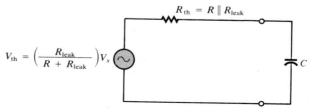

$$V_{th} = \left(\frac{R_{leak}}{R + R_{leak}} \right) V_s$$

(c)

resistance is comparable in value to the circuit resistance, R, the circuit response is drastically affected. The circuit looking from the capacitor toward the source can be Thevenized, as shown in Figure 16–57(b). The Thevenin equivalent resistance is R in parallel with R_{leak} (the source appears as a short), and the Thevenin equivalent voltage is determined by the voltage divider action of R and R_{leak}. The Thevenin equivalent circuit is shown in Figure 16–57(c).

$$R_{\text{th}} = R \| R_{\text{leak}}$$

$$R_{\text{th}} = \frac{RR_{\text{leak}}}{R + R_{\text{leak}}} \tag{16–41}$$

$$V_{\text{th}} = \frac{R_{\text{leak}}V_{\text{in}}}{R + R_{\text{leak}}} \tag{16–42}$$

As you can see, the voltage to which the capacitor will charge is reduced since $V_{\text{th}} < V_s$. Also, the circuit time constant is reduced, and the current is increased.

EXAMPLE 16–18

Assume that the capacitor in Figure 16–58 is degraded to a point where its leakage resistance is 10 kΩ. Determine the phase shift from input to output and the output voltage under the degraded condition.

FIGURE 16–58

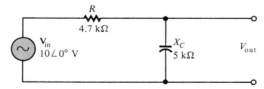

Solution:
The effective circuit resistance is

$$R_{\text{th}} = \frac{RR_{\text{leak}}}{R + R_{\text{leak}}} = \frac{(4.7\ \text{k}\Omega)(10\ \text{k}\Omega)}{14.7\ \text{k}\Omega} = 3.2\ \text{k}\Omega$$

The phase shift is

$$\phi = -90° + \tan^{-1}\left(\frac{X_C}{R_{\text{th}}}\right) = -90° + \tan^{-1}\left(\frac{5\ \text{k}\Omega}{3.2\ \text{k}\Omega}\right) = -32.62°$$

$$V_{th} = \left(\frac{R_{leak}}{R + R_{leak}}\right)V_{in} = \left(\frac{10 \text{ k}\Omega}{14.7 \text{ k}\Omega}\right)10 \text{ V} = 6.8 \text{ V}$$

$$V_{out} = \left(\frac{X_C}{\sqrt{R_{th}^2 + X_C^2}}\right)(V_{th})$$

$$= \left(\frac{5 \text{ k}\Omega}{\sqrt{(3.2 \text{ k}\Omega)^2 + (5 \text{ k}\Omega)^2}}\right)(6.8 \text{ V}) = 5.73 \text{ V}$$

SECTION REVIEW 16–9

1. Describe the effect of a leaky capacitor on the response of an *RC* circuit.

2. In a series *RC* circuit, all of the applied voltage appears across an open capacitor (T or F).

16–10 COMPUTER ANALYSIS

The program listed here provides for the computation of phase shift and normalized output voltage as functions of frequency for an *RC* lag network. Inputs required are the component values, the upper and lower frequency limits, and the frequency steps. A flowchart is shown in Figure 16–59.

```
10   CLS
20   PRINT "THIS PROGRAM COMPUTES THE PHASE SHIFT FROM INPUT
     TO"
30   PRINT "OUTPUT AND THE NORMALIZED OUTPUT VOLTAGE MAGNITUDE"
40   PRINT "AS FUNCTIONS OF FREQUENCY FOR AN RC LAG NETWORK."
50   PRINT:PRINT:PRINT
60   INPUT "TO CONTINUE PRESS 'ENTER'";X:CLS
70   INPUT "THE VALUE OF R IN OHMS";R
80   INPUT "THE VALUE OF C IN FARADS";C
90   INPUT "THE LOWEST NONZERO FREQUENCY IN HERTZ";FL
100  INPUT "THE HIGHEST FREQUENCY IN HERTZ";FH
110  INPUT "THE FREQUENCY INCREMENTS IN HERTZ";FI
120  CLS
130  PRINT "FREQUENCY(HZ)","PHASE SHIFT","VOUT"
140  FOR F=FL TO FH STEP FI
150  XC = 1/(2*3.1416*F*C)
160  PHI=-90+ATN(XC/R)*57.3
170  VO=XC/(SQR(R*R+XC*XC))
180  PRINT F,PHI,VO
190  NEXT
```

FIGURE 16–59

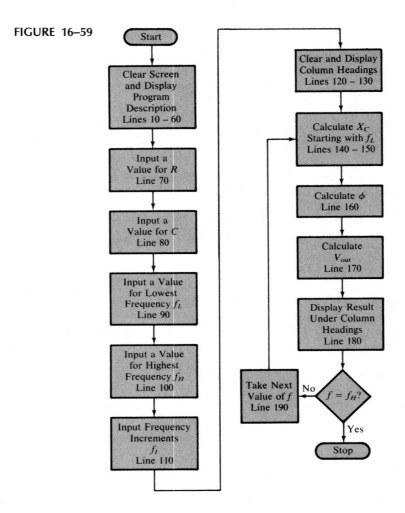

SECTION REVIEW 16–10

1. What is the purpose of the "STEP FI" portion of line 140?

2. Explain each line in the FOR/NEXT loop.

SUMMARY

1. A sinusoidal voltage applied to an *RC* circuit produces a sinusoidal current.

2. Current leads voltage in an *RC* circuit.

3. Impedance is the total opposition to sinusoidal current.

4. The unit of impedance is the ohm.

5. In a series *RC* circuit, the total impedance is the phasor sum of the resistance and the capacitive reactance.

6. The resistor voltage is always in phase with the current.

7. The capacitor voltage always lags the current by 90 degrees.

8. The phase angle between applied voltage and current is dependent on the relative values of R and X_C.

9. When $X_C = R$, the phase angle is 45 degrees.

10. The notation $\tan^{-1}$ for inverse tangent means "the angle whose tangent is."

11. The true power in a capacitor is zero.

12. When a sinusoidal source drives an *RC* circuit, part of the total power delivered by the source is resistive power (true power), and part of it is reactive power.

13. Reactive power is the rate at which energy is stored by a capacitor.

14. The total power being transferred between the source and the circuit is the combination of true power and reactive power and is called the apparent power; its unit is the volt-ampere (VA).

15. The power factor is the cosine of the phase angle.

16. In an *RC* lag network, the output voltage is across the capacitor and lags the input voltage.

17. In an *RC* lead network, the output voltage is across the resistor and leads the input voltage.

18. A low-pass filter passes lower frequencies and rejects higher frequencies.

19. A high-pass filter passes higher frequencies and rejects lower frequencies.

FORMULAS

Series *RC* Circuits

$$\mathbf{X}_C = -jX_C \qquad (16\text{--}1)$$

$$\mathbf{Z} = R - jX_C \qquad (16\text{--}2)$$

$$Z = \sqrt{R^2 + X_C^2} \tag{16-3}$$

$$\theta = -\tan^{-1}\left(\frac{X_C}{R}\right) \tag{16-4}$$

$$\mathbf{Z} = \sqrt{R^2 + X_C^2} \angle -\tan^{-1}\left(\frac{X_C}{R}\right) \tag{16-5}$$

$$\mathbf{V} = \mathbf{IZ} \tag{16-6}$$

$$\mathbf{I} = \frac{\mathbf{V}}{\mathbf{Z}} \tag{16-7}$$

$$\mathbf{Z} = \frac{\mathbf{V}}{\mathbf{I}} \tag{16-8}$$

$$\mathbf{V}_s = V_R - jV_C \tag{16-9}$$

$$\mathbf{V}_s = \sqrt{V_R^2 + V_C^2} \angle -\tan^{-1}\left(\frac{V_C}{V_R}\right) \tag{16-10}$$

$$V_s = \sqrt{V_R^2 + V_C^2} \tag{16-11}$$

$$\theta = -\tan^{-1}\left(\frac{V_C}{V_R}\right) \tag{16-12}$$

Parallel *RC* Circuits

$$\mathbf{Z} = \left(\frac{RX_C}{\sqrt{R^2 + X_C^2}}\right) \angle \left(-90° + \tan^{-1}\left(\frac{X_C}{R}\right)\right) \tag{16-13}$$

$$Z = \frac{RX_C}{\sqrt{R^2 + X_C^2}} \tag{16-14}$$

$$\theta = -90° + \tan^{-1}\left(\frac{X_C}{R}\right) \tag{16-15}$$

$$\mathbf{G} = \frac{1}{R\angle 0°} = G\angle 0° \tag{16-16}$$

$$\mathbf{B}_C = \frac{1}{X_C\angle -90°} = B_C\angle 90° = +jB_C \tag{16-17}$$

$$\mathbf{Y} = \frac{1}{Z\angle \pm\theta} = Y\angle \mp\theta \tag{16-18}$$

$$\mathbf{Y} = G + jB_C \tag{16-19}$$

$$\mathbf{V} = \frac{\mathbf{I}}{\mathbf{Y}} \tag{16-20}$$

$$\mathbf{I} = \mathbf{VY} \tag{16-21}$$

$$\mathbf{Y} = \frac{\mathbf{I}}{\mathbf{V}} \tag{16–22}$$

$$\mathbf{I}_T = I_R + jI_C \tag{16–23}$$

$$\mathbf{I}_T = \sqrt{I_R^2 + I_C^2}\angle\tan^{-1}\left(\frac{I_C}{I_R}\right) \tag{16–24}$$

$$I_T = \sqrt{I_R^2 + I_C^2} \tag{16–25}$$

$$\theta = \tan^{-1}\left(\frac{I_C}{I_R}\right) \tag{16–26}$$

Power in *RC* Circuits

$$P_{\text{true}} = I^2R \tag{16–27}$$

$$P_r = I^2X_C \tag{16–28}$$

$$P_a = I^2Z \tag{16–29}$$

$$P_{\text{true}} = P_a\cos\theta \tag{16–30}$$

$$P_{\text{true}} = VI\cos\theta \tag{16–31}$$

$$PF = \cos\theta \tag{16–32}$$

Lag Network

$$\phi = -90° + \tan^{-1}\left(\frac{X_C}{R}\right) \tag{16–33}$$

$$V_{\text{out}} = \left(\frac{X_C}{\sqrt{R^2 + X_C^2}}\right)V_{\text{in}} \tag{16–34}$$

$$V_{\text{out}} = IX_C \tag{16–35}$$

$$\mathbf{V}_{\text{out}} = V_{\text{out}}\angle\phi \tag{16–36}$$

Lead Network

$$\phi = \tan^{-1}\left(\frac{X_C}{R}\right) \tag{16–37}$$

$$V_{\text{out}} = \left(\frac{R}{\sqrt{R^2 + X_C^2}}\right)V_{\text{in}} \tag{16–38}$$

$$V_{\text{out}} = IR \tag{16–39}$$

$$\mathbf{V}_{\text{out}} = V_{\text{out}}\angle\phi \tag{16–40}$$

Troubleshooting

$$R_{th} = \frac{RR_{leak}}{R + R_{leak}} \tag{16-41}$$

$$V_{th} = \frac{R_{leak}V_{in}}{R + R_{leak}} \tag{16-42}$$

SELF-TEST

Solutions appear at the end of the book.

1. In a series *RC* circuit, explain how each voltage drop differs from the applied voltage.

2. Describe the phase relationships between the current in a series *RC* circuit and **(a)** the resistor voltage; **(b)** the capacitor voltage.

3. If the frequency of the voltage applied to an *RC* circuit is increased, what happens to the impedance? To the phase angle?

4. If the frequency is doubled and the resistance is doubled, how much does the value of the impedance change?

5. To reduce the current in a series *RC* circuit, should you increase or decrease the frequency?

6. In a series *RC* circuit, 10 V rms is measured across the resistor and 10 V rms across the capacitor. Is the rms source voltage
   (a) 20 V,   **(b)** 14.14 V,   **(c)** 28.28 V, or   **(d)** 10 V?

7. The voltages in Question 6 are measured at a certain frequency. To make the resistor voltage greater than the capacitor voltage, must the frequency be increased or reduced?

8. When the resistor voltage in Question 7 becomes greater than the capacitor voltage, does the phase angle increase or decrease?

9. Does the impedance of a parallel *RC* circuit increase or decrease with frequency?

10. In a parallel *RC* circuit, there is 1 A rms through the resistive branch and 1 A rms through the capacitive branch. Is the total rms current
    (a) 1 A,   **(b)** 2 A,   **(c)** 1.414 A, or   **(d)** 2.28 A?

11. What does a power factor close to 1 indicate?

12. If a load is purely resistive and the true power is 5 W, what is the apparent power?

13. For a certain load, the true power is 100 W and the reactive power is 100 VAR. Is the apparent power
    (a) 200 VA,   **(b)** 100 VA, or   **(c)** 141.4 VA?

14. Why is it important to rate sources according to volt-amperes rather than watts?

15. In a certain series RC circuit, $R = 2.7 \text{ k}\Omega$ and $C = 0.005 \ \mu\text{F}$. Express the total impedance in rectangular form when $f = 5 \text{ kHz}$. In polar form.

16. In Question 15, if the frequency is doubled, what is the magnitude of the total impedance? What is the phase angle?

17. In Figure 16–60, at what frequency does a 45-degree phase angle between the source voltage and the current occur?

18. In a given series RC circuit, the capacitance is $0.01 \ \mu\text{F}$. At 5 kHz, what value of resistance will produce a 60-degree phase angle?

19. In the circuit of Figure 16–60, what is the phase difference between the source voltage and the resistor voltage for $f = 100 \text{ kHz}$?

20. In a given series RC circuit, $R = 560 \ \Omega$ and $X_C = 1 \text{ k}\Omega$. What is the phase angle?

21. A series RC circuit has a resistance that is twice the capacitive reactance at a given frequency. By how many degrees does the voltage lag the current?

22. Determine the impedance in rectangular form for each circuit in Figure 16–61.

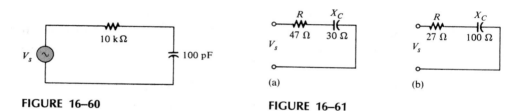

FIGURE 16–60

FIGURE 16–61

23. Determine the current and express it in polar form for each circuit in Figure 16–61, when $\mathbf{V}_s = 10\angle 0° \text{ V}$.

24. In a certain series RC circuit, the resistor voltage is 6 V, and the capacitor voltage is 4.5 V. What is the magnitude of the total voltage applied to the circuit?

25. Determine the impedance and express it in polar form for each circuit in Figure 16–62.

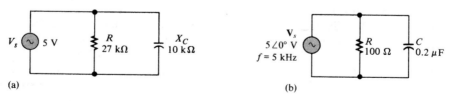

FIGURE 16–62

26. Determine the total current and each branch current in Figure 16–62 when $\mathbf{V}_s = 5\angle 0° \text{ V}$.

27. What is the phase angle between the applied voltage and the total current in Figure 16–62?

28. For the lag network in Figure 16–63, determine the angle by which the output voltage lags the input voltage (phase shift).

29. What is the magnitude of the output voltage in Figure 16–63?

30. Determine the phase shift from input to output for the lead network in Figure 16–64.

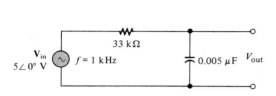

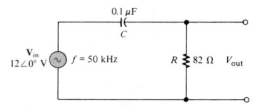

FIGURE 16–63   **FIGURE 16–64**

31. What is the magnitude of V_{out} in Figure 16–64?

32. For Figure 16–63, is the output voltage greater at 2 kHz or at 200 Hz?

33. For Figure 16–64, is the output voltage greater at 5 kHz or at 10 Hz?

34. Calculate the true power, the reactive power, and the apparent power for the circuits of Figures 16–63 and 16–64.

35. Calculate the power factors in Question 34.

PROBLEMS

Section 16–1

16–1   An 8-kHz sinusoidal voltage is applied to a series *RC* circuit. What is the frequency of the voltage across the resistor? The capacitor?

16–2   What is the wave shape of the current in the circuit of Problem 16–1?

Section 16–2

16–3   Express the total impedance of each circuit in Figure 16–65 in both polar and rectangular forms.

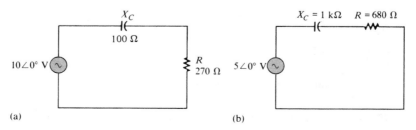

(a)   (b)

FIGURE 16–65

16–4 Determine the impedance magnitude and phase angle in each circuit in Figure 16–66.

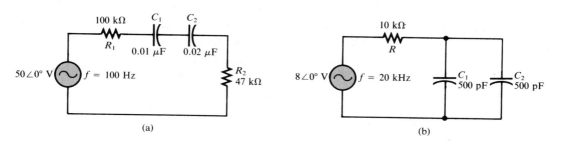

(a)

(b)

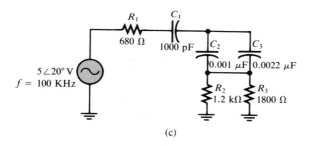

(c)

FIGURE 16–66

16–5 For the circuit of Figure 16–67, determine the impedance expressed in rectangular form for each of the following frequencies:
(a) 100 Hz (b) 500 Hz (c) 1 kHz (d) 2.5 kHz

FIGURE 16–67

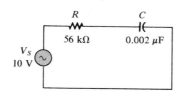

16–6 Repeat Problem 16–5 for $C = 0.005 \ \mu F$.

16–7 Determine the values of R and X_C in a series RC circuit for the following values of total impedance:
(a) $\mathbf{Z} = 33 \ \Omega - j50 \ \Omega$ (b) $\mathbf{Z} = 300\angle - 25° \ \Omega$
(c) $\mathbf{Z} = 1.8\angle - 67.2° \ k\Omega$ (d) $\mathbf{Z} = 789\angle - 45° \ \Omega$

Section 16–3

16–8 Express the current in polar form for each circuit of Figure 16–65.

16–9 Calculate the total current in each circuit of Figure 16–66, and express in polar form.

16–10 Determine the phase angle between the applied voltage and the current for each circuit in Figure 16–66.

16–11 Repeat Problem 16–10 for the circuit in Figure 16–67, using $f = 5$ kHz.

16–12 For the circuit in Figure 16–68, draw the phasor diagram showing all voltages and the total current. Indicate the phase angles.

16–13 For the circuit in Figure 16–69, determine the following in polar form:
(a) Z   **(b) I$_T$**   **(c) V$_R$**   **(d) V$_C$**

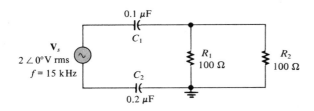

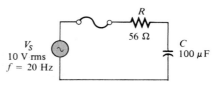

FIGURE 16–68   **FIGURE 16–69**

16–14 To what value must the rheostat be set in Figure 16–70 to make the total current 10 mA? What is the resulting phase angle?

16–15 Determine the series element or elements that must be installed in the block of Figure 16–71 to meet the following requirements:
(a) $P_{true} = 400$ W   **(b)** Leading power factor (I_T leads V_s)

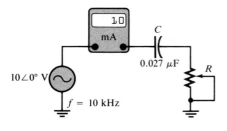

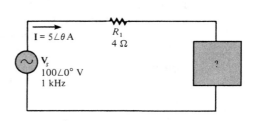

FIGURE 16–70   **FIGURE 16–71**

Section 16–4

16–16 Determine the impedance and express it in polar form for the circuit in Figure 16–72.

FIGURE 16–72

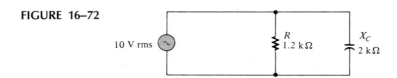

16–17 Determine the impedance magnitude and phase angle in Figure 16–73.

FIGURE 16–73

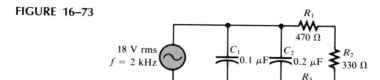

16–18 Repeat Problem 16–17 for the following frequencies:
(a) 1.5 kHz   (b) 3 kHz   (c) 5 kHz   (d) 10 kHz

Section 16–5
16–19 For the circuit in Figure 16–74, find all the currents and voltages in polar form.

16–20 For the parallel circuit in Figure 16–75, find the magnitude of each branch current and the total current. What is the phase angle between the applied voltage and the total current?

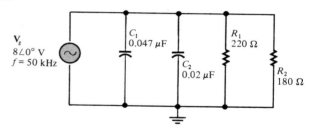

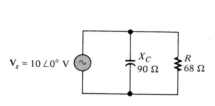

FIGURE 16–74

FIGURE 16–75

16–21 For the circuit in Figure 16–76, determine the following:
(a) Z (b) I_R (c) I_{CT} (d) I_T (e) θ

16–22 Repeat Problem 16–21 for $R = 5.6$ kΩ, $C_1 = 0.05$ μF, $C_2 = 0.022$ μF, and $f = 500$ Hz.

16–23 Convert the circuit in Figure 16–77 to an equivalent series form.

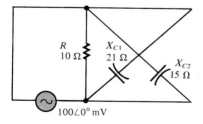

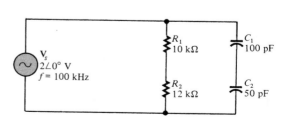

FIGURE 16–76

FIGURE 16–77

16–24 Determine the value to which R_1 must be adjusted to get a phase angle of 30 degrees between the source voltage and the total current in Figure 16–78.

FIGURE 16–78

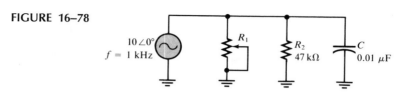

Section 16–6

16–25 Determine the voltages in polar form across each element in Figure 16–79. Sketch the voltage phasor diagram.

FIGURE 16–79

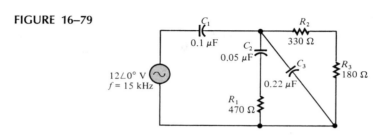

16–26 Is the circuit in Figure 16–79 predominantly resistive or predominantly capacitive?

16–27 Find the current through each branch and the total current in Figure 16–79. Express the currents in polar form. Sketch the current phasor diagram.

16–28 For the circuit in Figure 16–80, determine the following:
(a) I_T (b) θ (c) V_{R1} (d) V_{R2} (e) V_{R3} (f) V_C

16–29 Determine the value of C_2 in Figure 16–81 when $V_A = V_B$.

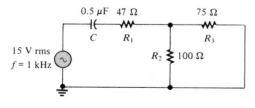

FIGURE 16–80

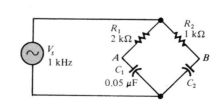

FIGURE 16–81

16–30 Determine the voltage and its phase angle at each point labeled in Figure 16–82.

16–31 Find the current through each component in Figure 16–82.

16–32 Sketch the voltage and current phasor diagram for Figure 16–82.

FIGURE 16–82

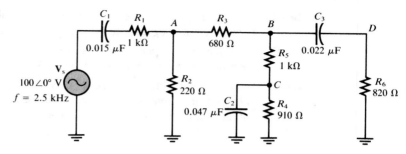

Section 16–7

16–33 In a certain series RC circuit, the true power is 2 W, and the reactive power is 3.5 VAR. Determine the apparent power.

16–34 In Figure 16–69, what is the true power and the reactive power?

16–35 What is the power factor for the circuit of Figure 16–77?

16–36 Determine P_{true}, P_r, P_a, and PF for the circuit in Figure 16–80. Sketch the power triangle.

16–37 A single 240-V, 60-Hz source drives two loads. Load A has an impedance of 50 Ω and a power factor of 0.85. Load B has an impedance of 72 Ω and a power factor of 0.95.
(a) How much current does each load draw?
(b) What is the reactive power in each load?
(c) What is the true power in each load?
(d) What is the apparent power in each load?
(e) Which load has more voltage drop along the lines connecting it to the source?

Section 16–8

16–38 For the lag network in Figure 16–83, determine the phase shift between the input voltage and the output voltage for each of the following frequencies:
(a) 1 Hz   (b) 100 Hz   (c) 1 kHz   (d) 10 kHz

16–39 The lag network in Figure 16–83 also acts as a low-pass filter. Draw a response curve for this circuit by plotting the output voltage versus frequency for 0 Hz to 10 kHz in 1-kHz increments.

16–40 Repeat Problem 16–38 for the lead network in Figure 16–84.

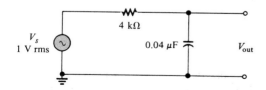

FIGURE 16–83

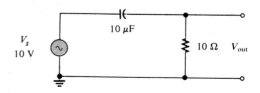

FIGURE 16–84

16–41 Plot the frequency response curve for the lead network in Figure 16–84 for a frequency range of 0 Hz to 10 kHz in 1-kHz increments.

16–42 Draw the voltage phasor diagram for each circuit in Figures 16–83 and 16–84 for a frequency of 5 kHz with $V_s = 1$ V rms.

16–43 What value of coupling capacitor is required in Figure 16–85 so that the signal voltage at the input of amplifier 2 is at least 70.7% of the signal voltage at the output of amplifier 1 when the frequency is 20 Hz?

FIGURE 16–85

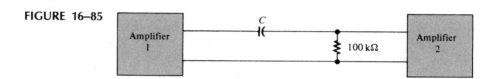

16–44 The rms value of the signal voltage out of amplifier *A* in Figure 16–86 is 50 mV. If the input resistance to amplifier *B* is 10 kΩ, how much of the signal is lost due to the coupling capacitor when the frequency is 3 kHz?

FIGURE 16–86

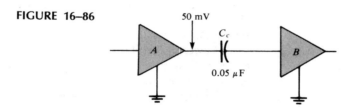

Section 16–9

16–45 Assume that the capacitor in Figure 16–87 is excessively leaky. Show how this degradation affects the output voltage and phase angle, assuming that the leakage resistance is 5 kΩ and the frequency is 10 Hz.

FIGURE 16–87

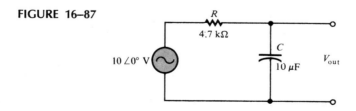

16–46 Each of the capacitors in Figure 16–88 has developed a leakage resistance of 2 kΩ. Determine the output voltages under this condition for each circuit.

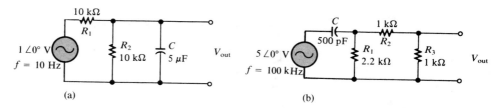

(a)   (b)

FIGURE 16–88

16–47 Determine the output voltage for the circuit in Figure 16–88(a) for each of the following failure modes, and compare it to the correct output:
(a) R_1 open **(b)** R_2 open **(c)** C open **(d)** C shorted

16–48 Determine the output voltage for the circuit in Figure 16–88(b) for each of the following failure modes, and compare it to the correct output:
(a) C open **(b)** C shorted **(c)** R_1 open
(d) R_2 open **(e)** R_3 open

Section 16–10
16–49 Modify the program in Section 16–10 to compute and tabulate current in addition to phase shift and output voltage. Modify the flowchart to reflect this.

16–50 Develop a computer program similar to the program in Section 16–10 for an *RC* lead network.

ANSWERS TO SECTION REVIEWS

Section 16–1
1. 60 Hz, 60 Hz. **2.** Closer to 0 degrees.

Section 16–2
1. $R = 150 \, \Omega, X_C = 220 \, \Omega$. **2.** $Z = 33 \, k\Omega - j50 \, k\Omega$. **3.** 59.9 Ω, $-56.58°$.

Section 16–3
1. 7.2 V. **2.** $-56.3°$. **3.** 90°. **4.** X_C decreases, Z decreases, θ decreases.

Section 16–4
1. Conductance is the reciprocal of resistance, capacitive susceptance is the reciprocal of capacitive reactance, and admittance is the reciprocal of impedance.
2. $Y = 0.01$ S. **3.** $Y = 0.025\angle32.1°$ S. **4.** 0.025 S, 32.1°.

Section 16–5
1. 21 mA. **2.** 18 mA, 56.3°, applied voltage. **3.** 90°.

Section 16–6
1. See Figure 16–89. **21.** $36.91\angle -51.56° \, \Omega$.

FIGURE 16–89

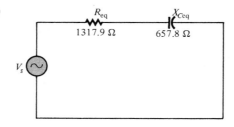

Section 16–7
1. Resistance.   **2.** 0.707.
3. P_{true} = 1320 W, P_r = 1840 VAR, P_a = 2264.51 VA.

Section 16–8
1. −62.84°.   **2.** 8.9 V rms.   **3.** Capacitor.   **4.** $X_C << R$.

Section 16–9
1. The leakage resistance acts in parallel with C, which alters the circuit time constant.
2. T.

Section 16–10
1. To advance the frequency value by FI each pass through the loop.
2. Line 150 calculates X_C; line 160 calculates ϕ; line 170 calculates V_{out}; and line 180 prints frequency, phase angle, and output voltage.

SEVENTEEN

RL
CIRCUIT ANALYSIS

An *RL* circuit contains both *resistance* and *inductance*. It is one of the basic types of reactive circuits to be studied. *RC* circuits were studied in the last chapter. Basic series and parallel *RL* circuits and their response to sinusoidal ac voltages are covered in this chapter. Series-parallel combinations are also analyzed. Power in *RL* circuits is studied, and basic applications are introduced. Troubleshooting and computer analysis are also introduced. The coverage in this chapter is basically parallel with the coverage of *RC* circuits in Chapter 16. This approach allows you to recognize the similarities and understand the differences in the analysis and response of *RC* and *RL* circuits.

In this chapter, you will learn:

☐   How to calculate the impedance of a series *RL* circuit.
☐   How to analyze a series *RL* circuit in terms of phase angle, current, and voltages using Ohm's law and Kirchhoff's voltage law.
☐   How to determine the effects of frequency on a series *RL* circuit.
☐   How to calculate the impedance of a parallel *RL* circuit.
☐   How to use the concepts of conductance, susceptance, and admittance in the analysis of parallel *RL* circuits.

☐ How to analyze parallel *RL* circuits in terms of phase angle, currents, and voltage using Ohm's law and Kirchhoff's current law.

☐ How to convert a parallel *RL* circuit to an equivalent series form.

☐ How to analyze circuits with combinations of series and parallel elements.

☐ How to determine true power, reactive power, apparent power, and the power factor in *RL* circuits.

☐ How to analyze *RL* lead and lag networks.

☐ How to analyze basic *RL* filters.

17–1   ## SINUSOIDAL RESPONSE OF *RL* CIRCUITS

As with the *RC* circuit, all currents and voltages in an *RL* circuit are sinusoidal when the input is sinusoidal. Phase shifts are introduced because of the inductance. As you will learn, the resistor voltage and current are in phase but lag the source voltage, and the inductor voltage leads the source voltage. The phase angle between the current and the inductor voltage is always 90 degrees. These generalized phase relationships are indicated in Figure 17–1. Notice that they are opposite from those of the *RC* circuit, as discussed in the last chapter.

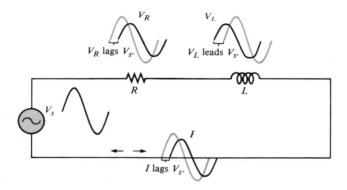

FIGURE 17–1
Illustration of sinusoidal response with general phase relationships of V_R, V_L, and I relative to V_s, V_R, and I are in phase, V_R lags V_s, and V_L leads V_s. V_R and V_L are 90 degrees out of phase with each other.

The amplitudes and the phase relationships of the voltages and current depend on the ohmic values of the resistance and the inductive reactance. When a circuit is purely inductive, the phase angle between the applied voltage and the total current is 90 degrees, with the current *lagging* the voltage. When there is a combination of both resistance and inductive reactance in a circuit, the phase angle is somewhere between zero and 90 degrees, depending on the relative values of R and X_L.

SECTION REVIEW 17–1

1. A 1-kHz sinusoidal voltage is applied to an *RL* circuit. What is the frequency of the resulting current?

2. When the resistance in an *RL* circuit is greater than the inductive reactance, do you think that the phase angle between the applied voltage and the total current is closer to zero or to 90 degrees?

17–2

IMPEDANCE OF SERIES *RL* CIRCUITS

As you know, impedance is the total opposition to sinusoidal current in a circuit and is expressed in ohms. The impedance of a series *RL* circuit is determined by the resistance and the inductive reactance.

Recall from Chapter 14 that inductive reactance is expressed as a phasor quantity in rectangular form as

$$\mathbf{X}_L = jX_L \tag{17–1}$$

In the series *RL* circuit of Figure 17–2, the total impedance is the phasor sum of *R* and jX_L and is expressed as

$$\mathbf{Z} = R + jX_L \tag{17–2}$$

FIGURE 17–2
Series *RL* circuit.

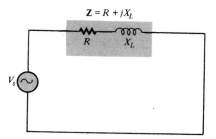

THE IMPEDANCE TRIANGLE

In ac analysis, both *R* and X_L are treated as phasor quantities, as shown in the phasor diagram of Figure 17–3(a), with X_L appearing at a +90-degree angle with respect to *R*. This relationship comes from the fact that the inductor voltage leads the current, and thus the resistor voltage, by 90 degrees. Since **Z** is the phasor sum of *R* and jX_L, its phasor representation is shown in Figure 17–3(b). A repositioning of the phasors, as shown in Part (c), forms a right triangle. This is called the *impedance triangle*. The length of each phasor represents the magnitude of the quantity, and θ is the phase angle between the applied voltage and the current in the *RL* circuit.

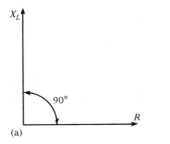

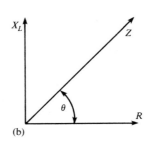

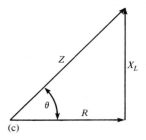

FIGURE 17–3
Development of the impedance triangle for a series *RL* circuit.

The impedance magnitude of the series *RL* circuit can be expressed in terms of the resistance and reactance as

$$Z = \sqrt{R^2 + X_L^2} \qquad (17–3)$$

The magnitude of the impedance is expressed in ohms.
The phase angle, θ, is expressed as

$$\theta = \tan^{-1}\left(\frac{X_L}{R}\right) \qquad (17–4)$$

Combining the magnitude and the angle, the impedance can be expressed in polar form as

$$\mathbf{Z} = \sqrt{R^2 + X_L^2}\angle\tan^{-1}\left(\frac{X_L}{R}\right) \qquad (17–5)$$

EXAMPLE 17–1

For each circuit in Figure 17–4, express the impedance in both rectangular and polar forms.

FIGURE 17–4

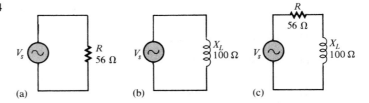

Solution:

(a) $\mathbf{Z} = R + j0 = R = 56\ \Omega$ in rectangular form ($X_L = 0$)

$\mathbf{Z} = R\angle 0° = 56\angle 0°\ \Omega$ in polar form

The impedance is simply equal to the resistance, and the phase angle is zero because pure resistance does not introduce a phase shift.

(b) $\mathbf{Z} = 0 + jX_L = j100 \ \Omega$ in rectangular form ($R = 0$)

$\mathbf{Z} = X_L \angle 90° = 100 \angle 90° \ \Omega$ in polar form

The impedance equals the inductive reactance in this case, and the phase angle is +90 degrees because the inductance causes the current to lag the voltage by 90 degrees.

(c) $\mathbf{Z} = R + jX_L = 56 \ \Omega + j100 \ \Omega$ in rectangular form

$$\mathbf{Z} = \sqrt{R^2 + X_L^2} \angle \tan^{-1}\left(\frac{X_L}{R}\right)$$

$$= \sqrt{(56 \ \Omega)^2 + (100 \ \Omega)^2} \angle \tan^{-1}\left(\frac{100}{56}\right) = 114.6 \angle 60.8° \ \Omega$$

The impedance is the phasor sum of the resistance and the inductive reactance. The phase angle is fixed by the relative values of X_L and R.

    Calculator sequences for Z and θ in part (c) are

Z: $\boxed{5}\ \boxed{6}\ \boxed{x^2}\ \boxed{+}\ \boxed{1}\ \boxed{0}\ \boxed{0}\ \boxed{x^2}\ \boxed{=}\ \boxed{\sqrt{x}}$

θ: $\boxed{1}\ \boxed{0}\ \boxed{0}\ \boxed{\div}\ \boxed{5}\ \boxed{6}\ \boxed{=}\ \boxed{\text{2nd F}}\ \boxed{\tan^{-1}}$

SECTION REVIEW 17–2

1. The impedance of a certain *RL* circuit is $150 \ \Omega + j220 \ \Omega$. What is the value of the resistance? The inductive reactance?

2. A series *RL* circuit has a total resistance of 33 kΩ and an inductive reactance of 50 kΩ. Write the expression for the impedance in rectangular form. Convert the impedance to polar form.

17–3

ANALYSIS OF SERIES *RL* CIRCUITS

OHM'S LAW

The application of Ohm's law to series *RL* circuits involves the use of the phasor quantities of **Z**, **V**, and **I**. The three equivalent forms of Ohm's law were stated in Chapter 16 for *RC* circuits. They apply also to *RL* circuits and are restated here for convenience: $\mathbf{V} = \mathbf{IZ}$, $\mathbf{I} = \mathbf{V/Z}$, and $\mathbf{Z} = \mathbf{V/I}$.

    Recall that since Ohm's law calculations involve multiplication and division operations, the voltage, current, and impedance should be expressed in polar form.

**EXAMPLE
17–2**

The current in Figure 17–5 is expressed as $\mathbf{I} = 0.2\angle 0°$ mA. Determine the source voltage and express it also in polar form.

FIGURE 17–5

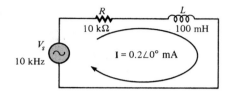

Solution:
The inductive reactance is

$$X_L = 2\pi fL = 2\pi(10 \text{ kHz})(100 \text{ mH}) = 6.28 \text{ k}\Omega$$

The impedance is

$$\mathbf{Z} = R + jX_L = 10 \text{ k}\Omega + j6.28 \text{ k}\Omega$$

Converting to polar form:

$$\mathbf{Z} = \sqrt{(10 \text{ k}\Omega)^2 + (6.28 \text{ k}\Omega)^2}\angle\tan^{-1}\left(\frac{6.28}{10}\right)$$

$$= 11.81\angle 32.13° \text{ k}\Omega$$

Applying Ohm's law:

$$\mathbf{V}_s = \mathbf{IZ} = (0.2\angle 0° \text{ mA})(11.81\angle 32.13° \text{ k}\Omega) = 2.36\angle 32.13° \text{ V}$$

The magnitude of the source voltage is 2.36 V at an angle of 32.13 degrees with respect to the current; that is, the voltage leads the current by 32.13 degrees, as shown in the phasor diagram of Figure 17–6.

FIGURE 17–6

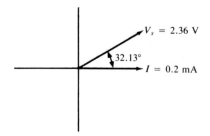

RELATIONSHIPS OF THE CURRENT AND VOLTAGES IN A SERIES *RL* CIRCUIT

In a series *RL* circuit, the current is the same through both the resistor and the inductor. Thus, the resistor voltage is in phase with the current, and the inductor voltage leads the current by 90 degrees. Therefore, there is a phase difference of 90 degrees between the resistor voltage, V_R, and the inductor voltage, V_L, as shown in the waveform diagram of Figure 17–7.

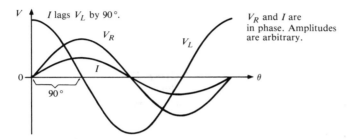

FIGURE 17–7
Phase relation of voltages and current in a series *RL* circuit.

From Kirchhoff's voltage law, the sum of the voltage drops must equal the applied voltage. However, since V_R and V_L are not in phase with each other, they must be added as phasor quantities with V_L leading V_R by 90 degrees, as shown in Figure 17–8(a). As shown in Part (b), $\mathbf{V}_s$ is the phasor sum of V_R and V_L.

$$\mathbf{V}_s = V_R + jV_L \tag{17–6}$$

This equation can be expressed in polar form as

$$\mathbf{V}_s = \sqrt{V_R^2 + V_L^2} \angle \tan^{-1}\left(\frac{V_L}{V_R}\right) \tag{17–7}$$

where the magnitude of the source voltage is

$$V_s = \sqrt{V_R^2 + V_L^2} \tag{17–8}$$

and the phase angle between the resistor voltage and the source voltage is

$$\theta = \tan^{-1}\left(\frac{V_L}{V_R}\right) \tag{17–9}$$

θ is also the phase angle between the source voltage and the current. Figure 17–9 shows a voltage and current phasor diagram that represents the waveform diagram of Figure 17–7.

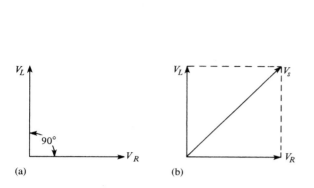

(a)

(b)

FIGURE 17–8
Voltage phasor diagram for a series *RL* circuit.

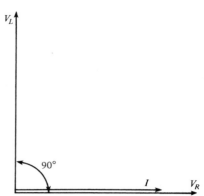

FIGURE 17–9
Voltage and current phasor diagram for the waveforms in Figure 17–7.

VARIATION OF IMPEDANCE AND PHASE ANGLE WITH FREQUENCY

The impedance triangle is useful in visualizing how the frequency of the applied voltage affects the *RL* circuit response. As you know, inductive reactance varies *directly* with frequency. When X_L increases, the magnitude of the total impedance also increases; and when X_L decreases, the magnitude of the total impedance decreases. Thus, *Z is directly dependent on frequency.*

The phase angle θ also varies *directly* with frequency, because $\theta = \tan^{-1}(X_L/R)$. As X_L increases with frequency, so does θ, and vice versa.

The impedance triangle is used in Figure 17–10 to illustrate the variations in X_L, Z, and θ as the frequency changes. Of course, R remains constant. The main point is that *because X_L varies directly as the frequency, so also do the magnitude of the total impedance and the phase angle.* Example 17–3 illustrates this.

FIGURE 17–10
Effect of frequency on impedance and phase angle.

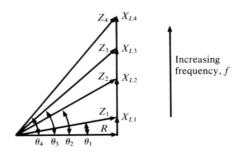

EXAMPLE
17–3

For the series *RL* circuit in Figure 17–11, determine the magnitude of the total impedance and the phase angle for each of the following frequencies: 10 kHz, 20 kHz, and
30 kHz.

FIGURE 17–11

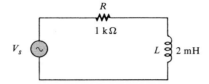

Solution:
For f = 10 kHz:

$$X_L = 2\pi fL = 2\pi (10 \text{ kHz})(20 \text{ mH}) = 1.26 \text{ k}\Omega$$

$$\mathbf{Z} = \sqrt{(1 \text{ k}\Omega)^2 + (1.26 \text{ k}\Omega)^2} \angle \tan^{-1}\left(\frac{1.26}{1}\right)$$

$$= 1.61\angle 51.56° \text{ k}\Omega$$

For f = 20 kHz:

$$X_L = 2\pi (20 \text{ kHz})(20 \text{ mH}) = 2.52 \text{ k}\Omega$$

$$\mathbf{Z} = \sqrt{(1 \text{ k}\Omega)^2 + (2.52 \text{ k}\Omega)^2} \angle \tan^{-1}\left(\frac{2.52}{1}\right)$$

$$= 2.71\angle 68.36° \text{ k}\Omega$$

For f = 30 kHz:

$$X_L = 2\pi (30 \text{ kHz})(20 \text{ mH}) = 3.77 \text{ k}\Omega$$

$$\mathbf{Z} = \sqrt{(1 \text{ k}\Omega)^2 + (3.77 \text{ k}\Omega)^2} \angle \tan^{-1}\left(\frac{3.77}{1}\right)$$

$$= 3.9\angle 75.14° \text{ k}\Omega$$

Notice that as the frequency increases, X_L, Z, and θ also increase.

SECTION REVIEW 17–3

1. In a certain series *RL* circuit, V_R = 2 V and V_L = 3 V. What is the magnitude of
the total voltage?

2. In Question 1, what is the phase angle between the total voltage and the current?

3. When the frequency of the applied voltage in a series *RL* circuit is increased, what happens to the inductive reactance? What happens to the magnitude of the total impedance? What happens to the phase angle?

17–4

IMPEDANCE OF PARALLEL *RL* CIRCUITS

A basic parallel *RL* circuit is shown in Figure 17–12. The expression for the total impedance is developed as follows.

$$\mathbf{Z} = \frac{(R\angle 0°)(X_L\angle 90°)}{R + jX_L}$$

$$= \frac{RX_L\angle(0° + 90°)}{\sqrt{R^2 + X_L^2}\angle\tan^{-1}\left(\dfrac{X_L}{R}\right)}$$

$$\mathbf{Z} = \left(\frac{RX_L}{\sqrt{R^2 + X_L^2}}\right)\angle\left(90° - \tan^{-1}\left(\frac{X_L}{R}\right)\right) \qquad \textbf{(17–10)}$$

Equation (17–10) is the expression for the total parallel impedance where the magnitude is

$$Z = \frac{RX_L}{\sqrt{R^2 + X_L^2}} \qquad \textbf{(17–11)}$$

and the phase angle between the applied voltage and the total current is

$$\theta = 90° - \tan^{-1}\left(\frac{X_L}{R}\right) \qquad \textbf{(17–12)}$$

This can also be expressed equivalently as $\theta = \tan^{-1}(R/X_L)$.

FIGURE 17–12
Parallel *RL* circuit.

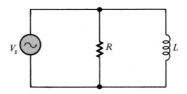

EXAMPLE 17–4

For each circuit in Figure 17–13, determine the magnitude of the total impedance and the phase angle.

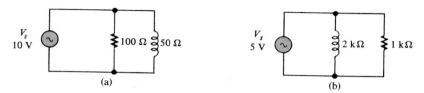

FIGURE 17–13

Solution:

(a) $Z = \left(\dfrac{RX_L}{\sqrt{R^2 + X_L^2}} \right) \angle \left(90° - \tan^{-1}\left(\dfrac{X_L}{R} \right) \right)$

$= \left[\dfrac{(100\ \Omega)(50\ \Omega)}{\sqrt{(100\ \Omega)^2 + (50\ \Omega)^2}} \right] \angle \left(90° - \tan^{-1}\left(\dfrac{50}{100} \right) \right)$

$= 44.72 \angle 63.43°\ \Omega$

Thus, $Z = 44.72\ \Omega$ and $\theta = 63.43$ degrees.

(b) $Z = \left[\dfrac{(1\ k\Omega)(2\ k\Omega)}{\sqrt{(1\ k\Omega)^2 + (2\ k\Omega)^2}} \right] \angle \left(90° - \tan^{-1}\left(\dfrac{2\ k\Omega}{1\ k\Omega} \right) \right)$

$= 894.4 \angle 26.57°\ \Omega$

Thus, $Z = 894.4\ \Omega$ and $\theta = 26.57$ degrees.

Notice that the positive angle indicates that the voltage leads the current, as opposed to the *RC* case where the voltage lags the current.

The calculator sequences for Part (a) are

Z: $\boxed{1}\ \boxed{0}\ \boxed{0}\ \boxed{\times}\ \boxed{5}\ \boxed{0}\ \boxed{\div}\ \boxed{(}\ \boxed{(}\ \boxed{1}\ \boxed{0}\ \boxed{0}\ \boxed{x^2}$
$\boxed{+}\ \boxed{5}\ \boxed{0}\ \boxed{x^2}\ \boxed{)}\ \boxed{\sqrt{x}}\ \boxed{=}$

θ: $\boxed{1}\ \boxed{0}\ \boxed{0}\ \boxed{\div}\ \boxed{5}\ \boxed{0}\ \boxed{=}\ \boxed{2\text{nd F}}\ \boxed{\tan^{-1}}$

SUSCEPTANCE AND ADMITTANCE

As you know from the previous chapter, conductance is the reciprocal of resistance, susceptance is the reciprocal of reactance, and admittance is the reciprocal of impedance.

For parallel *RL* circuits, inductive susceptance is expressed as

$$\mathbf{B}_L = \frac{1}{X_L\angle 90°} = B_L\angle - 90° = -jB_L \tag{17-13}$$

and the admittance is

$$\mathbf{Y} = \frac{1}{Z\angle \pm \theta} = Y\angle \mp \theta \tag{17-14}$$

As with the *RC* circuit, the unit for *G*, B_L, and *Y* is the siemen (S). In the basic parallel *RL* circuit shown in Figure 17–14, the total admittance is the phasor sum of the conductance, and the susceptance is expressed as

$$\mathbf{Y} = G - jB_L \tag{17-15}$$

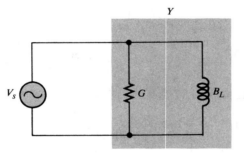
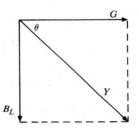

FIGURE 17–14
Admittance in a parallel *RL* circuit.

EXAMPLE 17–5

Determine the admittance in Figure 17–15.

FIGURE 17–15

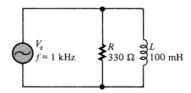

Solution:
$R = 330\ \Omega$; thus

$$G = \frac{1}{R} = \frac{1}{330\ \Omega} = 0.003\ \text{S}$$

$$X_L = 2\pi(1000\ \text{Hz})(100\ \text{mH}) = 628.3\ \Omega$$

$$B_L = \frac{1}{X_L} = 0.00159 \text{ S}$$

$$\mathbf{Y} = G - jB_L = 0.003 \text{ S} - j0.00159 \text{ S}$$

Y can be expressed in polar form as

$$\mathbf{Y} = \sqrt{(0.003 \text{ S})^2 + (0.00159 \text{ S})^2} \angle -\tan^{-1}\left(\frac{0.00159}{0.003}\right)$$

$$= 0.0034 \angle -27.92° \text{ S}$$

Converting to impedance, we get

$$\mathbf{Z} = \frac{1}{Y} = \frac{1}{0.0034 \angle -27.92° \text{ S}} = 294.12 \angle 27.92° \text{ } \Omega$$

Again, the positive phase angle in the impedance expression indicates that the voltage leads the current. The admittance phasor diagram is shown in Figure 17–16.

FIGURE 17–16

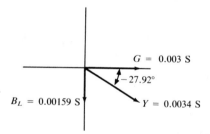

$G = 0.003 \text{ S}$

$-27.92°$

$B_L = 0.00159 \text{ S}$

$Y = 0.0034 \text{ S}$

SECTION REVIEW 17–4

1. If $Z = 500 \text{ } \Omega$, what is the value of Y?

2. In a certain parallel *RL* circuit, $R = 47 \text{ } \Omega$ and $X_L = 75 \text{ } \Omega$. Determine the magnitude of the admittance.

3. In the circuit of Question 2, does the total current lead or lag the applied voltage? By what phase angle?

17–5

ANALYSIS OF PARALLEL *RL* CIRCUITS

The following example applies Ohm's law to the analysis of a parallel *RL* circuit.

**EXAMPLE
17–6**

Determine the total current and the phase angle in the circuit of Figure 17–17.

FIGURE 17–17

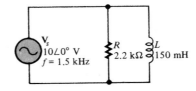

Solution:

$$X_L = 2\pi(1.5 \text{ kHz})(150 \text{ mH}) = 1.41 \text{ k}\Omega$$

The susceptance magnitude is

$$B_L = \frac{1}{X_L} = \frac{1}{1.41 \text{ k}\Omega} = 0.709 \text{ mS}$$

The conductance magnitude is

$$G = \frac{1}{R} = \frac{1}{2.2 \text{ k}\Omega} = 0.455 \text{ mS}$$

The total admittance is

$$\mathbf{Y} = G - jB_L = 0.455 \text{ mS} - j0.709 \text{ mS}$$

Converting to polar form:

$$\mathbf{Y} = \sqrt{(0.455 \text{ mS})^2 + (0.709 \text{ mS})^2} \angle -\tan^{-1}\left(\frac{0.709}{0.455}\right)$$

$$= 0.842 \angle -57.31° \text{ mS}$$

Applying Ohm's law:

$$\mathbf{I}_T = \mathbf{VY} = (10\angle 0° \text{ V})(0.842\angle -57.31° \text{ mS}) = 8.42\angle -57.31° \text{ mA}$$

The magnitude of the total current is 8.42 mA, and it lags the applied voltage by 57.31 degrees, as indicated by the negative angle associated with it. The phasor diagram in Figure 17–18 shows the voltage/current relationship.

FIGURE 17–18

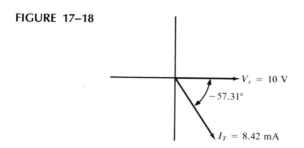

RELATIONSHIPS OF THE CURRENTS AND VOLTAGES IN A PARALLEL *RL* CIRCUIT

Figure 17–19(a) shows all the currents and voltages in a basic parallel *RL* circuit. As you can see, the applied voltage, V_s, appears across both the resistive and the inductive branches, so V_s, V_R, and V_L are all in phase and of the same magnitude. The total current, I_T, divides at the junction into the two branch currents, I_R and I_L.

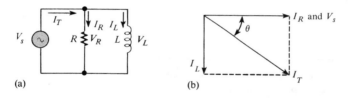

(a)   (b)

FIGURE 17–19
Currents and voltages in a parallel *RL* circuit.

The current through the resistor is in phase with the voltage. The current through the inductor lags the voltage and the resistor current by 90 degrees. By Kirchhoff's current law, the total current is the phasor sum of the two branch currents, as shown by the phasor diagram in Figure 17–19(b). The total current is expressed as

$$\mathbf{I}_T = I_R - jI_L \tag{17–16}$$

This equation can be expressed in polar form as

$$\mathbf{I}_T = \sqrt{I_R^2 + I_L^2} \angle -\tan^{-1}\left(\frac{I_L}{I_R}\right) \tag{17–17}$$

where the magnitude of the total current is

$$I_T = \sqrt{I_R^2 + I_L^2} \qquad\qquad (17\text{–}18)$$

and the phase angle between the resistor current and the total current is

$$\theta = -\tan^{-1}\left(\frac{I_L}{I_R}\right) \qquad\qquad (17\text{–}19)$$

Since the resistor current and the applied voltage are in phase, θ also represents the phase angle between the total current and the applied voltage. Figure 17–20 shows a complete current and voltage phasor diagram.

FIGURE 17–20
Current and voltage phasor diagram for a parallel *RL* circuit (amplitudes are arbitrary).

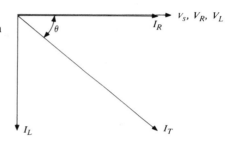

**EXAMPLE
17–7**

Determine the value of each current in Figure 17–21, and describe the phase relationship of each with the applied voltage.

FIGURE 17–21

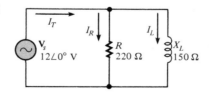

Solution:

$$I_R = \frac{V_s}{R} = \frac{12\angle 0° \text{ V}}{220\angle 0° \text{ }\Omega} = 54.55\angle 0° \text{ mA}$$

$$I_L = \frac{V_s}{X_L} = \frac{12\angle 0° \text{ V}}{150\angle 90° \text{ }\Omega} = 80\angle -90° \text{ mA}$$

$$I_T = I_R - jI_L = 54.55 \text{ mA} - j80 \text{ mA}$$

Converting I_T to polar form:

$$I_T = \sqrt{(54.55 \text{ mA})^2 + (80 \text{ mA})^2}\angle -\tan^{-1}\left(\frac{80}{54.55}\right)$$

$$= 96.83\angle -55.71° \text{ mA}$$

As the results show, the resistor current is 54.55 mA and is in phase with the applied voltage. The inductor current is 80 mA and lags the applied voltage by 90 degrees. The total current is 96.83 mA and lags the voltage by 55.71 degrees. The phasor diagram in Figure 17–22 shows these relationships.

FIGURE 17–22

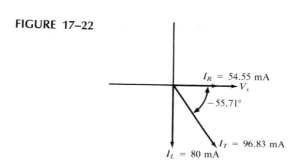

I_R = 54.55 mA
V_s
−55.71°
I_T = 96.83 mA
I_L = 80 mA

SECTION REVIEW 17–5

1. The admittance of an *RL* circuit is 0.004 S, and the applied voltage is 8 V. What is the total current?

2. In a certain parallel *RL* circuit, the resistor current is 12 mA, and the inductor current is 20 mA. Determine the magnitude and phase angle of the total current. This phase angle is measured with respect to what?

3. What is the phase angle between the inductor current and the applied voltage in a parallel *RL* circuit?

17–6

SERIES-PARALLEL ANALYSIS

In this section, *RL* circuits with combinations of both series and parallel *R* and *L* elements are analyzed. Examples are used to illustrate the procedures.

**EXAMPLE
17–8**

In the circuit of Figure 17–23, determine the following values:
(a) Z   (b) I_T (c) θ

FIGURE 17–23

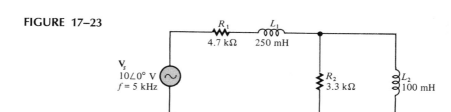

R_1
4.7 kΩ

L_1
250 mH

V_s
10∠0° V
f = 5 kHz

R_2
3.3 kΩ

L_2
100 mH

Solution:

(a) First, the inductive reactances are calculated:

$$X_{L1} = 2\pi(5 \text{ kHz})(250 \text{ mH}) = 7.85 \text{ k}\Omega$$
$$X_{L2} = 2\pi(5 \text{ kHz})(100 \text{ mH}) = 3.14 \text{ k}\Omega$$

R_1 and L_1 are in series with each other. R_2 and L_2 are in parallel. The series portion and the parallel portion are in series with each other.

$$\mathbf{Z}_1 = R_1 + jX_{L1} = 4.7 \text{ k}\Omega + j7.85 \text{ k}\Omega$$

$$G_2 = \frac{1}{R_2} = \frac{1}{3.3 \text{ k}\Omega} = 303 \ \mu\text{S}$$

$$B_{L2} = \frac{1}{X_{L2}} = \frac{1}{3.14 \text{ k}\Omega} = 318.5 \ \mu\text{S}$$

$$\mathbf{Y}_2 = G_2 - jB_L = 303 \ \mu\text{S} - j318.5 \ \mu\text{S}$$

Converting to polar form:

$$\mathbf{Y}_2 = \sqrt{(303 \ \mu\text{S})^2 + (318.5 \ \mu\text{S})^2} \angle -\tan^{-1}\left(\frac{318.5}{303}\right)$$

$$= 439.6 \angle -46.43° \ \mu\text{S}$$

Then

$$\mathbf{Z}_2 = \frac{1}{\mathbf{Y}_2} = \frac{1}{439.6 \angle -46.43° \ \mu\text{S}} = 2.28 \angle 46.43° \text{ k}\Omega$$

Converting to rectangular form:

$$\mathbf{Z}_2 = (2.28 \text{ k}\Omega)\cos(46.43°) + j(2.28 \text{ k}\Omega)\sin(46.43°)$$

$$= 1.57 \text{ k}\Omega + j1.65 \text{ k}\Omega$$

Combining $\mathbf{Z}_1$ and $\mathbf{Z}_2$, we get

$$\mathbf{Z}_T = \mathbf{Z}_1 + \mathbf{Z}_2$$

$$= (4.7 \text{ k}\Omega + j7.85 \text{ k}\Omega) + (1.57 \text{ k}\Omega + j1.65 \text{ k}\Omega)$$

$$= 6.27 \text{ k}\Omega + j9.5 \text{ k}\Omega$$

Expressing $\mathbf{Z}_T$ in polar form:

$$\mathbf{Z}_T = \sqrt{(6.27 \text{ k}\Omega)^2 + (9.5 \text{ k}\Omega)^2} \angle \tan^{-1}\left(\frac{9.5}{6.27}\right)$$

$$= 11.38 \angle 56.58° \text{ k}\Omega$$

(b) The total current is found using Ohm's law:

$$\mathbf{I}_T = \frac{\mathbf{V}_s}{\mathbf{Z}_T} = \frac{10\angle 0° \text{ V}}{11.38\angle 56.58° \text{ k}\Omega}$$

$$= 0.879\angle -56.58° \text{ mA}$$

(c) The total current lags the applied voltage by 56.58 degrees.

EXAMPLE 17–9

Determine the voltage across each element in Figure 17–24. Sketch a voltage phasor diagram.

FIGURE 17–24

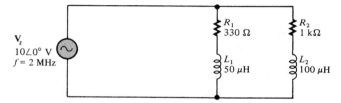

Solution:
First calculate X_{L1} and X_{L2}:

$$X_{L1} = 2\pi f L_1 = 2\pi(2 \text{ MHz})(50 \text{ } \mu\text{H}) = 628.3 \text{ } \Omega$$

$$X_{L2} = 2\pi f L_2 = 2\pi(2 \text{ MHz})(100 \text{ } \mu\text{H}) = 1.257 \text{ k}\Omega$$

Now, determine the impedance of each branch:

$$\mathbf{Z}_1 = R_1 + jX_{L1} = 330 \text{ } \Omega + j628.3 \text{ } \Omega$$

$$\mathbf{Z}_2 = R_2 + jX_{L2} = 1 \text{ k}\Omega + j1.257 \text{ k}\Omega$$

Converting to polar form:

$$\mathbf{Z}_1 = \sqrt{(330 \text{ } \Omega)^2 + (628.3 \text{ } \Omega)^2}\angle\tan^{-1}\left(\frac{628.3}{330}\right)$$

$$= 709.69\angle 62.29° \text{ } \Omega$$

$$\mathbf{Z}_2 = \sqrt{(1 \text{ k}\Omega)^2 + (1.257 \text{ k}\Omega)^2}\angle\tan^{-1}\left(\frac{1.257}{1}\right)$$

$$= 1.606\angle 51.5° \text{ k}\Omega$$

Calculate each branch current:

$$\mathbf{I}_1 = \frac{\mathbf{V}_s}{\mathbf{Z}_1} = \frac{10\angle 0° \text{ V}}{709.69\angle 62.29° \text{ } \Omega} = 14.09\angle -62.29° \text{ mA}$$

$$\mathbf{I}_2 = \frac{\mathbf{V}_s}{\mathbf{Z}_2} = \frac{10\angle 0° \text{ V}}{1.606\angle 51.5° \text{ k}\Omega} = 6.23\angle -51.5° \text{ mA}$$

Now, Ohm's law is used to get the voltage across each element:

$$\mathbf{V}_{R1} = \mathbf{I}_1\mathbf{R}_1 = (14.09\angle -62.2° \text{ mA})(330\angle 0° \ \Omega) = 4.65\angle -62.2° \text{ V}$$

$$\mathbf{V}_{L1} = \mathbf{I}_1\mathbf{X}_{L1} = (14.09\angle -62.2 \text{ mA})(628.3\angle 90° \ \Omega) = 8.85\angle 27.8° \text{ V}$$

$$\mathbf{V}_{R2} = \mathbf{I}_2\mathbf{R}_2 = (6.23\angle -51.5 \text{ mA})(1\angle 0° \text{ k}\Omega) = 6.23\angle -51.5° \text{ V}$$

$$\mathbf{V}_{L2} = \mathbf{I}_2\mathbf{X}_{L2} = (6.23\angle -51.5° \text{ mA})(1.257\angle 90° \text{ k}\Omega) = 7.83\angle 38.5° \text{ V}$$

The voltage phasor diagram is shown in Figure 17–25, and the current phasor diagram is shown in Figure 17–26.

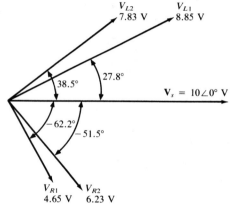

FIGURE 17–25

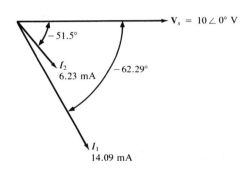

FIGURE 17–26

The two previous examples should give you a feel for how to approach the analysis of complex *RL* networks.

SECTION REVIEW 17–6

1. What is the total impedance of the circuit in Figure 17–24?

2. Determine the total current for the circuit in Figure 17–24.

17–7   POWER IN *RL* CIRCUITS

As you know, in a purely resistive ac circuit, all of the energy that is delivered by the source is dissipated in the form of heat by the resistance. In a purely inductive ac circuit, all of the energy delivered by the source is stored by the inductor in its magnetic field during a portion of the voltage cycle and then returned to the source during another portion of the cycle so that there is no net energy loss.

When there is both resistance and inductance in a circuit, some of the energy is alternately stored and returned by the inductance, and some is dissipated by the resistance. The amount of energy loss is determined by the relative values of resistance and inductive reactance.

When the resistance is greater than the inductive reactance, more of the total energy delivered by the source is dissipated by the resistance than is stored by the inductor, and when the reactance is greater than the resistance, more of the total energy is stored and returned than is lost.

As you know, the power loss in a resistance is called the *true power*. The power in an inductor is reactive power and is expressed as

$$P_r = I^2 X_L \qquad \text{(17–20)}$$

THE POWER TRIANGLE

The generalized power triangle for the *RL* circuit is shown in Figure 17–27. The apparent power, P_a, is the resultant of the average power, P_{true}, and the reactive power, P_r.

Recall that the power factor equals the cosine of θ ($PF = \cos \theta$). As the phase angle between the applied voltage and the total current increases, the power factor decreases, indicating an increasingly reactive circuit. The smaller the power factor, the smaller the true power is compared to the reactive power.

FIGURE 17–27
Power triangle for an *RL* circuit.

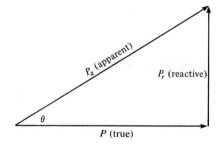

EXAMPLE 17–10

Determine the power factor, the true power, the reactive power, and the apparent power in Figure 17–28.

FIGURE 17–28

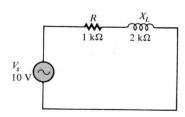

Solution:

The total impedance of the circuit is

$$\mathbf{Z} = R + jX_L$$

$$= \sqrt{(1\ k\Omega)^2 + (2\ k\Omega)^2}\angle\tan^{-1}\left(\frac{2000}{1000}\right)$$

$$= 2236\angle63.44°\ \Omega$$

The current magnitude is

$$I = \frac{V_s}{Z} = \frac{10\ V}{2236\ \Omega} = 4.47\ mA$$

The phase angle is indicated in the expression for **Z**.

$$\theta = 63.44°$$

The power factor is therefore

$$PF = \cos\theta = \cos(63.44°) = 0.447$$

The true power is

$$P_{\text{true}} = V_sI\cos\theta = (10\ V)(4.47\ mA)(0.447) = 19.99\ mW$$

The reactive power is

$$P_r = I^2X_L = (4.47\ mA)^2(2\ k\Omega) = 39.96\ mVAR$$

The apparent power is

$$P_a = I^2Z = (4.47\ mA)^2(2236\ \Omega) = 44.68\ mVA$$

SIGNIFICANCE OF THE POWER FACTOR

As you learned in Chapter 16, the power factor (*PF*) is very important in determining how much useful power (true power) is transferred to a load. The highest power factor is 1, which indicates that all of the current to a load is in phase with the voltage (resistive). When the power factor is 0, all of the current to a load is 90 degrees out of phase with the voltage (reactive).

Generally, a power factor as close to 1 as possible is desirable, because then most of the power transferred from the source to the load is useful or true power. True power goes only one way—from source to load—and performs work on the load in terms of energy dissipation. Reactive power simply goes back and forth between the source and the load with no net work being done. Energy must be used in order for work to be done.

Many practical loads have inductance as a result of their particular function, and it is essential for their proper operation. Examples are transformers, electric motors, and speakers, to name a few. Therefore, inductive (and capacitive) loads are a fact of life, and we must live with them.

To see the effect of the power factor on system requirements, refer to Figure 17–29 which shows a representation of a typical inductive load consisting effectively of inductance and resistance in parallel. Part (a) shows a load with a relatively low power factor (0.75), and Part (b) shows a load with a relatively high power factor (0.95). Both

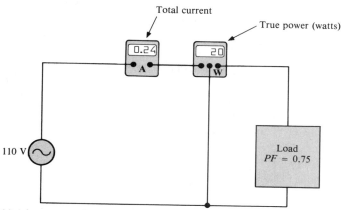

(a) A lower power factor means more total current for a given power dissipation (watts). A larger source is required to deliver the watts.

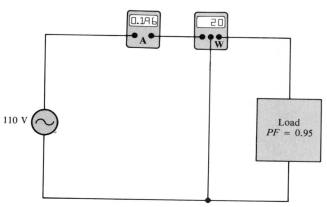

(b) A higher power factor means less total current for a given power dissipation. A smaller source can deliver the same true power (watts).

FIGURE 17–29

Illustration of the effect of the power factor on system requirements such as source rating (VA) and conductor size.

loads dissipate equal amounts of power as indicated by the wattmeters. Thus, an equal amount of work is done on both loads.

Although both loads are equivalent in terms of the amount of work done (true power), the low-power factor load in Part (a) draws more current from the source than does the high-power factor load in Part (b), as indicated by the ammeters. Therefore, the source in Part (a) must have a higher VA rating than the one in Part (b). Also, the lines connecting the source to the load must be a larger wire gage than those in Part (b), a condition that becomes significant when very long transmission lines are required, such as in power distribution.

Figure 17–29 has demonstrated that a higher power factor is an advantage in delivering power more efficiently to a load.

POWER FACTOR CORRECTION

The power factor of an inductive load can be increased by the addition of a capacitor in parallel, as shown in Figure 17–30. The capacitor compensates for the phase lag of the total current by creating a capacitive component of current that is 180 degrees out of phase with the inductive component. This has a canceling effect and reduces the phase angle (and power factor) as well as the total current, as illustrated in the figure.

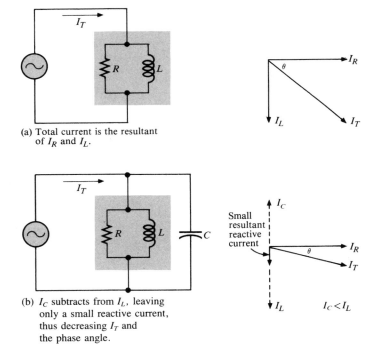

(a) Total current is the resultant of I_R and I_L.

(b) I_C subtracts from I_L, leaving only a small reactive current, thus decreasing I_T and the phase angle.

FIGURE 17–30

Example of how the power factor can be increased by the addition of a compensating capacitor.

SECTION REVIEW 17–7

1. To which component in an *RL* circuit is the energy loss due?
2. Calculate the power factor when $\theta = 50°$.
3. A certain *RL* circuit consists of a 470-Ω resistor and an inductive reactance of 620 Ω at the operating frequency. Determine P_{true}, P_r, and P_a when $I = 100$ mA.

17–8

BASIC APPLICATIONS

Two basic applications of *RL* circuits are introduced in this section: *phase shift networks* (*lead* and *lag*) and *frequency selective networks* (*filters*).

THE *RL* LEAD NETWORK

The first type of phase shift network is the *RL* lead network, in which the output voltage leads the input voltage by a specified amount. Figure 17–31(a) shows a series *RL* circuit with the *output voltage taken across the inductor* (note that in the *RC* lead network, the output was taken across the resistor). The source voltage is the input, V_{in}. As you know, θ is the angle between the current and the input voltage; it is also the angle between the resistor voltage and the input voltage, because V_R and I are in phase.

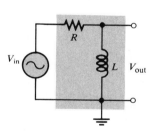

(a) A basic *RL* lead network

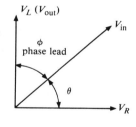

(b) Phasor voltage diagram showing phase lead between V_{in} and V_{out}

FIGURE 17–31
The *RL* lead network ($V_{\text{out}} = V_L$).

Since V_L *leads* V_R by 90 degrees, the phase angle between the inductor voltage and the input voltage is the difference between 90 degrees and θ, as shown in Figure 17–31(b). The inductor voltage is the *output;* it leads the input, thus creating a basic *lead* network.

When the input and output waveforms of the lead network are displayed on an oscilloscope, a relationship similar to that in Figure 17–32 is observed. The amount of

FIGURE 17–32
Input and output waveforms.

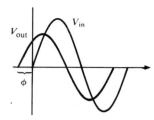

phase difference between the input and the output is dependent on the relative values of the inductive reactance and the resistance, as is the magnitude of the output voltage.

Phase Difference Between Input and Output The angle between V_{out} and V_{in} is designated ϕ (phi) and is developed as follows:

The polar expressions for the input voltage and the current are $V_{in}\angle 0°$ and $I\angle -\theta$, respectively. The output voltage is

$$\mathbf{V}_{out} = (I\angle -\theta)(X_L\angle 90°) = IX_L\angle(90° - \theta)$$

This shows that the output voltage is at an angle of $90° - \theta$ with respect to the input voltage. Since $\theta = \tan^{-1}(X_L/R)$, the angle ϕ between the input and output is

$$\phi = 90° - \tan^{-1}\left(\frac{X_L}{R}\right) \tag{17–21}$$

This angle can equivalently be expressed as

$$\phi = \tan^{-1}\left(\frac{R}{X_L}\right)$$

This angle is always positive, indicating that the output voltage *leads* the input voltage, as indicated in Figure 17–33.

FIGURE 17–33

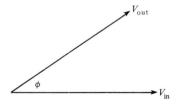

EXAMPLE 17–11

Determine the amount of phase lead from input to output in each lead network in Figure 17–34.

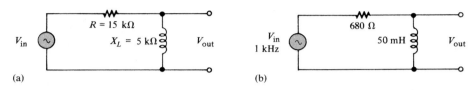

(a)   (b)

FIGURE 17–34

Solution:

(a) $\phi = 90° - \tan^{-1}\left(\dfrac{X_L}{R}\right) = 90° - \tan^{-1}\left(\dfrac{5000}{15,000}\right)$

$= 90° - 18.44° = 71.56°$

The output leads the input by 71.56 degrees.

(b) $X_L = 2\pi fL = 2\pi(1 \text{ kHz})(50 \text{ mH}) = 314.2 \ \Omega$

$\phi = 90° - \tan^{-1}\left(\dfrac{X_L}{R}\right) = 90° - \tan^{-1}\left(\dfrac{314.2}{680}\right) = 65.20°$

The output leads the input by 65.20 degrees.

Magnitude of the Output Voltage   To evaluate the output voltage in terms of its magnitude, visualize the RL lead network as a voltage divider. A portion of the total input voltage is dropped across R and a portion across L. Since the output voltage is V_L, it can be calculated as

$$V_{out} = \left(\frac{X_L}{\sqrt{R^2 + X_L^2}}\right)V_{in} \qquad\qquad \textbf{(17–22)}$$

Or it can be found using Ohm's law as follows:

$$V_{out} = IX_L \qquad\qquad \textbf{(17–23)}$$

The total phasor expression for the output voltage of an *RL* lead network is

$$\mathbf{V}_{out} = V_{out} \angle \phi \qquad\qquad (17\text{--}24)$$

EXAMPLE 17–12

For the lead network in Figure 17–34(b), determine the output voltage in phasor form when the input voltage has an rms value of 5 V. Sketch the input and output voltage waveforms showing the proper relationships.

Solution:

ϕ was found to be 65.20 degrees in Example 17–11.

$$\mathbf{V}_{out} = \left(\frac{X_L}{\sqrt{R^2 + X_L^2}} \right) V_{in} \angle \phi$$

$$= \left[\frac{314.2\ \Omega}{\sqrt{(680\ \Omega)^2 + (314.2\ \Omega)^2}} \right] 5 \angle 65.20°\ \text{V}$$

$$= 2.1 \angle 65.20°\ \text{V}$$

The waveforms are shown in Figure 17–35. Notice that the output voltage leads the input voltage by 65.20 degrees.

FIGURE 17–35

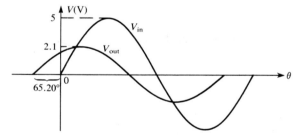

THE *RL* LAG NETWORK

The second basic type of phase shift network is the *RL* lag network. When the output of a series *RL* circuit is taken across the resistor rather than the inductor, as shown in Figure 17–36(a), it becomes a *lag* network. Lag networks cause the phase of the output voltage to *lag* the input by a specified amount.

Phase Difference Between Input and Output In a series *RL* circuit, the current lags the input voltage. Since the output voltage is taken across the resistor, the output *lags* the input, as indicated by the phasor diagram in Figure 17–36(b). The waveforms are shown in Part (c).

As in the lead network, the amount of phase difference between the input and output and the magnitude of the output voltage are dependent on the relative values of

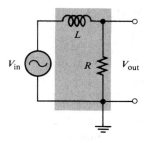

(a) A basic *RL* lag network

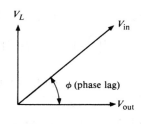

(b) Phasor voltage diagram showing phase lag between V_{in} and V_{out}

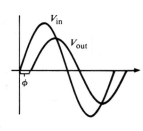

(c) Input and output waveforms

FIGURE 17–36
The *RL* lag network ($V_{out} = V_R$).

the resistance and the inductive reactance. When the input voltage is assigned a reference angle of 0 degrees, the angle of the output voltage (ϕ) with respect to the input voltage equals θ, because the resistor voltage (output) and the current are in phase with each other. The expression for the angle between the input voltage and the output voltage is

$$\phi = -\tan^{-1}\left(\frac{X_L}{R}\right) \qquad (17\text{–}25)$$

This angle is negative because the output lags the input.

EXAMPLE 17–13

Calculate the output phase angle for each circuit in Figure 17–37.

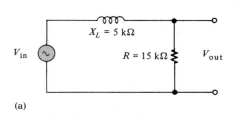

(a)

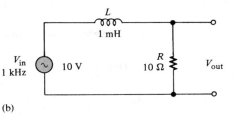

(b)

FIGURE 17–37

Solution:

(a) $\phi = -\tan^{-1}\left(\frac{X_L}{R}\right) = -\tan^{-1}\left(\frac{5 \text{ k}\Omega}{15 \text{ k}\Omega}\right) = -18.43°$

The output lags the input by 18.43 degrees.

(b) $X_L = 2\pi fL = 2\pi(1 \text{ kHz})(1 \text{ mH}) = 6.28 \ \Omega$

$$\phi = -\tan^{-1}\left(\frac{X_L}{R}\right) = -\tan^{-1}\left(\frac{6.28 \ \Omega}{10 \ \Omega}\right) = -32.13°$$

The output lags the input by 32.13 degrees.

Magnitude of the Output Voltage Since the output voltage of an *RL* lag network is taken across the resistor, the magnitude can be calculated using either the voltage divider approach or Ohm's law.

$$V_{out} = \left(\frac{R}{\sqrt{R^2 + X_L^2}}\right)V_{in} \qquad (17\text{--}26)$$

$$V_{out} = IR \qquad (17\text{--}27)$$

The expression for the output voltage in phasor form is

$$\mathbf{V}_{out} = V_{out}\angle -\phi \qquad (17\text{--}28)$$

EXAMPLE 17–14

The input voltage in Figure 17–37(b) (Example 17–13) has an rms value of 10 V. Determine the phasor expression for the output voltage. Sketch the waveform relationships for the input and output voltages.

Solution:
The phase angle was found to be -32.13 degrees in Example 17–13.

$$\mathbf{V}_{out} = \left(\frac{R}{\sqrt{R^2 + X_L^2}}\right)V_{in}\angle -\phi$$

$$= \left(\frac{10 \ \Omega}{11.81 \ \Omega}\right)10\angle -32.13° \text{ V} = 8.47\angle -32.13° \text{ V rms}$$

The waveforms are shown in Figure 17–38.

FIGURE 17–38

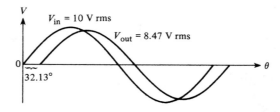

THE *RL* CIRCUIT AS A FILTER

As with *RC* circuits, series *RL* circuits also exhibit a frequency-selective characteristic and therefore act as basic filters. Filters are introduced here as an application example and will be covered in depth in a later chapter.

Low-Pass Filter   You have seen what happens to the output magnitude and phase angle in the lag network. In terms of the filtering action, the variation of the magnitude of the output voltage as a function of frequency is of primary importance.

　　Figure 17–39 shows the filtering action of a series *RL* circuit using specific values for purposes of illustration. In Part (a) of the figure, the input is *zero* frequency (dc). Since the inductor ideally acts as a short to constant direct current, the output voltage equals the full value of the input voltage (neglecting the winding resistance). Therefore, the circuit passes *all* of the input voltage to the output (10 V in, 10 V out).

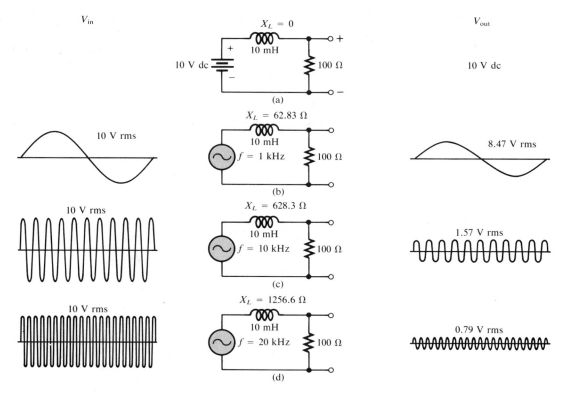

FIGURE 17–39
Low-pass filter action of an *RL* circuit (phase shift from input to output is not indicated).

In Figure 17–39(b), the frequency of the input voltage has been increased to 1 kHz, causing the inductive reactance to increase to 62.83 Ω. For an input voltage of 10 V rms, the output voltage is approximately 8.47 V rms, which can be calculated using the voltage divider approach or Ohm's law.

In Figure 17–39(c), the input frequency has been increased to 10 kHz, causing the inductive reactance to increase further to 628.3 Ω. For a constant input voltage of 10 V rms, the output voltage is now 1.57 V rms.

As the input frequency is increased further, the output voltage continues to decrease and approaches zero as the frequency becomes very high, as shown in Figure 17–39(d) for $f = 20$ kHz. A summary of the circuit action is as follows: As the frequency of the input increases, the inductive reactance increases. Because the resistance is constant and the inductive reactance increases, the voltage across the inductor increases, and that across the resistor (output voltage) decreases. The input frequency can be increased until it reaches a value at which the reactance is so large compared to the resistance that the output voltage can be neglected, because it becomes very small compared to the input voltage.

As shown in Figure 17–39, the circuit passes dc (zero frequency) completely. As the frequency of the input increases, less of the input voltage is passed through to the output. That is, the output voltage decreases as the frequency increases. It is apparent that the lower frequencies pass through the circuit much better than the higher frequencies. This *RL* circuit is therefore a very basic form of *low-pass filter*.

Figure 17–40 shows a response curve for a low-pass filter.

FIGURE 17–40
Low-pass filter response curve.

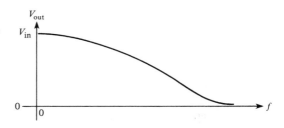

High-Pass Filter   In Figure 17–41(a), the output is taken across the inductor. When the input voltage is dc (zero frequency), the output is zero volts, because the inductor ideally appears as a short across the output.

In Figure 17–41(b), the frequency of the input signal has been increased to 100 Hz with an rms value of 10 V. The output voltage is 0.63 V rms. Thus, only a small percentage of the input voltage appears at the output at this frequency.

In Figure 17–41(c), the input frequency is increased further to 1 kHz, causing more voltage to be developed as a result of the increase in the inductive reactance. The output voltage at this frequency is 5.32 V rms. As you can see, the output voltage increases as the frequency increases. A value of frequency is reached at which the reactance is very large compared to the resistance and most of the input voltage appears across the inductor, as shown in Figure 17–41(d).

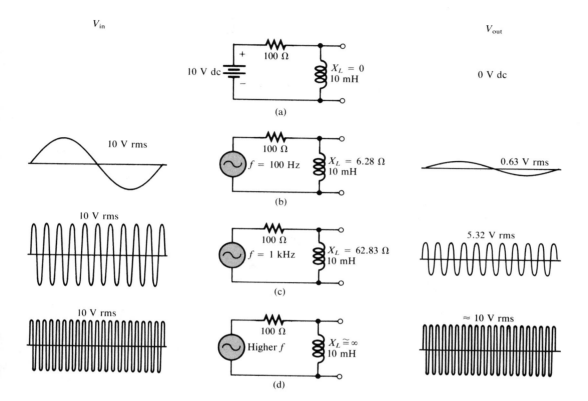

FIGURE 17–41
High-pass filter action of an *RL* circuit (phase shift from input to output is not indicated).

This circuit tends to prevent lower frequency signals from appearing on the output but permits higher frequency signals to pass through from input to output; thus, it is a very basic form of *high-pass filter*.

The response curve in Figure 17–42 shows that the output voltage increases and then levels off as it approaches the value of the input voltage as the frequency increases.

FIGURE 17–42
High-pass filter response curve.

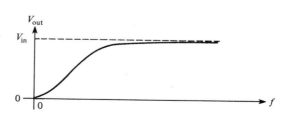

SECTION REVIEW 17-8

1. A certain *RL* lead network consists of a 3.3-kΩ resistor and a 15-mH inductor. Determine the phase shift between input and output at a frequency of 5 kHz.

2. An *RL* lag network has the same component values as the lead network in Question 1. What is the magnitude of the output voltage at 5 kHz when the input is 10 V rms?

3. When an *RL* circuit is used as a low-pass filter, across which component is the output taken?

17-9   TROUBLESHOOTING

EFFECTS OF AN OPEN INDUCTOR

The most common failure mode for inductors occurs when the winding opens as a result of excessive current or a mechanical contact failure. It is easy to see how an open coil affects the operation of a basic series *RL* circuit, as shown in Figure 17–43. Obviously, there is no current path—so the resistor voltage is zero, and the total applied voltage appears across the inductor.

FIGURE 17–43
Effect of an open coil.

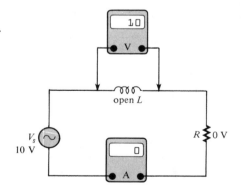

EFFECTS OF AN OPEN RESISTOR

When the resistor is open, there is no current, and the inductor voltage is zero. The total input voltage is across the open resistor, as shown in Figure 17–44.

FIGURE 17–44
Effect of an open resistor.

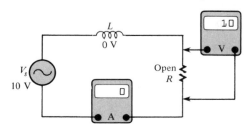

OPEN COMPONENTS IN PARALLEL CIRCUITS

In a parallel *RL* circuit, an open resistor or inductor will cause the total current to decrease, because the total impedance will increase. Obviously, the branch with the open component will have zero current. Figure 17–45 illustrates these conditions.

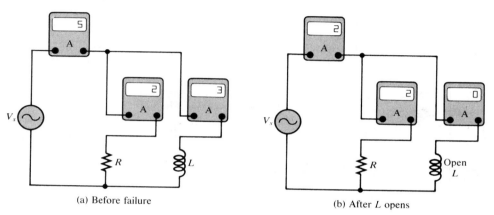

(a) Before failure

(b) After *L* opens

FIGURE 17–45
Effect of an open component in a parallel circuit with V_s constant.

EFFECTS OF AN INDUCTOR WITH SHORTED WINDINGS

It is possible for some of the windings of coils to short together as a result of damaged insulation. This failure mode is much less likely than the open coil. Shorted windings result in a reduction in inductance, since the inductance of a coil is proportional to the square of the number of turns.

SECTION REVIEW 17–9

1. Describe the effect of an inductor with shorted windings on the response of a series *RL* circuit.

2. In the circuit of Figure 17–46, indicate whether I_T, V_{R1}, and V_{R2} increase or decrease as a result of *L* opening.

FIGURE 17–46

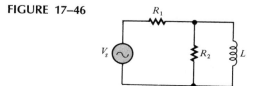

17–10   COMPUTER ANALYSIS

The following program provides for the computation of phase shift and output voltage as a function of frequency for an *RL* lead network. Inputs required are the component values, the frequency limits, and the increments of frequency.

```
 10 CLS
 20 PRINT "THIS PROGRAM COMPUTES THE PHASE SHIFT FROM INPUT
    TO"
 30 PRINT "OUTPUT AND THE NORMALIZED OUTPUT VOLTAGE MAGNITUDE"
 40 PRINT "AS FUNCTIONS OF FREQUENCY FOR AN RL LEAD NETWORK."
 50 PRINT:PRINT:PRINT
 60 INPUT "TO CONTINUE PRESS 'ENTER'";X:CLS
 70 INPUT "RESISTANCE IN OHMS";R
 80 INPUT "INDUCTANCE IN HENRIES";L
 90 INPUT "THE LOWEST FREQUENCY IN HERTZ";FL
100 INPUT "THE HIGHEST FREQUENCY IN HERTZ";FH
110 INPUT "THE FREQUENCY INCREMENTS IN HERTZ";FI
120 CLS
130 PRINT "FREQUENCY(HZ)","PHASE SHIFT","VOUT"
140 FOR F=FL TO FH STEP FI
150 XL=2*3.1416*F*L
160 PHI=90-ATN(XL/R)*57.3
170 VO=XL/(SQR(R*R+XL*XL))
180 PRINT F,PHI,VO
190 NEXT
```

SECTION REVIEW 17–10

1. List all of the variables in the program.

2. What values are displayed after the program runs?

SUMMARY

1. A sinusoidal voltage applied to an *RL* circuit produces a sinusoidal current.

2. Current lags voltage in an *RL* circuit.

3. In a series *RL* circuit, the total impedance is the phasor sum of the resistance and the inductive reactance.

4. The resistor voltage is always in phase with the current.

5. The inductor voltage always leads the current by 90 degrees.

6. The phase angle between the applied voltage and the current is dependent on the relative values of R and X_L.

7. When $X_L = R$, the phase angle is 45 degrees.

8. The true power in an ideal inductor is zero. The power loss is due only to the winding resistance in a practical inductor.

9. An inductor stores energy on the positive half of the ac cycle and then returns that stored energy to the source on the negative half-cycle.
10. When a sinusoidal source drives an *RL* circuit, part of the total power delivered by the source is resistive power (true power), and part of it is reactive power.
11. Reactive power is the rate at which energy is stored by an inductor.
12. The total power being transferred between the source and the circuit is the combination of true power and reactive power and is called the *apparent power*. Its unit is the volt-ampere (VA).
13. The power factor is the cosine of the phase angle between the source voltage and the total current.
14. In an *RL* lead network, the output voltage is across the inductor.
15. In an *RL* lag network, the output voltage is across the resistor.

FORMULAS

Series *RL* Circuits

$$\mathbf{X}_L = jX_L \tag{17-1}$$

$$\mathbf{Z} = R + jX_L \tag{17-2}$$

$$Z = \sqrt{R^2 + X_L^2} \tag{17-3}$$

$$\theta = \tan^{-1}\left(\frac{X_L}{R}\right) \tag{17-4}$$

$$\mathbf{Z} = \sqrt{R^2 + X_L^2}\angle\tan^{-1}\left(\frac{X_L}{R}\right) \tag{17-5}$$

$$\mathbf{V}_s = V_R + jV_L \tag{17-6}$$

$$\mathbf{V}_s = \sqrt{V_R^2 + V_L^2}\angle\tan^{-1}\left(\frac{V_L}{V_R}\right) \tag{17-7}$$

$$V_s = \sqrt{V_R^2 + V_L^2} \tag{17-8}$$

$$\theta = \tan^{-1}\left(\frac{V_L}{V_R}\right) \tag{17-9}$$

Parallel *RL* Circuits

$$\mathbf{Z} = \left(\frac{RX_L}{\sqrt{R^2 + X_L^2}}\right)\angle\left(90° - \tan^{-1}\left(\frac{X_L}{R}\right)\right) \tag{17-10}$$

$$Z = \frac{RX_L}{\sqrt{R^2 + X_L^2}} \tag{17-11}$$

$$\theta = 90° - \tan^{-1}\left(\frac{X_L}{R}\right) \tag{17-12}$$

$$\mathbf{B}_L = \frac{1}{X_L\angle 90°} = B_L\angle -90° = -jB_L \tag{17-13}$$

$$\mathbf{Y} = \frac{1}{Z\angle \pm\theta} = Y\angle \mp\theta \tag{17-14}$$

$$\mathbf{Y} = G - jB_L \tag{17-15}$$

$$\mathbf{I}_T = I_R - jI_L \tag{17-16}$$

$$\mathbf{I}_T = \sqrt{I_R^2 + I_L^2}\angle -\tan^{-1}\left(\frac{I_L}{I_R}\right) \tag{17-17}$$

$$I_T = \sqrt{I_R^2 + I_L^2} \tag{17-18}$$

$$\theta = -\tan^{-1}\left(\frac{I_L}{I_R}\right) \tag{17-19}$$

Power in *RL* Circuits

$$P_r = I^2 X_L \tag{17-20}$$

Lead Network

$$\phi = 90° - \tan^{-1}\left(\frac{X_L}{R}\right) \tag{17-21}$$

$$V_{\text{out}} = \left(\frac{X_L}{\sqrt{R^2 + X_L^2}}\right)V_{\text{in}} \tag{17-22}$$

$$V_{\text{out}} = IX_L \tag{17-23}$$

$$\mathbf{V}_{\text{out}} = V_{\text{out}}\angle \phi \tag{17-24}$$

Lag Network

$$\phi = -\tan^{-1}\left(\frac{X_L}{R}\right) \tag{17-25}$$

$$V_{\text{out}} = \left(\frac{R}{\sqrt{R^2 + X_L^2}}\right)V_{\text{in}} \tag{17-26}$$

$$V_{\text{out}} = IR \tag{17-27}$$

$$\mathbf{V}_{\text{out}} = V_{\text{out}}\angle -\phi \tag{17-28}$$

SELF-TEST

Solutions appear at the end of the book.

1. In a series *RL* circuit, in what way does each voltage drop differ from the applied voltage?

2. Describe the phase relationships between the current in a series *RL* circuit and
   (a) the resistor voltage   **(b)** the inductor voltage

3. If the frequency of the voltage applied to an *RL* circuit is increased, what happens to the impedance? To the phase angle?

4. If the frequency is doubled and the resistance is halved, how much does the value of the impedance change?

5. To reduce the current in a series *RL* circuit, should you increase or decrease the frequency?

6. In a series *RL* circuit, 10 V rms is measured across the resistor, and 10 V rms is measured across the inductor. What is the peak value of the source voltage?

7. The voltages in Question 6 are measured at a certain frequency. To make the resistor voltage greater than the inductor voltage, should you increase or decrease the frequency?

8. When the resistor voltage in Question 7 becomes greater than the inductor voltage, does the phase angle increase or decrease?

9. Does the impedance of a parallel *RL* circuit increase or decrease with frequency?

10. In a parallel *RL* circuit, there are 2 A rms through the resistive branch and 2 A rms through the inductive branch. What is the total rms current?

11. You are observing two voltage waveforms on an oscilloscope. The time base of the scope is adjusted so that one half-cycle of the waveforms covers the ten divisions on the horizontal axis of the screen. The positive-going zero crossing of one waveform is at the left-most division, and the positive-going zero crossing of the other is three divisions to the right. What is the phase shift between these two waveforms?

12. In general, which is more desirable, a power factor of 0.9 or a power factor of 0.78?

13. If a load is purely inductive and the reactive power is 10 VAR, what is the apparent power?

14. For a certain load, the true power is 10 W and the reactive power is 10 VAR. Which of the following does the apparent power equal?
    (a) 5 VA   **(b)** 20 VA   **(c)** 14.14 VA   **(d)** 100 VA

15. In a certain series *RL* circuit, $R = 3.3$ kΩ and $L = 20$ mH. Express the total impedance in rectangular form when $f = 10$ kHz. In polar form.

16. In Question 15, if the frequency is halved, what is the magnitude of the total impedance? What is the phase angle?

17. In Figure 17–47, at what frequency does a 45-degree phase angle between the source voltage and the current occur?

18. In a given series *RL* circuit, the inductance is 200 μH. At 5 kHz, what value of resistance will produce a 60-degree phase angle?

19. In a certain series *RL* circuit, $R = 560 \; \Omega$ and $X_L = 100 \; \Omega$. What is the phase angle?

20. Determine the current, resistor voltage, and inductor voltage in Figure 17–48. Express each in polar form.

FIGURE 17–47   FIGURE 17–48

21. Determine the impedance in polar form for each parallel *RL* circuit in Figure 17–49.

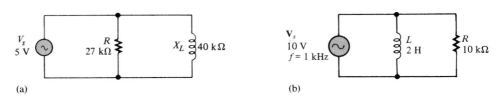

(a)   (b)

FIGURE 17–49

22. Find the total current and each branch current expressed in polar form for Figure 17–49 when $\mathbf{V}_s = 5\angle0° \; V$.

23. Determine the total impedance in Figure 17–50.

24. By what angle does the output lead the input in Figure 17–51?

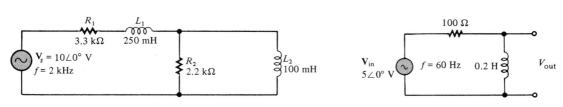

FIGURE 17–50   FIGURE 17–51

25. What is the magnitude of the output voltage in Figure 17–51?

26. For the lag network in Figure 17–52, determine the magnitude of the output voltage and the angle by which the output lags the input.

FIGURE 17–52

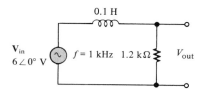

27. For Figure 17–51, is the output voltage greater at 2 kHz or at 1 kHz?

28. Determine each of the following quantities for Figure 17–52 when the frequency is 1 kHz:

    (a) P_{true} **(b)** P_r **(c)** P_a **(d)** power factor

---

PROBLEMS

Section 17–1

17–1   A 15-kHz sinusoidal voltage is applied to a series *RL* circuit. What is the frequency of I, V_R, and V_L?

17–2   What are the wave shapes of I, V_R, and V_L in Problem 17–1?

Section 17–2

17–3   Express the total impedance of each circuit in Figure 17–53 in both polar and rectangular forms.

FIGURE 17–53

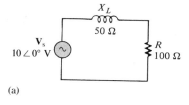

(a)

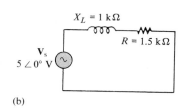

(b)

17–4   Determine the impedance magnitude and phase angle in each circuit in Figure 17–54. Sketch the impedance diagrams.

FIGURE 17–54

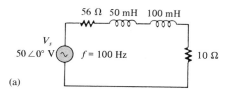

(a)

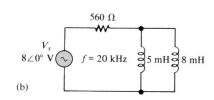

(b)

17–5   In Figure 17–55, determine the impedance at each of the following frequencies:
   (a) 100 Hz   **(b)** 500 Hz   **(c)** 1 kHz   **(d)** 2 kHz

17–6   Determine the values of R and X_L in a series *RL* circuit for the following values of total impedance:
   (a) $\mathbf{Z} = 20\ \Omega + j45\ \Omega$ **(b)** $\mathbf{Z} = 500\angle35°\ \Omega$
   (c) $\mathbf{Z} = 2.5\angle72.5°\ \text{k}\Omega$ **(d)** $\mathbf{Z} = 998\angle45°\ \Omega$

17–7   Reduce the circuit in Figure 17–56 to a single resistance and inductance in series.

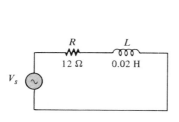

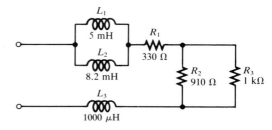

FIGURE 17–55   **FIGURE 17–56**

Section 17–3

17–8   Express the current in polar form for each circuit of Figure 17–53.

17–9   Calculate the total current in each circuit of Figure 17–54, and express in polar form.

17–10   Determine θ for the circuit in Figure 17–57.

17–11   If the inductance in Figure 17–57 is doubled, does θ increase or decrease, and by how many degrees?

17–12   Sketch the waveforms for $\mathbf{V}_s$, $\mathbf{V}_R$, and $\mathbf{V}_L$ in Figure 17–57. Show the proper phase relationships.

17–13   For the circuit in Figure 17–58, find $\mathbf{V}_R$ and $\mathbf{V}_L$ for each of the following frequencies:
   (a) 60 Hz   **(b)** 200 Hz   **(c)** 500 Hz   **(d)** 1 kHz

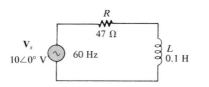

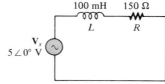

FIGURE 17–57   **FIGURE 17–58**

17–14   Determine the magnitude and phase angle of the source voltage in Figure 17–59.

FIGURE 17–59

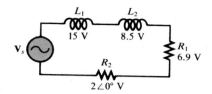

Section 17–4

17–15 What is the impedance expressed in polar form for the circuit in Figure 17–60?

FIGURE 17–60

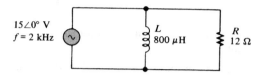

17–16 Repeat Problem 17–15 for the following frequencies:
(a) 1.5 kHz (b) 3 kHz (c) 5 kHz (d) 10 kHz

17–17 At what frequency does X_L equal R in Figure 17–60?

Section 17–5

17–18 Find the total current and each branch current in Figure 17–61.

17–19 Determine the following quantities in Figure 17–62:
(a) $\mathbf{Z}$ (b) $\mathbf{I}_R$ (c) $\mathbf{I}_L$ (d) $\mathbf{I}_T$ (e) θ

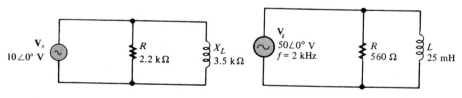

FIGURE 17–61   **FIGURE 17–62**

17–20 Repeat Problem 17–19 for $R = 56\ \Omega$ and $L = 330\ \mu H$.

17–21 Convert the circuit in Figure 17–63 to an equivalent series form.

17–22 Find the magnitude and phase angle of the total current in Figure 17–64.

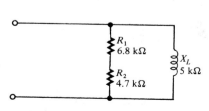

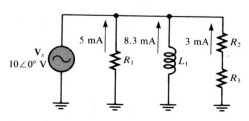

FIGURE 17–63   **FIGURE 17–64**

Section 17-6

17-23 Determine the voltages in polar form across each element in Figure 17-65. Sketch the voltage phasor diagram.

FIGURE 17-65

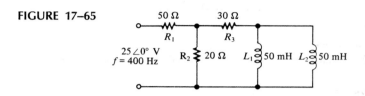

17-24 Is the circuit in Figure 17-65 predominantly resistive or predominantly inductive?

17-25 Find the current in each branch and the total current in Figure 17-65. Express the currents in polar form. Sketch the current phasor diagram.

17-26 For the circuit in Figure 17-66, determine the following:
(a) $\mathbf{I}_T$ (b) θ (c) $\mathbf{V}_{R1}$ (d) $\mathbf{V}_{R2}$
(e) $\mathbf{V}_{R3}$ (f) $\mathbf{V}_{L1}$ (g) $\mathbf{V}_{L2}$

17-27 For the circuit in Figure 17-67, determine the following:
(a) $\mathbf{I}_T$ (b) $\mathbf{V}_{L1}$ (c) $\mathbf{V}_{ab}$

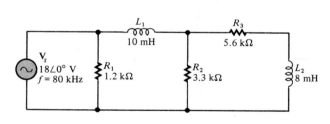

FIGURE 17-66

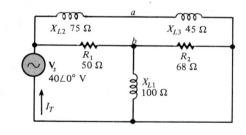

FIGURE 17-67

17-28 Draw the phasor diagram of all voltages and currents in Figure 17-67.

17-29 Determine the phase shift and attenuation (ratio of V_{out} to V_{in}) from the input to the output for the network in Figure 17-68.

FIGURE 17-68

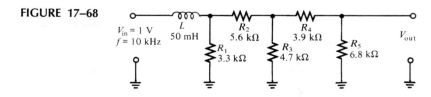

17-30 Determine the phase shift and attenuation from the input to the output for the ladder network in Figure 17-69.

FIGURE 17–69

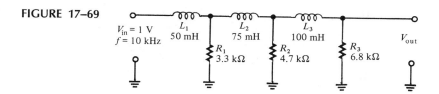

17–31 Design an ideal inductive switching circuit that will provide a momentary voltage of 2.5 kV from a 12-V dc source when a switch is thrown instantaneously from one position to another. The drain on the source must not exceed 1 A.

Section 17–7

17–32 In a certain *RL* circuit, the true power is 100 mW, and the reactive power is 340 mVAR. What is the apparent power?

17–33 Determine the true power and the reactive power in Figure 17–57.

17–34 What is the power factor in Figure 17–61?

17–35 Determine P_{true}, P_r, P_a, and *PF* for the circuit in Figure 17–66. Sketch the power triangle.

17–36 Find the true power for the circuit in Figure 17–67.

Section 17–8

17–37 For the lag network in Figure 17–70, determine the phase lag of the output voltage with respect to the input for the following frequencies:
  (a) 1 Hz   **(b)** 100 Hz   **(c)** 1 kHz   **(d)** 10 kHz

17–38 Draw the response curve for the circuit in Figure 17–70. Show the output voltage versus frequency in 1-kHz increments from 0 Hz to 5 kHz.

17–39 Repeat Problem 17–37 for the lead network in Figure 17–71.

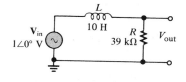

FIGURE 17–70

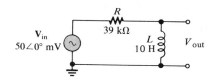

FIGURE 17–71

17–40 Using the same procedure as in Problem 17–38, draw the response curve for Figure 17–71.

17–41 Sketch the voltage phasor diagram for each circuit in Figures 17–70 and 17–71 for a frequency of 8 kHz.

Section 17–9

17–42 Determine the voltage across each element in Figure 17–66 if L_1 were open.

17–43 Determine the output voltage in Figure 17–72 for each of the following failure modes:
  (a) L_1 open   **(b)** L_2 open   **(c)** R_1 open   **(d)** a short across R_2

FIGURE 17–72

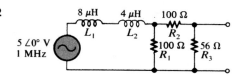

Section 17–10

17–44 Develop a flowchart for the program in Section 17–10.

17–45 Modify the program in Section 17–10 to compute and tabulate current and true power in addition to phase shift and output voltage.

17–46 Write a program similar to the program in Section 17–10 for an *RL* lag network.

ANSWERS TO SECTION REVIEWS

Section 17–1
1. 1 kHz.   **2.** Closer to zero.

Section 17–2
1. $R = 150 \ \Omega$, $X_L = 220 \ \Omega$.   **2.** 33 kΩ + j50 kΩ, 59.9$\angle$56.58° kΩ.

Section 17–3
1. 3.61 V.   **2.** 56.31°.   **3.** X_L increases; Z increases; θ increases.

Section 17–4
1. 0.002 S.   **2.** 0.025 S.   **3.** Lags, 33.69°

Section 17–5
1. 32 mA.   **2.** 23.32$\angle$ − 59.04° mA; the input voltage.   **3.** − 90°.

Section 17–6
1. $Z = 494.6\angle59.86°\ \Omega$.   **2.** $I_T = 20.22\angle - 59.86°$ mA.

Section 17–7
1. Resistor.   **2.** 0.643.   **3.** $P_{\text{true}} = 4.7$ W; $P_r = 6.2$ VAR; $P_a = 7.78$ VA.

Section 17–8
1. 81.87°.   **2.** 9.9 V.   **3.** Resistor.

Section 17–9
1. Shorted windings reduce L and thereby reduce X_L at any given frequency.
2. I_T decreases, V_{R1} decreases, V_{R2} increases.

Section 17–10
1. R, L, FL, FH, FI.
2. Frequency, phase shift, and output voltage.

EIGHTEEN

RLC CIRCUITS AND RESONANCE

In this chapter, the analysis methods learned in the last two chapters are extended to the coverage of circuits with combinations of resistive, inductive, and capacitive elements. Both series and parallel *RLC* circuits, plus combinations of both, are studied.

The concept of resonance is introduced in this chapter. Resonance is an important aspect of electronic applications. It is the basis for frequency selectivity in communications. For example, the ability of a radio or television receiver to select a certain frequency transmitted by a particular station and to eliminate frequencies from other stations is based on the principle of resonance.

Resonance occurs under certain specific conditions in circuits that have combinations of inductance and capacitance. The conditions that produce resonance and the characteristics of resonant circuits are covered in this chapter.

In this chapter, you will learn:

☐ How to determine the impedance of a series *RLC* circuit.
☐ How to analyze a series *RLC* circuit for current and voltage relationships, as well as for power.
☐ The meaning of *series resonance*.
☐ The conditions that produce resonance in series *RLC* circuits.

☐ How to determine the frequency, current, and impedance at resonance in a parallel *RLC* circuit.

☐ How to determine the impedance of a parallel *RLC* circuit.

☐ How to analyze a parallel *RLC* circuit for currents and voltage relationships.

☐ The meaning of parallel resonance, and how it differs from series resonance.

☐ The conditions that produce parallel resonance.

☐ How to determine the frequency, current, and impedance at resonance in a parallel *RLC* circuit.

☐ How to determine the bandwidth and *Q* of both series and parallel resonant circuits.

18–1   IMPEDANCE OF SERIES *RLC* CIRCUITS

A series *RLC* circuit is shown in Figure 18–1. It contains resistance, inductance, and capacitance. As you know, $\mathbf{X}_L$ causes the total current to lag the applied voltage. $\mathbf{X}_C$ has the opposite effect: It causes the current to lead the voltage. Thus $\mathbf{X}_L$ and $\mathbf{X}_C$ tend to offset each other. When they are equal, they cancel, and the total reactance is zero. In any case, the magnitude of the total reactance in the series circuit is

$$X_T = |X_L - X_C| \tag{18–1}$$

The term $|X_L - X_C|$ means the *absolute value* of the difference of the two reactances. That is, the sign of the result is considered positive no matter which reactance is greater. For example, $3 - 7 = -4$, but the absolute value is

$$|3 - 7| = 4$$

When $X_L > X_C$, the circuit is predominantly inductive, and when $X_C > X_L$, the circuit is predominantly capacitive.

FIGURE 18–1
Series *RLC* circuit.

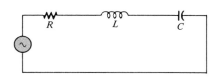

The total impedance for the series *RLC* circuit is stated in rectangular form in Equation (18–2) and in polar form in Equation (18–3).

$$\mathbf{Z} = R + jX_L - jX_C \tag{18–2}$$

$$\mathbf{Z} = \sqrt{R^2 + (X_L - X_C)^2} \angle \tan^{-1}\left(\frac{X_T}{R}\right) \tag{18–3}$$

In Equation (18–3), $\sqrt{R^2 + (X_L - X_C)^2}$ is the magnitude and $\tan^{-1}(X_T/R)$ is the phase angle between the total current and the applied voltage.

**EXAMPLE
18–1**

Determine the total impedance in Figure 18–2. Express it in both rectangular and polar forms.

FIGURE 18–2

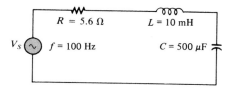

Solution:
First find X_C and X_L:

$$X_C = \frac{1}{2\pi fC} = \frac{1}{2\pi(100 \text{ Hz})(500 \text{ }\mu\text{F})} = 3.18 \text{ }\Omega$$

$$X_L = 2\pi fL = 2\pi(100 \text{ Hz})(10 \text{ mH}) = 6.28 \text{ }\Omega$$

In this case, X_L is greater than X_C, and thus the circuit is more inductive than capacitive. The magnitude of the total reactance is

$$X_T = |X_L - X_C| = |6.28 \text{ }\Omega - 3.18 \text{ }\Omega| = 3.1 \text{ }\Omega \quad \text{inductive}$$

The impedance in rectangular form is

$$\mathbf{Z} = R + (jX_L - jX_C) = 5.6 \text{ }\Omega + (j6.28 \text{ }\Omega - j3.18 \text{ }\Omega)$$
$$= 5.6 \text{ }\Omega + j3.1 \text{ }\Omega$$

The impedance in polar form is

$$\mathbf{Z} = \sqrt{R^2 + X_T^2} \angle \tan^{-1}\left(\frac{X_T}{R}\right)$$

$$= \sqrt{(5.6 \text{ }\Omega)^2 + (3.1 \text{ }\Omega)^2} \angle \tan^{-1}\left(\frac{3.1 \text{ }\Omega}{5.6 \text{ }\Omega}\right)$$

$$\doteq 6.40 \angle 28.97° \text{ }\Omega$$

The calculator sequence for conversion from the rectangular to the polar form is

Magnitude:
Angle: | b |

As you have seen, when the inductive reactance is greater than the capacitive reactance, the circuit appears inductive; so the current lags the applied voltage. When the capacitive reactance is greater, the circuit appears capacitive, and the current leads the applied voltage.

SECTION REVIEW 18–1

1. In a given series *RLC* circuit, X_C is 150 Ω and X_L is 80 Ω. What is the total reactance in ohms? Is it inductive or capacitive?
2. Determine the impedance in polar form for the circuit in Question 1 when $R = 47$ Ω. What is the magnitude of the impedance? What is the phase angle? Is the current leading or lagging the applied voltage?

18–2   ANALYSIS OF SERIES *RLC* CIRCUITS

Recall that capacitive reactance varies inversely with frequency and inductive reactance varies directly. As shown in Figure 18–3, you can see that for a typical series *RLC* circuit the total reactance behaves as follows: Starting at a very low frequency, X_C is high, and X_L is low, and the circuit is predominantly capacitive. As the frequency is increased, X_C decreases and X_L increases until a value is reached where $X_C = X_L$ and the two reactances cancel, making the circuit purely resistive. This condition is *series resonance* and will be studied in a later section. As the frequency is increased further, X_L becomes greater than X_C, and the circuit is predominantly inductive. The following example illustrates how the impedance and phase angle change as the source frequency is varied.

FIGURE 18–3
How X_C and X_L vary with frequency.

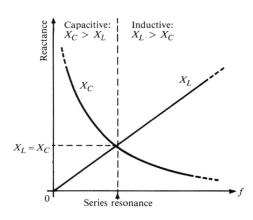

**EXAMPLE
18–2**

For each of the following input frequencies, find the impedance in polar form for the circuit in Figure 18–4. Note the change in magnitude and phase angle.
    (a) $f = 1$ kHz   **(b)** $f = 2$ kHz
    (c) $f = 3.5$ kHz   **(d)** $f = 5$ kHz

FIGURE 18–4

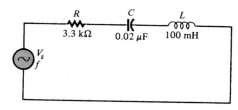

Solution:
(a) At $f = 1$ kHz:

$$X_C = \frac{1}{2\pi(1 \text{ kHz})(0.02 \ \mu\text{F})} = 7.96 \text{ k}\Omega$$

$$X_L = 2\pi(1 \text{ kHz})(100 \text{ mH}) = 628.3 \ \Omega$$

The circuit is clearly capacitive, and the impedance is

$$\mathbf{Z} = \sqrt{(3.3 \text{ k}\Omega)^2 + (628.3 \ \Omega - 7.96 \text{ k}\Omega)^2} \angle -\tan^{-1}\left(\frac{7.33 \text{ k}\Omega}{3.3 \text{ k}\Omega}\right)$$

$$= 8.04 \angle -65.76° \text{ k}\Omega$$

(b) At $f = 2$ kHz:

$$X_C = \frac{1}{2\pi(2 \text{ kHz})(0.02 \ \mu\text{F})} = 3.98 \text{ k}\Omega$$

$$X_L = 2\pi(2 \text{ kHz})(100 \text{ mH}) = 1.26 \text{ k}\Omega$$

The circuit is still capacitive, and the impedance is

$$\mathbf{Z} = \sqrt{(3.3 \text{ k}\Omega)^2 + (1.26 \text{ k}\Omega - 3.98 \text{ k}\Omega)^2} \angle -\tan^{-1}\left(\frac{2.72 \text{ k}\Omega}{3.3 \text{ k}\Omega}\right)$$

$$= 4.28 \angle -39.5° \text{ k}\Omega$$

(c) At $f = 3.5$ kHz:

$$X_C = \frac{1}{2\pi(3.5 \text{ kHz})(0.02 \ \mu\text{F})} = 2.27 \text{ k}\Omega$$

$$X_L = 2\pi(3.5 \text{ kHz})(100 \text{ mH}) = 2.20 \text{ k}\Omega$$

The circuit is very close to being purely resistive but is still slightly capacitive.

$$\mathbf{Z} = \sqrt{(3.3 \text{ k}\Omega)^2 + (2.20 \text{ k}\Omega - 2.27 \text{ k}\Omega)^2} \angle -\tan^{-1}\left(\frac{0.07 \text{ k}\Omega}{3.3 \text{ k}\Omega}\right)$$

$$\cong 3.3 \angle -1.2° \text{ k}\Omega$$

(d) At $f = 5$ kHz:

$$X_C = \frac{1}{2\pi(5 \text{ kHz})(0.02 \text{ }\mu\text{F})} = 1.59 \text{ k}\Omega$$

$$X_L = 2\pi(5 \text{ kHz})(100 \text{ mH}) = 3.14 \text{ k}\Omega$$

The circuit is now predominantly inductive.

$$\mathbf{Z} = \sqrt{(3.3 \text{ k}\Omega)^2 + (3.14 \text{ k}\Omega - 1.59 \text{ k}\Omega)^2} \angle \tan^{-1}\left(\frac{1.55 \text{ k}\Omega}{3.3 \text{ k}\Omega}\right)$$

$$= 3.64\angle 25.16° \text{ k}\Omega$$

Notice how the circuit changed from capacitive to inductive as the frequency increased. The phase condition changed from the current leading to the current lagging as indicated by the sign of the angle. It is interesting to note that the impedance magnitude decreased to a minimum approximately equal to the resistance and then began increasing again.

In a series *RLC* circuit, the capacitor voltage and the inductor voltage are *always* 180 degrees out of phase with each other. For this reason, V_C and V_L subtract from each other, and thus the voltage across L and C combined is always less than the larger individual voltage across either element, as illustrated in Figure 18–5 and in the waveform diagram of Figure 18–6.

FIGURE 18–5
The voltage across the series combination of C and L is always less than the larger individual voltage.

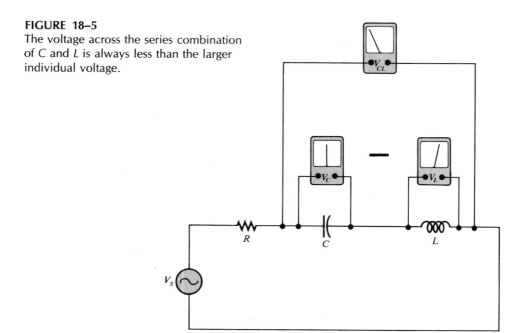

FIGURE 18–6

V_L and V_C effectively subtract.

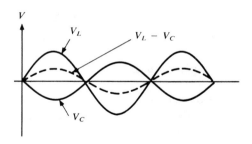

In the next example, Ohm's law is used to find the current and voltages in the series *RLC* circuit.

EXAMPLE 18–3

Find the current and the voltages across each element in Figure 18–7. Express each quantity in polar form, and draw a complete voltage phasor diagram.

FIGURE 18–7

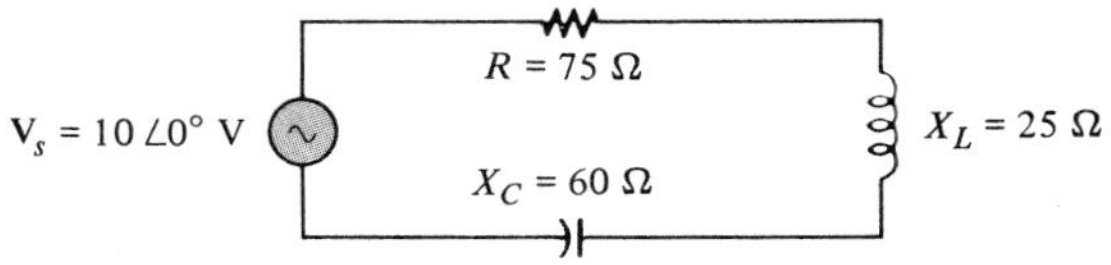

$\mathbf{V}_s = 10 \angle 0°$ V

$R = 75\ \Omega$

$X_C = 60\ \Omega$

$X_L = 25\ \Omega$

Solution:
First find the total impedance:

$$\mathbf{Z} = R + jX_L - jX_C = 75\ \Omega + j25\ \Omega - j60\ \Omega$$
$$= 75\ \Omega - j35\ \Omega$$

Convert to polar form for convenience in applying Ohm's law:

$$\mathbf{Z} = \sqrt{(75\ \Omega)^2 + (35\ \Omega)^2} \angle -\tan^{-1}\left(\frac{35}{75}\right)$$
$$= 82.76 \angle -25°\ \Omega$$

Apply Ohm's law to find the current:

$$\mathbf{I} = \frac{\mathbf{V}_s}{\mathbf{Z}} = \frac{10 \angle 0°\ \text{V}}{82.76 \angle -25°\ \Omega} = 0.121 \angle 25°\ \text{A}$$

Now apply Ohm's law to find the voltages across *R*, *L*, and *C*:

$$\mathbf{V}_R = \mathbf{I}R = (0.121 \angle 25°\ \text{A})(75 \angle 0°\ \Omega)$$
$$= 9.075 \angle 25°\ \text{V}$$
$$\mathbf{V}_L = \mathbf{I}X_L = (0.121 \angle 25°\ \text{A})(25 \angle 90°\ \Omega)$$
$$= 3.025 \angle 115°\ \text{V}$$

$$\mathbf{V}_C = \mathbf{IX}_C = (0.121 \angle 25° \text{ A})(60 \angle -90° \text{ }\Omega)$$

$$= 7.26 \angle -65° \text{ V}$$

The phasor diagram is shown in Figure 18–8. The magnitudes represent rms values. Notice that $\mathbf{V}_L$ is leading $\mathbf{V}_R$ by 90 degrees, and $\mathbf{V}_C$ is lagging $\mathbf{V}_R$ by 90 degrees. Also, there is a 180-degree phase difference between $\mathbf{V}_L$ and $\mathbf{V}_C$. If the current phasor were shown, it would be at the same angle as $\mathbf{V}_R$. The current is leading $\mathbf{V}_s$, the source voltage, by 25 degrees, indicating a capacitive circuit ($X_C > X_L$).

FIGURE 18–8

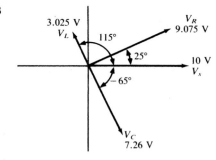

SECTION REVIEW 18–2

1. The following voltages occur in a certain series *RLC* circuit. Determine the source voltage: $\mathbf{V}_R = 24 \angle 30°$ V, $\mathbf{V}_L = 15 \angle 120°$ V, and $\mathbf{V}_C = 45 \angle -60°$ V.
2. When $R = 10$ Ω, $X_C = 18$ Ω, and $X_L = 12$ Ω, does the current lead or lag the applied voltage?
3. Determine the total reactance in Question 2.

18–3 SERIES RESONANCE

In a series *RLC* circuit, *series resonance* occurs when $X_L = X_C$. The frequency at which resonance occurs is called the *resonant frequency*, f_r. Figure 18–9 illustrates the series resonant condition.

FIGURE 18–9
Series resonance.

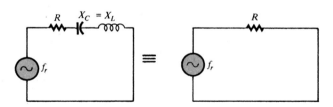

In a series resonant circuit, the total impedance is

$$\mathbf{Z} = R + jX_L - jX_C \qquad \text{(18–4)}$$

Since $X_L = X_C$, the j terms cancel, and the *impedance is purely resistive*. These resonant conditions are stated in the following equations.

$$X_L = X_C \qquad \text{(18–5)}$$

$$Z_r = R \qquad \text{(18–6)}$$

EXAMPLE 18–4

For the series *RLC* circuit in Figure 18–10, determine X_C and $\mathbf{Z}$ at resonance.

FIGURE 18–10

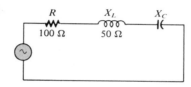

Solution:

$$X_L = X_C \quad \text{at the resonant frequency}$$

Thus,

$$X_C = 50\ \Omega$$

$$\mathbf{Z}_r = R + jX_L - jX_C = 100\ \Omega + j50\ \Omega - j50\ \Omega$$

$$= 100\angle 0°\ \Omega$$

The impedance is equal to the resistance because the reactances are equal in magnitude and therefore cancel.

WHY X_L AND X_C EFFECTIVELY CANCEL AT RESONANCE

At the series resonant frequency, the voltages across *C* and *L* are equal in magnitude because the reactances are equal and because each has the same current since they are in series ($IX_C = IX_L$). Also, V_L and V_C are always 180 degrees out of phase with each other.

During any given cycle, the polarities of the voltages across *C* and *L* are opposite, as shown in Figures 18–11(a) and 18–11(b). The equal and opposite voltages across *C* and *L* cancel, leaving zero volts from point *A* to point *B* as shown in the figure. Since there is no voltage drop from *A* to *B* but there is still current, the total reactance must be zero, as indicated in Part (c) of the figure. Also, the voltage phasor diagram in Part (d) shows that V_C and V_L are equal in magnitude and 180 degrees out of phase with each other.

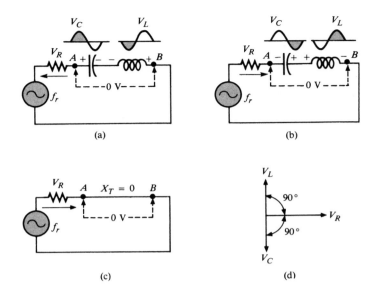

FIGURE 18–11

At the resonant frequency, f_r, the voltages across C and L are equal in magnitude. Since they are 180 degrees out of phase with each other, they cancel, leaving 0 V across the LC combination (point A to point B). The section of the circuit from A to B effectively looks like a short at resonance.

SERIES RESONANT FREQUENCY

For a given series *RLC* circuit, resonance happens at only one specific frequency. A formula for this resonant frequency is developed as follows:

$$X_L = X_C$$

Substituting the reactance formulas, we have

$$2\pi f_r L = \frac{1}{2\pi f_r C}$$

Solving for f_r:

$$(2\pi f_r L)(2\pi f_r C) = 1$$

$$4\pi^2 f_r^2 LC = 1$$

$$f_r^2 = \frac{1}{4\pi^2 LC}$$

Taking the square root of both sides:

$$f_r = \frac{1}{2\pi\sqrt{LC}} \tag{18–7}$$

EXAMPLE 18–5

Find the series resonant frequency for the circuit in Figure 18–12.

FIGURE 18–12

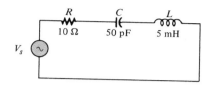

Solution:

$$f_r = \frac{1}{2\pi\sqrt{LC}} = \frac{1}{2\pi\sqrt{(5\ \text{mH})(50\ \text{pF})}}$$

$$= 318\ \text{kHz}$$

SERIES *RLC* IMPEDANCE

At frequencies below f_r, $X_C > X_L$; thus, the circuit is capacitive. At the resonant frequency, $X_C = X_L$, so the circuit is purely resistive. At frequencies above f_r, $X_L > X_C$; thus, the circuit is inductive.

   The impedance magnitude is minimum at resonance $(Z = R)$ and increases in value above and below the resonant point. The graph in Figure 18–13 illustrates how impedance changes with frequency. At zero frequency, both X_C and Z are infinitely large and X_L is zero, because the capacitor looks like an open at 0 Hz and the inductor looks like a short. As the frequency increases, X_C decreases and X_L increases. Since X_C is larger than X_L at frequencies below f_r, Z decreases along with X_C. At f_r, $X_C = X_L$ and $Z = R$. At frequencies above f_r, X_L becomes increasingly larger than X_C, causing Z to increase.

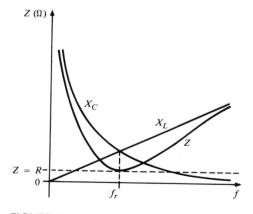

FIGURE 18–13
Series *RLC* impedance as a function of frequency.

EXAMPLE
18–6

For the circuit in Figure 18–14, determine the impedance magnitude at resonance, at 1000 Hz below resonance, and at 1000 Hz above resonance.

FIGURE 18–14

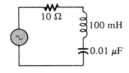

10 Ω
100 mH
0.01 μF

Solution:

$$f_r = \frac{1}{2\pi\sqrt{LC}} = \frac{1}{2\pi\sqrt{(100\ \text{mH})(0.01\ \mu\text{F})}}$$

$$= 5.03\ \text{kHz}$$

At 1000 Hz below f_r:

$$f_r - 1\ \text{kHz} = 4.03\ \text{kHz}$$

At 1000 Hz above f_r:

$$f_r + 1\ \text{kHz} = 6.03\ \text{kHz}$$

The impedance at resonance is equal to R:

$$Z = 10\ \Omega$$

The impedance at $f_r - 1\ \text{kHz}$:

$$Z = \sqrt{R^2 + (X_L - X_C)^2}$$
$$= \sqrt{(10\ \Omega)^2 + (2.53\ \text{k}\Omega - 3.95\ \text{k}\Omega)^2}$$
$$= 1.42\ \text{k}\Omega$$

Notice that X_C is greater than X_L; so Z is capacitive.
The impedance at $f_r + 1\ \text{kHz}$:

$$Z = \sqrt{R^2 + (X_L - X_C)^2}$$
$$= \sqrt{(10\ \Omega)^2 + (3.79\ \text{k}\Omega - 2.64\ \text{k}\Omega)^2}$$
$$= 1.15\ \text{k}\Omega$$

X_L is greater than X_C; so Z is inductive.

CURRENT AND VOLTAGES IN A SERIES *RLC* CIRCUIT

At the series resonant frequency, *the current is maximum* ($I_{max} = V_s/R$). Above and below resonance, the current decreases because the impedance increases. A *response curve* showing the plot of current versus frequency is shown in Figure 18–15(a).

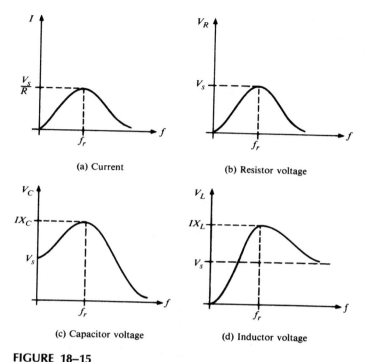

(a) Current

(b) Resistor voltage

(c) Capacitor voltage

(d) Inductor voltage

FIGURE 18–15
Current and voltage magnitudes as a function of frequency in a series *RLC* circuit. (Note that the peak magnitudes are not shown to scale with respect to each other.)

The resistor voltage, V_R, follows the current and is maximum (equal to V_s) at resonance and zero at $f = 0$ and at $f = \infty$, as shown in Figure 18–15(b). The general shapes of the V_C and V_L curves are indicated in Figure 18–15(c) and (d). Notice that $V_C = V_s$ when $f = 0$, because the capacitor appears open. Also notice that V_L approaches V_s as f approaches infinity, because the inductor appears open.

The voltages are maximum at resonance but drop off above and below f_r. The voltages across L and C at resonance are exactly *equal in magnitude but 180 degrees out of phase; so they cancel.* Thus, the total voltage across both L and C is zero, and $V_R = V_s$ at resonance, as indicated in Figure 18–16. Individually, V_L and V_C can be much greater than the source voltage, as you will see later. Keep in mind that V_L and V_C are always opposite in polarity regardless of the frequency, but only at resonance are their magnitudes equal.

FIGURE 18–16
Series *RLC* circuit at resonance.

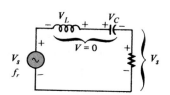

**EXAMPLE
18–7**

Find I, V_R, V_L, and V_C at resonance in Figure 18–17. The resonant values of X_L and X_C are shown.

FIGURE 18–17

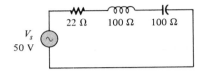

Solution:
At resonance, I is maximum and equal to V_s/R:

$$I = \frac{V_s}{R} = \frac{50 \text{ V}}{22 \text{ } \Omega} = 2.27 \text{ A}$$

Applying Ohm's law, the following voltage magnitudes are obtained:

$$V_R = IR = (2.27 \text{ A})(22 \text{ } \Omega) = 49.94 \text{ V}$$
$$V_L = IX_L = (2.27 \text{ A})(100 \text{ } \Omega) = 227 \text{ V}$$
$$V_C = IX_C = (2.27 \text{ A})(100 \text{ } \Omega) = 227 \text{ V}$$

Notice that all of the source voltage is dropped across the resistor. Also, of course, V_L and V_C are equal in magnitude but opposite in phase. This causes these voltages to cancel, making the *total* reactive voltage zero.

THE PHASE ANGLE OF A SERIES *RLC* CIRCUIT

At frequencies below resonance, $X_C > X_L$, and the current *leads* the source voltage, as indicated in Figure 18–18(a). The phase angle decreases as the frequency approaches the resonant value and is 0 degrees at resonance, as indicated in Part (b). At frequencies above resonance, $X_L > X_C$, and the current *lags* the source voltage, as indicated in Part (c). As the frequency goes higher, the phase angle approaches 90 degrees. A plot of phase angle versus frequency is shown in Part (d) of the figure.

SECTION REVIEW 18–3

1. What is the condition for series resonance?
2. Why is the current maximum at the resonant frequency?
3. Calculate the resonant frequency for $C = 1000$ pF and $L = 1000$ μH.
4. In Question 3, is the circuit inductive or capacitive at 50 kHz?

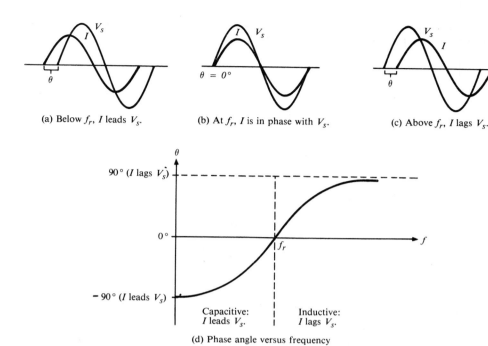

(a) Below f_r, I leads V_s. (b) At f_r, I is in phase with V_s. (c) Above f_r, I lags V_s.

(d) Phase angle versus frequency

FIGURE 18–18
The phase angle as a function of frequency in a series *RLC* circuit.

18–4

IMPEDANCE OF PARALLEL *RLC* CIRCUITS

A parallel *RLC* circuit is shown in Figure 18–19. The total impedance can be calculated using the sum-of-reciprocals method, just as was done for circuits with resistors in parallel.

$$\frac{1}{\mathbf{Z}} = \frac{1}{R\angle 0°} + \frac{1}{X_L\angle 90°} + \frac{1}{X_C\angle -90°}$$

or

$$\mathbf{Z} = \frac{1}{\dfrac{1}{R\angle 0°} + \dfrac{1}{X_L\angle 90°} + \dfrac{1}{X_C\angle -90°}} \qquad (18\text{–}8)$$

FIGURE 18–19
Parallel *RLC* circuit.

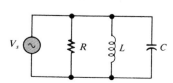

EXAMPLE 18–8

Find **Z** in polar form in Figure 18–20.

FIGURE 18–20

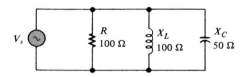

Solution:

$$\frac{1}{\mathbf{Z}} = \frac{1}{R\angle 0°} + \frac{1}{X_L\angle 90°} + \frac{1}{X_C\angle -90°}$$

$$= \frac{1}{100\angle 0° \ \Omega} + \frac{1}{100\angle 90° \ \Omega} + \frac{1}{50\angle -90° \ \Omega}$$

Applying the rule for division of polar numbers:

$$\frac{1}{\mathbf{Z}} = 0.01\angle 0° \ \text{S} + 0.01\angle -90° \ \text{S} + 0.02\angle 90° \ \text{S}$$

Recall that the sign of the denominator angle changes when dividing. Now, converting each term to its rectangular equivalent gives

$$\frac{1}{\mathbf{Z}} = 0.01 \ \text{S} - j0.01 \ \text{S} + j0.02 \ \text{S} = 0.01 \ \text{S} + j0.01 \ \text{S}$$

Taking the reciprocal to obtain **Z,** we get

$$\mathbf{Z} = \frac{1}{0.01 \ \text{S} + j0.01 \ \text{S}}$$

$$= \frac{1}{\sqrt{(0.01 \ \text{S})^2 + (0.01 \ \text{S})^2}\angle \tan^{-1}\left(\frac{0.01}{0.01}\right)}$$

$$= \frac{1}{0.01414\angle 45° \ \text{S}} = 70.7\angle -45° \ \Omega$$

The negative angle shows that the circuit is capacitive. This may surprise you, since $X_L > X_C$. However, in a parallel circuit, the smaller quantity has the greater effect on the total current. Just as in the case of all resistances in parallel, the smaller draws more current and has the greater effect on the total *R*.

In this circuit, the total current leads the total voltage by a phase angle of 45 degrees.

CONDUCTANCE, SUSCEPTANCE, AND ADMITTANCE

You have already learned the concepts of conductance, susceptance, and admittance. The formulas are restated here for convenience.

$$\mathbf{G} = \frac{1}{R\angle 0°} = G\angle 0° \qquad (18\text{--}9)$$

$$\mathbf{B}_C = \frac{1}{X_C\angle -90°} = B_C\angle 90° = jB_C \qquad (18\text{--}10)$$

$$\mathbf{B}_L = \frac{1}{X_L\angle 90°} = B_L\angle -90° = -jB_L \qquad (18\text{--}11)$$

$$\mathbf{Y} = \frac{1}{Z\angle \pm\theta} = Y\angle \mp\theta = G + jB_C - jB_L \qquad (18\text{--}12)$$

As you know, the unit of each of these quantities is the siemen (S).

EXAMPLE 18–9

Determine the conductance, capacitive susceptance, inductive susceptance, and total admittance in Figure 18–21.

FIGURE 18–21

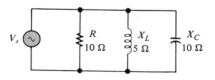

Solution:

$$\mathbf{G} = \frac{1}{R\angle 0°} = \frac{1}{10\angle 0° \ \Omega} = 0.1\angle 0° \ \text{S}$$

$$\mathbf{B}_C = \frac{1}{X_C\angle -90°} = \frac{1}{10\angle -90° \ \Omega} = 0.1\angle 90° \ \text{S}$$

$$\mathbf{B}_L = \frac{1}{X_L\angle 90°} = \frac{1}{5\angle 90° \ \Omega} = 0.2\angle -90° \ \text{S}$$

$$\mathbf{Y} = G + jB_C - jB_L = 0.1 \ \text{S} + j0.1 \ \text{S} - j0.2 \ \text{S}$$

$$= 0.1 \ \text{S} - j0.1 \ \text{S} = 0.1414\angle -45° \ \text{S}$$

From **Y**, we can get **Z**:

$$\mathbf{Z} = \frac{1}{\mathbf{Y}} = \frac{1}{0.1414\angle -45° \ \text{S}} = 7.07\angle 45° \ \Omega$$

SECTION REVIEW 18–4

1. In a certain parallel *RLC* circuit, the capacitive reactance is 60 Ω, and the inductive reactance is 100 Ω. Is the circuit predominantly capacitive or inductive?

2. Determine the admittance of a parallel circuit in which $R = 1$ kΩ, $X_C = 500$ Ω, and $X_L = 1.2$ kΩ.

3. In Question 2, what is the impedance?

18–5   ANALYSIS OF PARALLEL AND SERIES-PARALLEL *RLC* CIRCUITS

As you have seen, the smaller reactance in parallel dominates, because it results in the larger branch current. At low frequencies, the inductive reactance is less than the capacitive reactance; therefore, the circuit is inductive. As the frequency is increased, X_L increases and X_C decreases until a value is reached where $X_L = X_C$. This is the point of parallel resonance. As the frequency is increased further, X_C becomes smaller than X_L, and the circuit becomes capacitive.

CURRENT RELATIONSHIPS

In a parallel *RLC* circuit, the current in the capacitive branch and the current in the inductive branch are *always* 180 degrees out of phase with each other (neglecting any coil resistance). For this reason, I_C and I_L subtract from each other, and thus the total current into the parallel branches of L and C is always less than the largest individual branch current, as illustrated in Figure 18–22 and in the waveform diagram of Figure 18–23. Of course, the current in the resistive branch is always 90 degrees out of phase with both reactive currents, as shown in the current phasor diagram of Figure 18–24.

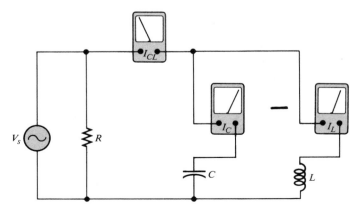

FIGURE 18–22
The total current into the parallel combination of C and L is the difference of the two branch currents.

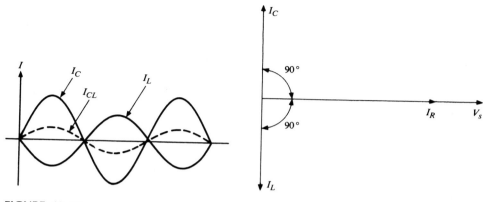

FIGURE 18–23
I_C and I_L effectively subtract.

FIGURE 18–24
Current phasor diagram for a parallel *RLC* circuit.

The total current can be expressed as

$$\mathbf{I}_T = \sqrt{I_R^2 + (I_C - I_L)^2} \angle \tan^{-1}\left(\frac{I_{CL}}{I_R}\right) \qquad (18\text{–}13)$$

where I_{CL} is $I_C - I_L$, the total current into the *L* and *C* branches.

EXAMPLE 18–10

Find each branch current and the total current in Figure 18–25.

FIGURE 18–25

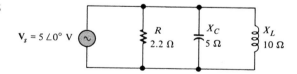

Solution:
Each branch current can be found using Ohm's law:

$$\mathbf{I}_R = \frac{\mathbf{V}_s}{\mathbf{R}} = \frac{5\angle0° \text{ V}}{2.2\angle0° \text{ }\Omega}$$

$$= 2.27\angle0° \text{ A}$$

$$\mathbf{I}_C = \frac{\mathbf{V}_s}{\mathbf{X}_C} = \frac{5\angle0° \text{ V}}{5\angle-90° \text{ }\Omega}$$

$$= 1\angle90° \text{ A}$$

$$\mathbf{I}_L = \frac{\mathbf{V}_s}{\mathbf{X}_L} = \frac{5\angle0° \text{ V}}{10\angle90° \text{ }\Omega}$$

$$= 0.5\angle-90° \text{ A}$$

The total current is the phasor sum of the branch currents. By Kirchhoff's law:

$$\mathbf{I}_T = \mathbf{I}_R + \mathbf{I}_C + \mathbf{I}_L$$

$$= 2.27\angle 0° \text{ A} + 1\angle 90° \text{ A} + 0.5\angle -90° \text{ A}$$

$$= 2.27 \text{ A} + j1 \text{ A} - j0.5 \text{ A} = 2.27 \text{ A} + j0.5 \text{ A}$$

$$= \sqrt{(2.27 \text{ A})^2 + (0.5 \text{ A})^2}\angle \tan^{-1}\left(\frac{0.5}{2.27}\right)$$

$$= 2.32\angle 12.42° \text{ A}$$

The total current is 2.32 A leading V_s by 12.42 degrees. Figure 18–26 is the current phasor diagram for the circuit.

FIGURE 18–26

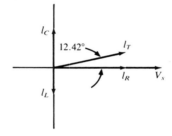

SERIES-PARALLEL ANALYSIS

The analysis of series-parallel combinations basically involves the application of the methods previously covered. The following two examples illustrate typical approaches to these more complex circuits.

EXAMPLE 18–11

In Figure 18–27, find the voltage across the capacitor in polar form. Is this circuit predominantly inductive or capacitive?

FIGURE 18–27

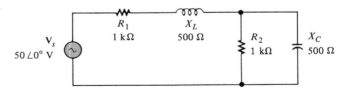

Solution:

We use the voltage-divider formula in this analysis. The impedance of the series combination of R_1 and X_L is called $\mathbf{Z}_1$.

$$\mathbf{Z}_1 = R_1 + jX_L$$

$$= 1000 \ \Omega + j500 \ \Omega \quad \text{in rectangular form}$$

$$\mathbf{Z}_1 = \sqrt{(1000 \ \Omega)^2 + (500 \ \Omega)^2} \angle \tan^{-1}\left(\frac{500}{1000}\right)$$

$$= 1118 \angle 26.57° \ \Omega \quad \text{in polar form}$$

The impedance of the parallel combination of R_2 and X_C is called $\mathbf{Z}_2$.

$$\mathbf{Z}_2 = \left(\frac{R_2 X_C}{\sqrt{R_2^2 + X_C^2}}\right) \angle \left(-90° + \tan^{-1}\left(\frac{X_C}{R_2}\right)\right)$$

$$= \left[\frac{(1000 \ \Omega)(500 \ \Omega)}{\sqrt{(1000 \ \Omega)^2 + (500 \ \Omega)^2}}\right] \angle \left(-90° + \tan^{-1}\left(\frac{500}{1000}\right)\right)$$

$$= 447.2 \angle -63.4° \ \Omega \quad \text{in polar form}$$

or

$$\mathbf{Z}_2 = 447.2 \cos(-63.4°) + j447.2 \sin(-63.4°)$$

$$= 200.24 \ \Omega - j399.87 \ \Omega \quad \text{in rectangular form}$$

The total impedance $\mathbf{Z}_T$ is

$$\mathbf{Z}_T = \mathbf{Z}_1 + \mathbf{Z}_2$$

$$= (1000 \ \Omega + j500 \ \Omega) + (200.24 \ \Omega - j399.87 \ \Omega)$$

$$= 1200.24 \ \Omega + j100.13 \ \Omega \quad \text{in rectangular form}$$

or

$$\mathbf{Z}_T = \sqrt{(1200.24 \ \Omega)^2 + (100.13 \ \Omega)^2} \angle \tan^{-1}\left(\frac{100.13}{1200.24}\right)$$

$$= 1204.41 \angle 4.77° \ \Omega \quad \text{in polar form}$$

Applying the voltage divider formula to get $\mathbf{V}_C$ yields

$$\mathbf{V}_C = \left(\frac{\mathbf{Z}_2}{\mathbf{Z}_T}\right)\mathbf{V}_s$$

$$= \left(\frac{447.2\angle-63.4°\ \Omega}{1204.41\angle 4.77°\ \Omega}\right)50\angle 0°\ \text{V}$$

$$= 18.57\angle-68.17°\ \text{V}$$

Therefore, V_C is 18.57 V and lags V_s by 68.17 degrees.

The $+j$ term in $\mathbf{Z}_T$, or the positive angle in its polar form, indicates that the circuit is more inductive than capacitive. However, it is just slightly more inductive, because the angle is small. This result is surprising, because $X_C = X_L = 500\ \Omega$. However, the capacitor is in *parallel* with a resistor, so the capacitor actually has less effect on the total impedance than does the inductor. Figure 18–28 shows the phasor relationship of $\mathbf{V}_C$ and $\mathbf{V}_s$. Although $X_C = X_L$, this circuit is not at resonance, since the j term of the total impedance is not zero due to the parallel combination of R_2 and X_C. You can see this by noting that the phase angle associated with $\mathbf{Z}_T$ is not zero.

FIGURE 18–28

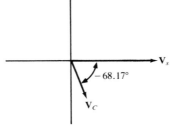

EXAMPLE
18–12

For the reactive circuit in Figure 18–29, find the voltage at point B with respect to ground.

FIGURE 18–29

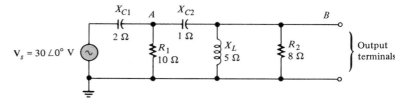

Solution:
The voltage at point B is the voltage across the open output terminals. Let's use the voltage divider approach. To do so, we must know the voltage at point A first; so we need to find the impedance from point A to ground as a starting point.

The parallel combination of X_L and R_2 is in series with X_{C2}. This combination is in parallel with R_1. We call this impedance from point A to ground, $\mathbf{Z}_A$. To find $\mathbf{Z}_A$, the following steps are taken:

R_2 and X_L in parallel ($\mathbf{Z}_1$):

$$\mathbf{Z}_1 = \left(\frac{R_2 X_L}{\sqrt{R_2^2 + X_L^2}}\right) \angle \left(90° - \tan^{-1}\left(\frac{X_L}{R}\right)\right)$$

$$= \left(\frac{(8\ \Omega)(5\ \Omega)}{\sqrt{(8\ \Omega)^2 + (5\ \Omega)^2}}\right) \angle \left(90° - \tan^{-1}\left(\frac{5}{8}\right)\right)$$

$$= \left(\frac{40}{9.43}\right) \angle (90° - 32°) = 4.24 \angle 58°\ \Omega$$

$\mathbf{Z}_1$ in series with $\mathbf{X}_{C2}$ ($\mathbf{Z}_2$):

$$\mathbf{Z}_2 = \mathbf{X}_{C2} + \mathbf{Z}_1$$
$$= 1 \angle -90°\ \Omega + 4.24 \angle 58°\ \Omega = -j1\ \Omega + 2.25\ \Omega + j3.6\ \Omega$$
$$= 2.25\ \Omega + j2.6\ \Omega$$
$$= \sqrt{(2.25\ \Omega)^2 + (2.6\ \Omega)^2} \angle \tan^{-1}\left(\frac{2.6}{2.25}\right)$$
$$= 3.44 \angle 49.13°\ \Omega$$

$\mathbf{R}_1$ in parallel with $\mathbf{Z}_2$ ($\mathbf{Z}_A$):

$$\mathbf{Z}_A = \frac{(10 \angle 0°)(3.44 \angle 49.13°)}{10 + 2.25 + j2.6}$$

$$= \frac{34.4 \angle 49.13°}{12.25 + j2.6}$$

$$= \frac{34.4 \angle 49.13°}{12.52 \angle 11.98°}$$

$$= 2.75 \angle 37.15°\ \Omega$$

The simplified circuit is shown in Figure 18–30.

FIGURE 18–30

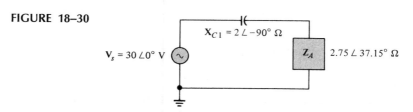

Now, the voltage divider principle can be applied to find $\mathbf{V}_A$. The total impedance is

$$
\begin{aligned}
\mathbf{Z}_T &= \mathbf{X}_{C1} + \mathbf{Z}_A \\
&= 2\angle -90° \ \Omega + 2.75\angle 37.15° \ \Omega = -j2 \ \Omega + 2.19 \ \Omega + j1.66 \ \Omega \\
&= 2.19 \ \Omega - j0.34 \ \Omega \\
&= \sqrt{(2.19 \ \Omega)^2 + (0.34 \ \Omega)^2} \angle -\tan^{-1}\left(\frac{0.34}{2.19}\right) \\
&= 2.22\angle -8.82° \ \Omega
\end{aligned}
$$

$$
\begin{aligned}
\mathbf{V}_A &= \left(\frac{\mathbf{Z}_A}{\mathbf{Z}_T}\right)\mathbf{V}_s \\
&= \left(\frac{2.75\angle 37.15° \ \Omega}{2.22\angle -8.82° \ \Omega}\right) 30\angle 0° \ \text{V} \\
&= 37.16\angle 45.97° \ \text{V}
\end{aligned}
$$

Next, $\mathbf{V}_B$ is found by dividing $\mathbf{V}_A$ down, as indicated in Figure 18–31. $\mathbf{V}_B$ is the open terminal output voltage.

$$
\begin{aligned}
\mathbf{V}_B &= \left(\frac{\mathbf{Z}_1}{\mathbf{Z}_2}\right)\mathbf{V}_A \\
&= \left(\frac{4.24\angle 58° \ \Omega}{3.44\angle 49.13° \ \Omega}\right) 37.16\angle 45.97° \ \text{V} \\
&- 45.8\angle 54.84° \ \text{V}
\end{aligned}
$$

Surprisingly, V_A is greater than V_s, and V_B is greater than V_A! This result is possible because of the out-of-phase relationship of the reactive voltages. Remember that X_C and X_L tend to cancel each other.

FIGURE 18–31

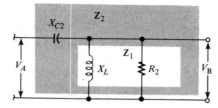

CONVERSION OF SERIES-PARALLEL TO PARALLEL

The particular series-parallel configuration shown in Figure 18–32 is important because it represents a circuit having parallel L and C branches, with the winding resistance of the coil taken into account as a series resistance in the L branch.

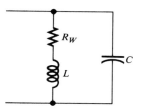

FIGURE 18–32

An important series-parallel *RLC* circuit ($Q = X_L/R_W$).

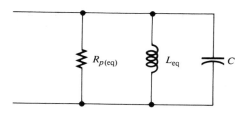

FIGURE 18–33

Parallel equivalent form of the circuit in Figure 18–32.

It is helpful to view the series-parallel circuit in Figure 18–32 in an *equivalent* purely parallel form, as indicated in Figure 18–33. This form will simplify our analysis of parallel resonant characteristics in the next section.

The equivalent inductance L_{eq} and the equivalent parallel resistance $R_{p(eq)}$ are given by the following formulas:

$$L_{eq} = L\left(\frac{Q^2 + 1}{Q^2}\right) \tag{18–14}$$

$$R_{p(eq)} = R_W(Q^2 + 1) \tag{18–15}$$

where Q is the quality factor of the coil, X_L/R_W. Derivations of these formulas are quite involved and thus are not given here. Notice in the equations that for a $Q \geq 10$, the value of L_{eq} is approximately the same as the original value of L. For example, if $L = 10$ mH, then

$$L_{eq} = 10 \text{ mH}\left(\frac{10^2 + 1}{10^2}\right) = 10 \text{ mH}(1.01) = 10.1 \text{ mH}$$

The equivalency of the two circuits means that at a given frequency, when the same value of voltage is applied to both circuits, the same total current flows in both circuits and the phase angles are the same. Basically, an equivalent circuit simply makes circuit analysis more convenient.

EXAMPLE 18–13

Convert the series-parallel circuit in Figure 18–34 to an equivalent parallel form at the given frequency.

FIGURE 18–34

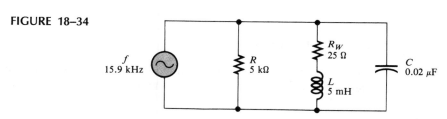

Solution: $\qquad X_L = 2\pi fL = 2\pi(15.9 \text{ kHz})(5 \text{ mH}) = 500 \ \Omega$

The Q of the coil is

$$Q = \frac{X_L}{R_W} = \frac{500 \ \Omega}{25 \ \Omega} = 20$$

Since $Q > 10$, then $L_{eq} \cong L = 5 \text{ mH}$.
    The equivalent parallel resistance is

$$R_{p(eq)} = R_W(Q^2 + 1) = (25 \ \Omega)(20^2 + 1) = 10.025 \text{ k}\Omega$$

This equivalent resistance appears in parallel with R as shown in Figure 18–35(a).
When combined, they give a total parallel resistance (R_{pT}) of 3.2 kΩ, as indicated in
Figure 18–35(b).

FIGURE 18–35

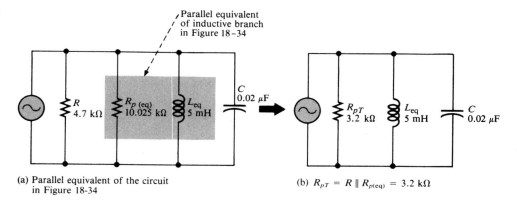

(a) Parallel equivalent of the circuit
in Figure 18-34

(b) $R_{pT} = R \parallel R_{p(eq)} = 3.2 \text{ k}\Omega$

SECTION REVIEW 18–5

1. In a three-branch parallel circuit, $R = 150 \ \Omega$, $X_C = 100 \ \Omega$, and $X_L = 50 \ \Omega$.
   Determine the current in each branch when $V_s = 12 \text{ V}$.

2. The impedance of a parallel RLC circuit is $2.8\angle -38.9° \text{ k}\Omega$. Is the circuit capacitive
   or inductive?

3. Find the equivalent parallel inductance and resistance for a 20-mH coil with a winding resistance of 10 Ω at a frequency of 1 kHz.

18–6   PARALLEL RESONANCE

First we will look at the resonant condition in an *ideal* parallel LC circuit. Then we will
progress to the more realistic case where the resistance of the coil is taken into account.

CONDITION FOR IDEAL PARALLEL RESONANCE

Ideally, parallel resonance occurs when $X_C = X_L$. The frequency at which resonance occurs is called the *resonant frequency,* just as in the series case. When $X_C = X_L$, the two branch currents, I_C and I_L, are equal in magnitude, and, of course, they are always 180 degrees out of phase with each other. Thus, the two currents cancel and the total current is zero, as shown in Figure 18–36.

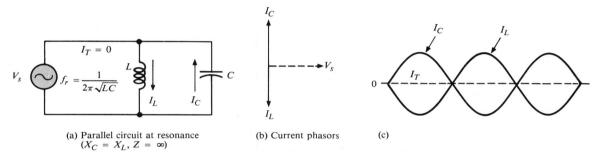

(a) Parallel circuit at resonance   (b) Current phasors   (c)
$(X_C = X_L, Z = \infty)$

FIGURE 18–36
An ideal parallel *LC* circuit at resonance.

Since the total current is zero, the impedance of the parallel *LC* circuit is infinitely large (∞). These *ideal* resonant conditions are stated as follows:

$$X_L = X_C$$
$$Z_r = \infty$$

THE IDEAL PARALLEL RESONANT FREQUENCY

For an ideal parallel resonant circuit, the frequency at which resonance occurs is determined by the same formula as in series resonant circuits:

$$f_r = \frac{1}{2\pi\sqrt{LC}}$$

WHY THE PARALLEL RESONANT *LC* CIRCUIT IS OFTEN CALLED A *TANK* CIRCUIT

The term *tank circuit* refers to the fact that the parallel resonant circuit stores energy in the magnetic field of the coil and in the electric field of the capacitor. The stored energy is transferred back and forth between the capacitor and the coil on alternate half-cycles as the current goes first one way and then the other when the inductor deenergizes and the capacitor charges, and vice versa, as illustrated in Figure 18–37.

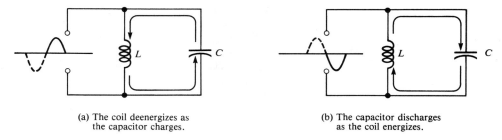

(a) The coil deenergizes as
the capacitor charges.

(b) The capacitor discharges
as the coil energizes.

FIGURE 18–37
Energy storage in an ideal parallel resonant tank circuit.

PARALLEL RESONANT CONDITIONS IN A NONIDEAL CIRCUIT

So far, the resonance of an ideal parallel *LC* circuit has been examined. Now we will consider resonance in a tank circuit with the resistance of the coil taken into account. Figure 18–38 shows a nonideal tank circuit and its parallel *RLC* equivalent.

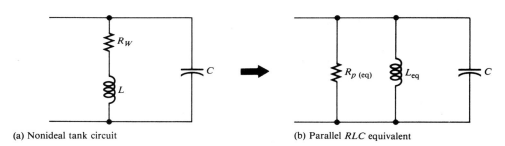

(a) Nonideal tank circuit

(b) Parallel *RLC* equivalent

FIGURE 18–38
A practical treatment of parallel resonant circuits must include the coil resistance.

The Q of the circuit at resonance is simply the Q of the coil:

$$Q = \frac{X_L}{R_W}$$

The expressions for the equivalent parallel resistance and the equivalent inductance were given in the previous section as

$$R_{p(eq)} = R_W(Q^2 + 1)$$

$$L_{eq} = L\left(\frac{Q^2 + 1}{Q^2}\right)$$

Recall that for $Q \geq 10$, $L_{eq} \cong L$.

At parallel resonance,

$$X_{L(eq)} = X_C$$

In the parallel equivalent circuit, we have $R_{p(eq)}$ in parallel with an ideal coil and a capacitor, so the L and C branches act as an ideal tank circuit which has an infinite impedance at resonance as shown in Figure 18–39. Therefore, the total impedance of the nonideal tank circuit at resonance can be expressed as simply the equivalent parallel resistance:

$$Z_r = R_W(Q^2 + 1) \qquad (18\text{–}16)$$

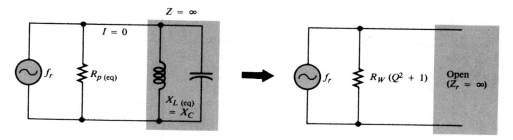

FIGURE 18–39
At resonance, the parallel LC portion appears open and the source sees only $R_{p(eq)}$.

A rigorous mathematical derivation of the impedance formula is given in the following steps:

$$\frac{1}{Z} = \frac{1}{-jX_C} + \frac{1}{R_W + jX_L}$$

$$= j\left(\frac{1}{X_C}\right) + \frac{R_W - jX_L}{(R_W + jX_L)(R_W - jX_L)}$$

$$= j\left(\frac{1}{X_C}\right) + \frac{R_W - jX_L}{R_W^2 + X_L^2}$$

The first term plus splitting the numerator of the second term yields

$$\frac{1}{Z} = j\left(\frac{1}{X_C}\right) - j\left(\frac{X_L}{R_W^2 + X_L^2}\right) + \frac{R_W}{R_W^2 + X_L^2}$$

At resonance, Z is purely resistive; so it has no j part (the j terms in the last expression cancel). Thus, only the real part is left, as stated in the following equation for Z at resonance:

$$Z_r = \frac{R_W^2 + X_L^2}{R_W} \qquad (18\text{–}17)$$

Splitting the denominator, we get

$$Z_r = \frac{R_W^2}{R_W} + \frac{X_L^2}{R_W} = R_W + \frac{X_L^2}{R_W}$$

Factoring out R_W gives

$$Z_r = R_W\left(1 + \frac{X_L^2}{R_W^2}\right)$$

Since $X_L^2/R_W^2 = Q^2$, then

$$Z_r = R_W(Q^2 + 1)$$

**EXAMPLE
18–14**

Determine the impedance of the circuit in Figure 18–40 at the resonant frequency ($f_r \cong 17,794$ Hz).

FIGURE 18–40

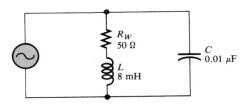

Solution:

$$X_L = 2\pi f_r L = 2\pi(17,794 \text{ Hz})(8 \text{ mH}) = 894 \text{ }\Omega$$

$$Q = \frac{X_L}{R_W} = \frac{894 \text{ }\Omega}{50 \text{ }\Omega} = 17.9$$

$$Z_r = R_W(Q^2 + 1) = 50 \text{ }\Omega(17.9^2 + 1) = 16.1 \text{ k}\Omega$$

The calculator sequence is

VARIATION OF THE IMPEDANCE WITH FREQUENCY

The impedance of a parallel resonant circuit is maximum at the resonant frequency and decreases at lower and higher frequencies, as indicated by the curve in Figure 18–41.

At very low frequencies, X_L is very small and X_C is very high, so the total impedance is essentially equal to that of the inductive branch. As the frequency goes up, the impedance also increases, and the inductive reactance dominates (because it is less than X_C) until the resonant frequency is reached. At this point, of course, $X_L \cong X_C$ (for $Q > 10$) and the impedance is at its maximum. As the frequency goes above resonance, the capacitive reactance dominates (because it is less than X_L) and the impedance decreases.

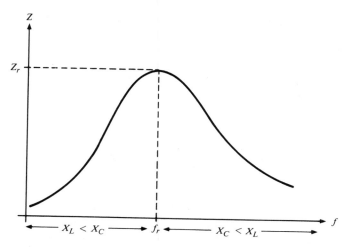

FIGURE 18–41
Generalized impedance curve for a parallel resonant circuit. The circuit is inductive below f_r, resistive at f_r, and capacitive above f_r.

CURRENT AND PHASE ANGLE AT RESONANCE

In the ideal tank circuit, the total current from the source at resonance is zero because the impedance is infinite. In the nonideal case, there is some total current at the resonant frequency, and it is determined by the impedance at resonance:

$$I_T = \frac{V_s}{Z_r} \qquad \text{(18–18)}$$

The phase angle of the parallel resonant circuit is 0 degrees because the impedance is purely resistive at the resonant frequency.

PARALLEL RESONANT FREQUENCY IN A NONIDEAL CIRCUIT

As you know, when the coil resistance is considered, the resonant condition is

$$X_{L(eq)} = X_C$$

which can be expressed as

$$2\pi f_r L\left(\frac{Q^2 + 1}{Q^2}\right) = \frac{1}{2\pi f_r C}$$

Solving for f_r, we get

$$f_r = \frac{1}{2\pi\sqrt{LC}}\sqrt{\frac{Q^2}{Q^2 + 1}} \qquad \text{(18–19)}$$

When $Q \geq 10$, the term with the Q factors is approximately 1:

$$\sqrt{\frac{Q^2}{Q^2 + 1}} = \sqrt{\frac{100}{101}} = 0.995 \cong 1$$

Therefore, the parallel resonant frequency is approximately the same as the series resonant frequency as long as Q is equal to or greater than 10:

$$f_r \cong \frac{1}{2\pi\sqrt{LC}} \quad \text{for } Q \geq 10$$

Equation (18–19) was given only to show the effect of Q on the resonant frequency. You cannot use it to calculate the f_r of a given circuit, because you must first know the value of Q. Since $Q = X_L/R_W$, you must know X_L at the resonant frequency. In order to get X_L, you must know f_r. Since f_r is what we are looking for in the first place, there is no way to proceed with only Equation (18–19) at our disposal.

A more precise expression for f_r in terms of the circuit component values is developed as follows:

The j terms in the equation preceding Equation (18–17) are equal.

$$\frac{1}{X_C} = \frac{X_L}{R_W^2 + X_L^2}$$

Thus,

$$R_W^2 = X_L^2 = X_L X_C$$

$$R_W^2 + (2\pi f_r L)^2 = \frac{2\pi f_r L}{2\pi f_r C}$$

$$R_W^2 + 4\pi^2 f_r^2 L^2 = \frac{L}{C}$$

$$4\pi^2 f_r^2 L^2 = \frac{L}{C} - R_W^2$$

Solving for f_r^2,

$$f_r^2 = \frac{\left(\dfrac{L}{C}\right) - R_W^2}{4\pi^2 L^2}$$

Multiply both numerator and denominator by C:

$$f_r^2 = \frac{L - R_W^2 C}{4\pi^2 L^2 C}$$

$$= \frac{L - R_W^2 C}{L(4\pi^2 LC)}$$

Factoring an L out of the numerator and canceling gives

$$f_r^2 = \frac{1 - (R_W^2 C/L)}{4\pi^2 LC}$$

Taking the square root of both sides yields f_r:

$$f_r = \frac{\sqrt{1 - (R_W^2 C/L)}}{2\pi\sqrt{LC}} \tag{18–20}$$

Now, f_r can be found from component values alone.

EXAMPLE 18–15

Find the frequency, impedance, and total current at resonance for the circuit in Figure 18–42.

FIGURE 18–42

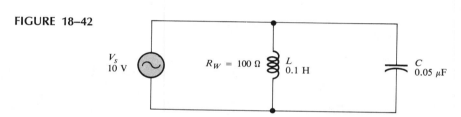

Solution:

$$f_r = \frac{\sqrt{1 - (R_W^2 C/L)}}{2\pi\sqrt{LC}} = \frac{\sqrt{1 - [(100 \text{ }\Omega)^2(0.05 \text{ } \mu\text{F})/0.1 \text{ H}]}}{2\pi\sqrt{(0.05 \text{ } \mu\text{F})(0.1 \text{ H})}} = 2.25 \text{ kHz}$$

$$X_L = 2\pi f_r L = 2\pi(2.25 \text{ kHz})(0.1 \text{ H}) = 1.4 \text{ k}\Omega$$

$$Q = \frac{X_L}{R_W} = \frac{1.4 \text{ k}\Omega}{100 \text{ }\Omega} = 14$$

$$Z_r = R_W(Q^2 + 1) = 100 \text{ }\Omega(14^2 + 1) = 19.7 \text{ k}\Omega$$

$$I_T = \frac{V_s}{Z_r} = \frac{10 \text{ V}}{19.7 \text{ k}\Omega} = 0.51 \text{ mA}$$

Note that since $Q > 10$, the approximate formula, $f_r \cong 1/2\pi\sqrt{LC}$, could be used. The calculator sequence for f_r is

HOW AN EXTERNAL PARALLEL LOAD RESISTANCE AFFECTS A TANK CIRCUIT

There are many practical situations in which an external load resistance appears in parallel with a tank circuit as shown in Figure 18–43(a). Obviously, the external resistor (R_L) will dissipate more of the energy delivered by the source and thus will lower the *overall Q* of the circuit. The external resistor effectively appears in parallel with the equivalent parallel resistance of the coil, $R_{p(eq)}$, and both are combined to determine a total parallel resistance, R_{pT}, as indicated in Figure 18–43(b):

$$R_{pT} = R_L \| R_{p(eq)}$$

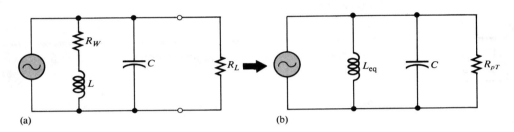

(a)   (b)

FIGURE 18–43
Tank circuit with a parallel load resistor and its equivalent circuit.

The overall parallel *RLC* circuit (Q_O) is expressed differently from the Q of a series circuit:

$$Q_O = \frac{R_{pT}}{X_{L(eq)}} \qquad \text{(18–21)}$$

As you can see, the effect of loading the tank circuit is to reduce its overall Q (which is equal to the coil Q when unloaded).

SECTION REVIEW 18–6

1. Is the impedance minimum or maximum at parallel resonance?
2. Is the current minimum or maximum at parallel resonance?
3. At ideal parallel resonance, $X_L = 1500 \ \Omega$. What is X_C?
4. A parallel tank circuit has the following values: $R_W = 4 \ \Omega$, $L = 50$ mH, and $C = 10$ pF. Calculate f_r and Z at resonance.
5. If $Q = 25$, $L = 50$ mH, and $C = 1000$ pF, what is f_r?
6. In Question 3, if $Q = 2.5$, what is f_r?
7. In a certain tank circuit, the coil resistance is $20 \ \Omega$. What is the total impedance at resonance if $Q = 20$?

18–7 BANDWIDTH OF RESONANT CIRCUITS

SERIES RESONANT CIRCUITS

As you learned, the current in a series RLC circuit is maximum at the resonant frequency and drops off on either side of this frequency. *The bandwidth is defined to be the range of frequencies for which the current is equal to or greater than 70.7% of its resonant value.* Bandwidth, sometimes abbreviated BW, is an important characteristic of a resonant circuit.

Figure 18–44 illustrates bandwidth on the response curve of a series RLC circuit. Notice that the frequency f_1 below f_r is the point at which the current is $0.707I_{max}$ and is commonly called the *lower cutoff frequency*. The frequency f_2 above f_r, where the current is again $0.707I_{max}$, is the *upper cutoff frequency*. Other names for f_1 and f_2 are *–3 dB frequencies, critical frequencies*, and *half-power frequencies*. The significance of the latter term is discussed later in the chapter.

FIGURE 18–44

Bandwidth on series resonant response curve for I.

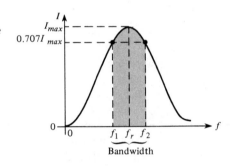

EXAMPLE 18–16

A certain series resonant circuit has a maximum current of 100 mA at the resonant frequency. What is the value of the current at the cutoff frequencies?

Solution:
Current at the cutoff frequencies is 70.7% of maximum:

$$I_{f1} = I_{f2} = 0.707I_{max}$$

$$= 0.707(100 \text{ mA}) = 70.7 \text{ mA}$$

PARALLEL RESONANT CIRCUITS

For a parallel resonant circuit, *the impedance is maximum at the resonant frequency;* so the total current is minimum. The bandwidth can be defined in relation to the impedance curve in the same manner that the current curve was used in the series circuit. Of course, f_r is the frequency at which Z is maximum; f_1 is the lower cutoff frequency at which $Z = 0.707Z_{max}$; and f_2 is the upper cutoff frequency at which again $Z = 0.707Z_{max}$. The bandwidth is the range of frequencies between f_1 and f_2, as shown in Figure 18–45.

FIGURE 18–45
Bandwidth of the parallel resonant
response curve for Z_T.

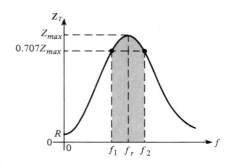

FORMULA FOR BANDWIDTH

The bandwidth for either series or parallel resonant circuits is the range of frequencies between the cutoff frequencies for which the response curve (I or Z) is 0.707 of the maximum value. Thus, the bandwidth is actually the *difference* between f_2 and f_1:

$$BW = f_2 - f_1 \qquad (18\text{--}22)$$

Ideally, f_r is the *center frequency* and can be calculated as follows:

$$f_r = \frac{f_1 + f_2}{2} \qquad (18\text{--}23)$$

**EXAMPLE
18–17**

A resonant circuit has a lower cutoff frequency of 8 kHz and an upper cutoff frequency of 12 kHz. Determine the bandwidth and center (resonant) frequency.

Solution:
$$BW = f_2 - f_1 = 12 \text{ kHz} - 8 \text{ kHz} = 4 \text{ kHz}$$

$$f_r = \frac{f_1 + f_2}{2} = \frac{12 \text{ kHz} + 8 \text{ kHz}}{2} = 10 \text{ kHz}$$

HALF-POWER FREQUENCIES

As previously mentioned, the upper and lower cutoff frequencies are sometimes called the *half-power frequencies*. This term is derived from the fact that the power from the source at these frequencies is one-half the power delivered at the resonant frequency. The following steps show that this is true for a series circuit. The same result applies to a parallel circuit.

At resonance,

$$P_{max} = I^2_{max}R$$

The power at f_1 or f_2 is

$$P_{f1} = I^2_{f1}R = (0.707I_{max})^2R = (0.707)^2I^2_{max}R$$
$$= 0.5I^2_{max}R = 0.5P_{max}$$

SELECTIVITY

The response curves in Figures 18–44 and 18–45 are also called *selectivity curves*. Selectivity defines how well a resonant circuit responds to a certain frequency and discriminates against all others. *The smaller the bandwidth, the greater the selectivity.*

We normally assume that a resonant circuit accepts frequencies within its bandwidth and completely eliminates frequencies outside the bandwidth. Such is not actually the case, however, because signals with frequencies outside the bandwidth are not completely eliminated. Their magnitudes, however, are greatly reduced. The further the frequencies are from the cutoff frequencies, the greater is the reduction, as illustrated in Figure 18–46(a). An ideal selectivity curve is shown in Figure 18–46(b).

FIGURE 18–46
Selectivity curve.

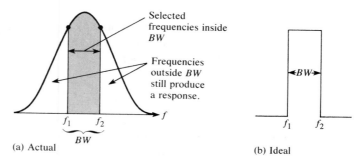

As you can see in Figure 18–46, another factor that influences selectivity is the *sharpness* of the slopes of the curve. The faster the curve drops off at the cutoff frequencies, the more selective the circuit is, because it responds only to the frequencies within the bandwidth. Figure 18–47 shows a general comparison of three response curves with varying degrees of selectivity.

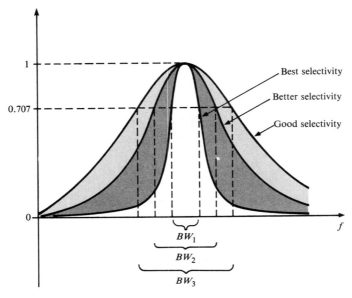

FIGURE 18–47
Comparative selectivity curves.

HOW *Q* AFFECTS BANDWIDTH

A higher value of circuit *Q* results in a smaller bandwidth. A lower value of *Q* causes a larger bandwidth. A formula for the bandwidth of a resonant circuit in terms of *Q* is stated in the following equation:

$$BW = \frac{f_r}{Q} \tag{18–24}$$

EXAMPLE 18–18

What is the bandwidth of each circuit in Figure 18–48?

FIGURE 18–48

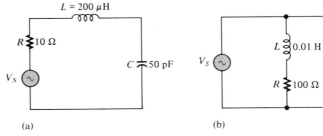

(a)   (b)

Solution:

(a)

$$f_r = \frac{1}{2\pi\sqrt{LC}} = 1.59 \text{ MHz}$$

$$Q = \frac{X_L}{R} = \frac{2 \text{ k}\Omega}{10 \text{ }\Omega} = 200$$

$$BW = \frac{f_r}{Q} = \frac{1.59 \text{ MHz}}{200} = 7.95 \text{ kHz}$$

(b)

$$f_r = \frac{\sqrt{1 - (R_W^2 C/L)}}{2\pi\sqrt{LC}} \cong \frac{1}{2\pi\sqrt{LC}} = 22.5 \text{ kHz}$$

$$Q = \frac{X_L}{R} = \frac{1.41 \text{ k}\Omega}{100 \text{ }\Omega} = 14.1$$

$$BW = \frac{f_r}{Q} = \frac{22.5 \text{ kHz}}{14.1} = 1.6 \text{ kHz}$$

SECTION REVIEW 18–7

1. What is the bandwidth when $f_2 = 2.2$ MHz and $f_1 = 1.8$ MHz?

2. For a resonant circuit with the cutoff frequencies in Question 1, what is the center frequency?

3. The power at resonance is 1.8 W. What is the power at the upper cutoff frequency?

4. Does a larger Q mean a smaller or a larger bandwidth?

18–8   SYSTEM APPLICATIONS

Resonant circuits are used in a wide variety of applications, particularly in communication systems. In this section, we will look briefly at a few common communication system applications. The purpose of this coverage is not to explain the systems that are used as examples but rather to show the importance of filters in electronic communication.

TUNED AMPLIFIERS

A tuned amplifier is a circuit that amplifies signals within a specified band. Typically, a parallel resonant circuit is used in conjunction with an amplifier to achieve the selectivity. In terms of the general operation, input signals with frequencies that range over a wide band are accepted on the amplifier's input and are amplified. The function of the resonant circuit is to allow only a relatively narrow band of those frequencies to be passed on. The variable capacitor allows tuning over the range of input frequencies so that a desired frequency can be selected, as indicated in Figure 18–49.

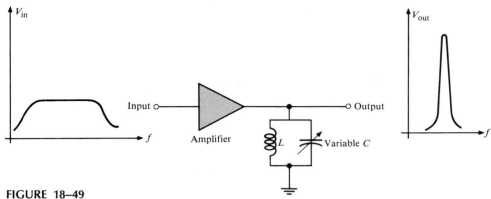

FIGURE 18–49
A basic tuned band-pass amplifier.

ANTENNA INPUT TO A RECEIVER

Radio signals are sent out from a transmitter via electromagnetic waves which propagate through the atmosphere. When the electromagnetic waves cut across the receiving antenna, small voltages are induced. Out of all the wide range of electromagnetic frequencies, only one frequency or a limited band of frequencies must be extracted. Figure 18–50 shows a typical arrangement of an antenna coupled to the receiver input by a transformer. A variable capacitor is connected across the transformer secondary to form a parallel resonant circuit.

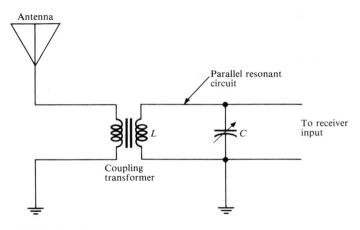

FIGURE 18–50
Resonant coupling from an antenna.

DOUBLE-TUNED TRANSFORMER COUPLING IN A RECEIVER

In some types of communication receivers, tuned amplifiers are transformer-coupled together to increase the amplification. Capacitors can be placed in parallel with the primary and secondary of the transformer, effectively creating two parallel resonant band-pass filters that are coupled together. This technique, illustrated in Figure 18–51, can result in a wider bandwidth and steeper slopes on the response curve, thus increasing the selectivity for a desired band of frequencies.

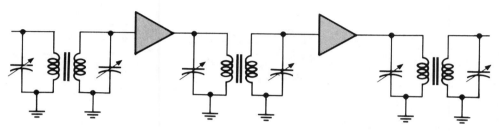

FIGURE 18–51
Double-tuned amplifiers.

SIGNAL RECEPTION AND SEPARATION IN A TV RECEIVER

A television receiver must handle both video (picture) signals and audio (sound) signals. Each TV transmitting station is allotted a 6-MHz bandwidth. Channel 2 is allotted a band from 54 MHz through 59 MHz, channel 3 is allotted a band from 60 MHz through 65 MHz, on up to channel 13 which has a band from 210 MHz through 215 MHz. You can tune the front end of the TV receiver to select any one of these channels by using tuned amplifiers. The signal output of the front end of the receiver has a bandwidth from 41 MHz through 46 MHz, regardless of the channel that is tuned in. This band, called the *intermediate frequency* (IF) band, contains both video and audio. Amplifiers tuned to the IF band boost the signal and feed it to the video amplifier.

Before the output of the video amplifier is applied to the picture tube, the audio signal is removed by a 4.5-MHz band-stop filter (called a *wave trap*), as shown in Figure 18–52. This trap keeps the sound signal from interfering with the picture. The video amplifier output is also applied to band-pass circuits that are tuned to the sound carrier frequency of 4.5 MHz. The sound signal is then processed and applied to the speaker as indicated in Figure 18–52.

SUPERHETERODYNE RECEIVER

Another good example of filter applications is in the common AM (amplitude modulation) receiver. The AM broadcast band ranges from 535 kHz to 1605 kHz. Each AM station is assigned a certain narrow bandwidth within that range. A simplified block diagram of a superheterodyne AM receiver is shown in Figure 18–53.

There are basically three parallel resonant band-pass filters in the front end of the receiver. Each of these filters is gang-tuned by capacitors; that is, the capacitors are mechanically or electronically linked together so that they change together as the tuning

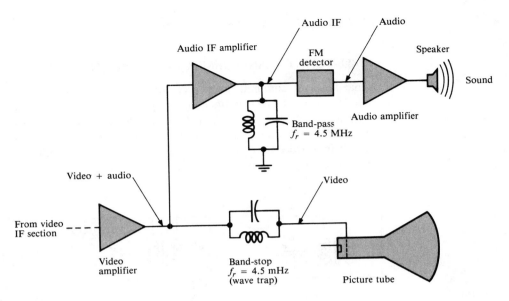

FIGURE 18–52

A simplified portion of a TV receiver showing filter usage.

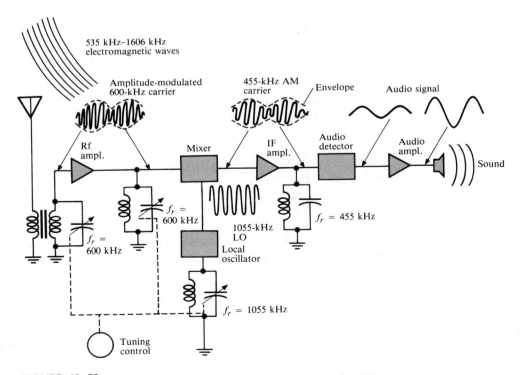

FIGURE 18–53

A simplified diagram of a superhterodyne AM radio broadcast receiver showing the application of tuned resonant circuits.

knob is turned. The front end is tuned to receive a desired station, for example, one that transmits at 600 kHz. The input filter from the antenna and the RF (radio frequency) amplifier filter select only a frequency of 600 kHz out of all the frequencies crossing the antenna. The actual audio (sound) signal is carried by the 600-kHz carrier frequency by modulating the amplitude of the carrier so that it follows the audio signal as indicated. The variation in the amplitude of the carrier corresponding to the audio signal is called the *envelope*. The 600 kHz is then applied to a circuit called the *mixer*. The *local oscillator* (LO) is tuned to a frequency that is 455 kHz above the selected frequency (1055 kHz, in this case). By a process called *heterodyning* or *beating*, the AM signal and the local oscillator signal are mixed together, and the 600-kHz AM signal is converted to a 455-kHz AM signal (1055 kHz − 600 kHz = 455 kHz). The 455 kHz is the intermediate frequency (IF) for standard AM receivers. No matter which station within the broadcast band is selected, its frequency is always converted to the 455-kHz IF. The amplitude-modulated IF is applied to an *audio detector* which removes the IF, leaving only the envelope or audio signal. The audio signal is then amplified and applied to the speaker.

SECTION REVIEW 18–8

1. Generally, why is a tuned filter necessary when a signal is coupled from an antenna to the input of a receiver?

2. What is a *wave trap*?

3. What is meant by *ganged tuning*?

18–9   COMPUTER ANALYSIS

The following program computes the impedance and phase angle over a specified frequency range for a series *RLC* circuit with specified parameters. The resonant frequency and bandwidth are also computed. A flowchart for this program appears in Figure 18–54.

```
10   CLS
20   PRINT "THE FOLLOWING PARAMETERS ARE COMPUTED FOR A SERIES
     RLC CIRCUIT:"
30   PRINT :PRINT"IMPEDANCE"
40   PRINT "PHASE ANGLE"
50   PRINT "RESONANT FREQUENCY"
60   PRINT "BANDWIDTH"
70   PRINT:PRINT:PRINT
80   INPUT "TO CONTINUE PRESS 'ENTER'";X:CLS
90   INPUT "THE VALUE OF R IN OHMS";R
100  INPUT "THE VALUE OF C IN FARADS";C
110  INPUT "THE VALUE OF L IN HENRIES";L
120  FR=1/(2*3.1416*SQR(L*C))
130  Q=2*3.1416*FR*L/R
140  BW=FR/Q
```

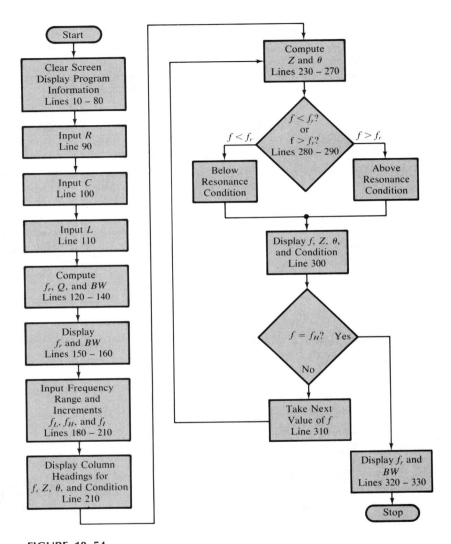

FIGURE 18–54

```
150 PRINT "RESONANT FREQUENCY = ";FR;"HZ"
160 PRINT "BANDWIDTH = ";BW;"HZ"
170 PRINT:PRINT:PRINT
180 INPUT "FOR IMPEDANCE, PHASE ANGLE, AND RESPONSE CONDITION,
    PRESS 'ENTER'";X
190 INPUT "THE MINIMUM FREQUENCY IN HERTZ";FL
200 INPUT "THE MAXIMUM FREQUENCY IN HERTZ";FH
210 INPUT "THE INCREMENTS OF FREQUENCY IN HERTZ";FI
220 PRINT "FREQUENCY(HZ)", "IMPEDANCE", "PHASE ANGLE",
    "CONDITION"
230 FOR F = FL TO FH STEP FI
240 XC = 1/(2*3.1416*F*C)
```

```
250 XL = 2*3.1416*F*L
260 Z=SQR(R*R+(XL-XC)*(XL-XC))
270 THETA = ATN((XL-XC)/R)
280 IF F<FR THEN C$ = "BELOW RESONANCE"
290 IF F>FR THEN C$ = "ABOVE RESONANCE"
300 PRINT F, Z, THETA, C$
310 NEXT
320 PRINT "RESONANT FREQUENCY = ";FR;"HZ"
330 PRINT "BANDWIDTH = ";BW;"HZ"
```

SECTION REVIEW 18–9

1. How are below-resonance and above-resonance conditions determined?

2. In line 260, can the positions of X_L and X_C be exchanged without affecting the outcome?

SUMMARY

1. X_L and X_C have opposing effects in an *RLC* circuit.

2. In a series *RLC* circuit, the larger reactance determines the net reactance of the circuit.

3. At series resonance, the inductive and capacitive reactances are equal.

4. The impedance of a series *RLC* circuit is purely resistive at resonance.

5. In a series *RLC* circuit, the current is maximum at resonance.

6. The reactive voltages V_L and V_C cancel at resonance in a series *RLC* circuit, because they are equal in magnitude and 180 degrees out of phase.

7. In a parallel *RLC* circuit, the smaller reactance determines the net reactance of the circuit.

8. In a parallel resonant circuit, the impedance is maximum at the resonant frequency.

9. A parallel resonant circuit is commonly called a *tank circuit*.

10. The impedance of a parallel resonant circuit is purely resistive at resonance.

11. The bandwidth of a series resonant circuit is the range of frequencies for which the current is $0.707I_{max}$ or greater.

12. The bandwidth of a parallel resonant circuit is the range of frequencies for which the impedance is $0.707Z_{max}$ or greater.

13. The cutoff frequencies are the frequencies above and below resonance where the circuit response is 70.7% of the maximum response.

14. The quality factor, Q, of a circuit is a ratio of stored energy to lost energy.

15. A higher Q produces a narrower bandwidth.

FORMULAS

Series *RLC* Circuits

$$X_T = |X_L - X_C| \tag{18–1}$$

$$\mathbf{Z} = R + jX_L - jX_C \tag{18–2}$$

$$\mathbf{Z} = \sqrt{R^2 + (X_L - X_C)^2} \angle \tan^{-1}\left(\frac{X_T}{R}\right) \tag{18–3}$$

Series Resonance

$$\mathbf{Z} = R + jX_L - jX_C \tag{18–4}$$

$$X_L = X_C \tag{18–5}$$

$$Z_r = R \tag{18–6}$$

$$f_r = \frac{1}{2\pi\sqrt{LC}} \tag{18–7}$$

Parallel *RLC* Circuits

$$\mathbf{Z} = \frac{1}{\dfrac{1}{R\angle 0°} + \dfrac{1}{X_L\angle 90°} + \dfrac{1}{X_C\angle -90°}} \tag{18–8}$$

$$\mathbf{G} = \frac{1}{R\angle 0°} = G\angle 0° \tag{18–9}$$

$$\mathbf{B}_C = \frac{1}{X_C\angle -90°} = B_C\angle 90° = jB_C \tag{18–10}$$

$$\mathbf{B}_L = \frac{1}{X_L\angle 90°} = B_L\angle -90° = -jB_L \tag{18–11}$$

$$\mathbf{Y} = \frac{1}{Z\angle \pm\theta} = Y\angle \mp\theta = G + jB_C - jB_L \tag{18–12}$$

$$\mathbf{I}_T = \sqrt{I_R^2 + (I_C - I_L)^2} \angle \tan^{-1}\left(\frac{I_{CL}}{I_R}\right) \tag{18–13}$$

$$L_{eq} = L\left(\frac{Q^2 + 1}{Q^2}\right) \tag{18–14}$$

$$R_{p(eq)} = R_W(Q^2 + 1) \tag{18–15}$$

Parallel Resonance

$$Z_r = R_W(Q^2 + 1) \tag{18-16}$$

$$Z_r = \frac{R_W^2 + X_L^2}{R_W} \tag{18-17}$$

$$I_T = \frac{V_s}{Z_r} \tag{18-18}$$

$$f_r = \frac{1}{2\pi\sqrt{LC}}\sqrt{\frac{Q^2}{Q^2 + 1}} \tag{18-19}$$

$$f_r = \frac{\sqrt{1 - (R_W^2 C/L)}}{2\pi\sqrt{LC}} \tag{18-20}$$

$$Q_O = \frac{R_{pT}}{X_{L(eq)}} \tag{18-21}$$

$$BW = f_2 - f_1 \tag{18-22}$$

$$f_r = \frac{f_1 + f_2}{2} \tag{18-23}$$

$$BW = \frac{f_r}{Q} \tag{18-24}$$

SELF-TEST

Solutions appear at the end of the book.

1. What is the total reactance of a series RLC circuit at resonance?

2. What is the phase angle of a series RLC circuit at resonance?

3. What is the impedance of a series LC circuit at the resonant frequency? The component values are $L = 15$ mH, $C = 0.015$ μF, and $R_W = 80$ Ω.

4. In a series RLC circuit operating below the resonant frequency, does the current lead or lag the applied voltage? Why?

5. In a series RLC circuit, if C is increased, what happens to the resonant frequency?

6. In a certain series resonant circuit, $V_C = 150$ V, $V_L = 150$ V, and $V_R = 50$ V. What is the value of the source voltage? What is the voltage across L and C combined?

7. A certain series resonant band-pass filter has a bandwidth of 1000 Hz. If the existing coil is replaced by a coil with a lower Q, what happens to the bandwidth?

8. In a parallel *RLC* circuit, why does the current lag the source voltage at frequencies below resonance?

9. Ideally, what is the total current into the *L* and *C* branches at resonance?

10. In order to tune a parallel resonant circuit to a lower frequency, should you increase or decrease the capacitance?

11. Under what condition is the parallel resonant frequency the same as for series resonance?

12. What happens to the bandwidth of a parallel resonant filter if the parallel resistance is reduced?

13. Determine the impedance in polar form in Figure 18–55.

14. Find each of the following in Figure 18–55:
    (a) I   **(b) V_R**   **(c) V_L**   **(d) V_C**   **(e) P_{true}**   **(f) P_r (net)**

15. A certain series *RLC* circuit has $R_W = 15 \ \Omega$, $L = 20 \ mH$, and $C = 12 \ pF$. Determine the resonant frequency. What is the total impedance at resonance?

16. In Question 5, what are the impedance and phase angle 10 kHz above resonance?

17. Find the following in Figure 18–56 at the resonant frequency:
    (a) I   **(b) f_r**   **(c) V_R**   **(d) V_L**   **(e) V_C**   **(f) θ**

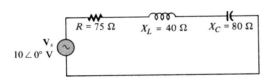

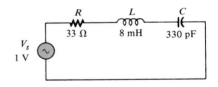

FIGURE 18–55

FIGURE 18–56

18. Determine the impedance in polar form for the circuit in Figure 18–57.

19. Find the current in each branch of Figure 18–57.

20. Determine the following in Figure 18–58:
    (a) Z_T   **(b) I_T**   **(c) I_L**   **(d) I_C**

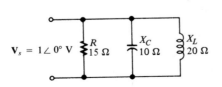

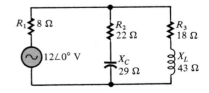

FIGURE 18–57

FIGURE 18–58

21. Determine f_r and *Z* at resonance for the tank circuit in Figure 18–59.

22. In Figure 18–59, how much current is drawn from the source at resonance?

FIGURE 18–59

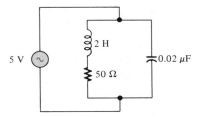

23. Determine Q and the bandwidth of the circuit in Figure 18–60.

FIGURE 18–60

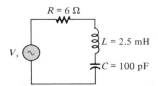

24. A tank circuit has a lower cutoff frequency of 8 kHz and a bandwidth of 2 kHz. What is the resonant frequency?

PROBLEMS

Section 18–1

18–1 A certain series RLC circuit has the following values: $R = 10\ \Omega$, $C = 0.05\ \mu\text{F}$, and $L = 5$ mH. Determine the impedance in polar form. What is the net reactance? The source frequency is 5 kHz.

18–2 Find the impedance in Figure 18–61, and express it in polar form.

FIGURE 18–61

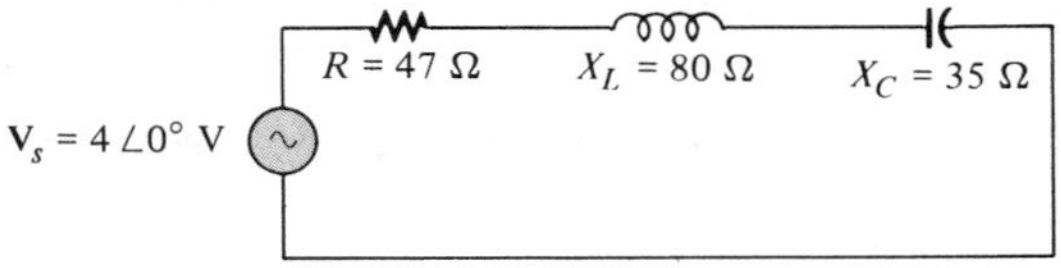

$R = 47\ \Omega \qquad X_L = 80\ \Omega \qquad X_C = 35\ \Omega$

$V_s = 4 \angle 0°$ V

18–3 If the frequency of the source voltage in Figure 18–61 is doubled from the value that produces the indicated reactances, how does the magnitude of the impedance change?

18–4 For the circuit of Figure 18–61, determine the net reactance that will make the impedance magnitude equal to 100 Ω.

Section 18–2

18–5 For the circuit in Figure 18–61, find $\mathbf{I}_T$, $\mathbf{V}_R$, $\mathbf{V}_L$, and $\mathbf{V}_C$ in polar form.

18–6 Sketch the voltage phasor diagram for the circuit in Figure 18–61

18–7 Analyze the circuit in Figure 18–62 for the following ($f = 25$ kHz):

   (a) I_T (b) P_{true} (c) P_r (d) P_a

FIGURE 18–62

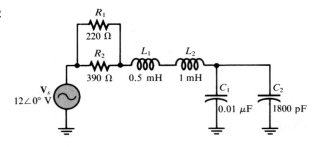

Section 18–3

18–8 Find X_L, X_C, Z, and I at the resonant frequency in Figure 18–63.

18–9 A certain series resonant circuit has a maximum current of 50 mA and a V_L of 100 V. The applied voltage is 10 V. What is Z? What are X_L and X_C?

18–10 For the RLC circuit in Figure 18–64, determine the resonant frequency and the cutoff frequencies.

FIGURE 18–63   **FIGURE 18–64**

18–11 What is the value of the current at the half-power points in Figure 18–64?

18–12 Determine the phase angle between the applied voltage and the current at the cutoff frequencies in Figure 18–64. What is the phase angle at resonance?

18–13 Design a circuit in which the following series resonant frequencies are switch-selectable:

   (a) 500 kHz   (b) 1000 kHz   (c) 1500 kHz   (d) 2000 kHz

Section 18–4

18–14 Express the impedance of the circuit in Figure 18–65 in polar form.

FIGURE 18–65

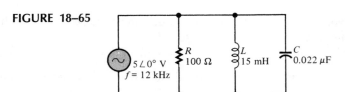

18–15 Is the circuit in Figure 18–65 capacitive or inductive? Explain.

18–16 At what frequency does the circuit in Figure 18–65 change its reactive characteristic (from inductive to capacitive or vice versa)?

Section 18–5

18–17 For the circuit in Figure 18–65, find all the currents and voltages in polar form.

18–18 Find the total impedance for each circuit in Figure 18–66.

18–19 For each circuit in Figure 18–66, determine the phase angle between the source voltage and the total current.

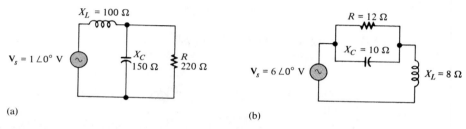

(a)   (b)

FIGURE 18–66

18–20 Determine the voltage across each element in Figure 18–67, and express each in polar form.

18–21 Convert the circuit in Figure 18–67 to an equivalent series form.

18–22 What is the current through R_2 in Figure 18–68?

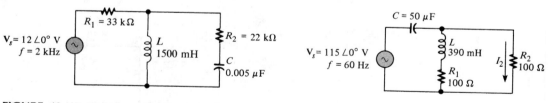

FIGURE 18–67   **FIGURE 18–68**

18–23 In Figure 18–68, what is the phase angle between I_2 and the source voltage?

18–24 Determine the total resistance and the total reactance in Figure 18–69.

18–25 Find the current through each component in Figure 18–69. Find the voltage across each component.

FIGURE 18–69

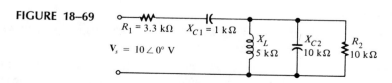

18–26 Determine if there is a value of C that will make $V_{ab} = 0$ V in Figure 18–70. If not, explain.

FIGURE 18–70

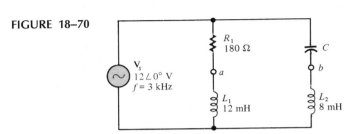

18–27 If the value of C is 0.02 μF, what is the current through a 100-Ω resistor connected from a to b in Figure 18–70?

Section 18–6

18–28 What is the impedance of an ideal parallel resonant circuit (no resistance in either branch)?

18–29 Find Z at resonance and f_r for the tank circuit in Figure 18–71.

18–30 How much current is drawn from the source in Figure 18–71 at resonance? What are the inductive current and the capacitive current at the resonant frequency?

18–31 Determine the resonant frequencies and the output voltage at each frequency in Figure 18–72.

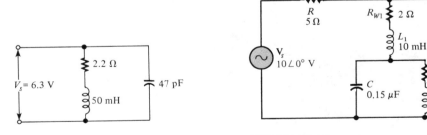

FIGURE 18–71   FIGURE 18–72

18–32 Design a parallel-resonant network using a single coil and switch-selectable capacitors to produce the following resonant frequencies: 8 MHz, 9 MHz, 10 MHz, and 11 MHz. Assume a 10-μH coil with a winding resistance of 5 Ω.

Section 18–7

18–33 At resonance, $X_L = 2$ kΩ and $R_W = 25$ Ω in a parallel RLC circuit. The resonant frequency is 5 kHz. Determine the bandwidth.

18–34 If the lower cutoff frequency is 2400 Hz and the upper cutoff frequency is 2800 Hz, what is the bandwidth? What is the resonant frequency?

18–35 In a certain RLC circuit, the power at resonance is 2.75 W. What is the power at the lower cutoff frequency?

18–36 What values of L and C should be used in a tank circuit to obtain a resonant frequency of 8 kHz? The bandwidth must be 800 Hz. The winding resistance of the coil is 10 Ω.

18–37 A parallel resonant circuit has a Q of 50 and a BW of 400 Hz. If Q is doubled, what is the bandwidth for the same f_r?

Section 18–9

18–38 Write a computer program to compute f_r, Q, and BW of a specified tank circuit with L, C, and R_W as the input variables.

18–39 Write a program to compute the impedance and phase shift of a specified parallel tank circuit over a specified frequency range.

ANSWERS TO SECTION REVIEWS

Section 18–1
1. 70 Ω, capacitive.   **2.** $84.31\angle -56.12° \Omega$, 84.31 Ω, $-56.12°$, leading.

Section 18–2
1. $38.42\angle -21.34°$ V.   **2.** Leads.   **3.** 6 Ω.

Section 18–3
1. $X_L = X_C$.   **2.** The impedance is minimum.   **3.** 159 kHz.   **4.** Capacitive.

Section 18–4
1. Capacitive.   **2.** $0.00154\angle 49.4°$ S.   **3.** $649.4\angle -49.4° \Omega$.

Section 18–5
1. $I_R = 80$ mA, $I_C = 120$ mA, $I_L = 240$ mA.   **2.** Capacitive.
3. 20.13 mH, 1589 Ω.

Section 18–6
1. Maximum   **2.** Minimum   **3.** 1500 Ω.   **4.** $f_r = 225$ kHz, $Z = 1250$ MΩ.   **5.** 22.51 kHz   **6.** 20.9 kHz   **7.** 8020 Ω.

Section 18–7
1. 400 kHz.   **2.** 2 MHz.   **3.** 0.9 W.   **4.** Smaller BW.

Section 18–8
1. To select a narrow band of frequencies.   **2.** A band-stop filter.
3. Several capacitors (or inductors) whose values can be varied simultaneously with a common control.

Section 18–9
1. In lines 280 and 290 using comparisons.   **2.** Yes.

NINETEEN

FILTERS

19–1   LOW-PASS FILTERS
19–2   HIGH-PASS FILTERS
19–3   BAND-PASS FILTERS
19–4   BAND-STOP FILTERS
19–5   COMPUTER ANALYSIS

Filters were introduced briefly in Chapters 16, 17, and 18 to illustrate applications of *RC, RL,* and *RLC* circuits and are covered more thoroughly here because of their importance in electronic systems. In this chapter, *passive filters* are discussed. Passive filters use various combinations of resistors, capacitors, and inductors. You have already seen how basic *RC, RL,* and *RLC* circuits can be used as filters. In this chapter, you will see that passive filters can be placed in four general categories according to their function: *low-pass, high-pass, band-pass,* and *band-stop.* Within each functional category, there are several common types to be examined.

In this chapter, you will learn:

☐ The definition of *band pass* in relation to low-pass filters.
☐ How various combinations of L and C form low-pass filters.
☐ How various combinations of L and C form high-pass filters.
☐ How basic band-pass filters are formed using high-pass, low-pass, and resonant circuits.
☐ How basic band-stop filters are formed using high-pass, low-pass, and resonant circuits.
☐ How to use the decibel (dB) measurement.
☐ How to generate a Bode plot of a filter's response.

19–1 LOW-PASS FILTERS

As you already know, a low-pass filter allows signals with lower frequencies to pass from input to output while rejecting higher frequencies. A block diagram and a general response curve for a low-pass filter appear in Figure 19–1.

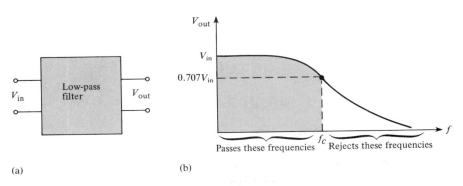

(a)   (b)

FIGURE 19–1
Low-pass filter block diagram and general response curve.

The range of low frequencies passed by a low-pass filter within a specified limit is called the *pass band* of the filter. The point considered to be the *upper end* of the band pass is at the *critical frequency*, f_c, as illustrated in Figure 19–1(b). The critical frequency is the frequency at which the filter's output voltage is 70.7% of the maximum. The filter's critical frequency is also called the *break frequency* or *cutoff frequency*. The output is often said to be *down 3 dB* at this frequency. The term *dB* (*decibel*) is a commonly used one that you should understand.

DECIBELS

Because of the importance of the decibel unit in filter measurements, additional coverage of this topic is necessary before going any further.

The basis for the decibel unit stems from the logarithmic response of the human ear to the intensity of sound. The decibel is a logarithmic measurement of the ratio of one power to another or one voltage to another, which can be used to express the input-to-output relationship of a filter. The following equation expresses a *voltage* ratio in decibels:

$$dB = 20 \log\left(\frac{V_{out}}{V_{in}}\right) \qquad \textbf{(19–1)}$$

The following equation is the decibel formula for a *power* ratio:

$$dB = 10 \log\left(\frac{P_{out}}{P_{in}}\right) \qquad \textbf{(19–2)}$$

EXAMPLE
19–1

At a certain frequency, the output voltage of a filter is 5 V and the input is 10 V. Express the voltage ratio in decibels.

Solution:

$$20 \log\left(\frac{V_{out}}{V_{in}}\right) = 20 \log\left(\frac{5 \text{ V}}{10 \text{ V}}\right) = 20 \log(0.5)$$

$$= -6.02 \text{ dB}$$

The calculator sequence is

$\boxed{5}\ \boxed{\div}\ \boxed{10}\ \boxed{=}\ \boxed{\log}\ \boxed{\times}\ \boxed{20}\ \boxed{=}$

RC LOW-PASS FILTER

A basic *RC* low-pass filter is shown in Figure 19–2. Notice that the output voltage is taken across the *capacitor*.

FIGURE 19–2

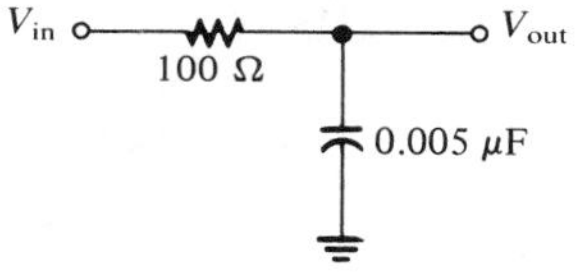

When the input is dc (0 Hz), the output voltage equals the input voltage because X_C is infinitely large. As the input frequency is increased, X_C decreases and, as a result, V_{out} gradually decreases until a frequency is reached where $X_C = R$. This is the *critical frequency*, f_c, of the filter.

$$X_C = R$$

$$\frac{1}{2\pi f_c C} = R$$

$$f_c = \frac{1}{2\pi RC} \tag{19–3}$$

At the critical frequency, the output voltage magnitude is

$$V_{out} = \left(\frac{X_C}{\sqrt{R^2 + X_C^2}}\right) V_{in}$$

by application of the voltage divider formula. Since $X_C = R$ at f_c, the output voltage can be expressed as

$$V_{out} = \left(\frac{R}{\sqrt{R^2 + R^2}}\right) V_{in} = \left(\frac{R}{\sqrt{2R^2}}\right) V_{in}$$

$$= \left(\frac{R}{R\sqrt{2}}\right) V_{in} = \left(\frac{1}{\sqrt{2}}\right) V_{in}$$

$$= 0.707 V_{in}$$

These calculations show that the output is 70.7% of the input when $X_C = R$. The frequency at which this occurs is, by definition, the critical frequency.

The ratio of output voltage to input voltage at the critical frequency can be expressed in dB as follows.

$$V_{out} = 0.707V_{in}$$

$$\frac{V_{out}}{V_{in}} = 0.707$$

$$20 \log\left(\frac{V_{out}}{V_{in}}\right) = 20 \log(0.707) = -3 \text{ dB}$$

EXAMPLE 19–2

Determine the critical frequency for the low-pass RC filter in Figure 19–2.

Solution:

$$f_c = \frac{1}{2\pi RC} = \frac{1}{2\pi(100\ \Omega)(0.005\ \mu\text{F})}$$

$$= 318.3 \text{ kHz}$$

The output voltage is 3 dB below V_{in} at this frequency (V_{out} has a maximum value of V_{in}).

"ROLL-OFF" OF THE RESPONSE CURVE

The dashed lines in Figure 19–3 show an actual response curve for a low-pass filter. The maximum output is defined to be 0 dB as a reference. Zero decibels corresponds to $V_{out} = V_{in}$, because $20 \log(V_{out}/V_{in}) = 20 \log 1 = 0$ dB. The output drops from 0 dB to -3 dB at the critical frequency and then continues to decrease at a *fixed* rate. This pattern of decrease is called the *roll-off* of the frequency response. The solid line shows an ideal output response that is considered to be "flat" out to f_c. The output then decreases at the fixed rate.

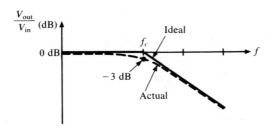

FIGURE 19–3
Actual and ideal response curves for a low-pass filter.

As you have seen, the output voltage of a low-pass filter decreases by 3 dB when the frequency is increased to the critical value f_c. As the frequency continues to increase above f_c, the output voltage continues to decrease. In fact, for each tenfold increase in frequency above f_c, there is a 20-dB reduction in the output, as shown in the following steps.

Let's take a frequency that is ten times the critical frequency ($f = 10f_c$). Since $R = X_C$ at f_c, then $R = 10X_C$ at $10f_c$ because of the inverse relationship of X_C and f.

The *attenuation* of the RC circuit is the ratio V_{out}/V_{in} and is developed as follows:

$$\frac{V_{out}}{V_{in}} = \frac{X_C}{\sqrt{R^2 + X_C^2}} = \frac{X_C}{\sqrt{(10X_C)^2 + X_C^2}}$$

$$= \frac{X_C}{\sqrt{100X_C^2 + X_C^2}} = \frac{X_C}{\sqrt{X_C^2(100 + 1)}}$$

$$= \frac{X_C}{X_C\sqrt{101}} = \frac{1}{\sqrt{101}} \cong \frac{1}{10} = 0.1$$

The dB attenuation is

$$20 \log\left(\frac{V_{out}}{V_{in}}\right) = 20 \log(0.1) = -20 \text{ dB}$$

A tenfold change in frequency is called a *decade*. So, for the RC network, the output voltage is reduced by 20 dB for each decade increase in frequency. A similar result can be derived for the high-pass network. The roll-off is a constant 20 dB/decade for a basic RC or RL filter. Figure 19–4 shows a frequency response plot on a semilog scale, where each interval on the horizontal axis represents a tenfold increase in frequency. This response curve is called a *Bode plot*.

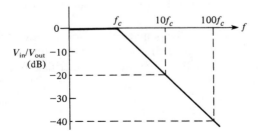

FIGURE 19–4
Frequency roll-off for a low-pass RC filter (Bode plot).

**EXAMPLE
19–3**

Make a Bode plot for the filter in Figure 19–5 for three decades of frequency. Use semilog graph paper.

FIGURE 19–5

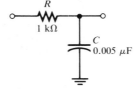

Solution:
The critical frequency for this high-pass filter is

$$f_c = \frac{1}{2\pi RC} = \frac{1}{2\pi(1\ k\Omega)(0.005\ \mu F)}$$
$$= 31.8\ kHz$$

The idealized Bode plot is shown with the solid line on the semilog graph in Figure 19–6. The approximate actual response curve is shown with the dashed line. Notice first that the horizontal scale is logarithmic and the vertical scale is linear. The frequency is on the logarithmic scale, and the filter output in decibels is on the vertical.

The output is flat below f_c (31.8 kHz). As the frequency is increased above f_c, the output drops at a 20 dB/decade rate. Thus, for the ideal curve, every time the frequency is increased by ten, the output is reduced by 20 dB. A slight variation from this occurs in actual practice. The output is actually at -3 dB rather than 0 dB at the critical frequency.

RL LOW-PASS FILTER

A basic *RL* low-pass filter is shown in Figure 19–7. Notice that the output voltage is taken across the *resistor*.

When the input is dc (0 Hz), the output voltage ideally equals the input voltage because X_L is a short (if R_W is neglected). As the input frequency is increased, X_L increases and, as a result, V_{out} gradually decreases until the critical frequency is reached. At this point, $X_L = R$ and the frequency is

$$2\pi f_c L = R$$

$$f_c = \frac{R}{2\pi L}$$

$$f_c = \frac{1}{2\pi(L/R)} \qquad \textbf{(19–4)}$$

Just as in the *RC* low-pass filter, $V_{out} = 0.707 V_{in}$ and, thus, the output voltage is down 3 dB at the critical frequency.

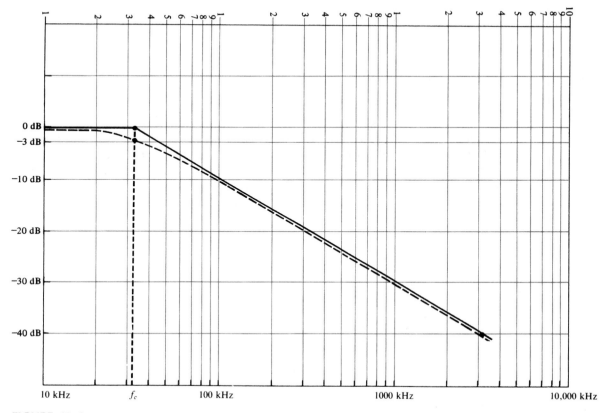

FIGURE 19–6
Bode plot for Example 19–3.

FIGURE 19–7
RL low-pass filter.

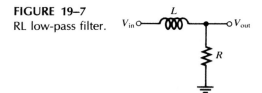

EXAMPLE 19–4

Make a Bode plot for the filter in Figure 19–8 for three decades of frequency. Use semilog graph paper.

FIGURE 19–8

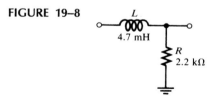

Solution:
The critical frequency for this high-pass filter is

$$f_c = \frac{1}{2\pi(L/R)} = \frac{1}{2\pi(4.7 \text{ mH}/2.2 \text{ k}\Omega)}$$

$$= 74.5 \text{ kHz}$$

The idealized Bode plot is shown with the solid line on the semilog graph in Figure 19–9. The approximate actual response curve is shown with the dashed line. Notice first that the horizontal scale is logarithmic and the vertical scale is linear. The

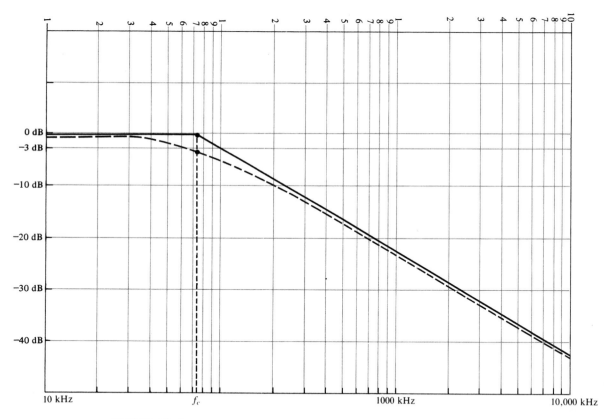

FIGURE 19–9
Bode plot for Example 19– 4.

frequency is on the logarithmic scale, and the filter output in decibels is on the vertical.

The output is flat below f_c (74.5 kHz). As the frequency is increased above f_c, the output drops at a 20 dB/decade rate. Thus, for the ideal curve, every time the frequency is increased by ten, the output is reduced by 20 dB. A slight variation from this occurs in actual practice. The output is actually at -3 dB rather than 0 dB at the critical frequency.

PHASE SHIFT IN A LOW-PASS FILTER

The RC low-pass filter acts as a lag network. Recall from Chapter 16 that the phase shift from input to output is expressed as

$$\phi = -90° + \tan^{-1}\left(\frac{X_C}{R}\right)$$

At the critical frequency, $X_C = R$ and, therefore, $\phi = -45$ degrees. As the input frequency is reduced, ϕ decreases and approaches 0 degrees as the frequency approaches zero, as shown in Figure 19–10.

FIGURE 19–10
Phase characteristic of a low-pass filter.

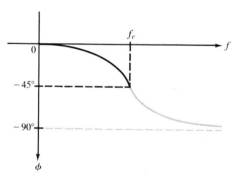

The RL low-pass filter also acts as a lag network. Recall from Chapter 17 that the phase shift is expressed as

$$\phi = -\tan^{-1}\left(\frac{X_L}{R}\right)$$

As in the RC filter, the phase shift from input to output is -45 degrees at the critical frequency and decreases for frequencies below f_c.

SECTION REVIEW 19–1

1. In a certain low-pass filter, f_c = 2.5 kHz. What is its band pass?
2. In a certain low-pass filter, R = 100 Ω and X_C = 2 Ω at a frequency, f_1. Determine $\mathbf{V}_{\text{out}}$ at f_1 when $\mathbf{V}_{\text{in}}$ = 5∠0° V rms.
3. V_{out} = 400 mV, and V_{in} = 1.2 V. Express the ratio $V_{\text{out}}/V_{\text{in}}$ in dB.

19–2

HIGH-PASS FILTERS

A high-pass filter allows signals with higher frequencies to pass from input to output while rejecting lower frequencies. A block diagram and a general response curve for a high-pass filter are shown in Figure 19–11.

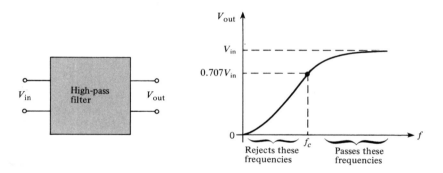

FIGURE 19–11
High-pass filter block diagram and response curve.

The frequency considered to be the *lower* end of the band is called the *critical frequency*. Just as in the low-pass filter, it is the frequency at which the output is 70.7% of the maximum, as indicated in the figure.

RC HIGH-PASS FILTER

A basic *RC* high-pass filter is shown in Figure 19–12. Notice that the output voltage is taken across the *resistor*.

FIGURE 19–12
RC high-pass filter.

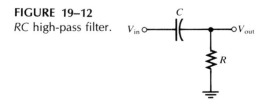

When the input frequency is at its critical value, $X_C = R$ and the output voltage is $0.707V_{in}$, just as in the case of the low-pass filter. As the input frequency increases above f_c, X_C decreases and, as a result, the output voltage increases and approaches a value equal to V_{in}. The expression for the critical frequency of the high-pass filter is the same as for the low-pass,

$$f_c = \frac{1}{2\pi RC}$$

Below f_c, the output voltage decreases (rolls off) at a rate of 20 dB/decade. Figure 19–13 shows an actual and an ideal response curve for a high-pass filter.

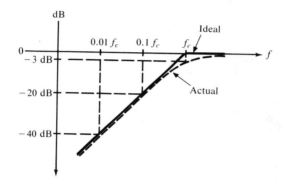

FIGURE 19–13
Actual and ideal response curves for a high-pass filter.

EXAMPLE 19–5

Make a Bode plot for the filter in Figure 19–14 for three decades of frequency. Use semilog graph paper.

FIGURE 19–14

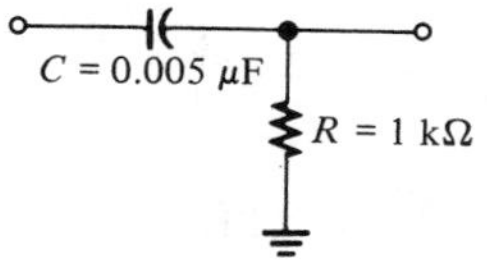

$C = 0.005\ \mu F$

$R = 1\ k\Omega$

Solution:
The critical frequency for this high-pass filter is

$$f_c = \frac{1}{2\pi RC} = \frac{1}{2\pi(100\ \Omega)(0.159\ \mu F)}$$

$$= 10\ kHz$$

The idealized Bode plot is shown with the solid line on the semilog graph in Figure 19–15. The approximate actual response curve is shown with the dashed line. Notice first that the horizontal scale is logarithmic and the vertical scale is linear. The frequency is on the logarithmic scale, and the filter output in decibels is on the vertical.

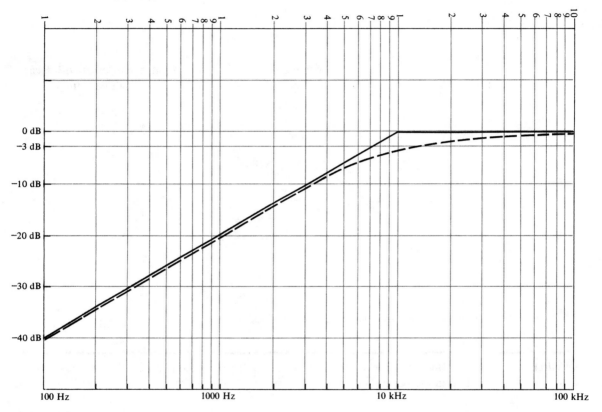

FIGURE 19–15
Bode plot for Example 19–5.

The output is flat beyond f_c (10 kHz). As the frequency is reduced below f_c, the output drops at a 20 dB/decade rate. Thus, for the ideal curve, every time the frequency is reduced by ten, the output is reduced by 20 dB. A slight variation from this occurs in actual practice. The output is actually at -3 dB rather than 0 dB at the critical frequency.

RL HIGH-PASS FILTER

A basic *RL* high-pass filter is shown in Figure 19–16. Notice that the output is taken across the *inductor*.

FIGURE 19–16
RL high-pass filter.

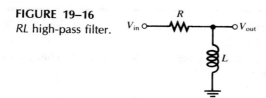

When the input frequency is at its critical value, $X_L = R$, and the output voltage is $0.707V_{in}$. As the frequency increases above f_c, X_L increases and, as a result, the output voltage increases until it equals V_{in}. The expression for the critical frequency of the high-pass filter is the same as for the low-pass,

$$f_c = \frac{1}{2\pi(L/R)}$$

PHASE SHIFT IN A HIGH-PASS FILTER

Both the *RC* and the *RL* high-pass filters act as lead networks. Recall from Chapters 16 and 17 that the phase shift from input to output for the *RC* lead network is

$$\phi = \tan^{-1}\left(\frac{X_C}{R}\right)$$

and for the *RL* lead network is

$$\phi = 90° - \tan^{-1}\left(\frac{X_L}{R}\right)$$

At the critical frequency, $X_L = R$ and, therefore, $\phi = 45$ degrees. As the frequency is increased, ϕ decreases toward 0 degrees as shown in Figure 19–17.

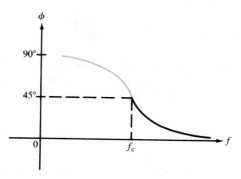

FIGURE 19–17
Phase characteristic of a high-pass filter.

EXAMPLE
19–6

(a) In Figure 19–18, find the value of C so that X_C is ten times less than R at an input frequency of 10 kHz.

(b) If a 5-V sine wave with a dc level of 10 V is applied, what are the output voltage magnitude and the phase shift?

FIGURE 19–18

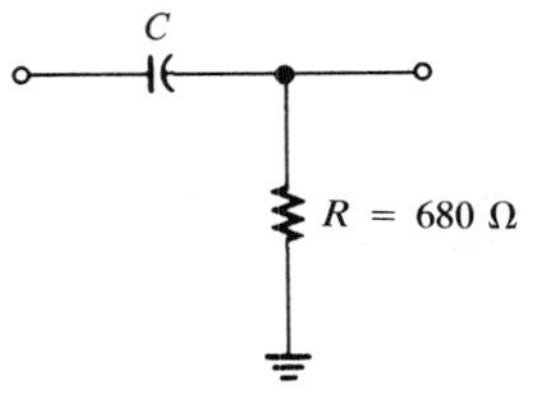

$R = 680\ \Omega$

Solution:

(a) The value of C is determined as follows:

$$X_C = 0.1R$$
$$= 60\ \Omega$$

$$C = \frac{1}{2\pi f X_C} = \frac{1}{2\pi (10\ \text{kHz})(60\ \Omega)}$$
$$= 0.265\ \mu\text{F}$$

(b) The magnitude of the sine wave output is determined as follows:

$$V_{\text{out}} = \left(\frac{R}{\sqrt{R^2 + X_C^2}}\right) V_{\text{in}} = \left(\frac{680\ \Omega}{\sqrt{(680\ \Omega)^2 + (68\ \Omega)^2}}\right) 5\ \text{V}$$
$$= 4.98\ \text{V}$$

The phase shift is

$$\phi = \tan^{-1}\left(\frac{X_C}{R}\right) = \tan^{-1}\left(\frac{68\ \Omega}{680\ \Omega}\right) = 5.7°$$

At $f = 10$ kHz, which is a decade above the critical frequency, the sinusoidal output is almost equal to the input in magnitude, and the phase shift is very small. The 10-V dc level has been filtered out and does not appear at the output.

SECTION REVIEW 19–2

1. The input voltage of a high-pass filter is 1 V. What is V_{out} at the critical frequency?

2. In a certain high-pass filter, $\mathbf{V}_{\text{in}} = 10\angle 0°$ V, $R = 1$ kΩ, and $X_L = 15$ kΩ. Determine $\mathbf{V}_{\text{out}}$.

19–3 BAND-PASS FILTERS

A band-pass filter allows a certain *band* of frequencies to pass and attenuates or rejects all frequencies below and above the band pass. A typical band-pass response curve is shown in Figure 19–19.

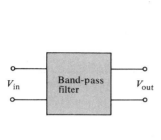

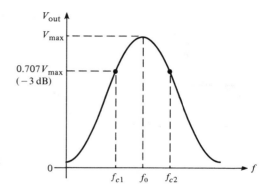

FIGURE 19–19
Typical band-pass response curve.

LOW-PASS/HIGH-PASS FILTER

A combination of a low-pass and a high-pass filter can be used to form a band-pass filter, as illustrated in Figure 19–20.

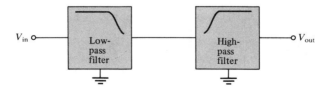

FIGURE 19–20
Low-pass and high-pass filters used to form a band-pass filter.

If the critical frequency of the low-pass ($f_{c(l)}$) is higher than the critical frequency of the high-pass ($f_{c(h)}$), the responses overlap. Thus, all frequencies except those between $f_{c(h)}$ and $f_{c(l)}$ are eliminated, as shown in Figure 19–21.

> **The bandwidth of a band-pass filter is** *the range of frequencies for which the current (or output voltage) is equal to or greater than 70.7% of its value at the resonant frequency.*

As you know, bandwidth is often abbreviated *BW*.

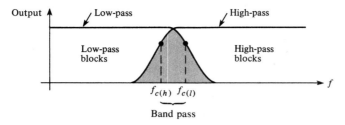

FIGURE 19–21
Overlapping response curves of a high-pass/low-pass filter.

**EXAMPLE
19–7**

A high-pass filter with f_c = 2 kHz and a low-pass filter with f_c = 2.5 kHz are used to construct a band-pass filter. What is the bandwidth of the band pass?

Solution:
$$BW = f_{c(l)} - f_{c(h)} = 2.5 \text{ kHz} - 2 \text{ kHz}$$
$$= 500 \text{ Hz}$$

SERIES RESONANT BAND-PASS FILTER

A type of series resonant band-pass filter is shown in Figure 19–22. As you learned in the last chapter, a series resonant circuit has *minimum impedance* and *maximum current* at the resonant frequency, f_r. Thus, most of the input voltage is dropped across the resistor at the resonant frequency. Therefore, the output across R has a band-pass characteristic with a maximum output at the frequency of resonance. The resonant frequency is called the *center frequency*, f_0. The bandwidth is determined by the circuit Q, as was discussed in Chapter 18.

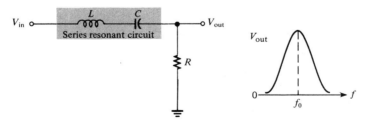

FIGURE 19–22
Series resonant band-pass filter.

A higher value of circuit Q results in a smaller bandwidth. A lower value of Q causes a larger bandwidth. A formula for the bandwidth of a resonant circuit in terms of Q is stated in the following equation:

$$BW = \frac{f_0}{Q} \qquad \textbf{(19–5)}$$

**EXAMPLE
19–8**

Determine the output voltage magnitude at the center frequency (f_0) and the bandwidth for the filter in Figure 19–23.

FIGURE 19–23

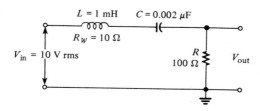

Solution:
At f_0, the impedance of the resonant circuit is equal to the winding resistance, R_W. By the voltage divider formula,

$$V_{out} = \left(\frac{R}{R + R_W}\right)V_{in} = \left(\frac{100\ \Omega}{110\ \Omega}\right)10\ \text{V}$$

$$= 9.09\ \text{V}$$

The center frequency is

$$f_0 = \frac{1}{2\pi\sqrt{LC}}$$

$$= 112.5\ \text{kHz}$$

The circuit Q is

$$Q = \frac{X_L}{R_T} = \frac{707\ \Omega}{110\ \Omega}$$

$$= 6.4$$

The bandwidth is

$$BW = \frac{f_0}{Q} = \frac{112.5\ \text{kHz}}{6.4}$$

$$= 17.6\ \text{kHz}$$

PARALLEL RESONANT BAND-PASS FILTER

A type of band-pass filter using a parallel resonant circuit is shown in Figure 19–24. Recall that a parallel resonant circuit has *maximum impedance* at resonance. The circuit in Figure 19–24 acts as a voltage divider. At resonance, the impedance of the tank is much greater than R. Thus, most of the input voltage is across the tank, producing a maximum output voltage at the resonant (center) frequency.

FIGURE 19–24
Parallel resonant band-pass filter.

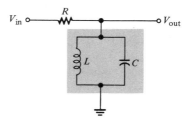

For frequencies above or below resonance, the tank impedance drops off, and more of the input voltage is across R. As a result, the output voltage across the tank drops off, creating a band-pass characteristic.

EXAMPLE 19–9

What is the center frequency of the filter in Figure 19–25? Assume $R_W = 0\ \Omega$.

FIGURE 19–25

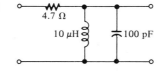

Solution:
The center frequency of the filter is its resonant frequency:

$$f_0 = \frac{1}{2\pi\sqrt{LC}} = \frac{1}{2\pi\sqrt{(10\ \mu\text{H})(100\ \text{pF})}}$$

$$= 5.03\ \text{MHz}$$

EXAMPLE 19–10

Determine the center frequency and bandwidth for the band-pass filter in Figure 19–26 if the inductor has a winding resistance of 15 Ω.

FIGURE 19–26

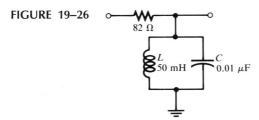

Solution:
Recall from Chapter 18 that the resonant (center) frequency of a nonideal tank circuit is

$$f_0 = \frac{\sqrt{1 - (R_W^2 C/L)}}{2\pi\sqrt{LC}}$$

$$= \frac{\sqrt{1 - (15^2\ \Omega)(0.01\ \mu F)/50\ mH}}{2\pi\sqrt{(50\ mH)(0.01\ \mu F)}}$$

$$= 7.12\ kHz$$

The Q of the coil at resonance is

$$Q = \frac{X_L}{R_W} = \frac{2\pi f_0 L}{R_W} = \frac{2\pi(7.12\ kHz)(50\ mH)}{15\ \Omega}$$

$$= 149$$

The bandwidth of the filter is

$$BW = \frac{f_0}{Q} = \frac{7.12\ kHz}{149} = 47.79\ Hz$$

Note that since $Q > 10$, the ideal formula could have been used to calculate f_0. However, we did not know this in advance.

SECTION REVIEW 19–3

1. For a band-pass filter, $f_{c(h)} = 29.8$ kHz and $f_{c(l)} = 30.2$ kHz. What is the bandwidth?
2. A parallel resonant band-pass filter has the following values: $R_W = 15\ \Omega$, $L = 50\ \mu H$, and $C = 470$ pF. Determine the approximate center frequency.

19–4 BAND-STOP FILTERS

A band-stop filter is essentially the opposite of a band-pass filter in terms of the responses. It allows *all* frequencies to pass except those lying within a certain *stop band*. A general band-stop response curve is shown in Figure 19–27.

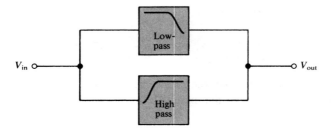

FIGURE 19–27
General band-stop response curve.

LOW-PASS/HIGH-PASS FILTER

A band-stop filter can be formed from a low-pass and a high-pass filter, as shown in Figure 19–28.

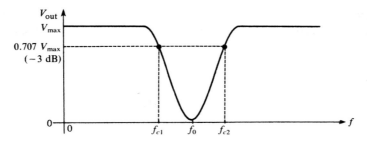

FIGURE 19–28
Low-pass and high-pass filters used to form a band-stop filter.

If the low-pass critical frequency $f_{c(l)}$ is set lower than the high-pass critical frequency $f_{c(h)}$, a band-stop characteristic is formed as illustrated in Figure 19–29.

FIGURE 19–29
Band-stop response curve.

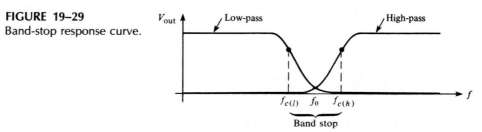

SERIES RESONANT BAND-STOP FILTER

A series resonant circuit used in a band-stop configuration is shown in Figure 19–30. Basically, it works as follows: At the resonant frequency, the impedance is minimum,

FIGURE 19–30
Series resonant band-stop filter.

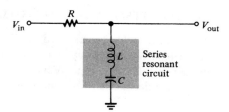

and therefore the output voltage is minimum. Most of the input voltage is dropped across R. At frequencies above and below resonance, the impedance increases, causing more voltage across the output.

EXAMPLE 19–11

Find the output voltage magnitude at f_0 and the bandwidth in Figure 19–31.

FIGURE 19–31

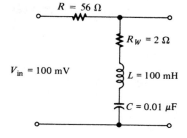

Solution:
Since $X_L = X_C$ at resonance, then

$$V_{out} = \left(\frac{R_W}{R + R_W}\right)V_{in} = \left(\frac{2\ \Omega}{58\ \Omega}\right)100\ mV$$

$$= 3.45\ mV$$

$$f_0 = \frac{1}{2\pi\sqrt{LC}} = \frac{1}{2\pi\sqrt{(100\ mH)(0.01\ \mu F)}}$$

$$= 5.03\ kHz$$

$$Q = \frac{X_L}{R} = \frac{3160\ \Omega}{58\ \Omega}$$

$$= 54.5$$

$$BW = \frac{f_0}{Q} = \frac{5.03\ kHz}{54.5}$$

$$= 92.3\ Hz$$

PARALLEL RESONANT BAND-STOP FILTER

A parallel resonant circuit used in a band-stop configuration is shown in Figure 19–32. At the resonant frequency, the tank impedance is maximum, and so most of the input voltage appears across it. Very little voltage is across R at resonance. As the tank impedance decreases above and below resonance, the output voltage increases.

FIGURE 19–32
Parallel resonant band-stop filter.

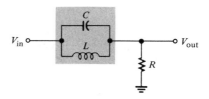

EXAMPLE 19–12

Find the center frequency of the filter in Figure 19–33. Sketch the output response curve showing the minimum and maximum voltages.

FIGURE 19–33

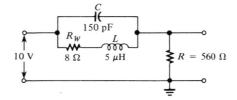

Solution:

$$f_0 = \frac{\sqrt{1 - R_W^2 C/L}}{2\pi\sqrt{LC}} = \frac{\sqrt{1 - (64)(150 \text{ pF})/5 \ \mu\text{H}}}{2\pi\sqrt{(5 \ \mu\text{H})(150 \text{ pF})}}$$

$$= 5.79 \text{ MHz}$$

At the center (resonant) frequency,

$$X_L = 2\pi f_0 L = 2\pi(5.79 \text{ MHz})(5 \ \mu\text{H}) = 181.9 \ \Omega$$

$$Q = \frac{X_L}{R_W} = \frac{181.9 \ \Omega}{8 \ \Omega} = 22.74$$

$$Z_r = R_W(Q^2 + 1) = 8 \ \Omega(22.74^2 + 1)$$

$$= 4.145 \text{ k}\Omega \text{ (resistive)}$$

Now, we use the voltage divider formula to find the *minimum* output voltage magnitude:

$$V_{\text{out(min)}} = \left(\frac{R}{R + Z_r}\right)V_{\text{in}} = \left(\frac{560 \ \Omega}{4.705 \text{ k}\Omega}\right)10 \text{ V}$$

$$= 1.19 \text{ V}$$

At zero frequency, the impedance of the tank is R_W because $X_C = \infty$ and $X_L = 0\ \Omega$. Therefore, the maximum output voltage is

$$V_{out(max)} = \left(\frac{R}{R + R_W}\right)V_{in} = \left(\frac{560\ \Omega}{568\ \Omega}\right)10\ \text{V}$$

$$= 9.86\ \text{V}$$

As the frequency increases much higher than f_0, X_C approaches $0\ \Omega$, and V_{out} approaches V_{in}. Figure 19–34 is the response curve.

FIGURE 19–34

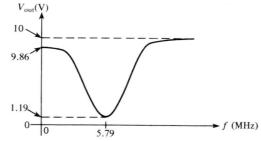

SECTION REVIEW 19–4

1. How does a band-stop filter differ from a band-pass filter?

2. Name three basic ways to construct a band-stop filter.

19–5   COMPUTER ANALYSIS

With this program, the response of low-pass RC and high-pass RC filters can be computed. The critical frequency is calculated, and the output voltages are displayed for a specified range of input frequencies. Other inputs required are the component values and the input voltage.

```
10   CLS
20   PRINT "PLEASE SELECT AN OPTION BY NUMBER."
30   PRINT
40   PRINT "(1) RC LOW-PASS ANALYSIS"
50   PRINT "(2) RC HIGH-PASS ANALYSIS"
60   INPUT X
70   ON X GOTO 80,290
80   CLS
90   PRINT "RC LOW-PASS ANALYSIS"
100  PRINT
110  INPUT "RESISTANCE IN OHMS";R
120  INPUT "CAPACITANCE IN FARADS";C
130  INPUT "INPUT VOLTAGE IN VOLTS";VIN
140  PRINT
150  INPUT "LOWEST FREQUENCY IN HERTZ";FL
```

```
160 INPUT "HIGHEST FREQUENCY IN HERTZ";FH
170 INPUT "FREQUENCY INCREMENTS IN HERTZ";FI
180 CLS
190 PRINT "FREQUENCY(HZ)","VOUT"
200 FOR F=FL TO FH STEP FI
210 XC=1/(2*3.1416*F*C)
220 V=(XC/(SQR(R*R+XC*XC)))*VIN
230 PRINT F, V
240 NEXT
250 FC=1/(2*3.1416*R*C)
260 PRINT "THE CRITICAL FREQUENCY IS";FC;"HZ"
270 PRINT:PRINT:INPUT "TO RETURN TO MENU, ENTER 1";X
280 ON X GOTO 10
290 CLS
300 PRINT "RC HIGH-PASS ANALYSIS"
310 PRINT
320 INPUT "RESISTANCE IN OHMS";R
330 INPUT "CAPACITANCE IN FARADS";C
340 INPUT "INPUT VOLTAGE IN VOLTS";VIN
350 PRINT
360 INPUT "LOWEST FREQUENCY IN HERTZ";FL
370 INPUT "HIGHEST FREQUENCY IN HERTZ";FH
380 INPUT "FREQUENCY INCREMENTS IN HERTZ";FI
390 CLS
400 PRINT "FREQUENCY(HZ)","VOUT(V)"
410 FOR F=FL TO FH STEP FI
420 XC=1/(2*3.1416*F*C)
430 V=(R/(SQR(R*R+XC*XC)))*VIN
440 PRINT F, V
450 NEXT
460 FC=1/(2*3.1416*R*C)
470 PRINT "THE CRITICAL FREQUENCY IS";FC;"HZ"
480 PRINT:PRINT:INPUT "TO RETURN TO MENU, ENTER 1";X
490 ON X GOTO 10
```

SECTION REVIEW 19–5

1. On what circuit principle is the calculation in line 430 based?

2. What is the purpose of line 490?

SUMMARY

1. In an *RC* low-pass filter, the output voltage is taken across the capacitor and the output lags the input.

2. In an *RL* low-pass filter, the output voltage is taken across the resistor and the output lags the input.

3. In an *RC* high-pass filter, the output is taken across the resistor and the output leads the input.

4. In an RL high-pass filter, the output is taken across the inductor and the output leads the input.

5. The roll-off rate of a basic RC or RL filter is 20 dB per decade.

6. A band-pass filter passes frequencies between the lower and upper cutoff frequencies and rejects all others.

7. A band-stop filter rejects frequencies between its lower and upper cutoff frequencies and passes all others.

8. The bandwidth of a resonant filter is determined by the quality factor (Q) of the circuit and the resonant frequency.

9. Cutoff frequencies are also called -3 dB frequencies.

10. The output voltage is 70.7% of its maximum at the cutoff frequencies.

FORMULAS

$$dB = 20 \log\left(\frac{V_{out}}{V_{in}}\right) \tag{19–1}$$

$$dB = 10 \log\left(\frac{P_{out}}{P_{in}}\right) \tag{19–2}$$

$$f_c = \frac{1}{2\pi RC} \tag{19–3}$$

$$f_c = \frac{1}{2\pi(L/R)} \tag{19–4}$$

$$BW = \frac{f_0}{Q} \tag{19–5}$$

SELF-TEST

Solutions appear at the end of the book.

1. The maximum output voltage of a given low-pass filter is 10 V. What is the output voltage at the cutoff frequency?

2. A sine wave with a peak-to-peak value of 15 V is applied to an RC low-pass filter. The average value of the input is 10 V. If the reactance at the sine-wave frequency is assumed to be zero, what is the output voltage?

3. The same signal in Question 2 is applied to an RC high-pass filter. If the reactance is zero at the signal frequency, what is the output voltage of the filter?

4. Calculate the value of C used in a low-pass filter if its reactance must be at least ten times smaller than the resistance of 2.2 kΩ at 1 kHz.

5. A low-pass RL filter has X_L = 100 Ω and R = 4.7 Ω at the input frequency. What is the exact output voltage if the input is a 5-V peak-to-peak sine wave riding on an 8-V constant dc level? Assume a winding resistance of 0 Ω.

6. Sketch a series resonant band-pass filter and a series resonant band-stop filter.

7. Sketch a parallel resonant band-pass filter and a parallel resonant band-stop filter.

8. The resonant frequency of a band-pass filter is 5 kHz. The circuit Q is 20. What is the bandwidth?

9. The output voltage of a filter is 4 V, and the input is 12 V. Express this relationship in dB.

10. The output power of a circuit is 5 W, and the input power is 10 W. What is the dB power ratio?

11. What is the cutoff frequency of an RC filter with R = 1 kΩ and C = 0.02 μF?

12. What is the cutoff frequency of an RL filter with R = 5.6 kΩ and L = 0.01 mH?

PROBLEMS

Section 19–1

19–1   In a certain low-pass filter, X_C = 500 Ω and R = 2.2 kΩ. What is the output voltage when the input is 10 V rms?

19–2   A certain low-pass filter has a cutoff frequency of 3 kHz. Determine which of the following frequencies are passed and which are rejected:
  (a) 100 Hz   **(b)** 1 kHz   **(c)** 2 kHz   **(d)** 3 kHz   **(e)** 5 kHz

19–3   Determine the output voltage of each filter in Figure 19–35 at the specified frequency when V_{in} = 10 V.

FIGURE 19–35

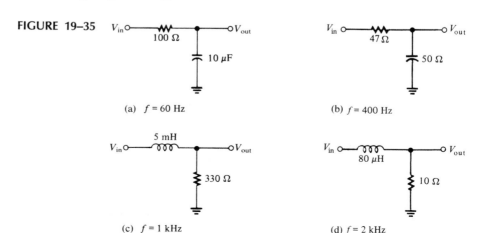

(a) f = 60 Hz

(b) f = 400 Hz

(c) f = 1 kHz

(d) f = 2 kHz

19–4 What is f_c for each filter in Figure 19–35? Determine the output voltage at f_c in each case when $V_{in} = 5$ V.

19–5 For the filter in Figure 19–36, calculate the value of C required for each of the following cutoff frequencies:
(a) 60 Hz   (b) 500 Hz   (c) 1 kHz   (d) 5 kHz

FIGURE 19–36

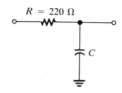

19–6 Determine the critical frequency for each switch position on the switched filter network of Figure 19–37.

FIGURE 19–37

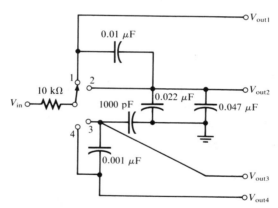

19–7 Sketch a Bode plot for each part of Problem 19–5.

19–8 For each following case, express the voltage ratio in dB:
(a) $V_{in} = 1$ V, $V_{out} = 1$ V   (b) $V_{in} = 5$ V, $V_{out} = 3$ V
(c) $V_{in} = 10$ V, $V_{out} = 7.07$ V   (d) $V_{in} = 25$ V, $V_{out} = 5$ V

19–9 The input voltage to a low-pass RC filter is 8 V rms. Find the output voltage at the following dB levels:
(a) -1 dB   (b) -3 dB   (c) -6 dB   (d) -20 dB

19–10 For a basic RC low-pass filter, find the output voltage in dB relative to a 0-dB input for the following frequencies ($f_c = 1$ kHz):
(a) 10 kHz   (b) 100 kHz   (c) 1 MHz

Section 19–2

19–11 In a high-pass filter, $X_C = 500$ Ω and $R = 2.2$ kΩ. What is the output voltage when $V_{in} = 10$ V rms?

19–12 A high-pass filter has a cutoff frequency of 50 Hz. Determine which of the following frequencies are passed and which are rejected:
(a) 1 Hz   (b) 20 Hz   (c) 50 Hz   (d) 60 Hz   (e) 30 kHz

19–13 Determine the output voltage of each filter in Figure 19–38 at the specified frequency when $V_{in} = 10$ V.

FIGURE 19–38

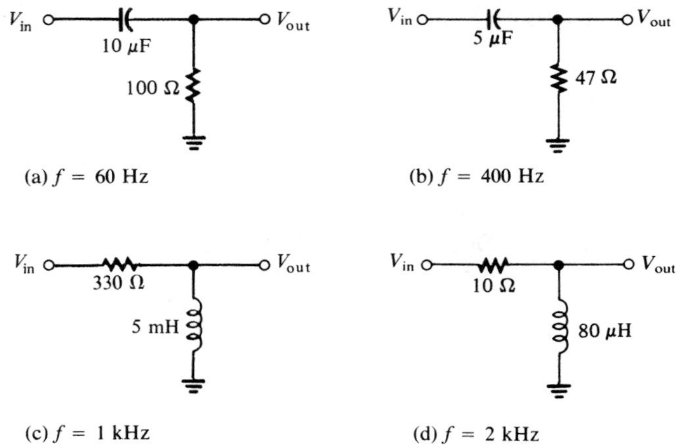

(a) $f = 60$ Hz

(b) $f = 400$ Hz

(c) $f = 1$ kHz

(d) $f = 2$ kHz

19–14 What is f_c for each filter in Figure 19–38? Determine the output voltage at f_c in each case ($V_{in} = 10$ V).

19–15 Sketch the Bode plot for each filter in Figure 19–38.

19–16 Determine f_c for each switch position in Figure 19–39.

FIGURE 19–39

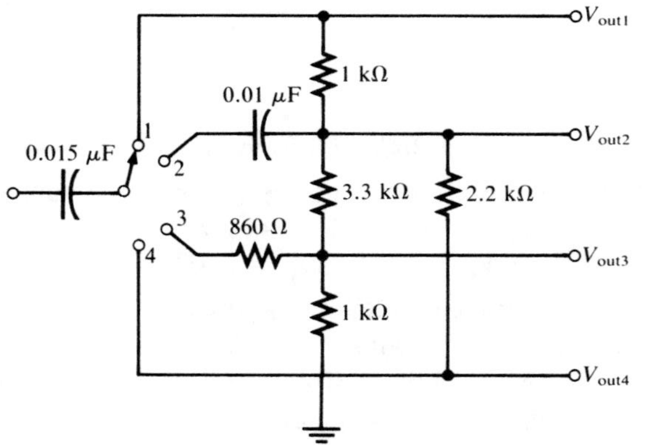

Section 19–3

19–17 Determine the center frequency for each filter in Figure 19–40.

19–18 Assuming that the coils in Figure 19–40 have a winding resistance of 10 Ω, find the bandwidth for each filter.

FIGURE 19–40

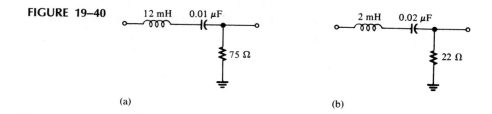

(a)   (b)

19–19 What are the upper and lower cutoff frequencies for each filter in Figure 19–40? Assume the response is symmetrical about f_0. Neglect R_W.

19–20 For each filter in Figure 19–41, find the center frequency of the band pass.

FIGURE 19–41

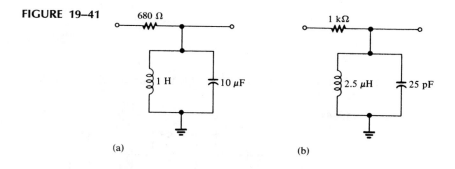

(a)   (b)

19–21 If the coils in Figure 19–41 have a winding resistance of 4 Ω, what is the output voltage at resonance when $V_{in} = 120$ V?

19–22 Determine the separation of center frequencies for all switch positions in Figure 19–42. Do any of the responses overlap? Assume $R_W = 0$ Ω for each coil.

FIGURE 19–42

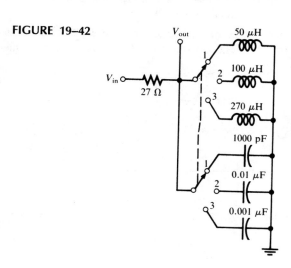

19–23 Design a band-pass filter using a parallel resonant circuit to meet the following specifications:
(a) $BW = 500$ Hz
(b) $Q = 40$
(c) $I_{C(max)} = 20$ mA, $V_{C(max)} = 2.5$ V

Section 19–4
19–24 Determine the center frequency for each filter in Figure 19–43.

FIGURE 19–43

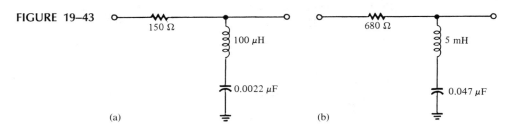

(a)   (b)

19–25 For each filter in Figure 19–44, find the center frequency of the stop band.

FIGURE 19–44

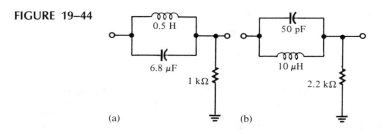

(a)   (b)

19–26 If the coils in Figure 19–44 have a winding resistance of 8 Ω, what is the output voltage at resonance when $V_{in} = 50$ V?

19–27 Determine the values of L_1 and L_2 in Figure 19–45 to pass a signal with a frequency of 1200 kHz and stop (reject) a signal with a frequency of 456 kHz.

FIGURE 19–45

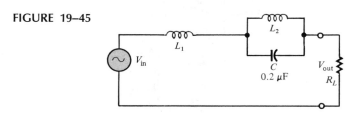

Section 19–5

19–28 Develop a program to compute the output voltage for a low-pass *RL* circuit over a specified range of frequencies and for a specified input voltage. The winding resistance is to be treated as an input variable along with values for *R* and *L*.

19–29 Modify the program in Problem 19–28 to compute the attenuation of the filter, and express it in dB for a specified range of frequencies.

19–30 Develop a flowchart for the program in Section 19–5.

ANSWERS TO SECTION REVIEWS

Section 19–1
1. 0 Hz to 2.5 kHz.   **2.** $0.1\angle -88.85°$ V rms.   **3.** -9.54 dB.

Section 19–2
1. 0.707 V.   **2.** $9.97\angle 3.81°$ V.

Section 19–3
1. 400 Hz.   **2.** 1.04 MHz.

Section 19–4
1. It *rejects* a certain band of frequencies.
2. High-pass/low-pass combination, series resonant circuit, and parallel resonant circuit.

Section 19–5
1. Voltage divider principle.
2. When a key is pressed, the program returns to line 10.

TWENTY

CIRCUIT THEOREMS IN ac ANALYSIS

Several important circuit theorems were introduced in Chapter 8 with emphasis on their applications in the analysis of dc circuits. In this chapter, these theorems are covered in relation to the analysis of ac circuits with reactive elements.

In this chapter, you will learn:

☐ How to use the superposition theorem to analyze multiple-source ac circuits.
☐ How to apply Thevenin's theorem to circuits with ac sources and reactive components.
☐ How to apply Norton's theorem to circuits with ac sources and reactive components.
☐ How to use Millman's theorem to reduce parallel voltage sources and impedances to a single equivalent voltage source and impedance.
☐ How the maximum power transfer theorem applies to reactive circuits.

20–1 THE SUPERPOSITION THEOREM

The *superposition theorem* was introduced in Chapter 8 for dc circuit analysis. In this section, the superposition theorem is applied to circuits with ac sources and reactive elements. The theorem can be stated as follows:

> **The current in any given branch of a multiple-source circuit can be found by determining the currents in that particular branch produced by each source acting alone, with all other sources replaced by their internal impedances. The total current in the given branch is the phasor sum of the individual source currents in that branch.**

The procedure for the application of the superposition theorem is:

1. Leave one of the sources in the circuit, and reduce all others to zero. Reduce voltage sources to zero by placing a theoretical short between the terminals; any internal series impedance remains.

2. Find the current in the branch of interest produced by the one remaining source.

3. Repeat steps 1 and 2 for each source in turn. When complete, you will have a number of current values equal to the number of sources in the circuit.

4. Add the individual current values as phasor quantities.

The following examples illustrate this procedure.

EXAMPLE 20–1

Find the current in R of Figure 20–1 using the superposition theorem. Assume the internal source impedance is zero.

FIGURE 20–1

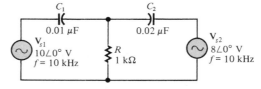

Solution:
First, by zeroing V_{s2}, find the current in R due to V_{s1}, as indicated in Figure 20–2.

FIGURE 20–2

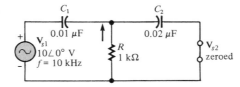

Looking from V_{s1}, we know that the impedance is

$$\mathbf{Z} = \mathbf{X}_{C1} + \frac{R\mathbf{X}_{C2}}{R + \mathbf{X}_{C2}}$$

$$X_{C1} = \frac{1}{2\pi(10\ \text{kHz})(0.01\ \mu\text{F})} = 1.59\ \text{k}\Omega$$

$$X_{C2} = \frac{1}{2\pi(10\ \text{kHz})(0.02\ \mu\text{F})} = 796\ \Omega$$

$$\mathbf{Z} = 1.59\angle-90°\ \text{k}\Omega + \frac{(1\angle 0°\ \text{k}\Omega)(795\angle-90°\ \Omega)}{1\ \text{k}\Omega - j795\ \Omega}$$

$$= 1.59\angle-90°\ \text{k}\Omega + 622.3\angle-51.51°\ \Omega$$

$$= -j1.59\ \text{k}\Omega + 387.3\ \Omega - j487.1\ \Omega$$

$$= 387.3\ \Omega - j2077\ \Omega = 2112.8\angle-79.44°\ \Omega$$

The total current from source 1 is:

$$\mathbf{I}_{s1} = \frac{\mathbf{V}_{s1}}{\mathbf{Z}} = \frac{10\angle 0°\ \text{V}}{2112.8\angle-79.44°\ \Omega} = 4.73\angle 79.44\ \text{mA}$$

Using the current divider formula, the current through R due to V_{s1} is

$$\mathbf{I}_{R1} = \left(\frac{X_{C2}\angle-90°}{R - jX_{C2}}\right)\mathbf{I}_{s1}$$

$$= \left(\frac{796\angle-90°\ \Omega}{1\ \text{k}\Omega - j796\ \Omega}\right)(4.73\angle 79.44\ \text{mA})$$

$$= (0.663\angle-51.48°\ \Omega)(4.73\angle 79.44°\ \text{mA})$$

$$= 3.14\angle 27.96°\ \text{mA}$$

Next, find the current in R due to source V_{s2} by zeroing V_{s1}, as shown in Figure 20–3.

FIGURE 20–3

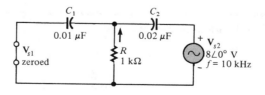

Looking from V_{s2}, we see that the impedance is

$$\mathbf{Z} = \mathbf{X}_{C2} + \frac{R\mathbf{X}_{C1}}{R + \mathbf{X}_{C1}}$$

$X_{C1} = 1.59 \text{ k}\Omega$

$X_{C2} = 796 \text{ }\Omega$

$$\mathbf{Z} = 796\angle -90° \text{ }\Omega + \frac{(1\angle 0° \text{ k}\Omega)(1.59\angle -90° \text{ k}\Omega)}{1 \text{ k}\Omega - j1.59 \text{ k}\Omega}$$

$$= 796\angle -90° \text{ }\Omega + 846.5\angle -32.17° \text{ }\Omega$$

$$= -j796 \text{ }\Omega + 716.54 \text{ }\Omega - j450.71 \text{ }\Omega$$

$$= 716.54 \text{ }\Omega - j1246.71 \text{ }\Omega = 1437.95\angle -60.11° \text{ }\Omega$$

The total current from source 2 is

$$\mathbf{I}_{s2} = \frac{\mathbf{V}_{s2}}{\mathbf{Z}} = \frac{8\angle 0° \text{ V}}{1437.95\angle -60.11° \text{ }\Omega}$$

$$\doteq 5.56\angle 60.11° \text{ mA}$$

Using the current divider formula, the current through R due to V_{s2} is

$$\mathbf{I}_{R2} = \left(\frac{\mathbf{X}_{C1}\angle -90°}{R - jX_{C1}} \right) \mathbf{I}_{s2}$$

$$= \frac{1.59\angle -90° \text{ k}\Omega}{1.88\angle -57.83° \text{ k}\Omega} (5.56\angle 60.11° \text{ mA})$$

$$= 4.70\angle 27.94° \text{ mA}$$

The two individual resistor currents are now added in phasor form:

$$\mathbf{I}_{R1} = 3.14\angle 27.96° \text{ mA} = 2.77 \text{ mA} + j1.47 \text{ mA}$$

$$\mathbf{I}_{R2} = 4.70\angle 27.94° \text{ mA} = 4.15 \text{ mA} + j2.20 \text{ mA}$$

$$\mathbf{I}_R = \mathbf{I}_{R1} + \mathbf{I}_{R2} = 6.92 \text{ mA} + j3.67 \text{ mA} = 7.83\angle 27.94° \text{ mA}$$

EXAMPLE 20–2

Find the coil current in Figure 20–4.

FIGURE 20–4

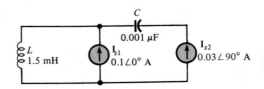

Solution:
First, find the current through the inductor due to current source I_{s1} by replacing source I_{s2} with an open, as shown in Figure 20–5. As you can see, the entire 0.1 A from the current source I_{s1} is through the coil.

FIGURE 20–5

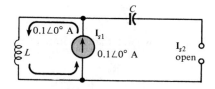

Next, find the current through the inductor due to current source I_{s2} by replacing source I_{s1} with an open, as indicated in Figure 20–6. Notice that all of the 0.03 A from source I_{s2} is through the coil.

FIGURE 20–6

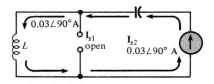

To get the total inductor current, the two individual currents are superimposed and added as phasor quantities.

$$\mathbf{I}_L = \mathbf{I}_{L1} + \mathbf{I}_{L2}$$
$$= 0.1\angle 0° \text{ A} + 0.03\angle 90° \text{ A} = 0.1 \text{ A} + j0.03 \text{ A}$$
$$= 0.104\angle 16.7° \text{ A}$$

EXAMPLE 20–3

Find the total current in the resistor R_L in Figure 20–7.

FIGURE 20–7

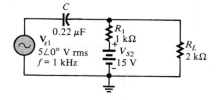

Solution:
First, find the current through R_L due to source V_{S1} by zeroing source V_{S2}, as shown in Figure 20–8.

FIGURE 20–8

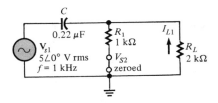

Looking from V_{s1}, we see that the impedance is

$$\mathbf{Z} = \mathbf{X}_C + \frac{\mathbf{R}_1\mathbf{R}_L}{\mathbf{R}_1 + \mathbf{R}_L}$$

$$X_C = \frac{1}{2\pi(1\ \text{kHz})(0.22\ \mu\text{F})} = 723\ \Omega$$

$$\mathbf{Z} = 723\angle-90°\ \Omega + \frac{(1\angle0°\ \text{k}\Omega)(2\angle0°\ \text{k}\Omega)}{3\angle0°\ \text{k}\Omega}$$

$$= -j723\ \Omega + 667\ \Omega = 983.68\angle-47.31°\ \Omega$$

The total current from source 1 is

$$\mathbf{I}_{s1} = \frac{\mathbf{V}_{s1}}{\mathbf{Z}} = \frac{5\angle0°\ \text{V}}{983.68\angle-47.31°\ \Omega} = 5.08\angle47.31°\ \text{mA}$$

Using the current divider approach, the current in R_L due to V_{s1} is

$$\mathbf{I}_{L1} = \left(\frac{R_1}{R_1 + R_L}\right)\mathbf{I}_{s1}$$

$$= \left(\frac{1\ \text{k}\Omega}{3\ \text{k}\Omega}\right)(5.08\angle47.31°\ \text{mA})$$

$$= 1.69\angle47.31°\ \text{mA}$$

Next, find the current in R_L due to the dc source V_{S2} by zeroing V_{s1}, as shown in Figure 20–9.

FIGURE 20–9

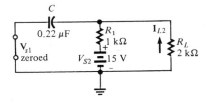

The impedance as seen by V_{S2} is

$$\mathbf{Z} = R_1 + R_L = 3\ \text{k}\Omega$$

The current produced by V_{S2} is

$$I_{L2} = \frac{V_{S2}}{Z} = \frac{15 \text{ V}}{3 \text{ k}\Omega}$$

$$= 5 \text{ mA dc}$$

By superposition, the total current in R_L is $1.69\angle 47.31°$ mA riding on a dc level of 5 mA, as indicated in Figure 20–10.

FIGURE 20–10

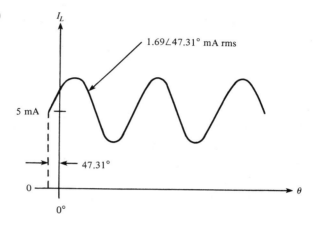

SECTION REVIEW 20–1

1. If two equal currents are in opposing directions at any instant of time in a given branch of a circuit, what is the net current at that instant?
2. Why is the superposition theorem useful in the analysis of multiple-source circuits?
3. Using the superposition theorem, find the magnitude of the current through R in Figure 20–11.

FIGURE 20–11

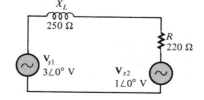

20–2   THEVENIN'S THEOREM

Thevenin's theorem, as applied to ac circuits, provides a method for reducing any circuit to an equivalent form that consists of an *equivalent ac voltage source* in series with an *equivalent impedance*.

The form of Thevenin's equivalent circuit is shown in Figure 20–12. Regardless of how complex the original circuit is, it can always be reduced to this equivalent form. The equivalent voltage source is designated $\mathbf{V}_{th}$; the equivalent impedance is designated $\mathbf{Z}_{th}$ (lowercase subscript denotes ac quantity). Notice that the impedance is represented by a block in the circuit diagram. This is because the equivalent impedance can be of several forms: purely resistive, purely capacitive, purely inductive, or a combination of resistance and a reactance.

FIGURE 20–12
Thevenin's equivalent circuit.

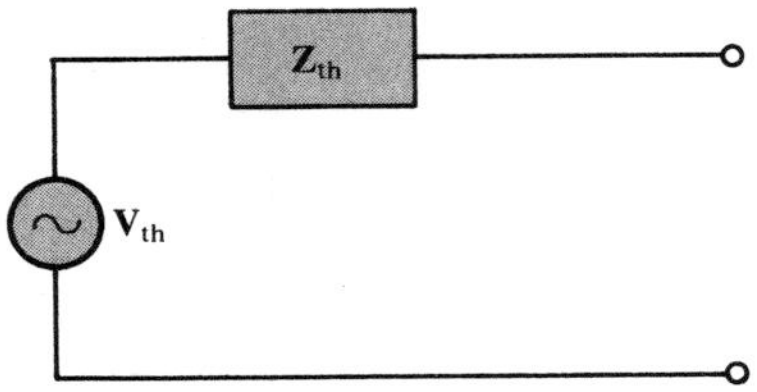

EQUIVALENCY

Figure 20–13(a) shows a block diagram that represents an ac circuit of any given complexity. This circuit has two output terminals, A and B. A load impedance, $\mathbf{Z}_L$, is connected to the terminals. The circuit produces a certain voltage, $\mathbf{V}_L$, and a certain current, $\mathbf{I}_L$, as illustrated.

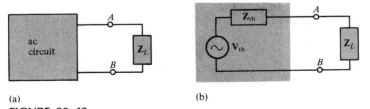

(a)   (b)

FIGURE 20–13
An ac circuit of any complexity can be reduced to a Thevenin equivalent for analysis purposes.

By Thevenin's theorem, the circuit in the block can be reduced to an *equivalent* form, as indicated by the dashed lines of Figure 20–13(b). The term *equivalent* means that when the same value of load is connected to both the original circuit and Thevenin's equivalent circuit, the load voltages and currents are equal for both. Therefore, as far as the load is concerned, there is no difference between the original circuit and Thevenin's equivalent circuit. The load "sees" the same current and voltage regardless of whether it is connected to the original circuit or to the Thevenin equivalent.

THEVENIN'S EQUIVALENT VOLTAGE (V_{th})

As you have seen, the equivalent voltage, $\mathbf{V}_{th}$, is one part of the complete Thevenin equivalent circuit. $\mathbf{V}_{th}$ is defined as *the open circuit voltage between two specified points in a circuit*.

To illustrate, assume that an ac circuit of some type has a resistor connected between two defined points, *A* and *B*, as shown in Figure 20–14(a). We wish to find the Thevenin equivalent circuit for the circuit as "seen" by *R*. $\mathbf{V}_{th}$ is the voltage across the points *A* and *B*, *with R removed*, as shown in Part (b) of the figure. The circuit is viewed from the open terminals *AB*, and *R* is considered external to the circuit for which the Thevenin equivalent is to be found. The following three examples show how to find $\mathbf{V}_{th}$.

FIGURE 20–14

How $\mathbf{V}_{th}$ is determined.

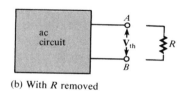

(a) Circuit   (b) With *R* removed

EXAMPLE 20–4

Determine $\mathbf{V}_{th}$ for the circuit external to R_L in Figure 20–15.

Solution:

FIGURE 20–15

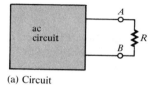

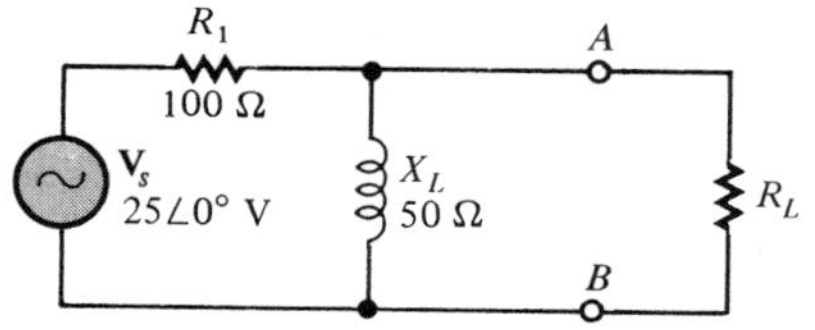

Remove R_L and determine the voltage from *A* to *B* ($\mathbf{V}_{th}$). In this case, the voltage from *A* to *B* is the same as the voltage across X_L. This is determined using the voltage divider method:

$$\mathbf{V}_L = \left(\frac{X_L\angle 90°}{R_1 + jX_L}\right)\mathbf{V}_s$$

$$= \left(\frac{50\angle 90° \ \Omega}{111.8\angle 26.57° \ \Omega}\right)25\angle 0° \ \text{V}$$

$$= 11.18\angle 63.43° \ \text{V}$$

$$\mathbf{V}_{th} = \mathbf{V}_{AB} = \mathbf{V}_L = 11.18\angle 63.43° \ \text{V}$$

**EXAMPLE
20-5**

For the circuit in Figure 20-16, determine the Thevenin voltage as seen by R_L.

FIGURE 20-16

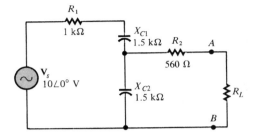

Solution:
Thevenin's voltage for the circuit between terminals A and B is the voltage that appears across A and B with R_L removed from the circuit.

There is no voltage drop across R_2 because the open terminals AB prevent current through it. Thus, $\mathbf{V}_{AB}$ is the same as $\mathbf{V}_{C2}$ and can be found by the voltage divider formula:

$$\mathbf{V}_{AB} = \mathbf{V}_{C2} = \left(\frac{X_{C2}\angle -90°}{R_1 - jX_{C1} - jX_{C2}}\right)\mathbf{V}_s$$

$$= \left(\frac{1.5\angle -90° \text{ k}\Omega}{1 \text{ k}\Omega - j3 \text{ k}\Omega}\right)10\angle 0° \text{ V}$$

$$= \left(\frac{1.5\angle -90° \text{ k}\Omega}{3.16\angle -71.57° \text{ k}\Omega}\right)10\angle 0° \text{ V}$$

$$= 4.75\angle -18.43° \text{ V}$$

$$\mathbf{V}_{th} = \mathbf{V}_{AB} = 4.75\angle -18.43° \text{ V}$$

**EXAMPLE
20-6**

For Figure 20-17, find $\mathbf{V}_{th}$ for the circuit external to R_L.

FIGURE 20-17

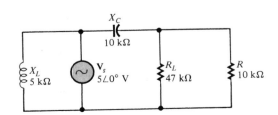

Solution:

First remove R_L and determine the voltage across the resulting open terminals, which is $\mathbf{V}_{th}$. We find $\mathbf{V}_{th}$ by applying the voltage divider formula to X_C and R:

$$\mathbf{V}_{th} = \mathbf{V}_R = \left(\frac{R\angle 0°}{R - jX_C}\right)\mathbf{V}_s$$

$$= \left(\frac{10\angle 0° \text{ k}\Omega}{10 \text{ k}\Omega - j10 \text{ k}\Omega}\right)5\angle 0° \text{ V}$$

$$= \left(\frac{10\angle 0° \text{ k}\Omega}{14.14\angle -45° \text{ k}\Omega}\right)5\angle 0° \text{ V}$$

$$= 3.54\angle 45° \text{ V}$$

Notice that L has no effect on the result, since the 5-V source appears across C and R in combination.

THEVENIN'S EQUIVALENT IMPEDANCE (Z_{th})

The previous examples illustrated how to find only one part of a Thevenin equivalent circuit. Now, we turn our attention to determining the Thevenin equivalent impedance, $\mathbf{Z}_{th}$. As defined by Thevenin's theorem, $\mathbf{Z}_{th}$ *is the total impedance appearing between two specified terminals in a given circuit with all sources replaced by their internal impedances (ideally zeroed).* Thus, when we wish to find $\mathbf{Z}_{th}$ between any two terminals in a circuit, all the voltage sources are shorted (ideally, neglecting any internal impedance), and all the current sources are opened (ideally, to represent infinite impedance). Then the total impedance between the two terminals is determined. The following three examples illustrate how to find $\mathbf{Z}_{th}$.

EXAMPLE 20–7

Find $\mathbf{Z}_{th}$ for the part of the circuit in Figure 20–18 that is external to R_L. This is the same circuit used in Example 20–4.

FIGURE 20–18

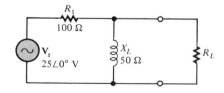

Solution:

First, reduce V_s to zero by shorting it, as shown in Figure 20–19.

FIGURE 20–19

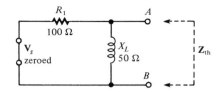

Looking in at terminals A and B, R and X_L are in parallel. Thus,

$$\mathbf{Z}_{th} = \frac{(R\angle 0°)(X_L \angle 90°)}{R + jX_L}$$

$$= \frac{(100\angle 0° \ \Omega)(50\angle 90° \ \Omega)}{100 \ \Omega + j50 \ \Omega}$$

$$= \frac{(100\angle 0° \ \Omega)(50\angle 90° \ \Omega)}{111.8\angle 26.57° \ \Omega}$$

$$= 44.72\angle 63.43° \ \Omega$$

**EXAMPLE
20–8**

For the circuit in Figure 20–20, determine $\mathbf{Z}_{th}$ as seen by R_L. This is the same circuit used in Example 20–5.

FIGURE 20–20

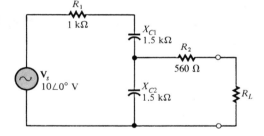

Solution:
First, zero the voltage source, as shown in Figure 20–21.

FIGURE 20–21

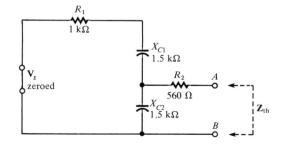

Looking from terminals A and B, C_2 appears in parallel with the series combination of R_1 and C_1. This entire combination is in series with R_2. The calculation for Z_{th} is as follows:

$$Z_{th} = R_2\angle 0° + \frac{(X_{C2}\angle -90°)(R_1 - jX_{C1})}{R_1 - jX_{C1} - jX_{C2}}$$

$$= 560\angle 0° \ \Omega + \frac{(1.5\angle -90° \ k\Omega)(1 \ k\Omega - j1.5 \ k\Omega)}{1 \ k\Omega - j3 \ k\Omega}$$

$$= 560\angle 0° \ \Omega + \frac{(1.5\angle -90° \ k\Omega)(1.8\angle -56.31° \ k\Omega)}{3.16\angle -71.57° \ k\Omega}$$

$$= 560\angle 0° \ \Omega + 854\angle -74.74° \ \Omega$$

$$= 560 \ \Omega + 225 \ \Omega - j824 \ \Omega$$

$$= 785 \ \Omega - j824 \ \Omega$$

$$= 1138\angle -46.39° \ \Omega$$

**EXAMPLE
20–9**

For the circuit in Figure 20–22, determine Z_{th} for the portion of the circuit external to R_L. This is the same circuit as in Example 20–6.

FIGURE 20–22

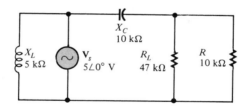

Solution:
With the voltage source zeroed, X_L is effectively out of the circuit. R and C appear in parallel when viewed from the open terminals, as indicated in Figure 20–23. Z_{th} is calculated as follows:

$$Z_{th} = \frac{(R\angle 0°)(X_C\angle -90°)}{R - jX_C}$$

$$= \frac{(10\angle 0° \ k\Omega)(10\angle -90° \ k\Omega)}{14.14\angle -45° \ k\Omega}$$

$$= 7.07\angle -45° \ k\Omega$$

FIGURE 20–23

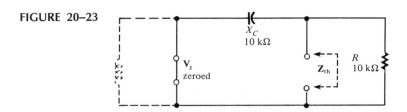

THEVENIN'S EQUIVALENT CIRCUIT

The previous examples have shown how to find the two equivalent components of a Thevenin circuit, V_{th} and Z_{th}. Keep in mind that V_{th} and Z_{th} can be found for any circuit. Once these equivalent values are determined, they must be connected in *series* to form the Thevenin equivalent circuit. The following examples use the previous examples to illustrate this final step.

EXAMPLE 20–10

Draw the Thevenin equivalent circuit for the original circuit in Figure 20–24 that is external to R_L.

FIGURE 20–24

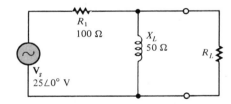

Solution:
We found in Examples 20–4 and 20–7 that $V_{th} = 11.18\angle 63.43°$ V and $Z_{th} = 44.72\angle 63.43°$ Ω.
    In rectangular form, the impedance is

$$Z_{th} = 20 \ \Omega + j40 \ \Omega$$

This form indicates that the impedance is a 20-Ω resistor in series with a 40-Ω inductive reactance. The Thevenin equivalent circuit is shown in Figure 20–25.

FIGURE 20–25

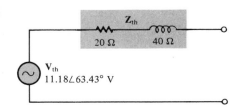

EXAMPLE 20–11

For the original circuit in Figure 20–26, sketch the Thevenin equivalent circuit external to R_L. This is the circuit used in Examples 20–5 and 20–8.

FIGURE 20–26

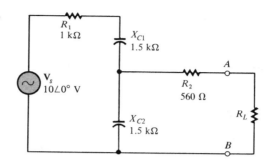

Solution:

In Examples 20–5 and 20–8, $\mathbf{V}_{th}$ was found to be $4.75\angle -18.43°$ V, and $\mathbf{Z}_{th}$ was found to be $1138\angle -46.39°$ Ω. In rectangular form, $\mathbf{Z}_{th} = 785\ \Omega - j824\ \Omega$. The Thevenin equivalent circuit is shown in Figure 20–27.

FIGURE 20–27

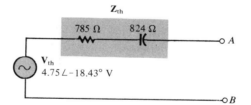

EXAMPLE 20–12

For the original circuit in Figure 20–28, determine the Thevenin equivalent circuit as seen by R_L.

FIGURE 20–28

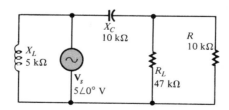

Solution:

From Examples 20–6 and 20–9, $\mathbf{V}_{th} = 3.54\angle 45°$ V, and $\mathbf{Z}_{th} = 7.07\angle -45°$ kΩ. The impedance in rectangular form is

$$\mathbf{Z}_{th} = 5\ k\Omega - j5\ k\Omega$$

Thus, the Thevenin equivalent circuit is as shown in Figure 20–29.

FIGURE 20–29

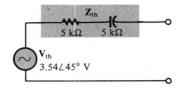

SUMMARY OF THEVENIN'S THEOREM

Remember that the Thevenin equivalent circuit is always of the series form regardless of the original circuit that it replaces. The significance of Thevenin's theorem is that the equivalent circuit can replace the original circuit as far as any external load is concerned. Any load connected between the terminals of a Thevenin equivalent circuit experiences the same current and voltage as if it were connected to the terminals of the original circuit.

A summary of steps for applying Thevenin's theorem follows:

1. Open the two terminals between which you want to find the Thevenin circuit. This is done by removing the component from which the circuit is to be viewed.
2. Determine the voltage across the two open terminals.
3. Determine the impedance viewed from the two open terminals with all sources zeroed (voltage sources replaced with shorts and current sources replaced with opens).
4. Connect $\mathbf{V}_{th}$ and $\mathbf{Z}_{th}$ in series to produce the complete Thevenin equivalent circuit.

SECTION REVIEW 20–2

1. What are the two basic components of a Thevenin equivalent ac circuit?
2. For a certain circuit, $\mathbf{Z}_{th} = 25\ \Omega - j50\ \Omega$, and $\mathbf{V}_{th} = 5\angle 0°$ V. Sketch the Thevenin equivalent circuit.
3. For the circuit in Figure 20–30, find the Thevenin equivalent looking from terminals AB.

FIGURE 20–30

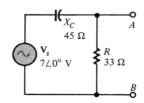

20–3 NORTON'S THEOREM

Like Thevenin's theorem, Norton's theorem provides a method of reducing a more complex circuit to a simpler, more manageable form. The basic difference is that Norton's theorem gives an equivalent current source (rather than a voltage source) in parallel (rather than series) with an equivalent impedance. The form of Norton's equivalent circuit is shown in Figure 20–31. Regardless of how complex the original circuit is, it can be reduced to this equivalent form. The equivalent current source is designated I_n, and the equivalent impedance is Z_n (lowercase subscript denotes ac quantity).

FIGURE 20–31
Norton equivalent circuit.

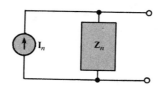

Norton's theorem shows you how to find I_n and Z_n. Once they are known, simply connect them in parallel to get the complete Norton equivalent circuit.

NORTON'S EQUIVALENT CURRENT SOURCE (I_n)

I_n is one part of the Norton equivalent circuit; Z_n is the other part. I_n is defined as *the short circuit current between two specified points in a given circuit*. Any load connected between these two points effectively "sees" a current source I_n in parallel with Z_n.

To illustrate, suppose that the circuit shown in Figure 20–32 has a load resistor connected to points A and B, as indicated in Part (a). We wish to find the Norton equivalent for the circuit external to R_L. To find I_n, calculate the current between points A and B with those terminals shorted, as shown in Part (b). The following example shows how to find I_n.

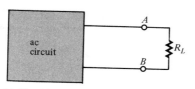

(a) Circuit with load resistor   (b) Short circuit current is I_{in}.

FIGURE 20–32
How I_n is determined.

**EXAMPLE
20–13**

In Figure 20–33, determine $\mathbf{I}_n$ for the circuit as "seen" by the load resistor. The shaded area identifies the portion of the circuit to be Nortonized.

FIGURE 20–33

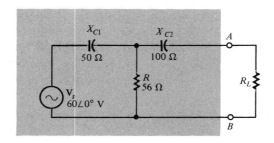

Solution:
Short the terminals *A* and *B,* as shown in Figure 20–34. $\mathbf{I}_n$ is the current through the short and is calculated as follows:

FIGURE 20–34

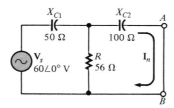

First, the total impedance viewed from the source is

$$\mathbf{Z} = \mathbf{X}_{C1} + \frac{\mathbf{R}\mathbf{X}_{C2}}{\mathbf{R} + \mathbf{X}_{C2}}$$

$$= 50\angle -90°\ \Omega + \frac{(56\angle 0°\ \Omega)(100\angle -90°\ \Omega)}{56\ \Omega - j100\ \Omega}$$

$$= 50\angle -90°\ \Omega + 48.86\angle -29.25°\ \Omega$$

$$= -j50\ \Omega + 42.63\ \Omega - j23.87\ \Omega$$

$$= 42.63\ \Omega - j73.87\ \Omega$$

$$= 85.29\angle -60.01°\ \Omega$$

The total current from the source is

$$\mathbf{I}_s = \frac{\mathbf{V}_s}{\mathbf{Z}} = \frac{60\angle 0°\ \text{V}}{85.29\angle -60.01°\ \Omega} = 0.703\angle 60.01°\ \text{A}$$

Applying the current divider formula to get $\mathbf{I}_n$ (the current through the short):

$$\mathbf{I}_n = \left(\frac{\mathbf{R}}{\mathbf{R} + \mathbf{X}_{C2}}\right)\mathbf{I}_s$$

$$= \left(\frac{56\angle 0° \ \Omega}{56 \ \Omega - j100 \ \Omega}\right)(0.703\angle 60.01° \ A)$$

$$= 0.343\angle 120.76° \ A$$

This is the value for the equivalent Norton current source.

NORTON'S EQUIVALENT IMPEDANCE ($\mathbf{Z}_n$)

$\mathbf{Z}_n$ is defined the same as $\mathbf{Z}_{\text{th}}$. It is the total impedance appearing between two specified terminals of a given circuit viewed from the open terminals with all sources zeroed.

EXAMPLE 20–14

Find $\mathbf{Z}_n$ for the circuit in Figure 20–33 (Example 20–13) viewed from the open terminals AB.

Solution:
First reduce $\mathbf{V}_s$ to zero, as indicated in Figure 20–35.
    Looking in at terminals AB, C_2 is in series with the parallel combination of R and C_1. Thus,

$$\mathbf{Z}_n = \mathbf{X}_{C2} + \frac{\mathbf{R}\mathbf{X}_{C1}}{\mathbf{R} + \mathbf{X}_{C1}}$$

$$= 100\angle -90° \ \Omega + \frac{(56\angle 0° \ \Omega)(50\angle -90° \ \Omega)}{56 \ \Omega - j50 \ \Omega}$$

$$= 100\angle -90° \ \Omega + 37.30\angle -48.24° \ \Omega$$

$$= -j100 \ \Omega + 24.84 \ \Omega - j27.82 \ \Omega$$

$$= 24.84 \ \Omega - j127.82 \ \Omega$$

The Norton equivalent impedance is a 24.84-Ω resistance in series with a 127.82-Ω capacitive reactance.

FIGURE 20–35

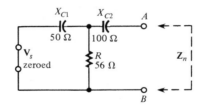

The previous two examples have shown how to find the two equivalent components of a Norton equivalent circuit. Keep in mind that these values can be found for any given ac circuit. Once these values are known, they are connected in parallel to form the Norton equivalent circuit, as the following example illustrates.

EXAMPLE 20–15

Sketch the complete Norton equivalent circuit for the original circuit in Figure 20–33 (Example 20–13).

Solution:
From Examples 20–13 and 20–14:

$$\mathbf{I}_n = 0.343\angle120.76° \text{ A}$$

$$\mathbf{Z}_n = 24.84 \ \Omega - j127.82 \ \Omega$$

The Norton equivalent circuit is shown in Figure 20–36.

FIGURE 20–36

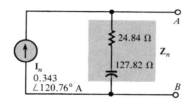

SUMMARY OF NORTON'S THEOREM

Any load connected between the terminals of a Norton equivalent circuit will have the same current through it and the same voltage across it as it would when connected to the terminals of the original circuit. A summary of steps for theoretically applying Norton's theorem is as follows:

1. Short the two terminals between which the Norton circuit is to be determined.
2. Determine the current through the short. This is $\mathbf{I}_n$.
3. Determine the impedance between the two open terminals with all sources zeroed. This is $\mathbf{Z}_n$.
4. Connect $\mathbf{I}_n$ and $\mathbf{Z}_n$ in parallel.

SECTION REVIEW 20–3

1. For a given circuit, $\mathbf{I}_n = 5\angle0°$ mA, and $\mathbf{Z}_n = 150 \ \Omega + j100 \ \Omega$. Draw the Norton equivalent circuit.
2. Find the Norton circuit as seen by R_L in Figure 20–37.

FIGURE 20–37

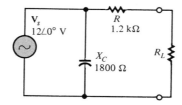

20–4

MILLMAN'S THEOREM

Millman's theorem permits any number of parallel branches consisting of voltage sources and impedances to be reduced to a single equivalent voltage source and equivalent impedance. It is an alternative to Thevenin's theorem for the case of all parallel voltage sources. The Millman conversion is illustrated in Figure 20–38.

FIGURE 20–38
The Millman conversion.

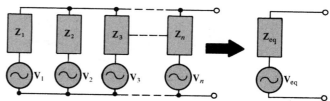

MILLMAN'S EQUIVALENT VOLTAGE (V_{eq}) AND EQUIVALENT IMPEDANCE (Z_{eq})

Millman's theorem provides a formula for calculating the equivalent voltage, V_{eq}. To find V_{eq}, convert each of the parallel voltage sources into current sources, as shown in Figure 20–39.

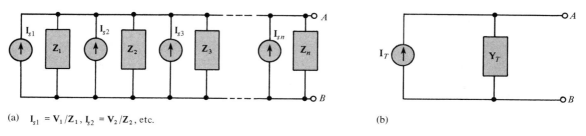

(a) $I_{s1} = V_1/Z_1$, $I_{s2} = V_2/Z_2$, etc.

(b)

FIGURE 20–39

In Figure 20–39(b), the total current from the parallel current sources is

$$I_T = I_{s1} + I_{s2} + I_{s3} + \cdots + I_{sn}$$

The total admittance between terminals AB is

$$\mathbf{Y}_T = \mathbf{Y}_1 + \mathbf{Y}_2 + \mathbf{Y}_3 + \cdot \cdot \cdot + \mathbf{Y}_n$$

where $\mathbf{Y}_T = 1/\mathbf{Z}_T$, $\mathbf{Y}_1 = 1/\mathbf{Z}_1$, and so on. (Note that n in this equation represents the number of elements and does not represent a Norton quantity.) Remember that current sources are effectively open (ideally). Therefore, by Millman's theorem, the *equivalent impedance* is the total impedance, $\mathbf{Z}_T$.

$$\mathbf{Z}_{eq} = \frac{1}{\mathbf{Y}_T} = \frac{1}{\dfrac{1}{\mathbf{Z}_1} + \dfrac{1}{\mathbf{Z}_2} + \cdot \cdot \cdot + \dfrac{1}{\mathbf{Z}_n}} \qquad (20\text{–}1)$$

By Millman's theorem, the *equivalent voltage* is $\mathbf{I}_T\mathbf{Z}_{eq}$, where $\mathbf{I}_T$ is expressed as follows:

$$\mathbf{I}_T = \frac{\mathbf{V}_1}{\mathbf{Z}_1} + \frac{\mathbf{V}_2}{\mathbf{Z}_2} + \cdot \cdot \cdot + \frac{\mathbf{V}_n}{\mathbf{Z}_n}$$

The following is the formula for the equivalent voltage:

$$\mathbf{V}_{eq} = \frac{\mathbf{V}_1/\mathbf{Z}_1 + \mathbf{V}_2/\mathbf{Z}_2 + \cdot \cdot \cdot + \mathbf{V}_n/\mathbf{Z}_n}{1/\mathbf{Z}_1 + 1/\mathbf{Z}_2 + \cdot \cdot \cdot + 1/\mathbf{Z}_n}$$

$$\mathbf{V}_{eq} = \frac{\mathbf{V}_1/\mathbf{Z}_1 + \mathbf{V}_2/\mathbf{Z}_2 + \cdot \cdot \cdot + \mathbf{V}_n/\mathbf{Z}_n}{\mathbf{Y}_{eq}} \qquad (20\text{–}2)$$

Equations (20–1) and (20–2) are the two Millman formulas.

EXAMPLE 20–16

Use Millman's theorem to find the voltage across R_L and the current through R_L in Figure 20–40.

FIGURE 20–40

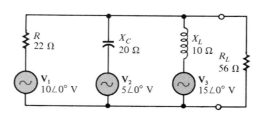

Solution:
Apply Millman's theorem as follows:

$$\mathbf{Z}_{eq} = \cfrac{1}{\cfrac{1}{R\angle 0°} + \cfrac{1}{X_C \angle -90°} + \cfrac{1}{X_L \angle 90°}}$$

$$= \cfrac{1}{\cfrac{1}{22\angle 0°\ \Omega} + \cfrac{1}{20\angle -90°\ \Omega} + \cfrac{1}{10\angle 90°\ \Omega}}$$

$$= \cfrac{1}{0.045\angle 0° + 0.05\angle 90° + 0.1\angle -90°}$$

$$= \cfrac{1}{0.045 + j0.05 - j0.1}$$

$$= \cfrac{1}{0.045 - j0.05}$$

$$= \cfrac{1}{0.0673\angle -48°}$$

$$= 14.86\angle 48°\ \Omega = 9.94\ \Omega + j11.04\ \Omega$$

$$\mathbf{Y}_{eq} = 0.0673\angle -48°\ S$$

$$\mathbf{V}_{eq} = \cfrac{\cfrac{V_1 \angle 0°}{R\angle 0°} + \cfrac{V_2 \angle 0°}{X_C \angle -90°} + \cfrac{V_3 \angle 0°}{X_L \angle 90°}}{\mathbf{Y}_{eq}}$$

$$= \cfrac{\cfrac{10\angle 0°\ V}{22\angle 0°\ \Omega} + \cfrac{5\angle 0°\ V}{22\angle -90°\ \Omega} + \cfrac{15\angle 0°\ V}{10\angle 90°\ \Omega}}{0.0707\angle -45°\ S}$$

$$= \cfrac{0.45\angle 0° + 0.23\angle 90° + 1.5\angle -90°}{0.0673\angle -48°\ S}$$

$$= \cfrac{1.35\angle -70.5°}{0.0673\angle -48°\ S}$$

$$= 20.06\angle -22.5°\ V$$

The single equivalent voltage source is shown in Figure 20–41.

FIGURE 20–41

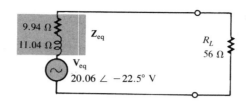

$\mathbf{I}_L$ and $\mathbf{V}_L$ are calculated as follows:

$$\mathbf{I}_L = \frac{\mathbf{V}_{eq}}{\mathbf{Z}_{eq} + \mathbf{R}_L}$$

$$= \frac{20.06\angle-22.5°\ \text{V}}{9.94\ \Omega + j11.04\ \Omega + 56\ \Omega}$$

$$= \frac{20.06\angle-22.5°\ \text{V}}{66.86\angle9.5°\ \Omega}$$

$$= 0.3\angle-32°\ \text{A}$$

$$\mathbf{V}_L = \mathbf{I}_L\mathbf{R}_L = (0.3\angle-32°\ \text{A})(56\angle0°\ \Omega)$$

$$= 16.8\angle-32°\ \text{V}$$

SECTION REVIEW 20–4

1. To what type of circuit does Millman's theorem apply?

2. Find the load current in Figure 20–42 using Millman's theorem.

FIGURE 20–42

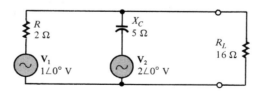

20–5   MAXIMUM POWER TRANSFER THEOREM

The maximum power transfer theorem as applied to ac circuits states:

> **When a load is connected to a circuit, maximum power is transferred to the load when the load impedance is the complex conjugate of the circuit's output impedance.**

The complex conjugate of $R - jX_C$ is $R + jX_L$, where the resistances and the reactances are equal in magnitude. The output impedance is effectively Thevenin's equivalent impedance viewed from the output terminals. When $\mathbf{Z}_L$ is the complex conjugate of $\mathbf{Z}_{out}$, maximum power is transferred from the circuit to the load with a power factor of 1. An equivalent circuit with its output impedance and load is shown in Figure 20–43.

FIGURE 20–43
Equivalent circuit with load.

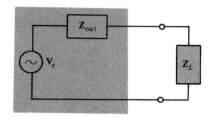

Example 20–17 shows that maximum power occurs when the impedances are *conjugately* matched.

EXAMPLE 20–17

The circuit to the left of terminals A and B in Figure 20–44 provides power to the load, $\mathbf{Z}_L$. It is the Thevenin equivalent of a more complex circuit. Calculate the power delivered to the load for each of the following frequencies: 10 kHz, 30 kHz, 50 kHz, 80 kHz, and 100 kHz.

FIGURE 20–44

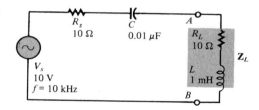

Solution:
For $f = 10$ kHz:

$$X_C = \frac{1}{2\pi(10 \text{ kHz})(0.01 \ \mu\text{F})} = 1.59 \text{ k}\Omega$$

$$X_L = 2\pi(10 \text{ kHz})(1 \text{ mH}) = 62.8 \ \Omega$$

$$Z_T = \sqrt{(R_s + R_L)^2 + (X_L - X_C)^2}$$

$$= \sqrt{(20 \ \Omega)^2 + (1.53 \text{ k}\Omega)^2}$$

$$= 1.53 \text{ k}\Omega$$

$$I = \frac{V_s}{Z_T} = \frac{10 \text{ V}}{1.53 \text{ k}\Omega} = 6.54 \text{ mA}$$

$$P_L = I^2 R_L = (6.54 \text{ mA})^2(10 \ \Omega)$$

$$= 0.428 \text{ mW}$$

For $f = 30$ kHz:

$$X_C = \frac{1}{2\pi(30 \text{ kHz})(0.01 \ \mu\text{F})} = 530.5 \ \Omega$$

$$X_L = 2\pi(30 \text{ kHz})(1 \text{ mH}) = 188.5 \ \Omega$$

$$Z_T = \sqrt{(20 \ \Omega)^2 + (342 \ \Omega)^2}$$

$$= 342.6 \ \Omega$$

$$I = \frac{V_s}{Z_T} = \frac{10 \text{ V}}{342.6 \ \Omega} = 29.2 \text{ mA}$$

$$P_L = I^2 R_L = (29.2 \text{ mA})^2(10 \ \Omega)$$

$$= 8.53 \text{ mW}$$

For $f = 50$ kHz:

$$X_C = \frac{1}{2\pi(50 \text{ kHz})(0.01 \ \mu\text{F})} = 318.3 \ \Omega$$

$$X_L = 2\pi(50 \text{ kHz})(1 \text{ mH}) = 314.2 \ \Omega$$

Note that X_C and X_L are very close to being equal and make the impedances complex conjugates. The actual frequency at which $X_L = X_C$ is 50.33 kHz.

$$Z_T = \sqrt{(20 \ \Omega)^2 + (4.1 \ \Omega)^2}$$

$$= 20.42 \ \Omega$$

$$I = \frac{V_s}{Z_T} = \frac{10 \text{ V}}{20.42 \ \Omega} = 489.7 \text{ mA}$$

$$P_L = I^2 R_L = (489.7 \text{ mA})^2(10 \ \Omega)$$

$$= 2.5 \text{ W}$$

For $f = 80$ kHz:

$$X_C = \frac{1}{2\pi(80 \text{ kHz})(0.01 \ \mu\text{F})} = 198.9 \ \Omega$$

$$X_L = 2\pi(80 \text{ kHz})(1 \text{ mH}) = 502.7 \ \Omega$$

$$Z_T = \sqrt{(20 \ \Omega)^2 + (303.8 \ \Omega)^2}$$

$$= 304.5 \ \Omega$$

$$I = \frac{V_s}{Z_T} = \frac{10 \text{ V}}{304.5 \ \Omega} = 32.8 \text{ mA}$$

$$P_L = I^2 R_L = (32.8 \text{ mA})^2(10 \ \Omega)$$

$$= 10.76 \text{ mW}$$

For $f = 100$ kHz:

$$X_C = \frac{1}{2\pi(100 \text{ kHz})(0.01 \ \mu\text{F})} = 159.2 \ \Omega$$

$$X_L = 2\pi(100 \text{ kHz})(1 \text{ mH}) = 628.3 \ \Omega$$

$$Z_T = \sqrt{(20 \ \Omega)^2 + (469.1 \ \Omega)^2}$$

$$= 469.5 \ \Omega$$

$$I = \frac{V_s}{Z_T} = \frac{10 \text{ V}}{469.5 \ \Omega} = 21.3 \text{ mA}$$

$$P_L = I^2 R_L = (21.3 \text{ mA})^2 (10 \ \Omega)$$

$$= 4.54 \text{ mW}$$

As you can see from the results, the power to the load peaks at the frequency at which the load impedance is the complex conjugate of the output impedance (when the reactances are equal in magnitude). A graph of the load power versus frequency is shown in Figure 20–45.

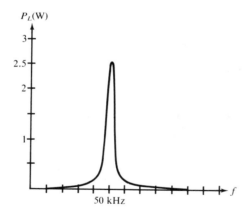

FIGURE 20–45

EXAMPLE 20–18

(a) Determine the frequency at which maximum power is transferred from the amplifier to the speaker in Figure 20–46(a). The amplifier and coupling capacitor are the source, and the speaker is the load, as shown in the equivalent circuit of Figure 20–46(b).

(b) How many watts of power are delivered to the speaker at this frequency if $V_s = 3.8$ V rms?

FIGURE 20–46

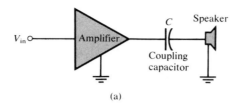

(a)

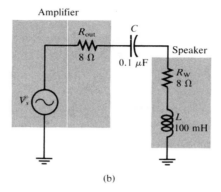

(b)

Solution:

(a) When the power to the speaker is maximum, the source impedance $(R_{out} - jX_C)$ and the load impedance $(R_W + jX_L)$ are complex conjugates, so

$$X_C = X_L$$

$$\frac{1}{2\pi fC} = 2\pi fL$$

Solving for f:

$$f^2 = \frac{1}{4\pi^2 LC}$$

$$f = \frac{1}{2\pi\sqrt{LC}}$$

$$= \frac{1}{2\pi\sqrt{(100 \text{ mH})(0.1 \text{ }\mu\text{F})}} \cong 1592 \text{ Hz}$$

(b)

$$Z_T = R_{out} + R_W = 16 \text{ }\Omega$$

$$I = \frac{V_s}{Z_T} = \frac{3.8 \text{ V}}{16 \text{ }\Omega} = 0.238 \text{ A}$$

$$P_{max} = I^2 R_W = (0.238 \text{ A})^2 8 \text{ }\Omega = 0.453 \text{ W}$$

EXAMPLE 20–19

To what value must C be adjusted in the transmitter of Figure 20–47 in order to deliver maximum power to the antenna at a frequency of 15 kHz? Assume the antenna and its coupling transformer can be represented by a resistance of 300 Ω in series with 0.001 H. The output resistance of the transmitter is 600 Ω.

FIGURE 20–47

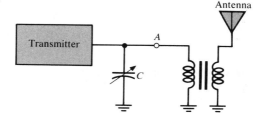

Solution:
The impedance of the load is

$$\mathbf{Z}_L = 300 \ \Omega + j94.2 \ \Omega$$

Looking back toward the transmitter from point A, the equivalent circuit is shown in Figure 20–48. By Thevenin's theorem, the magnitude of the source impedance is

$$Z_s = \frac{R_{out}X_C}{\sqrt{R_{out}^2 + X_C^2}} = \frac{(600 \ \Omega)X_C}{\sqrt{(600 \ \Omega)^2 + X_C^2}}$$

In order to provide maximum power to the load, the source impedance and the load impedance must be complex conjugates at a frequency of 15 kHz. Therefore, the magnitudes of Z_s and Z_L must be equal:

$$\mathbf{Z}_L = 300 \ \Omega - j94.2 \ \Omega = 314.4\angle -17.4° \ \Omega$$

Thus,

$$Z_L = 314.4 \ \Omega$$

$$Z_s = Z_L \quad \text{for maximum power transfer}$$

$$\frac{(600 \ \Omega)X_C}{\sqrt{(600 \ \Omega)^2 + X_C^2}} = 314.4 \ \Omega$$

FIGURE 20–48

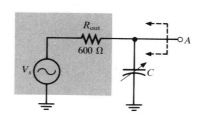

Solving for X_C,

$$600X_C = 314.4\sqrt{600^2 + X_C^2}$$

$$600^2 X_C^2 = 314.4^2(600^2 + X_C^2)$$

$$(600^2 - 314.4^2)X_C^2 = (600^2)(314.4^2)$$

$$X_C = \sqrt{\frac{(600^2)(314.4^2)}{600^2 - 314.4^2}} = 369 \ \Omega$$

The value to which the capacitor must be set to achieve maximum power to the antenna at 15 kHz is

$$\frac{1}{2\pi fC} = 369 \ \Omega$$

$$C = \frac{1}{2\pi(15 \text{ kHz})(369 \ \Omega)} = 0.028 \ \mu\text{F}$$

SECTION REVIEW 20–5

1. If the output impedance of a certain driving circuit is $50 \ \Omega - j10 \ \Omega$, what value of load impedance will result in the maximum power to the load?

2. For the circuit in Question 1, how much power is delivered to the load when the load impedance is the complex conjugate of the output impedance and when the load current is 2 A?

20–6   COMPUTER ANALYSIS

The following program permits the computation and tabulation of power transferred to a load for specified output and load impedances over a defined range of frequencies. The maximum possible power and the frequency at which it is realized are computed and displayed also.

```
10   CLS
20   PRINT "POWER TRANSFER":PRINT
30   PRINT "YOU SPECIFY THE OUTPUT IMPEDANCE OF A CIRCUIT AND
     THE"
40   PRINT "LOAD IMPEDANCE BOTH IN TERMS OF RESISTANCE AND THE"
50   PRINT "REACTIVE COMPONENT VALUES AND A RANGE OF
     FREQUENCIES."
60   PRINT "THE COMPUTER WILL DETERMINE THE POWER TO THE LOAD"
70   PRINT "FOR EACH FREQUENCY INCREMENT IN THE RANGE. ALSO,
     THE"
80   PRINT "MAXIMUM LOAD POWER AND THE FREQUENCY AT WHICH IT"
```

```
90   PRINT "OCCURS IS DETERMINED."
100  PRINT:PRINT:PRINT
110  INPUT "TO CONTINUE PRESS 'ENTER'";X:CLS
120  INPUT "OUTPUT RESISTANCE IN OHMS";RO
130  INPUT "OUTPUT CAPACITANCE IN FARADS";CO
140  INPUT "OUTPUT INDUCTANCE IN HENRIES";LO
150  INPUT "LOAD RESISTANCE IN OHMS";RL
160  INPUT "LOAD CAPACITANCE IN FARADS";CL
170  INPUT "LOAD INDUCTANCE IN HENRIES";LL
180  INPUT "SOURCE VOLTAGE IN VOLTS";VS
190  PRINT
200  INPUT "MINIMUM FREQUENCY IN HERTZ";FL
210  INPUT "MAXIMUM FREQUENCY IN HERTZ";FH
220  INPUT "FREQUENCY INCREMENTS IN HERTZ";FI
230  CLS
240  PRINT "FREQUENCY(HZ)", "LOAD POWER(W)"
250  FOR F=FL TO FH STEP FI
260  IF CO=0 THEN GOTO 290
270  XO=2*3.1416*F*LO-(1/(2*3.1416*F*CO))
280  GOTO 300
290  XO=2*3.1416*F*LO
300  IF CL=0 THEN GOTO 330
310  XL=2*3.1416*F*LL-(1/2*3.1416*F*CL))
320  GOTO 340
330  XL=2*3.1416*F*LL
340  XT=XO+XL
350  R=RO+RL
360  I=VS/(SQR(R*R+XT*XT))
370  PL=I*I*RL
380  PRINT F,PLA
390  NEXT
400  IMAX=VS/R
410  PLMAX=IMAX*IMAX*RL
420  IF LO+LL=0 OR CO+CL=0 GOTO 470
430  IF CO>0 AND CL>0 GOTO 460
440  FM=1/(2*3.1416*(SQR((LO+LL)*(CO+CL))))
450  GOTO 470
460  FM=1/(2*3.1416*(SQR((LO+LL)*(CO*CL/(CO+CL)))))
470  PRINT "THE MAXIMUM LOAD POWER IS ";PLMAX;"W"
480  IF (LO>0 OR LL>0)AND(CO>0 OR CL>0) THEN 520
490  IF (LO=0 AND LL=0)AND(CO=0 AND CL=0) THEN 530
500  IF LO=0 AND LL=0 THEN 540
510  IF CO=0 AND CL=0 THEN 550
520  PRINT "THE FREQUENCY AT WHICH MAXIMUM POWER OCCURS
     IS";FM;"HZ":END
530  PRINT "POWER IS NOT A FUNCTION OF FREQUENCY.":END
540  PRINT "THE FREQUENCY AT WHICH MAXIMUM POWER OCCURS IS
     INFINITY.":END
550  PRINT "THE FREQUENCY AT WHICH MAXIMUM POWER OCCURS IS
     ZERO.":END
```

SECTION REVIEW 20–6

1. Explain the purpose of line 260.

2. Explain the purpose of line 490.

SUMMARY

1. The superposition theorem is useful for the analysis of multiple-source circuits.

2. Thevenin's theorem provides a method for the reduction of any ac circuit to an equivalent form consisting of an equivalent voltage source in series with an equivalent impedance.

3. The term *equivalency,* as used in Thevenin's and Norton's theorems, means that when a given load impedance is connected to the equivalent circuit, it will have the same voltage across it and the same current through it as when it is connected to the original circuit.

4. Norton's theorem provides a method for the reduction of any ac circuit to an equivalent form consisting of an equivalent current source in parallel with an equivalent impedance.

5. Millman's theorem provides a method for the reduction of parallel voltage sources and impedances to a single equivalent voltage source and an equivalent impedance.

6. Maximum power is transferred to a load when the load impedance is the complex conjugate of the output impedance of the driving circuit.

FORMULAS

Millman's Theorem

$$\mathbf{Z}_{eq} = \frac{1}{\mathbf{Y}_T} = \frac{1}{\dfrac{1}{\mathbf{Z}_1} + \dfrac{1}{\mathbf{Z}_2} + \cdots + \dfrac{1}{\mathbf{Z}_n}} \tag{20-1}$$

$$\mathbf{V}_{eq} = \frac{\mathbf{V}_1/\mathbf{Z}_1 + \mathbf{V}_2/\mathbf{Z}_2 + \cdots + \mathbf{V}_n/\mathbf{Z}_n}{\mathbf{Y}_{eq}} \tag{20-2}$$

SELF-TEST

Solutions appear at the end of the book.

1. In Figure 20–49, use the superposition theorem to find the total current in *R*.

2. In Figure 20–49, what is the total current in the *C* branch?

3. Reduce the circuit external to R_L in Figure 20–50 to its Thevenin equivalent.

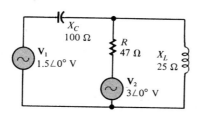

FIGURE 20–49

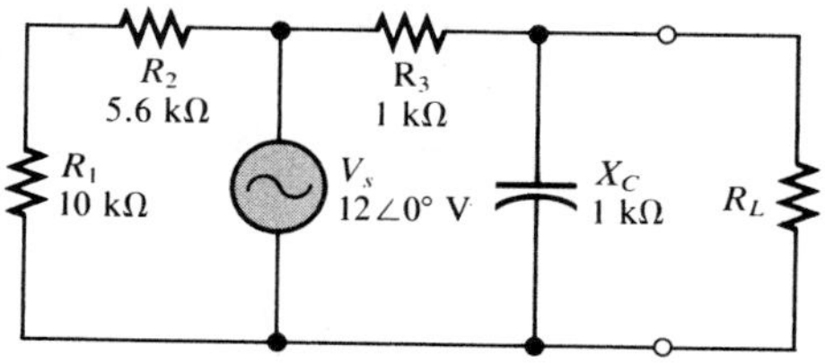

FIGURE 20–50

4. Using Thevenin's theorem, find the current in R_L for the circuit in Figure 20–51.

5. Reduce the circuit in Figure 20–51 to its Norton equivalent as seen by R_L.

6. Use Millman's theorem to simplify the circuit (external to $\mathbf{Z}_L$) in Figure 20–52 to a single voltage source and impedance.

7. What value of $\mathbf{Z}_L$ in Figure 20–52 is required for maximum power transfer to the load?

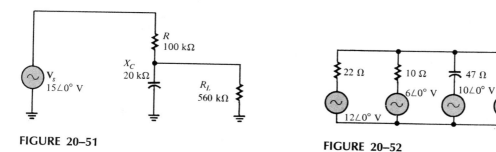

FIGURE 20–51

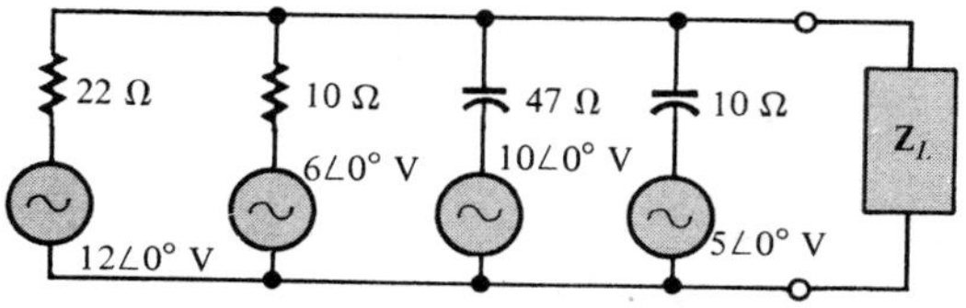

FIGURE 20–52

PROBLEMS

Section 20–1

20–1 Using the superposition method, calculate the current in the right-most branch of Figure 20–53.

FIGURE 20–53

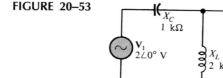

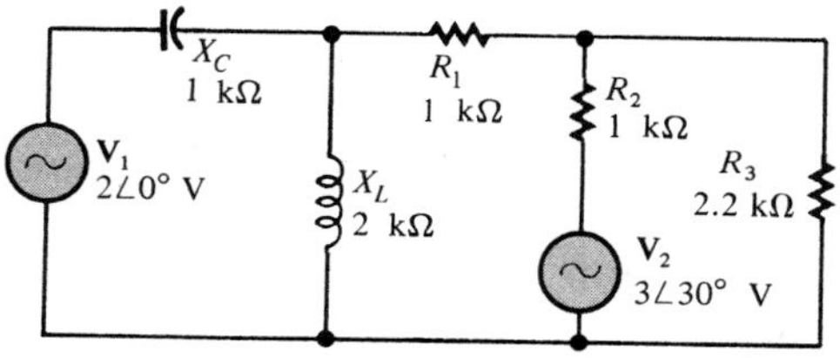

20–2 Use the superposition theorem to find the current in and the voltage across the R_2 branch of Figure 20–53.

20–3 Using the superposition theorem, solve for the current through R_1 in Figure 20–54.

FIGURE 20–54

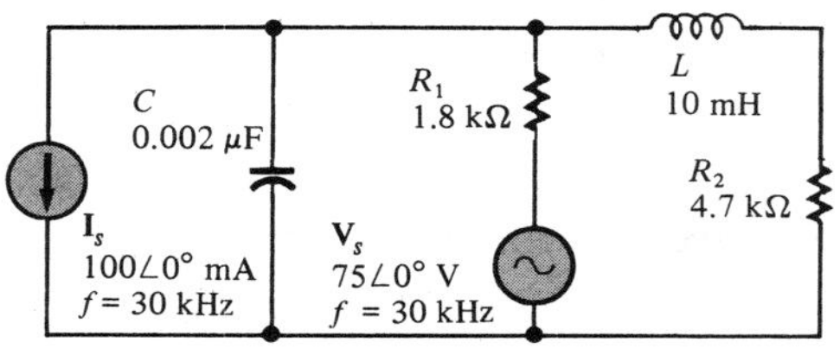

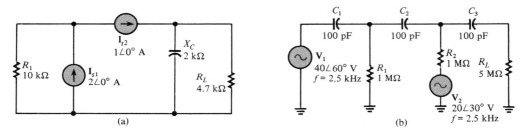

20–4 Using the superposition theorem, find the load current in each circuit of Figure 20–55.

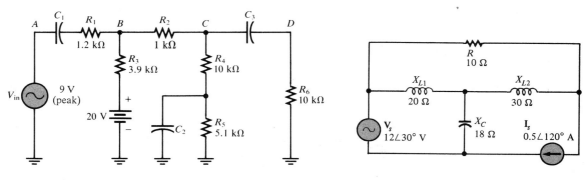

FIGURE 20–55

20–5 Determine the voltage at each point (A, B, C, D) in Figure 20–56. Assume $X_C = 0$ for all capacitors. Sketch the voltage waveforms at each of the points.

20–6 Use the superposition theorem to find the capacitor current in Figure 20–57.

FIGURE 20–56   **FIGURE 20–57**

Section 20–2

20–7   For each circuit in Figure 20–58, determine the Thevenin equivalent circuit for the portion of the circuit viewed by R_L.

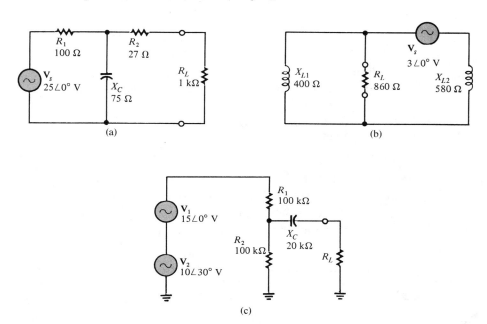

FIGURE 20–58

20–8   Using Thevenin's theorem, determine the current through the load R_L in Figure 20–59.

20–9   Using Thevenin's theorem, find the voltage across R_4 in Figure 20–60.

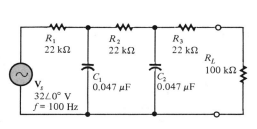

FIGURE 20–59

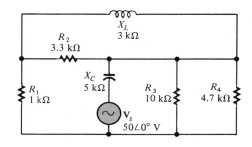

FIGURE 20–60

20–10   Simplify the circuit external to R_3 in Figure 20–61 to its Thevenin equivalent.

FIGURE 20-61

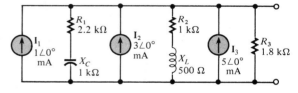

Section 20-3

20-11 For each circuit in Figure 20-58, determine the Norton equivalent as seen by R_L.

20-12 Using Norton's theorem, find the current through the load resistor R_L in Figure 20-59.

20-13 Using Norton's theorem, find the voltage across R_4 in Figure 20-60.

Section 20-4

20-14 Apply Millman's theorem to the circuit of Figure 20-62.

20-15 Use Millman's theorem and source conversions to reduce the circuit in Figure 20-63 to a single voltage source and impedance.

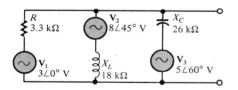

FIGURE 20-62

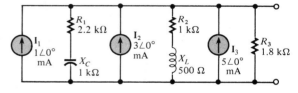

FIGURE 20-63

Section 20-5

20-16 For each circuit in Figure 20-64, maximum power is to be transferred to the load R_L. Determine the appropriate value for the load impedance in each case.

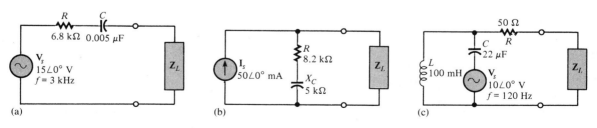

FIGURE 20-64

20-17 Determine $\mathbf{Z}_L$ for maximum power in Figure 20-65.

FIGURE 20–65

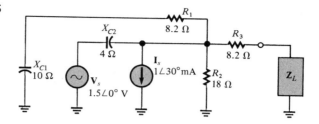

20–18 Find the load impedance required for maximum power transfer to Z_L in Figure 20–66. Determine the maximum true power.

FIGURE 20–66

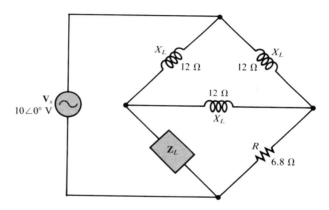

20–19 A load is to be connected in the place of R_2 in Figure 20–61 to achieve maximum power transfer. Determine the type of load, and express it in rectangular form.

Section 20–6

20–20 Write a computer program to convert a specified Thevenin equivalent circuit to a Norton equivalent.

20–21 Write a computer program implementing Millman's theorem.

ANSWERS TO SECTION REVIEWS

Section 20–1

1. 0.   **2.** The circuit can be analyzed one source at a time.   **3.** 12 mA.

Section 20–2

1. Equivalent voltage source and equivalent series impedance.
2. See Figure 20–67.   **3.** $Z_{th} = 21.45\ \Omega - j15.73\ \Omega$, $V_{th} = 4.14\angle53.75°$ V.

FIGURE 20–67

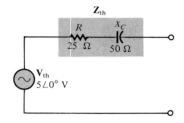

Section 20–3
1. See Figure 20–68.   **2.** $\mathbf{Z}_n = R\angle 0° = 1.2\angle 0°$ kΩ, $\mathbf{I}_n = 10\angle 0°$ mA.

FIGURE 20–68

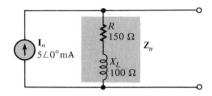

Section 20–4
1. Parallel voltage sources with series impedances.   **2.** $\mathbf{I}_L = 66.5\angle 19.1°$ mA.

Section 20–5
1. $\mathbf{Z}_L = 50\ \Omega + j10\ \Omega$.   **2.** 200 W.

Section 20–6
1. If the output capacitance is 0, then only X_L need be calculated in line 290.
2. If all inductances and capacitances are zero, the source and load are pure resistances. Thus, the load power is not a function of frequency as line 530 indicates.

TWENTY-ONE

PULSE RESPONSE OF REACTIVE CIRCUITS

In Chapters 16 and 17, the frequency response of *RC* and *RL* circuits was covered. In this chapter, the response of *RC* and *RL* circuits to pulse inputs is examined. With pulse inputs, the *time* response of the circuits is of primary importance. In the areas of pulse and digital circuits, the technician is often concerned with how a circuit responds over an interval of time to rapid changes in voltage or current. The relationship of the circuit *time constant* to the input pulse characteristics, such as pulse width and period, determines the shapes of the circuit voltages.

Before starting this chapter, you should review the material in Sections 13–5 and 14–5. Understanding exponential changes in voltages and currents in inductors and capacitors is crucial to the study of pulse response. Exponential formulas that were given in Chapter 13 are used throughout this chapter.

Integrator and *differentiator,* terms used throughout this coverage, refer to mathematical functions that are approximated by these circuits under certain conditions. Integration is an averaging process, and differentiation is a process for establishing a rate of change of a function.

In this chapter, you will learn:

☐ How to recognize an *RC* and an *RL* integrating circuit.
☐ How to recognize an *RC* and an *RL* differentiating circuit.
☐ The meaning of *transient time*.

☐ The meaning of the term *steady state*.
☐ How the output voltage of integrators and differentiators changes with changes in input pulse width and/or frequency.
☐ How the output voltage of integrators and differentiators is a function of the circuit time constant.
☐ How time response and frequency response are related.

21–1 THE *RC* INTEGRATOR

Figure 21–1 shows a series *RC* circuit with a pulse input. When the output is taken across the capacitor, this circuit is known as an *integrator* in terms of its pulse response. Recall that in terms of frequency response, it is a basic *low-pass* filter. The term *integrator* is derived from a mathematical function which this circuit approximates under certain conditions.

FIGURE 21–1
An *RC* integrating circuit.

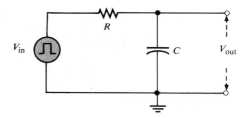

HOW THE CAPACITOR CHARGES AND DISCHARGES WITH A PULSE INPUT

When a pulse generator is connected to the input, as symbolized in Figure 21–1, the capacitor will charge and discharge in response to the pulses. When the input goes from its low level to its high level, the capacitor *charges* toward the high level of the pulse through the resistor. This action is analogous to connecting a battery through a switch to the *RC* network, as illustrated in Figure 21–2(a). When the pulse goes from its high level back to its low level, the capacitor *discharges* back through the source. The resistance of the source is assumed to be negligible compared to *R*. This action is analogous to replacing the source with a closed switch, as illustrated in Figure 21–2(b).

The capacitor will charge and discharge following an *exponential curve*. Its rate of charging and discharging, of course, depends on the *RC* time constant ($\tau = RC$).

For an ideal pulse, both edges are considered to occur instantaneously. Two basic rules of capacitor behavior aid in understanding *RC* circuit responses to pulse inputs:

1. The capacitor appears as a *short* to an instantaneous change in current and as an open to dc.
2. The voltage across the capacitor cannot change instantaneously—it can change only exponentially.

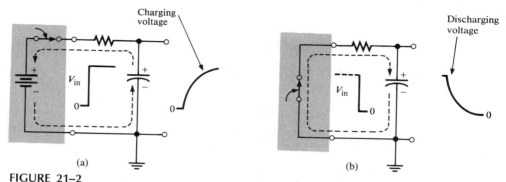

FIGURE 21–2

The equivalent action when a pulse source charges and discharges the capacitor.

THE CAPACITOR VOLTAGE

In an *RC* integrator, *the output is the capacitor voltage*. The capacitor charges during the time that the pulse is high. If the pulse is at its high level long enough, the capacitor will fully charge to the voltage amplitude of the pulse, as illustrated in Figure 21–3. The capacitor discharges during the time that the pulse is low. If the low time between pulses is long enough, the capacitor will fully discharge to zero, as shown in the figure. Then when the next pulse occurs, it will charge again.

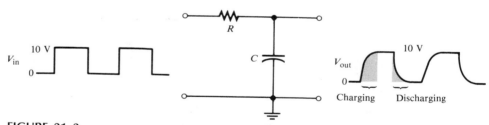

FIGURE 21–3

Illustration of a capacitor fully charging and discharging in response to a pulse input.

SECTION REVIEW 21–1

1. Define the term *integrator* in relation to an *RC* circuit.

2. What causes a capacitor in an *RC* circuit to charge and discharge?

21–2

RESPONSE OF AN *RC* INTEGRATOR TO A SINGLE PULSE

From the previous section, you have a general idea of how an *RC* integrator responds to a pulse input. In this section, we examine the response to a single pulse in detail. In the next section, we will expand this basic knowledge to include the general case of a series of pulses.

Two conditions of pulse response must be considered:

1. when the input pulse width (t_W) is equal to or greater than five time constants ($t_W \geq 5\tau$)
2. when the input pulse width is less than five time constants ($t_W < 5\tau$)

Recall that five time constants is accepted as the time for a capacitor to fully charge or fully discharge; this time is often called the *transient time*.

WHEN THE PULSE WIDTH IS EQUAL TO OR GREATER THAN FIVE TIME CONSTANTS

The capacitor will *fully charge* if the pulse width is equal to or greater than five time constants (5τ). This condition is expressed as $t_W \geq 5\tau$. At the end of the pulse, the capacitor *fully discharges* back through the source.

Figure 21–4 illustrates the output waveforms for various values of time constant and a fixed input pulse width. Notice that the shape of the output pulse approaches that of the input as the transient time is made small compared to the pulse width. In each case the output reaches the full amplitude of the input.

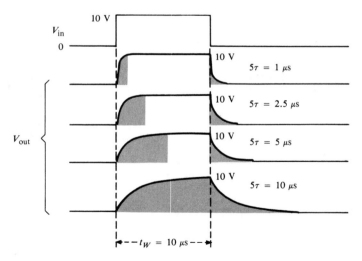

FIGURE 21–4
Variation of an integrator's output pulse shape with time constant. The shaded areas indicate when the capacitor is charging and discharging.

Figure 21–5 shows how a fixed time constant and a variable input pulse width affect the integrator output. Notice that as the pulse width is increased, the shape of the output pulse approaches that of the input. Again, this means that the transient time is short compared to the pulse width.

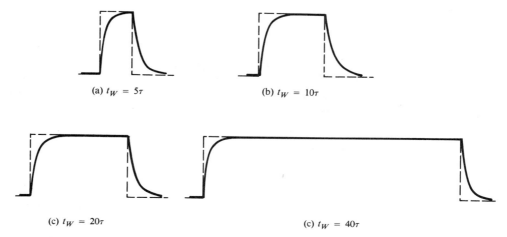

(a) $t_W = 5\tau$ (b) $t_W = 10\tau$

(c) $t_W = 20\tau$ (c) $t_W = 40\tau$

FIGURE 21–5
Variation of an integrator's output pulse shape with input pulse width (the time constant is fixed). Dashed is input and black is output.

WHEN THE PULSE WIDTH IS LESS THAN FIVE TIME CONSTANTS

Now let us examine the case in which the width of the input pulse is less than five time constants of the integrator. This condition is expressed as $t_W < 5\tau$.

As before, the capacitor charges for the duration of the pulse. However, because the pulse width is less than the time it takes the capacitor to fully charge (5τ), the output voltage will *not* reach the full input voltage before the end of the pulse. *The capacitor only partially charges,* as illustrated in Figure 21–6 for several values of *RC* time constants. Notice that for longer time constants, the output reaches a lower voltage because the capacitor cannot charge as much. Of course, in our examples with a single pulse input, the capacitor fully discharges after the pulse ends.

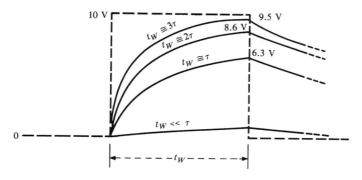

FIGURE 21–6
Capacitor voltage for various time constants that are longer than the input pulse width. Dashed is input and black is output.

When the time constant is much greater than the input pulse width, the capacitor charges very little, and, as a result, the output voltage becomes almost negligible, as indicated in Figure 21–6.

Figure 21–7 illustrates the effect of reducing the input pulse width for a *fixed* time constant value. As the width is reduced, the output voltage becomes smaller because the capacitor has less time to charge. However, it takes the capacitor the same length of time (5τ) to discharge back to zero for each condition after the pulse is removed.

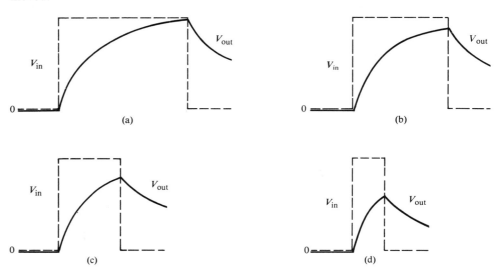

FIGURE 21–7
The capacitor charges less and less as the input pulse width is reduced. The time constant is fixed.

EXAMPLE 21–1

A single 10-V pulse with a width of 100 μs is applied to the integrator in Figure 21–8.

(a) To what voltage will the capacitor charge?

(b) How long will it take the capacitor to discharge if the internal resistance of the pulse source is 50 Ω?

(c) Sketch the output voltage.

FIGURE 21–8

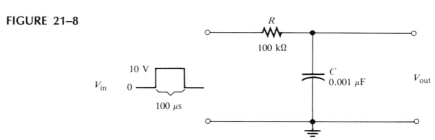

Solution:

(a) The circuit time constant is

$$\tau = RC = (100 \text{ k}\Omega)(0.001 \ \mu\text{F}) = 100 \ \mu\text{s}$$

Notice that the pulse width is exactly equal to the time constant. Thus, the capacitor will charge 63% of the full input amplitude in one time constant, so the output will reach a maximum voltage of 6.3 V.

(b) The capacitor discharges back through the source when the pulse ends. We can neglect the 50-Ω source resistance in series with 100 kΩ. The total discharge time therefore is

$$5\tau = 5(100 \ \mu\text{s}) = 500 \ \mu\text{s}$$

(c) The output charging and discharging curve is shown in Figure 21–9.

FIGURE 21–9 V_{out} (V)

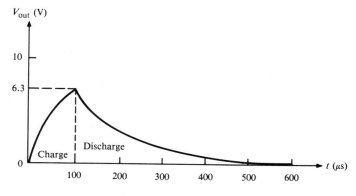

EXAMPLE
21–2

Determine how much the capacitor in Figure 21–10 will charge when the single pulse is applied to the input.

FIGURE 21–10

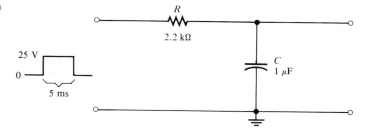

Solution:

Calculate the time constant:

$$\tau = RC = (2.2 \text{ k}\Omega)(1 \ \mu\text{F}) = 2.2 \text{ ms}$$

Because the pulse width is 5 ms, the capacitor charges for 2.27 time constants (5 ms/2.2 ms = 2.27). We use the exponential formula from Chapter 13 to find the voltage to which the capacitor will charge. The calculation is done as follows:

$$v = V_F(1 - e^{-t/RC}) \quad \text{where } V_F = 25 \text{ V} \quad \text{and} \quad t = 5 \text{ ms}$$

$$= (25 \text{ V})(1 - e^{-5/2.2}) = (25 \text{ V})(1 - e^{-2.27})$$

$$= (25 \text{ V})(1 - 0.103) = (25 \text{ V})(0.897)$$

$$= 22.43\text{V}$$

These calculations show that the capacitor charges to 22.43 V during the 5-ms duration of the input pulse. It will discharge back to zero when the pulse goes back to zero.

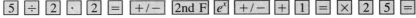

SECTION REVIEW 21–2

1. When an input pulse is applied to an *RC* integrator, what condition must exist in order for the output voltage to reach full amplitude?

2. For the circuit in Figure 21–11, which has a single input pulse, find the maximum output voltage and determine how long the capacitor will discharge.

FIGURE 21–11

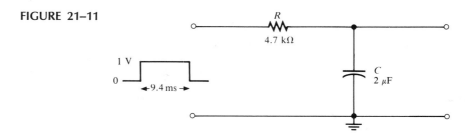

3. For Figure 21–11, sketch the approximate shape of the output voltage with respect to the input pulse.

4. If the integrator time constant equals the input pulse width, will the capacitor fully charge?

5. Describe the condition under which the output voltage has the approximate shape of a rectangular input pulse.

21–3

RESPONSE OF AN *RC* INTEGRATOR TO REPETITIVE PULSES

In the last section you learned how an *RC* integrator responds to a *single* pulse input. These basic ideas will be extended in this section to include the integrator response to *repetitive* pulses. In electronic systems, you will encounter repetitive pulse waveforms much more often than single pulses. However, an understanding of the *RC* integrator response to single pulses is essential to learning how these circuits act with repeated pulses.

If a periodic pulse waveform is applied to an *RC* integrator, as shown in Figure 21–12, *the output waveshape depends on the relationship of the circuit time constant and the frequency (period) of the input pulses*. The capacitor, of course, charges and discharges in response to a pulse input. *The amount of charge and discharge of the capacitor depends both on the circuit time constant and on the input frequency*, as mentioned.

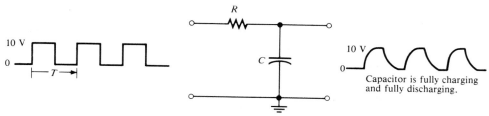

FIGURE 21–12
RC integrator with a repetitive pulse waveform input.

If the pulse width and the time between pulses are each equal to or greater than five time constants, the capacitor will fully charge and fully discharge during each period of the input waveform. This case is shown in Figure 21–12.

WHEN THE CAPACITOR DOES NOT FULLY CHARGE AND DISCHARGE

When the pulse width and the time between pulses are *shorter than five time constants*, as illustrated in Figure 21–13 for a square wave, the capacitor will *not* completely charge or discharge. We will now examine the effects of this situation on the output voltage of the *RC* integrator.

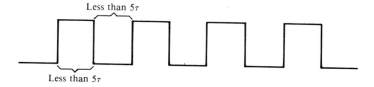

FIGURE 21–13
Waveform that does not allow full charge or discharge of integrator capacitor.

For illustration, let us take an example of an *RC* integrator with a charging and discharging time constant equal to the pulse width of a 10-V square wave input, as in Figure 21–14. This choice will simplify the analysis and will demonstrate the basic

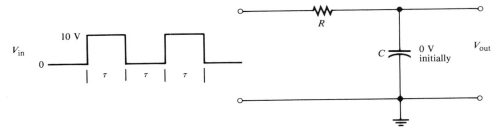

FIGURE 21-14
Integrator with a square wave input having a period equal to two time constants ($T = 2\tau$).

action of the integrator under these conditions. At this point, we really do not care what the exact time constant value is because we know from Chapter 10 that an RC circuit charges 63% during one time constant interval.

We will assume that the capacitor in Figure 21–14 begins initially uncharged and examine the output voltage on a pulse-by-pulse basis. The results of this analysis are shown in Figure 21–15.

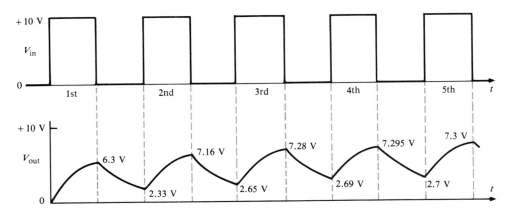

FIGURE 21-15
Input and output for the initially uncharged integrator in Figure 21–14.

First Pulse   During the first pulse, the capacitor charges. The output voltage reaches 6.3 V (63% of 10 V), as shown in Figure 21–15.

Between First and Second Pulses   The capacitor discharges, and the voltage decreases to 37% of the voltage at the beginning of this interval: 0.37(6.3 V) = 2.33 V.

Second Pulse   The capacitor voltage begins at 2.33 V and increases 63% of the way to 10 V. This calculation is as follows: The total charging range is 10 V − 2.33 V = 7.67 V. The capacitor voltage will increase an additional 63% of 7.67 V, which is 4.83 V. Thus, at the end of the second pulse, the output voltage is 2.33 V + 4.83 V = 7.16 V, as shown in Figure 21–15. Notice that the *average* is building up.

Between Second and Third Pulses   The capacitor discharges during this time, and therefore the voltage decreases to 37% of the initial voltage by the end of the second pulse: 0.37(7.16 V) = 2.65 V.

Third Pulse   At the start of the third pulse, the capacitor voltage begins at 2.65 V. The capacitor charges 63% of the way from 2.65 V to 10 V: 0.63(10 V − 2.65 V) = 4.63 V. Therefore, the voltage at the end of the third pulse is 2.65 V + 4.63 V = 7.28 V.

Between Third and Fourth Pulses   The voltage during this interval decreases due to capacitor discharge. It will decrease to 37% of its value by the end of the third pulse. The final voltage in this interval is 0.37(7.28 V) = 2.69 V.

Fourth Pulse   At the start of the fourth pulse, the capacitor voltage is 2.69 V. The voltage increases by 0.63(10 V − 2.69 V) = 4.605 V. Therefore, at the end of the fourth pulse, the capacitor voltage is 2.69 V + 4.61 V = 7.295 V. Notice that the values are leveling off as the pulses continue.

Between Fourth and Fifth Pulses   Between these pulses, the capacitor voltage drops to 0.37(7.295 V) = 2.7 V.

Fifth Pulse   During the fifth pulse, the capacitor charges 0.63(10 V − 2.7 V) = 4.6 V. Since it started at 2.7 V, the voltage at the end of the pulse is 2.7 V + 4.6 V = 7.3 V.

STEADY STATE RESPONSE

In the preceding discussion, the output voltage gradually built up and then began leveling off. It takes approximately 5τ for the output voltage to build up to a constant *average* value. This interval is the *transient time* of the circuit. Once the output voltage reaches *the average value of the input voltage, a steady state condition* is reached which continues as long as the periodic input continues. This state is illustrated in Figure 21–16 based on the values obtained in the preceding discussion.

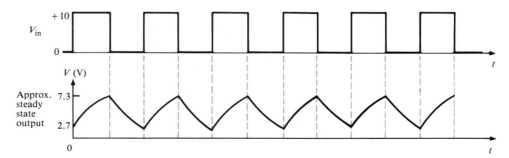

FIGURE 21–16
Output reaches steady state after 5τ.

The transient time for our example circuit is the time from the beginning of the first pulse to the end of the third pulse. The reason for this interval is that the capacitor voltage at the end of the third pulse is 7.28 V, which is about 99% of the final voltage.

THE EFFECT OF AN INCREASE IN TIME CONSTANT

What happens to the output voltage if the *RC* time constant of the integrator is increased with a variable resistor, as indicated in Figure 21–17? As the time constant is increased, the capacitor charges *less* during a pulse and discharges *less* between pulses. The result is a *smaller* fluctuation in the output voltage for increasing values of time constant, as shown in Figure 21–18.

FIGURE 21–17
Integrator with a variable time constant.

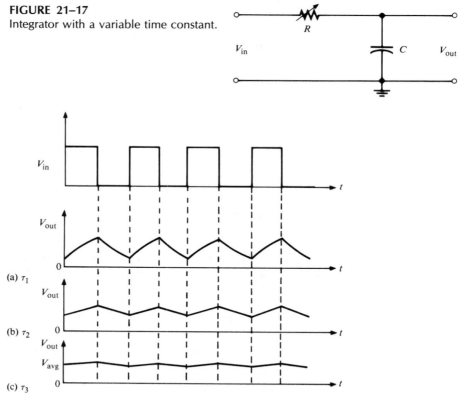

(a) τ_1

(b) τ_2

(c) τ_3

FIGURE 21–18
Effect of longer time constants on the output of an integrator ($\tau_3 > \tau_2 > \tau_1$).

As the time constant becomes extremely long compared to the pulse width, the output voltage approaches a constant dc voltage, as shown in Figure 21–18(c). This value is the *average value* of the input. For a square wave, it is one-half the amplitude.

**EXAMPLE
21–3**

Determine the output voltage waveform for the first *two* pulses applied to the integrator circuit in Figure 21–19. Assume that the capacitor is initially uncharged.

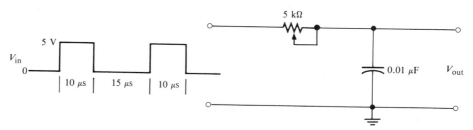

FIGURE 21–19

Solution:
First calculate the circuit time constant:

$$\tau = RC = (5 \text{ k}\Omega)(0.01 \text{ }\mu\text{F}) = 50 \text{ }\mu\text{s}$$

Obviously, the time constant is much longer than the input pulse width or the interval between pulses (notice that the input is not a square wave). Thus, in this case, the exponential formulas must be applied, and the analysis is relatively difficult. Follow the solution carefully.

Step 1: *Calculation for first pulse:* Use the equation for an increasing exponential because *C* is charging. Note that V_F is 5 V, and *t* equals the pulse width of 10 μs. Therefore,

$$v_C = V_F(1 - e^{-t/RC}) = (5 \text{ V})(1 - e^{-10\mu s/50\mu s})$$
$$= (5 \text{ V})(1 - 0.819) = 0.906 \text{ V}$$

This result is plotted in Figure 21–20(a).

Step 2: *Calculation for interval between first and second pulse:* Use the equation for a decreasing exponential because *C* is discharging. Note that V_i is 0.906 V because *C* begins to discharge from this value at the end of the first pulse. The discharge time is 15μs. Therefore,

$$v_C = V_i e^{-t/RC} = 0.906e^{-15\mu s/50\mu s}$$
$$= 0.906(0.741) = 0.671 \text{ V}$$

This result is shown in Figure 21–20(b).

Step 3: *Calculation for second pulse:* At the beginning of the second pulse, the output voltage is 0.671 V. During the second pulse, the capacitor will again charge. In this

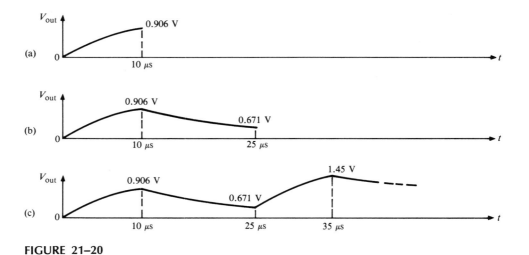

FIGURE 21–20

case it does not begin at zero volts. It already has 0.671 V from the previous charge and discharge. To handle this situation, we must use the general exponential formula:

$$v = V_F + (V_i - V_F)e^{-t/\tau}$$

Using this equation, we can calculate the voltage across the capacitor at the end of the second pulse as follows:

$$
\begin{aligned}
v_C &= V_F + (V_i - V_F)e^{-t/RC} \\
&= 5\text{ V} + (0.671\text{ V} - 5\text{ V})e^{-10\mu s/50\mu s} \\
&= 5\text{ V} + (-4.33\text{ V})(0.819) = 5\text{ V} - 3.55\text{ V} \\
&= 1.45\text{ V}
\end{aligned}
$$

This result is shown in Figure 21–20(c).

Notice that the output waveform builds up on successive input pulses. After approximately 5τ, it will reach its steady state and will fluctuate between a constant maximum and a constant minimum, with an average equal to the average value of the input. You can see this pattern by carrying the analysis in this example further.

SECTION REVIEW 21–3

1. What conditions allow an *RC* integrator capacitor to fully charge and discharge when a periodic pulse waveform is applied to the input?

2. What will the output waveform look like if the circuit time constant is extremely small compared to the pulse width of a square wave input?

3. When 5τ is greater than the pulse width of an input square wave, the time required for the output voltage to build up to a constant average value is called _____.

4. Define *steady state response*.

5. What does the average value of the output voltage of an integrator equal during steady state?

21–4   RESPONSE OF AN *RC* DIFFERENTIATOR TO A SINGLE PULSE

Figure 21–21 shows a series *RC* circuit. Notice that the output is taken across the *resistor*, thereby distinguishing the *RC* differentiator from the *RC* integrator. This circuit is called a *differentiator* in terms of its pulse response. Recall that in terms of its frequency response, it is a basic *high-pass* filter.

FIGURE 21–21
An *RC* differentiating circuit.

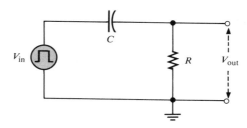

The same action occurs in the differentiator as in the integrator, except the output voltage is taken across the resistor rather than the capacitor. The capacitor charges exponentially at a rate depending on the *RC* time constant. The shape of the differentiator's resistor voltage is determined by the charging and discharging action of the capacitor.

To understand how the output voltage is shaped by a differentiator, we must consider the following:

1. the response to the rising pulse edge,
2. the response between the rising and falling edges, and
3. the response to the falling pulse edge.

RESPONSE TO THE RISING EDGE OF THE INPUT PULSE

Assume that the capacitor is initially uncharged prior to the rising pulse edge. Prior to the pulse, the input is zero volts. Thus, there are zero volts across the capacitor and also zero volts across the resistor, as indicated in Figure 21–22(a).

Now assume that a 10-V pulse is applied to the input. When the rising edge occurs, point *A* goes to +10 V. Recall that *the voltage across a capacitor cannot*

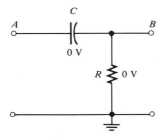

(a) Before pulse is applied

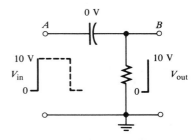

(b) At rising edge of input pulse

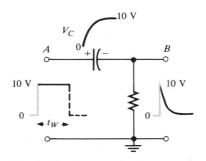

(c) During level part of pulse when $t_W \geqslant 5\tau$

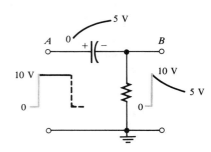

(d) During level part of pulse when $t_W < 5\tau$

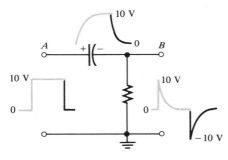

(e) At falling edge of pulse when $t_W \geqslant 5\tau$

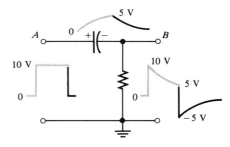

(f) At falling edge of pulse when $t_W < 5\tau$

FIGURE 21–22

Examples of the response of a differentiator to a single input pulse under two conditions: $t_W \geq 5\tau$ and $t_W < 5\tau$.

change instantaneously, and thus the capacitor appears instantaneously as a short. Therefore, if point A goes instantly to $+10$ V, then point B *must* also go instantly to $+10$ V, keeping the capacitor voltage zero for the instant of the rising edge. The capacitor voltage is the voltage from point A to point B.

    The voltage at point B with respect to ground is the voltage across the resistor (and the output voltage). Thus, the output voltage suddenly goes to $+10$ V in response to the rising pulse edge, as indicated in Figure 21–22(b).

RESPONSE DURING PULSE WHEN $t_W \geq 5\tau$

While the pulse is at its high level between the rising edge and the falling edge, the capacitor is charging. When the pulse width is equal to or greater than five time constants ($t_W \geq 5\tau$), the capacitor has time to *fully charge*.

As the voltage across the capacitor builds up exponentially, the voltage across the resistor *decreases* exponentially until it reaches zero volts at the time the capacitor reaches full charge ($+10$ V in this case). This decrease in the resistor voltage occurs because the sum of the capacitor voltage and the resistor voltage at any instant must be equal to the applied voltage, in compliance with Kirchhoff's voltage law ($v_C + v_R = v_{in}$). This part of the response is illustrated in Figure 21–22(c).

RESPONSE DURING PULSE WHEN $t_W < 5\tau$

When the pulse width is less than five time constants ($t_W < 5\tau$), *the capacitor does not have time to fully charge*. Its partial charge depends on the relation of the time constant and the pulse width.

Because the capacitor does not reach the full $+10$ V, *the resistor voltage will not reach zero volts* by the end of the pulse. For example, if the capacitor charges to $+5$ V during the pulse interval, the resistor voltage will decrease to $+5$ V, as illustrated in Figure 21–22(d).

RESPONSE TO FALLING EDGE WHEN $t_W \geq 5\tau$

Let us first examine the case in which the capacitor is *fully charged* at the end of the pulse ($t_W \geq 5\tau$). Refer to Figure 21–22(e). On the falling edge, the input pulse suddenly goes from $+10$ V back to zero. An instant before the falling edge, the capacitor is charged to 10 V, so point A is $+10$ V and point B is 0 V. Since the voltage across a capacitor cannot change instantaneously, when point A makes a transition from $+10$ V to zero on the falling edge, point B *must* also make a 10-V transition from zero to -10 V. This keeps the voltage across the capacitor at 10 V for the instant of the falling edge.

The capacitor now begins to *discharge* exponentially. As a result, the resistor voltage goes from -10 V to zero in an exponential curve, as indicated in Figure 21–22(e).

RESPONSE TO FALLING EDGE WHEN $t_W < 5\tau$

Next, let us examine the case in which the capacitor is only partially charged at the end of the pulse ($t_W < 5\tau$). For example, if the capacitor charges to $+5$ V, the resistor voltage at the instant before the falling edge is also $+5$ V, because the capacitor voltage plus the resistor voltage must add up to $+10$ V, as illustrated in Figure 21–22(d).

When the falling edge occurs, point A goes from $+10$ V to zero. As a result, point B goes from $+5$ V to -5 V, as illustrated in Figure 21–22(f). This decrease occurs, of course, because the capacitor voltage cannot change at the instant of the falling edge. Immediately after the falling edge, the capacitor begins to discharge to zero. As a result, the resistor voltage goes from -5 V to zero, as shown.

SUMMARY OF DIFFERENTIATOR RESPONSE TO A SINGLE PULSE

Perhaps a good way to summarize this section is to look at the general output waveforms of a differentiator as the time constant is varied from one extreme, when 5τ is much less than the pulse width, to the other extreme, when 5τ is much greater than the pulse width. These situations are illustrated in Figure 21–23. In Part (a) of the figure, the output consists of narrow positive and negative "spikes." In Part (f), the output approaches the shape of the input. Various conditions between these extremes are illustrated in Parts (b) through (e).

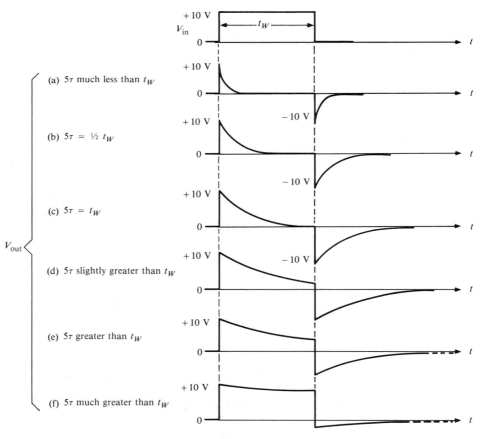

FIGURE 21–23
Effects of a change in time constant on the shape of the output voltage of a differentiator.

**EXAMPLE
21–4**

Sketch the output voltage for the circuit in Figure 21–24.

FIGURE 21–24

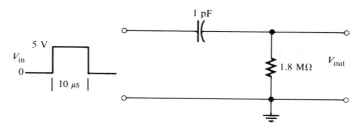

Solution:
First calculate the time constant:

$$\tau = RC = (1.8 \text{ M}\Omega)(1 \text{ pF}) = 1.8 \text{ } \mu s$$

In this case, $t_W > 5\tau$, so the capacitor reaches full charge before the end of the pulse.
On the rising edge, the resistor voltage jumps to $+5$ V and then decreases exponentially to zero by the end of the pulse. On the falling edge, the resistor voltage jumps to -5 V and then goes back to zero exponentially. The resistor voltage is, of course, the output, and its shape is shown in Figure 21–25.

FIGURE 21–25

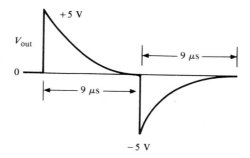

**EXAMPLE
21–5**

Determine the output voltage waveform for the differentiator in Figure 21–26.

FIGURE 21–26

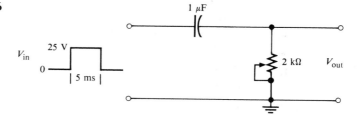

Solution:
First calculate the time constant:

$$\tau = (2 \text{ k}\Omega)(1 \text{ } \mu\text{F}) = 2 \text{ ms}$$

On the rising edge, the resistor voltage immediately jumps to $+25$ V. Because the pulse width is 5 ms, the capacitor charges for 2.5 time constants and therefore does not reach full charge. Thus, we must use the formula for a decreasing exponential in order to calculate to what voltage the output decreases by the end of the pulse:

$$v_{\text{out}} = V_i e^{-t/RC} \quad \text{where } V_i = 25 \text{ V} \quad \text{and} \quad t = 5\text{ms}$$

$$= 25e^{-5\text{ms}/2\text{ms}} = 25(0.082)$$

$$= 2.05 \text{ V}$$

This calculation gives us the resistor voltage at the end of the 5-ms pulse width interval.

On the falling edge, the resistor voltage immediately jumps from $+2.05$ V down to -22.95 V (a 25-V transition). The resulting waveform of the output voltage is shown in Figure 21–27.

FIGURE 21–27

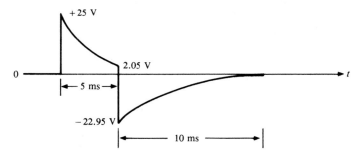

SECTION REVIEW 21–4

1. Sketch the output of a differentiator for a 10-V input pulse when $5\tau = 0.5t_W$.

2. Under what condition does the output pulse shape most closely resemble the input pulse for a differentiator?

3. What does the differentiator output look like when 5τ is much less than the pulse width of the input?

4. If the resistor voltage in a differentiating circuit is down to $+5$ V at the end of a 15-V input pulse, to what negative value will the resistor voltage go in response to the *falling* edge of the input?

21–5

RESPONSE OF AN *RC* DIFFERENTIATOR TO REPETITIVE PULSES

The differentiator response to a single pulse, covered in the preceding section, is extended in this section to repetitive pulses.

If a periodic pulse waveform is applied to an *RC* differentiating circuit, two conditions again are possible: $t_W \geq 5\tau$ or $t_W < 5\tau$. Figure 21–28 shows the output when $t_W = 5\tau$. As the time constant is reduced, both the positive and the negative portions of the output become narrower. Notice that the *average* value of the output is *zero*.

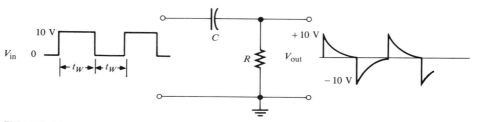

FIGURE 21–28
Example of differentiator response when $t_W = 5\tau$.

Figure 21–29 shows the *steady state* output when $t_W < 5\tau$. As the time constant is increased, the positively and negatively sloping portions become flatter. For a very long time constant, the output approaches the shape of the input, but with an average value of zero. An average value of zero means that the waveform has equal positive and negative portions. The average value of a waveform is its *dc* component. Because a capacitor blocks dc, the dc component of the input is prevented from passing through to the output.

Like the integrator, the differentiator output takes time (5τ) to reach steady state. To illustrate the response, let us take an example in which the time constant *equals* the input pulse width.

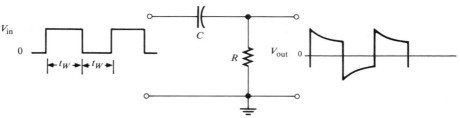

FIGURE 21–29
Example of differentiator response when $t_W < 5\tau$.

ANALYSIS OF A REPETITIVE WAVEFORM

At this point, we do not care what the circuit time constant is, because we know that the resistor voltage will decrease to 37% of its maximum value during one pulse (1τ). We will assume that the capacitor in Figure 21–30 begins initially uncharged, and then we will examine the output voltage on a pulse-by-pulse basis. The results of the analysis to follow are shown in Figure 21–31.

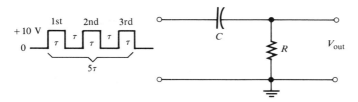

FIGURE 21–30
RC differentiator with $\tau = t_W$.

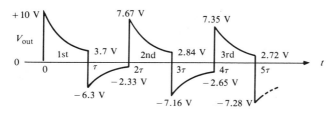

FIGURE 21–31
Differentiator output waveform during transient time for the circuit in Figure 21–30.

First Pulse   On the rising edge, the output instantaneously jumps to $+10$ V. Then the capacitor partially charges to 63% of 10 V, which is 6.3 V. Thus, the output voltage must decrease to 3.7 V, as shown in Figure 21–31.

On the falling edge, the output instantaneously makes a negative-going 10-V transition to -6.3 V (-10 V $+ 3.7$ V $= -6.3$ V).

Between First and Second Pulses   The capacitor discharges to 37% of 6.3 V, which is 2.33 V. Thus, the resistor voltage, which starts at -6.3 V, must increase to -2.33 V. Why? Because at *the instant prior to* the next pulse, the input voltage is zero. Therefore, the sum of v_C and v_R must be zero (2.33 V $-$ 2.33 V $= 0$). Remember that $v_C + v_R = v_{in}$ at all times.

Second Pulse   On the rising edge, the output makes an instantaneous, positive-going, 10-V transition from -2.33 V to 7.67 V. Then the capacitor charges $0.63 \times$ (10 V $-$ 2.33 V) $= 4.83$ V by the end of the pulse. Thus, the capacitor voltage increases from 2.33 V to 2.33 V $+ 4.83$ V $= 7.16$ V. The output voltage drops to $0.37 \times$ (7.67 V) $= 2.84$ V.

On the falling edge, the output instantaneously makes a negative-going transition from 2.84 V to −7.16 V, as shown in Figure 21–31.

Between Second and Third Pulses The capacitor discharges to 37% of 7.16 V, which is 2.65 V. Thus, the output voltage starts at −7.16 V and increases to −2.65 V, because the capacitor voltage and the resistor voltage must add up to zero at the instant prior to the third pulse (the input is zero).

Third Pulse On the rising edge, the output makes an instantaneous 10-V transition from −2.65 V to +7.35 V. Then the capacitor charges 0.63 × (10 V − 2.65 V) = 4.63 V to 2.65 V + 4.63 V = 7.28 V. As a result, the output voltage drops to 0.37 × 7.35 V = 2.72 V.

On the falling edge, the output instantly goes from +2.72 V down to −7.28 V.

After the third pulse, five time constants have elapsed, and the output voltage is close to its steady state. Thus, it will continue to vary from a positive maximum of about +7.3 V to a negative maximum of about −7.3 V, with an average value of zero.

SECTION REVIEW 21–5

1. What conditions allow an *RC* differentiator to fully charge and discharge when a periodic pulse waveform is applied to the input?

2. What will the output waveform look like if the circuit time constant is extremely small compared to the pulse width of a square wave input?

3. What does the average value of the differentiator output voltage equal during steady state?

21–6

THE *RL* INTEGRATOR

Figure 21–32 shows a series *RL* circuit with *the output taken across the resistor*. This circuit is also known as an *integrator*. Its output waveform under equivalent conditions is the same as that for the *RC* integrator. Recall that in the *RC* case, the output was across the capacitor.

FIGURE 21–32
An *RL* integrating circuit.

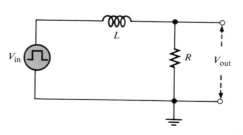

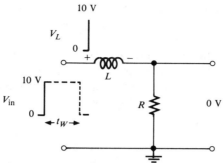

(a) At rising edge of pulse ($i = 0$)

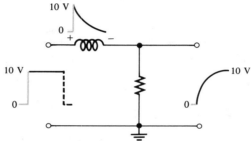

(b) During flat portion of pulse

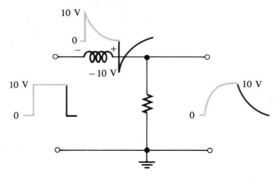

(c) At falling edge of pulse and after

FIGURE 21–33
Illustration of the pulse response of an *RL* integrator *($t_W > 5\tau$)*.

As you know, both edges of an ideal pulse are considered to occur instantaneously. Two basic rules for inductor behavior will aid in analyzing *RL* circuit responses to pulse inputs:

1. The inductor appears as an open to an instantaneous change in current and as a short (ideally) to dc.
2. The current in an inductor cannot change instantaneously—it can change only exponentially.

RESPONSE OF THE INTEGRATOR TO A SINGLE PULSE

When a pulse generator is connected to the input of the integrator and the voltage pulse goes from its low level to its high level, the inductor prevents a sudden change in current. As a result, the inductor acts as an open, and all of the input voltage is across it at the instant of the rising pulse edge. This situation is indicated in Figure 21–33(a).

After the rising edge, the current builds up, and the output voltage follows the current as it increases exponentially, as shown in Figure 21–33(b). The current can reach a maximum of V_p/R if the transient time is shorter than the pulse width ($V_p = 10$ V in this example).

When the pulse goes from its high level to its low level, an induced voltage with reversed polarity is created across the coil in an effort to keep the current equal to V_p/R. The output voltage begins to decrease exponentially, as shown in Figure 21–33(c).

The exact shape of the output depends on the L/R time constant as summarized in Figure 21–34 for various relationships between the time constant and the pulse width. You should note that the response of this *RL* circuit *in terms of the shape of the output* is identical to that of the *RC* integrator. The relationship of the L/R time constant to the input pulse width has the same effect as the *RC* time constant that we discussed earlier in this chapter. For example, when $t_W < 5\tau$, the output voltage will not reach its maximum possible value.

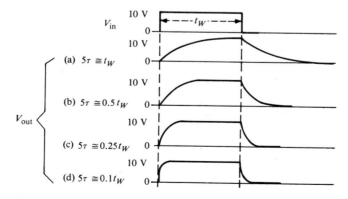

FIGURE 21–34
Illustration of the variation in integrator output pulse shape with time constant.

**EXAMPLE
21–6**

Determine the maximum output voltage for the integrator in Figure 21–35 when a single pulse is applied as shown.

FIGURE 21–35

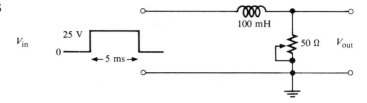

Solution:
Calculate the time constant:

$$\tau = \frac{L}{R} = \frac{100 \text{ mH}}{50 \text{ }\Omega} = 2 \text{ ms}$$

Because the pulse width is 5 ms, the inductor charges for 2.5τ. We must use the exponential formula to calculate the voltage as follows:

$$v_{out} = V_F(1 - e^{-t/\tau}) = 25(1 - e^{-5/2})$$
$$= 25(1 - e^{-2.5}) = 25(1 - 0.082)$$
$$= 25(0.918) = 22.95 \text{ V}$$

**EXAMPLE
21–7**

A pulse is applied to the *RL* integrator circuit in Figure 21–36. Determine the complete waveshapes and the values for I, V_R, and V_L.

FIGURE 21–36

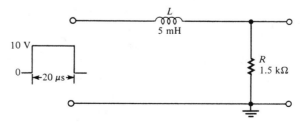

Solution:
The circuit time constant is

$$\tau = \frac{L}{R} = \frac{5 \text{ mH}}{1.5 \text{ k}\Omega} = 3.33 \ \mu s$$

Since $5\tau = 16.65 \ \mu s$ is less than t_W, the current will reach its maximum value and remain there until the end of the pulse.

At the rising edge of the pulse:

$$i = 0 \text{ A}$$
$$v_R = 0 \text{ V}$$
$$v_L = 10 \text{ V}$$

Since the inductor is initially an open, all of the input voltage appears across *L*.

During the pulse:

i increases exponentially to $V_p/R = 10 \text{ V}/1.5 \text{ k}\Omega$
$$= 6.67 \text{ mA in } 16.65 \ \mu s$$

v_R increases exponentially to 10 V in 16.65 μs.
v_L decreases exponentially to zero in 16.65 μs.

At the falling edge of the pulse:

$$i = 6.67 \text{ mA}$$
$$v_R = 10 \text{ V}$$
$$v_L = -10 \text{ V}$$

After the pulse:

i decreases exponentially to zero in 16.65 μs.
v_R decreases exponentially to zero in 16.65 μs.
v_L decreases exponentially to zero in 16.65 μs.

The waveforms are shown in Figure 21–37.

FIGURE 21–37

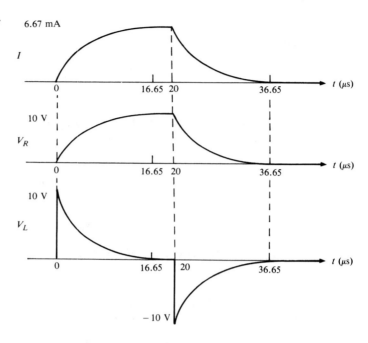

**EXAMPLE
21–8**

A 10-V pulse with a width of 1 ms is applied to the integrator in Figure 21–38. Determine the voltage level that the output will reach during the pulse. If the source has an internal resistance of 30 Ω, how long will it take the output to decay to zero? Sketch the output voltage waveform.

FIGURE 21–38

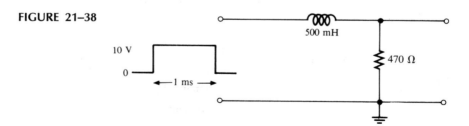

Solution:
The coil charges through the 30-Ω source resistance plus the 470-Ω resistor. The time constant is,

$$\tau = \frac{L}{R_T} = \frac{500 \text{ mH}}{470 \text{ Ω} + 30 \text{ Ω}} = \frac{500 \text{ mH}}{0.5 \text{ kΩ}} = 1 \text{ ms}$$

Notice that the pulse width is exactly equal to τ. Thus, the output V_R will reach 63% of the full input amplitude in 1τ. Therefore, the output voltage gets to 6.3 V at the end of the pulse.

   After the pulse is gone, the inductor discharges back through the 30-Ω source and the 470-Ω resistor. The source takes 5τ to completely discharge; $5\tau = 5(1 \text{ ms}) = 5 \text{ ms}$.

   The output voltage is shown in Figure 21–39.

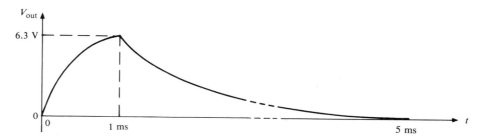

FIGURE 21–39

SECTION REVIEW 21–6

1. In an *RL* integrator, across which component is the output voltage taken?

2. When a pulse is applied to an *RL* integrator, what condition must exist in order for the output voltage to reach the amplitude of the input?

3. Under what condition will the output voltage have the approximate shape of the input pulse?

21–7

THE *RL* DIFFERENTIATOR

Figure 21–40 shows a series *RL* circuit with *the output across the inductor*. Compare this circuit to the *RC* differentiator in which the output is the resistor voltage.

FIGURE 21–40
An *RL* differentiating circuit.

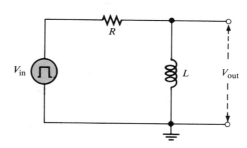

RESPONSE OF THE DIFFERENTIATOR TO A SINGLE PULSE

A pulse generator is connected to the input of the differentiator. Initially, before the pulse, there is no current in the circuit. When the input pulse goes from its low level to its high level, the inductor prevents a sudden change in current. It does so, as you know, with an induced voltage equal and opposite to the input. As a result, L looks like an open, and *all* of the input voltage appears across it at the instant of the rising edge, as shown in Figure 21–41(a) with a 10-V pulse.

During the pulse, the current exponentially builds up. As a result, the inductor voltage decreases, as shown in Figure 21–41(b). The rate of decrease, as you know,

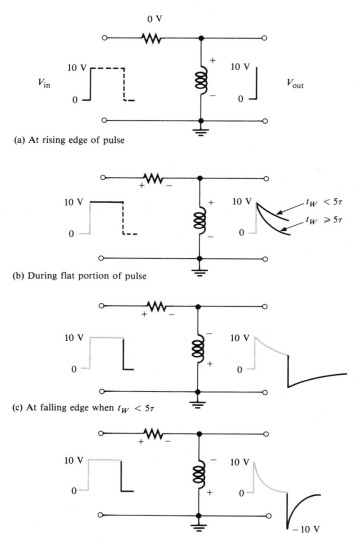

(a) At rising edge of pulse

(b) During flat portion of pulse

(c) At falling edge when $t_W < 5\tau$

(d) At falling edge when $t_W \geqslant 5\tau$.

FIGURE 21–41
Illustration of the response of an *RL* differentiator for both time constant conditions.

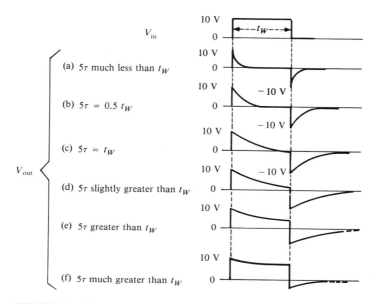

FIGURE 21–42
Illustration of the variation in output pulse shape with the time constant.

depends on the *L/R* time constant. When the falling edge of the input appears, the inductor reacts to keep the current as is, by creating an induced voltage in a direction as indicated in Part (c). This reaction is seen as a sudden negative-going transition of the inductor voltage, as indicated in Parts (c) and (d).

Two conditions are possible, as indicated in Parts (c) and (d). In Part (c), 5τ is greater than the input pulse width, and the output voltage does not have time to decay to zero. In Part (d), 5τ is less than the pulse width, and so the output decays to zero before the end of the pulse. In this case a full, negative, 10-V transition occurs at the trailing edge.

Keep in mind that as far as the input and output waveforms are concerned, the *RL* integrator and differentiator perform the same as their *RC* counterparts.

A summary of the *RL* differentiator response for relationships of various time constants and pulse widths is shown in Figure 21–42.

EXAMPLE 21–9

Sketch the output voltage for the circuit in Figure 21–43.

FIGURE 21–43

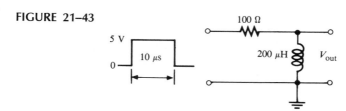

Solution:
First calculate the time constant:

$$\tau = \frac{L}{R} = \frac{200 \ \mu\text{H}}{100 \ \Omega} = 2 \ \mu\text{s}$$

In this case, $t_W = 5\tau$, so the output will decay to zero at the end of the pulse.

On the rising edge, the inductor voltage jumps to $+5$ V and then decays exponentially to zero. It reaches approximately zero at the instant of the falling edge. On the falling edge of the input, the inductor voltage jumps to -5 V and then goes back to zero. The output waveform is shown in Figure 21–44.

FIGURE 21–44

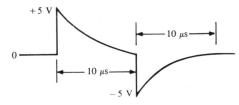

EXAMPLE 21–10

Determine the output voltage waveform for the differentiator in Figure 21–45.

FIGURE 21–45

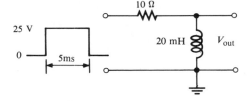

Solution:
First calculate the time constant:

$$\tau = \frac{L}{R} = \frac{20 \ \text{mH}}{10 \ \Omega} = 2 \ \text{ms}$$

On the rising edge, the inductor voltage immediately jumps to $+25$ V. Because the pulse width is 5 ms, the inductor charges for only 2.5τ, so we must use the formula for a decreasing exponential:

$$\begin{aligned} v_L &= V_i e^{-t/\tau} = 25e^{-5/2} \\ &= 25e^{-2.5} = 25(0.082) \\ &= 2.05 \ \text{V} \end{aligned}$$

This result is the inductor voltage at the end of the 5-ms input pulse.

On the falling edge, the output immediately jumps from $+2.05$ V down to -22.95 V (a 25-V negative-going transition). The complete output waveform is sketched in Figure 21–46.

FIGURE 21–46

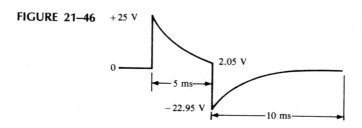

SECTION REVIEW 21–7

1. In an *RL* differentiator, across which component is the output taken?
2. Under what condition does the output pulse shape most closely resemble the input pulse?
3. If the inductor voltage in an *RL* differentiator is down to $+2$ V at the end of a $+10$-V input pulse, to what negative voltage will the output voltage go in response to the falling edge of the input?

21–8

RELATIONSHIP OF TIME RESPONSE TO FREQUENCY RESPONSE

There is a definite relationship between time response and frequency response: *The fast rising and falling edges of a pulse waveform contain the higher frequency components in that waveform. The "top" or flatter portions of the pulses represent the slow changes or lower frequency components. The average value of the pulse waveform is its dc component.* These relationships are indicated in Figure 21–47.

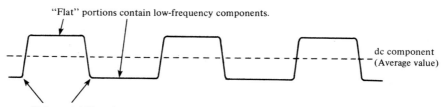

"Flat" portions contain low-frequency components.

dc component
(Average value)

Rising and falling edges contain high-frequency components.

FIGURE 21–47
General relationship of a pulse waveform to frequency content.

THE *RC* INTEGRATOR ACTS AS A LOW-PASS FILTER

As you learned, the integrator tends to exponentially "round off" the edges of the applied pulses. This rounding off occurs to varying degrees, depending on the relationship of the time constant to the pulse width and period. The rounding off of the edges indicates that the integrator tends to reduce the higher frequency components of the pulse waveform, as illustrated in Figure 21–48.

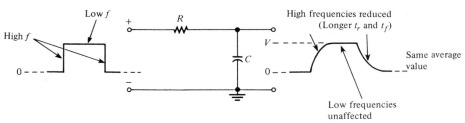

FIGURE 21–48
Time and frequency response relationship in an integrator (one pulse in a repetitive waveform shown).

THE *RL* INTEGRATOR ACTS AS A LOW-PASS FILTER

Like the *RC* integrator, the *RL* integrator also acts as a basic low-pass filter, because *L* is in series between input and output. X_L is small for low frequencies and offers little opposition. It increases with frequency, so at higher frequencies most of the total voltage is dropped across *L* and very little across *R*, the output. If the input is dc, *L* is like a short ($X_L = 0$). At high frequencies, *L* becomes like an open, as illustrated in Figure 21–49.

FIGURE 21–49
Low-pass filter action.

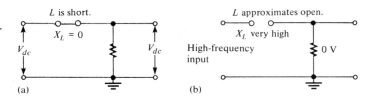

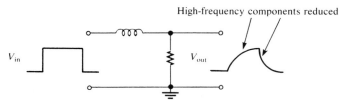

(c) High-frequency components of input pulse reduced

THE *RC* DIFFERENTIATOR ACTS AS A HIGH-PASS FILTER

As you know, the differentiator tends to introduce *tilt* to the flat portion of a pulse. That is, it tends to reduce the lower frequency components of a pulse waveform. Also, it completely eliminates the dc component of the input and produces a zero average-value output. This action is illustrated in Figure 21–50.

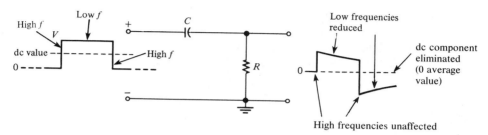

FIGURE 21–50
Time and frequency response relationship in a differentiator (one pulse in a repetitive waveform shown).

THE *RL* DIFFERENTIATOR ACTS AS A HIGH-PASS FILTER

Again like the *RC* differentiator, the *RL* differentiator also acts as a basic high-pass filter. Because L is connected across the output, less voltage is developed across it at lower frequencies than at higher ones. There are zero volts across the output for dc. For high frequencies, most of the input voltage is dropped across the output coil ($X_L = 0$ for dc; $X_L \cong$ open for high frequencies). Figure 21–51 shows high-pass filter action.

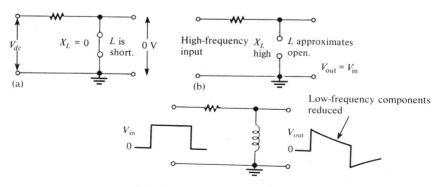

(c) Low-frequency components of input pulse reduced

FIGURE 21–51
High-pass filter action.

FORMULA RELATING RISE TIME AND FALL TIME TO FREQUENCY

It can be shown that the fast transitions of a pulse (rise and fall times) are related to the *highest frequency* component, f_h, in that pulse by the following formula:

$$t_r = \frac{0.35}{f_h} \qquad \qquad \text{(21–1)}$$

This formula also applies to fall time, and the fastest transition determines the highest frequency in the pulse waveform.

Equation (21–1) can be rearranged to give the highest frequency as follows:

$$f_h = \frac{0.35}{t_r} \qquad \qquad \text{(21–2)}$$

or

$$f_h = \frac{0.35}{t_f} \qquad \qquad \text{(21–3)}$$

EXAMPLE 21–11

What is the highest frequency contained in a pulse that has rise and fall times equal to 10 nanoseconds (10 ns)?

Solution:

$$f = \frac{0.35}{t_r} = \frac{0.35}{10 \times 10^{-9} \text{ s}} = 0.035 \times 10^9 \text{ Hz}$$

$$= 35 \times 10^6 \text{ Hz} = 35 \text{ MHz}$$

SECTION REVIEW 21–8

1. What type of filter is an integrator?
2. What type of filter is a differentiator?
3. What is the highest frequency component in a pulse waveform having t_r and t_f equal to 1 μs?

SUMMARY

1. In an *RC* integrating circuit, the output voltage is taken across the capacitor.
2. In an *RC* differentiating circuit, the output voltage is taken across the resistor.
3. In an *RL* integrating circuit, the output voltage is taken across the resistor.
4. In an *RL* differentiating circuit, the output voltage is taken across the inductor.

5. In an integrator, when the pulse width (t_W) of the input is much less than the transient time, the output voltage approaches a constant level equal to the average value of the input.

6. In an integrator, when the pulse width of the input is much greater than the transient time, the output voltage approaches the shape of the input.

7. In a differentiator, when the pulse width of the input is much less than the transient time, the output voltage approaches the shape of the input but with an average value of zero.

8. In a differentiator, when the pulse width of the input is much greater than the transient time, the output voltage consists of narrow, positive- and negative-going spikes occurring on the leading and trailing edges of the input pulses.

9. The *transient time* of an *RC* or *RL* circuit equals five time constants; $5RC$ or $5L/R$.

10. The *steady state* is the condition of a circuit after the transient time has elapsed.

11. The *dc component* of a pulse waveform is equal to its average value.

12. The rising and falling edges of a pulse waveform contain the higher frequency components.

13. The flat portion of the pulse contains the lower frequency components.

FORMULAS

$$t_r = \frac{0.35}{f_h} \qquad (21\text{--}1)$$

$$f_h = \frac{0.35}{t_r} \qquad (21\text{--}2)$$

$$f_h = \frac{0.35}{t_f} \qquad (21\text{--}3)$$

SELF-TEST

Solutions appear at the end of the book.

1. How do you distinguish an *RC* differentiating circuit from an integrating circuit?

2. In an *RC* integrator, to what voltage does the capacitor charge if the 10-V input pulse has a width equal to one time constant?

3. Repeat Question 2 for an *RC* differentiator.

4. In an *RC* integrator, under what condition does the output voltage waveform closely resemble that of the input waveform?

5. Repeat Question 4 for an *RC* differentiator.

6. Under what condition are the positive and negative portions of the differentiator output voltage of equal amplitude?

7. How do you distinguish an *RL* differentiating circuit from an integrating circuit?

8. Express the maximum possible current in an *RL* circuit.

9. Under what condition in an *RL* differentiating circuit does the current reach its maximum possible value?

10. You have two differentiating circuits sitting side-by-side. One is an *RC* and the other an *RL*. Both have equal time constants. If you apply the same pulse waveform to each circuit, how can you tell the *RC* from the *RL* by observing only their output voltages?

11. A 10-V pulse is applied to a series *RC* circuit. The pulse width equals *one* time constant. To what voltage does the capacitor charge during the pulse? Assume that it is initially uncharged.

12. Repeat Question 11 for the following values of t_W:
    (a) 2τ (b) 3τ (c) 4τ (d) 5τ

13. A 5-V pulse with a width of 5 μs is applied to an *RC* circuit with $\tau = 10$ μs. If the capacitor is initially uncharged, to how many volts does it charge by the end of the pulse?

14. The capacitor in an *RC* circuit is charged to 15 V. If it is allowed to discharge for an interval equal to one time constant, what is its voltage?

15. Sketch the approximate shape of an integrator output where 5τ is much less than the pulse width of a 10-V square-wave input. Repeat for the case in which 5τ is much larger than the pulse width.

16. Repeat Question 15 for a differentiator.

17. Determine the output voltage for an integrator with a single input pulse, as shown in Figure 21–52. For repetitive pulses, how long will it take this circuit to reach steady state?

FIGURE 21–52

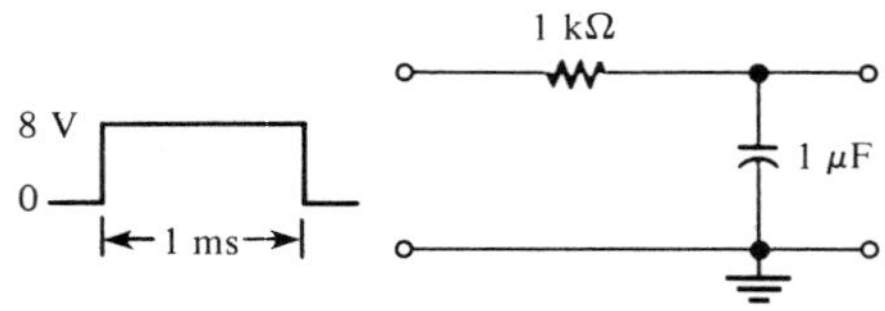

18. Make a differentiator from the circuit in Figure 21–52, and repeat the requirements of Question 17.

19. Sketch the approximate shape of an integrator output for a single input pulse when $5\tau = 0.5t_W$. Repeat for $5\tau = 5t_W$.

20. Repeat Question 19 for a differentiator.

21. Determine the output voltage for the circuit in Figure 21–53. A single input pulse is applied as shown.

FIGURE 21–53

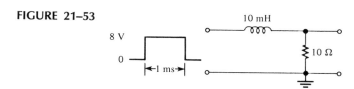

22. Convert the circuit in Figure 21–53 to a differentiator, and repeat Question 21.

23. What is the highest frequency for a fall time of 250 ns?

PROBLEMS

Section 21–1

21–1 An integrating circuit has $R = 2.2\ k\Omega$ in series with $C = 0.05\ \mu F$. What is the time constant?

21–2 Determine how long it takes the capacitor in an integrating circuit to reach full charge for each of the following series RC combinations:
(a) $R = 56\ \Omega$, $C = 50\ \mu F$ (b) $R = 3300\ \Omega$, $C = 0.015\ \mu F$
(c) $R = 22\ k\Omega$, $C = 100\ pF$ (d) $R = 5.6\ M\Omega$, $C = 10\ pF$

Section 21–2

21–3 A 20-V pulse is applied to an RC integrator. The pulse width equals *one* time constant. To what voltage does the capacitor charge during the pulse? Assume that it is initially uncharged.

21–4 Repeat Problem 21–3 for the following values of t_w:
(a) 2τ (b) 3τ (c) 4τ (d) 5τ

21–5 Sketch the approximate shape of an integrator output voltage where 5τ is much less than the pulse width of a 10-V square-wave input. Repeat for the case in which 5τ is much larger than the pulse width.

21–6 Determine the output voltage for an integrator with a single input pulse, as shown in Figure 21–54. For repetitive pulses, how long will it take this circuit to reach steady state?

FIGURE 21–54

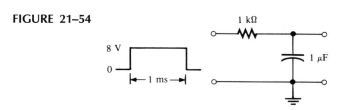

21–7 **(a)** What is τ in Figure 21–55? **(b)** Sketch the output voltage.

FIGURE 21–55

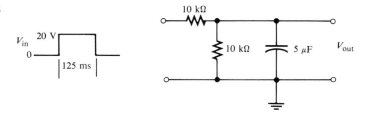

Section 21–3

21–8 Sketch the integrator output in Figure 21–56, showing maximum voltages.

FIGURE 21–56

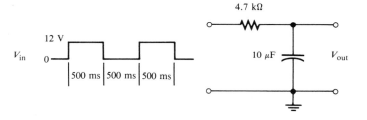

21–9 A 1-V, 10-kHz pulse waveform with a duty cycle of 25% is applied to an integrator with $\tau = 25$ μs. Graph the otuput voltage for three initial pulses. C is initially uncharged.

21–10 What is the steady state output voltage of the integrator with a square-wave input shown in Figure 21–57?

FIGURE 21–57

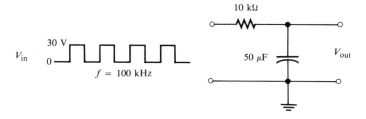

Section 21–4

21–11 Repeat Problem 21–5 for an RC differentiator.

21–12 Redraw the circuit in Figure 21–54 to make it a differentiator, and repeat Problem 21–6.

21–13 **(a)** What is τ in Figure 21–58? **(b)** Sketch the output voltage.

FIGURE 21–58

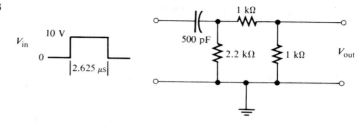

Section 21–5

21–14 Sketch the differentiator output in Figure 21–59, showing maximum voltages.

FIGURE 21–59

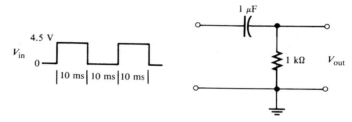

21–15 What is the steady state output voltage of the differentiator with the square-wave input shown in Figure 21–60?

FIGURE 21–60

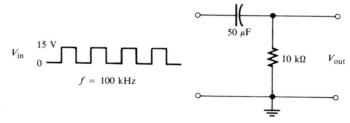

Section 21–6

21–16 Determine the output voltage for the circuit in Figure 21–61. A single input pulse is applied as shown.

FIGURE 21–61

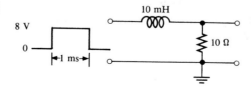

21–17 Sketch the integrator output in Figure 21–62, showing maximum voltages.

FIGURE 21–62

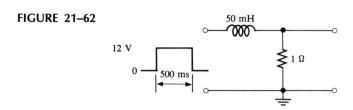

21–18 Determine the time constant in Figure 21–63. Is this circuit an integrator or a differentiator?

FIGURE 21–63

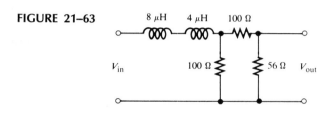

Section 21–7
21–19 **(a)** What is τ in Figure 21–64? **(b)** Sketch the output voltage.

FIGURE 21–64

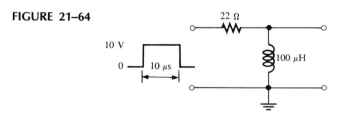

21–20 Draw the output waveform if a periodic pulse waveform with $t_W = 25$ μs and $T = 60$ μs is applied to the circuit in Figure 21–64.

Section 21–8
21–21 What is the highest frequency component in the output of an integrator with $\tau = 10$ μs? Assume that $5\tau < t_W$.

21–22 A certain pulse waveform has a rise time of 55 ns and a fall time of 42 ns. What is the highest frequency component in the waveform?

ANSWERS TO SECTION REVIEWS

Section 21–1
1. A series RC circuit in which the output is across the capacitor.
2. A voltage applied to the input causes the capacitor to charge. A short across the input causes the capacitor to discharge.

Section 21–2
1. $5\tau \le t_W$.   **2.** 0.63 V, 47 ms.   **3.** See Figure 21–65.   **4.** No.
5. $5\tau \ll t_W$ (5τ much less than t_W).

FIGURE 21–65

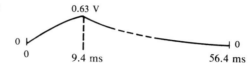

Section 21–3
1. $5\tau \le t_W$, and $5\tau \le$ time between pulses.   **2.** Like the input.
3. *transient time*.   **4.** The response after the transient time has passed.
5. The average value of the input voltage.

Section 21–4
1. See Figure 21–66.   **2.** $5\tau \gg t_W$.   **3.** Positive and negative spikes.
4. -10 V.

FIGURE 21–66

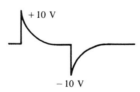

Section 21–5
1. $5\tau \le t_W$ and $5\tau \le$ time between pulses.   **2.** Positive and negative spikes.
3. 0 V.

Section 21–6
1. Resistor.   **2.** $5\tau \le t_W$.   **3.** $5\tau \ll t_W$.

Section 21–7
1. Inductor.   **2.** $5\tau \gg t_W$.   **3.** -8 V.

Section 21–8
1. Low-pass.   **2.** High-pass.   **3.** 350 kHz.

TWENTY-TWO

POLYPHASE SYSTEMS IN POWER APPLICATIONS

In our coverage of ac analysis up to this point, only *single-phase* sinusoidal sources have been considered. In Chapter 11, it was discussed how a sinusoidal voltage can be generated by rotating a conductor at a constant velocity in a magnetic field, and the basic concepts of ac generators were introduced.

In this chapter, the basic generation of polyphase sinusoidal waveforms is examined. The advantages of polyphase systems in power applications are covered, and various types of three-phase connections and power measurement are introduced.

In this chapter, you will learn:

☐ The differences among single-phase, two-phase, and three-phase generators.
☐ The advantages of polyphase voltages over single-phase.
☐ The various types of three-phase generator configurations.
☐ The relationships of currents and voltages in a three-phase system.
☐ How to analyze three-phase systems.
☐ The difference between a balanced and an unbalanced load.
☐ Two methods for measuring power in three-phase systems.
☐ How a basic power distribution system works.

22–1   BASIC POLYPHASE MACHINES

Polyphase generators produce simultaneous multiple sine waves that are separated by differing phase angles. This is accomplished by multiple loops rotating through the magnetic field. Similarly, polyphase motors operate with multiple-phase sinusoidal inputs.

A BASIC TWO-PHASE GENERATOR

Figure 22–1(a) shows a second separate conductor loop added to a basic single-phase generator with the two loops separated by 90 degrees. Both loops are mounted on the same rotor and therefore rotate at the same speed. Loop A is 90 degrees ahead of loop B in the direction of rotation. As they rotate, two induced sinusoidal voltages are produced that are 90 degrees apart in phase, as shown in Part (b) of the figure.

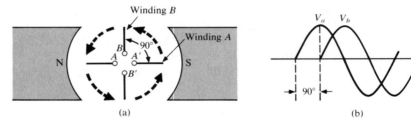

(a)   (b)

FIGURE 22–1
Basic two-phase generator.

A BASIC THREE-PHASE GENERATOR

Figure 22–2(a) shows a generator with three separate conductor loops placed at 120-degree intervals around the rotor. This configuration generates three sinusoidal voltages

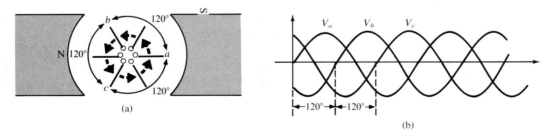

(a)   (b)

FIGURE 22–2
Basic three-phase generator.

that are separated from each other by phase angles of 120 degrees, as shown in Part (b).

A PRACTICAL THREE-PHASE GENERATOR

The purpose of using the simple ac generators (alternators) presented up to this point was to illustrate the basic concept of the generation of sinusoidal voltages from the rotation of a conductor in a magnetic field. Most practical ac generators produce three-phase ac using a configuration that is somewhat different from that previously discussed. The basic principle, however, is the same. The reasons for the use of three-phase ac are examined in the next section.

A basic two-pole, three-phase alternator is shown in Figure 22–3. Most practical alternators are of this basic form. Rather than using a permanent magnet in a fixed position, a rotating electromagnet is used. The electromagnet is created by passing direct current (I_F) through a winding around the rotor, as shown. These windings are called *field windings*. The direct current is applied through a brush and slip ring assembly. The stationary outer portion of the alternator is called the *stator*. Three separate windings are placed 120 degrees apart around the stator; three-phase voltages are induced in these windings as the magnetic field rotates, as indicated in Part (b) of Figure 22–2.

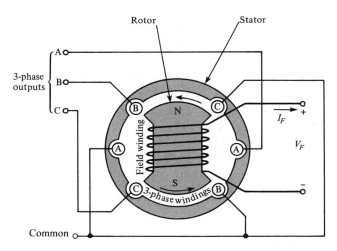

FIGURE 22–3
Basic two-pole, three-phase alternator.

A BASIC THREE-PHASE MOTOR

The most common type of ac motor is the three-phase induction motor. Basically, it consists of a stator with stator windings and a rotor assembly constructed as a cylindrical frame of metal bars arranged in a ''squirrel-cage'' type configuration. A basic end-view diagram is shown in Figure 22–4.

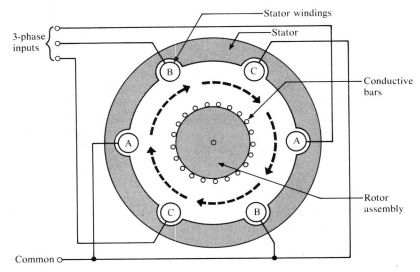

FIGURE 22–4
Basic three-phase induction motor.

When the three-phase voltages are applied to the stator windings, a rotating magnetic field is established. As the magnetic field rotates, currents are *induced* in the conductors of the squirrel-cage rotor. The interaction of the induced currents and the magnetic field produces forces that cause the rotor to also rotate.

SECTION REVIEW 22–1

1. Describe the basic principle used in ac generators.

2. A two-pole, single-phase generator rotates at 400 rps. What frequency is produced?

3. How many separate armature windings are required in a three-phase alternator?

22–2   ADVANTAGES OF POLYPHASE IN POWER APPLICATIONS

There are several advantages of using polyphase alternators to deliver power to a load over using a single-phase system to deliver power to a given load. These considerations are now discussed.

THE COPPER ADVANTAGE

A Single-Phase System   Figure 22–5 is a simplified representation of a single-phase alternator connected to a resistive load. The coil symbol represents the alternator winding.

FIGURE 22–5
Simplified representation of a single-phase alternator connected to a resistive load.

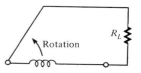

For example, a single-phase sinusoidal voltage is induced in the winding and applied to a 60-Ω load, as indicated in Figure 22–6. The resulting load current is

$$\mathbf{I}_L = \frac{120\angle0° \text{ V}}{60\angle0° \text{ }\Omega} = 2\angle0° \text{ A}$$

FIGURE 22–6
Single-phase example.

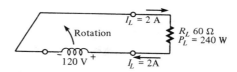

The total current that must be delivered by the alternator to the load is $2\angle0°$ A. This means that the two conductors carrying current to and from the load must *each* be capable of handling 2 A; thus, the *total copper cross section* must handle 4 A. (The copper cross section is a measure of the *total* amount of wire required based on its physical size as related to its diameter.) The total load power is $I_L^2R_L = 240$ W.

A Two-Phase System Figure 22–7 shows a simplified representation of a two-phase alternator connected to two 120-Ω load resistors. The coils drawn at a 90-degree angle represent the armature windings spaced 90 degrees apart. An equivalent single-phase system would be required to feed two 120-Ω resistors in parallel, thus creating a 60-Ω load.

FIGURE 22–7
Simplified representation of a two-phase
alternator with each phase connected to a
120-Ω load.

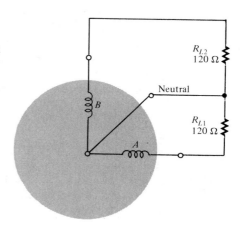

The voltage across load R_{L1} is $120\angle0°$ V, and the voltage across load R_{L2} is $120\angle90°$ V, as indicated in Figure 22–8(a). The current in R_{L1} is

$$\mathbf{I}_{L1} = \frac{120\angle0° \text{ V}}{120\angle0° \text{ }\Omega} = 1\angle0° \text{ A}$$

and the current in R_{L2} is

$$\mathbf{I}_{L2} = \frac{120\angle90° \text{ V}}{120\angle0° \text{ }\Omega} = 1\angle90° \text{ A}$$

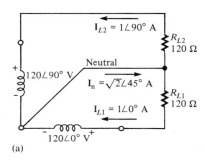
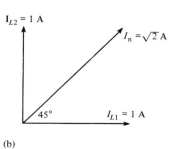

(a)   (b)

FIGURE 22–8
Two-phase example.

Notice that the two load resistors are connected to a common or neutral conductor, which provides a return path for the currents. Since the two load currents are 90 degrees out of phase with each other, the current in the neutral conductor (I_n) is the *phasor sum* of the load currents.

$$I_n = \sqrt{I_{L1}^2 + I_{L2}^2} = \sqrt{2} \text{ A} = 1.414 \text{ A}$$

Figure 22–8(b) shows the phasor diagram for the currents in the two-phase system.

Three conductors are required in this system to carry current to and from the loads. Two of the conductors must be capable of handling 1 A each, and the neutral conductor must handle 1.414 A. The *total* copper cross section must handle 1 A + 1 A + 1.414 A = 3.414 A. This is *less* copper cross section than was required in the single-phase system to deliver the same amount of total load power:

$$P_{L1} + P_{L2} = I_{L1}^2 R_{L1} + I_{L2}^2 R_{L2} = 120 \text{ W} + 120 \text{ W} = 240 \text{ W}$$

A Three-Phase System   Figure 22–9 shows a simplified representation of a three-phase alternator connected to three 180-Ω resistive loads. An equivalent single-phase system would be required to feed three 180-Ω resistors in parallel, thus creating an effective load resistance of 60 Ω. The coils represent the alternator windings separated by 120 degrees.

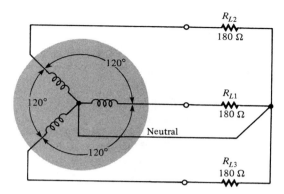

FIGURE 22–9
A simplified representation of a three-phase alternator with each phase connected to a
180-Ω load.

The voltage across R_{L1} is $120\angle0°$ V, the voltage across R_{L2} is $120\angle120°$ V,
and the voltage across R_{L3} is $120\angle-120°$ V, as indicated in Figure 22–10(a). The
current from each winding to its respective load is as follows:

$$\mathbf{I}_{L1} = \frac{120\angle0° \text{ V}}{180\angle0° \text{ }\Omega} = 0.667\angle0° \text{ A}$$

$$\mathbf{I}_{L2} = \frac{120\angle120° \text{ V}}{180\angle0° \text{ }\Omega} = 0.667\angle120° \text{ A}$$

$$\mathbf{I}_{L3} = \frac{120\angle-120° \text{ V}}{180\angle0° \text{ }\Omega} = 0.667\angle-120° \text{ A}$$

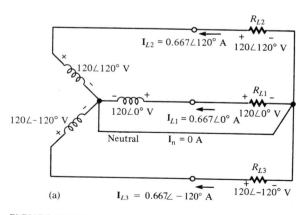

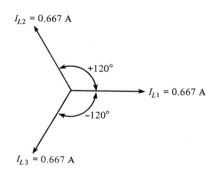

FIGURE 22–10
Three-phase example.

The total load power is

$$P_{tot} = I_{L1}^2 R_{L1} + I_{L2}^2 R_{L2} + I_{L3}^2 R_{L3} = 240 \text{ W}$$

This is the same total load power as delivered by both the single-phase and the two-phase systems previously discussed.

Notice that four conductors, including the neutral, are required to carry the currents to and from the loads. The current in each of the three conductors is 0.667 A, as indicated in the figure. The current in the neutral conductor is the phasor sum of the three load currents and is equal to *zero*, as shown in the following equation, with reference to the phasor diagram in Figure 22–10(b). This condition, where all loads are equal and the neutral current is zero, is called a *balanced load* condition.

$$\begin{aligned}
\mathbf{I}_{L1} + \mathbf{I}_{L2} + \mathbf{I}_{L3} &= 0.667\angle 0° \text{ A} + 0.667\angle 120° \text{ A} + 0.667\angle -120° \text{ A} \\
&= 0.667 \text{ A} - 0.3335 \text{ A} + j0.578 \text{ A} - 0.3335 \text{ A} - j0.578 \text{ A} \\
&= 0.667 \text{ A} - 0.667 \text{ A} \\
&= 0 \text{ A}
\end{aligned}$$

The total copper cross section must handle 0.667 A + 0.667 A + 0.667 A + 0 A = 2 A. This result shows that considerably less copper is required to deliver the same load power with a three-phase system than is required for either the single-phase or the two-phase systems. The amount of copper is an important consideration in power distribution systems.

EXAMPLE 22–1

Compare the total copper cross sections in terms of current-carrying capacity for single-phase and three-phase 120-V systems with effective load resistances of 12 Ω.

Solution:
Single-phase system: The load current is

$$I_L = \frac{120 \text{ V}}{12 \text{ Ω}} = 10 \text{ A}$$

The conductor to the load must carry 10 A, and the conductor from the load must also carry 10 A. *The total copper cross section, therefore, must be sufficient to handle 20 A.*

Three-phase system: For an effective load resistance of 12 Ω, the three-phase alternator feeds three load resistors of 36 Ω each. The current in each load resistor is

$$I_L = \frac{120 \text{ V}}{36 \text{ Ω}} = 3.33 \text{ A}$$

Each of the three conductors feeding the balanced load must carry 3.33 A, and the neutral current is zero. *Therefore, the total copper cross section must be sufficient to handle 9.99 A.* This is significantly less than for the single-phase system with an equivalent load.

THE ADVANTAGE OF CONSTANT POWER

A second advantage of polyphase systems over single-phase is that polyphase systems produce a constant amount of power in the load.

A Single-Phase System   As shown in Figure 22–11, the load power fluctuates as the square of the sinusoidal voltage divided by the resistance. It changes from a maximum of $V^2_{R(max)}/R_L$ to a minimum of zero at a frequency equal to twice that of the voltage.

FIGURE 22–11
Single-phase load power (sin² curve).

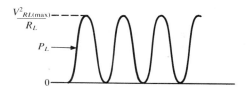

A Polyphase System   The power waveform across one of the load resistors in a two-phase system is 90 degrees out of phase with the power waveform across the other, as shown in Figure 22–12. When the instantaneous power to one load is maximum, the other is minimum. In between the maximum and minimum points, one power is increasing while the other is decreasing. Careful examination of the power waveforms shows that when two instantaneous values are added, the sum is always constant and equal to $V^2_{R(max)}/R_L$.

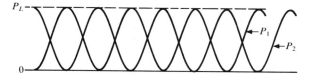

FIGURE 22–12
Two-phase power ($P_L = V^2_{RL(max)}/R_L$).

In a three-phase system, the total power delivered to the load resistors is also constant for the same basic reason as for the two-phase system. The sum of the instantaneous voltages is always the same; therefore, the power has a constant value. A constant load power means a uniform conversion of mechanical to electrical energy, which is an important consideration in many power applications.

THE ADVANTAGE OF A CONSTANT, ROTATING MAGNETIC FIELD

In many applications, ac generators are used to drive ac motors for conversion of electrical energy to mechanical energy in the form of shaft rotation in the motor. The original energy for operation of the generator can come from any of several sources such as hydroelectric, steam, and so forth. Figure 22–13 illustrates the basic concept.

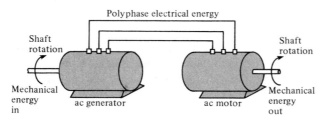

FIGURE 22–13
Simple example of mechanical-to-electrical-to-mechanical energy conversion.

When a polyphase generator is connected to the motor windings as depicted in Figure 22–14, where two-phase is used for purposes of illustration, a magnetic field is created within the motor that has a constant flux density and that rotates at the frequency of the two-phase sine wave. The motor's rotor is pulled around at a constant rotational velocity by the rotating magnetic field, producing a constant shaft rotation.

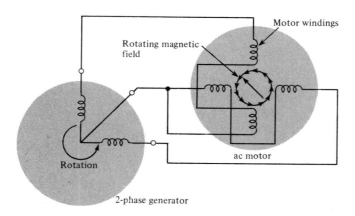

FIGURE 22–14
A two-phase generator producing a constant rotating magnetic field in an ac motor.

Three-phase, of course, has the same advantage and is used universally in the field of power distribution. Single-phase is unsuitable for such applications, because it produces a magnetic field that fluctuates in flux density and reverses direction during each cycle without providing the advantage of constant rotation.

SECTION REVIEW 22–2

1. List three advantages of polyphase systems over single-phase systems.

2. Which advantage(s) are most important in mechanical-to-electrical energy conversions?

3. Which advantage(s) are most important in electrical-to-mechanical energy conversions?

22–3 THREE-PHASE GENERATOR CONFIGURATIONS

THE Y-CONNECTED GENERATOR

In the introduction of three-phase systems in previous sections, the Y-connection was used for illustration. We now examine this configuration further. A second configuration, called the *delta* (Δ), is introduced later.

A Y-connected system can be either a *three-wire* or, when the neutral is used, a *four-wire* system, as shown in Figure 22–15, connected to a generalized load represented by the block. Recall that when the loads are perfectly balanced, the neutral current is zero; therefore, the neutral conductor is unnecessary. However, in cases where the loads are not equal (unbalanced), a neutral wire is essential to provide a return current path, because the neutral current has a nonzero value.

FIGURE 22–15
Y-connected generator.

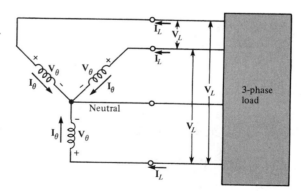

The voltages across the generator windings are called *phase voltages* (V_θ), and the currents through the windings are called *phase currents* (I_θ). Also, the currents in the lines connecting the generator windings to the load are called *line currents* (I_L), and the voltages across the lines are called the *line voltages* (V_L). Note that the magnitude of each line current is equal to the corresponding phase current in the Y-connected circuit.

$$I_L = I_\theta \tag{22–1}$$

In Figure 22–16, the line terminations of the windings are designated *a*, *b*, and *c*, and the neutral point is designated "n." These letters are added as subscripts to the

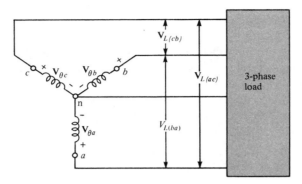

FIGURE 22–16
Phase voltages and line voltages in a Y-connected system.

phase and line currents to indicate the phase with which each is associated. The phase voltages are also designated in the same manner. Notice that the phase voltages are always positive at the terminal end of the winding and are negative at the neutral point. The line voltages are from one winding terminal to another, as indicated by the double-letter subscripts. For example, $V_{L(ba)}$ is the line voltage from b to a.

Figure 22–17(a) shows a phasor diagram for the phase voltages. By rotation of the phasors, as shown in Part (b), $V_{\theta a}$ is given a reference angle of zero, and the polar expressions for the phasor voltages are as follows.

$$\mathbf{V}_{\theta a} = V_{\theta a}\angle 0°$$

$$\mathbf{V}_{\theta b} = V_{\theta b}\angle 120°$$

$$\mathbf{V}_{\theta c} = V_{\theta c}\angle -120°$$

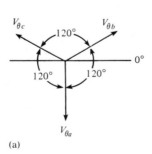

(a)

(b)

FIGURE 22–17
Phase voltage diagram.

There are three line voltages: one between a and b, one between a and c, and another between b and c. It can be shown that the magnitude of each line voltage is equal to $\sqrt{3}$ times the magnitude of the phase voltage and that there is a phase angle of 30 degrees between each line voltage and the nearest phase voltage.

$$V_L = \sqrt{3} \, V_\theta \tag{22-2}$$

Since all phase voltages are equal in magnitude,

$$\mathbf{V}_{L(ba)} = \sqrt{3} \, V_\theta \angle 150°$$

$$\mathbf{V}_{L(ac)} = \sqrt{3} \, V_\theta \angle 30°$$

$$\mathbf{V}_{L(cb)} = \sqrt{3} \, V_\theta \angle -90°$$

The line voltage phasor diagram is shown in Figure 22–18 superimposed on the phasor diagram for the phase voltages. Notice that there is a phase angle of 30 degrees between each line voltage and the nearest phase voltage and that the line voltages are 120 degrees apart.

FIGURE 22–18

Phasor diagram for the phase voltages and line voltages in a Y-connected, three-phase system.

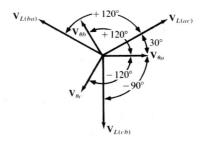

EXAMPLE 22–2

The instantaneous position of a certain Y-connected ac generator is shown in Figure 22–19. If each phase voltage has a magnitude of 120 V rms, determine the magnitude of each line voltage, and sketch the phasor diagram.

FIGURE 22–19

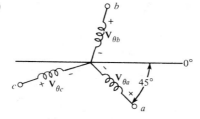

Solution:
The magnitude of each line voltage is

$$V_L = \sqrt{3} \, V_\theta = \sqrt{3} \, (120 \text{ V}) = 207.85 \text{ V}$$

The phasor diagram for the given instantaneous generator position is shown in Figure 22–20.

FIGURE 22–20

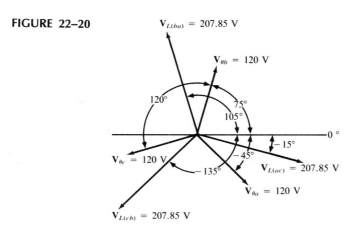

$V_{L(ba)} = 207.85$ V

$V_{\theta b} = 120$ V

$V_{\theta c} = 120$ V

$V_{\theta a} = 120$ V

$V_{L(ac)} = 207.85$ V

$V_{L(cb)} = 207.85$ V

THE Δ-CONNECTED GENERATOR

In the Y-connected generator, two voltage magnitudes are available at the terminals in the four-wire system: the *phase voltage* and the *line voltage*. Also, in the Y-connected generator, the line current is equal to the phase current. Keep these characteristics in mind as you examine the Δ-connected generator.

The windings of a three-phase generator can be rearranged to form a Δ-connected generator, as shown in Figure 22–21. By examination of this diagram, it is apparent that the magnitudes of the line voltages and phase voltages are equal, but the line currents do not equal the phase currents.

FIGURE 22–21
Δ-connected generator.

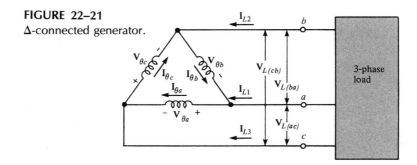

Since this is a three-wire system, only a single voltage *magnitude* is available, expressed as:

$$V_L = V_\theta$$

(22–3)

All of the phase voltages are equal in magnitude; thus, the line voltages are expressed in polar form as follows:

$$\mathbf{V}_{L(ac)} = V_\theta \angle 0°$$

$$\mathbf{V}_{L(ba)} = V_\theta \angle 120°$$

$$\mathbf{V}_{L(cb)} = V_\theta \angle -120°$$

The phasor diagram for the phase currents is shown in Figure 22–22, and the polar expressions for each current are as follows:

$$\mathbf{I}_{\theta a} = I_{\theta a} \angle 0°$$

$$\mathbf{I}_{\theta b} = I_{\theta b} \angle 120°$$

$$\mathbf{I}_{\theta c} = I_{\theta c} \angle -120°$$

FIGURE 22–22
Phase current diagram for the Δ-connected system.

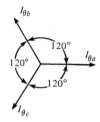

It can be shown that the magnitude of each line current is equal to $\sqrt{3}$ times the magnitude of the phase current and that there is a phase angle of 30 degrees between each line current and the nearest phase current.

$$I_L = \sqrt{3}\, I_\theta \qquad\qquad (22\text{--}4)$$

Since all phase currents are equal in magnitude,

$$\mathbf{I}_{L1} = \sqrt{3}\, I_\theta \angle -30°$$

$$\mathbf{I}_{L2} = \sqrt{3}\, I_\theta \angle 90°$$

$$\mathbf{I}_{L3} = \sqrt{3}\, I_\theta \angle -150°$$

The current phasor diagram is shown in Figure 22–23.

FIGURE 22–23
Phasor diagram of phase currents and line currents.

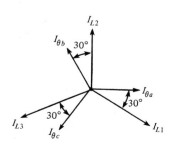

EXAMPLE 22–3

The three-phase Δ-connected generator represented in Figure 22–24 is driving a balanced load such that each phase current is 10 A in magnitude. When $\mathbf{I}_{\theta a} = 10\angle30°$ A, determine the following:

(a) The polar expressions for the other phase currents.

(b) The polar expressions for each of the line currents.

(c) The complete current phasor diagram.

FIGURE 22–24

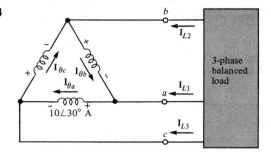

Solution:

(a) The phase currents are separated by 120 degrees; therefore

$$\mathbf{I}_{\theta b} = 10\angle(30° + 120°) = 10\angle150° \text{ A}$$

$$\mathbf{I}_{\theta c} = 10\angle(30° - 120°) = 10\angle-90° \text{ A}$$

(b) The line currents are separated from the nearest phase current by 30 degrees; therefore

$$\mathbf{I}_{L1} = \sqrt{3}\, I_{\theta a}\angle(30° - 30°) = 17.32\angle0° \text{ A}$$

$$\mathbf{I}_{L2} = \sqrt{3}\, I_{\theta b}\angle(150° - 30°) = 17.32\angle120° \text{ A}$$

$$\mathbf{I}_{L3} = \sqrt{3}\, I_{\theta c}\angle(-90° - 30°) = 17.32\angle-120° \text{ A}$$

(c) The phasor diagram is shown in Figure 22–25.

FIGURE 22–25

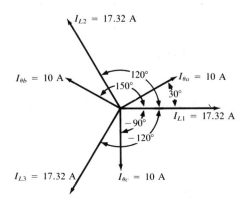

SECTION REVIEW 22–3

1. In a certain three-wire, Y-connected generator, the phase voltages are 1 kV. Determine the magnitude of the line voltages.

2. In the Y-connected generator mentioned in Question 1, all the phase currents are 5 A. What are the line current magnitudes?

3. In a Δ-connected generator, the phase voltages are 240 V. What are the line voltages?

4. In a Δ-connected generator, a phase current is 2 A. Determine the magnitude of the line current.

22–4

ANALYSIS OF THREE-PHASE SOURCE/LOAD CONFIGURATIONS

As with the generator connections, the load can be either in a Y or a Δ configuration. A Y-connected load is shown in Figure 22–26(a), and a Δ-connected load is shown in Part (b). The blocks represent the load impedances, which can be resistive, reactive, or both.

FIGURE 22–26
Three-phase loads.

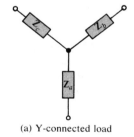

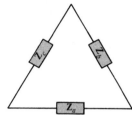

       (a) Y-connected load   (b) Δ-connected load

In this section, four source/load configurations are examined: a Y-connected source driving a Y-connected load (Y-Y system), a Y-connected source driving a Δ-connected load (Y-Δ system), a Δ-connected source driving a Δ-connected load (Δ-Δ system), and a Δ-connected source driving a Y-connected load (Δ-Y system).

THE Y-Y SYSTEM

Figure 22–27 shows a Y-connected source driving a Y-connected load. The load can be a balanced load, such as a three-phase motor where $Z_a = Z_b = Z_c$, or it can be three independent single-phase loads where, for example, Z_a is a lighting circuit, Z_b is a heater, and Z_c is an air-conditioning compressor.

    An important feature of a Y-connected source is that two different values of three-phase voltage are available: the phase voltage and the line voltage. For example, in the standard power distribution system, a three-phase transformer can be considered a source of three-phase voltage supplying 120 V and 208 V. In order to utilize a phase

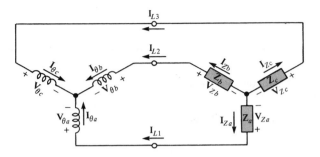

FIGURE 22–27
A Y-connected source feeding a Y-connected load.

voltage of 120 V, the loads are connected in the Y configuration. Later, you will see that a Δ-connected load is used for the 208-V line voltages.

Notice in the Y-Y system in Figure 22–27 that the phase current, the line current, and the load current are all equal in each phase. Also, each load voltage equals the corresponding phase voltage. These relationships are expressed as follows and are true for either a balanced or an unbalanced load.

$$I_\theta = I_L = I_Z \qquad \text{(22–5)}$$

$$V_\theta = V_Z \qquad \text{(22–6)}$$

where V_Z and I_Z are the load voltage and current.

For a balanced load, all the phase currents are equal, and the neutral current is zero. For an unbalanced load, each phase current is different, and the neutral current is, therefore, nonzero.

EXAMPLE 22–4

In the Y-Y system of Figure 22–28, determine the following:

(a) Each load current   (b) Each line current
(c) Each phase current   (d) Neutral current
(e) Each load voltage

FIGURE 22–28

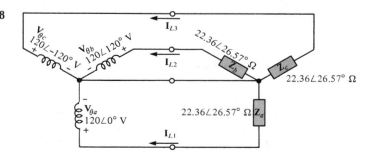

Solution:
This system has a balanced load.

(a) $\mathbf{Z}_a = \mathbf{Z}_b = \mathbf{Z}_c = 20\ \Omega + j10\ \Omega = 22.36\angle 26.57°\ \Omega$

$$\mathbf{I}_{Za} = \frac{\mathbf{V}_{\theta a}}{\mathbf{Z}_a} = \frac{120\angle 0°\ \text{V}}{22.36\angle 26.57°\ \Omega} = 5.37\angle -26.57°\ \text{A}$$

$$\mathbf{I}_{Zb} = \frac{\mathbf{V}_{\theta b}}{\mathbf{Z}_b} = \frac{120\angle 120°\ \text{V}}{22.36\angle 26.57°\ \Omega} = 5.37\angle 93.43°\ \text{A}$$

$$\mathbf{I}_{Zc} = \frac{\mathbf{V}_{\theta c}}{\mathbf{Z}_c} = \frac{120\angle -120°\ \text{V}}{22.36\angle 26.57°\ \Omega} = 5.37\angle -146.57°\ \text{A}$$

(b) $\mathbf{I}_{L1} = 5.37\angle -26.57°\ \text{A}$

$\mathbf{I}_{L2} = 5.37\angle 93.43°\ \text{A}$

$\mathbf{I}_{L3} = 5.37\angle -146.57°\ \text{A}$

(c) $\mathbf{I}_{\theta a} = 5.37\angle -26.57°\ \text{A}$

$\mathbf{I}_{\theta b} = 5.37\angle 93.43°\ \text{A}$

$\mathbf{I}_{\theta c} = 5.37\angle -146.57°\ \text{A}$

(d) $\mathbf{I}_n = \mathbf{I}_{Za} + \mathbf{I}_{Zb} + \mathbf{I}_{Zc}$

$= 5.37\angle -26.57°\ \text{A} + 5.37\angle 93.43°\ \text{A} + 5.37\angle -146.57°\ \text{A}$

$= (4.8\ \text{A} - j2.4\ \text{A}) + (-4.48\ \text{A} - j2.96\ \text{A})$

$\quad + (-0.32\ \text{A} + j5.36\ \text{A})$

$= 0\ \text{A}$

If the load impedances were not equal (balanced load), the neutral current would have a *nonzero* value.

(e) The load voltages are equal to the corresponding source phase voltages:

$$\mathbf{V}_{Za} = 120\angle 0°\ \text{V}$$

$$\mathbf{V}_{Zb} = 120\angle 120°\ \text{V}$$

$$\mathbf{V}_{Zc} = 120\angle -120°\ \text{V}$$

THE Y-Δ SYSTEM

Figure 22–29 shows a Y-connected source feeding a Δ-connected load. An important feature of this configuration is that each phase of the load has the full *line voltage* across it.

$$V_Z = V_L \tag{22-7}$$

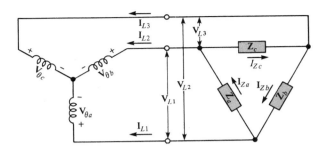

FIGURE 22–29
A Y-connected source feeding a Δ-connected load.

The line currents equal the corresponding phase currents, and each line current divides into two load currents, as indicated. For a *balanced* load ($Z_a = Z_b = Z_c$), the expression for the current in each load is

$$I_L = \sqrt{3}\, I_Z \qquad\qquad\qquad\qquad \textbf{(22–8)}$$

EXAMPLE 22–5

Determine the load voltages and load currents in Figure 22–30, and show their relationship in a phasor diagram.

FIGURE 22–30

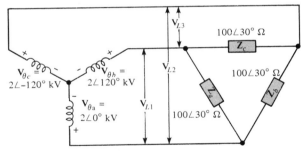

Solution:
$$\mathbf{V}_{Za} = \mathbf{V}_{L1} = 2\sqrt{3}\angle 150°\text{ kV} = 3.46\angle 150°\text{ kV}$$
$$\mathbf{V}_{Zb} = \mathbf{V}_{L2} = 2\sqrt{3}\angle 30°\text{ kV} = 3.46\angle 30°\text{ kV}$$
$$\mathbf{V}_{Zc} = \mathbf{V}_{L3} = 2\sqrt{3}\angle -90°\text{ kV} = 3.46\angle -90°\text{ kV}$$

The load currents are
$$\mathbf{I}_{Za} = \frac{\mathbf{V}_{Za}}{\mathbf{Z}_a} = \frac{3.46\angle 150°\text{ kV}}{100\angle 30°\text{ }\Omega} = 34.6\angle 120°\text{ A}$$

$$\mathbf{I}_{Zb} = \frac{\mathbf{V}_{Zb}}{\mathbf{Z}_b} = \frac{3.46\angle 30°\text{ kV}}{100\angle 30°\text{ }\Omega} = 34.6\angle 0°\text{ A}$$

$$\mathbf{I}_{Zc} = \frac{\mathbf{V}_{Zc}}{\mathbf{Z}_c} = \frac{3.46\angle -90°\text{ kV}}{100\angle 30°\text{ }\Omega} = 34.6\angle -120°\text{ A}$$

The phasor diagram is shown in Figure 22–31.

FIGURE 22–31

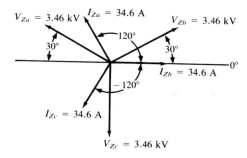

THE Δ-Y SYSTEM

Figure 22–32 shows a Δ-connected source feeding a Y-connected balanced load. By examination of the figure, you can see that the line voltages are equal to the corresponding phase voltages of the source. Also, each phase voltage equals the difference of the corresponding load voltages, as you can see by the polarities.

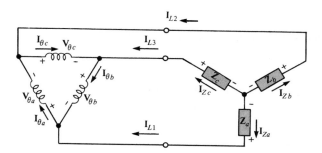

FIGURE 22–32
A Δ-connected source feeding a Y-connected load.

Each load current equals the corresponding line current. The sum of the load currents is zero, because the load is balanced; thus, there is no need for a neutral return.

The relationship between the load voltages and the corresponding phase voltages (and line voltages) is

$$V_\theta = \sqrt{3} \, V_Z \qquad\qquad (22\text{–}9)$$

The line currents and corresponding load currents are equal, and for a balanced load, the sum of the load currents is zero.

$$\mathbf{I}_L = \mathbf{I}_Z \qquad\qquad (22\text{–}10)$$

As you can see in Figure 22–32, each line current is the difference of two phase currents.

$$\mathbf{I}_{L1} = \mathbf{I}_{\theta a} - \mathbf{I}_{\theta b}$$

$$\mathbf{I}_{L2} = \mathbf{I}_{\theta c} - \mathbf{I}_{\theta a}$$

$$\mathbf{I}_{L3} = \mathbf{I}_{\theta b} - \mathbf{I}_{\theta c}$$

EXAMPLE 22–6

Determine the currents and voltages in the balanced load and the magnitude of the line voltages in Figure 22–33.

FIGURE 22–33

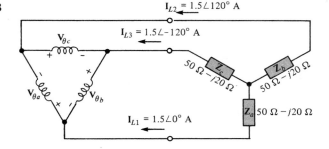

Solution:
The load currents equal the specified line currents.

$$\mathbf{I}_{Za} = \mathbf{I}_{L1} = 1.5\angle 0° \text{ A}$$

$$\mathbf{I}_{Zb} = \mathbf{I}_{L2} = 1.5\angle 120° \text{ A}$$

$$\mathbf{I}_{Zc} = \mathbf{I}_{L3} = 1.5\angle -120° \text{ A}$$

The load voltages are

$$
\begin{aligned}
\mathbf{V}_{Za} &= \mathbf{I}_{Za}\mathbf{Z}_a \\
&= (1.5\angle 0° \text{ A})(50 \text{ } \Omega - j20 \text{ } \Omega) \\
&= (1.5\angle 0° \text{ A})(53.85\angle -21.8° \text{ } \Omega) \\
&= 80.78\angle -21.8° \text{ V} \\
\mathbf{V}_{Zb} &= \mathbf{I}_{Zb}\mathbf{Z}_b \\
&= (1.5\angle 120° \text{ A})(53.85\angle -21.8° \text{ } \Omega) \\
&= 80.78\angle 98.2° \text{ V} \\
\mathbf{V}_{Zc} &= \mathbf{I}_{Zc}\mathbf{Z}_c \\
&= (1.5\angle -120° \text{ A})(53.85\angle -21.8° \text{ } \Omega) \\
&= 80.78\angle -141.8° \text{ V}
\end{aligned}
$$

The magnitude of the line voltages is

$$V_L = V_\theta = \sqrt{3}\, V_Z = \sqrt{3}\,(80.78\ \text{V}) = 139.92\ \text{V}$$

THE Δ-Δ SYSTEM

Figure 22–34 shows a Δ-connected source driving a Δ-connected load. Notice that the load voltage, line voltage, and source phase voltage are all equal for a given phase.

$$V_{\theta a} = V_{L1} = V_{Za}$$
$$V_{\theta b} = V_{L2} = V_{Zb}$$
$$V_{\theta c} = V_{L3} = V_{Zc}$$

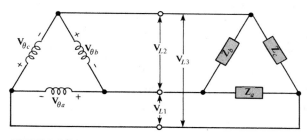

FIGURE 22–34
A Δ-connected source feeding a Δ-connected load.

Of course, when the load is balanced, all the voltages are equal, and a general expression can be written.

$$V_\theta = V_L = V_Z \qquad\qquad (22\text{–}11)$$

For a balanced load and equal source phase voltages, it can be shown that

$$I_L = \sqrt{3}\, I_Z \qquad\qquad (22\text{–}12)$$

EXAMPLE 22–7

Determine the magnitude of the load currents and the line currents in Figure 22–35.

FIGURE 22–35

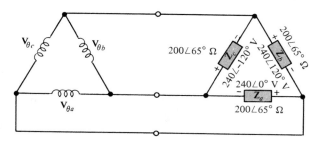

Solution:
$$V_{Za} = V_{Zb} = V_{Zc} = 240 \text{ V}$$

The magnitude of the load currents is

$$I_{Za} = I_{Zb} = I_{Zc} = \frac{V_{Za}}{Z_a}$$

$$= \frac{240 \text{ V}}{200 \ \Omega} = 1.2 \text{ A}$$

The magnitude of the line currents is

$$I_r = \sqrt{3} \, I_Z = \sqrt{3} \, (1.2 \text{ A}) = 2.08 \text{ A}$$

SECTION REVIEW 22–4

1. List the four types of three-phase source/load configurations.

2. In a certain Y-Y system, the source phase currents each have a magnitude of 3.5 A. What is the magnitude of each load current for a balanced load condition?

3. In a given Y-Δ system, $V_L = 220$ V. Determine V_Z.

4. Determine the line voltages in a balanced Δ-Y system when the magnitude of the source phase voltages is 60 V.

5. Determine the magnitude of the load currents in a balanced Δ-Δ system having a line current magnitude of 3.2 A.

22–5

POWER IN THREE-PHASE SYSTEMS

Each phase of a *balanced* three-phase load has an equal amount of power. Therefore, the total true load power is three times the power in each phase of the load.

$$P_T = 3V_Z I_Z \cos \theta \qquad (22\text{–}13)$$

where V_Z and I_Z are the voltage and current associated with each phase of the load, and $\cos \theta$ is the power factor.

Recall that in a balanced Y-connected system,

$$V_L = \sqrt{3} \, V_Z \quad \text{and} \quad I_L = I_Z$$

and in a balanced Δ-connected system,

$$V_L = V_Z \quad \text{and} \quad I_L = \sqrt{3} \, I_Z$$

When either of these relationships is substituted into Equation (22–13), the total true power for both Y- and Δ-connected systems is

$$P_T = \sqrt{3} \, V_L I_L \cos \theta \qquad (22\text{–}14)$$

**EXAMPLE
22–8**

In a certain Δ-connected balanced load, the line voltages are 250 V and the imped-ances are $50∠30°$ Ω. Determine the total load power.

Solution:
In a Δ-connected system, $V_Z = V_L$ and $I_L = \sqrt{3} I_Z$. The load current magnitudes are

$$I_Z = \frac{V_Z}{Z} = \frac{250 \text{ V}}{50 \text{ Ω}} = 5 \text{ A}$$

and

$$I_L = \sqrt{3} I_Z = \sqrt{3} (5) \text{ A} = 8.66 \text{ A}$$

The power factor is

$$\cos \theta = \cos 30° = 0.866$$

The total power is

$$P_T = \sqrt{3} V_L I_L \cos \theta$$
$$= \sqrt{3} (250 \text{ V})(8.66 \text{ A})(0.866) = 3247 \text{ W}$$

POWER MEASUREMENT

Power is measured in three-phase systems using wattmeters. The wattmeter uses a basic electrodynamometer-type movement consisting of two coils. One coil is used to measure the current, and the other is used to measure the voltage. The needle of the meter is deflected proportionally to the current through a load and the voltage across the load, thus indicating power. Figure 22–36 shows a basic wattmeter symbol and the connec-tions for measuring power in a load. The resistor in series with the voltage coil limits the current through the coil to a small amount proportional to the voltage across the coil.

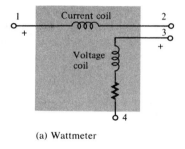

(a) Wattmeter

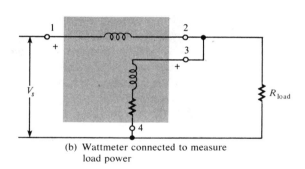

(b) Wattmeter connected to measure load power

FIGURE 22–36

Three-Wattmeter Method Power can be measured easily in a balanced or unbalanced three-phase load of either the Y or the Δ type by using three wattmeters connected as shown in Figure 22–37. This is sometimes known as the *three-wattmeter method*.

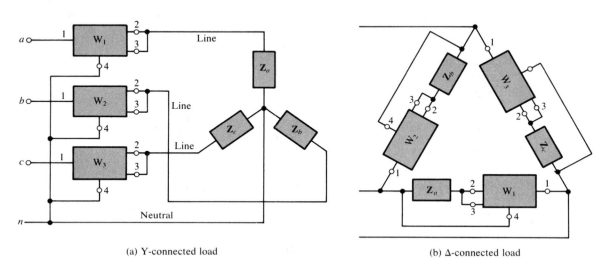

(a) Y-connected load

(b) Δ-connected load

FIGURE 22–37
Three-wattmeter method of power measurement.

The total power is determined by summing the three wattmeter readings.

$$P_T = P_1 + P_2 + P_3 \qquad (22\text{–}15)$$

If the load is balanced, the total power is simply three times the reading on any one wattmeter.

In many three-phase loads, particularly the Δ configuration, it is difficult to connect a wattmeter such that the voltage coil is across the load or such that the current coil is in series with the load because of inaccessibility of points within the load.

Two-Wattmeter Method Another method of three-phase power measurement uses only two wattmeters. The connections for this *two-wattmeter method* are shown in Figure 22–38. Notice that the voltage coil of each wattmeter is connected across a line voltage and that the current coil has a line current flowing through it. It can be shown that the *algebraic* sum of the two wattmeter readings equals the total power in the Y- or Δ-connected load.

$$P_T = P_1 \pm P_2 \qquad (22\text{–}16)$$

FIGURE 22–38
Two-wattmeter method.

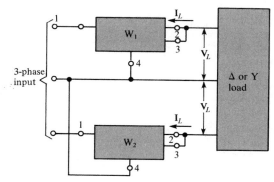

SECTION REVIEW 22–5

1. $V_L = 30$ V, $I_L = 1.2$ A, and the power factor is 0.257. What is the total power in a balanced Y-connected load? In a balanced Δ-connected load?

2. Three wattmeters connected to measure the power in a certain balanced load indicate a total of 2678 W. How much power does each meter measure?

22–6   ## POWER TRANSMISSION AND DISTRIBUTION SYSTEMS

Three-phase systems are used for the transmission and distribution of power to industrial and residential users. In this section, a typical power distribution system and its components are examined.

THREE-PHASE TRANSFORMERS

Three-phase transformers are used to step up voltages for energy transmission to a customer and to step down voltages to a required level at the point of energy use.

The three-phase windings can be connected in any one of the four configurations discussed in the last section: Y-Y, Y-Δ, Δ-Y, and Δ-Δ. In Figure 22–39, the left group of windings represents the primary, and the right group of windings represents the secondary.

In a typical power distribution system, a Y-connected generator is connected to a step-up transformer with a Δ-connected primary and a Y-connected secondary. The Y-connected secondary acts as a source to feed a high-voltage transmission line. Often, several step-up transformers are used to raise the voltage to the transmission line level. At the end of the transmission line, step-down transformers are used to reduce the high voltage for local distribution. For example, a Y-connected secondary can supply a user with 120-V single-phase and 208-V three-phase as illustrated in Figure 22–40.

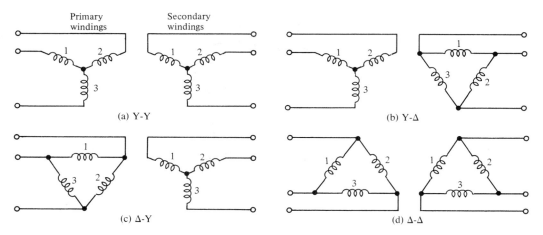

FIGURE 22–39
Three-phase transformers.

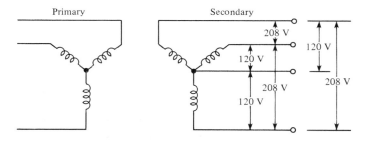

FIGURE 22–40
Example of three-phase transformer voltages.

A typical complete power generation and distribution system is shown in Figure 22–41.

SECTION REVIEW 22–6

1. List the major components of a power distribution system.

2. Does the secondary of a power transformer feeding a transmission line act as a source or a load to the transmission line?

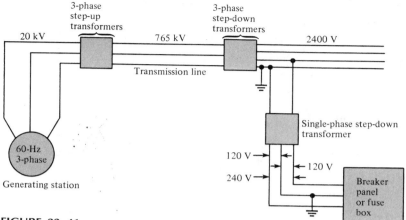

FIGURE 22–41
Simplified power generation and distribution system. Voltages shown are approximate magnitudes for a typical system.

SUMMARY

1. A simple two-phase generator consists of two conductive loops, separated by 90 degrees, rotating in a magnetic field.

2. A simple three-phase generator consists of three conductive loops separated by 120 degrees.

3. Three advantages of polyphase systems over single-phase systems are a smaller copper cross section for the same power delivered to the load, constant power delivered to the load, and a constant, rotating magnetic field.

4. In a Y-connected generator, $I_L = I_\theta$ and $V_L = \sqrt{3}\, V_\theta$.

5. In a Y-connected generator, there is a 30-degree difference between each line voltage and the nearest phase voltage.

6. In a Δ-connected generator, $V_L = V_\theta$ and $I_L = \sqrt{3}\, I_\theta$.

7. In a Δ-connected generator, there is a 30-degree difference between each line current and the nearest phase current.

8. A balanced load is one in which all the impedances are equal.

9. Power is measured in a three-phase load using either the three-wattmeter method or the two-wattmeter method.

FORMULAS

Y Generator

$$I_L = I_\theta \tag{22-1}$$

$$V_L = \sqrt{3}\, V_\theta \tag{22-2}$$

Δ Generator

$$V_L = V_\theta \tag{22-3}$$

$$I_L = \sqrt{3}\, I_\theta \tag{22-4}$$

Y-Y System

$$I_\theta = I_L = I_Z \tag{22-5}$$

$$V_\theta = V_Z \tag{22-6}$$

Y-Δ System

$$V_Z = V_L \tag{22-7}$$

$$I_L = \sqrt{3}\, I_Z \tag{22-8}$$

Δ-Y System

$$V_\theta = \sqrt{3}\, V_Z \tag{22-9}$$

$$\mathbf{I_L = I_Z} \tag{22-10}$$

Δ-Δ System

$$V_\theta = V_L = V_Z \tag{22-11}$$

$$I_L = \sqrt{3}\, I_Z \tag{22-12}$$

Three-Phase Power

$$P_T = 3V_Z I_Z \cos\theta \tag{22-13}$$

$$P_T = \sqrt{3}\, V_L I_L \cos\theta \tag{22-14}$$

Three-Wattmeter Method

$$P_T = P_1 + P_2 + P_3 \tag{22-15}$$

Two-Wattmeter Method

$$P_T = P_1 \pm P_2 \tag{22-16}$$

SELF-TEST

1. Describe the purpose of the field windings in an ac generator.

2. A single-phase voltage of $30\angle0°$ V is applied across a load impedance of $15\angle22°$ Ω. Determine the load current in polar form.

3. The phase current produced by a certain Y-connected generator is 12 A. What is the corresponding line current?

4. If each phase voltage in a Y-connected generator is 180 V, what is the magnitude of the line voltages?

5. A certain Δ-connected generator produces phase voltages of 30 V. What are the line voltages in terms of their magnitude?

6. A loaded Δ-connected generator produces phase currents of 5 A. Determine the line currents.

7. A certain Y-Y system produces phase currents of 15 A. Determine the line current and load current magnitudes.

8. The phase voltages produced by the generator in a certain Y-Y system have a magnitude of 1.5 kV. Determine the magnitudes of the line voltages and the load voltages.

9. In a particular Y-Δ system, the load voltages are each 75 V in magnitude, and the load currents are 3.3 A. Determine the line voltages and currents.

10. The source phase voltages of a Δ-Y system are 220 V. Find the load voltage magnitude.

11. A balanced three-phase system exhibits a value of load voltage and load current of 145 V and 2.5 A, respectively. Determine the total load power when $\theta = 30°$.

PROBLEMS

Section 22–1

22–1 The output of an ac generator has a maximum value of 250 V. At what angle is the instantaneous value equal to 75 V?

22–2 A certain two-pole generator has a speed of rotation of 3600 rpm. What is the frequency of the voltage produced by this generator?

Section 22–2

22–3 A single-phase generator feeds a load consisting of a 200-Ω resistor and a capacitor with a reactance of 175 Ω. The generator produces a voltage of 100 V. Determine the magnitude of the load current and its phase relation to the generator voltage.

22–4 In a two-phase system, the two currents through the lines connecting the generator to the load are 3.8 A. Determine the current in the neutral line.

22–5 A certain three-phase unbalanced load in a four-wire system has currents of $2\angle20°$ A, $3\angle140°$ A, and $1.5\angle-100°$ A. Determine the current in the neutral line.

Section 22–3

22–6 Determine the line voltages in Figure 22–42.

22–7 Determine the line currents in Figure 22–43.

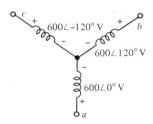

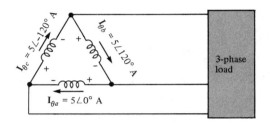

FIGURE 22–42   **FIGURE 22–43**

22–8 Develop a complete current phasor diagram for Figure 22–43.

Section 22–4

22–9 Determine the following quantities for the Y-Y system in Figure 22–44:
(a) Line voltages   (b) Phase currents   (c) Line currents
(d) Load currents   (e) Load voltages

22–10 Repeat Problem 22–9 for the system in Figure 22–45, and also find the neutral current.

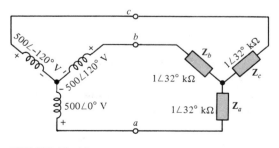

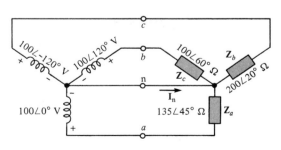

FIGURE 22–44   **FIGURE 22–45**

22–11 Repeat Problem 22–9 for the system in Figure 22–46.

22–12 Repeat Problem 22–9 for the system in Figure 22–47.

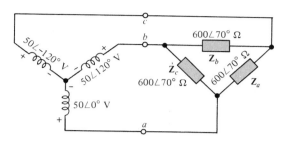

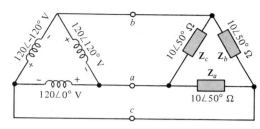

FIGURE 22–46

FIGURE 22–47

22–13 Determine the line voltages and load currents for the system in Figure 22–48.

FIGURE 22–48

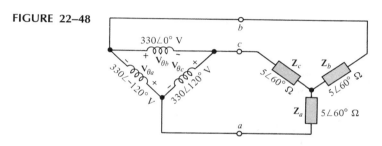

Section 22–5

22–14 The power in each phase of a balanced three-phase system is 1200 W. What is the total power?

22–15 Determine the load power in Figures 22–44 through 22–48.

22–16 Find the total load power in Figure 22–49.

FIGURE 22–49

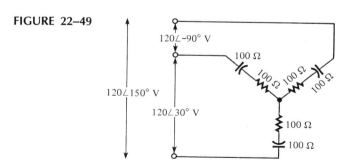

22–17 Using the three-wattmeter method for the system in Figure 22–49, how much power does each wattmeter indicate?

22–18 Repeat Problem 22–17 using the two-wattmeter method.

Section 22–6

22–19 A certain Y-Y three-phase step-up transformer has a turns ratio of 10 for each phase. If the input line voltages are 100 V, what voltages are available on the outputs?

22–20 Convert the Y-connected load in Figure 22–50 to an equivalent Δ-connected load.

FIGURE 22–50

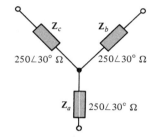

ANSWERS TO SECTION REVIEWS

Section 22–1

1. A sinusoidal voltage is induced when a conductive loop is rotated in a magnetic field at a constant speed.
2. 400 Hz. **3.** Three.

Section 22–2

1. Less copper cross section to conduct current, constant power to load, and constant, rotating magnetic field.
2. Constant power. **3.** Constant field.

Section 22–3

1. 1.73 kV. **2.** 5 A. **3.** 240 V. **4.** 3.46 A.

Section 22–4

1. Y-Y, Y-Δ, Δ-Y, and Δ-Δ. **2.** 3.5 A. **3.** 220 V. **4.** 60 V.
5. 1.85 A.

Section 22–5

1. 16.025 W, 16.025 W. **2.** 892.67 W.

Section 22–6

1. Generator, three-phase transformers, transmission line, load. **2.** Source.

APPENDIX A

WIRE SIZES

Wires are the most common form of conductive material used in electrical applications. They vary in diameter size and are arranged according to standard *gage numbers,* called *American Wire Gage* (AWG) sizes. The larger the gage number is, the smaller the wire diameter is. The AWG sizes are listed in Table A–1.

TABLE A–1
American Wire Gage (AWG) sizes for solid round copper.

AWG #	Area (CM)	Ω/1000 ft at 20°C	AWG #	Area (CM)	Ω/1000 ft at 20°C
0000	211,600	0.0490	19	1,288.1	8.051
000	167,810	0.0618	20	1,021.5	10.15
00	133,080	0.0780	21	810.10	12.80
0	105,530	0.0983	22	642.40	16.14
1	83,694	0.1240	23	509.45	20.36
2	66,373	0.1563	24	404.01	25.67
3	52,634	0.1970	25	320.40	32.37
4	41,742	0.2485	26	254.10	40.81
5	33,102	0.3133	27	201.50	51.47
6	26,250	0.3951	28	159.79	64.90
7	20,816	0.4982	29	126.72	81.83
8	16,509	0.6282	30	100.50	103.2
9	13,094	0.7921	31	79.70	130.1
10	10,381	0.9989	32	63.21	164.1
11	8,234.0	1.260	33	50.13	206.9
12	6,529.0	1.588	34	39.75	260.9
13	5,178.4	2.003	35	31.52	329.0
14	4,106.8	2.525	36	25.00	414.8
15	3,256.7	3.184	37	19.83	523.1
16	2,582.9	4.016	38	15.72	659.6
17	2,048.2	5.064	39	12.47	831.8
18	1,624.3	6.385	40	9.89	1049.0

As the table shows, the size of a wire is also specified in terms of its *cross-sectional area*, as illustrated also in Figure A–1. The unit of cross-sectional area is the *circular mil*, abbreviated CM. One circular mil is the area of a wire with a diameter of 0.001 inch (1 mil). We find the cross-sectional area by expressing the diameter in thousandths of an inch (mils) and squaring it, as follows:

$$A = d^2$$

where A is the cross-sectional area in circular mils and d is the diameter in mils.

FIGURE A–1
Cross-sectional area of a wire.

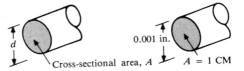

Cross-sectional area, A 0.001 in. A = 1 CM

EXAMPLE A–1

What is the cross-sectional area of a wire with a diameter of 0.005 inch?

Solution: $d = 0.005$ in. $= 5$ mils
$$A = d^2 = 5^2 = 25 \text{ CM}$$

WIRE RESISTANCE

Although copper wire conducts electricity extremely well, it still has some resistance, as do all conductors. The resistance of a wire depends on four factors: (1) type of material, (2) length of wire, (3) cross-sectional area, and (4) temperature.

Each type of conductive material has a characteristic called its *resistivity, ρ*. For each material, ρ is a constant value at a given temperature. The formula for the resistance of a wire of length l and cross-sectional area A is

$$R = \frac{\rho l}{A}$$

This formula tells us that resistance increases with resistivity and length and decreases with cross-sectional area. For resistance to be calculated in ohms, the length must be in feet, the cross-sectional area in circular mils, and the resistivity in CM-Ω/ft.

EXAMPLE A–2

Find the resistance of a 100-ft length of copper wire with a cross-sectional area of 810.1 CM. The resistivity of copper is 10.4 CM-Ω/ft.

Solution: $$R = \frac{\rho l}{A} = \frac{(10.4 \text{ CM-}\Omega\text{/ft})(100 \text{ ft})}{810.1 \text{ CM}} = 1.284 \ \Omega$$

Table A–1 lists the resistance of the various standard wire sizes in ohms per 1000 feet at 20°C. For example, a 1000-ft length of 14-gage copper wire has a resistance of 2.525 Ω. A 1000-ft length of 22-gage wire has a resistance of 16.14 Ω. For a given length, the smaller wire has more resistance. Thus, for a given voltage, larger wires can carry more current than smaller ones.

APPENDIX B

TABLE OF
STANDARD
RESISTOR VALUES

Resistance Tolerance (±%)

0.1% 0.25% 0.5%	1%	2% 5%	10%	0.1% 0.25% 0.5%	1%	2% 5%	10%	0.1% 0.25% 0.5%	1%	2% 5%	10%	0.1% 0.25% 0.5%	1%	2% 5%	10%	0.1% 0.25% 0.5%	1%	2% 5%	10%	0.1% 0.25% 0.5%	1%	2% 5%	10%
10.0	10.0	10	10	14.7	14.7	—	—	21.5	21.5	—	—	31.6	31.6	—	—	46.4	46.4	—	—	68.1	68.1	68	68
10.1	—	—	—	14.9	—	—	—	21.8	—	—	—	32.0	—	—	—	47.0	—	47	47	69.0	—	—	—
10.2	10.2	—	—	15.0	15.0	15	15	22.1	22.1	22	22	32.4	32.4	—	—	47.5	47.5	—	—	69.8	69.8	—	—
10.4	—	—	—	15.2	—	—	—	22.3	—	—	—	32.8	—	—	—	48.1	—	—	—	70.6	—	—	—
10.5	10.5	—	—	15.4	15.4	—	—	22.6	22.6	—	—	33.2	33.2	33	33	48.7	48.7	—	—	71.5	71.5	—	—
10.6	—	—	—	15.6	—	—	—	22.9	—	—	—	33.6	—	—	—	49.3	—	—	—	72.3	—	—	—
10.7	10.7	—	—	15.8	15.8	—	—	23.2	23.2	—	—	34.0	34.0	—	—	49.9	49.9	—	—	73.2	73.2	—	—
10.9	—	—	—	16.0	—	16	—	23.4	—	—	—	34.4	—	—	—	50.5	—	—	—	74.1	—	—	—
11.0	11.0	11	—	16.2	16.2	—	—	23.7	23.7	—	—	34.8	34.8	—	—	51.1	51.1	51	—	75.0	75.0	75	—
11.1	—	—	—	16.4	—	—	—	24.0	—	24	—	35.2	—	—	—	51.7	—	—	—	75.9	—	—	—
11.3	11.3	—	—	16.5	16.5	—	—	24.3	24.3	—	—	35.7	35.7	—	—	52.3	52.3	—	—	76.8	76.8	—	—
11.4	—	—	—	16.7	—	—	—	24.6	—	—	—	36.1	—	36	—	53.0	—	—	—	77.7	—	—	—
11.5	11.5	—	—	16.9	16.9	—	—	24.9	24.9	—	—	36.5	36.5	—	—	53.6	53.6	—	—	78.7	78.7	—	—
11.7	—	—	—	17.2	—	—	—	25.2	—	—	—	37.0	—	—	—	54.2	—	—	—	79.6	—	—	—
11.8	11.8	—	—	17.4	17.4	—	—	25.5	25.5	—	—	37.4	37.4	—	—	54.9	54.9	—	—	80.6	80.6	—	—
12.0	—	12	12	17.6	—	—	—	25.8	—	—	—	37.9	—	—	—	55.6	—	—	—	81.6	—	—	—
12.1	12.1	—	—	17.8	17.8	—	—	26.1	26.1	—	—	38.3	38.3	—	—	56.2	56.2	56	56	82.5	82.5	82	82
12.3	—	—	—	18.0	—	18	18	26.4	—	—	—	38.8	—	—	—	56.9	—	—	—	83.5	—	—	—
12.4	12.4	—	—	18.2	18.2	—	—	26.7	26.7	—	—	39.2	39.2	39	39	57.6	57.6	—	—	84.5	84.5	—	—
12.6	—	—	—	18.4	—	—	—	27.1	—	27	27	39.7	—	—	—	58.3	—	—	—	85.6	—	—	—
12.7	12.7	—	—	18.7	18.7	—	—	27.4	27.4	—	—	40.2	40.2	—	—	59.0	59.0	—	—	86.6	86.6	—	—
12.9	—	—	—	18.9	—	—	—	27.7	—	—	—	40.7	—	—	—	59.7	—	—	—	87.6	—	—	—
13.0	13.0	13	—	19.1	19.1	—	—	28.0	28.0	—	—	41.2	41.2	—	—	60.4	60.4	—	—	88.7	88.7	—	—
13.2	—	—	—	19.3	—	—	—	28.4	—	—	—	41.7	—	—	—	61.2	—	—	—	89.8	—	—	—
13.3	13.3	—	—	19.6	19.6	—	—	28.7	28.7	—	—	42.2	42.2	—	—	61.9	61.9	62	—	90.9	90.9	91	—
13.5	—	—	—	19.8	—	—	—	29.1	—	—	—	42.7	—	—	—	62.6	—	—	—	92.0	—	—	—
13.7	13.7	—	—	20.0	20.0	20	—	29.4	29.4	—	—	43.2	43.2	43	—	63.4	63.4	—	—	93.1	93.1	—	—
13.8	—	—	—	20.3	—	—	—	29.8	—	—	—	43.7	—	—	—	64.2	—	—	—	94.2	—	—	—
14.0	14.0	—	—	20.5	20.5	—	—	30.1	30.1	30	—	44.2	44.2	—	—	64.9	64.9	—	—	95.3	95.3	—	—
14.2	—	—	—	20.8	—	—	—	30.5	—	—	—	44.8	—	—	—	65.7	—	—	—	96.5	—	—	—
14.3	14.3	—	—	21.0	21.0	—	—	30.9	30.9	—	—	45.3	45.3	—	—	66.5	66.5	—	—	97.6	97.6	—	—
14.5	—	—	—	21.3	—	—	—	31.2	—	—	—	45.9	—	—	—	67.3	—	—	—	98.8	—	—	—

Note: These values are generally available in multiples of 0.1, 1, 10, 100, 1 k, and 1 M.

APPENDIX C

BATTERIES

Batteries are an important source of dc voltage. They are available in two basic categories: the *wet cell* and the *dry cell*. A battery generally is made up of several individual cells.

A cell consists basically of two *electrodes* immersed in an *electrolyte*. A voltage is developed between the electrodes as a result of the *chemical* action between the electrodes and the electrolyte. The electrodes typically are two dissimilar metals, and the electrolyte is a chemical solution.

SIMPLE WET CELL

Figure C–1 shows a simple copper-zinc (Cu-Zn) chemical cell. One electrode is made of copper, the other of zinc. These electrodes are immersed in a solution of water and hydrochloric acid (HCl), which is the electrolyte.

FIGURE C–1
Simple chemical cell.

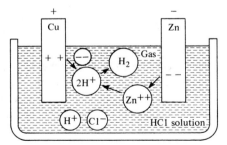

Positive hydrogen ions (H^+) and negative chlorine ions (Cl^-) are formed when the HCl ionizes in the water. Since zinc is more active than hydrogen, zinc atoms leave the zinc electrode and form zinc ions (Zn^{++}) in the solution. When a zinc ion is formed, two excess electrons are left on the zinc electrode, and two hydrogen ions are displaced from the solution. These two hydrogen ions will migrate to the copper electrode, take two electrons from the copper, and form a molecule of hydrogen gas (H_2). As a result of this reaction, a negative charge develops on the zinc electrode, and a positive charge develops on the copper electrode, creating a potential difference or voltage between the two electrodes.

In this copper-zinc cell, the hydrogen gas given off at the copper electrode tends to form a layer of bubbles around the electrodes, insulating the copper from the electrolyte. This effect, called *polarization,* results in a reduction in the voltage produced by the cell. Polarization can be remedied by the addition of an agent to the electrolyte to remove hydrogen gas or by the use of an electrolyte that does not form hydrogen gas.

Lead-Acid Cell   The positive electrode of a lead-acid cell is lead peroxide (PbO_2), and the negative electrode is spongy lead (Pb). The electrolyte is sulfuric acid (H_2SO_4) in water. Thus, the lead-acid cell is classified as a wet cell.

Two positive hydrogen ions ($2H^+$) and one negative sulfate ion (SO_4^{--}) are formed when the sulfuric acid ionizes in the water. Lead ions (Pb^{++}) from both electrodes displace the hydrogen ions in the electrolyte solution. When the lead ion from the spongy lead electrode enters the solution, it combines with a sulfate ion (SO_4^{--}) to form lead sulfate ($PbSO_4$), and it leaves two excess electrons on the electrode.

When a lead ion from the lead peroxide electrode enters the solution, it also leaves two excess electrons on the electrode and forms lead sulfate in the solution. However, because this electrode is lead peroxide, two free oxygen atoms are created when a lead atom leaves and enters the solution as a lead ion. These two oxygen atoms take four electrons from the lead peroxide electrode and become oxygen ions (O^{--}). This process creates a deficiency of two electrons on this electrode (there were initially two excess electrons).

The two oxygen ions ($2O^{--}$) combine in the solution with four hydrogen ions ($4H^+$) to produce two molecules of water ($2H_2O$). This process dilutes the electrolyte over a period of time. Also, there is a buildup of lead sulfate on the electrodes. These two factors result in a reduction in the voltage produced by the cell and necessitate periodic recharging.

As you have seen, for each departing lead ion, there is an excess of two electrons on the spongy lead electrode, and there is a deficiency of two electrons on the lead peroxide electrode. Therefore, the lead peroxide electrode is positive, and the spongy lead electrode is negative. This chemical reaction is pictured in Figure C–2.

As mentioned, the dilution of the electrolyte by the formation of water and lead sulfate requires that the lead-acid cell be recharged to *reverse* the chemical process. A chemical cell that can be recharged is called a *secondary cell.* One that cannot be recharged is called a *primary cell.*

The cell is recharged by connection of an external voltage source to the electrodes, as shown in Figure C–3. The formula for the chemical reaction in a lead-acid cell is as follows:

$$Pb + PbO_2 + 2H_2SO_4 \longrightarrow 2PbSO_4 + 2H_2O$$

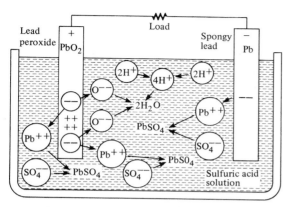

FIGURE C–2
Chemical reaction in a discharging lead-acid cell.

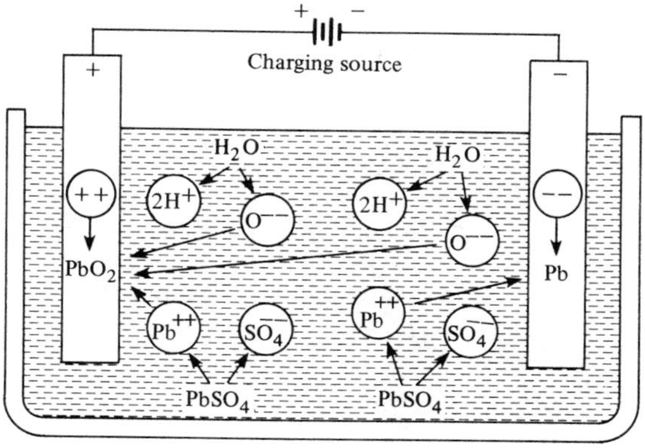

FIGURE C–3
Recharging a lead-acid cell.

DRY CELL

In the dry cell, some of the disadvantages of a liquid electrolyte are overcome. Actually, the electrolyte in a typical dry cell is not dry but rather is in the form of a moist paste. This electrolyte is a combination of granulated carbon, powdered manganese dioxide, and ammonium chloride solution.

A typical carbon-zinc dry cell is illustrated in Figure C–4. The zinc container or can is dissolved by the electrolyte. As a result of this reaction, an excess of electrons accumulates on the container, making it the negative electrode.

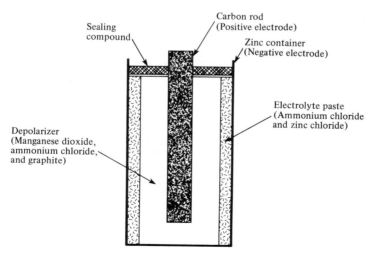

FIGURE C–4
Simplified construction of a dry cell.

The hydrogen ions in the electrolyte take electrons from the carbon rod, making it the positive electrode. Hydrogen gas is formed near the carbon electrode, but this gas is eliminated by reaction with manganese dioxide (called a *depolarizing agent*). This depolarization prevents bursting of the container due to gas formation. Because the chemical reaction is not reversible, the carbon-zinc cell is a primary cell.

TYPES OF CHEMICAL CELLS

Although only two common types of battery cells have been discussed, there are several types, listed in Table C–1.

TABLE C–1
Types of battery cells.

Type	+ electrode	− electrode	Electrolyte	Volts	Comments
Carbon-zinc	Carbon	Zinc	Ammonium and zinc chloride	1.5	Dry, primary
Lead-acid	Lead peroxide	Spongy lead	Sulfuric acid	2.0	Wet, secondary
Manganese-alkaline	Manganese dioxide	Zinc	Potassium hydroxide	1.5	Dry, primary or secondary
Mercury	Zinc	Mercuric oxide	Potassium hydroxide	1.3	Dry, primary
Nickel-cadmium	Nickel	Cadmium hydroxide	Potassium hydroxide	1.25	Dry, secondary
Nickel-iron (Edison cell)	Nickel oxide	Iron	Potassium hydroxide	1.36	Wet, secondary

APPENDIX D

DERIVATIONS

rms (EFFECTIVE) VALUE OF A SINE WAVE

The abbreviation "rms" stands for the *root mean square* process by which this value is derived. In the process, we first square the equation of a sine wave:

$$v^2 = V_p^2 \sin^2\theta$$

Next, we obtain the mean or average value of v^2 by dividing the area under a half-cycle of the curve by π (see Figure D–1). The area is found by integration and trigonometric identities:

$$V_{\text{avg}}^2 = \frac{\text{area}}{\pi}$$

$$= \frac{1}{\pi} \int_0^\pi V_p^2 \sin^2\theta \, d\theta$$

$$= \frac{V_p^2}{2\pi} \int_0^\pi (1 - \cos 2\theta) \, d\theta$$

$$= \frac{V_p^2}{2\pi} \int_0^\pi 1 \, d\theta - \frac{V_p^2}{2\pi} \int_0^\pi (-\cos 2\theta) \, d\theta$$

$$= \frac{V_p^2}{2\pi} (\theta - \tfrac{1}{2} \sin 2\theta)_0^\pi$$

$$= \frac{V_p^2}{2\pi} (\pi - 0)$$

$$= \frac{V_p^2}{2}$$

FIGURE D–1

Finally, the square root of V_{avg}^2 is V_{rms}:

$$V_{\text{rms}} = \sqrt{V_{\text{avg}}^2}$$
$$= \sqrt{V_p^2/2}$$
$$= \frac{V_p}{\sqrt{2}}$$
$$= 0.707V_p$$

AVERAGE VALUE OF A HALF-CYCLE SINE WAVE

The average value of a sine wave is determined for a half-cycle because the average over a full cycle is zero.

The equation for a sine wave is

$$v = V_p \sin \theta$$

The average value of the half-cycle is the area under the curve divided by the distance of the curve along the horizontal axis (see Figure D–2):

$$V_{\text{avg}} = \frac{\text{area}}{\pi}$$

FIGURE D–2

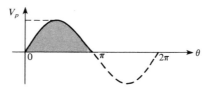

To find the area, we use integral calculus:

$$V_{\text{avg}} = \frac{1}{\pi} \int_0^\pi V_p \sin \theta \, d\theta$$
$$= \frac{V_p}{\pi} (-\cos \theta) \Big|_0^\pi$$

$$= \frac{V_p}{\pi} [-\cos \pi - (-\cos 0)]$$

$$= \frac{V_p}{\pi} [-(-1) - (-1)]$$

$$= \frac{V_p}{\pi} (2)$$

$$= \frac{2}{\pi} V_p$$

$$= 0.637 V_p$$

REACTANCE DERIVATIONS

DERIVATION OF CAPACITIVE REACTANCE

$$\theta = 2\pi f t = \omega t$$

$$i = C \frac{dV}{dt} = C \frac{d(V_p \sin \theta)}{dt}$$

$$= C \frac{d(V_p \sin \omega t)}{dt}$$

$$= \omega C (V_p \cos \omega t)$$

$$I_{rms} = \omega C V_{rms}$$

$$X_C = \frac{V_{rms}}{I_{rms}} = \frac{V_{rms}}{\omega C V_{rms}} = \frac{1}{\omega C}$$

$$= \frac{1}{2\pi f C}$$

DERIVATION OF INDUCTIVE REACTANCE

$$v = L \frac{di}{dt} = L \frac{d(I_p \sin \omega t)}{dt}$$

$$= \omega L (I_p \cos \omega t)$$

$$V_{rms} = \omega L I_{rms}$$

$$X_L = \frac{V_{rms}}{I_{rms}} = \frac{\omega L I_{rms}}{I_{rms}} = \omega L$$

$$= 2\pi f L$$

SOLUTIONS TO SELF-TESTS

CHAPTER ONE

1. Current: amperes. Voltage: volts. Resistance: ohms. Power: watts. Energy: joules.

2. Amperes: A. Volts: V. Ohms: Ω. Watts: W. Joules: J.

3. Current: I. Voltage: V. Resistance: R. Power: P. Energy: W.

4. **(a)** 10 mA   **(b)** 5 kV   **(c)** 15 μW   **(d)** 20 MΩ

5. **(a)** 0.000008 A   **(b)** 25,000,000 W   **(c)** 0.100 V

CHAPTER TWO

1. A neutral atom with an atomic number of 3 has 3 electrons.

2. Shells are the orbits in which electrons revolve.

3. Semiconductors have fewer free electrons than conductors, but more than insulators.

4. The presence of a net positive charge or a net negative charge in a material.

5. They will be pulled toward each other because opposite charges attract.

6. A single electron has a charge of 1.6×10^{-19} coulomb.

7. A positive ion is created when a valence electron escapes from its orbit.

8. Potential difference is voltage, and its unit is the volt.

9. The unit of energy is the joule.

10. Voltage sources.

11. Voltage can exist between two points when there is no current.

12. The repulsive force between the negative voltage source terminal and the free electrons produces movement of the electrons through the material toward the positive terminal, which attracts them.

13. One ampere is the amount of electrical current equal to one coulomb of charge passing a point in one second (1 C/s).

14. Resistors limit current and produce heat.

15. Potentiometers and rheostats are variable resistors.

16. In a linear potentiometer, the resistance varies in direct proportion to the wiper movement. In a tapered device, the resistance is not linearly proportional to the wiper movement.

17. The circuit is open.

18. A current measurement is more difficult than a voltage measurement because the meter must be connected in series, a connection that requires a line to be broken.

19. $I = Q/t = 50$ C/5 s $= 10$ A

20. $Q = It = (2$ A$)(10$ s$) = 20$ C

21. $V = W/Q = 500$ J/100 C $= 5$ V

22. $G = 1/R = 1/10\ \Omega = 0.1\ \text{S}$

23. Blue: 6, 1st digit
Green: 5, 2nd digit
Red: 2, number of zeros
Gold: 5%, tolerance
$R = 6500\ \Omega \pm 5\%$

24. 500 Ω. Center position is one-half the resistance.

CHAPTER THREE

1. Ohm's law states that voltage, current, and resistance are linearly related: $V = IR$.

2. **(a)** The current triples when the voltage is tripled.
(b) The current decreases by 75% when the voltage is reduced by 75%.
(c) The current decreases by one-half when the resistance is doubled.
(d) The current increases by 35% when the resistance is reduced by 35%.
(e) The current increases by four times when the voltage is doubled and the resistance is cut in half.
(f) The current does not change when the voltage is doubled and the resistance is doubled.

3. $I = V/R$

4. $V = IR$

5. $R = V/I$

6. **(a)** Volts divided by kilohms gives milliamperes: $V/k\Omega = mA$.
(b) Volts divided by megohms gives microamperes: $V/M\Omega = \mu A$.
(c) Milliamperes times kilohms gives volts: $mA \times k\Omega = V$.
(d) Milliamperes times megohms gives kilovolts: $mA \times M\Omega = kV$.
(e) Volts divided by milliamperes gives kilohms: $V/mA = k\Omega$.
(f) Volts divided by microamperes gives megohms: $V/\mu A = M\Omega$.

7. $I = V/R = 10\ \text{V}/5.6\ \Omega = 1.79\ \text{A}$

8. $I = V/R = 75\ \text{V}/10\ \text{k}\Omega = 7.5\ \text{mA}$

9. $V = IR = (2.5\ \text{A})(22\ \Omega) = 55\ \text{V}$

10. $V = IR = (30\ \text{mA})(2.2\ \text{k}\Omega) = 66\ \text{V}$

11. $R = V/I = 12\ \text{V}/4.44\ \text{A} = 2.7\ \Omega$

12. $R = V/I = 9\ \text{V}/109.8\ \mu\text{A} = 0.082\ \text{M}\Omega$
$= 82\ \text{k}\Omega$

13. $I = V/R = 50\ \text{kV}/2.7\ \text{k}\Omega = 18.52\ \text{A}$

14. $I = V/R = 15\ \text{mV}/10\ \text{k}\Omega = 1.5\ \mu\text{A}$

15. $V = IR = (4\ \mu\text{A})(100\ \text{k}\Omega) = 400\ \text{mV}$

16. $V = IR = (100\ \text{mA})(1\ \text{M}\Omega) = 100\ \text{kV}$

17. $R = V/I = 5\ \text{V}/1.85\ \text{mA} = 2.7\ \text{k}\Omega$

18. $I = V/R = 1.5\ \text{V}/4.7\ \text{k}\Omega = 0.319\ \text{mA}$

CHAPTER FOUR

1. Power is the rate at which energy is used.

2. **(a)** 1000 watts in a kilowatt.
(b) 1,000,000 watts in a megawatt.

3. Connect the voltmeter across the resistor, and connect the ammeter in series with the resistor. Multiply the voltage across the resistor and the current through it to find the power.

4. **(a)** $P = VI$
(b) $P = I^2R$
(c) $P = V^2/R$

5. Use a 2-W resistor to handle 1.1 W.

6. An open resistor shows an infinite (∞) reading on the ohmmeter.

7. $P = W/t = 200\ \text{J}/10\ \text{s} = 20\ \text{W}$

8. $P = W/t = 10,000\ \text{J}/300\ \text{ms}$
$= 33.33\ \text{kW}$

9. $50\ \text{kW} = 50 \times 10^3\ \text{W} = 50,000\ \text{W}$

10. $0.045\ \text{W} = 45 \times 10^{-3}\ \text{W} = 45\ \text{mW}$

11. $P_R = VI = (10\ \text{V})(350\ \text{mA})$
$= 3500\ \text{mW} = 3.5\ \text{W}$
$P_{BATT} = 3.5\ \text{W}$

12. $P = V^2/R = (50\ \text{V})^2/1000\ \Omega = 2.5\ \text{W}$

13. $P = I^2R = (0.5\ \text{A})^2(5.6\ \text{k}\Omega)$
$= 1400\ \text{W}$

14. $P = VI = (120\ \text{V})(2\ \text{A}) = 240\ \text{W}$

15. $W = Pt = (15\ \text{W})(60\ \text{s}) = 900\ \text{J}$

16. $P = 500\ \text{W} = 0.5\ \text{kW}$
$t = 24\ \text{h}$
$W = Pt = (0.5\ \text{kW})(24\ \text{h}) = 12\ \text{kWh}$

17. $(75\ \text{W})(10\ \text{h}) = 750\ \text{Wh}$

18. Power supply 2 has a greater load because it must supply more current than power supply 1. Number 2 load has a smaller resistive value because it supplies more current for the same voltage.

19. 1: $P = VI = (25 \text{ V})(0.1 \text{ A}) = 2.5 \text{ W}$
   2: $P = VI = (25 \text{ V})(0.5 \text{ A}) = 12.5 \text{ W}$

20. $I_L = 12 \text{ V}/600 \text{ Ω} = 0.02 \text{ A}$
   50 Ah/0.02 A = 2500 h

21. (8 A)(2.5 h) = 20 Ah

22. $P_{OUT} = P_{IN} = P_{LOST} = 2 \text{ W} - 0.25 \text{ W}$
   $= 1.75 \text{ W}$

23. Efficiency $= (P_{OUT}/P_{IN}) \times 100\%$
   $= (0.5 \text{ W}/0.6 \text{ W}) \times 100\%$
   $= 83.33\%$

CHAPTER FIVE

1. See Figure S–1.

FIGURE S–1

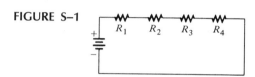

2. Since current is the same at *all* points in a series circuit, there are 5 A out of the sixth and tenth resistors as well as out of all of the others.

3. See Figure S–2.

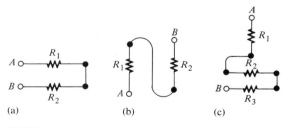

(a)   (b)   (c)

FIGURE S–2

4. $I_{AB} = I_{BC} = I_{CD} = I_{DE} = I_{EF} = I_{FG}$
   $= I_{GH} = I_{HA} = 3 \text{ A}$

5. $R_T = 82 \text{ Ω} + 470 \text{ Ω} = 552 \text{ Ω}$

6. $R_T = (8)(56 \text{ Ω}) = 448 \text{ Ω}$

7. $R = 22 \text{ kΩ}$
   $- (1 \text{ kΩ} + 1 \text{ kΩ} + 5.6 \text{ kΩ} + 3.3 \text{ kΩ})$
   $= 11.1 \text{ kΩ}$

8. $R_T = (3)(1 \text{ kΩ}) = 3 \text{ kΩ}$
   $I = 5 \text{ V}/3 \text{ kΩ} = 1.67 \text{ mA}$

9. (a) $R_T = 100 \text{ Ω} + 47 \text{ Ω} + 68 \text{ Ω}$
   $= 215 \text{ Ω}$
   $I = 5 \text{ V}/215 \text{ Ω} = 23.26 \text{ mA}$
   (b) $R_T = 56 \text{ Ω} + 180 \text{ Ω} + 33 \text{ Ω}$
   $= 269 \text{ Ω}$
   $I = 8 \text{ V}/269 \text{ Ω} = 29.74 \text{ mA}$
   Circuit (b) has more current.

10. $R_T = V/I = 12 \text{ V}/2 \text{ mA} = 6 \text{ kΩ}$
   $R_{EACH} = 6 \text{ kΩ}/6 = 1 \text{ kΩ}$

11. $V_T = 8 \text{ V} + 5 \text{ V} + 1.5 \text{ V} = 14.5 \text{ V}$

12. $V_T = 9 \text{ V} - 9 \text{ V} = 0 \text{ V}$

13. See Figure S–3.

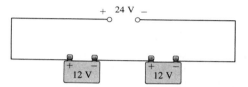

FIGURE S–3

14. The light is dimmer because the total voltage is reduced; therefore, there is less current through the bulb.

15. (a) $V_T = 50 \text{ V} - 30 \text{ V} - 10 \text{ V} = 10 \text{ V}$
   See Figure S–4.

FIGURE S–4

   (b) $V_T = 1.5 \text{ V} + 3 \text{ V} + 4.5 \text{ V} - 9 \text{ V}$
   $- 1.5 \text{ V} = -1.5 \text{ V}$
   See Figure S–5.

FIGURE S–5

16. By Kirchhoff's voltage law, the sum of the voltages is zero.

17. $V_S = 6(5 \text{ V}) = 30 \text{ V}$

18. Since the resistors are all of equal value, they each drop 2 V.
$$V_T = (5)(2 \text{ V}) = 10 \text{ V}$$

19. The 10-kΩ resistor has the greatest voltage drop because it has the largest value.

20. $V_S = V_1 + V_2 + V_3$
$V_3 = V_S - V_1 - V_2$
$\quad = 50 \text{ V} - 10 \text{ V} - 15 \text{ V} = 25 \text{ V}$

21. $R_T = V/I = 30 \text{ V}/10 \text{ mA} = 3 \text{ k}\Omega$
$R_T = R_1 + R_2 + R_3 + R_4 + R_5$
$R_3 = R_T - (R_1 + R_2 + R_4 + R_5)$
$\quad = 3 \text{ k}\Omega - 2.1 \text{ k}\Omega = 900 \text{ }\Omega$

22. $V_x = (R_x/R_T)(V_T)$
$V_{47\Omega} = (47 \text{ }\Omega/196 \text{ }\Omega)(20 \text{ V}) = 4.8 \text{ V}$

23. $V_x = (R_x/R_T)(V_T)$
$V_1 = (33 \text{ }\Omega/102 \text{ }\Omega)(90 \text{ V}) = 29.12 \text{ V}$
$V_2 = (22 \text{ }\Omega/102 \text{ }\Omega)(90 \text{ V}) = 19.41 \text{ V}$
$V_3 = (47 \text{ }\Omega/102 \text{ }\Omega)(90 \text{ V}) = 41.47 \text{ V}$

24. $V_x = (R_x/R_T)(V_T)$
$3 \text{ V} = (R_x/100 \text{ k}\Omega)(9 \text{ V})$
$R_{x1} = (3 \text{ V})(100 \text{ k}\Omega)/9 \text{ V} = 33.33 \text{ k}\Omega$
$R_{x2} = R_T - R_{x1} = 100 \text{ k}\Omega - 33.33 \text{ k}\Omega$
$\quad = 66.67 \text{ k}\Omega$

25. $P_T = P_1 + P_2 + P_3$
$\quad = 2.5 \text{ W} + 5 \text{ W} + 1.2 \text{ W} = 8.7 \text{ W}$

26. One 100-Ω resistor dissipates the most power $(P = V^2/R)$.

27. $P_{\text{EACH}} = 10 \text{ W}/5 = 2 \text{ W}$

28. Check for an open. The voltage across the open equals the source voltage.

29. Check for a short.

CHAPTER SIX

1. See Figure S–6.

FIGURE S–6

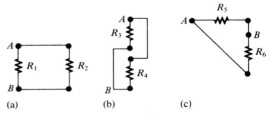

2. $I_{BR} = I_T/2 = 5 \text{ A}/2 = 2.5 \text{ A}$

3. See Figure S–7.

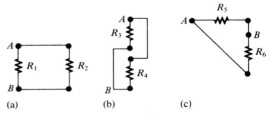

(a)   (b)   (c)

FIGURE S–7

4. $V_{AB} = V_{AC} = V_{EC} = V_{DC} = 5 \text{ V}$
(because all of the resistors are in parallel)

5. $R_T = \dfrac{(82 \text{ }\Omega)(150 \text{ }\Omega)}{82 \text{ }\Omega + 150 \text{ }\Omega} = 53 \text{ }\Omega$

6. $1/R_T = (1/1000 \text{ }\Omega) + (1/820 \text{ }\Omega)$
$\quad + (1/560 \text{ }\Omega) + (1/220 \text{ }\Omega)$
$\quad + (1/100 \text{ }\Omega) = 0.01855 \text{ S}$
$R_T = 1/0.01855 \text{ S} = 53.91 \text{ }\Omega$

7. $R_T = R/n = 56 \text{ }\Omega/8 = 7 \text{ }\Omega$

8. See Figure S–8.

FIGURE S–8

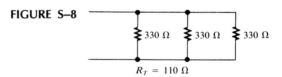

330 Ω 330 Ω 330 Ω

$R_T = 110 \text{ }\Omega$

9. $R_T = R/n = 680 \text{ }\Omega/3 = 226.67 \text{ }\Omega$
$I_T = V/R_T = 5 \text{ V}/226.67 \text{ }\Omega = 0.022 \text{ A}$
$\quad = 22 \text{ mA}$

10. (a) $R_T = (56 \text{ }\Omega)(27 \text{ }\Omega)/83 \text{ }\Omega$
$\quad = 18.22 \text{ }\Omega$
$I_T = 10 \text{ V}/18.22 \text{ }\Omega = 0.549 \text{ A}$
$\quad = 549 \text{ mA}$
(b) $R_T = (15 \text{ }\Omega)(100 \text{ }\Omega)/115 \text{ }\Omega$
$\quad = 13.04 \text{ }\Omega$
$I_T = 8 \text{ V}/13.04 \text{ }\Omega = 0.613 \text{ A}$
$\quad = 613 \text{ mA}$
Circuit (b) has the greater total current.

11. $R_T = V/I_T = 12 \text{ V}/3 \text{ mA} = 4 \text{ k}\Omega$
$R_T = R/n$
$4 \text{ k}\Omega = R_{\text{EACH}}/6$
$R_{\text{EACH}} = (6)(4 \text{ k}\Omega) = 24 \text{ k}\Omega$

12. $I_T = 1 \text{ A} - 2 \text{ A} + 4 \text{ A} = 3 \text{ A}$

13. $I_T = 1 \text{ A} - 1 \text{ A} = 0 \text{ A}$

14. $I_T = (5)(25 \text{ mA}) = 125 \text{ mA}$

15. $I_T = I_1 + I_2 + I_3$
$$I_3 = I_T - (I_1 + I_2)$$
$$= 0.5 \text{ A} - (0.1 \text{ A} + 0.2 \text{ A})$$
$$= 0.5 \text{ A} - 0.3 \text{ A} = 0.2 \text{ A}$$

16. $V_S = V_1 = V_2 = V_3$
$$V_S = V_1 = I_1 R_1 = (1 \text{ A})(10 \ \Omega) = 10 \text{ V}$$
$$R_2 = V_s/I_2 = 10 \text{ V}/0.5 \text{ A} = 20 \ \Omega$$
$$I_3 = I_T - (I_1 + I_2) = 5 \text{ A} - 1.5 \text{ A}$$
$$= 3.5 \text{ A}$$
$$R_3 = V_s/I_3 = 10 \text{ V}/3.5 \text{ A} = 2.86 \ \Omega$$

17. $1/R_T = (1/2200 \ \Omega) + (1/6800 \ \Omega)$
$$+ (1/3300 \ \Omega) + (1/1000 \ \Omega)$$
$$= 0.0019 \text{ S}$$
$$R_T = 1/0.0019 \text{ S} = 526 \ \Omega$$
$$I_x = (R_T/R_x)(I_T)$$
$$I_{2.2k\Omega} = (526 \ \Omega/2200 \ \Omega)(1 \text{ A}) = 0.239 \text{ A}$$
$$I_{6.8k\Omega} = (526 \ \Omega/6800 \ \Omega)(1 \text{ A})$$
$$= 0.077 \text{ A}$$
$$I_{3.3k\Omega} = (526 \ \Omega/3300 \ \Omega)(1 \text{ A})$$
$$= 0.159 \text{ A}$$
$$I_{1k\Omega} = (500 \ \Omega/1000 \ \Omega)(1 \text{ A}) = 0.5 \text{ A}$$

18. $I_1 = \left(\dfrac{R_2}{R_1 + R_2}\right)I_T = \left(\dfrac{22 \text{ k}\Omega}{26.7 \text{ k}\Omega}\right)750 \text{ mA}$
$$= 617.98 \text{ mA}$$
$$I_2 = \left(\dfrac{R_1}{R_1 + R_2}\right)I_T = \left(\dfrac{4.7 \text{ k}\Omega}{26.7 \text{ k}\Omega}\right)750 \text{ mA}$$
$$= 132.02 \text{ mA}$$

19. $P_1 = I_1^2 R_1 = (617.98 \text{ mA})^2(4.7 \text{ k}\Omega) = 1794.9 \text{ W}$
$$P_2 = I_2^2 R_2 = (132.02 \text{ mA})^2(22 \text{ k}\Omega)$$
$$= 383.4 \text{ W}$$
$$P_T = P_1 + P_2 = 1794.9 \text{ W} + 383.4 \text{ W}$$
$$= 2178.3 \text{ W}$$

20. $I_{220\Omega} = 10 \text{ V}/220 \ \Omega = 0.045 \text{ A}$
$$= 45 \text{ mA}$$
$$I_{560\Omega} = 10 \text{ V}/560 \ \Omega = 0.01786 \text{ A}$$
$$= 17.86 \text{ mA}$$
The total current should be 70 mA, but it is equal to the current in the 560-Ω branch only. This finding indicates that the 220-Ω branch is *open*.

21. $R_x = \dfrac{R_A R_T}{R_A - R_T} = \dfrac{(100 \ \Omega)(45 \ \Omega)}{100 \ \Omega - 45 \ \Omega}$
$$= 82 \ \Omega$$

CHAPTER SEVEN

1. See Figure S–9.

FIGURE S–9

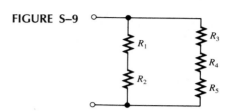

2. $R_T = \dfrac{(R_1 + R_2)(R_3 + R_4 + R_5)}{R_1 + R_2 + R_3 + R_4 + R_5}$

3. The most current goes through R_1 and R_2.

4. Since there are 6 V across one 1-kΩ resistor, there are also 6 V across the other because they are in series. The voltage across the two 1-kΩ resistors is 12 V, which is also the voltage across the parallel 2.2-kΩ resistor.

5. The 330-Ω resistor has the total current through it. Since 330 Ω is greater than the resistance of the parallel combination of four 1-kΩ resistors (250 Ω), the 330-Ω resistor has the largest voltage drop.

6. Each of the four 1-kΩ resistors in parallel carries 25% of the total current.

7. $V_{BA} = V_B - V_A = 8 \text{ V} - 5 \text{ V} = 3 \text{ V}$
$$V_{CA} = V_C - V_A = 12 \text{ V} - 5 \text{ V} = 7 \text{ V}$$

8. The voltage will decrease.

9. The 100-kΩ load resistance has the greatest effect.

10. More current will be drawn from the source when a load is connected.

11. There is 0 V across the output of a balanced bridge.

12. A device that indicates current in either direction.

13. The voltmeter probably is loading the circuit.

14. R_2, R_3, and R_4 are in parallel, and this parallel combination is in series with both R_1 and R_5.

15. (a) $R_T = R_1 + R_5 + (R_2/3)$;
$$R_2 = R_3 = R_4 = 33 \ \Omega.$$
$$R_T = 10 \ \Omega + 10 \ \Omega + 11 \ \Omega$$
$$= 31 \ \Omega$$
(b) $I_T = V_S/R_T = 30 \text{ V}/31 \ \Omega = 0.968 \text{ A}$

(c) $I_3 = I_T/3 = 0.968$ A$/3 = 0.323$ A
(d) $V_4 = I_4R_4; I_4 = I_3 = 0.323$ A
$V_4 = (0.323$ A$)(30$ $\Omega) = 9.69$ V

16. (a) $R_1\|R_2 = 2.2$ k$\Omega/2 = 1.1$ kΩ
$R_3\|R_4 = (330$ $\Omega)(680$ $\Omega)/1010$ Ω
$= 222$ Ω
$R_5\|R_6 = 1.5$ k$\Omega/2 = 750$ Ω
$R_1\|R_2 + R_3\|R_4 + R_5\|R_6$
$= 1100$ $\Omega + 222$ Ω
$+ 750$ $\Omega = 2.072$ kΩ
(b) $I_T = 10$ V$/2.072$ k$\Omega = 4.83$ mA
(c) $I_1 = I_T/2 = 4.83$ mA$/2 = 2.415$ mA
(because I_T splits equally between R_1 and R_2)
(d) $I_6 = I_1 = 2.415$ mA
$V_6 = (2.415$ mA$)(1.5$ k$\Omega) = 3.62$ V

17. (a) $R_4\|R_5 = 4.7$ k$\Omega/2 = 2.35$ kΩ
$R_4\|R_5 + R_3 = 2.35$ k$\Omega + 3.3$ kΩ
$= 5.65$ kΩ
5.65 k$\Omega\|R_2 = (5.65$ k$\Omega)(3.3$ k$\Omega)/8.95$ kΩ
$= 2.08$ kΩ
$R_T = 2.08$ k$\Omega\|R_1$
$= (2.08$ k$\Omega)(10$ k$\Omega)/12.08$ kΩ
$= 1.72$ kΩ
(b) $I_T = V_S/R_T = 6$ V$/1.72$ kΩ
$= 3.49$ mA
(c) The resistance to the right of AB is 2.08 kΩ.
The current through this part of the circuit is
$I = 6$ V$/2.08$ k$\Omega = 2.88$ mA.
$I_3 = \left(\dfrac{R_2}{R_2 + 5.65\text{ k}\Omega}\right)2.88$ mA
$= \left(\dfrac{3.3\text{ k}\Omega}{8.95\text{ k}\Omega}\right)2.88$ mA
$= 1.06$ mA
$I_5 = I_3/2 = 1.06$ mA$/2 = 0.53$ mA
(d) $V_2 = V_S = 6$ V

18. $V_A = 25$ V
R from point B to gnd $= 84.7$ Ω
$V_B = (84.7$ $\Omega/131.7$ $\Omega)(25$ V$) = 16.08$ V
$V_C = (60$ $\Omega/160$ $\Omega)(16.08$ V$)$
$= 6.03$ V

19. $V_A = 15$ V
$V_B = 0$ V
$V_C = 0$ V

20. $V_{3.3\text{k}\Omega} = (1.62$ k$\Omega/2.62$ k$\Omega)(10$ V$) = 6.18$ V
The 7.62-V reading is incorrect.
$V_{2.2\text{k}\Omega} = (2.2$ k$\Omega/3.2$ k$\Omega)(6.18$ V$) = 4.23$ V
The 5.24-V reading is incorrect. The 3.3-kΩ resistor must be open to produce these incorrect voltages.

21. $V_{\text{OUT}} = (22$ k$\Omega/32$ k$\Omega)(30$ V$)$
$= 20.63$ V unloaded
22 k$\Omega\|220$ k$\Omega = 20$ kΩ
$V_{\text{OUT}} = (20$ k$\Omega/30$ k$\Omega)(30$ V$)$
$= 20$ V loaded with 220 kΩ

22. $V_A = (6.6$ k$\Omega/9.9$ k$\Omega)(10$ V$)$
$= 6.67$ V switch open
6.6 k$\Omega\|10$ k$\Omega = 3.98$ kΩ
$V_A = (3.98$ k$\Omega/7.28$ k$\Omega)(10$ V$)$
$= 5.47$ V switch closed

23. (a) $R_3 + R_4 = 75$ $\Omega + 27$ $\Omega = 102$ Ω
$(R_3 + R_4)\|R_2 = 102$ $\Omega\|100$ $\Omega = 50.5$ Ω
$R_T = (R_3 + R_4)\|R_2 + R_1$
$= 50.5$ $\Omega + 47$ $\Omega = 97.5$ Ω
(b) $I_T = 50$ V$/97.5$ $\Omega = 0.513$ A
(c) $I_3 = \left(\dfrac{R_2}{R_2 + R_3 + R_4}\right)I_T$
$= \left(\dfrac{100\text{ }\Omega}{202\text{ }\Omega}\right)0.513$ A $= 0.254$ A
(d) $I_4 = I_3 = 0.254$ A
(e) $V_A = I_3(R_3 + R_4)$
$= (0.254$ A$)(102$ $\Omega) = 25.9$ V
(f) $V_B = I_4R_4 = (0.254$ A$)(27$ $\Omega)$
$= 6.86$ V

24. $R_{C \text{ to gnd}} = 8.2$ kΩ
$R_{B \text{ to gnd}} = 15$ k$\Omega\|16.4$ k$\Omega = 7.83$ kΩ
$R_{A \text{ to gnd}} = 15$ k$\Omega\|(8.2$ k$\Omega + 7.83$ k$\Omega)$
$= 15$ k$\Omega\|16.03$ k$\Omega = 7.75$ kΩ
$V_A = \left(\dfrac{7.75\text{ k}\Omega}{15.95\text{ k}\Omega}\right)10$ V $= 4.86$ V
$V_B = \left(\dfrac{7.83\text{ k}\Omega}{16.03\text{ k}\Omega}\right)4.86$ V $= 2.37$ V
$V_C = \left(\dfrac{8.2\text{ k}\Omega}{16.4\text{ k}\Omega}\right)2.37$ V $= 1.185$ V

25. $R_{\text{UNK}} = R_V(R_2/R_4)$
$= (10$ k$\Omega)(1$ k$\Omega/8.2$ k$\Omega) = 1.22$ kΩ

CHAPTER EIGHT

1. $I_S = V_S/R_S = 25$ V$/5$ $\Omega = 5$ A
$R_S = 5$ Ω

2. $I_S = 30 \text{ V}/10 \ \Omega = 3 \text{ A}; R_S = 10 \ \Omega$
See Figure S-10.

FIGURE S-10

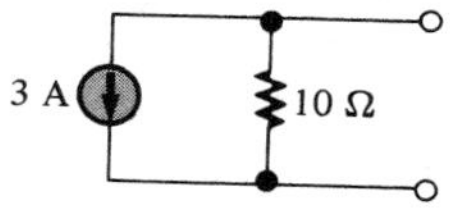

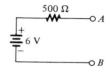

3. $V_S = (5 \text{ mA})(25 \text{ k}\Omega) = 125 \text{ V};$
$R_S = 25 \text{ k}\Omega$
See Figure S-11.

FIGURE S-11

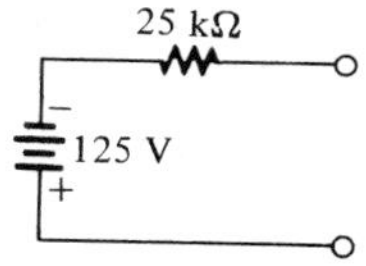

25 kΩ

125 V

4. Replace all but one source with their internal resistances; then find the currents or voltages due to the remaining source. Repeat for each source, and combine the results.

5. $I = 10 \text{ mA} - 8 \text{ mA} = 2 \text{ mA}$ (in the direction of the larger).

6. For 1-V source:
$R_T = 117.15 \ \Omega$
$I_T = 8.54 \text{ mA}$
$I_3 = \left(\dfrac{R_2}{R_2 + R_3}\right)I_T = \left(\dfrac{47 \ \Omega}{74 \ \Omega}\right)8.54 \text{ mA}$
$= 5.42 \text{ mA down}$
For 1.5-V source:
$R_T = 68.26 \ \Omega$
$I_T = 21.97 \text{ mA}$
$I_3 = \left(\dfrac{R_1}{R_1 + R_3}\right)I_T = \left(\dfrac{100 \ \Omega}{127 \ \Omega}\right)21.97 \text{ mA}$
$= 17.3 \text{ mA down}$
$I_3 \text{ (total)} = 5.42 \text{ mA} + 17.3 \text{ mA}$
$= 22.72 \text{ mA}$

7. For 1-V source:
$I_2 = \left(\dfrac{R_3}{R_2 + R_3}\right)I_T = \left(\dfrac{27 \ \Omega}{74 \ \Omega}\right)8.54 \text{ mA}$
$= 3.12 \text{ mA down}$
For 1.5-V source:
$I_2 = I_T = 21.97 \text{ mA up}$
$I_2 \text{ (total)} = 21.97 \text{ mA} - 3.12 \text{ mA}$
$= 18.85 \text{ mA}$

8. See Figure S-12.

FIGURE S-12

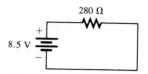

280 Ω

8.5 V

9. $R_{TH} = 1 \text{ k}\Omega/2 = 500 \ \Omega$ (5.6-kΩ and 10-kΩ resistors are "shorted" by source)
$V_{TH} = (1 \text{ k}\Omega/2 \text{ k}\Omega)(12 \text{ V}) = 6 \text{ V}$
See Figure S-13.

FIGURE S-13

500 Ω

6 V

A

B

10. Looking in at *open* terminals *AB*:
$R_{TH} = (22 \text{ k}\Omega)(100 \text{ k}\Omega)/122 \text{ k}\Omega$
$= 18.03 \text{ k}\Omega$
$V_{TH} = (22 \text{ k}\Omega/122 \text{ k}\Omega)(15 \text{ V}) = 2.7 \text{ V}$
With R_L connected:
$I_L = V_{TH}/(R_{TH} + R_L)$
$= 2.7 \text{ V}/578.03 \text{ k}\Omega = 4.67 \ \mu\text{A}$

11. $I_N = V_{TH}/R_{TH} = 2.7 \text{ V}/18.03 \text{ k}\Omega$
$= 0.15 \text{ mA}$
$R_N = R_{TH} = 18.03 \text{ k}\Omega$
See Figure S-14.

FIGURE S-14

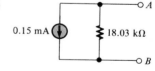

0.15 mA

18.03 kΩ

A

B

12. $R_{EQ} = \dfrac{1}{(1/22 \ \Omega) + (1/10 \ \Omega) + (1/56 \ \Omega) + (1/10 \ \Omega)}$
$= 3.8 \ \Omega$
$V_{EQ} = \dfrac{(12 \text{ V}/22 \ \Omega) + (6 \text{ V}/10 \ \Omega) - (10 \text{ V}/56 \ \Omega) - (5 \text{ V}/10 \ \Omega)}{0.263}$
$= 0.47/0.263 = 1.79 \text{ V}$
See Figure S-15.

FIGURE S-15

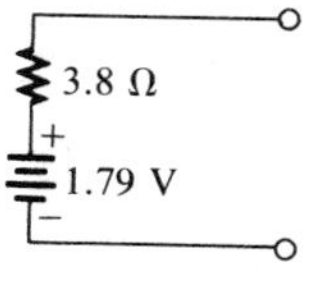

3.8 Ω

1.79 V

13. $R_L = R_{EQ} = 3.8\ \Omega$

14. **(a)** $R_1 = R_A R_C / (R_A + R_B + R_C)$
$= (1\ k\Omega)(560\ \Omega)/2.56\ k\Omega$
$= 218.75\ \Omega$
$R_2 = R_B R_C / (R_A + R_B + R_C)$
$= (1\ k\Omega)(560\ \Omega)/2.56\ k\Omega$
$= 218.75\ \Omega$
$R_3 = R_A R_B / (R_A + R_B + R_C)$
$= (1\ k\Omega)(1\ k\Omega)/2.56\ k\Omega$
$= 391\ \Omega$
See Figure S-16.

FIGURE S-16

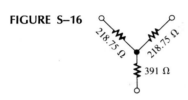

(b) $R_A = \dfrac{R_1 R_2 + R_1 R_3 + R_2 R_3}{R_2}$

$= \dfrac{(47\ \Omega)(47\ \Omega) + (47\ \Omega)(100\ \Omega) + (47\ \Omega)(100\ \Omega)}{47\ \Omega}$

$= 247\ \Omega$
$R_B = 11{,}609/R_1 = 11{,}609/47$
$= 247\ \Omega$
$R_C = 11{,}609/R_3 = 11{,}609/100$
$= 116.09\ \Omega$
See Figure S-17.

FIGURE S-17

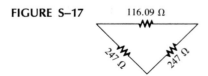

116.09 Ω

247 Ω   247 Ω

CHAPTER NINE

1. 2 loops, 5 nodes

2. The currents are shown in Figure S-18.

FIGURE S-18

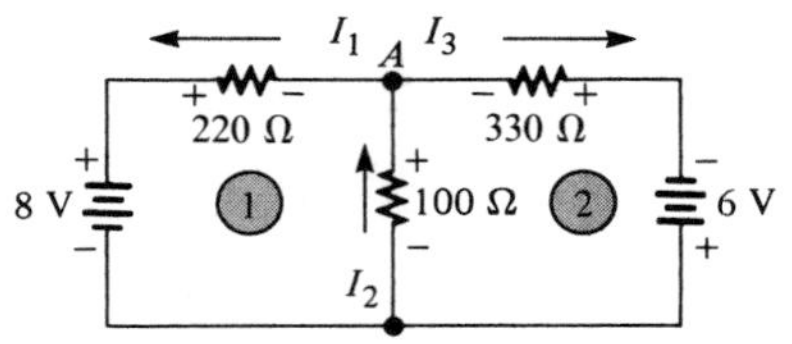

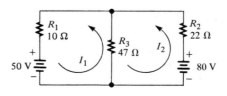

The calculations are as follows.
Loop 1: $220I_1 + 100I_2 - 8 = 0$
Loop 2: $100I_2 + 330I_3 + 6 = 0$
Node A: $I_1 - I_2 + I_3 = 0$
Solving, we obtain
$I_1 = I_2 - I_3$
$220(I_2 - I_3) + 100I_2 = 8$
$320I_2 - 220I_3 = 8$
$I_2 = (-6 - 330I_3)/100$
$320\left(\dfrac{-6 - 330I_3}{100}\right) - 220I_3 = 8$
$-19.2 - 1056I_3 - 220I_3 = 8$
$1276I_3 = -27.2$
$I_3 = -0.021\ A$
$100I_2 + 330(-0.021) = -6$
$I_2 = (-6 + 6.93)/100 = 0.0093\ A$
$I_1 = I_2 - I_3$
$= 0.0093\ A - (-0.021\ A)$
$= 0.0303\ A$

3. $I_1 = (15 - 5I_2)/2$
$6(15 - 5I_2)/2 - 8I_2 = 10$
$45 - 15I_2 - 8I_2 = 10$
$23I_2 = 35$
$I_2 = 35/23 = 1.52\ A$
Substituting, we have
$2I_1 + 5(1.52) = 15$
$I_1 = (15 - 7.6)/2 = 3.7\ A$

4. $I_1 = \dfrac{\begin{vmatrix} 25 & -12 \\ 18 & 3 \end{vmatrix}}{\begin{vmatrix} 15 & -12 \\ 9 & 3 \end{vmatrix}} = \dfrac{75 - (-216)}{45 - (-108)} = \dfrac{291}{153}$

$= 1.9\ A$

5. Third-order determinants are the limit for expansion method.

6. Fourth-order determinants are required for four simultaneous equations. The cofactor method must be used.

7. The currents are shown in Figure S-19.

FIGURE S-19

The calculations are as follows:

$$57I_1 - 47I_2 = 50$$
$$-47I_1 + 69I_2 = -80$$

$$I_1 = \frac{\begin{vmatrix} 50 & -47 \\ -80 & 69 \end{vmatrix}}{\begin{vmatrix} 57 & -47 \\ -47 & 69 \end{vmatrix}} = \frac{3450 - 3760}{3933 - 2209}$$

$$= \frac{-310}{1724} = -0.18 \text{ A}$$

$$I_2 = \frac{\begin{vmatrix} 57 & 50 \\ -47 & -80 \end{vmatrix}}{1724} = \frac{-4560 - (-2350)}{1724}$$

$$= \frac{-2210}{1724} = -1.28 \text{ A}$$

$$I_{R3} = I_2 - I_1 = 1.28 \text{ A} - 0.18 \text{ A}$$
$$= 1.1 \text{ A up}$$

8. The calculations are as follows:

$$I_1 + I_2 = I_3$$

$$\frac{-15 - V_A}{100} + \frac{-15 - V_A}{330} - \frac{V_A}{680} = 0$$

$$\frac{-V_A}{100} - \frac{V_A}{330} - \frac{V_A}{680} = \frac{15}{100} + \frac{15}{330}$$

$$\frac{-6.8V_A - 2.06V_A - V_A}{680} = \frac{49.5 + 15}{330}$$

$$\frac{-9.86V_A}{680} = \frac{64.5}{330}$$

$$V_A = -\frac{(680)(64.5)}{(9.86)(330)} = -13.48 \text{ V}$$

Also see Figure S-20.

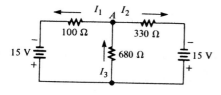

FIGURE S-20

CHAPTER TEN

1. Like poles repel.

2. A magnetic field consists of the lines of force from the north pole to the south pole of a magnet.

3. Flux density: teslas. Magnetic flux: webers. Magnetomotive force: ampere-turns.

4. Reluctance is the opposition to the establishment of a magnetic field. It is analogous to resistance.

5. Current in the coil of a recording head sets a magnetic field that varies with the direction and magnitude of the current. The magnetic field magnetizes the recording surface proportional to its direction and strength.

6. In a solenoid, the electromagnetic action produces a mechanical movement of a plunger or core. In a relay, the electromagnetic action produces the opening or closing of an electrical contact.

7. Movement of a conductor relative to a magnetic field.

8. True.

9. The induced voltage increases when the number of turns is increased.

10. When the tab cuts through and alters the magnetic field.

CHAPTER ELEVEN

1. Alternating current periodically reverses direction. Direct current is always in one direction.

2. A sine wave has one positive peak and one negative peak during each cycle.

3. A cycle is one repetition of a sine wave, consisting of a positive alternation and a negative alternation.

4. A periodic waveform repeats itself at regular, fixed intervals. A nonperiodic wave does not repeat at regular intervals.

5. The sine wave with the highest frequency, 20 kHz, is changing at the fastest rate.

6. The sine wave with the shortest period, 2 ms, is changing at the fastest rate.

7. 60 Hz = 60 cycles per second.
   Total cycles in 10 s = (60 cycles/s)(10 s) = 600 cycles.

8. $f = 1/T = 1/200 \text{ ms} = 5 \text{ Hz}$

9. $T = 1/f = 1/25 \text{ kHz} = 40 \ \mu s$

10. $T = 5 \ \mu s$
    $f = 1/T = 1/5 \ \mu s = 0.2 \text{ MHz}$

11. $V_p = 25 \text{ V}$
    $V_{pp} = 50 \text{ V}$
    $V_{rms} = 0.707V_p = 17.68 \text{ V}$
    $V_{avg} = 0 \text{ V}$

12. $V_p = 1.414V_{rms} = 1.414(115 \text{ V})$
$= 162.6$ *V*

13. $I_{rms} = V_{rms}/R = 10 \text{ V}/100 \text{ }\Omega = 0.1 \text{ A}$

14. 180 degrees.

15. $(57.3°/\text{rad})(2 \text{ rad}) = 114.6°$

16. $\theta = 45° - 10° = 35°$

17. $v = 15 \sin 32° = 7.95 \text{ V}$

18. $V_R = IR = (5 \text{ mA})(10 \text{ k}\Omega) = 50 \text{ V}$

19. $V_s = 6.5 \text{ V} + 3.2 \text{ V} = 9.7 \text{ V}$

20. The one with 50-μs pulses.

21. 50%.

22. A sawtooth has two ramps.

CHAPTER TWELVE

1. Phasors can represent quantities that have both magnitude and direction (angle).

2. $20° \equiv -340°$
$60° \equiv -300°$
$135° \equiv -225°$
$200° \equiv -160°$
$315° \equiv -45°$
$330° \equiv -30°$

3. $\omega = 2\pi f$
$f = \omega/(2\pi)$
$= (1000 \text{ rad/s})/(2\pi \text{ rad/cycle})$
$= 159 \text{ Hz}$

4. See Figure S–21.

5. See Figure S–22.

6. **(a)** $5 + j5 = \sqrt{5^2 + 5^2}\angle\tan^{-1}(5/5)$
$= 7.07\angle45°$

   (b) $12 + j9$
$= \sqrt{12^2 + 9^2}\angle\tan^{-1}(9/12)$
$= 15\angle36.87°$

   (c) $8 - j10$
$= \sqrt{8^2 + 10^2}\angle\tan^{-1}(-10/8)$
$= 12.8\angle-51.34°$

   (d) $100 - j50$
$= \sqrt{100^2 + 50^2}\angle\tan^{-1}(-50/100)$
$= 111.8\angle-26.57°$

7. **(a)** $1\angle45° = 1 \cos 45° + j1 \sin 45°$
$= 0.707 + j0.707$

   (b) $12\angle60° = 12 \cos 60° + j12 \sin 60°$
$= 6 + j10.39$

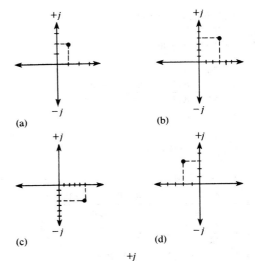

(a)　　　(b)

(c)　　　(d)

(e)

FIGURE S–21

   (c) $100\angle-80° = 100 \cos(-80°)$
$+ j100 \sin(-80°)$
$= 17.36 - j98.48$

   (d) $40\angle125° = 40 \cos 125°$
$+ j40 \sin 125°$
$= -22.94 + j32.77$

8. **(a)** First quadrant　　**(b)** First quadrant
　(c) Fourth quadrant　**(d)** Second quadrant

9. **(a)** $(5\angle20°)(2\angle45°)$
$= (5)(2)\angle(20° + 45°)$
$= 10\angle65°$

   (b) $(3.5\angle-15°)(6\angle120°)$
$= (3.5)(6)\angle(-15° + 120°)$
$= 21\angle105°$

10. **(a)** $(8 + j7) + (12 + j4)$
$= (8 + 12) + j(7 + 4)$
$= 20 + j11$

   (b) $(12 - j10) + (-8 + j5)$
$= (12 - 8) + j(-10 + 5)$
$= 4 - j5$

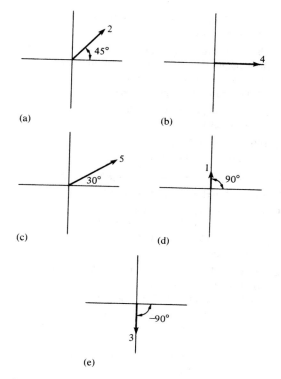

FIGURE S–22

CHAPTER THIRTEEN

1. (a) True   (b) True   (c) False
   (d) False

2. (a) False   (b) True   (c) False
   (d) True

3. $0.00001 \text{ F} = 10 \ \mu\text{F}$ and is larger than $0.01 \ \mu\text{F}$.

4. $0.0001 \ \mu\text{F} = 100 \text{ pF}$ and is smaller than 1000 pF.

5. Since $Q = CV$, an increase in V produces an increase in Q.

6. Since $W = (1/2)CV^2$, doubling V increases the stored energy by four times.

7. The voltage rating of a capacitor can be increased by increasing the plate separation or by using a dielectric with a greater dielectric constant.

8. All of the following increase the capacitance value:
   (c) Move plates closer;
   (d) Increase plate area;
   (e) Decrease the dielectric thickness.

9. $C_T < 0.05 \ \mu\text{F}$

10. $C_T = 4(0.02 \ \mu\text{F}) = 0.08 \ \mu\text{F}$

11. (a) $v_C = 0 \text{ V}$ at instant of switch closure.
    (b) The capacitor is fully charged at five time constants from switch closure.
    (c) $v_C = V_S$ when the capacitor is fully charged.
    (d) $I = 0 \text{ A}$ when the capacitor is fully charged.

12. The current increases because the capacitive reactance decreases with an increase in frequency.

13. When the frequency is decreased, the capacitor has the greatest voltage because X_C becomes larger than R.

14. The capacitor is very leaky (the leakage resistance is low).

15. $C = Q/V = (50 \times 10^{-6} \text{ C})/5 \text{ V}$
    $= 10 \times 10^{-6} \text{ F} = 10 \ \mu\text{F}$

16. $V = Q/C = (5 \times 10^{-8} \text{ C})/0.01 \ \mu\text{F}$
    $= (5 \times 10^{-8} \text{ C})/(0.01 \times 10^{-6} \text{ F})$
    $= 500 \times 10^{-2} \text{ V} = 5 \text{ V}$

17. $(0.005 \ \mu\text{F})(10^6 \text{ pF}/\mu\text{F}) = 5000 \text{ pF}$

18. $(2000 \text{ pF})(10^{-6} \ \mu\text{F}/\text{pF}) = 0.002 \ \mu\text{F}$

19. $(1500 \ \mu\text{F})(0.000001 \text{ F}/\mu\text{F}) = 0.0015 \text{ F}$

20. $W = \frac{1}{2}CV^2 = \frac{1}{2}(10 \times 10^{-6} \text{ F})(100 \text{ V})^2$
    $= 0.05 \text{ J}$

21. $\epsilon = \epsilon_r\epsilon_0 = (1200)(8.85 \times 10^{-12} \text{ F/m})$
    $= 1.062 \times 10^{-8} \text{ F/m}$

22. $C = \dfrac{A\epsilon_r(8.85 \times 10^{-12})}{d}$
    $= \dfrac{(0.009 \text{ m}^2)(2)(8.85 \times 10^{-12} \text{ F/m})}{0.0015 \text{ m}}$
    $= 106.2 \text{ pF}$

23. $\tau = RC = (2 \text{ k}\Omega)(0.05 \ \mu\text{F})$
    $= 0.1 \times 10^{-3} \text{ s} = 0.1 \text{ ms}$

24. $v_C = (0.37)(15 \text{ V}) = 5.55 \text{ V}$

25. $v = 10 \text{ V}(1 - e^{-15\text{ms}/10\text{ms}})$
    $= 10 \text{ V}(0.777)$
    $= 7.77 \text{ V}$

26. (a) $1/C_T = (1/0.05 \ \mu\text{F}) + (1/0.01 \ \mu\text{F})$
    $+ (1/0.02 \ \mu\text{F}) + (1/0.05 \ \mu\text{F})$
    $= 1.9 \times 10^8$
    $C_T = 0.00526 \ \mu\text{F}$
    (b) $C_T = (300 \text{ pF})(100 \text{ pF})/400 \text{ pF}$
    $= 75 \text{ pF}$
    (c) $C_T = 12 \text{ pF}/3 = 4 \text{ pF}$

27. $1/C_T = (1/0.025 \ \mu F)$
$\qquad + (1/0.04 \ \mu F) + (1/0.1 \ \mu F)$
$\qquad = 7.5 \times 10^7$
$\quad C_T = 0.0133 \ \mu F$
$\quad V_1 = (C_T/C_1)V_T$
$\qquad = (0.0133 \ \mu F/0.025 \ \mu F)250 \ V$
$\qquad = 133 \ V$
$\quad V_2 = (C_T/C_2)V_T$
$\qquad = (0.0133 \ \mu F/0.04 \ \mu F)250 \ V$
$\qquad = 83.13 \ V$
$\quad V_3 = (C_T/C_3)V_T$
$\qquad = (0.0133 \ \mu F/0.1 \ \mu F)250 \ V$
$\qquad = 33.25 \ V$

28. (a) $C_T = 100 \ pF + 47 \ pF + 33 \ pF$
$\qquad + 10 \ pF = 190 \ pF$
(b) $C_T = (5)(0.008 \ \mu F) = 0.04 \ \mu F$

29. $X_C = \dfrac{1}{2\pi f C}$

$\qquad = \dfrac{1}{2\pi (2 \times 10^6 \ Hz)(1 \times 10^{-6} \ F)}$
$\qquad = 0.0796 \ \Omega$

30. $X_C = V_{rms}/I_{rms} = 25 \ V/50 \ mA$
$\qquad = 0.5 \ k\Omega$

$\quad X_C = \dfrac{1}{2\pi f C}$

$\quad f = \dfrac{1}{2\pi X_C C}$

$\qquad = \dfrac{1}{2\pi (500 \ \Omega)(0.3 \times 10^{-6} \ F)}$
$\qquad = 1.06 \ kHz$

31. $P_{true} = 0$
$\quad P_r = \dfrac{V_{rms}^2}{X_C} = \dfrac{(25 \ V)^2}{500 \ \Omega} = 1.25 \ VAR$

32. The capacitor has a very low resistance, which indicates it is faulty.

CHAPTER FOURTEEN

1. $0.000005 \ H = 5 \ \mu H$ and is larger than $0.05 \ \mu H$.

2. Since $0.33 \ mH = 330 \ \mu H$, the 33-μH inductance is the smaller.

3. Since $W = (1/2)LI^2$, the stored energy increases when the current increases.

4. Since $W = (1/2)LI^2$, a doubling of the current increases the stored energy by four times.

5. The winding resistance can be decreased by using larger-sized wire.

6. The following result in an increase in inductance:
(b) Change from an air core to an iron core;
(c) Increase the number of turns;
(e) Decrease the length of the core.

7. $v_{ind} = N(d\phi/dt) = 50(25 \ Wb/s)$
$\qquad = 1250 \ V$

8. $1500 \ \mu H = 1.5 \ mH,$
$\quad 20 \ mH = 20,000 \ \mu H$

9. $\tau = L/R = 10 \ mH/2 \ k\Omega = 5 \ \mu s$

10. $L_T = 75 \ \mu H + 20 \ \mu H = 95 \ \mu H$

11. $1/L_T = (1/1 \ H) + (1/1 \ H)$
$\qquad + (1/0.5 \ H) + (1/0.5 \ H)$
$\qquad + (1/2 \ H) = 6.5$
$\quad L_T = 0.154 \ H$

12. (a) $v_L = V_S$ at the instant of switch closure.
(b) $i = 0 \ A$ at the instant of switch closure.
(c) The current reaches maximum at five time constants from switch closure.
(d) $v_L = 0 \ V$ at five time constants after switch closure.
(e) $v_R = 0 \ V$ at instant of switch closure and $v_R = V_S$ at five time constants.

13. The current decreases because X_L increases with an increase in frequency.

14. The resistor voltage is greatest because X_L decreases with a decrease in frequency.

15. $f = 2 \ MHz$
$\quad X_L = 2\pi f L = 2\pi (2 \ MHz)(15 \ \mu H)$
$\qquad = 188.5 \ \Omega$

16. $X_L = V_{rms}/I_{rms} = 25 \ V/50 \ mA$
$\qquad = 0.5 \ k\Omega$
$\quad X_L = 2\pi f L$
$\quad f = \dfrac{X_L}{2\pi L} = \dfrac{500 \ \Omega}{2\pi (8 \ \mu H)} = 9.95 \ MHz$

17. $L = \dfrac{X_L}{2\pi f} = \dfrac{10 \ M\Omega}{2\pi (50 \ MHz)} = 31.83 \ mH$

18. The inductor is open.

CHAPTER FIFTEEN

1. $k = 0.9$

2. $L_M = k\sqrt{L_1 L_2} = 0.9\sqrt{(10 \ \mu H)(5 \ \mu H)}$
$\qquad = 6.36 \ \mu H$

3. $N_s/N_p = 36/12 = 3$. It is a step-up transformer.

4. See Figure S–23.

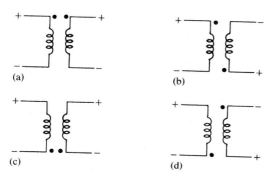

(a)

(b)

(c)

(d)

FIGURE S–23

5. $N_s/N_p = 15/10 = 1.5$
$V_s = 1.5V_p = 1.5(120 \text{ V}) = 180 \text{ V}$

6. $V_s = 0.5V_p = 0.5(50 \text{ V}) = 25 \text{ V}$

7. $N_s/N_p = V_s/V_p = 60 \text{ V}/240 \text{ V} = 0.25$

8. $I_s = (N_p/N_s)I_p = (1/0.25)(0.25 \text{ A})$
$= 1 \text{ A}$

9. $Z_p = (N_p/N_s)^2 Z_L = (1/5)^2(1 \text{ k}\Omega)$
$= 40 \ \Omega$

10. $R_s = R_{\text{reflected}}$
$= 10 \ \Omega$ for maximum power to R_L
$R_{\text{reflected}} = R_p = (N_p/N_s)^2(R_L)$
$(N_p/N_s)^2(R_L) = 10 \ \Omega$
$R_L = 10 \ \Omega/0.04 = 250 \ \Omega$

11. 1: $V_1 = (N_s/N_p)V_s = (100/10)120 \text{ V}$
$= 1200 \text{ V}$
2: $V_2 = (10/10)120 \text{ V} = 120 \text{ V}$
3: $V_3 = (5/10)\ 120 \text{ V} = 60 \text{ V}$

12. 1200 V/2 = 600 V. The polarities are opposite.

13. $V_s = 0$ V because dc voltage cannot be coupled in a transformer.

14. Transformer losses reduce the ideal output voltage.

15. 95% of the primary magnetic flux lines pass through the secondary.

CHAPTER SIXTEEN

1. Both V_C and V_R are less than V_s in magnitude and are different in phase.

2. (a) V_R is in phase with I.
(b) V_C lags I by 90 degrees.

3. Both Z and θ decrease as frequency increases.

4. Z remains the same because X_C is halved and R is doubled.

5. Decrease the frequency to decrease the current.

6. $V_s = \sqrt{V_R^2 + V_C^2} = \sqrt{(10 \text{ V})^2 + (10 \text{ V})^2}$
$= 14.14 \text{ V}$

7. Frequency must be increased to reduce X_C so that $R > X_C$.

8. The phase angle decreases.

9. The impedance decreases due to a decrease in X_C.

10. $I_T = \sqrt{I_R^2 + I_C^2} = \sqrt{(1 \text{ A})^2 + (1 \text{ A})^2}$
$= 1.414 \text{ A}$

11. $PF = 1$ means a purely resistive circuit.

12. $P_a = 5 \text{ VA}$

13. $P_a = \sqrt{(100 \text{ W})^2 + (100 \text{ VAR})^2} = 141.4 \text{ VA}$

14. To avoid exceeding the maximum allowable current.

15. $Z_T = R - jX_C$
$= 2.7 \text{ k}\Omega - j6.37 \text{ k}\Omega$
$= 6.92\angle -67° \text{ k}\Omega$

16. $Z_T = 2.7 \text{ k}\Omega - j3.19 \text{ k}\Omega$
$= 4.18\angle -49.8° \text{ k}\Omega$

17. $X_C = R$ at 45°
$X_C = 10 \text{ k}\Omega$
$1/(2\pi fC) = 10 \text{ k}\Omega$
$f = \dfrac{1}{2\pi(10 \text{ k}\Omega)(100 \text{ pF})} = 159 \text{ kHz}$

18. $\tan \theta = X_C/R$
$X_C = \dfrac{1}{2\pi fC} = \dfrac{1}{2\pi(5 \text{ kHz})(0.01 \ \mu\text{F})}$
$= 3.18 \text{ k}\Omega$
$X_C/R = \tan \theta$
$R = X_C/\tan \theta = 3.18 \text{ k}\Omega/\tan 60°$
$= 3.18 \text{ k}\Omega/1.732$
$= 1.84 \text{ k}\Omega$

19. $\theta = \tan^{-1}(X_C/R)$
$X_C = \dfrac{1}{2\pi fC} = \dfrac{1}{2\pi(100 \text{ kHz})(100 \text{ pF})}$
$= 15.9 \text{ k}\Omega$
$\theta = \tan^{-1}(15.9 \text{ k}\Omega/10 \text{ k}\Omega) = 57.83°$

20. $\theta = \tan^{-1}(1000 \ \Omega/560 \ \Omega) = 60.75°$

21. $\theta = \tan^{-1}(X_C/R) = \tan^{-1}(X_C/2X_C)$
$= \tan^{-1} 0.5 = 26.57°$

22. (a) $\mathbf{Z} = 47\,\Omega - j30\,\Omega$
(b) $\mathbf{Z} = 27\,\Omega - j100\,\Omega$

23. (a) $\mathbf{Z} = 55.76\angle -32.55°\,\Omega$
$\mathbf{I} = \dfrac{\mathbf{V}_s}{\mathbf{Z}} = \dfrac{10\angle 0°\text{ V}}{55.76\angle -32.55°\,\Omega}$
$= 179\angle 32.55°\text{ mA}$
(b) $\mathbf{Z} = 103.6\angle -74.89°\,\Omega$
$\mathbf{I} = \dfrac{10\angle 0°\text{ V}}{103.6\angle -74.89°\,\Omega}$
$= 96.5\angle 74.89°\text{ mA}$

24. $V = \sqrt{(6\text{ V})^2 + (4.5\text{ V})^2}$
$= 7.5\text{ V}$

25. (a) $\mathbf{Y} = G + jB_C$
$= 3.7 \times 10^{-5}\text{ S} + j1 \times 10^{-4}\text{ S}$
$= 1.07 \times 10^{-4}\angle 69.7°\text{ S}$
$\mathbf{Z} = 1/\mathbf{Y} = 9.38\angle -69.7°\text{ k}\Omega$
(b) $X_C = 159\,\Omega$
$\mathbf{Y} = 0.01\text{ S} + j6.28 \times 10^{-3}\text{ S}$
$= 1.18 \times 10^{-2}\angle 32.13°\text{ S}$
$\mathbf{Z} = 1/\mathbf{Y} = 84.7\angle -32.13°\,\Omega$

26. (a) $\mathbf{I}_T = \dfrac{\mathbf{V}_s}{\mathbf{Z}} = \dfrac{5\angle 0°\text{ V}}{9.38\angle -68.7°\text{ k}\Omega}$
$= 0.533\angle 69.7°\text{ mA}$
$\mathbf{I}_C = \dfrac{\mathbf{V}_s}{\mathbf{X}_C} = \dfrac{5\angle 0°\text{ V}}{10\angle -90°\text{ k}\Omega}$
$= 0.5\angle 90°\text{ mA}$
$\mathbf{I}_R = \dfrac{\mathbf{V}_s}{\mathbf{R}} = \dfrac{5\angle 0°\text{ V}}{27\angle 0°\text{ k}\Omega}$
$= 0.185\angle 0°\text{ mA}$
(b) $\mathbf{I}_T = \dfrac{\mathbf{V}_s}{\mathbf{Z}} = \dfrac{5\angle 0°\text{ V}}{84.7\angle -32.13°\,\Omega}$
$= 59\angle 32.13°\text{ mA}$
$\mathbf{I}_C = \dfrac{5\angle 0°\text{ V}}{159\angle -90°\,\Omega}$
$= 31.45\angle 90°\text{ mA}$
$\mathbf{I}_R = \dfrac{5\angle 0°\text{ V}}{100\angle 0°\,\Omega} = 50\angle 0°\text{ mA}$

27. (a) $\theta = 69.7°$ V lags I
(b) $\theta = 32.13°$ V lags I

28. $X_C = \dfrac{1}{2\pi(1\text{ kHz})(0.005\ \mu\text{F})}$
$= 31.83\text{ k}\Omega$
$\phi = -90° + \tan^{-1}(31.83\text{ k}\Omega/33\text{ k}\Omega)$
$= -90° + 43.97° = -46.03°$

29. $V_{\text{out}} = \dfrac{X_C}{\sqrt{R^2 + X_C^2}}\,V_{\text{in}}$
$= \dfrac{31.83\text{ k}\Omega}{\sqrt{(33\text{ k}\Omega)^2 + (31.83\text{ k}\Omega)^2}}\,5\text{ V}$
$= 3.47\text{ V}$

30. $X_C = \dfrac{1}{2\pi(50\text{ kHz})(0.1\ \mu\text{F})} = 31.83\,\Omega$
$\theta = \tan^{-1}(X_C/R) = \tan^{-1}(31.83/82)$
$= 21.22°$

31. $V_{\text{out}} = \dfrac{R}{\sqrt{R^2 + X_C^2}}\,V_{\text{in}}$
$= \dfrac{82\,\Omega}{\sqrt{(82\,\Omega)^2 + (31.83\,\Omega)^2}}\,12\text{ V}$
$= 11.19\text{ V}$

32. 200 Hz. This filter is a low-pass.

33. 5 kHz. This filter is a high-pass.

34. Figure 16–63:
$I = 5\text{ V}/Z = 5\text{ V}/45.85\text{ k}\Omega$
$= 0.109\text{ mA}$
$P_{\text{true}} = I^2R = (0.109\text{ mA})^2(33\text{ k}\Omega)$
$= 0.392\text{ mW}$
$P_r = I^2X_C = (0.109\text{ mA})^2(31.83\text{ k}\Omega)$
$= 0.378\text{ mVAR}$
$P_a = I^2Z = (0.109\text{ mA})^2(45.85\text{ k}\Omega)$
$= 0.545\text{ mVA}$
Figure 16–64:
$I = 12\text{ V}/Z = 12\text{ V}/87.96\,\Omega$
$= 0.136\text{A}$
$P_{\text{true}} = I^2R = (0.136\text{ A})^2(82\,\Omega)$
$= 1.52\text{ W}$
$P_r = I^2X_C = (0.136\text{ A})^2(31.83\,\Omega)$
$= 0.589\text{ VAR}$
$P_a = I^2Z = (0.136\text{ A})^2(87.96\,\Omega)$
$= 1.63\text{ VA}$

35. Figure 16–63:
$\theta = \tan^{-1}(X_C/R)$
$= \tan^{-1}(31.83\text{ k}\Omega/33\text{ k}\Omega)$
$= 43.97°$
$PF = \cos\theta = \cos 43.97° = 0.72$
Figure 16–64:
$\theta = \tan^{-1}(X_C/R)$
$= \tan^{-1}(31.83\,\Omega/82\,\Omega)$
$= 21.22°$
$PF = \cos\theta = \cos 21.22° = 0.93$

CHAPTER SEVENTEEN

1. Both V_R and V_L are less than V_s in magnitude and differ in phase angle.

2. **(a)** V_R is in phase with I.
   (b) V_L leads I by 90 degrees.

3. Both Z and θ increase as frequency increases.

4. Z does not change.

5. Increase the frequency to reduce the current.

6. $V_s = \sqrt{(10\text{ V})^2 + (10\text{ V})^2} = 14.14$ V rms
   $V_s = (1.414)(14.14\text{ V}) = 19.99$ V peak

7. The frequency must be decreased to reduce X_L so that $R > X_L$.

8. The phase angle decreases.

9. Z increases due to an increase in X_L.

10. $I_T = \sqrt{(2\text{ A})^2 + (2\text{ A})^2} = 2.83$ A

11. 3 div × 18°/div = 54°

12. 0.9

13. $P_a = 10$ VA

14. $P_a = \sqrt{(10\text{ W})^2 + (10\text{ VAR})^2} = 14.14$ VA

15. $X_L = 2\pi f L = 1.26$ kΩ
   $\mathbf{Z} = R + jX_L = 3.3\text{ kΩ} + j1.26\text{ kΩ}$
   $= 3.53\angle 20.9°$ kΩ

16. $X_L = 2\pi f L = 2\pi(5\text{ kHz})(20\text{ mH})$
   $= 628$ Ω
   $Z = \sqrt{R^2 + X_L^2}$
   $= \sqrt{(3.3\text{ kΩ})^2 + (628\text{ Ω})^2} = 3.36$ kΩ
   $\theta = \tan^{-1}(X_L/R) = 10.9°$

17. $X_L = R = 10$ kΩ at $\theta = 45°$
   $f = X_L/2\pi L = 10\text{ kΩ}/2\pi(100\text{ mH})$
   $= 15.9$ kHz

18. $X_L = 2\pi f L = 2\pi(5\text{ kHz})(200\text{ μH})$
   $= 6.28$ Ω
   $\tan 60° = 6.28\text{ Ω}/R$
   $R = 6.28\text{ Ω}/\tan 60° = 6.28\text{ Ω}/1.73$
   $= 3.63$ Ω

19. $\theta = \tan^{-1}(X_L/R)$
   $= \tan^{-1}(100\text{ Ω}/560\text{ Ω})$
   $= 10.13°$

20. $\mathbf{Z} = 47\text{ kΩ} + j20\text{ kΩ}$
   $= 51.1\angle 23.05°$ kΩ
   $\mathbf{I} = \dfrac{120\angle 0°\text{ V}}{51.1\angle 23.05°\text{ kΩ}}$
   $= 2.35\angle -23.05°$ mA

$\mathbf{V}_R = \mathbf{IR}$
   $= (2.35\angle -23.05°\text{ mA})(47\angle 0°\text{ kΩ})$
   $= 110.5\angle -23.05°$ V
$\mathbf{V}_L = \mathbf{I}X_L$
   $= (2.35\angle -23.05\text{ mA})(20\angle 90°\text{ kΩ})$
   $= 47\angle 66.95°$ V

21. **(a)** $\mathbf{Y}_T = (1/27\text{ kΩ}) - j(1/40\text{ kΩ})$
   $= 3.7 \times 10^{-5}\text{ S} - j2.5 \times 10^{-5}$ S
   $= 4.47 \times 10^{-5}\angle -34°$ S
   $\mathbf{Z}_T = 1/\mathbf{Y}_T = 22.4\angle 34°$ kΩ
   (b) $X_L = 2\pi f L = 2\pi(1\text{ kHz})(2\text{ H})$
   $= 12.57$ kΩ
   $\mathbf{Y}_T = \dfrac{1}{10\text{ kΩ}} - j\dfrac{1}{12.57\text{ kΩ}}$
   $= 0.1 \times 10^{-3}$ S
   $\quad - j0.0796 \times 10^{-3}$ S
   $= 10 \times 10^{-5}$ S
   $\quad -j7.96 \times 10^{-5}$ S
   $= 12.78 \times 10^{-5}\angle -38.52°$ S
   $\mathbf{Z}_T = 1/\mathbf{Y}_T = 7.82\angle 38.52°$ kΩ

22. **(a)** $\mathbf{I}_T = \dfrac{5\angle 0°\text{ V}}{22.4\angle 34°\text{ kΩ}}$
   $= 0.223\angle -34°$ mA
   $\mathbf{I}_R = \dfrac{5\angle 0°\text{ V}}{27\angle 0°\text{ kΩ}} = 0.185\angle 0°$ mA
   $\mathbf{I}_L = \dfrac{5\angle 0°\text{ V}}{40\angle 90°\text{ kΩ}}$
   $= 0.125\angle -90°$ mA
   (b) $\mathbf{I}_T = \dfrac{10\angle 0°\text{ V}}{7.82\angle 38.52°\text{ kΩ}}$
   $= 1.28\angle -38.52°$ mA
   $\mathbf{I}_R = \dfrac{10\angle 0°\text{ V}}{10\angle 0°\text{ kΩ}} = 1\angle 0°$ mA
   $\mathbf{I}_L = \dfrac{10\angle 0°\text{ V}}{12.57\angle 90°\text{ kΩ}}$
   $= 0.796\angle -90°$ mA

23. $X_{L1} = 2\pi f L_1 = 3.14$ kΩ
   $X_{L2} = 2\pi f L_2 = 1.26$ kΩ
   $\mathbf{Y}_2 = \dfrac{1}{2.2\text{ kΩ}} - j\dfrac{1}{1.26\text{ kΩ}}$
   $= 0.91 \times 10^{-3}\angle -60.2°$ S
   $\mathbf{Z}_2 = 1/\mathbf{Y}_2 = 1.1\angle 60.2°$ kΩ
   $= 547\text{ Ω} + j955$ Ω
   $\mathbf{Z}_T = \mathbf{Z}_1 + \mathbf{Z}_2$
   $= (3.3\text{ kΩ} + j3.14\text{ kΩ})$
   $\quad + (547\text{ Ω} + j955\text{ Ω})$
   $= 3.85\text{ kΩ} + j4.1\text{ kΩ}$
   $= 5.62\angle 46.8°$ kΩ

24. $X_L = 2\pi fL = 2\pi(60 \text{ Hz})(0.2 \text{ H})$
$= 75.4 \text{ }\Omega$
$\phi = 90° - \tan^{-1}(75.4/100) = 52.98°$

25. $V_{out} = V_L = (X_L/\sqrt{R^2 + X_L^2})V_{in}$
$= (75.4 \text{ }\Omega/\sqrt{(100 \text{ }\Omega)^2 + (75.4 \text{ }\Omega)^2})5 \text{ V}$
$= 3.01 \text{ V}$

26. $V_{out} = V_R = (R/\sqrt{R^2 + X_L^2})V_{in}$
$= (1.2 \text{ k}\Omega/\sqrt{(1.2 \text{ k}\Omega)^2 + (628.3 \text{ }\Omega)^2})6 \text{ V}$
$= 5.32 \text{ V}$
$\phi = -\tan^{-1}(628.3 \text{ }\Omega/1.2 \text{ k}\Omega)$
$= -27.6°$

27. At 2 kHz, because X_L is greater.

28. $I = 6 \text{ V}/1354.5 \text{ }\Omega = 4.4 \text{ mA}$
(a) $P_{true} = I^2R = (4.4 \text{ mA})^2(1.2 \text{ k}\Omega)$
$= 23.23 \text{ mW}$
(b) $P_r = I^2X_L = (4.4 \text{ mA})^2(628.3 \text{ }\Omega)$
$= 12.16 \text{ mVAR}$
(c) $P_a = I^2Z = (4.4 \text{ mA})^2(1354.5 \text{ }\Omega)$
$= 26.2 \text{ mVA}$
(d) $PF = \cos \theta = \cos(27.6°) = 0.886$

CHAPTER EIGHTEEN

1. $X_T = 0$

2. $\theta = 0°$

3. $Z = R_W = 80 \text{ }\Omega$

4. I leads V_s because the circuit is capacitive ($X_C > X_L$).

5. Since $f_r = 1/2\pi\sqrt{LC}$, an increase in C causes a decrease in f_r.

6. $V_s = V_R = 50 \text{ V}$
$V_L + V_C = 150 \text{ V} - 150 \text{ V} = 0 \text{ V}$

7. BW increases when the Q is lowered.

8. Because the circuit is predominantly inductive.

9. $I_T = 0 \text{ A}$ at parallel resonance.

10. Increase C to decrease f_r.

11. $Q \geq 10$, as a rule of thumb.

12. A reduction in R_{Tp} reduces the Q and thereby increases BW.

13. $Z = R + j(X_L - X_C)$
$= 75 \text{ }\Omega + j(40 \text{ }\Omega - 80 \text{ }\Omega)$
$= 75 \text{ }\Omega - j40 \text{ }\Omega$
$= 85\angle -28.1° \text{ }\Omega$

14. (a) $I = \dfrac{V_s}{Z} = \dfrac{10\angle 0° \text{ V}}{85\angle -28.1° \text{ }\Omega}$
$= 117.6\angle 28.1° \text{ mA}$
(b) $V_R = IR$
$= (117.6\angle 28.1° \text{ mA})(75\angle 0° \text{ }\Omega)$
$= 8.82\angle 28.1° \text{ V}$
(c) $V_L = IX_L$
$= (117.6\angle 28.1° \text{ mA})(40\angle 90° \text{ }\Omega)$
$= 4.7\angle 118.1° \text{ V}$
(d) $V_C = IX_C$
$= (117.6\angle 28.1° \text{ mA})(80\angle -90° \text{ }\Omega)$
$= 9.4\angle -61.9° \text{ V}$
(e) $P_{true} = I^2R = (117.6 \text{ mA})^2(75 \text{ }\Omega)$
$= 1.03 \text{ W}$
(f) $P_{r(net)} = I^2(X_L - X_C)$
$= (117.6 \text{ mA})^2(40 \text{ }\Omega)$
$= 0.553 \text{ VAR}$

15. $f_r = \dfrac{1}{2\pi\sqrt{LC}} = \dfrac{1}{2\pi\sqrt{(20 \text{ mH})(12 \text{ pF})}}$
$= 324.9 \text{ kHz}$
$Z_r = R = 15 \text{ }\Omega$

16. $f_r + 10 \text{ kHz} = 334.9 \text{ kHz}$
$X_L = 2\pi fL = 2\pi(334.9 \text{ kHz})(20 \text{ mH})$
$= 42.1 \text{ k}\Omega$
$X_C = \dfrac{1}{2\pi fC} = \dfrac{1}{2\pi(334.9 \text{ kHz})(12 \text{ pF})}$
$= 39.6 \text{ k}\Omega$
$Z = \sqrt{R^2 + (X_L - X_C)^2} \approx 2.5 \text{ k}\Omega$
$\theta = \tan^{-1}(2.5 \text{ k}\Omega/15) = 89.66°$

17. (a) $I = V_s/R = 1 \text{ V}/33 \text{ }\Omega = 30.3 \text{ mA}$
(b) $f_r = \dfrac{1}{2\pi\sqrt{LC}} = 97.95 \text{ kHz}$
(c) $V_R = V_s = 1 \text{ V}$
(d) $X_L = 2\pi f_rL = 4.92 \text{ k}\Omega$
$Q = X_L/R = 4.92 \text{ k}\Omega/33 \text{ }\Omega$
$= 149$
$V_L = QV_s = 149 \text{ V}$
(e) $V_C = V_L = 149 \text{ V}$ (180 degrees out-of-phase with V_L)
(f) $\theta = 0°$ since $Z = R$

18. $Y = 1/R + j(1/X_C) - j(1/X_L)$
$= 1/15 \text{ }\Omega + j(1/10 \text{ }\Omega) - j(1/20 \text{ }\Omega)$
$= 0.067 \text{ S} + j0.05 \text{ S}$
$= 0.084\angle 36.7° \text{ S}$
$Z = 1/Y = 11.9\angle -36.7° \text{ }\Omega$

19. $\mathbf{I}_R = \mathbf{V}_s/R = 1\angle 0°\ \text{V}/15\angle 0°\ \Omega$
$= 67\angle 0°\ \text{mA}$
$= \mathbf{V}_s/\mathbf{X}_C = 1\angle 0°\ \text{V}/10\angle -90°\ \Omega$
$= 100\angle 90°\ \text{mA}$
$\mathbf{I}_L = \mathbf{V}_s/\mathbf{X}_L = 1\angle 0°\ \text{V}/20\angle 90°\ \Omega$
$= 50\angle -90°\ \text{mA}$

20. (a) $\mathbf{Z}_A = R_2 - jX_C = 22\ \Omega - j29\ \Omega$
$= 36.4\angle -52.8°\ \Omega$
$\mathbf{Z}_B = R_3 + jX_L = 18\ \Omega + j43\ \Omega$
$= 46.6\angle 67.3°\ \Omega$

$$\mathbf{Z}_C = \frac{\mathbf{Z}_A\mathbf{Z}_B}{\mathbf{Z}_A + \mathbf{Z}_B}$$

$$= \frac{(36.4\angle -52.8°\ \Omega)(46.6\angle 67.3°\ \Omega)}{42.4\angle 19.3°\ \Omega}$$

$= 40\angle -4.8°\ \Omega$
$\mathbf{Z}_T = R_1 + \mathbf{Z}_C$
$= 8\ \Omega + 39.8\ \Omega - j3.35$
$= 47.8 - j3.35$
$= 47.9\angle -4°\ \Omega$

(b) $\mathbf{I}_T = 12\angle 0°\ \text{V}/47.9\angle -4°\ \Omega$
$= 0.25\angle 4°\ \text{A}$

(c) $\mathbf{I}_L = \left(\dfrac{\mathbf{Z}_T}{\mathbf{Z}_B}\right)\mathbf{I}_T$

$= \left(\dfrac{47.9\angle -4°\ \Omega}{46.6\angle 67.3°\ \Omega}\right)0.25\angle 4°\ \text{A}$

$= 0.26\angle -67.3°\ \text{A}$

(d) $\mathbf{I}_C = \left(\dfrac{\mathbf{Z}_T}{\mathbf{Z}_A}\right)\mathbf{I}_T$

$= \left(\dfrac{47.9\angle -4°\ \Omega}{36.4\angle -52.8°\ \Omega}\right)0.25\angle 4°\ \text{A}$

$= 0.33\angle 52.8°\ \text{A}$

21. $f_r = \dfrac{\sqrt{1 - (R_W^2 C/L)}}{2\pi\sqrt{LC}}$

$= \dfrac{\sqrt{1 - [(50\ \Omega)^2(0.02\ \mu\text{F})/2\ \text{H}]}}{2\pi\sqrt{(2\ \text{H})(0.02\ \mu\text{F})}}$

$\cong 795.8\ \text{Hz}$
$X_L = 2\pi f_r L = 2\pi(795.8\ \text{Hz})(2\ \text{H})$
$= 10\ \text{k}\Omega$
$Q = X_L/R_W = 10\ \text{k}\Omega/50\ \Omega = 200$
$Z_r = R_W(Q^2 + 1) = (50\ \Omega)(200^2 + 1)$
$\cong 2\ \text{M}\Omega$

22. $I_r = 5\ \text{V}/Z_r = 5\ \text{V}/2\ \text{M}\Omega = 2.5\ \mu\text{A}$

23. $f_r = \dfrac{\sqrt{1 - (R_W^2 C/L)}}{2\pi\sqrt{LC}} \cong 318.3\ \text{kHz}$

$X_L = 2\pi f_r L = 5\ \text{k}\Omega;\ Q = X_L/R$
$= 5\ \text{k}\Omega/6\ \Omega = 833$
$BW = f_r/Q = 318.3\ \text{kHz}/833$
$= 382\ \text{Hz}$

24. $f_2 = 8\ \text{kHz} + 2\ \text{kHz} = 10\ \text{kHz}$
$f_r = (f_1 + f_2)/2 = (8\ \text{kHz} + 10\ \text{kHz})/2$
$= 9\ \text{kHz}$

CHAPTER NINETEEN

1. $V_{\text{out}} = 0.707V_{\text{max}} = (0.707)(10\ \text{V})$
$= 7.07\ \text{V}$

2. $V_{\text{out}} = 10\ \text{V dc}$; the sine wave frequency is shorted to ground by the capacitor.

3. V_{out} is a 15-V peak-to-peak sine wave with a *zero* average value; the dc component is blocked by the capacitor.

4. $X_C = (1/10)2.2\ \text{k}\Omega = 220\ \Omega$

$$C = \frac{1}{2\pi f X_C} = \frac{1}{2\pi(1\ \text{kHz})(220\ \Omega)}$$

$= 0.723\ \mu\text{F}$
Use next highest standard value.

5. $V_{\text{out}} = \left(\dfrac{R}{\sqrt{R^2 + X_L^2}}\right)V_{\text{in}}$

$$= \left(\frac{5\ \Omega}{\sqrt{(4.7\ \Omega)^2 + (100\ \Omega)^2}}\right)5\ \text{V}$$

$= 0.25\ \text{V peak-to-peak}$
$V_{\text{avg}} = 8\ \text{V}$
The output is a 0.25-V peak-to-peak sine wave riding on an 8-V dc level.

6. See Figure S–24.

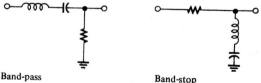

Band-pass   Band-stop

FIGURE S–24

7. See Figure S–25.

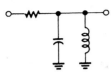

Band-pass   Band-stop

FIGURE S–25

8. $BW = f_r/Q = 5 \text{ kHz}/20 = 250 \text{ Hz}$

9. $20 \log(4/12) = -9.54 \text{ dB}$

10. $10 \log(5/10) = -3 \text{ dB}$

11. $f_c = \dfrac{1}{2\pi RC} = \dfrac{1}{2\pi(1 \text{ k}\Omega)(0.02 \ \mu\text{F})}$
$= 7.96 \text{ kHz}$

12. $f_c = \dfrac{1}{2\pi(L/R)} = \dfrac{1}{2\pi(0.01 \text{ mH}/5.6 \text{ k}\Omega)}$
$= 89.13 \text{ MHz}$

CHAPTER TWENTY

1. With V_2 zeroed:
$$\mathbf{Z}_{V1} = \mathbf{X}_C + \mathbf{R} \| \mathbf{X}_L$$
$$= 100\angle -90° \ \Omega + 47\angle 0° \ \Omega \| 25\angle 90° \ \Omega$$
$$= 100\angle -90° \ \Omega + 22.07\angle 62° \ \Omega$$
$$= 10 \ \Omega - j80.51 \ \Omega$$
$$= 81.13\angle -82.92° \ \Omega$$
$$\mathbf{I}_{V1} = \frac{1.5\angle 0° \text{ V}}{81.13\angle -82.92° \ \Omega}$$
$$= 18.5\angle 82.92° \text{ mA}$$
$$\mathbf{I}_{R(V1)} = \left(\frac{\mathbf{X}_L}{\mathbf{R} + \mathbf{X}_L}\right)\mathbf{I}_{V1}$$
$$= \left(\frac{25\angle 90° \ \Omega}{53.24\angle 28° \ \Omega}\right)18.5\angle 82.92° \text{ mA}$$
$$= 8.69\angle 144.9° \text{ mA}$$
With V_1 zeroed:
$$\mathbf{Z}_{V2} = \mathbf{R} + \mathbf{X}_C \| \mathbf{X}_L$$
$$= 47\angle 0° \ \Omega + 100\angle -90° \ \Omega \| 25\angle 90° \ \Omega$$
$$= 47 \ \Omega + j33.3 \ \Omega = 57.6\angle 35.3° \ \Omega$$
$$\mathbf{I}_{R(V2)} = \mathbf{I}_{V2} = \frac{3\angle 0° \text{ V}}{57.6\angle 35.3° \ \Omega}$$
$$= 52\angle -35.3° \text{ mA}$$
$$\mathbf{I}_R = \mathbf{I}_{R(V1)} + \mathbf{I}_{R(V2)}$$
$$= 8.69\angle 144.9° \text{ mA} + 52\angle -35.3° \text{ mA}$$
$$= 35.33 \text{ mA} - j25 \text{ mA}$$
$$= 43.28\angle -35.3° \text{ mA}$$

2. $\mathbf{I}_{C(V1)} = \mathbf{I}_{V1} = 18.5\angle 82.92° \text{ mA}$
$$\mathbf{I}_{C(V2)} = \left(\frac{\mathbf{X}_L}{\mathbf{X}_C + \mathbf{X}_L}\right)\mathbf{I}_{V2}$$
$$= \left(\frac{25\angle 90° \ \Omega}{75\angle -90° \ \Omega}\right)52\angle -35.3° \text{ A}$$
$$= 17.33\angle 144.7° \text{ mA}$$
$$\mathbf{I}_C = \mathbf{I}_{C(V1)} + \mathbf{I}_{C(V2)}$$
$$= 30.75\angle 112.69° \text{ mA}$$

3. $\mathbf{V}_{th} = \left(\dfrac{\mathbf{X}_C}{\mathbf{R}_3 + \mathbf{X}_C}\right)\mathbf{V}_s$
$$= \left(\frac{1\angle -90° \text{ k}\Omega}{1 \text{ k}\Omega - j1 \text{ k}\Omega}\right)12\angle 0° \text{ V}$$
$$= 8.49\angle -45° \text{ V}$$
$$\mathbf{Z}_{th} = \mathbf{R}_3 \| \mathbf{X}_C$$
$$= \frac{(1\angle 0° \text{ k}\Omega)(1\angle -90° \text{ k}\Omega)}{1 \text{ k}\Omega - j1 \text{ k}\Omega}$$
$$= 707\angle -45° \ \Omega$$
$$= 500 \ \Omega - j500 \ \Omega$$
See Figure S–26.

FIGURE S–26

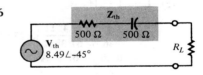

4. $\mathbf{V}_{th} = \left(\dfrac{\mathbf{X}_C}{\mathbf{R} + \mathbf{X}_C}\right)\mathbf{V}_s$
$$= \left(\frac{20\angle -90° \text{ k}\Omega}{100 \text{ k}\Omega - j20 \text{ k}\Omega}\right)15\angle 0° \text{ V}$$
$$= \left(\frac{20\angle -90° \text{ k}\Omega}{102\angle -11.3° \text{ k}\Omega}\right)15\angle 0° \text{ V}$$
$$= 2.94\angle -78.7° \text{ V}$$
$$\mathbf{Z}_{th} = \mathbf{R} \| \mathbf{X}_C$$
$$= \frac{(100\angle 0° \text{ k}\Omega)(20\angle -90° \text{ k}\Omega)}{102\angle -11.3° \text{ k}\Omega}$$
$$= 19.6\angle -78.7° \text{ k}\Omega$$
$$\mathbf{I}_L = \frac{\mathbf{V}_{th}}{\mathbf{Z}_{th} + \mathbf{R}_L} = \frac{2.94\angle -78.7° \text{ V}}{563.8 \text{ k}\Omega - j19.2 \text{ k}\Omega}$$
$$= \frac{2.94\angle -78.7° \text{ V}}{564\angle -1.95° \text{ k}\Omega}$$
$$= 5.21\angle -76.75° \ \mu\text{A}$$

5. $\mathbf{Z}_n = \mathbf{Z}_{th} = 19.6\angle -78.7° \text{ k}\Omega$
$$= 3.84 \text{ k}\Omega - j19.22 \text{ k}\Omega$$

$$\mathbf{I}_n = \frac{\mathbf{V}_{th}}{\mathbf{Z}_{th}} = \frac{2.94\angle -78.7° \text{ V}}{19.6\angle -78.7° \text{ k}\Omega}$$
$$= 0.15\angle 0° \text{ mA}$$

See Figure S–27.

FIGURE S–27

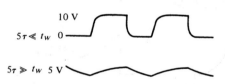

6. $\mathbf{I}_{eq} = \dfrac{12\angle 0° \text{ V}}{22\angle 0° \ \Omega} + \dfrac{6\angle 0° \text{ V}}{10\angle 0° \ \Omega}$

$\qquad + \dfrac{10\angle 0° \text{ V}}{47\angle -90° \ \Omega} + \dfrac{5\angle 0° \text{ V}}{10\angle -90° \ \Omega}$

$\qquad = 0.545 \text{ A} + 0.6 \text{ A} + j0.213 \text{ A} + j0.5 \text{ A}$

$\qquad = 1.35\angle 31.9° \text{ A}$

$\mathbf{Z}_{eq} = \dfrac{1}{0.045 \text{ S} + 0.1 \text{ S} + j0.021 \text{ S} + j0.1 \text{ S}}$

$\qquad = \dfrac{1}{0.145 \text{ S} + j0.121 \text{ S}}$

$\qquad = 5.3\angle -39.8° \ \Omega$

$\qquad = 4.07 \ \Omega - j3.39 \ \Omega$

$\mathbf{V}_{eq} = \mathbf{I}_{eq}\mathbf{Z}_{eq}$

$\qquad = (1.35\angle 31.9° \text{ A})(5.3\angle -39.8° \ \Omega)$

$\qquad = 7.16\angle -7.9° \text{ V}$

See Figure S–28.

FIGURE S–28

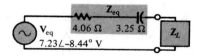

7. $\mathbf{Z}_L = 4.07 \ \Omega + j3.39 \ \Omega$

CHAPTER TWENTY-ONE

1. The output of an RC differentiator is taken across R. The output of an RC integrator is taken across C.

2. $v_C = 0.63(10 \text{ V}) = 6.3 \text{ V}$

3. Same as Problem 2.

4. When the input pulse width is much greater than 5τ.

5. When the input pulse width is much less than 5τ.

6. For $t_w \geq 5\tau$.

7. The output of an RL differentiator is taken across L. The output of an RL integrator is taken across R.

8. $I_{max} = V_s/R$

9. When the input pulse width is equal to or greater than 5τ.

10. You can't tell the difference by observing the output voltages.

11. $v_C = (0.63)(10 \text{ V}) = 6.3 \text{ V}$

12. **(a)** $v_C = (0.86)(10 \text{ V}) = 8.6 \text{ V}$
    (b) $v_C = (0.95)(10 \text{ V}) = 9.5 \text{ V}$
    (c) $v_C = (0.98)(10 \text{ V}) = 9.8 \text{ V}$
    (d) $v_C \cong (1)(10 \text{ V}) = 10 \text{ V}$

13. $v_C = V_F(1 - e^{-t/\tau}) = 5 \text{ V}(1 - e^{-5/10})$
    $\quad = 5 \text{ V}(1 - 0.607)$
    $\quad = 1.97 \text{ V}$

14. $v_C = (0.37)(15 \text{ V}) = 5.55 \text{ V}$

15. See Figure S–29.

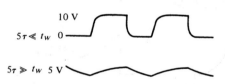

FIGURE S–29

16. See Figure S–30.

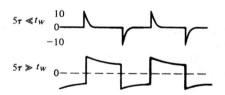

FIGURE S–30

17. $\tau = RC = (1 \text{ k}\Omega)(1 \ \mu\text{F}) = 1 \text{ ms}$. See Figure S–31. The time to reach steady state is $5\tau = 5 \text{ ms}$ for repetitive pulses.

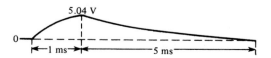

FIGURE S–31

18. See Figure S–32.

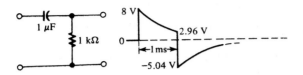

FIGURE S–32

19. See Figure S–33.

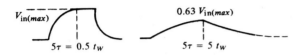

FIGURE S–33

20. See Figure S–34.

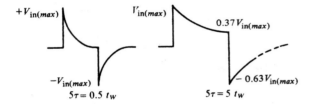

FIGURE S–34

21. $\tau = 10 \text{ mH}/10 \ \Omega = 1 \text{ ms}$
$5\tau = 5 \text{ ms}$
$v_{out(max)} = (0.63)8 \text{ V} = 5.04 \text{ V}$
See Figure S–35.

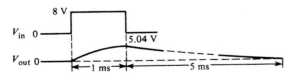

FIGURE S–35

22. See Figure S–36.

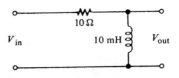

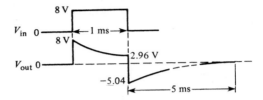

FIGURE S–36

23. $f = 0.35/t_r = 0.35/250 \text{ ns} = 1.4 \text{ MHz}$

CHAPTER TWENTY-TWO

1. Current through the field windings produces a magnetic field for rotor coil to cut in order to induce a voltage in the rotor windings.

2. $I_L = \dfrac{30\angle 0° \text{ V}}{15\angle 22° \ \Omega} = 2\angle -22° \text{ A}$

3. $I_L = I_\theta = 12 \text{ A}$

4. $V_L = \sqrt{3} \ V_\theta = \sqrt{3} \ (180 \text{ V}) = 311.8 \text{ V}$

5. $V_L = V_\theta = 30 \text{ V}$

6. $I_L = \sqrt{3} \ I_\theta = \sqrt{3} \ (5 \text{ A}) = 8.66 \text{ A}$

7. $I_\theta = I_L = I_Z = 15 \text{ A}$

8. $V_L = \sqrt{3} \ V_\theta = \sqrt{3} \ (1.5 \text{ kV}) = 2.6 \text{ kV}$
$V_Z = V_\theta = 1.5 \text{ kV}$

9. $V_L = V_Z = 75 \text{ V}$
$I_L = \sqrt{3} \ I_Z = \sqrt{3} \ (3.3 \text{ A}) = 5.7 \text{ A}$

10. $V_Z = V_\theta/\sqrt{3} = 220 \text{ V}/\sqrt{3} = 127 \text{ V}$

11. $P_T = 3V_Z I_Z \cos \theta$
$= 3(145 \text{ V})(2.5 \text{ A})(0.866)$
$= 941.8 \text{ W}$

ANSWERS TO SELECTED ODD-NUMBERED PROBLEMS

CHAPTER ONE

1–1 (a) 3×10^3 (b) 75×10^3
(c) 2×10^6

1–3 (a) $8.4 \times 10^3 = 0.84 \times 10^4$
$= 0.084 \times 10^5$
(b) $99 \times 10^3 = 9.9 \times 10^4$
$= 0.99 \times 10^5$
(c) $200 \times 10^3 = 20 \times 10^4$
$= 2 \times 10^5$

1–5 (a) 0.0000025 (b) 5000 (c) 0.39

1–7 (a) 126×10^6 (b) 5.00085×10^3
(c) 6060×10^{-9}

1–9 (a) 20×10^8 (b) 36×10^{14}
(c) 15.4×10^{-15}

1–11 (a) 2370×10^{-6} (b) 18.56×10^{-6}
(c) $0.00574389 \times 10^{-6}$
(d) $100,000,000,000 \times 10^{-6}$

1–13 (a) 4.95×10^7 (b) 5×10^3
(c) 10.575×10^{-6} (d) 22,511.31

1–15 (a) $3 \ \mu F$ (b) $3.3 \ M\Omega$
(c) 350 nA or $0.35 \ \mu A$

1–17 (a) $24 \ \mu A$ (b) $9.7 \ M\Omega$ (c) 3 pW

1–19 (a) $82 \ \mu W$ (b) 450 W

1–21 (a) 1000 (b) 50,000 (c) 2×10^{-5}
(d) 1.55×10^{-4}

CHAPTER TWO

2–1 $80 \times 10^{12} \ C$

2–3 (a) 10 V (b) 2.5 V (c) 4 V

2–5 20 V

2–7 33.33 V

2–9 0.2 A

2–11 0.15 C

2–13 (a) $68 \ k\Omega \ \pm \ 5\%$ (b) $4.7 \ k\Omega \ \pm \ 10\%$
(c) $390 \ \Omega \ \pm \ 5\%$

2–15 Red, violet, brown

2–17 (a) 0.2 S (b) 0.04 S (c) 0.01 S

2–19 AWG #27

2–21 Through lamp 2

2–23 Circuit (b)

2–25 See Figure P–1.

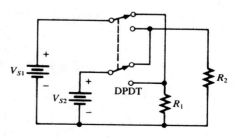

FIGURE P–1

2–27 See Figure P–2.

FIGURE P–2

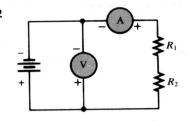

2–29 Position 1: $V1 = 0$ V, $V2 = V_S$
Position 2: $V1 = V_S$, $V2 = 0$ V

2–31 See Figure P–3.

2–33 **(a)** 12.5 V   **(b)** 0.7 V

2–35 **(a)** 200 Ω   **(b)** 150 MΩ   **(c)** 4500 Ω

CHAPTER THREE

3–1 **(a)** 5 A   **(b)** 1.5 A   **(c)** 0.5 A
(d) 2 mA   **(e)** 44.64 μA

3–3 1.2 A

3–5 **(a)** 0.455 mA, 4.55 mA
(b) 1.79 mA, 17.9 mA
(c) 10.9 μA, 109 μA

3–7 **(a)** 36 V   **(b)** 280 V   **(c)** 1700 V
(d) 28.2 V   **(e)** 56 V

3–9 81 V

3–11 **(a)** 59.9 mA   **(b)** 5.99 V   **(c)** 0.005 V

3–13 **(a)** 2 kΩ   **(b)** 3.5 kΩ   **(c)** 2 kΩ
(d) 100 kΩ   **(e)** 1 MΩ

3–15 150 Ω

3–17 133 Ω, 100 Ω

3–19 The graph is a straight line, indicating a linear relationship between V and I.

3–21 $R_1 = 0.5$ Ω, $R_2 = 1$ Ω, $R_3 = 2$ Ω

3–23 4 V, 6 V

3–27 See Figure P–4 for flowchart.

```
10   CLS
20   PRINT "THIS PROGRAM
     COMPUTES CURRENT FOR A
     SPECIFIED RANGE"
30   PRINT "OF VOLTAGES AND
     A SPECIFIED RESISTANCE
     VALUE."
40   FOR T=0 TO 3000:NEXT:CLS
50   INPUT "RESISTANCE IN
     OHMS";R
60   INPUT "THE LOWEST
     VOLTAGE";VL
70   INPUT "THE HIGHEST
     VOLTAGE";VH
80   INPUT "VOLTAGE
     INTERVAL";VI
90   FOR V=VL TO VH STEP VI
100  I=V/R
110  PRINT "V=";V;"V, I=";I;
     "A"
120  NEXT
```

CHAPTER FOUR

4–1 350 W

4–3 20,000 W

FIGURE P–3

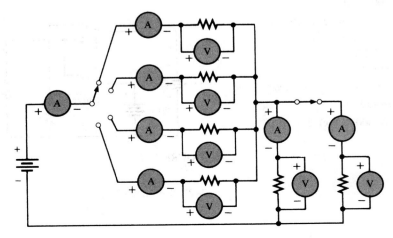

FIGURE P-4

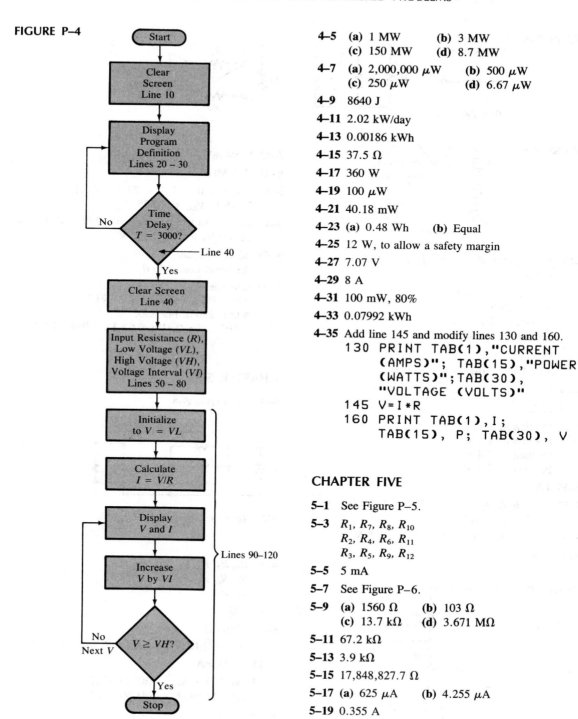

4-5 (a) 1 MW   (b) 3 MW
     (c) 150 MW   (d) 8.7 MW

4-7 (a) 2,000,000 μW   (b) 500 μW
     (c) 250 μW   (d) 6.67 μW

4-9 8640 J

4-11 2.02 kW/day

4-13 0.00186 kWh

4-15 37.5 Ω

4-17 360 W

4-19 100 μW

4-21 40.18 mW

4-23 (a) 0.48 Wh   (b) Equal

4-25 12 W, to allow a safety margin

4-27 7.07 V

4-29 8 A

4-31 100 mW, 80%

4-33 0.07992 kWh

4-35 Add line 145 and modify lines 130 and 160.
```
130 PRINT TAB(1),"CURRENT
    (AMPS)"; TAB(15),"POWER
    (WATTS)";TAB(30),
    "VOLTAGE (VOLTS)"
145 V=I*R
160 PRINT TAB(1),I;
    TAB(15), P; TAB(30), V
```

CHAPTER FIVE

5-1 See Figure P-5.

5-3 R_1, R_7, R_8, R_{10}
R_2, R_4, R_6, R_{11}
R_3, R_5, R_9, R_{12}

5-5 5 mA

5-7 See Figure P-6.

5-9 (a) 1560 Ω (b) 103 Ω
     (c) 13.7 kΩ (d) 3.671 MΩ

5-11 67.2 kΩ

5-13 3.9 kΩ

5-15 17,848,827.7 Ω

5-17 (a) 625 μA   (b) 4.255 μA

5-19 0.355 A

FIGURE P–5

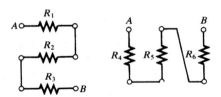

FIGURE P–6

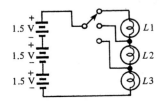

5–21 $R_1 = 330\ \Omega$, $R_2 = 220\ \Omega$, $R_3 = 100\ \Omega$, $R_4 = 470\ \Omega$

5–23 Position A: 5.45 mA
Position B: 6.06 mA
Position C: 7.95 mA
Position D: 12 mA

5–25 14 V

5–27 (a) 23 V   (b) 35 V   (c) 0 V

5–29 4 V

5–31 22 Ω

5–33 Position A: 4 V
Position B: 4.5 V
Position C: 5.4 V
Position D: 7.2 V

5–35 4.82%

5–37 $V_R = 6$ V, $V_{2R} = 12$ V, $V_{3R} = 18$ V, $V_{4R} = 24$ V, $V_{5R} = 30$ V

5–39 $V_2 = 1.79$ V, $V_3 = 1$ V, $V_4 = 17.9$ V

5–41 See Figure P–7.

FIGURE P–7

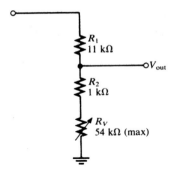

5–43 54.94 mW

5–45 12.5 MΩ

5–47 $V_{AG} = 100$ V, $V_{BG} = 57.575$ V, $V_{CG} = 15.15$ V, $V_{DG} = 7.575$ V

5–49 $V_{AG} = 14.815$ V, $V_{BG} = 12.967$ V, $V_{CG} = 12.637$ V, $V_{DG} = 9.337$ V

5–51 (a) R_4 is open.
(b) Short from A to B

5–53 Add line:
245 P(X)=V(X)↑2/R(X)
Add to end of line 250:
,"P"; X; "="; P(X); "WATTS"

CHAPTER SIX

6–1 See Figure P–8.

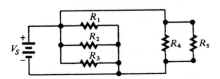

FIGURE P–8

6–3 R_1, R_2, R_5, R_9, R_{10}, R_{12}
R_4, R_6, R_7, R_8
R_3, R_{11}

6–5 100 V

6–7 1350 mA

6–9 $R_2 = 22\ \Omega$, $R_3 = 100\ \Omega$, $R_4 = 33\ \Omega$

6–11 11.36 mA

6–13 (a) 359.97 Ω (b) 25.55 Ω
(c) 818.86 Ω (d) 996.51 Ω

6–15 567 Ω

6–17 22.65 Ω

6–19 (a) 510 kΩ (b) 244.59 kΩ
       (c) 510 kΩ (d) 192.78 kΩ

6–21 10 A

6–23 50 mA; When one bulb burns out the others remain on.

6–25 53.68 Ω

6–27 $I_1 = 0.253$ A, $I_2 = 0.147$ A, $I_3 = 0.1$ A, $R = 395$ Ω

6–29 Position A: 2.25 A
       Position B: 4.75 A
       Position C: 7 A

6–31 (a) $I_1 = 6.875$ μA, $I_2 = 3.125$ μA
       (b) $I_1 = 5.25$ mA, $I_2 = 2.39$ mA, $I_3 = 1.59$ mA, $I_4 = 0.772$ mA

6–33 $R_1 = 3.3$ kΩ, $R_2 = 1.8$ kΩ, $R_3 = 5.6$ kΩ, $R_4 = 3.9$ kΩ

6–35 0.05 Ω

6–37 (a) 68.75 μW (b) 52.5 mW

6–39 $P_1 = 1.25$ W, $I_2 = 74.91$ mA, $I_1 = 125.1$ mA, $V_S = 10$ V, $R_1 = 79.9$ Ω, $R_2 = 133.5$ Ω

6–41 0.682 A, 3.41 A

6–43 The 8.2-kΩ resistor is open.

6–45 Checks for a single open resistor:
       Ohmmeter connected to pins 1 and 2: 767 Ω is correct. A 1-kΩ or a 3.3-kΩ reading indicates an open resistor.

FIGURE P–9

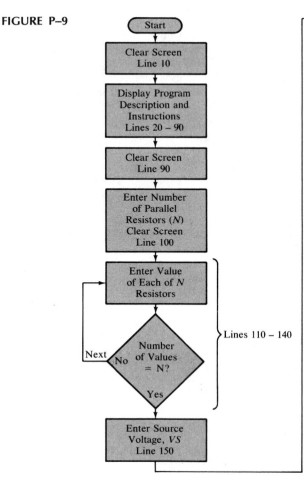

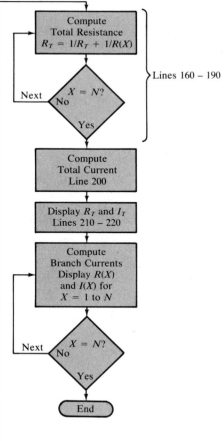

Ohmmeter connected to pins 3 and 4: 159.5 Ω is correct. A 270-Ω or a 390-Ω reading indicates an open resistor.

Ohmmeter connected to pins 5 and 6: 198 kΩ is correct. A 247-kΩ reading indicates that R_5 is open; a 227-kΩ reading indicates that R_6 is open. A 279-kΩ reading indicates that R_7 is open; a 323-kΩ reading indicates that R_8 is open.

6–47
```
10   CLS
20   INPUT "TOTAL RESISTANCE
     (OHMS)";RT
30   INPUT "HOW MANY
     RESISTORS IN
     PARALLEL";N
40   FOR X=1 TO N-1
50   PRINT "THE VALUE OF
     R";X;"IN OHMS"
60   INPUT R(X)
70   NEXT
80   RS=R(1)
90   FOR X=2 TO N-1
100  RS=1/(1/(RS)+(1/R(X)))
110  NEXT
120  CLS
130  RN=RT*RS/(RS-RT)
140  PRINT "UNKNOWN RESISTOR
     VALUE IS";RN;"OHMS."
```

6–49 See Figure P–9.

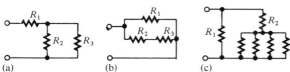

(a)   (b)   (c)

FIGURE P–10
FIGURE P–11

CHAPTER SEVEN

7–1 See Figure P–10.

7–3 (a) R_1 and R_4 are in series with the parallel combination of R_2 and R_3.

(b) R_1 is in series with the parallel combination of R_2, R_3, and R_4.

(c) The parallel combination of R_2 and R_3 is in series with the parallel combination of R_4 and R_5. This is all in parallel with R_1.

7–5 See Figure P–11. R_8, R_9, and R_{11} can be removed.

7–7 See Figure P–12.

7–9 (a) 133 Ω   (b) 779.44 Ω   (c) 852 Ω

7–11 (a) $I_1 = I_4 = 11.28$ mA, $I_2 = I_3 = 5.64$ mA, $V_1 = 0.63$ V, $V_2 = V_3 = 0.564$ V, $V_4 = 0.305$ V

(b) $I_1 = 3.85$ mA, $I_2 = 0.563$ mA, $I_3 = 1.16$ mA, $I_4 = 2.13$ mA, $V_1 = 2.62$ V, $V_2 = V_3 = V_4 = 0.383$ V

(c) $I_1 = 5$ mA, $I_2 = 0.303$ mA, $I_3 = 0.568$ mA, $I_4 = 0.313$ mA, $I_5 = 0.558$ mA, $V_1 = 5$ V, $V_2 = V_3 = 1.879$ V, $V_4 = V_5 = 3.13$ V

7–13 S_1 closed, S_2 open: 220 Ω
S_1 closed, S_2 closed: 200 Ω
S_1 open, S_2 open: 320 Ω
S_1 open, S_2 closed: 300 Ω

7–15 $V_{AG} = 100$ V, $V_{BG} = 61.52$ V, $V_{CG} = 15.73$ V, $V_{DG} = 7.87$ V

7–17 Measure the voltage at A with respect to ground and the voltage at B with respect to ground. The difference is V_{R2}.

7–19 109.9 Ω

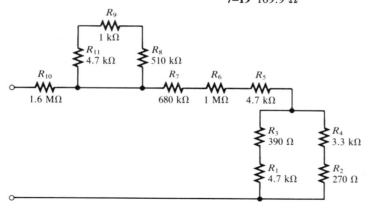

FIGURE P–12

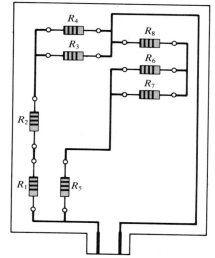

7–21 $R_{AB} = 1.65 \text{ k}\Omega$
$R_{BC} = 1.65 \text{ k}\Omega$
$R_{CD} = 0 \, \Omega$

7–23 7.5 V unloaded, 7.29 loaded

7–25 47 kΩ

7–27 $R_1 = 1000 \, \Omega$; $R_2 = R_3 = 500 \, \Omega$;
$V_{5V} = 3.12$ V (loaded);
$V_{2.5V} = 1.25$ V (loaded)

7–29 (a) $V_G = 1.75$ V, $V_S = 0.25$ V
(b) $I_1 = I_2 = 6.48 \, \mu A$, $I_D = I_S = 166.67 \, \mu A$
(c) $V_{DS} = 7.92$ V, $V_{DG} = 6.42$ V

7–31 (a) 270.8 Ω (b) 0.22 A
(c) 0.059 A (d) 12.03 V

7–33 620.8 Ω, $I_1 = I_9 = 16.11$ mA, $I_2 = 8.27$ mA,
$I_3 = I_8 = 7.84$ mA, $I_4 = 4.06$ mA, $I_5 = I_6 = I_7 = 3.78$ mA

7–35 0.97 A, 0 V

7–37 (a) 9 V (b) 3.75 V (c) 11.25 V

7–39 2184 Ω

7–41 No, it should be 4.39 V.

7–43 The 2.2-kΩ resistor is open.

7–45 The 3.3-kΩ resistor is open.

7–47 $V_A = 15$ V, $V_B = 0$ V, $V_C = 0$ V

7–49
```
10    CLS
20    INPUT "WHAT IS THE
      INPUT VOLTAGE";VIN
30    PRINT:PRINT
```

```
40    FOR X=1 TO 8
50    PRINT "VALUE OF
      R";X;"IN OHMS"
60    INPUT R(X)
70    NEXT
80    CLS
90    RC=(R(6)*(R(7)+R(8)))/(R(6)
      +R(7)+R(8))
100   RB=(R(4)*(R(5)+RC))/(R(4)
      +R(5)+RC)
110   RA=(R(2)*(R(3)+RB))/(R(2)
      +R(3)+RB)
120   RT=RA+R(1)
130   IT=VIN/RT
140   VA=IT*RA
150   I(1)=IT
160   I(2)=VA/R(2)
170   I(3)=IT-I(2)
180   VB=I(3)*RB
190   I(4)=VB/R(4)
200   I(5)=I(3)-I(4)
210   VC=I(5)*RC
220   I(6)=VC/R(6)
230   I(7)=I(5)-I(6)
240   I(8)=I(7)
250   VD=I(8)*R(8)
260   PRINT
      "VA=";VA;"V","VB=";VB;
      "V","VC=";VC;"V","VD=";VD;
      "V"
270   PRINT
280   FOR Y=1 TO 8
290   PRINT
      "I";Y;"=";I(Y);"AMP"
300   NEXT
```

CHAPTER EIGHT

8–1 $I_S = 6$ A, $R_S = 50 \, \Omega$

8–3 $V_S = 720$ V, $R_S = 1.2 \text{ k}\Omega$

8–5 0.845 mA

8–7 1.6 mA

8–9 −79.7 V

8–11 $I_{S1} = 137.5 \, \mu A$, $I_{S2} = 205.3 \, \mu A$

8–13 0.116 mA

8–15 $R_{TH} = 88.6 \, \Omega$, $V_{TH} = 1.086$ V

8–17 0.1 mA

8–19 (a) $I_N = 0.11$ A, $R_N = 76.66$ Ω
(b) $I_N = 0.011$ A, $R_N = 72.97$ Ω
(c) $I_N = 50$ μA, $R_N = 35.9$ kΩ
(d) $I_N = 0.069$ A, $R_N = 1.3$ kΩ

8–21 17.89 V

8–23 $I_N = 0.99$ mA, $R_N = 1174.6$ Ω

8–25 $R_{EQ} = 56.9$ Ω, $V_{EQ} = -2.74$ V

8–27 S_1, S_2 closed: $I_L = 0.448$ mA
S_1, S_3 closed: $I_L = 0.506$ mA
S_2, S_3 closed: $I_L = 0.315$ mA
S_1, S_2, S_3 closed: $I_L = 0.459$ mA

8–29 11.14 Ω

8–31 $R_{TH} = 48$ Ω, $R_4 = 159.56$ Ω

8–33 (a) $R_A = 39.82$ Ω, $R_B = 73$ Ω, $R_C = 48.67$ Ω
(b) $R_A = 21.18$ kΩ, $R_B = 10.28$ kΩ, $R_C = 14.87$ kΩ

8–35
```
10   CLS
20   PRINT "PLEASE PROVIDE
     THE DELTA RESISTOR
     VALUES WHEN PROMPTED."
30   PRINT:PRINT
40   INPUT "RA IN
     KILOHMS";RA
50   INPUT "RB IN
     KILOHMS";RB
60   INPUT "RC IN
     KILOHMS";RC
70   CLS
80   RT=RA+RB+RC
90   R1=RA*RC/RT
100  R2=RB*RC/RT
110  R3=RA*RB/RT
120  PRINT "THE VALUES FOR
     THE WYE NETWORK ARE AS
     FOLLOWS:"
130  PRINT:PRINT
     "R1=";R1;"KOHMS"
140  PRINT "R2=";R2;"KOHMS"
150  PRINT "R3=";R3;"KOHMS"
```

8–37
```
10   CLS
20   PRINT "THIS PROGRAM
     CONVERTS A SPECIFIED
     WYE NETWORK TO A
     DELTA."
30   PRINT "R1 IS UPPER LEFT
     ARM OF WYE. R2 IS UPPER
     RIGHT ARM OF
40   PRINT "WYE. R3 IS
     BOTTOM ARM OF WYE."
50   PRINT
60   INPUT "R1, R2, AND
     R3";R1,R2,R3
70   CLS
80   RS=R1*R2+R1*R3+R2*R3
90   RA=RS/R2
100  RB=RS/R1
110  RC=RS/R3
120  PRINT "RA=";RA
130  PRINT "RB=";RB
140  PRINT "RC=";RC
```

CHAPTER NINE

9–1 Six possible loops

9–3 $I_1 - I_2 - I_3 = 0$

9–5 $V_1 = 5.662$ V, $V_2 = 6.325$ V, $V_3 = 0.325$ V

9–7 $I_1 = 0.738$ A, $I_2 = -0.527$ A, $I_3 = -0.469$ A

9–9 -1.837 V

9–11 $I_1 = 0$ A, $I_2 = 2$ A

9–13 (a) $-16,470$ (b) -1.591

9–15 $I_1 = 1.24$ A, $I_2 = 2.05$ A, $I_3 = 1.89$ A

9–17 $I_1 = -5.11$ mA, $I_2 = -3.52$ mA

9–19 $V_{1k} = 5.11$ V, $V_{560} = 0.89$ V, $V_{820} = 2.89$ V

9–21 $I_1 = 0.0123$ A, $I_2 = -0.0797$ A, $I_3 = -0.0952$ A

9–23 8 mV

9–25 7.89 mA

9–27 $I_{82} = 20.61$ mA, $I_{47} = 192.59$ mA, $I_{68} = -171.91$ mA

9–29 $V_A = 1.5$ V, $V_B = -5.65$ V

9–31 $I_{R1} = 193.37$ μA, $I_{R2} = 370$ μA, $I_{R3} = 177$ μA, $I_{R4} = 328.13$ μA, $I_{R5} = 1.462$ mA, $I_{R6} = 521.5$ μA, $I_{R7} = 2.16$ mA, $I_{R8} = 1.639$ mA

$V_A = -3.70$ V, $V_B = -5.85$ V, $V_C = -15.68$ V

9-33
```
10   CLS
20   PRINT "THIS PROGRAM
     WILL EVALUATE THIRD
     ORDER DETERMINANTS"
30   PRINT "     A1
       B1     C1"
40   PRINT
50   PRINT "     A2
       B2     C2"
60   PRINT
70   PRINT "     A3
       B3     C3"
80   INPUT "A1";A1
90   INPUT "A2";A2
100  INPUT "A3";A3
110  INPUT "B1";B1
120  INPUT "B2";B2
130  INPUT "B3";B3
140  INPUT "C1";C1
150  INPUT "C2";C2
160  INPUT "C3";C3
170  PRINT:PRINT:PRINT:PRINT
180  D=A1*(B2*C3-B3*C2)-A2*
     (B1*C3-B3*C1)+A2*(B1*
     C2-B2*C1)
190  PRINT "D=";D
```

CHAPTER TEN

10-1 Decreases

10-3 37.5 μWb

10-5 596.8

10-7 150 At

10-9 Change the current.

10-11 Material A

10-13 1 mA

10-15 Lenz's law defines the *polarity* of the induced voltage.

10-17 The commutator and brush arrangement electri-ᶜally connects the loop to the external circuit.

10-19 See Figure P-13.

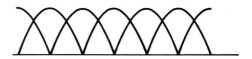

FIGURE P-13

CHAPTER ELEVEN

11-1 (a) 1 Hz   (b) 5 Hz
      (c) 20 Hz   (d) 1000 Hz
      (e) 2 kHz   (f) 100 kHz

11-3 2 μs

11-5 (a) 8.48 V   (b) 24 V
      (c) 0 V

11-7 (a) 7.07 mA   (b) 0 mA
      (c) 10 mA   (d) 20 mA
      (e) 10 mA

11-9 512.6 W

11-11 120 Hz

11-13 (a) 0.524 rad   (b) 0.785 rad
      (c) 1.361 rad   (d) 2.356 rad
      (e) 3.491 rad   (f) 5.236 rad

11-15 15 degrees, A leading

11-17 See Figure P-14.

FIGURE P-14

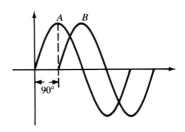

11-19 (a) 57.36 mA   (b) 99.62 mA
       (c) −17.36 mA   (d) −57.36 mA
       (e) −99.62 mA   (f) 0 mA

11-21   30°:   12.99 V
       45°:   14.49 V
       90°:   12.99 V
     180°:   −7.5 V
     200°:   −11.49 V
     300°:   −7.5 V

11–23 23.08 V

11–25 0.2%

11–27 **(a)** 25% **(b)** 66.67%

11–29 **(a)** 250 kHz **(b)** 33.33 Hz

11–31 3 V

11–33 25 kHz

11–35
```
10 CLS
20 PRINT "THIS PROGRAM
   CONVERTS DEGREES TO
   RADIANS"
30 PRINT
40 INPUT "DEGREES";D
50 CLS
60 R=(3.1416/180)*D
70 PRINT
   D"DEGREES=";R;"RADIANS"
```

CHAPTER TWELVE

12–1 See Figure P–15.

FIGURE P–15

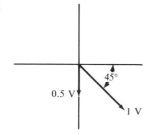

0.5 V 45° 1 V

12–3 **(a)** 9.55 Hz **(b)** 57.3 Hz
       (c) 0.318 Hz **(d)** 199.9 Hz

12–5 29.5 μs

12–7 See Figure P–16.

12–9 **(a)** -5, $+j3$ and 5, $-j3$
       (b) -1, $-j7$ and 1, $+j7$
       (c) -10, $+j10$ and 10, $-j10$

12–11 18.03

12–13 **(a)** $642.8 - j766$ **(b)** $-14.1 + j5.13$
        (c) $-17.68 - j17.68$ **(d)** $-3 + j0$

12–15 **(a)** Fourth **(b)** Fourth **(c)** Fourth
        (d) First

12–17 **(a)** $12\angle 115°$ **(b)** $20\angle 230°$
        (c) $100\angle 190°$ **(d)** $50\angle 160°$

FIGURE P–16

12–19 **(a)** $1.1 + j0.7$ **(b)** $-81 - j35$
        (c) $5.3 - j5.27$ **(d)** $-50.4 + j62.5$

12–21 **(a)** $3.2\angle 11°$ **(b)** $7\angle -101°$
        (c) $1.52\angle 70.6°$ **(d)** $2.79\angle -63.5°$

12–23 $9.87\angle -8.8°$ V, $1.97\angle -8.8°$ mA

CHAPTER THIRTEEN

13–1 **(a)** 5 μF **(b)** 1 μC **(c)** 10 V

13–3 **(a)** 0.001 μF **(b)** 0.0035 μF
      (c) 0.00025 μF

13–5 2.3×10^{-22}

13–7 **(a)** 8.9×10^{-12} F/m
      (b) 35.4×10^{-12} F/m
      (c) 66.375×10^{-12} F/m
      (d) 17.7×10^{-12} F/m

13–9 8.9 pF

13–11 12.5-pF increase

13–13 Ceramic

13–15 Aluminum, tantalum; they are polarized.

13–17 **(a)** 0.02 μF **(b)** 0.047 μF
        (c) 0.001 μF **(d)** 220 pF

13–19 **(a)** 0.67 μF **(b)** 68.97 pF **(c)** 2.7 μF

13–21 2 μF

13–23 **(a)** 1057 pF **(b)** 0.121 μF

13–25 **(a)** 2.5 μF **(b)** 716.66 pF **(c)** 1.6 μF

13–27 $\Delta V_5 = +1.39$ V, $\Delta V_6 = -2.96$ V

13–29 (a) 14 ms   (b) 247.5 μs
       (c) 11 μs   (d) 280 μs

13–31 (a) 9.2 V   (b) 1.24 V
       (c) 0.46 V   (d) 0.168 V

13–33 (a) 17.97 V   (b) 12.79 V   (c) 6.61 V

13–35 7.62 μs

13–37 2.94 μs

13–39 See Figure P–17.

13–41 (a) 31.83 Ω (b) 111 kΩ (c) 49.7 Ω

13–43 200 Ω

13–45 0 W, 3.39 mVAR

13–47 0.0052 μF

13–49 X_C of the bypass capacitor should ideally be 0 Ω.

13–51 The capacitor is shorted.

13–53
```
10   CLS
20   PRINT "THIS PROGRAM
     COMPUTES AND TABULATES
     THE CAPACITIVE"
30   PRINT "REACTANCE FOR
     SPECIFIED R AND C OVER
     A SPECIFIED"
40   PRINT "FREQUENCY
     RANGE.":PRINT
```

```
50   INPUT "R IN OHMS";R
60   INPUT "C IN FARADS";C
70   INPUT "MINIMUM
     FREQUENCY IN HERTZ";FL
80   INPUT "MAXIMUM
     FREQUENCY IN HERTZ";FH
90   INPUT "FREQUENCY
     INCREMENTS IN
     HERTZ";FI:CLS
100  PRINT "FREQUENCY (HZ)",
     "CAPACITIVE REACTANCE
     (OHMS)"
110  FOR F=FL TO FH STEP FI
120  XC=1/(2*3.1416*F*C)
130  PRINT F,XC
140  NEXT
```

CHAPTER FOURTEEN

14–1 (a) 1000 mH   (b) 0.25 mH
       (c) 0.01 mH   (d) 0.5 mH

14–3 50 mV

14–5 0.02 V

14–7 1 A

14–9 155 μH

14–11 50.51 mH

14–13 7.14 μH

FIGURE P–17

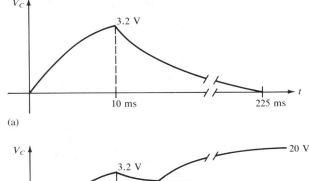

(a)

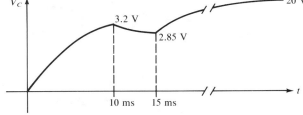

(b)

14–15 (a) 4.33 H   (b) 50 mH   (c) 0.57 μH

14–17 (a) 1 μs   (b) 2.13 μs   (c) 2 μs

14–19 (a) 5.52 V   (b) 2.03 V   (c) 0.75 V
       (d) 0.27 V   (e) 0.1 V

14–21 (a) 12.28 V   (b) 9.1 V   (c) 3.35 V

14–23 10.99 μs

14–25 3.18 ms

14–27 240 V

14–29 (a) 144 Ω (b) 10.1 Ω (c) 13.4 Ω

14–31 (a) 55.5 Hz   (b) 796 Hz   (c) 597 Hz

14–33 $26.1\angle -90°$ mA

14–35 (a) Infinite resistance   (b) Zero resistance
       (c) Lower R_W

14–37
```
10  CLS
20  INPUT "INDUCTANCE IN
    HENRIES";L
30  INPUT "WINDING
    RESISTANCE IN OHMS";RW
40  INPUT "DIRECT CURRENT IN
    AMPS"; I
50  E=.5*L*I*I
60  P=I*I*RW
70  CLS
80  PRINT "ENERGY STORED BY
    INDUCTOR=";E;"JOULES"
90  PRINT "POWER DISSIPATED
    BY WINDING
    RESISTANCE=";P;"WATTS"
```

CHAPTER FIFTEEN

15–1 1.5 μH

15–3 4; 0.25

15–5 (a) 100 V rms; same polarity
       (b) 100 V rms; opposite polarity
       (c) 20 V rms; opposite polarity

15–7 600 V

15–9 0.25 (4:1)

15–11 60 V

15–13 (a) 25 mA   (b) 50 mA
        (c) 15 V   (d) 0.75 W

15–15 1.83

15–17 39.1 W

15–19 (a) 6 V   (b) 0 V   (c) 40 V

15–21 94.5 W

15–23 0.98

15–25 25 kVA

15–27 $V_1 = 11.5$ V, $V_2 = 23$ V, $V_3 = 23$ V,
       $V_4 = 46$ V

15–29 (a) 48 V   (b) 25 V

15–31 (a) $V_{RL} = 35\angle 0°$ V, $I_{RL} = 2.92\angle 0°$ A,
         $V_C = 15\angle 0°$ V, $I_C = 1.5\angle 90°$ A
       (b) $34.5\angle -12.6°$ Ω

CHAPTER SIXTEEN

16–1 8 kHz, 8 kHz

16–3 (a) 270 Ω $- j100$ Ω, $287.9\angle -20.3°$ Ω
       (b) 680 Ω $- j1000$ Ω, $1209\angle -55.8°$ Ω

16–5 (a) 56 kΩ $- j796$ kΩ
       (b) 56 kΩ $- j159$ kΩ
       (c) 56 kΩ $- j79.6$ kΩ
       (d) 56 kΩ $- j31.8$ kΩ

16–7 (a) $R = 33$ Ω, $X_C = 50$ Ω
       (b) $R = 271.9$ Ω, $X_C = 126.8$ Ω
       (c) $R = 697.5$ Ω, $X_C = 1.66$ kΩ
       (d) $R = 557.9$ Ω, $X_C = 557.9$ Ω

16–9 (a) $0.179\angle 58.2°$ mA   (b) $0.625\angle 38.5°$ mA
       (c) $1.99\angle 76.2°$ mA

16–11 15.85°

16–13 (a) $97.3\angle -54.9°$ Ω (b) $103\angle 54.9°$ mA
       (c) $5.76\angle 54.9°$ V   (d) $8.18\angle -35.1°$ V

16–15 $R_X = 12$ Ω, $C_X = 13.3$ μF in series.

16–17 261 Ω, $-79.83°$

16–19 $\mathbf{V}_C = \mathbf{V}_R = 10\angle 0°$ V
       $\mathbf{I}_T = 184.2\angle 37.1°$ mA
       $\mathbf{I}_R = 147\angle 0°$ mA
       $\mathbf{I}_C = 111.1\angle 90°$ mA

16–21 (a) $6.58\angle -48.81°$ Ω (b) $10\angle 0°$ mA
       (c) $11.43\angle 90°$ mA   (d) $15.2\angle 48.81°$ mA
       (e) $-48.81°$ (I_T leading V_s)

16–23 18.16-kΩ resistor in series with 190.6-pF capacitor.

16–25 $\mathbf{V}_{C1} = 8.43\angle -3.1°$ V, $\mathbf{V}_{C2} = 1.48\angle -58.9°$ V
       $\mathbf{V}_{C3} = 3.6\angle 6.8°$ V, $\mathbf{V}_{R1} = 3.28\angle 31.1°$ V
       $\mathbf{V}_{R2} = 2.33\angle 6.8°$ V, $\mathbf{V}_{R3} = 1.27\angle 6.8°$ V

16–27 $\mathbf{I}_T = 79.5\angle 87°$ mA, $\mathbf{I}_{C2R1} = 6.98\angle 31.1°$ mA
       $\mathbf{I}_{C3} = 75\angle 96.8°$ mA, $\mathbf{I}_{R2R3} = 7.1\angle 6.8°$ mA

16–29 0.1 μF

16–31 $I_{C1} = I_{R1} = 22.56\angle74.4°$ mA
$I_{R2} = 20.3\angle72°$ mA
$I_{R3} = 2.44\angle94.9°$ mA
$I_{R4} = 1.47\angle40.2°$ mA
$I_{R5} = 1.78\angle74.3°$ mA
$I_{R6} = I_{C3} = 1\angle133.8°$ mA
$I_{C2} = 0.99\angle130.2°$ mA

16–33 4.03 VA

16–35 0.91

16–37 (a) $I_{LA} = 4.8$ A, $I_{LB} = 3.33$ A
(b) $P_{rA} = 606$ VAR, $P_{rB} = 249$ VAR
(c) $P_{trueA} = 979.2$ W, $P_{trueB} = 758.5$ W
(d) $P_{aA} = 1151.6$ VA, $P_{aB} = 798.3$ VA
(e) Load A

16–39

0 Hz	1 V
1 kHz	0.71 V
2 kHz	0.45 V
3 kHz	0.32 V
4 kHz	0.24 V
5 kHz	0.20 V
6 kHz	0.16 V
7 kHz	0.14 V
8 kHz	0.12 V
9 kHz	0.11 V
10 kHz	0.10 V

16–41

0 Hz	0 V
1 kHz	5.32 V
2 kHz	7.82 V
3 kHz	8.83 V
4 kHz	9.29 V
5 kHz	9.53 V
6 kHz	9.66 V
7 kHz	9.76 V
8 kHz	9.80 V
9 kHz	9.84 V
10 kHz	9.87 V

16–43 0.0796 μF

16–45 Reduces V_{out} to 2.83 V and θ to -56.7 degrees.

16–47 (a) No output voltage
(b) $0.3\angle-72.34°$ V
(c) $0.5\angle0°$ V
(d) 0 V

16–49
```
10   CLS
20   INPUT "INPUT VOLTAGE IN
     VOLTS";VIN
30   INPUT "THE VALUE OF R
     IN OHMS";R
40   INPUT "THE VALUE OF C
     IN FARADS";C
50   INPUT "THE LOWEST
     NONZERO FREQUENCY IN
     HERTZ";FL
60   INPUT "THE HIGHEST
     FREQUENCY IN HERTZ";FH
70   INPUT "THE FREQUENCY
     INCREMENTS IN HERTZ";FI
80   CLS
90   PRINT "FREQUENCY(HZ)",
     "PHASE SHIFT","VOUT", "I"
100  FOR F=FL TO FH STEP FI
110  XC=1/(2*3.1416*F*C)
120  PHI=-90+ATN(XC/R)*57.3
130  VO=VIN*XC/(SQR(R*R+XC*
     XC))
140  I=VIN/(SQR(R*R+XC*XC))
150  PRINT F,PHI,VO,I
160  NEXT
```

CHAPTER SEVENTEEN

17–1 15 kHz

17–3 (a) $100 \ \Omega + j50 \ \Omega$;
$111.8\angle26.6° \ \Omega$
(b) $1.5 \ k\Omega + j1 \ k\Omega$;
$1.8\angle33.7° \ k\Omega$

17–5 (a) $17.38\angle46.4° \ \Omega$
(b) $63.97\angle79.2° \ \Omega$
(c) $126.23\angle84.5° \ \Omega$
(d) $251.61\angle87.3° \ \Omega$

17–7 $806.4 \ \Omega$, 4.11 mH

17–9 (a) $435\angle-55°$ mA
(b) $11.75\angle-34.7°$ mA

17–11 θ increases from 38.7 degrees to 58 degrees.

17–13 (a) $V_R = 4.85\angle-14.1°$ V
$V_L = 1.22\angle75.9°$ V
(b) $V_R = 3.83\angle-39.96°$ V
$V_L = 3.21\angle50.04°$ V
(c) $V_R = 2.16\angle-64.5°$ V
$V_L = 4.51\angle25.5°$ V
(d) $V_R = 1.16\angle-76.6°$ V
$V_L = 4.86\angle13.4°$ V

17–15 $7.69\angle50.3° \ \Omega$

17–17 2387 Hz

17–19 (a) $274\angle 60.7° \ \Omega$
    (b) $89.3\angle 0°$ mA
    (c) $159\angle -90°$ mA
    (d) $182.5\angle -60.7°$ mA
    (e) 60 7° (I_T lagging V_s)

17–21 1.83-kΩ resistor in series with 4.21-kΩ inductive reactance

17–23 $\mathbf{V}_{R1} = 18.63\angle -3.4°$ V
    $\mathbf{V}_{R2} = 6.52\angle 9.7°$ V
    $\mathbf{V}_{R3} = 2.81\angle -54.8°$ V
    $\mathbf{V}_{L1} = \mathbf{V}_{L2} = 5.88\angle 35.2°$ V

17–25 $\mathbf{I}_{R1} = 0.373\angle -3.4°$ A
    $\mathbf{I}_{R2} = 0.326\angle 9.7°$ A
    $\mathbf{I}_{R3} = 0.094\angle -54.8°$ A
    $\mathbf{I}_{L1} = \mathbf{I}_{L2} = 0.047\angle -54.8°$ A

17–27 (a) $0.57\angle -51.2°$ A   (b) $20.95\angle 17.1°$ V
    (c) $8.01\angle -129.7°$ V

17–29 $\theta = 52.5°$ (V_{out} lags V_{in}), 0.14

17–31 See Figure P–18.

FIGURE P–18

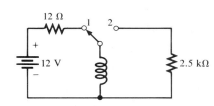

17–33 1.29 W, 1.04 VAR

17–35 $P_{true} = 0.29$ W; $P_r = 0.05$ VAR; $P_a = 0.297$ VA; $PF = 0.99$

17–37 (a) $-0.092°$ (b) $-9.15°$
    (c) $-58.17°$ (d) $-86.45°$

17–39 (a) $89.91°$ (b) $80.85°$
    (c) $31.83°$ (d) $3.55°$

17–41 See Figure P–19.

17–43 (a) 0 V   (b) 0 V   (c) $1.62\angle -25.79°$ V
    (d) $2.15\angle -64.5°$ V

17–45
```
10   CLS
20   PRINT "THIS PROGRAM
     COMPUTES THE PHASE
     SHIFT FROM INPUT TO"
30   PRINT "OUTPUT,
     NORMALIZED OUTPUT
     VOLTAGE MAGNITUDE,"
35   PRINT "NORMALIZED
     CURRENT, AND TRUE
     POWER"
40   PRINT "AS FUNCTIONS OF
     FREQUENCY FOR AN RL
     LEAD NETWORK."
50   PRINT:PRINT:PRINT
60   INPUT "TO CONTINUE
     PRESS 'ENTER'";X:CLS
70   INPUT "RESISTANCE IN
     OHMS";R
80   INPUT "INDUCTANCE IN
     HENRIES";L
90   INPUT "THE LOWEST
     FREQUENCY IN HERTZ";FL
100  INPUT "THE HIGHEST
     FREQUENCY IN HERTZ";FH
110  INPUT "THE FREQUENCY
     INCREMENTS IN HERTZ";FI
120  CLS
130  PRINT
     TAB(5)"FREQ(HZ)";TAB(15)"PH
     SHFT";TAB(25)"VOUT";TAB(35)"
     (A)";TAB(45)"PTRU(W)"
140  FOR F=FL TO FH STEP FI
150  XL=2*3.1416*F*L
160  PHI=90-ATN(XL/R)*57.3
```

FIGURE P–19

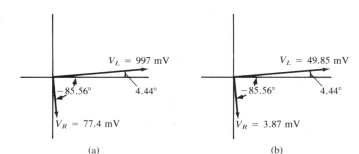

$V_L = 997$ mV $V_L = 49.85$ mV

$-85.56°$ $4.44°$ $-85.56°$ $4.44°$

$V_R = 77.4$ mV $V_R = 3.87$ mV

(a)   (b)

```
170  VO=XL/(SQR(R*R+XL*XL))
180  I=VO/XL
190  PTRU=I*I*R
200  PRINT
     TAB(5)F;TAB(15)PHI;
     TAB(25)VO;TAB(35)I;TAB(45)PTRU
210  NEXT
```

CHAPTER EIGHTEEN

18–1 $479.6\angle -88.8°$ Ω; 479.5-Ω capacitive

18–3 Impedance increases.

18–5 $I_T = 0.061\angle -43.75°$ A
$V_R = 2.87\angle -43.75°$ V
$V_L = 4.88\angle 46.25°$ V
$V_C = 2.14\angle -133.75°$ V

18–7 (a) $35.8\angle 65°$ mA   (b) 0.181 W
     (c) 0.39 VAR   (d) 0.43 VA

18–9 $Z = 200$ Ω, $X_C = X_L = 2$ kΩ

18–11 0.5 A

18–13 See Figure P–20.

FIGURE P–20

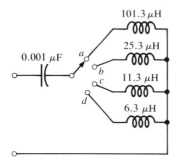

18–15 A phase angle of -4.69 degrees indicates a slightly capacitive circuit.

18–17 $I_T \cong 0.05\angle 4.5°$ A; $I_R = 0.05\angle 0°$ A
$I_L = 4.4\angle -90°$ mA;
$I_C = 8.3\angle 90°$ mA;
$V_R = V_L = V_C = 5\angle 0°$ V

18–19 (a) $2°$ (V_s leads I_T)
     (b) $23.2°$ (V_s leads I_T)

18–21 48.95-kΩ resistor in series with 1.33-H inductor.

18–23 $42.87°$ (I_2 leads V_s)

18–25 $I_{R1} = I_{C1} = 1.1\angle -25.7°$ mA
$I_{R2} = 0.768\angle 19.3°$ mA
$I_{C2} = 0.768\angle 109.3°$ mA
$I_L = 1.54\angle -70.7°$ mA
$V_{R2} = V_{C2} = V_L = 7.68\angle 19.3°$ V
$V_{R1} = 3.61\angle -25.7°$ V
$V_{C1} = 1.1\angle -115.7°$ V

18–27 $49\angle 130.9°$ mA

18–29 483 MΩ, 103.8 kHz

18–31 $f_{r(series)} = 4.1$ kHz
$V_{out} = 9.97\angle 3.26°$ V
$f_{r(parallel)} = 2.6$ kHz
$V_{out} \cong 10\angle 0°$ V

18–33 62.5 Hz

18–35 1.375 W

18–37 200 Hz

18–39
```
10   CLS
20   PRINT "THE FOLLOWING
     PARAMETERS ARE COMPUTED
     FOR A PARALLEL TANK
     CKT:"
30   PRINT :PRINT"IMPEDANCE"
40   PRINT "PHASE ANGLE"
50   PRINT:PRINT:PRINT
60   INPUT "TO CONTINUE
     PRESS 'ENTER'";X:CLS
70   INPUT "THE VALUE OF R
     IN OHMS";R
80   INPUT "THE VALUE OF C
     IN FARADS";C
90   INPUT "THE VALUE OF L
     IN HENRIES";L
100  INPUT "THE MINIMUM
     FREQUENCY IN HERTZ";FL
110  INPUT "THE MAXIMUM
     FREQUENCY IN HERTZ";FH
120  INPUT "THE INCREMENTS
     OF FREQUENCY IN
     HERTZ";FI
130  PRINT "FREQUENCY(HZ)",
     "IMPEDANCE", "PHASE
     ANGLE"
140  FOR F=FL TO FH STEP FI
150  XC=1/(2*3.1416*F*C)
160  XL=2*3.1416*F*L
170  Z=(SQR(R*R+XL*XL)*XC)/
     SQR(R*R+(XL-XC)*
     (XL-XC))
```

```
180  THETA=ATN((XL-XC)/R)
190  PRINT F, Z, THETA
200  NEXT
```

```
140  ATTN=V/VIN
150  PRINT F,V,ATTN
160  NEXT
```

CHAPTER NINETEEN

19–1 2.22 V rms

19–3 (a) 9.36 V   (b) 7.29 V
        (c) 9.96 V   (d) 9.95 V

19–5 (a) 12.1 μF   (b) 1.45 μF
        (c) 0.723 μF   (d) 0.144 μF

19–9 (a) 7.13 V   (b) 5.67 V
        (c) 4 V   (d) 0.8 V

19–11 9.75 V

19–13 (a) 3.53 V   (b) 5.08 V
         (c) 0.947 V   (d) 0.995 V

19–17 (a) 14.5 kHz   (b) 25.2 kHz

19–19 (a) 14.99 kHz, 14 kHz
         (b) 26.95 kHz, 23.45 kHz

19–21 (a) 116.8 V   (b) 115.4 V

19–23 $C = 0.064$ μF, $L = 0.989$ mH, $f_r = 20$ kHz

19–25 (a) 86.3 Hz   (b) 7.1 MHz

19–27 $L_1 = 0.087$ μH, $L_2 = 0.609$ μH

19–29
```
10   CLS
20   INPUT "VALUE OF R IN
     OHMS";R
30   INPUT "VALUE OF L IN
     HENRIES";L
40   INPUT "VALUE OF WINDING
     RESISTANCE IN OHMS";RW
50   INPUT "INPUT VOLTAGE IN
     VOLTS";VIN
60   INPUT "LOWEST FREQUENCY
     IN HERTZ";FL
70   INPUT "HIGHEST
     FREQUENCY IN HERTZ";FH
80   INPUT "FREQUENCY
     INCREMENTS IN HERTZ";FI
90   CLS
100  PRINT "FREQUENCY
     (HZ)","VOUT
     (VOLTS)","ATTN"
110  FOR F=FL TO FH STEP FI
120  XL=2*3.1416*F*L
130  V=(R/(SQR((R+RW)*(R+RW)+
     XL*XL)))*VIN
```

CHAPTER TWENTY

20–1 1.22∠28.9° mA

20–3 80.5∠−11.3° mA

20–5 $V_{A(dc)} = 0$ V, $V_{B(dc)} = 16.1$ V, $V_{C(dc)} = 15.1$ V,
        $V_{D(dc)} = 0$ V, $V_{A(peak)} = 9$ V, $V_{B(peak)} = 5.96$ V,
        $V_{C(peak)} = V_{D(peak)} = 4.96$ V

20–7 (a) $\mathbf{V}_{th} = 15∠−53.1°$ V
             $\mathbf{Z}_{th} = 63$ Ω $− j48$ Ω
        (b) $\mathbf{V}_{th} = 1.22∠0°$ V
             $\mathbf{Z}_{th} = j237$ Ω
        (c) $\mathbf{V}_{th} = 12.1∠11.9°$ V
             $\mathbf{Z}_{th} = 50$ kΩ $− j20$ kΩ

20–9 16.9∠88.24° V

20–11 (a) $\mathbf{I}_n = 0.19∠−15.83°$ A
              $\mathbf{Z}_n = 63$ Ω $− j48$ Ω
         (b) $\mathbf{I}_n = 5.2∠−90°$ A
              $\mathbf{Z}_n = j236.7$ Ω
         (c) $\mathbf{I}_n = 0.23∠33.7°$ mA
              $\mathbf{Z}_n = 50$ kΩ $− j20$ kΩ

20–13 17.8∠88.5° V

20–15 $\mathbf{V}_{eq} = 5.09∠7.6°$ V
         $\mathbf{Z}_{eq} = 0.565∠7.6°$ kΩ

20–17 8.52 Ω $+ j2.57$ Ω

20–19 95.2 Ω $+ j42.7$ Ω

20–21
```
10   CLS
20   PRINT "THIS PROGRAM
     COMPUTES THE MILLMAN
     EQUIVALENT COMPONENTS"
30   PRINT "FOR RESISTIVE
     IMPEDANCES ONLY. THE
     VOLTAGE AND IMPEDANCE"
40   PRINT "IN EACH BRANCH
     OF THE ORIGINAL CIRCUIT
     ARE REQUIRED AS"
50   PRINT "INPUTS."
60   PRINT
70   INPUT "NUMBER OF
     PARALLEL BRANCHES";N
80   CLS:FOR X=1 TO N
90   PRINT "BRANCH";X;
     "VOLTAGE"
```

```
100 INPUT V(X)
110 PRINT "BRANCH";X;
    "IMPEDANCE IN OHMS"
120 INPUT Z(X)
130 NEXT
140 CLS
150 FOR X=1 TO N
160 YEQ=YEQ+1/Z(X)
170 IEQ=IEQ+V(X)/Z(X)
180 NEXT
190 CLS
200 PRINT "ZEQ=";1/YEQ;
    "OHMS"
210 PRINT "VEQ=";IEQ/YEQ;
    "VOLTS"
```

CHAPTER TWENTY-ONE

21–1 110 μs

21–3 12.6 V

21–5 See Figure P–21.

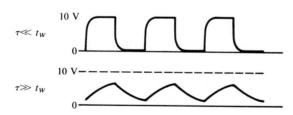

FIGURE P–21

FIGURE P–22

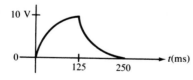

FIGURE P–23

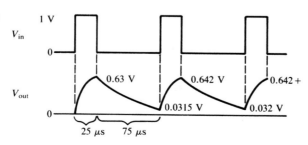

21–7 (a) 25 ms
       (b) See Figure P–22.

21–9 See Figure P–23.

21–11 See Figure P–24.

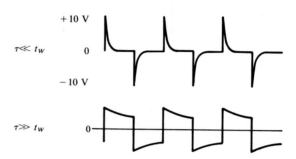

FIGURE P–24

21–13 (a) 0.525 μs
       (b) See Figure P–25.

FIGURE P–25

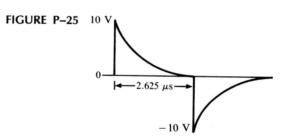

21–15 A square wave with an average value of zero.

21–17 See Figure P–26.

21–19 (a) 4.55 μs
       (b) See Figure P–27.

21–21 15.945 kHz

FIGURE P–26

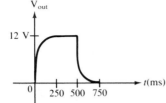

FIGURE P–27

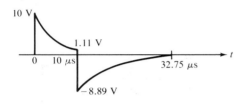

CHAPTER TWENTY-TWO

22–1 $17.46°$

22–3 $0.376∠41.2° \text{ A}$

22–5 $1.32∠121° \text{ A}$

22–7 $\mathbf{I}_{La} = 8.66∠ - 30° \text{ A}$
$\mathbf{I}_{Lb} = 8.66∠90° \text{ A}$
$\mathbf{I}_{Lc} = 8.66∠ - 150° \text{ A}$

22–9 **(a)** $\mathbf{V}_{L(ab)} = 866∠ - 30° \text{ V}$
$\mathbf{V}_{L(ca)} = 866∠ - 150° \text{ V}$
$\mathbf{V}_{L(bc)} = 866∠90° \text{ V}$
(b) $\mathbf{I}_{\theta a} = 0.5∠ - 32° \text{ A}$
$\mathbf{I}_{\theta b} = 0.5∠88° \text{ A}$
$\mathbf{I}_{\theta c} = 0.5∠ - 152° \text{ A}$
(c) $\mathbf{I}_{La} = 0.5∠ - 32° \text{ A}$
$\mathbf{I}_{Lb} = 0.5∠88° \text{ A}$
$\mathbf{I}_{Lc} = 0.5∠ - 152° \text{ A}$
(d) $\mathbf{I}_{Za} = 0.5∠ - 32° \text{ A}$
$\mathbf{I}_{Zb} = 0.5∠88° \text{ A}$
$\mathbf{I}_{Zc} = 0.5∠ - 152° \text{ A}$
(e) $\mathbf{V}_{Za} = 500∠0° \text{ V}$
$\mathbf{V}_{Zb} = 500∠120° \text{ V}$
$\mathbf{V}_{Zc} = 500∠ - 120° \text{ V}$

22–11 **(a)** $\mathbf{V}_{L(ab)} = 86.6∠ - 30° \text{ V}$
$\mathbf{V}_{L(ca)} = 86.6∠ - 150° \text{ V}$
$\mathbf{V}_{L(bc)} = 86.6∠90° \text{ V}$
(b) $\mathbf{I}_{\theta a} = 0.25∠110° \text{ A}$
$\mathbf{I}_{\theta b} = 0.25∠ - 130° \text{ A}$
$\mathbf{I}_{\theta c} = 0.25∠ - 10° \text{ A}$
(c) $\mathbf{I}_{La} = 0.25∠110° \text{ A}$
$\mathbf{I}_{Lb} = 0.25∠ - 130° \text{ A}$
$\mathbf{I}_{Lc} = 0.25∠ - 10° \text{ A}$
(d) $\mathbf{I}_{Za} = 0.144∠140° \text{ A}$
$\mathbf{I}_{Zb} = 0.144∠20° \text{ A}$
$\mathbf{I}_{Zc} = 0.144∠ - 100° \text{ A}$
(e) $\mathbf{V}_{Za} = 86.6∠ - 150° \text{ V}$
$\mathbf{V}_{Zb} = 86.6∠90° \text{ V}$
$\mathbf{V}_{Zc} = 86.6∠ - 30° \text{ V}$

22–13 $\mathbf{V}_{L(ab)} = 330∠ - 120° \text{ V}$
$\mathbf{V}_{L(ca)} = 330∠120° \text{ V}$
$\mathbf{V}_{L(bc)} = 330∠0° \text{ V}$
$\mathbf{I}_{Za} = 38.06∠ - 150° \text{ A}$
$\mathbf{I}_{Zb} = 38.06∠ - 30° \text{ A}$
$\mathbf{I}_{Zc} = 38.06∠90° \text{ A}$

22–15 Figure 22–44:   636 W
Figure 22–45:   149.36 W
Figure 22–46:   12.8 W
Figure 22–47:   2776.8 W
Figure 22–48:   10.9 kW

22–17 24 W

22–19 Output line voltages = 1 kV;
output phase voltages = 577.4 V.

GLOSSARY

Admittance A measure of the ability of a reactive circuit to permit current. The reciprocal of impedance.

Alternating current (ac) Current that reverses direction in response to a change in voltage polarity.

Ammeter An electrical instrument used to measure current.

Ampere (A or amp) The unit of electrical current.

Ampere-hour (Ah) The measure of the capacity of a battery to supply electrical current.

Amplification The process of increasing the power, voltage, or current of an electrical signal.

Amplifier An electronic circuit having the capability of amplification and designed specifically for that purpose.

Amplitude The voltage or current value of an electrical signal which, in some cases, implies the maximum value.

Analog Characterizing a linear process in which a variable takes on a *continuous* set of values within a given range.

Apparent power The power that *appears* to be delivered to a reactive circuit. The product of volts and amperes with units of VA (volt-amperes).

Arc tangent An inverse trigonometric function meaning "the angle whose tangent is." Also called *inverse tangent* as symbolized by $\tan^{-1}$.

Asynchronous Having no fixed time relationship, such as two waveforms that are not related to each other in terms of their time variations.

Atom The smallest particle of an element possessing the unique characteristics of that element.

Attenuation The process of reducing the power, voltage, or current value of an electrical signal. It can be thought of as negative amplification.

Audio Related to ability of the human ear to detect sound. Audio frequencies (af) are those that can be heard by the human ear, typically from 20 Hz to 20 kHz.

Autotransformer A transformer having only one coil for both its primary and its secondary.

AWG American Wire Gage, a standardization of wire sizes according to the diameter of the wire.

Band-pass The characteristic of a certain type of filter whereby frequencies within a certain range are passed through the circuit and all others are blocked.

Band-stop The characteristics of a certain type of filter whereby frequencies within a certain range are blocked from passing through the circuit.

Base One of the semiconductor regions in a bipolar transistor.

Baseline The normal level of a pulse waveform. The level in the absence of a pulse.

Battery An energy source that uses a chemical reaction to convert chemical energy into electrical energy.

Bias The application of a dc voltage to a diode, transistor, or other electronic device to produce a desired mode of operation.

Bode plot An idealized graph of the gain, in dB, versus frequency. Used to illustrate graphically the response of an amplifier or filter circuit.

Branch One of the current paths in a parallel circuit.

Capacitance The ability of a capacitor to store electrical charge.

Capacitor An electrical device possessing the property of capacitance.

Cascade An arrangement of circuits in which the output of one circuit becomes the input to the next.

Center tap (CT) A connection at the midpoint of the secondary of a transformer.

Charge An electrical property of matter in which an attractive force or a repulsive force is created between two particles. Charge can be positive or negative.

Chassis The metal framework or case upon which an electrical circuit or system is constructed.

Chip A tiny piece of semiconductor material upon which an integrated circuit is constructed.

Choke An inductor. The term is more commonly used in connection with inductors used to block or *choke off* high frequencies.

Circuit An interconnection of electrical components designed to produce a desired result.

Circuit breaker A resettable protective electrical device used for interrupting current in a circuit when the current has reached an excessive level.

Circular mil (CM) The unit of the cross-sectional area of a wire. A wire with a diameter of 0.001 inch has a cross-sectional area of one circular mil.

Closed circuit A circuit with a complete current path.

Coefficient of coupling A constant associated with transformers that specifies the magnetic field in the secondary as a result of that in the primary.

Coil A common term for an inductor or for the primary or secondary winding of a transformer.

Common A term sometimes used for the ground or reference point in a circuit.

Computer An electronic system that can process data and perform calculations at a very fast rate. It operates under the direction of a *stored program*.

Conductance The ability of a circuit to allow current. It is the reciprocal of resistance. The units are siemens (S).

Conductor A material that allows electrical current to flow with relative ease. An example is copper.

Core The material within the windings of an inductor that influences the electromagnetic characteristics of the inductor.

Coulomb (C) The unit of electrical charge.

CRT Cathode ray tube.

Current The rate of flow of electrons in a circuit.

Cycle The repetition of a pattern.

Decibel (dB) The unit of the logarithmic expression of a ratio, such as power or voltage gain.

Degree The unit of angular measure corresponding to 1/360 of a complete revolution.

Derivative The rate of change of a function, determined mathematically.

Dielectric The insulating material used between the plates of a capacitor.

Differentiator An *RC* or *RL* circuit that produces an output which approaches the mathematical derivative of the input.

Digital Characterizing a nonlinear process in which a variable takes on discrete values within a given range.

Diode An electronic device that permits current flow in only one direction. It can be a semiconductor or a tube.

Direct current (dc) Current that flows in only one direction.

DMM Digital multimeter.

Duty cycle A characteristic of a pulse waveform that indicates the percentage of time that a pulse is present during a cycle.

DVM Digital voltmeter.

Effective value A measure of the heating effect of a sine wave. Also known as the rms (root mean square) value.

Electrical Related to the use of electrical voltage and current to achieve desired results.

Electromagnetic Related to the production of a magnetic field by an electrical current in a conductor.

Electron The basic particle of electrical charge in matter. The electron possesses negative charge.

Electronic Related to the movement and control of free electrons in semiconductors or vacuum devices.

Element One of the unique substances that make up the known universe. Each element is characterized by a unique atomic structure.

Emitter One of the three regions in a bipolar transistor.

Energy The ability to do work.

Exponent The number of times a given number is multiplied by itself. The exponent is called the *power,* and the given number is called the *base.*

Falling edge The negative-going transition of a pulse.

Fall time The time interval required for a pulse to change from 90% of its amplitude to 10% of its amplitude.

Farad (F) The unit of capacitance.

Field The invisible forces that exist between oppositely charged particles (electric field) or between the north and south poles of a magnet or electromagnet (magnetic field).

Filter A type of electrical circuit that passes certain frequencies and rejects all others.

Flux The lines of force in a magnetic field.

Flux density The amount of flux per unit area in a magnetic field.

Free electron A valence electron that has broken away from its parent atom and is free to move from atom to atom within the atomic structure of a material.

Frequency A measure of the rate of change of a periodic function. The electrical unit of frequency is hertz (Hz).

Function generator An electronic test instrument that is capable of producing several types of electrical waveforms, such as sine, triangular, and square waves.

Fuse A protective electrical device that burns open when there is excessive current in a circuit.

Gain The amount by which an electrical signal is increased or amplified.

Generator A general term for any of several types of devices or instruments that are sources for electrical signals.

Germanium A semiconductor material.

Giga A prefix used to designate 10^9 (one thousand million).

Ground In electrical circuits, the common or reference point. It can be chassis ground or earth ground.

Harmonics The frequencies contained in a composite waveform that are integer multiples of the repetition frequency or fundamental frequency of the waveform.

Henry (H) The unit of inductance.

Hertz (Hz) The unit of frequency. One hertz is one cycle per second.

High-pass The characteristics of a certain type of filter whereby higher frequencies are passed and lower frequencies are rejected.

Hypotenuse The longest side of a right triangle.

Impedance The total opposition to current in a reactive circuit. The unit is ohms (Ω).

Induced voltage Voltage produced as a result of a changing magnetic field.

Inductance The property of an inductor whereby a change in current causes the inductor to produce an opposing voltage.

Inductor An electrical device having the property of inductance. Also known as a *coil* or a *choke.*

Infinite Having no bounds or limits.

Input The voltage, current, or power applied to an electrical circuit to produce a desired result.

Instantaneous value The value of a variable at a given instant in time.

Insulator A material that does not allow current under normal conditions.

Integrated circuit (IC) A type of circuit in which all of the components are constructed on a single, tiny piece of semiconductor material.

Integrator A type of *RC* or *RL* circuit that produces an output which approaches the mathematical integral of the input.

Joule (J) The unit of energy.

Kilo A prefix used to designate 10^3 (one thousand).

Kilowatt-hour (kWh) A common unit of energy used mainly by utility companies.

Kirchhoff's laws A set of circuit laws that describes certain voltage and current relationships in a circuit.

Lag A condition of the phase or time relationship of waveforms in which one waveform is behind the other in phase or time.

Lead A wire or cable connection to an electrical or electronic device or instrument. Also, a condition of the phase or time relationship of waveforms in which one waveform is ahead of the other in phase or time.

Leading edge   The first step or transition of a pulse.

Linear   Characterized by a straight-line relationship.

Load   The device upon which work is performed.

Logarithm   The exponent to which a base number must be raised to produce a given number. For example, the logarithm of 100 is 2 ($10^2 = 100$).

Loop   A closed path in a circuit.

Low-pass   The characteristics of a certain type of filter whereby lower frequencies are passed and higher frequencies are rejected.

Magnetic   Related to or possessing characteristics of magnetism. Having a north and a south pole with lines of force extending between the two.

Magnetomotive force (mmf)   The force produced by a current in a coiled wire in establishing a magnetic field. The unit is ampere-turns (At).

Magnitude   The value of a quantity, such as the number of amperes of current or the number of volts of voltage.

Mega   A prefix designating 10^6 (one million).

Mesh   An arrangement of loops in a circuit.

Micro   A prefix designating 10^{-6} (one-millionth).

Milli   A prefix designating 10^{-3} (one-thousandth).

Modulation   The process whereby a signal containing information (such as voice) is used to modify the amplitude (AM) or the frequency (FM) of a much higher frequency sine wave (carrier).

Multimeter   An instrument used to measure current, voltage, and resistance.

Mutual inductance   The inductance between two separate coils, such as in a transformer.

Nano   A prefix designating 10^{-9} (one thousand-millionth).

Network   A circuit.

Neutron   An atomic particle having no electrical charge.

Node   A point or junction in a circuit where two or more components connect.

Ohm (Ω)   The unit of resistance.

Ohmmeter   An instrument for measuring resistance.

Open circuit   A circuit in which there is not a complete current path.

Oscillator   An electronic circuit that internally produces a time-varying signal without an external signal input.

Oscilloscope   A measurement instrument that displays signal waveforms on a screen.

Output   The voltage, current, or power produced by a circuit in response to an input or to a particular set of conditions.

Overshoot   A short duration of excessive amplitude occurring on the positive-going transition of a pulse.

Parallel   The relationship in electric circuits in which two or more current paths are connected between the same two points.

Peak value   The maximum value of an electrical waveform, particularly in relation to sine waves.

Period   The time interval of one cycle of a periodic waveform.

Periodic   Characterized by a repetition at fixed intervals.

Permeability   A measure of the ease with which a magnetic field can be established within a material.

Phase   The relative displacement of a time-varying waveform in terms of its occurrence.

Pico   A prefix designating 10^{-12} (one-billionth).

Potentiometer   A three-terminal variable resistor.

Power   The rate of energy consumption. The unit is watts (W).

Power factor   The relationship between volt-amperes and true power or watts. Volt-amperes multiplied by the power factor equals true power.

Power supply   An electronic instrument that produces voltage, current, and power from the ac power line or batteries in a form suitable for use in various applications to power electronic equipment.

Primary   The input winding of a transformer.

Proton   A positively charged atomic particle.

Pulse   A type of waveform that consists of two equal and opposite steps in voltage or current, separated by a time interval.

Pulse width   The time interval between the opposite steps of an ideal pulse. Also, the time between the 50% points on the leading and trailing edges of a nonideal pulse.

Quality factor (Q)   The ratio of reactive power to true power in a coil or a resonant circuit.

Radian   A unit of angular measurement. There are 2π radians in a complete revolution. One radian equals $57.3°$.

Ramp A type of waveform characterized by a linear increase or decrease in voltage or current.

Reactance The opposition of a capacitor or an inductor to sinusoidal current. The unit is ohms (Ω).

Reactive power The rate at which energy is stored and alternately returned to the source by a reactive component.

Rectifier An electronic circuit that converts ac into pulsating dc.

Reluctance The opposition to the establishment of a magnetic field in an electromagnetic circuit.

Resistance Opposition to current. The unit is ohms (Ω).

Resistivity The resistance that is characteristic of a given material.

Resistor An electrical component possessing resistance.

Resonance In an *LC* circuit, the condition when the impedance is minimum (series) or maximum (parallel).

Response In electronic circuits, the reaction of a circuit to a given input.

Rheostat A two-terminal, variable resistor.

Right angle A 90° angle.

Ringing An unwanted oscillation on a waveform.

Rise time The time interval required for a pulse to change from 10% of its amplitude to 90% of its amplitude.

Rising edge The positive-going transition of a pulse.

rms Root mean square. The value of a sine wave that indicates its heating effect. Also known as *effective value*.

Sawtooth A type of electrical waveform composed of ramps.

Secondary The output winding of a transformer.

Semiconductor A material that has a conductance value between that of a conductor and that of an insulator. Silicon and germanium are examples.

Series In an electrical circuit, a relationship of components in which the components are connected such that they provide a single current path between two points.

Short A zero resistance connection between two points.

Siemen (S) The unit of conductance.

Signal A time-varying electrical waveform.

Silicon A semiconductor material used in transistors.

Sine wave A type of alternating electrical waveform.

Slope The vertical change in a line for a given horizontal change.

Source Any device that produces energy.

Steady state An equilibrium condition in a circuit.

Step A voltage or current transition from one level to another.

Susceptance The ability of a reactive component to permit current flow. The reciprocal of reactance.

Switch An electrical or electronic device for opening and closing a current path.

Synchronous Having a fixed time relationship.

Tangent A trigonometric function that is the ratio of the opposite side of a right triangle to the adjacent side.

Tank A parallel resonant circuit.

Tapered Nonlinear, such as a tapered potentiometer.

Temperature coefficient A constant specifying the amount of change in the value of a quantity for a given change in temperature.

Tesla (T) The unit of flux density. Also, webers per square meter.

Tetrode A vacuum tube that has four elements.

Thermistor A resistor whose resistance decreases with an increase in temperature.

Tilt A slope on the normally flat portion of a pulse.

Time constant A fixed time interval, set by *R*, *C*, and *L* values, that determines the time response of a circuit.

Tolerance The limits of variation in the value of an electrical component.

Trailing edge The last edge to occur in a pulse.

Transformer An electrical device that operates on the principle of electromagnetic induction. It is used for increasing or decreasing an ac voltage and for various other applications.

Transient A temporary or passing condition in a circuit. A sudden and temporary change in circuit conditions.

Transistor A semiconductor device used for amplification and switching applications in electronic circuits.

Triangular wave A type of electrical waveform that consists of ramps.

Trimmer A small, variable resistor or capacitor.

Troubleshooting The process and technique of identifying and locating faults in an electrical or electronic circuit.

True power   The average rate of energy consumption. In an electrical circuit, true power occurs only in the resistance and represents a net energy loss.

Turns ratio   The ratio of the number of secondary turns to the number of primary turns in the transformer windings.

Undershoot   The opposite of overshoot, occurring on the negative-going edge of a pulse.

Valence   Related to the outer shell or orbit of an atom.

Volt   The unit of voltage or electromotive force (emf).

Voltage   The amount of energy available to move a certain number of electrons from one point to another in an electrical circuit.

Watt (W)   The unit of power.

Waveform   The pattern of variations of a voltage or a current.

Wavelength   The length in space occupied by one cycle of an electromagnetic wave.

Weber   The unit of magnetic flux.

Winding   The loops of wire or coil in an inductor or transformer.

Wiper   The variable contact in a potentiometer or other device.

INDEX